Biology of Nonvascular Plants

Biology of

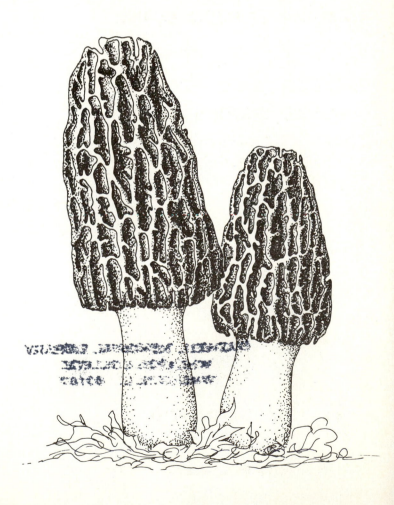

Nonvascular Plants

HAYDEN N. PRITCHARD, Ph.D.

Associate Professor of Biology,
Lehigh University, Bethlehem, Pennsylvania

PATRICIA T. BRADT, Ph.D.

Adjunct Associate Professor of Biology,
Lehigh University, Bethlehem, Pennsylvania

with 319 illustrations

original drawings by **AMY E. CONTI**

New York, New York

 TIMES MIRROR/MOSBY
COLLEGE PUBLISHING

ST. LOUIS • TORONTO • SANTA CLARA   1984

To

Mary, Paul, David, and Aunt Leila;

L. Jack, James, Julia, and George

Our families,
who have provided us
with encouragement when we most needed it

Editor: Diane L. Bowen
Assistant editor: Susan Dust Schapper
Editorial assistant: Jackie V. Yaiser
Manuscript editor: Suzanne L. Harrawood
Design: Jeanne Bush
Production: Carol O'Leary, Teresa Breckwoldt

Library of Congress Cataloging In Publication Data

Pritchard, Hayden N.
    Biology of nonvascular plants.

    Bibliography: p.
    Includes index.
    1. Cryptogams.   I.   Bradt, Patricia T.   II.   Title.
QK505.P84   1984   586   83-1018
ISBN   0-8016-4043-1

AC/VH/VH   9 8 7 6 5 4 3 2 1   02/D/240

Front cover: Near apical segment of the giant kelp, *Macrocystis* sp. (Phaeophyta), showing its gas-filled flotation bladders and its photosynthetic blades. These kelps are one of the few algae that possess cells and ducts for internal transport of metabolic products. (Photo by Robert B. Evans/Tom Stack and Associates.)

Back cover: Group of small delicate colorful mushrooms of the genus *Mycena* (Basidiomycotina) growing on a rotting tree stump in Washington's Olympic rain forest. There are over 230 species of *Mycena,* all of which are dainty, fragile saprophytes bearing variously colored cups. (Photo by Jim Brandenburg.)

PREFACE

The idea for development of this book grew from our long-standing relationship and our mutual work in botany, ecology, and aquatic biology. We saw a need to develop a useful, interesting, academically meaningful course that introduces students to the world of nonvascular plants. Agreeing that such a course should be both practical and intellectually challenging, we also recognized the need for a comprehensive text for the beginning student. The decision to write this book was based on our unsuccessful search for a textbook that emphasized the ecological and economic importance as well as the taxonomy and morphology of nonvascular plants.

The development of the text began with an outline for a one-semester course in algae and fungi. With this as a base we expanded and refined the original outline to include various physiological and ecological aspects of all nonvascular plants, including algae, fungi, lichens, and bryophytes. We have not included the prokaryotic organisms known as bacteria, since most colleges and universities have entire courses that deal exclusively with that group.

Each chapter includes an abbreviated and annotated classification for the division discussed in that chapter. The classification lists individual genera that are discussed in the text so that the student can relate each genus to its taxonomic position. Taxonomic details are purposely kept to a minimum. Chapters open with a description of cellular characteristics found in each division, including ultra-structural, microscopic, and gross morphological features. Representative reproductive mechanisms (asexual and sexual) as well as characteristic growth patterns are separately discussed, as are individual life histories.

The physiology section of each chapter presents selected important physiological aspects of each particular division. We have chosen those physiological phenomena that we consider most representative and characteristic of the group.

Following this, each chapter includes a section on the position of each group in an evolutionary scheme, the importance of each division in the biosphere, and the economic impact each group has on our society. We hope to demonstrate to students the reasons why nonvascular plants are important not only in the natural world but also as tools in the laboratory for the study of genetics, molecular biology, and biochemistry. The commercial use of nonvascular plants is also discussed, together with the multitude of uses humans make of these plants. Both algae and fungi can also be very troublesome: large populations of algae may render a lake or stream extremely unpleasant or even toxic, while certain fungi may cause disease epidemics in both plants or animals. However, algae are the major contributors of fixed carbon to the biosphere, while fungi are important decomposers of dead organic material, especially cellulose and lignin.

We think students will appreciate the link between organism identification and the role each plays in everyday life. In reading the text and following its exposition, the student will be able to recognize and identify each genus mentioned and place that organism in its proper evolutionary position with respect to its form, physiology, and biochemistry. Most genera mentioned are readily available as both live and preserved specimens from various biological supply houses. Thus the text is a good tool for an accompanying laboratory session. In fact, the book has been written with laboratory and field work in mind.

Our approach to nonvascular plants is very practical: it represents what we consider the easiest and most interesting method of learning about these organisms. The book is intended to be introductory in nature, suitable for sophomore or junior students of biology or even graduate students who have had minimal exposure to nonvascular plants. The text could also be divided and used as an introductory text in either psychology (Chapters 1 through 11) or mycology (Chapters 12 through 19 or 20). For those readers who wish to consider the groups in greater detail, we have included references at the rear of the book to specific works that have been cited in the text. Anticipating that readers may wish to read further about individual divisions, we have ended each chapter with a list of selected additional references that discuss each division in greater depth. We hope that all readers' curiosity will be piqued to the extent that they will continue investigating particular organisms in both the laboratory and the literature and that students find these organisms as fascinating and challenging as we did in gathering material for this book.

Hayden N. Pritchard
Patricia T. Bradt

ACKNOWLEDGMENTS

We would like to recognize the encouragement and assistance given us during the preparation of the manuscript. We include the academic and intellectual support of the following: Sidney S. Herman, Chairman of the Biology Department (during the writing of the text), Dean John W. Hunt, Barry Bean, J. Donald Ryan, Bland Montenecourt, Saul B. Barber, Roy C. Herrenkohl, Richard G. Malsberger, Jon I. Parker, and James M. Parks, all of Lehigh University; Robert F. Schmaltz of The Pennsylvania State University; Robert S. Chase of Lafayette College; and all undergraduate and graduate students of Lehigh who have contributed to our excitement about and knowledge of nonvascular plants.

For their help in photography we thank Diane Spess (who also contributed graphs and diagrams), Martin Berg, William Fisk, Sr., William Fisk, Jr., and Julia Bradt. For use of their original photographs we thank Vernon Ahmadjian, Margaret Goreau, Erika Hartweig, Jerome Jacobs, Jonathan King, George Knaphus, Andrew Knoll, Donald Marx, William Merrill, Jon I. Parker, Alastair Pringle, Rudd Scheffer, Karen Steidinger, L.S. Tester, Paul Wuest, and *The Express* of Easton, Pa.

We are also grateful to our typists, Miriam Fretzo, Geraldine Dettra, Anna Mae Stern, Marie Tracy, Judy Dreisbach, and Barbara Schmidt.

Special recognition is given to our major critic, proofreader, and indexer, Barbara K. Ryan, without whose timely help and humor our task would have been much more difficult and much less interesting. Gregory T. Lautenslager also helped proofread, and Sharon L. Siegler of Lehigh's Mart Library gave unselfishly of her time and talent, saving us from unproductive searches.

We would also like to give special acknowledgment to Howard and Myrt Whitcomb of Lehigh University for encouraging us to write about nonvascular plants and for their continued interest in our progress.

It was indeed a pleasure working with our gifted and knowledgeable artist, Amy Elizabeth Conti, who showed a wonderful ability to understand what was really desired in the illustrations; with our trusting editors, Diane Bowen and Susan Dust Schapper, both of whom were a constant source of encouragement, stimulation, and valuable information; and with our intrepid manuscript editor, Suzanne Harrawood, whose patience and knowledge made our work much easier.

In addition, we would like to acknowledge the two former Lehigh University biologists who guided us into the fascinating world of algae and fungi: Francis J. Trembley and Basil W. Parker.

Finally, we are indebted to our reviewers for constructive criticism and valuable advice. We recognize and extend our sincere thanks to the following:

Doyle E. Enderegg   *University of Idaho*
Robert G. Anderson   *University of Missouri, Kansas City*
John E. Averett   *University of Missouri, St. Louis*
John L. Blum   *University of Wisconsin, Milwaukee*

Edwin R. Florance   *Lewis and Clark College, Portland, Oregon*
J.R. Goodin   *Texas Tech University*
J. Herbert Graffius   *Ohio University*
Alan Graham   *Kent State University*
Harold J. Humm   *University of South Florida*
Deana T. Klein   *St. Michael's College, Winooski, Vermont*
Chester R. Leathers   *Arizona State University*
Alfred Loeblich III   *University of Houston*
David J. McLaughlin   *University of Minnesota*
Steven N. Murray   *California State University, Fullerton*
David K. Northington   *Texas Tech University*
Craig W. Schneider   *Trinity College, Hartford, Connecticut*
Daniel E. Wujek   *Central Michigan University*

Edwin Florance and Craig Schneider deserve special credit for reviewing and commenting on the entire draft. Doyle Anderegg provided exceptional assistance by sending us his private translation of the Henssen and Jahns monograph that otherwise would have been inaccessible. Though all of these people saved us from many pitfalls, we assume responsibility for any errors that remain.

CONTENTS

FUNGI

LICHENS AND BRYOPHYTES

The world of nonvascular plants includes organisms with and without chlorophyll that have no vascular or conducting tissue. The absence of conducting and supporting tissues restricts the distribution of these plants. They are either aquatic or terrestrial; if the latter, they are relatively small and usually prostrate in form. They rely solely upon physical diffusion for nutrient and water transport. In the aquatic environment, however, certain nonvascular plants such as the kelps (Phaeophyta—Chapter 7) may grow to be some of the largest plants on earth. These and other "seaweeds" are surrounded by water, so nutrients diffuse easily to all parts of the plant.

The plants discussed in this book include members of four of the five taxonomic kingdoms described by Whittaker (1969). The Cyanobacteria or blue-green algae (Chapter 3) are members of the prokaryotic kingdom Monera. The Euglenophyta or euglenoids (Chapter 6), the Chrysophyta or yellow-brown algae (Chapter 8), the Pyrrhophyta or fire algae (Chapter 9), and the Xanthophyta or yellow-green algae (Chapter 10) are all classified in the eukaryotic kingdom Protista. The Charophyta or stoneworts (Chapter 5), the Phaeophyta or brown algae (Chapter 7), and the Rhodophyta or red algae (Chapter 11) are members of the kingdom Plantae because they are multicellular and macroscopic. The Chlorophyta or green algae (Chapter 4) are difficult to classify because the division contains both microscopic single-celled organisms and macroscopic multicelled organisms. Whittaker (1969) places the Chlorophyta in the kingdom Plantae, but in this text the Chlorophyta are placed as overlapping the two kingdoms Protista and Plantae (Fig. 1-1).

The fungi are all placed in the kingdom Fungi, including the Myxomycota or slime molds (Chapter 14), the Mastigomycotina or flagellated fungi (Chapter 15), the Zygomycotina or zygospore-forming fungi (Chapter 16), the Ascomycotina or sac fungi (Chapter 17),

INTRODUCTION

1

the Basidiomycotina or club fungi (Chapter 18), and the Deuteromycotina or imperfect fungi (Chapter 19). The lichens (Chapter 20), representing symbiotic associations between an alga and a fungus, encompass members of two kingdoms and are therefore not placed in a separate kingdom. The Bryophyta or mosses (Chapter 21) are members of the kingdom Plantae.

Nonvascular plants have historically been studied as a unit because (1) they lack conducting tissue, (2) many are microscopic in size, and (3) their sex organs are relatively simple structurally. Linnaeus grouped nonvascular plants into the category **Cryptogamia,** which means "concealed" reproductive organs. "Cryptogams" include not only algae and fungi but also ferns, mosses, and lichens. The algae and fungi were often placed in a subdivision of Cryptogamia, the **Thallophyta,** which means having a thallus (i.e., plant body) lacking true roots, stems, and leaves. The Bryophyta were placed in their own subdivision of the Cryptogamia because their sex organs and zygote development are more complex than those of the Thallophyta. The terms "cryptogam" and "thallophyte" are still used occasionally, so the student should be familiar with them.

Classification of many nonvascular plants is not agreed upon by phycologists and mycologists; it is an ever-present problem for biologists, because they attempt to place in discreet taxonomic categories a widely diverse group of organisms whose characteristics developed along a continuum rather than in neatly separated sections. Organisms that bear characteristics of two or more taxonomic groups are always a problem, and their classification is debated constantly. For example, the kingdom Protista includes both plants (algae) and animals (protozoa). At this level of organization, primarily single-celled, it is often extremely difficult, if not impossible, to designate an organism as plant or animal. Certain euglenoids (Euglenophyta—Chapter 6) contain chlorophyll and photosynthesize under favorable conditions. However, if light is reduced and organic material abounds, these euglenoids may lose their chlorophyll, cease photosynthesizing, and use the abundant organic material as an energy source. Are these organisms algae or protozoa? Their mode of nutrition may be either autotrophic or heterotrophic, depending on conditions. The kingdom Protista was conceived to include such enigmas and to avoid such problems of "plant" or "animal." Students may discover that they are studying protists such as euglenophytes or pyrrhophytes in a course in protozoology or invertebrate zoology, because these two divisions have many "animal-like" members that can live heterotrophically.

The taxonomy presented in this book is widely used by many researchers, but it is by no means universally accepted; in fact, no taxonomy is universally accepted by all phycologists, mycologists, lichenologists, and bryologists. This book attempts to keep taxonomic details to a minimum and the discussion thereof as brief as possible. We wish to convey to the reader a sense of the importance of each group in the evolutionary scheme, in the functioning of ecological systems, and in the group's economic use.

As a beginning to understanding the role of nonvascular plants in the biosphere, the student is first introduced to their classification, cell structure, reproduction, growth, and physiology. These sections are followed by evolution, ecology, and economic importance.

The sections on classification and cell structure are designed to place the plants into a perspective comprehensible to all beginning biologists. Each classification scheme follows an evolutionary sequence so that the earliest organisms occur first and derived organisms occur later on. In general this means that the structurally simplest organisms are presented first, while the structurally more complex are presented later on. Cell structure should relate to the increasing complexity established by the classification section.

Descriptions of reproductive mechanisms and growth are meant to reinforce the classification and cellular structure already introduced. The life histories of the various organisms instruct students as to how these organisms fit into the overall environment, reflecting the cellular composition of the group as well as its ecological and evolutionary position.

The physiological discussion aids in the evaluation of the adaptations of these plants to their particular environments. In many harsh environments certain nonvascular plants may be found that are adapted in unique ways to prevailing conditions. For example, rocks in Antarctica may contain both algae and lichens, while rocks of the desert may support lichens on their surface and fungi inside.

The discussion of evolution enables the reader to place these organisms in an overall evolutionary scheme. Unfortunately, the fossil record of both algal and fungal development contains numerous gaps, so the evolutionary pathways along which nonvascular plants developed remain somewhat uncertain and subject to conjecture. The study of the various theories of evolution by themselves presents a challenging undertaking. Theories of the evolution of nonvascular plants will be debated for many, many years before the fossil record yields enough evidence to definitively outline the evolutionary course.

By understanding the relationship of nonvascular plants to their environment and their role in various ecosystems, the student will realize the importance of these plants, even though the majority are microscopic and therefore not obvious to the casual observer. Their size is by no means related to the significance of their ecological role.

Algae living in the sea contribute 30% to 40% of the total organic carbon fixed in the biosphere, while freshwater and terrestrial algae contribute an additional 1% to 5%. The vast amount of marine habitat (360×10^6 km^2) compared to terrestrial (150×10^6 km^2) that algae can and do populate ensures their very significant contribution to biosphere primary productivity.

The ecological role of fungi is primarily that of decomposers of organic material. They can synthesize an array of enzymes that can break down numerous naturally occurring organic molecules into inorganic components. Nutrients contained therein are recycled into inorganic forms and are then available for use by green plants. In the process of metabolizing organic molecules, fungi obtain energy for their own growth and reproduction. Various fungal groups inhabit different environments; for example, the Mastigomycotina are primarily aquatic and are therefore important decomposers in water. The Zygomycotina, Ascomycotina, Basidiomycotina, and Deuteromycotina are primarily terrestrial and may be isolated from a wide variety of habitats from the desert to the arctic.

In the economic importance sections, commercial use of algae as a food or energy source is discussed. In addition, the manner is examined in which large algal populations affect other aquatic organisms and human use of the waters. Many fungal species are grown commercially for such important organic products as antibiotics, vitamins, and organic acids. The biochemical ability of certain fungi to anaerobically convert glucose to ethyl alcohol is utilized in the commercial production of beer, wine, and other alcoholic beverages. Other fungi are used to enzymatically modify foods such as milk or soybeans and to enhance or change their taste (e.g., cheeses) or nutritive value (e.g., fermented soybeans). Still other fungi such as mushrooms are collected or cultivated commercially as a very desirable human food. Fungi are also being evaluated as an inexpensive source of both protein-rich food and renewable energy. Because of their role in the decomposition process, fungi are being used to dispose of waste products with resulting production of glucose and other organic compounds that can then be used in the synthesis of commercially valuable products.

Many fungi are opportunistic, switching from a saprophytic role to a parasitic role when suitable environmental conditions or susceptible hosts are present. Such parasitic fungi have caused widespread and devastating losses to both food and forests. Certain fungi may also parasitize animals, including humans, causing various diseases. Mycotoxins produced by other fungi during their growth may cause severe disease in animals that consume contaminated products.

The chapters on the individual algal divisions (Chapters 3-11) are preceded by two introductory chapters: Chapter 1 on algal structural and reproductive characteristics, and Chapter 2 on the relationship of algae to their environment. The chapters on the individual fungal subdivisions (Chapters 14-19) are preceded by Chapter 12, which discusses general fungal morphological characteristics and reproductive methods, and Chapter 13, which treats the role of fungi in the ecosystem.

The lichens (Chapter 20) are plants that represent a symbiotic association between an alga and a fungus. The lichen, however, has unique characteristics that are not exhibited by either the alga or the fungus alone. In the development of this association both the alga and fungus are modified to the extent that it is difficult to identify either organism. To be positively identified, each symbiotic partner must be cultured individually. In addition, these symbiotic organisms can live in harsh environments such as glaciers, high altitudes, or the polar regions, where neither the fungal or algal partner could survive alone.

The mosses (Chapter 21) are structurally more complex than both the algae and fungi. They are almost entirely terrestrial, but they still require moisture for reproduction. Most researchers consider the bryophytes the most evolutionarily advanced nonvascular plants because of their multicellular reproductive organs, their parenchymatous thalli, and their structures that closely resemble, but are not identical to, roots, leaves, and stems. Although they are small, the bryophytes are widely distributed over unpolluted areas of the earth, and some species have invaded aquatic habitats.

At the conclusion of the text there is a glossary for those terms used in this book to which the student may not have been previously exposed. Highly specialized language has been kept to a minimum, and as new terms are introduced they are defined within the text, in addition to being defined in the glossary. All the cited references may be found in the "Literature Cited" section. This list of sources is presented as a brief introduction to the very extensive and diverse literature available.

In conclusion, this book attempts to relate nonvascular plants to the experiences of college students so that they can begin to comprehend the importance of these plants to the biosphere. We sincerely hope that students will not only enjoy the material presented here but also will be stimulated to pursue the subject in greater depth.

ALGAE

1

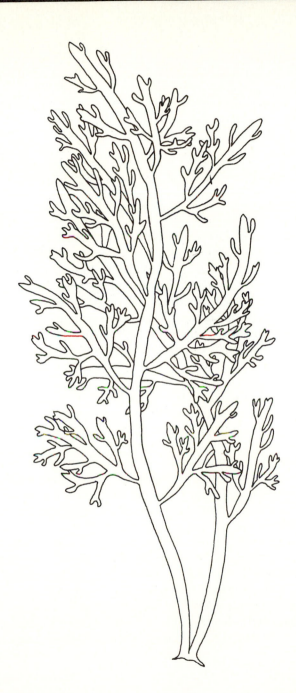

The algae (sing. alga) constitute a diversified, difficult-to-define group of organisms. Most students understand what is meant when they hear the word "algae," for they have seen the green scum associated with indoor aquaria, and they are familiar with the "seaweed" algae from the oceans. Scientists are more cautious in their use of the term algae, because they find it covers a range of organisms that is too broad to effectively isolate under a single term. Algae, therefore, is a common name rather than a scientific taxonomic category.

Generally the algae are considered to be a category of nonvascular, plantlike organisms whose biochemical composition includes a specific type of chlorophyll (chlorophyll a), and whose reproductive structures are relatively simple because they lack a layer of sterile, vegetative cells surrounding the eggs or sperm. There are, however, some exceptions to this definition, because not all algae contain chlorophyll, and some algae have structurally complex reproductive organs. Some brown algae (Phaeophyta) even contain vascular tissues.

Courses in protozoology usually include chapters that describe many of the organisms considered in this text. The phytoflagellates or photosynthetic flagellates, including organisms from Euglenophyta, Chlorophyta, Chrysophyta, Pyrrhophyta, and Xanthophyta, are all considered to be protozoans in these texts. Textbooks concerned primarily with bacteriology or microbiology often contain sections on the blue-green bacteria (Cyanobacteria).

ALGAL KINGDOMS

Further ambiguities become apparent when it is noted that the algal complex is composed of organisms that belong to three different kingdoms based upon the five-kingdom classification developed by Whittaker (1969). Algae, as we know them, are included as members of the kingdoms Monera, Protista, and Plantae (Fig. 1-1). Monera comprises all of the prokaryotic organisms, since they all lack membrane-bound nuclei and cellular organelles. Included in the Monera are two main groups, the bacteria and the blue-green algae. The latter group has historically been considered to be an alga, but recent evidence has produced a reevaluation of them; this text will treat them as more closely related to bacteria than to algae. Evidence for this is presented in Chapter 2, Cyanobacteria.

Bergey's Manual of Determinative Bacteriology (Buchanan and Gibbons, 1974) divides all prokaryotic organisms (Monera) in two divisions—Bacteria and Cyanobacteria. Bacteria comprise a large group of prokaryotic organisms that are not generally considered to be either algae or fungi, because of this, they will not be discussed in this text. Bacteria are distinct from eukaryotic algae and fungi in both structural and biochemical characteristics. They lack a membrane-bound nucleus. They also differ with respect to the complete absence of chlorophyll *a* (which occurs in algae), the lack of cellulose or chitin in their cell walls (which occurs in both algae and fungi), the absence of starch as an energy storage product (which occurs in some algae), and the occurrence of photosynthetic pigments that are quite distinct from those found in algae.

The distinction between the kingdom Monera and the rest of the biological world is based upon the presence or absence of a membrane-bound nucleus. If such a nuclear structure is absent, the cells are said to be **prokaryotic** (fr. Greek *pro* = before + *karyon* = kernel). When the cells contain a membrane-bound nucleus, they are called **eukaryotic** (fr. Greek *eu* = true + *karyon* = kernel). The kingdoms Protista,

Fungi, Plantae, and Animalia are composed of organisms whose cells contain nuclei, and they are therefore eukaryotic.

Eukaryotic cells throughout the various kingdoms all exhibit certain ultrastructural characteristics associated with their nuclei and cytoplasm. Within the nucleus is a group of chromosomes that carry the hereditary material deoxyribonucleic acid (DNA). The structure and chemical nature of the chromosome are remarkably constant in all eukaryotic cells, with each chromosome being formed by two chromatids joined at the centromere. Also within each nucleus is one or more nucleoli (sing. nucleolus),

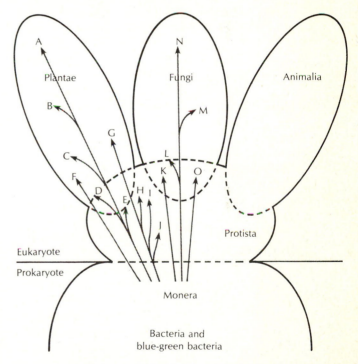

A. Tracheophyta (vascular plants)
B. Bryophyta
C. Charophyta
D. Chlorophyta
E. Euglenophyta
F. Rhodophyta
G. Phaeophyta
H. Chrysophyta
I. Pyrrhophyta
J. Xanthophyta
K. Mastigomycotina
L. Zygomycotina
M. Ascomycotina
N. Basidiomycotina
O. Myxomycota

FIG. 1-1 The five kingdoms.

large round structures that are associated with ribonucleic acid (RNA) storage and synthesis and that are formed by nucleolar organizers associated with specific chromosomes.

The cytoplasm contains numerous membrane-limited cellular organelles and is bound at its outer limit by the plasma membrane. Biological membranes are all complex structures whose multiple architectural forms and multiple functions are just beginning to be understood. Each membrane is composed of two layers of phospholipid molecules in which proteins and glycoproteins (combinations of proteins and sugars) are embedded (Singer and Nicolson, 1972). The precise chemical composition and molecular arrangement of the lipids, proteins, and glycoproteins varies from one membrane to another and from one cell to another. Physical openings or pores may be found in some membranes. This sort of model for biological membranes is known as the "fluid mosaic" model, since the lipid component and some of the proteins are able to flow freely from one part of the membrane to another, thus acting like a liquid interface between the exterior and interior of the cell. Every molecule that enters or leaves the cell must pass through the plasma membrane. The fluid mosaic model for biological membranes appears to be common to all organelles associated with living cells including the plasma membrane.

Most of the cytoplasmic organelles are limited by fluid mosaic membranes. The endoplasmic reticulum (ER) is a series of canals or vesicles that are membrane bound and form a network throughout the cytoplasm of most cells. The membranes that form the ER are similar to the plasma membrane in both structure and chemical composition. They are often continuous with the membranes that form the nuclear membrane and the plasma membrane.

Ribosomes are small ribonucleoprotein particles consisting of 30% to 45% protein (by weight); the remainder is ribonucleic acid (RNA). They are suspended in the fluid of the cytoplasm or may be attached to the membranes composing the ER. When ribosomes are attached, the ER is known as rough ER. Without ribosomes, the membranes of the ER are known as smooth ER. Rough ER is associated with protein synthesis. Both rough ER and smooth ER provide cells with a system of canals through which materials may be transported throughout the cell.

The cytoplasm also contains a system of membranes that are stacked in series forming one or more Golgi bodies or dictyosomes. The membranes of the Golgi bodies enclose spaces known as the cisternae. It is within the cisternae that secretory products are synthesized and encased within membranes. The Golgi bodies are a sort of packaging section of the cell, whose products are processed for future export. These structures are best seen in cells active in secretion. They are not always a prominent part of algal or fungal cells, although dictyosomes have been identified in both groups.

Mitochondria are cytoplasmic organelles bound by fluid mosaic membranes. They contain all the enzymes necessary for aerobic respiration and are the structures that generate all of the adenosine triphosphate (ATP) necessary to provide energy for the metabolic activity of the cell. The structure of the mitochondrion may vary to some extent among eukaryotic cells, but

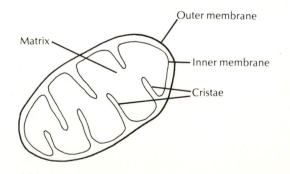

FIG. 1-2 Single mitochondrion from a eukaryotic cell. The mitochondrion is the energy-generating source of eukaryotes and is therefore the site of cellular respiration and formation of adenosine triphosphate (ATP).

it always has infoldings of the inner membrane forming cristae, which provide the site for the enzymatic activities of respiration (Fig. 1-2).

Eukaryotic cells, which are photosynthetic, all possess cytoplasmic inclusions called chloroplasts. On the ultrastructural level chloroplasts are complicated, membrane-bound structures with a distinctive architecture (see Fig. 1-7). They are the site of photosynthesis, and they contain all the photosynthetic, light-capturing pigments, including the various chlorophyll types. Chloroplasts, which are easily seen with the light microscope, are often green as a result of the presence of chlorophyll. They may be colorless, however, or may take on the color of the pigment that is present in the greatest concentration.

The cytoplasm may also contain a variety of other organelles such as microtubules, crystals, and vacuoles containing food storage products, pigments, or waste products. Extensions of flagella called basal bodies may be found in association with flagellate cells. Basal bodies sometimes are attached to similar pairs of structures, located just outside the nuclear membrane, which are called centrioles. Paired centrioles are a constant feature of animal cells, but in algae and fungi they are found only in some forms, most of which are motile flagellates. In some cases it appears that basal bodies and centrioles are identical structures that migrate from the nuclear membrane to the plasma membrane at different times during the life of a cell (Lee, 1980).

Protista constitute a group of organisms that are mostly unicellular. However, this kingdom also contains some organisms that have developed loose associations of cells, colonies, filaments, sheets, or even three-dimensional growth habits. Within the Protista are found the protozoa (unicellular animals) and a number of algal divisions, including some green algae (Chlorophyta), the euglenoids (Euglenophyta), the yellow-green algae (Xanthophyta), the fire algae (Pyrrhophyta), and the golden-brown algae (Chrysophyta).

The plant kingdom contains those photosynthetic multicellular plants that include all vascular plants and mosses (Bryophyta) as well as those algal groups with structurally complex three-dimensional habits. These include the three divisions made up of the red algae (Rhodophyta), the brown algae (Phaeophyta), and the stoneworts (Charophyta).

The kingdom Fungi contains organisms composed of eukaryotic cells, many of which form multinucleate filaments. The fungi do not possess chlorophyll and therefore lack the ability to photosynthesize. Their nutrition is absorptive. The similarities and differences between the algae and the fungi are discussed in more detail in later chapters.

The animal kingdom contains all of the multicellular animals. Their nutrition is ingestive. They lack cell walls, and they do not carry out photosynthesis.

STRUCTURAL CHARACTERISTICS OF ALGAE

In any description characterizing algal organisms, certain morphological, physiological, biochemical, and ecological qualities help investigators to focus upon the specimen in question. Ultimate decisions as to the scientific classification of any algal specimen are based upon characteristics that may or may not be readily apparent or easily discernible. These include the gross morphology, ultrastructural traits of particular cellular organelles, and the biochemical properties of various pigments, cell walls, and energy storage products (see Fig. 1-3 and Tables 1-1 and 1-3). However, certain generalizations can be made about the algae that are of great assistance to students in identifying and classifying specimens.

Structural characteristics are the simplest features to use when identifying unknown algal specimens. Algal organisms range in size from small unicells such as blue-green bacteria, most of which measure 2 to 3 μm in diameter, to large multicellular forms such as the giant kelps,

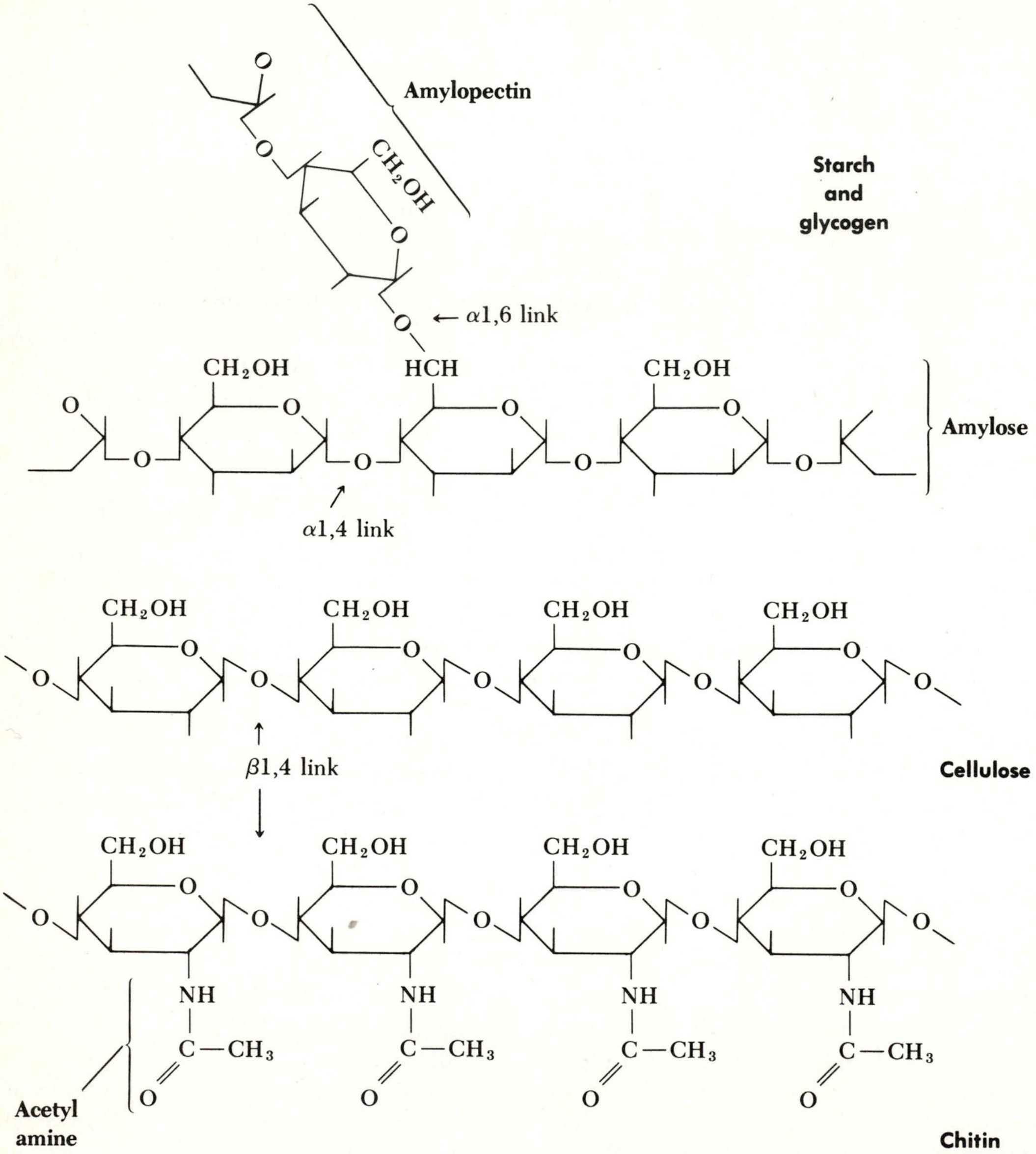

FIG. 1-3 Polysaccharides important to algae and fungi.

TABLE 1-1. Biochemical Characteristics of Algae and Fungi

Division	Chlorophyll types	Accessory pigments	Distinctive cell wall components
Cyanobacteria (blue-green bacteria)	*a*	Phycobilins: Phycocyanin Phycoerythrin Allophycocyanin Carotenes Xanthophylls	Muramic acid Diaminopimelic acid
Chlorophyta (green algae)	*a* and *b*	Carotenes Xanthophylls (including lutein)	Cellulose Sulfated polysaccharides $CaCO_3$, $MgCO_3$
Charophyta (stoneworts)	*a* and *b*	Carotenes Xanthophylls (including lutein)	Cellulose $CaCO_3$
Euglenophyta (euglenoids)	*a* and *b*	Carotenes Xanthophylls	Proteinaceous pellicle
Phaeophyta (brown algae)	*a* and *c*	Carotenes Xanthophylls (including fuco-xanthin)	Cellulose Alginic acid Sulfated mucopolysac-charides
Chrysophyta (golden-brown algae, including diatoms)	*a* and *c*	Carotenes Xanthophylls (including fuco-xanthin)	Silicates Cellulose Some naked
Pyrrhophyta (fire algae, including dinoflagel-lates)	*a* and *c*	Carotenes Xanthophylls (including fuco-xanthin)	Cellulose Some naked
Xanthophyta (grass-green algae)	*a* and *c*	Carotenes Xanthophylls	Cellulose Uronic acids
Rhodophyta (red algae)	*a* and *d*	Phycobilins: Phycocyanin Phycoerythrin Allophycocyanin Carotenes Xanthophylls (including lutein)	Cellulose Sulfated polysaccharides (agar, carrageenan) $CaCO_3$, $MgCO_3$
Kingdom Fungi	None	None	Chitin Some cellulose

which reach lengths up to 70 m (over 200 ft) and weigh up to 250 kg (over 500 lb).

The vegetative plant of the algae is made up of a wide variety of shapes and forms ranging from single cells to complex, three-dimensional cellular assemblages. Motile unicellular organisms are common among the more primitive eukaryotic algae (e.g., *Chlamydomonas*). All other eukaryotic forms are believed to be evolutionary derivatives of motile eukaryotic algae. For example, some motile unicellular forms have been observed settling on a substrate and resorbing their flagella, thus producing a unicellular, nonmotile algal cell. Some common algae are derivatives of this nonmotile condition (e.g., *Chlorella*). Motile cells also have been known to divide mitotically, without complete separation of individual daughter cells, thus forming a multicellular colonial type of organism (e.g., *Vol-*

vox; see Fig. 4-4). These cells usually retain **plasmodesmata** connecting adjacent cells. The cells of other organisms divide mitotically and produce completely separate individual cells. These cells become embedded in a secreted gelatinous matrix, forming a **palmella** type of organism (e.g., *Tetraspora,* Fig. 4-25). Such cells are completely autonomous, unlike the colonial organisms that exhibit some degree of coordination among individual cells.

Many multicellular algae are constructed on a body plan that superficially resembles the found in higher plants. That is, it looks as though the alga in question contains roots, stems, and leaves. However, the function and internal cellular pattern of these structures in the algae are quite different, and they are therefore given separate names: the holdfast, stipe, and blade. The **holdfast,** like the root, is an anchoring organ, but it does not absorb nutrients nor store food products more than any other segment of the plant. The **stipe,** like the stem, holds the organism in an erect position, but it generally does not act as a major transport organ, nor does it store food products. The **blade,** like the leaf, provides a flat surface for photo-

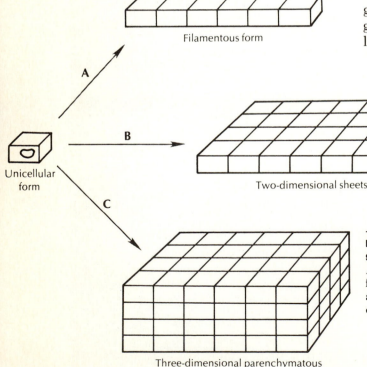

Filamentous form

Unicellular form

Two-dimensional sheets

Three-dimensional parenchymatous

FIG. 1-4 Growth patterns indicating mitotic cell divisions.
A, One dimension only, resulting in filamentous forms; **B,** two dimensions, resulting in sheets of cells; and **C,** three dimensions, resulting in parenchymatous construction.

synthetic activity, but it does not have a high degree of cellular differentiation, nor does it provide for gas exchange with its environment as in leaves of vascular plants.

When eukaryotic cells divide and the individual daughter cells remain attached to each other, the resultant structure may be a filament, a two-dimensional sheet of cells, or a three-dimensional, **parenchymatous** structure (Fig. 1-4). During the course of evolution algal cells have exhibited all four constructions. Increasing complexity from unicellular to the parenchymatous form apparently has accompanied evolutionary development. Flagellate unicells are considered therefore to be the primitive cell type, while all other forms are derived. Unicellular organisms are found in all algal divisions except the brown algae (Phaeophyta). Filamentous forms are also common; examples can be found in all divisions. Two-dimensional organisms whose cells are arranged in sheets are less common, but do occur in most divisions (e.g., *Ulva* from the green algae and *Porphyra* from the red algae). Three-dimensional parenchymatous organisms are not especially common among the algae, although they do appear among the most highly specialized forms from the brown algae (Phaeophyta), red algae (Rhodophyta), and green algae (Chlorophyta). One other growth pattern is found among algae and fungi: the **heterotrichous** habit. Heterotrichy is a term used to describe a filamentous form that is divided by branching into a prostrate filament and an erect portion.

Most freshwater algae produce a dormant or protective stage sometime during their life cycle. These stages may involve the construction of a thick cell wall around the vegetative cell, producing an encapsulated cyst. For example, motile flagellate unicells resorb their flagella and secrete a thickened, cellulosic cell wall in some dinoflagellates and euglenoids. Another type of dormant stage is found when motile algal forms resorb their flagella and produce a palmella stage consisting of nonmotile cells encased in a secreted gelatinous matrix. Members

of the Volvocales (Chlorophyta) often undergo temporary nonmotile periods in palmella stages.

The reproductive structures of algae are always exposed to the immediate environment. That is, eggs and sperm are formed without having any layer of sterile vegetative cells surrounding them (as is the case in higher plants). This means that the gametes are not protected and nurtured by the parent **thallus** (plant body). A notable exception to this is found in the division Charophyta, where both eggs and sperm develop within a complex, multicellular reproductive organ. Other exceptions are found among the red and brown algae.

When egg and sperm cell fuse to form a zygote, the process is called **syngamy** or fertilization. Zygotes thus formed do not form multicellular embryos while they are still attached to the parent plant, as is the case in higher plants. In fact, most algae release their eggs and sperm directly into an aquatic environment, and fertilization and embryonic development occur in open water. The gametes, zygote, and young embryo are independent of the parent thallus from the very beginning.

Another easily identifiable characteristic of algae is the presence or absence of motile flagellate cells. Many algae produce cells possessing flagella at some period during their life cycles. These motile cells may be vegetative (as in many unicellular forms) or reproductive. Flagellate reproductive cells may be the gametes (egg or sperm) or motile spores called zoospores. It is difficult to differentiate between a gamete and a zoospore based upon morphological structure alone. The difference is really physiological, whereby zoospores may develop directly into an adult, while gametes must fuse before subsequent development.

PHYSIOLOGICAL AND BIOCHEMICAL CHARACTERISTICS OF ALGAE

All algal divisions are primarily composed of photoautotrophic organisms that contain chlorophyll *a* and a second type of accessory chlorophyll, either chlorophyll *b*, *c*, or *d* (see Table 1-1). The cells also possess certain types of accessory photoreceptive pigments, including three special groups of light-sensitive biochemical compounds—the carotenes, the xanthophylls,* and the phycobilins. However, some algal divisions also contain colorless heterotrophic members that lack the ability to synthesize chlorophyll and other pigments. They are especially common in fresh water.

The cells of algae may possess a rigid cell wall that is usually composed of cellulose, but there are many exceptions to this rule. Some algae lack cell walls and exist as naked (having only a plasma membrane) protoplasts. Other algae may have cell coverings composed of a thickened plasma membrane called a pellicle (e.g., *Euglena*). Still others are enclosed within ge-

*Carotenes and xanthophylls are both types of relatively simple yellow, orange, or red **carotenoid** pigments. The carotenes are oxygen-free hydrocarbons, while the xanthophylls are their oxygenated derivatives.

latinous coverings called loricas (e.g., *Dinobryon*). Some cellulosic cell walls contain a wide variety of inorganic substances such as calcium carbonate, magnesium carbonate, iron, and silicon. The walls of diatoms (Chrysophyta) are completely siliceous without any trace of cellulose (see Table 1-1).

Calcium carbonate ($CaCO_3$) is a common mineral component of cell walls in many species of algae. It is easily determined from a glance at Table 1-2 that the ability to deposit calcium has arisen a number of times during the course of evolution, since all three lines of algal evolution have developed it. The various algal divisions indicated by Table 1-2 also show that calcium deposition has developed in both warm and cold water in both marine and freshwater environments. The chemical process of biologial calcium deposition is complicated and not entirely understood. It is generally agreed that the process is associated with photosynthesis, since calcification rates increase in light and decrease in the dark. It appears that the final deposition of

TABLE 1-2. Calcareous Algae

Algae	Habitat	CaCO₃ crystal	Example
Division Cyanobacteria	Alkaline fresh water	Calcite	*Rivularia*
Division Chlorophyta Order Caulerpales	Warm, marine shallow-water benthos	Aragonite	*Halimeda, Udotea, Penicillus*
Order Dasycladales	Warm, marine shallow-water benthos	Aragonite	*Cymopolia, Neomeris, Acetabularia*
Division Charophyta Order Charales	Alkaline fresh water and brackish water	Calcite	*Chara*
Division Phaeophyta Order Dictyotales	Warm, marine shallow-water benthos	Aragonite	*Padina*
Division Chrysophyta Order Prymnesiales	Warm water marine plankton	Calcite	Coccolithophorids
Division Rhodophyta Order Nemaliales	Warm water marine benthos	Aragonite	*Liagora, Galaxaura*
Order Crypto-nemiales	Warm to cool water marine benthos	Calcite	Coralline red algae, including *Corallina, Amphiroa, Porolithon, Lithothamnium, Goniolithon, Melobesia*

calcium results from a combination of biological activity (photosynthesis) in conjunction with the physical chemistry of waters already high in carbonate and calcium ions.

It is well known that many surficial waters in the marine environment have high concentrations of both calcium and carbon dioxide. Tropical surface waters are often supersaturated with calcium carbonate, giving way to undersaturation at depths ranging from a few hundred meters to nearly 1 kilometer. In contrast to many minerals, calcium carbonate (whether in the form of calcite or aragonite) decreases in solubility in warmer waters. This helps to explain the supersaturation of tropical seas and undersaturation in arctic regions (Davis, 1973). The solubility, however, will increase with increasing salinity, thus raising the concentration of carbon dioxide dissolved in the water and decreasing the pH. Supersaturation of these waters can be demonstrated by "seeding" the water with solid crystalline samples of calcium carbonate. Such a reaction is expressed by the following:

$$Ca^{++} \text{ (excess)} + CO_3^= \text{ (excess)} +$$
$$CaCO_3 \text{ (seed)} \rightleftharpoons 2\ CaCO_3 \text{ (solid)}$$

Actually this reaction involves a number of intermediate steps. The several reactions involved represent (1) the dissolution of carbon dioxide in water to form carbonic acid, (2a and 2b) the first and second ionization of carbonic acid, and (3) the dissolution of solid calcium carbonate.

(1) $H_2O + CO_2 \rightleftharpoons H_2CO_3$

(2a) $H_2CO_3 \rightleftharpoons H^+ + HCO_3^-$

(2b) $HCO_3^- \rightleftharpoons H^+ + CO_3^{2-}$

(3) $CaCO_3 \rightleftharpoons Ca^{2+} + CO_3^{2-}$

These reactions may be combined for convenience into either of two forms:

(4) $CaCO_3 + H_2O + CO_2 \rightleftharpoons Ca^{2+} + HCO_3^-$

(5) $CaCO_3 + H^+ \rightleftharpoons Ca^{2+} + HCO_3^+$

It is evident from reaction (4) that adding water or carbon dioxide to a solution will cause calcium carbonate to dissolve, while removing water or carbon dioxide will cause the reaction to move toward the left, precipitating calcium carbonate. The removal of water by evaporation leads to formation of evaporitic limestones and possibly some cave deposits, while spontaneous degassing as a result of increasing temperatures might induce inorganic precipitation of calcium carbonate. Reaction (5) clearly shows the importance of pH in carbonate solution-precipitation reactions. Increased hydrogen ion activity (lower pH) will cause reaction (5) to move to the right, dissolving calcium carbonate, while increased pH (reduced hydrogen ion activity) will favor precipitation of the solid (a move to the left).

Since marine waters are slightly alkaline (pH 8.0 to 8.4), the tendency is for most of the carbon dioxide in the oceans to exist as the carbonate ion (HCO_3^-), as seen in equation 2a. The hydrogen (H^+) is buffered in the ocean by a complex series of carbonate and silicate reactions (Garrels and McKenzie, 1971).

Digby (1977 a, b) has attempted to explain how biological deposition of calcium carbonate is accomplished through pH changes caused by photosynthesis. His theory is explained in a three-step process:

(1)   Free elections from photosynthetic reactions combine with bicarbonate ions.

$$2HCO_3^- + 2e^+ \rightleftharpoons 2H^+ + 2CO_3^=$$

(2)   The carbonate ions diffuse from the site of photosynthesis to the cell's surface, where they react with water, forming hydroxyl ions.

$$2CO_3^= + H_2O \rightleftharpoons 2HCO_3^- + 2\ [OH^-]$$

(3)   The unstable free hydroxyl ions cause a rise in pH at the cell's surface that in turn precipitates calcium carbonate from the sea water.

$$2CO_3^= + 2Ca^{++} \rightleftharpoons 2CaCO_3$$

The free hydroxyls quickly react with available hydrogen to form water. The precipitated calcium carbonate is held in place by organic compounds, which form a matrix around the mineral (Lewin, 1962).

While Digby's theory helps us to address the problem of calcification, it also presents some interesting questions: Why don't all algae that photosynthesize in marine waters also deposit calcium carbonate? How is it possible for different minerals and different compositions of carbonates to be deposited by various algal species? Are the mechanisms for calcium deposition the same in all freshwater and saltwater species? Could not the same pH changes, brought about by the electron transport system in Digby's theory, be accomplished by the uptake of CO_2 in equation (4), thus causing $CaCO_3$ to precipitate through a type of degassing process? And finally, what is the role of enzyme-mediated reactions in the process? It is highly likely that specific enzyme systems function in these reactions similar to the process found in some mammalian cells in which the enzyme carbonic anhydrase facilitates the carriage of carbon dioxide in blood.

$$CO_2 + H_2O \overset{\text{carbonic anhydrase}}{\rightleftharpoons} H_2CO_3$$

Specific enzymes would help to explain why certain algae have the ability to deposit calcium carbonate while others do not, and why only certain parts of some algae have this ability.

Two crystalline forms of calcium carbonate are deposited in algal cell walls (see Table 1-2). **Calcite** crystals belong to the rhombohedral (hexagonal system) class of crystals, while **aragonite** crystallizes in the orthorhombic class. They differ from each other not only in their geometry, but also in their physical properties such as melting point, solubility, hardness, and specific gravity. Calcite in algal walls is more common in cooler waters, while aragonite prevails in warmer tropical environments. Calcite is softer than aragonite and is less dense and less soluble in water. Mixtures of calcite and aragonite are not found among the algae as they are among calcifying invertebrates. However, minerals other than calcium often are found as carbonates in algal cell walls. Magnesium carbonate may comprise 7% to 30% of calcite deposits in some red algae. In aragonitic walls magnesium carbonate seldom exceeds more than 1% of the mineral deposition (Chave, 1954). Strontium carbonate is more common in aragonitic walls, although seldom reaching concentrations of more than 2%. Magnesium hydroxide may be present in high-magnesium calcite deposited by some algae.

The organic matrix within which the mineral carbonates are deposited is composed of gelatinous or mucilaginous substances that form an integral part of the cell wall. It is not known whether these organic molecules play a role in carbonate deposition. However, it has been demonstrated that when the cell dies or is killed, calcium carbonate loss to the environment is increased, thus demonstrating that metabolic processes are important in cell wall maintenance. The various polysaccharides associated with calcium deposition in cell walls are reviewed by Pentecost (1980).

Macromolecular substances in algae and fungi often provide important clues to the phylogenetic relationships of specific organisms. Energy storage products are always mentioned in characterizing the various divisions (Table 1-3). Most of the molecules are polymers (chains) of simple sugars that form polysaccharides of one type or another, although alcohols, lipids (fats and oils), and even proteins may be stored in some cells. Polysaccharides also form important structural molecules in most cells, since the cell walls of algae, fungi, and vascular plant cells are all formed from them.

The most important polysaccharides are starch, glycogen, cellulose, and chitin (Fig. 1-3). Each of these important sugars is formed by a linear arrangement of glucose molecules. However, the manner by which individual glucose units are joined is different—sometimes in the nature of the chemical bond and sometimes in the degree of branching. These variations alter the physical-chemical properties of the molecule, allowing glucose units to be used for various functions in the same cell.

Glycogen and starch are similar in structure.

TABLE 1-3. Energy Storage Molecules

Division Cyanobacteria (blue-green algae)	α 1-4, 1-6 Glucose polymers similar to glycogen (also called myxophycean starch)
Division Chlorophyta (green algae)	α 1-4, 1-6 Glucose polymers (starch)
Division Charophyta (stoneworts)	α 1-4, 1-6 Glucose polymers (starch)
Division Euglenophyta (euglenoids)	β 1-3 Glucose polymers (paramylon)
Division Phaeophyta (brown algae)	β 1-3, 1-6 Glucose polymers (laminarin)
Division Chrysophyta (golden-brown algae)	β 1-3 Glucose polymer with two β 1-6 glucose polymer chains (chrysolaminarin)
Division Pyrrhophyta (fire algae)	α 1-4, 1-6 Glucose polymers (starch)
Division Xanthophyta (grass-green algae)	(Both paramylon and chrysolaminarin reported)
Division Rhodophyta (red algae)	α 1-4, 1-6 Glucose polymers (floridean starch) similar to glycogen
Kingdom Fungi	α 1-4, 1-6 Glucose polymers (glycogen)

Both are water soluble. Generally glycogen is known as "animal starch," although it is also found in fungi, and starch is known as "plant starch" because it is found in all the higher plants and green algae. Both molecules function as energy storage products in cells. The different algal divisions each utilize a specific glucose polymer as their energy storage compound, such as paramylon, laminarin, chrysolaminarin, and floridean starch (Table 1-3).

Starch differs from glycogen in the degree of branching of glucose polymers. Starch is a high-molecular-weight molecule composed of subdivisions of lesser weight, amylose and amylopectin. Amylose is an unbranched polymer of glucose having α 1-4 linkages. **Amylopectin** is a branched polymer of glucose units similar to glycogen but not so highly branched. The amylopectin polymers are attached to the amylose polymers with α 1-6 linkages (Fig. 1-3).

Glycogen is a high-molecular-weight molecule of α 1-4 glucose units containing frequent α 1-6 glucose unit branches, each of which begins a new α 1-4 linked chain of 8-12 glucose units. Glycogen is not found in algal cells, although some blue-green algae contain molecules that are very similar. The red algae also contain an energy storage product, floridean starch, which has some similarity to glycogen. All the fungi store glycogen.

Cellulose and chitin are structural polysaccharides. They are both insoluble high-molecular-weight polymers of glucose. Most algal and higher plant cell walls are composed of cellulose, while the fungi contain chitin. The difference between these polysaccharides and soluble starches and glycogens is the nature of the chemical bond between glucose units. Cellulose and chitin contain a different stereochemical bond (β bond) than the others (an α bond). The result of this is the degree of twisting in the polymers. The β linkage produces a linear unbranched molecule, while the α link causes a helical arrangement with six glucose units per turn. The β link allows adjacent linear polymers to lie close to each other, leaving no intra-molecular spaces and closing off the free hydroxyl groups (-OH) of the glucose units from surrounding solvents. Such an arrangement imparts considerable structural rigidity to the molecule, allowing for these substances to be used in cell walls.

The α bond, on the other hand, with its attendant helix, exposes the glucose hydroxyl groups for reaction with surrounding solvents and enzymes. The looseness of this structure allows for open spaces within the molecule and accounts for the solubility and flexibility of starch and glycogen.

The distinction between the α and β linkages also has significance for humans and other organisms that attempt to eat plant materials. Humans produce enzymes capable of breaking α 1-4 linkages, but they have no effect on β 1-4 links. The hydrolysis (breaking) of β links requires a special enzyme, cellulase. Some fungi, protozoans, and bacteria synthesize cellulases and can therefore break down plant cell walls. Other organisms are incapable of this.

ENVIRONMENTAL CHARACTERISTICS OF ALGAE

Algae also possess some general ecological traits that are helpful in characterizing them as a coherent, related group of living organisms. First of all, a great majority of algae are aquatic. Some are capable of existing in both freshwater and saltwater habitats, although most are restricted to one or the other. There are also some terrestrial algae, which can be found growing in such diverse places as the soil, on or in rocks, or even as epiphytes existing on the bark of trees.

Second, algae have a wide range of environmental tolerance and therefore have the capability of invading most habitats. Algal cells have been found growing in temperatures ranging from zero to 90° C, in salinities ranging from zero to 60 °/oo (parts per thousand), in pH ranges from 2 to 10, in oxygen tensions ranging from

saturation to zero (some algae are anaerobic), in nitrate-deficient environments (some algae can fix nitrogen), and in waters polluted with organic substances (some algae are heterotrophic and are capable of extracting energy from the chemical bonds of organic molecules).

Third, algae are usually considered to be an **autotrophic** group, requiring only light and inorganic substances (mainly nitrogen, phosphorus, potassium, carbon dioxide, and water) in order to maintain themselves. There are, however, many exceptions to the autotrophic existence, and most algal divisions contain some species that live heterotrophically. **Heterotrophs,** unlike autotrophs, require an external source of energy that is usually in the form of organic molecules. They then derive energy from the chemical bonds of the organic molecules rather than from the sun, as autotrophs do. Some categories of organisms lie somewhere between the purely autotrophic and completely heterotrophic forms; these organisms are called **mixotrophs** (or facultative heterotrophs) and are capable of both photosynthesis and deriving energy from the chemical bonds of organic molecules. Another group of photosynthetic organisms is referred to as **auxotrophic,** because these organisms require small amounts of preformed organic molecules that act as coenzymes in their metabolism. These molecules are usually vitamins.

Other modes of nutrition among the algae are exhibited by specialized heterotrophs that obtain energy from dead organic matter. These organisms are called **saprotrophs.** The colorless algae, bacteria, protozoans, and fungi are often saprotrophic. Some other specialized heterotrophs exhibit symbiosis, whereby two unrelated organisms live in close association with each other. There are three types of symbiosis, all of which occur among nonvascular plants. The first, parasitism, occurs when one symbiont, the parasite, derives its energy and nutrients from the other, the host. A second type, mutualism, is encountered when both organisms derive equal benefit from their symbiotic association. The third type, commensalism, is found when one individual benefits from the relationship, while the other is physically unharmed. Symbiotic associations including parasitism, mutualism, and commensalism may be seen among the algae and fungi.

Some algae use other plants and algae as substrates on which to grow. These algae are called **epiphytes** (fr. Greek *epi* = upon + *phytes* = plant). If the alga uses a member of the animal kingdom on which to grow, it is called an **epizoic** (fr. Greek *epi* = upon + *zoon* = animal). Many of the smaller algae grow epiphytically upon the larger algae, and epizoics are common among aquatic vertebrates and invertebrates. A blue-green epizoic even grows on the fur of the South American sloth, causing the animal to appear green during the hot, rainy season.

ALGAL EVOLUTION

The algae may be divided into three separate lines of evolutionary ascent based upon considerations of their biochemical and ultrastructural characteristics. The green line of algal evolution has cells that contain **chlorophyll** *a* and *b*. The brown line organisms are characterized by **chlorophyll** *a* and *c,* and the red line by **chlorophyll** *a* and *d* (Table 1-1 and Fig. 1-5).

One of the more interesting problems associated with algal evolution is the derivation of

FIG. 1-5 Evolution of algae.

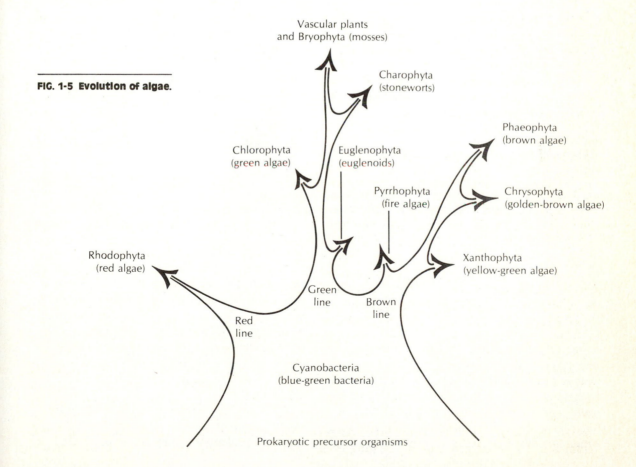

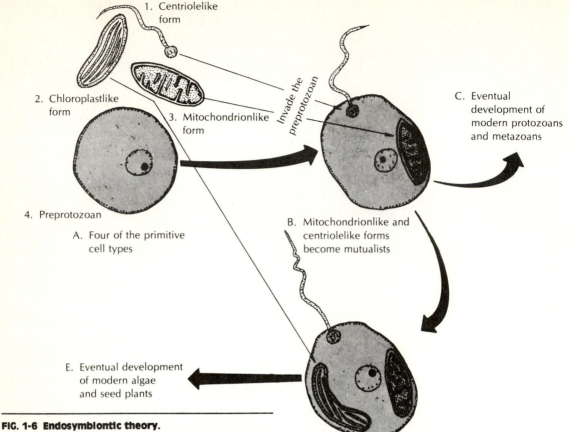

1. Centriolelike form

2. Chloroplastlike form

3. Mitochondrionlike form

Invade the preprotozoan

C. Eventual development of modern protozoans and metazoans

4. Preprotozoan

A. Four of the primitive cell types

B. Mitochondrionlike and centriolelike forms become mutualists

E. Eventual development of modern algae and seed plants

D. Prealgal-type develops after chloroplastlike form becomes mutualistic

FIG. 1-6 Endosymbiontic theory.
This concept of how complex (eukaryotic) cells may have arisen suggests that originally a number of simple types arose, *A*. These united in several combinations through development of mutualistic relations. Combination of the centriolelike and mitochondrionlike types with the preprotozoan resulted in the ancestral stock of protozoans and animals, *B* and *C*. Then, after a chloroplastlike organism had invaded this stock, *D*, algae and higher plants gradually evolved *E*. (From Dillon, L.S. 1978. Evolution: concepts and consequences, ed. 2, St. Louis, The C.V. Mosby Co.)

eukaryotic cells. Klein (1970) has proposed an evolutionary process whereby the outer membranes formed folds, developing internal membrane-bound organelles. Others, such as Margulis (1971, 1981), have proposed that eukaryotes evolved by prokaryotic cells ingesting other smaller prokaryotes. For example, a large

heterotrophic bacterium might ingest a small photosynthetic blue-green cell, thus establishing a mutually beneficial relationship. The blue-green cell would initially have been an endosymbiotic cell growing within the bacterium. Evidence in support of this endosymbiotic theory for eukaryotic evolution is found in the discovery of DNA within organelles such as chloroplasts and mitochondria (Dillon, 1978; see Fig. 1-6).

The search continues for a satisfactory intermediate type of organism that would fill the gap between prokaryotes and eukaryotes. One or-

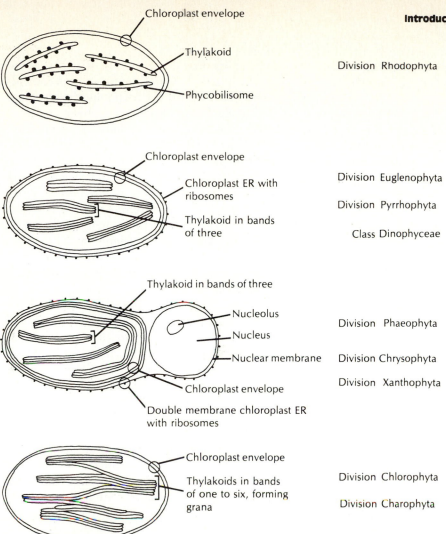

Division Rhodophyta

Division Euglenophyta

Division Pyrrhophyta

Class Dinophyceae

Division Phaeophyta

Division Chrysophyta

Division Xanthophyta

Division Chlorophyta

Division Charophyta

ganism, *Cyanidium caldarium*, resembles such an intermediary, since it combines characteristics of blue-green, red, and green algae. This organism is classified in this text as a blue-green bacterium because of the presence of characteristic pigments, the phycocyanins. However, electron micrographs of *Cyanidium* have revealed a membrane-bound nucleus, a chloroplast, and numerous mitochondria. Therefore it is likely that this organism represents a transitional form between the prokaryotic Cyanobacteria and the higher eukaryotic algae such as those found in the Rhodophyta (Fredrick, 1980).

FIG. 1-7 Chloroplast ultrastructural types.
Ultrastructural features of chloroplasts from the three lines of algal evolution. At top is a chloroplast from a red alga. Note presence of phycobilisomes (similar to those found in blue-green bacteria) and double-membrane chloroplast envelope. Bottom figure shows chloroplast from a green alga or charophyte. Note presence of thylakoid bands in numbers from one to six. Also note double-membrane chloroplast envelope. Middle two chloroplasts represent early and derived members of the brown line of evolution and euglenophytes from the green line. Note addition of single chloroplast ER (CER) membrane around each chloroplast in Euglenophyta and Pyrrhophyta; also note addition of two chloroplast ER membranes in derived divisions of the brown line, with outer membrane containing the nucleus.

The structure, biochemistry, and physiology of *Cyanidium* present a challenge to evolutionary biologists.

Characterization of each algal division is based upon distinctive biochemical features and the ultrastructural traits of chloroplasts, eyespots, and flagella. Table 1-1 lists some of the special biochemical components associated with cell walls and photosynthesis (chlorophyll types and accessory pigments). Each of these items will be discussed in succeeding chapters on individual algal divisions.

Electron microscope studies have revealed certain structural features that indicate evolutionary relationships among the algae (Dodge, 1973). The nature of the chloroplast structure and its position relative to the nucleus, pyrenoid, and endoplasmic reticulum have phylo-

genetic significance (Fig. 1-7). Flagellar ultrastructure (Fig. 1-8) and the position of individual flagella (Fig. 1-9) are both important features used in algal identification and phylogeny. And finally, the structure and location of eyespots provide an important consideration when establishing algal relationships and phylogeny.

Chloroplasts in eukaryotic algae are always bound by a double membrane. Within the inner membrane is a liquid **stroma** or matrix in which is suspended a series of flat, membranous disks called **thylakoids.** The arrangements of the thylakoids vary from one group of algae to another. In primitive prokaryotic algae, thylakoids occur as single membranes (e.g., Cyanobacteria). A progression may be followed in which thylakoids are arranged in bands of two, three, and finally into multiples up to six thylakoids per band, forming a **granum** (as in Chlorophyta and Charophyta). Associated with the thylakoids are the light-sensitive pigments, including the various chlorophylls, carotenes, xanthophylls, and phycobilins. In the blue-green and red algae the phycobilins are incorporated into proteinaceous structural units called **phycobilisomes** (Fig. 1-7).

Chloroplasts in the euglenoids and all divisions in the brown line of evolution possess an additional membrane or two around the outside of the plastid (Fig. 1-7). These membranes are called the **chloroplast endoplasmic reticulum** (**chloroplast ER** or **CER**). The outside CER

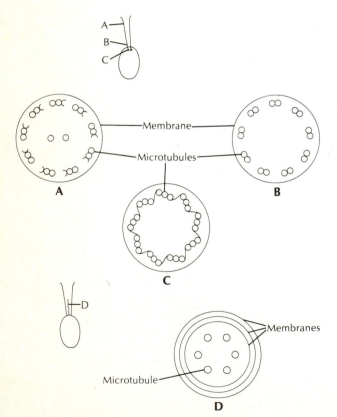

A

B

Membrane

Microtubules

C

D

Membranes

Microtubule

D

FIG. 1-8 Cell appendage ultrastructure.
Comparative view of flagella at different levels along long axis of its structure. **A,** Flagellum in cross section midway along its length. This is the classic view of the flagellum with its typical 9-2 arrangement of protein microtubules. Outer circumferential microtubules may vary from doublets (pairs) to triplets; central microtubules always occur singly. **B,** Flagellum in transition zone just outside cell wall. Note loss of the two central microtubules. **C,** Flagellum from the basal body region showing microtubules occurring in triplets. **D,** Haptonema cross section from Prymnesiophyceae. Note three outer membranes and six central microtubules.

membrane may have ribosomes attached to its external surface. In certain divisions the outer CER also surrounds the nucleus.

Within the stroma of the chloroplast may be found various other biochemical and structural features, including proteinaceous pyrenoids, ribosomes, strands of DNA, deposits of lipid material, and various polysaccharides.

The length, position, and ultrastructure of flagella also provide a significant taxonomic tool. Electron micrographs have revealed that flagella of eukaryotic cells differ significantly from those of prokaryotic cells (bacteria) because they have an internal organization composed of proteinaceous microtubules. These microtubules, referred to as the axoneme, form an axial complex that extends the length of the flagellum and into the cytoplasm of the cell. The precise arrangement of the microtubules varies along the axis of the flagellum (Fig. 1-8). At the base of each flagellum is an enlarged area called the basal body. Microtubules emerge from the basal body and extend into the cytoplasm, forming microtubular roots that are especially important in showing phylogenetic relationships among the green algae (Chlorophyta; see Fig. 4-8).

Another ultrastructural feature of flagella is the presence or absence of **mastigonemes** (hairs) and scales on the axial surface of the flagellar membrane. When mastigonemes are present, the flagellum is said to be of a tinsel type. When the surface is smooth, the flagellum is called a whiplash type. Some cells contain both whiplash- and tinsel-type flagella (e.g., the dinoflagellates and unicellular members of Xanthophyta). The presence or absence of mastigonemes is detectable only with the electron microscope.

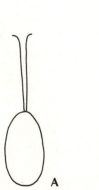

FIG. 1-9 Flagella positions.
Representative positions of emergent flagella among the algae. **A,** Two isokont flagella of *Chlamydomonas.* **B,** Two heterokont flagella of *Dinobryon;* note presence of a special cell covering, the lorica. **C,** Two isokont flagella and a special median appendage, the haptonema, in *Prymnesium.* **D,** Two heterokont flagella in a sperm cell from a brown alga such as *Fucus.* **E,** Stephanokont flagella from the sperm cells of *Oedogonium.* **F,** Two heterokont flagella of *Gonyaulax;* note that one flagellum is circumferential while the other trails behind. **G,** Single emergent flagellum of *Euglena. Euglena* possesses another much shorter flagellum that is nonemergent.

Flagellar features that can be seen with the light microscope, and are therefore more useful to most students, include the number of flagella present, the length of the flagella, and their position of emination from the cell surface (Fig. 1-9). Most motile algal cells are biflagellate (with two flagella; e.g., *Chlamydomonas*). Some, however, have a single emergent apical flagellum (e.g., *Euglena*); others may be quadriflagellate (with four flagella). This condition is especially common where diploid motile cells are concerned.

Flagella of equal length are said to be **isokont** (fr. Greek *iso* = same + *kontos* = pole). When the flagella are of unequal lengths, they are said to be **heterokont** (fr. Greek *hetero* = different). If many flagella form a ring around the cell, they are said to be **stephanokont** (fr. Greek *stephanos* = crown). Only a few genera produce stephanokont cells (e.g., *Oedogonium*).

Flagella may emerge from various positions on algal cells. They may eminate from an **apical** position at one end of the cells (e.g., *Chlamydomonas*), a subapical position as in the motile cells of the brown algae (Phaeophyta), or from a lateral position as in the dinoflagellates (Pyrrhophyta).

In addition to the flagella produced by motile cells, a different sort of appendage is formed in some algae belonging to the division Chrysophyta. The family Prymnesiophyceae (coccolithophorids) produces a special flagellum-like appendage called an haptonema (Fig. 1-8). Its internal structure differs from that of a flagellum in that it has only six or seven microtubules, which are surrounded by three external membranes, as indicated by Manton (1964).

Another ultrastructural feature that figures significantly in the development of phylogenetic relationships among the algae is the structure and position of the light-sensitive red eyespot (stigma). This structure is easily identified with the light microscope as a red-colored spot within the cells of many motile algae. However, to identify its precise position or its ultrastructure, electron micrographs are required. Dodge (1973) has classified and described the eyespots of the major algal divisions. In the green algae (Chlorophyta) the eyespot forms a part of the chloroplast but is not associated with the flagella. In the Chrysophyta, Phaeophyta, and some Xanthophyta, the eyespot is a part of the chloroplast and is closely associated with the basal part of one of the flagella. In the euglenoids (Euglenophyta) and some Xanthophyta, the eyespot is independent of the chloroplasts and is located in close apposition to the basal portion of one of the flagella. In the dinoflagellates (Pyrrhophyta), the eyespot structure ranges from small and within the chloroplast to large and external to the chloroplast. In some dinoflagellates, the structure forms a complex light receptor called an ocellus, which includes a lens (Dodge, 1973).

ALGAL REPRODUCTION

The study of reproduction among algae is largely a matter of mastering a few basic concepts and applying these to individual life cycles. The more primitive algae tend to be single-celled organisms whose sexual reproduction is **isogamous** (fr. Greek *iso* = equal + *gamete* = wife or husband), thus producing eggs and sperm that are morphologically identical (Fig. 1-10). Because males and females are identical in form at this level, they may be designated as plus (+) or minus (−).

As algae became more complex during the course of evolution they developed a type of sexual reproduction designated as **anisogamous** (fr. Greek *aniso* = unequal). The gametes produced were morphologically distinct, with the larger gamete being designated the female and the smaller gamete the male. In anisogamy both gametes are usually, but not necessarily, motile.

The most evolutionarily advanced condition is **oogamous** (fr. Greek *oon* = egg), with a very large, nonmotile egg cell and a small, motile sperm cell. The sperm cells are often so small that they are difficult to see with the light microscope.

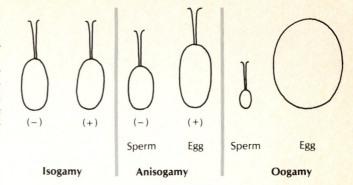

FIG. 1-10 Evolution of sex.
The three types of sexual gametic union are represented: isogamy, anisogamy, and oogamy. Sequence represents an evolutionary development from the earliest algae, which were isogamous, to the derived forms, culminating in the oogamous condition.

Isogamy, anisogamy, and oogamy (Fig. 1-10) are gametic conditions that are found in association with the three basic life cycle types—**haplontic, diplohaplontic,** and **diplontic** (Fig. 1-11). The most primitive life cycle is the haplontic one in which the organism spends most of its existence as a haploid adult. The only diploid phase is the zygote. This type of life cycle is characterized by **zygotic meiosis,** which means that the first two nuclear divisions of the zygotic nucleus are meiotic divisions, and the product is always four haploid nuclei. Nuclear division is usually followed by cytoplasmic division (**cytokinesis**). Haplontic individuals are common among all Moneran organisms (prokaryotic organisms) and many of the unicellular algae and the primitive multicellular algae. Fungi are also frequently haplontic.

Diplohaplontic and diplontic life cycles are derived from the haplontic cycle through a delay in the timing of meiosis, thus providing for a longer period of diploid existence. In diplohaplontic life cycles, the diploid zygotic nucleus does not undergo meiosis. Instead it divides by mitosis and by cytokinesis, producing diploid cells that form a diploid organism called the **sporophyte.** Its major reproductive function is the production of spores. **Spores** are differentiated from **gametes** by their ability to germinate and undergo further development without prior fusion with another cell (see the glossary). In diplohaplontic individuals, the spores produced

are haploid, because they are formed by meiotic cell divisions (and are therefore called **meiospores**). Diploid **mitospores** (produced by mitosis and cytokinesis) may also be produced as a vegetative means of reproduction (see Fig. 1-11). The meiospores germinate by mitotic cell divisions, producing a haploid organism called the **gametophyte.** Its major reproductive function is the production of gametes. This type of life cycle is characterized by two things: (1) **sporogenic meiosis,** which means that the meiotic cell divisions produce spores (meiospores) instead of gametes; and (2) the alternation of haploid and diploid adult generations. These cycles are common to all the higher plants, all the slime molds (Myxomycota), most of the brown algae (Phaeophyta), and red algae (Rhodophyta), and they are widespread among the rest of the algal and fungal divisions.

An added degree of complexity occurs in the diplohaplontic life cycles, when the haploid and diploid adult generations are identical in appearance. Such conditions are said to have **isomorphic** (fr. Greek *isos* = equal + *morph* = form) **alternation of generations** (e.g., some of the brown algae such as *Ectocarpus* or *Dictyota*). When the two generations are morphologically distinct it is termed a **heteromorphic** (fr. Greek *hetero* = other) **alternation of generations** (e.g., the giant kelps).

The diplontic life cycle (Fig. 1-11) is the most advanced condition. These organisms spend

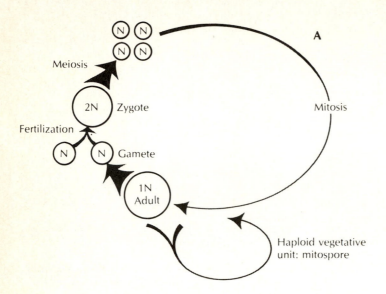

Meiosis

2N Zygote

Fertilization

N Gamete

1N
Adult

Mitosis

Haploid vegetative
unit: mitospore

A

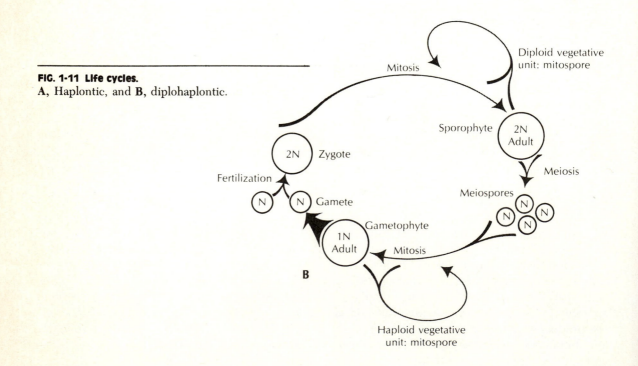

FIG. 1-11 Life cycles.
A, Haplontic, and **B,** diplohaplontic.

Mitosis

Diploid vegetative
unit: mitospore

Sporophyte

2N
Adult

Meiosis

Meiospores

2N Zygote

Fertilization

N Gamete

1N
Adult

Gametophyte

Mitosis

B

Haploid vegetative
unit: mitospore

most of their lives as diploid individuals, with the only haploid phase being the gamete. The major characteristic of this type of life cycle is **gametogenic meiosis,** which means the production of gametes by meiotic cell divisions. Diplontic life cycles are common to all animals, but rare among algae and fungi, although they can be found in some diatoms (Chrysophyta), some of the Fucales (Phaeophyta), and some Chlorophytes (e.g., *Codium*).

SELECTED REFERENCES

Bold, H.C. and M.J. Wynne. 1978. Introduction to the algae: structure and reproduction. Prentice-Hall, Inc., Englewood Cliffs, N.J. 706 pp.

Davis, R.A. 1973. Principles of oceanography. Addison-Wesley Publishing Co., Inc. Reading, Mass.

Dillon, L.S. 1978. Evolution, concepts and consequences. Ed. 2. The C.V. Mosby Co., St. Louis.

Dodge, J.D. 1973. The fine structure of algal cells. Academic Press, Inc., New York. 261 pp.

Fritsch, F.E. 1945. The structure and reproduction of the algae. Vol. II. Cambridge University Press, Cambridge, England. 939 pp.

Lee, R.E. 1980. Phycology. Cambridge University Press, Cambridge, England. 478 pp.

Margulis, L. 1981. Symbiosis in cell evolution. Life and its environment on the early earth. W.H. Freeman & Co., San Francisco. 419 pp.

Pentecost, A. 1980. Calcification in plants. Int. Rev. Cytol. **62:**1-28, G.H. Bourne and J.T. Danielli (eds.) Academic Press, Inc., New York.

Stewart, W.D.P. (ed.) 1974. Algal physiology and biochemistry. Botanical Monographs, Vol. 10. University of California Press, Berkeley. 989 pp.

FIG. 1-11, cont'd Life cycles.
C, Diplontic.

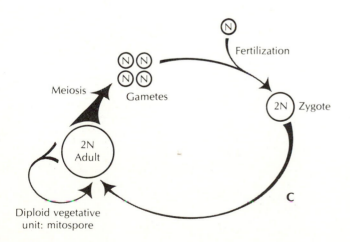

2

The distribution and abundance of algae in various habitats of the world are governed by the location of the particular ecosystem and many other physical, chemical, and biological factors. Numbers of algae may be controlled by either physical or biological processes: physical factors (wind, wave action, temperature, and light) generally influence algal populations more in temperate and arctic areas, while biological factors (competition, parasitism, and predation) usually regulate populations in the more physically stable tropical environments (Odum, 1971).

Ecosystems, whether marine, freshwater, or terrestrial, are composed of biotic (living) and abiotic (nonliving) components. The living components are (1) the **producers** of energy, the autotrophic green plants that convert solar energy to chemical energy; (2) the **macroconsumers,** the heterotrophic animals that eat the producers; and (3) the **decomposers** or **microconsumers** (primarily heterotrophic bacteria and fungi) that recycle or transform organic material into inorganic material that is usable by the producers. Nonliving components are (1) inorganic nutrients such as orthophosphate, nitrate, carbon dioxide, and water; (2) nonliving organic materials such as proteins, carbohydrates, lipids, and **humic** substances; and (3) climatic factors such as precipitation, temperature, and sunlight. The interaction of all six of these components makes each individual ecosystem unique in function, energy flow, nutrient distribution, flora, and fauna.

The role of algae in the ecosystem is that of producer: algae convert CO_2 by light energy from the sun via photosynthesis into the chemical energy of carbohydrates (and other organic materials) usable by the macro- and microconsumers of the ecosystem. The equation for photosynthesis shows that the process requires carbon dioxide, water, sunlight, and nutrients in order to make high-energy-containing molecules of carbohydrate.

$$1,300,000 \text{ kcal light energy} + 106 \text{ } CO_2 + 90 \text{ } H_2O +$$
$$16 \text{ } NO_3 + 1 \text{ } PO_4 + \text{minerals} =$$
$$13,000 \text{ kcal potential energy in } 3258 \text{ g}$$
$$\text{protoplasm}^* + 154 \text{ } O_2 +$$
$$1,287,000 \text{ kcal heat energy dispersed}$$

Note the need for water, carbon dioxide, trace minerals, and nutrients in the photosynthetic equation. A shortage of any one of these essential materials may inhibit the photosynthetic process and reduce growth or reproduction of the plant. Note also that only 1% of energy from the sun is "fixed" into chemical energy by photosynthesis, which is not a very efficient process.

Algae contain energy-rich organic molecules. The energy in these molecules is transferred either to animals grazing upon the algae or to the decomposers processing dead algal cells. The algae will also use a certain proportion of the energy for their own metabolism. However the transfer occurs, the energy from the organic molecules is used for the growth and reproduction of the **consumers** or decomposers. Ultimately the remaining energy is passed from the plant-eating **herbivores** or the **detritus**-eating decomposers to the meat-eating secondary or tertiary consumers (**carnivores**) at the top of the **food pyramid.** Such a food pyramid is seen in Fig. 2-1, where each block represents the kilocalories of available energy. Each step of the pyramid is called a **trophic level;** thus the producers are placed in the first trophic level, the primary consumers or herbivores on the second trophic level, the secondary consumers or carnivores on the third level, and so on.

The decomposers recycle energy from dead organic molecules at each trophic level. Decomposers may also be called **transformers,** because they transform organic molecules into inorganic molecules. In some ecosystems a fourth or fifth trophic level may be present, represented by secondary or tertiary carnivores. A loss of energy occurs because (1) the organisms of each

*The 3258 g of protoplasm contains 106 C, 180 H, 46 O, 16 N, 1 P + 815 g mineral ash. (Equation from Odum, 1971.)

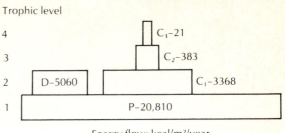

Trophic level

Energy flow: kcal/m²/year

FIG. 2-1 Energy pyramid for a spring in Florida.
The primary producers (*P*) are primarily attached algae and aquatic vascular plants. Primary consumers (C_1) (herbivores) are mainly aquatic invertebrates and herbivorous vertebrates (fish and turtles). Secondary consumers (C_2) are small carnivorous invertebrates and vertebrates; and tertiary consumers (C_3) are large carnivorous vertebrates (fish and birds). Decomposers (*D*) transform organic matter (dead plants and animals, waste products) into inorganic compounds utilized by producers. (From Odum, E.P.: Fundamentals of ecology, ed. 3. Copyright © 1971 by W.B. Saunders Co. Copyright 1953 and 1959 by W.B. Saunders Co. Reprinted by permission of Holt, Rinehart & Winston, CBS College Publishing, New York.)

level must use a substantial amount of energy for their growth and reproduction, and (2) some of this energy is dissipated as unusable heat energy. Note in the diagrammed ecosystem of a spring in Florida (Fig. 2-1) that considerable energy (24% = 5060 kcal) from the producers is channeled to the decomposers. The amount of producer energy channeled to the herbivores and decomposers varies with each ecosystem. Fig. 2-2 shows the components of an aquatic ecosystem—a small pond.

Depending on the ecosystem, there may be three, four, or even five trophic levels. The fewer the trophic levels, the greater the amount of energy available for the organisms at the top of the pyramid. Aquatic ecosystems may have up to five levels, but terrestrial ecosystems generally have fewer. (See Gates, 1971, for additional discussion of energy flow through ecosystems.)

LIMITING FACTORS

The amount of energy entering an ecosystem is determined by the amount of sunlight reaching the producers and by the nutrients available to the producers for photosynthesis (see photosynthetic equation, p. 33). The nonliving components of the ecosystem, therefore, often control the numbers of the biotic components. For example, algal populations often are limited by the availability of the inorganic nutrients orthophosphate and nitrate. In one body of water growth of a particular algal population may be limited by the concentration of orthophosphate (PO_4^{3-}) the soluble form of phosphorous usable by producers, while in another body of water the growth of another population may be controlled by the concentration of ammonia (NH_4^+) or nitrate (NO_3^-), the soluble forms of nitrogen utilized by producers. Diatoms (Chrysophyta) in fresh water may be limited by silica, which they need for their outer shells, but it is not always readily available in these environs. Algal populations may also be controlled by the amount of light reaching the systems, the ambient temperature, or various other physical factors. When a particular factor limits the growth of a population, it is called a **limiting factor.**

The concept of limiting factor is derived from a combination of Liebig's (1842) "Law of the Minimum" and Shelford's (1913) "Law of Tolerance." Liebig's Law says that the growth and success of an organism is dependent on the amount of that essential material that is present in the lowest concentration. Shelford's Law maintains that organisms live within a range of too much and too little of an essential factor and that the presence of the organism is limited by

FIG. 2-2 Diagram of a freshwater pond ecosystem.
Abiotic components *(A)* are inorganic and organic nutrients. Primary producers *(P)* are planktonic algae and attached vascular plants. Primary consumers *(C₁, herbivores)* are planktonic invertebrates (zooplankton) and benthic invertebrates. Secondary consumers *(C₂, carnivores)* are small fish and carnivorous invertebrates. Tertiary consumers *(C₃)* are large carnivorous fish or fish-eating birds or mammals. All levels of consumers may feed upon decomposers *(D)*. Decomposers (primarily bacteria and fungi) recycle organic materials. (From Odum, E.P.: Fundamentals of ecology, ed. 3. Copyright © 1971 by W.B. Saunders Co. Copyright 1953 and 1959 by W.B. Saunders Co. Reprinted by permission of Holt, Rinehart & Winston, CBS College Publishing, New York.)

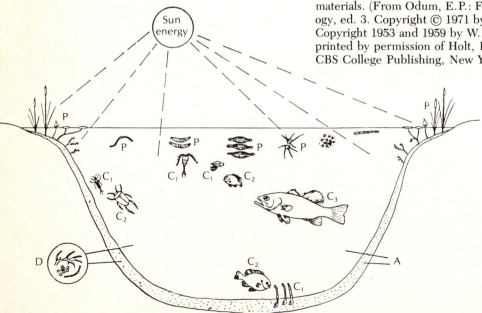

that factor for which it has the narrowest tolerance. The limiting factor concept states that the numbers of a particular organism and its presence or absence are controlled by a deficiency or excess of chemical and physical factors. (For further discussion see Odum, 1971). The various physical, chemical, and biological factors interact to control a species' numbers; it is often very difficult, if not impossible, to determine definitely which specific factor limits the abundance of a particular species. See O'Brien (1972) and Schindler (1978) for the relationship between limiting factors and phytoplankton populations.

In spite of the difficulty of determining which factor limits a species' numbers, much research has been devoted to this subject in both freshwater and saltwater systems. For example, the long debate over the presence of phosphate compounds in detergents centered around the question of whether phosphate was or was not a limiting factor in the growth of algae in fresh water. (See Likens, 1972, for a thorough discussion of this subject.) When a lake or river receives large quantities of nitrates and phosphates, the algal populations respond to these excess nutrients by increasing their numbers; the lake is then said to be **eutrophic**, or overly

---

FIG. 2-3 Electromagnetic spectrum.

Spectrum of electromagnetic radiation and forms of energy this radiation takes. Each type of radiation has characteristic wavelength and content of energy; these are inversely related. For example, cosmic rays contain greatest amount of energy but have shortest wavelength, while radio and television rays have least amount of energy and longest wavelengths. Radiation reaching earth's atmosphere is attenuated by atmospheric dust and gases, ozone layer, and cloud cover and moisture. Amount of visible solar radiation reaching earth's surface is estimated at $5.5 \times 10^5/$ m$_2$/yr (Whittaker, 1975). Only a small fraction of this visible solar radiation is absorbed by the photosensitive pigments of green plants and ultimately converted into chemical energy of organic materials.

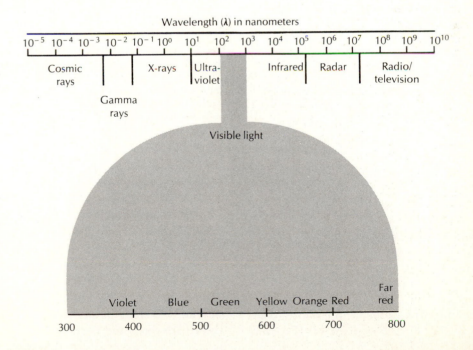

fertile. If the herbivores cannot increase as rapidly as the algae, the algal populations build up, are not eaten by the consumers, and eventually die and sink to the bottom of the lake. The decomposers increase in abundance, utilizing the chemical energy from the dead algae. Oxygen is required for the breakdown of the large numbers of dead algae by the decomposers, who transform the organic material of the dead algae into inorganic nutrients that are usable by subsequent algal populations (Fig. 2-2). In eutrophic waters much of the energy fixed by the algal producers, therefore, is directed to the decomposers of the ecosystem. When large populations of rapidly multiplying decomposers occur, they consume large quantities of oxygen with their respiration and breakdown of organic materials. The dissolved oxygen concentrations in the lake may drop below levels essential for fish life. A fish kill results; now even more dead organic matter requires oxygen for decomposition. The lake becomes a foul-smelling and undesirable body of water. Such eutrophic conditions have occurred in parts of Lake Erie and in Lake Washington in Seattle, but control of sewage inflows (which contain large amounts of nitrates and phosphates) into these lakes has substantially lessened the problem. Edmundson (1970) discusses the Lake Washington case in detail.

In studies of lake eutrophication phosphate is usually investigated first as the potential limiting factor. Nitrogen is not often considered as a limiting factor, because many blue-greens can fix nitrogen enzymatically—that is, they can convert atmospheric nitrogen (N_2) into ammonia, usable by algae. Populations of certain species of Cyanobacteria, therefore, can obtain their own usable nitrogen when supplies in the surrounding environment become low. In fresh water the limiting factors most often investigated are phosphate, carbon, light, occasionally nitrate, and, in the case of the diatoms, silica.

Photosynthesizing organisms utilize light energy within the visible spectrum (Fig. 2-3)

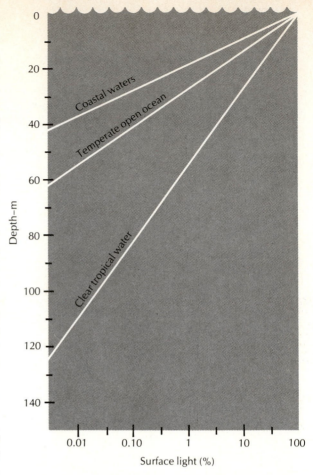

FIG. 2-4 Reduction in light intensity as light passes through marine waters.
Coastal waters are more turbid than open ocean waters because of suspended matter washed in from land. Temperate and arctic waters are more turbulent and less transparent than tropical waters because of higher winds. Most photosynthesis occurs above the 1% surface light level. In Antarctic, however, photosynthesis may proceed at light levels of 0.1% to 0.06% of surface light (Raymont, 1980). (Data from Jerlov, 1951, 1968, 1976.)

and convert this light energy to chemical energy by photosynthesis. Only energy from certain portions of the visible spectrum can be absorbed by the chlorophylls or by accessory photosynthetic pigments. The photosensitive pigments and the wavelengths of visible light that they absorb are discussed next.

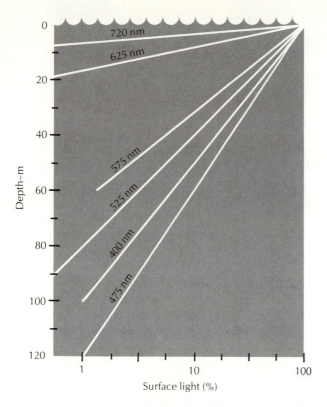

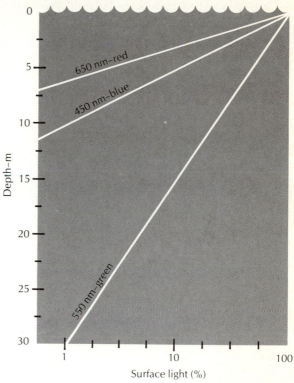

FIG. 2-5 Spectral composition of light as it passes through waters of the open ocean.
Blue-green light waves (475 nm) penetrate farthest into water column while red light waves (720 nm) are rapidly absorbed in the first few meters. (Data from Clarke, 1939, and Jerlov, 1951.)

FIG. 2-6 Changes in the spectral composition of light as it passes through turbid coastal waters.
Note that in this environment green light penetrates deepest into water column. Compare with Fig. 2-5, which shows changes in the open ocean. (Data from Levring, 1947, 1966.)

Light varies seasonally, and in both fresh and salt water the quantity (intensity) and the quality (wavelengths) may be altered by turbidity or turbulence in the water column. The intensity of light is reduced substantially as it passes through water (Fig. 2-4). The quality of the light also varies with the depth (Fig. 2-5). The blue-green wavelengths (425 to 500 nm) penetrate clear water most deeply. Green light (500 to 550 nm), however, penetrates most deeply into coastal waters, which usually have high turbidity (Fig. 2-6). The green algae (Chlorophyta)

contain chlorophyll a and b, which absorb the energy from light waves in both the 400 to 500 nm range and in the 625 to 675 nm range. Species of blue-green (Cyanobacteria), brown (Phaeophyta), and red (Rhodophyta) algae possess **accessory photosensitive pigments** (also called **antenna pigments**) that enable them to absorb light energy in the 475 to 600 nm range (Fig. 2-7). These accessory pigments enable Cyanobacteria and brown and red algae to live at greater depths where only blue and green light predominate. (See Govindjee and Braun,

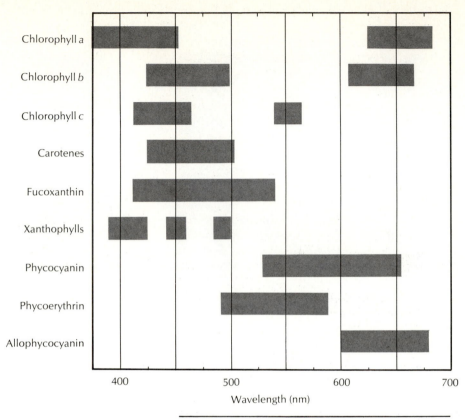

FIG. 2-7 Absorption of different wavelengths of light by various photosynthetic pigments.
Quality of light absorbed by different pigments may vary considerably, depending on alga, extraction process, and solvent used. (Data from Levring, 1966, Govindjee and Braun, 1974, Lehninger, 1975, Raymont, 1980, and Zilinskas et al, 1980.)

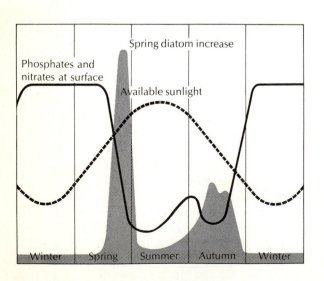

FIG. 2-8 Spring diatom increase (SDI) in temperate seas.
Diatom population increases when amounts of sunlight and nutrients are high. As diatom numbers increase, the nutrients (phosphates and nitrates or ammonia) are utilized by algae and concentration decreases. Sunlight remains adequate for diatom growth but growth is limited by nutrients. In fall a smaller peak in the diatom population occurs. (From Russell-Hunter, W.D.: Aquatic productivity. Reprinted with permission of Macmillan Publishing Co. Copyright © 1970 by W.D. Russell-Hunter.)

1974, for a discussion of pigments, light, and photosynthesis.)

In the marine environment factors similar to those in fresh water are limiting, such as light, temperature, and nutrients. Nitrate is thought to play a more important role in the marine ecosystem, because there are fewer nitrogen-fixing blue-green bacteria, and those present seldom become abundant. The factors that limit algal growth may also vary with the time of the year. A good example of this seasonal change in limiting factors is the **spring diatom increase** (SDI) that occurs in both salt and fresh waters of the temperate zone (Fig. 2-8). As spring approaches the amount of daylight increases; nutrients are plentiful because winter populations of diatoms are low. In winter diatoms are limited by low temperatures and available light. Given longer photoperiods, increasing temperatures, and ample nutrients, diatom populations build up, often peaking in late spring. Such large populations consume the available nutrients, herbivorous consumers graze upon the algae, and the diatom population is reduced. The available nutrients are used up by early summer, having been incorporated into the protoplasm of the producers or consumers. In spite of favorable light and temperatures in summer, the low concentrations of nutrients now control the numbers of diatoms. Nutrients continue to limit diatom numbers until the autumn, when nutrient concentrations gradually build up via turbulence to levels adequate to support a smaller increase in diatom numbers. However, in the autumn both available light and temperatures are dropping, and the autumn diatom peak drops off rapidly. Raymont (1980) discusses in detail phytoplankton populations in the oceans. The physiology of phytoplankton is treated in Morris (1981).

DENSITY OF WATER

Another factor must be considered when evaluating available nutrients: the maximum density of water occurs at 4° C, so as water approaches that temperature it will start to sink. Salinity also affects the specific gravity of water and therefore the density—the higher the salinity, the denser the water. In temperate zones maximum water density at 4° C leads to **thermal stratification.** In winter the colder water or ice is 1° C or 0° C and less dense than the 4° C water that has sunk to the bottom. In winter, therefore, the thermal stratification implies cold or frozen water on top and warmer water on the bottom (Fig. 2-9). In the spring, as the surface water warms to 4° C, it will sink, and the temperature will be the same from top to bottom. As the weather warms, the warmer and less-dense water stays on top, the cooler water on the bottom. Thus in summer the water is again thermally stratified, but with the warmer water on the top.

When a lake is thermally stratified, the temperature will drop or rise rapidly at the depth where the warm water meets the cooler water. This area, called the **thermocline,** can be felt when diving deeply into a body of water. The depth of the thermocline can also be measured with a thermometer on a long cable. Once the body of water has thermally stratified in either summer or winter, no exchange of nutrients or other solutes occurs between the upper (**epilimnion**) and lower (**hypolimnion**) layers of the lake. In a thermally stratified body of water, therefore, many nutrients are trapped in the lower layer and are not available to the floating algae, known as **phytoplankton,** in the upper layer. In the autumn when the epilimnion cools to 4° C, the water mass sinks, and nutrients from the bottom are evenly distributed throughout the water column. In the following spring nutrients are again made available to the algae in the epilimnion as the surface warming water sinks and the nutrients from the hypolimnion are evenly distributed in the water column.

The nutrients concentrate in the hypolimnion, because as plants and animals die they fall to the bottom. The bottom-dwelling decomposers then transform the organic materials into inorganic nutrients that are usable by the algae

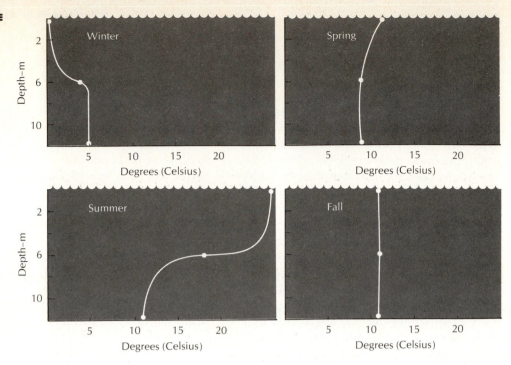

FIG. 2-9 Thermal stratification in a temperate fresh-water lake.
In winter the less-dense, colder water or ice (less than 4° C) remains on the surface. Temperature throughout water column is relatively constant during spring and autumn "overturn." In summer thermal stratification results in less-dense, warmer water staying on the surface.

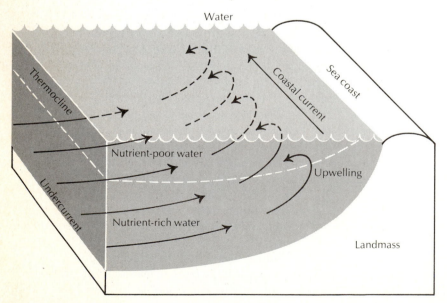

FIG. 2-10 Upwelling of nutrient-rich marine waters.
Undercurrents carry nutrient-laden deep waters toward the land mass. As currents contact land, deep waters rise along shallower coastal area and may be carried along coast by coastal current. Areas of upwelling are extremely productive because increased nutrients contribute to large populations of algae.

when mixing occurs in the water column. In lakes the process of thorough mixing of the water column is called the **spring** or **fall overturn** —depending on the season. Thermal stratification also occurs in oceans and estuaries, but tides, winds, and currents may counteract or reduce the stratification.

In the marine ecosystem, coastal areas have more nutrients than the open ocean as a result of the high nutrient content of land runoff, which contains nutrients from sewage treatment plants and agricultural fertilizers. The nutrient content of salt water is also influenced by a process known as **upwelling.** In certain areas of the world where ocean currents meet land masses, these currents rise from the ocean bottom (Fig. 2-10). These upwelling areas are found in the Arctic and Antarctic, the west coasts of North and South America, the northeast coast of North America, and the west coasts of Africa and Australia. Upwelling areas have large algal populations because of the ample concentrations of nutrients available. Since nutrients are not usually a problem, algal populations in upwelling areas are usually limited by light or by the grazing of herbivores. As a result of the large algal populations in upwelling areas, there are also large invertebrate and fish populations. Georges Bank off the eastern North American coast between Maine and Nova Scotia is one of the most productive fishing areas in the world. Likewise, upwelling areas off the coast of Peru are highly productive and heavily fished. (See Idyll, 1973, for a discussion of the herbivorous anchovy populations off the coast of Peru.)

LIGHT

The light available for photosynthesis depends on many factors: in clear tropical waters adequate light may penetrate as deeply as 200 m; in the more turbulent temperate and arctic waters, light penetration is reduced to about 30 m. Penetration of light is also reduced in coastal areas because of the turbidity from the suspended matter in land runoff. Even large al-

gal populations can reduce light penetration by shading algae beneath them. The zone through which adequate light for photosynthesis penetrates is called the **photic zone.** As previously mentioned, this zone is deep in clear tropical waters but shallower in the more turbulent temperate waters and in the more turbid coastal waters (Fig. 2-11). In fresh water the depth of light penetration may vary greatly, depending upon nutrients, dissolved organic matter, algal

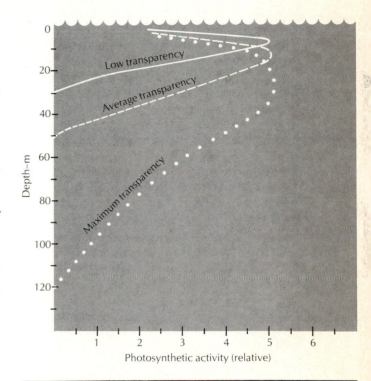

FIG. 2-11 Relationship of photosynthesis to depth. Amount of photosynthetic activity is related to intensity and spectral composition of light penetrating water column. Open ocean waters in the tropics usually have maximum transparency, while temperate and arctic coastal waters have low transparency and light penetration. Depth of photic zone will vary with turbidity, turbulence, and degree of stratification, in addition to many other factors. Note maximum photosynthetic activity takes place several meters below surface. In most algae photosynthesis is inhibited by high light intensities such as those found at water's surface.

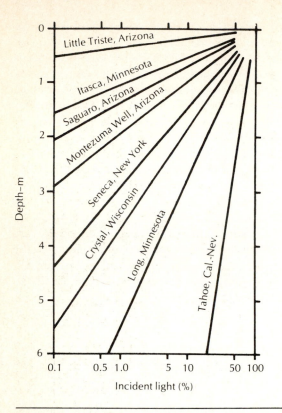

FIG. 2-12 Penetration of light into fresh water at different locations in the United States.
Light penetration into lakes or ponds depends on amount of suspended solids in the lake and its trophic status (oligotrophic, mesotrophic, or eutrophic; see p. 44). Light penetration into eutrophic lakes may be severely reduced by large algal populations or suspended solids. Lake Tahoe is an oligotrophic lake and very clear, while Little Triste Lake is either eutrophic or very turbid because of suspended solids. (From Cole, G.A. 1983. Textbook of limnology, ed. 3, St. Louis, The C.V. Mosby Co.)

population, and many other factors (Fig. 2-12). Remember that both quality and quantity of light are altered as the light passes through water.

The life of a planktonic alga depends on maintaining itself in the photic zone, where it can receive enough solar energy to efficiently photosynthesize. The alga must also respire in order to obtain energy from stored organic materials to support metabolic processes. A planktonic alga must constantly maintain a balance between the organic materials it produces via photosynthesis and the organic material it respires (i.e., oxidizes) for energy. The depth at which energy gained by photosynthesis equals that used in respiration is called the **compensation depth.** No photosynthesizing alga can live below this depth for very long, because if respiration exceeds production the alga will die from lack of energy. However, some algae may live heterotrophically below the compensation depth. Many algae can actively or passively migrate vertically in the water column, thus maintaining themselves at that depth where production exceeds respiration. Some algae, in fact, can even pass through or remain in the gut of herbivores and continue to photosynthesize (Epp and Lewis, 1981). Certain planktonic crustacea may contain metabolically active algae even after the invertebrates are dead. These undigested algae probably contribute oxygen and carbohydrates to the gut, at the same time removing carbon dioxide and other wastes.

Maximum photosynthesis does not usually take place at the surface, but rather many centimeters below the surface, because high light intensity at the surface inhibits photosynthesis in most algae (Fig. 2-11). Light intensity and light penetration will vary with the time of day and the season, so the ability to migrate vertically enables the alga to move to that depth where it can most efficiently photosynthesize.

Up to this point the discussion has centered primarily on the phytoplankton—those alga that live suspended in the water column—but these are not the only producers in fresh and salt water. Attached algae living on rocks, mud, and sand on the bottom of a body of water are called benthic algae, and their habitat is called the **benthos.** In shallow waters benthic vascular plants (seagrasses) may also contribute substantially to energy production.

In both marine and freshwater environments life does exist below the photic zone. Dead and decaying algae and animals constantly fall, creating a "rain of detritus" on those heterotrophic animals and decomposers that live below. The organisms below the photic zone break down the organic debris into inorganic substances such as carbon dioxide, nitrate, and orthophosphate, which may be returned to the zone via overturn or upwelling. The return of nutrients to this zone is not very efficient; therefore, deeper waters usually contain higher nutrient concentrations than the shallow photic zone, where nutrients are rapidly utilized by algae and the vascular plants.

PRIMARY PRODUCTIVITY

The algae (planktonic and benthic) are essential primary producers of energy, not only in the marine environment but also in fresh water. The amount of organic material fixed by green plants in an ecosystem can be measured in several ways: One way is to determine the amount of carbon in the organic material; a second way is to figure the dry weight of the total organic material; and a third way is to measure the amount of energy, in the form of calories, in the organic material. The weight of the organic material in the ecosystem (either as grams of carbon or grams of organic material—dry weight) can be estimated and reported as **biomass** or **standing crop** per unit of area. This measurement provides an estimate of how much living material is present in the ecosystem and may be used to compare various areas. For example, biomass (kg/m²) of the open ocean is 0.003 kg/m² (low), while the biomass of estuaries is 1.0 kg/m² (high) (Whittaker, 1975). Biomass measurements may be made either of all the organisms in an area or only one particular group of organisms (e.g., the algae). The biomass of algae during a spring diatom increase (SDI) in April may be as high as 0.08 kg/m² in the upwelling area of George Bank. Biomass may vary greatly over the year, depending on the latitude and many other factors.

Most ecologists prefer measuring the *rate* at which new organic matter is produced in the ecosystem rather than just determining the amount (biomass) of organic material present at one time. The rate at which organic matter is produced is called **primary productivity** and is expressed as weight (grams of carbon or total organic material) or energy content (calories in the organic material) per unit of area per unit of time. Primary productivity may be calculated by the amount of carbon dioxide taken in by the plants or by the amount of oxygen evolved during photosynthesis. These determinations can then be converted to grams of organic matter, grams of carbon, or kilocalories.

Measurements of primary productivity are very useful in comparing various areas of the world: ocean versus land, arctic versus tropics, agricultural land versus forests, and so on. Tropical coral reefs, for example, are highly productive (2500 g dry organic material/m²/yr). Two forms of primary productivity may be determined: **gross primary productivity** (GPP), which includes all the energy fixed by plants into organic materials, and **net primary productivity** (NPP), which includes the total energy content of the organic material produced by plants *minus* the energy used by the plants in respiration for growth, reproduction, and other functions. Therefore, net primary production equals gross primary production minus respiration: NPP = GPP − R. Net primary productivity values indicate how much energy is available to the next trophic level.

When comparing the primary productivity (gross or net) of different areas (Table 2-1), it is important to remember that in areas such as the polar seas primary productivity may be very high for several months, but for comparison with other latitudes, the productivity must be averaged over the year (Fig. 2-13). An additional factor to be considered is the size of the area in which the primary productivity occurs. For example, coral reefs are highly productive (2500 g dry organic material/m²/yr), but they cover only a small fraction of the earth's surface. On the other hand, the open ocean is low in productivity (125 g/m²/yr), but it covers about 70% of

TABLE 2-1. Annual Primary Productivity

Ecosystem	Area (10⁶ km²)	Annual mean NPP* (Grams dry organic material/m²/yr)	World NPP* (10⁹ dry ton/yr)
Tropical forest	24.5	2000	49.4
Desert and semidesert	42.0	40	1.7
Lake and stream	2.0	250	0.5
Eutrophic lake	†	325	†
Other	80.5	†	63.4
TOTAL CONTINENTAL	149.0	773	115.0
Open ocean	332.0	125	41.5
Upwelling areas	0.4	500	0.2
Algal beds and coral reefs	0.6	2500	1.6
Estuaries	1.4	1500	2.1
Other	26.6	†	9.6
TOTAL MARINE	361.0	152	55.0
TOTAL WORLD	510	333	170

From Whittaker and Likens, 1975; Likens, 1975.
*NPP = net primary productivity;
1 g dry organic matter = approximately 4.6 kcal energy;
g dry organic matter = 2.2 × g carbon
†Not available.

the earth's surface (Table 2-1). The open ocean, therefore, is responsible for more primary production than coral reefs. Primary production in estuaries is also very high (1500 g/m²/yr), but is responsible for only 1% of the world's primary productivity. The role of estuaries in the marine ecosystem is broader than just contributing primary productivity, however. Estuaries support the juvenile stages of many commercially and ecologically important invertebrates and vertebrates. The juvenile stages move out of the estuaries as they grow and populate the open ocean.

In freshwater eutrophic lakes primary productivity averages 325 g/m²/yr (Table 2-1). The production in a eutrophic lake may result in a large biomass of undesirable fish that have little sport or economic value or in many decomposers. Problems with oxygen deficiency may develop, resulting in fish kills (see Chapter 3). Over geological time lakes become eutrophic naturally; this is known as **natural eutrophication.** As sediments build up in the lake from land drainage, the lake becomes shallower and

organic materials increase, insuring an adequate supply of nutrients. Usually this process takes thousands of years; however, the activities of humans have accelerated the process of natural eutrophication by adding nutrients from sewage and fertilizers to the lake and by increasing runoff and siltation from the land. When human activities are involved in accelerating the eutrophication process, it is called **cultural eutrophication.** Likens (1972) discusses the role of nutrients in the eutrophication process.

Lakes that are low in primary productivity and are infertile are called **oligotrophic.** They are characterized by exceptionally clear waters, low numbers of phytoplankton, and low concentrations of nutrients. These oligotrophic lakes are usually located in relatively isolated mountain areas away from human activity and are highly prized for recreation. Lakes that are intermediate between eutrophic and oligotrophic are designated **mesotrophic,** depending on primary productivity and levels of nutrients, chiefly orthophosphate and nitrate.

FIG. 2-13 Seasonal variations in phytoplankton and herbivorous zooplankton populations at different latitudes.

Note summer peak in Arctic, two peaks in temperate zone, and relatively constant populations in tropics. (From Fogg, G.E. 1980. Phytoplanktonic primary productivity. In Barnes, R.S.K., and K.H. Mann, [eds.]: Fundamentals of aquatic ecosystems, Oxford, England, Blackwell Scientific Publications Ltd.)

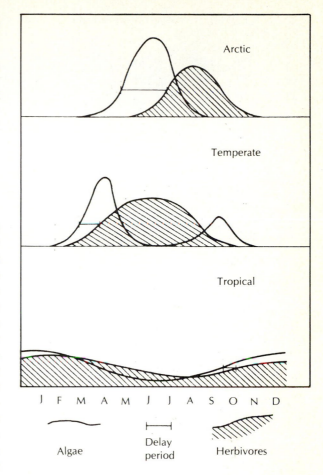

ZONATION

Some algae have narrow ranges of tolerance for various environmental factors and occupy very specific habitats where the environment fluctuates only slightly. Other algae can tolerate a broad range of environmental factors and are found in many different habitats. Algae may be (1) attached to rocks, sand, or mud—benthic algae; (2) living on the surface of larger algae or vascular plants—epiphytic algae; or (3) suspended in the water column as single cells or small colonies—phytoplankton. Benthic algae may be found anywhere light reaches the bottom of a body of water, epiphytic algae may be found anywhere large algae (**macroalgae**) or vascular plants are growing, and phytoplankton may be found in all waters to the depths of light penetration.

In the marine environment benthic algae, which live between the high and low tide marks, live in the **intertidal zone** (Fig. 2-14), also called the littoral zone. The area below the low tide mark where light penetrates is called the **sublittoral (subtidal)** zone. Many benthic algae live in this area where hard substrate is available. Shallow water areas above the **continental shelf** are called the **neritic zone,** while waters above the deeper **continental slope** and **abyssal plain** are called the **oceanic zone.** The neritic zone and the oceanic zone together form the **pelagic zone,** which includes the entire mass of ocean water.

Primary production is limited to those areas where enough light can penetrate (the photic

zone) to enable the algae to produce enough energy via photosynthesis to equal or exceed the energy consumed in respiration.

The photic zone is deeper in the open ocean and tropical waters than it is in coastal areas and temperate and arctic waters. In high latitudes turbulence resulting from winds and upwelling decreases light penetration, while in coastal areas both suspended solids from land runoff and turbulence from upwelling restrict light penetration. The photic zone in these areas, therefore, is not as deep as it is in the open ocean (Fig. 2-4).

In fresh water slightly different terms are used for the zones. The **littoral zone** refers to

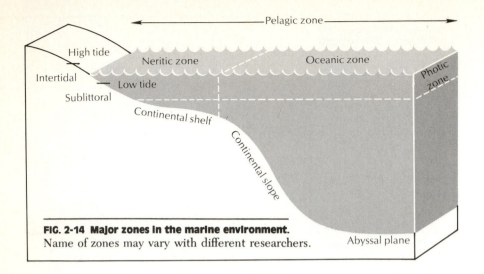

FIG. 2-14 Major zones in the marine environment.
Name of zones may vary with different researchers.

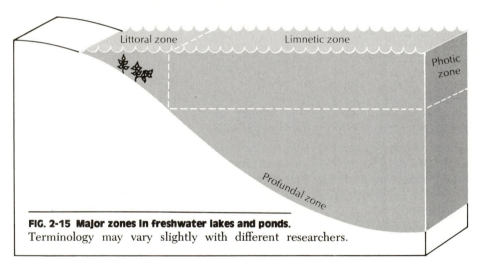

FIG. 2-15 Major zones in freshwater lakes and ponds.
Terminology may vary slightly with different researchers.

that area where light penetrates to the bottom and where aquatic vascular plants are found (Fig. 2-15), and the **profundal zone** is that deep area of lake water where little light penetrates. The area of open water beyond the littoral zone is called the **limnetic zone.** Those algae living at the lower depths of the photic zone must be adapted to using light of low intensity in the blue to green wavelengths (Fig. 2-5). Ice and snow cover may substantially reduce light transmission. Clear ice transmits approximately

the same amount of light as water, but snow on top of the ice can significantly reduce the amount of solar energy reaching the underlying water. For example, Cyanobacteria have been found growing on the bottom of Antarctic lakes, where they are adapted to extremely low light intensities. Approximately 97% of the visible light is absorbed by the ice and underlying water (Parker et al, 1981).

Distribution of benthic algae along a shore is governed by both biotic and abiotic factors.

FIG. 2-16 Zonation of marine algae on an intertidal rock face along Oregon (U.S.) coast.
Three divisions of algae are represented—Cyanobacteria, Phaeophyta, and Rhodophyta. Note highest "black zone" composed mainly of Cyanobacteria. Red algae are generally located in the high intertidal zone, and larger browns (kelps) are found low in the intertidal zone. Fifteen different genera of algae comprise the algal community on these rocks. (From McConnaughey, B.H. and R. Zottoli 1983. Introduction to marine biology, ed. 4, St. Louis, The C.V. Mosby Co.)

1	Cyanobacteria	
2	*Pelvetiopsis*	Phaeophyta
3	*Halosaccion*	Rhodophyta
4	*Endocladia*	Rhodophyta
5	*Fucus*	Phaeophyta
6	*Gigartina*	Rhodophyta
7	*Porphyra*	Rhodophyta
8	*Iridaea*	Rhodophyta
9	*Alaria*	Phaeophyta
10	*Cystoseira*	Phaeophyta
11	*Nereocystis*	Phaeophyta
12	*Laminaria*	Phaeophyta
13	*Postelsia*	Phaeophyta
14	*Odonthalia*	Rhodophyta
15	*Microcladia*	Rhodophyta

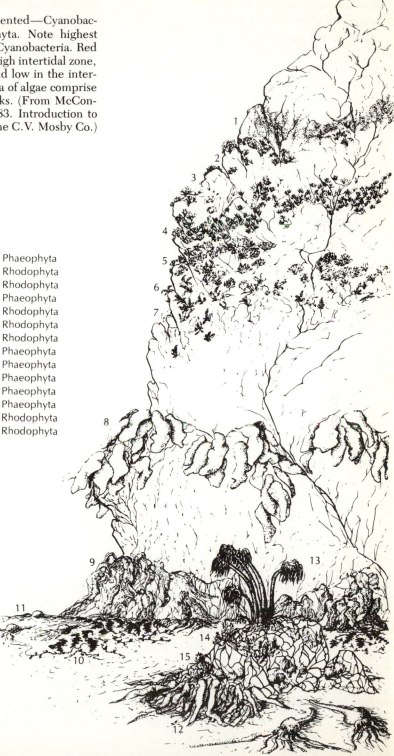

Various "zones" of algae may be observed on temperate rocky shores, for example. Species of algae are distributed in intertidal areas along environmental gradients such as salinity, light intensity, and moisture. The location of various algae and their invertebrate predators is determined both by their physiological tolerance for environmental variations and by their interactions with other species in the same zone. Such zonation may be seen in the intertidal rock community in Oregon (Fig. 2-16); fifteen genera of algae may be found in different zones on these rocks.

SUCCESSION AND CONDITIONING

Over the course of a year a fairly predictable **succession** of algal species may be observed in a

particular environment. Such successions are also documented for vascular plants in both aquatic and terrestrial ecosystems. Two probable reasons for succession are (1) the population of each species is controlled by different limiting factors, and (2) the preceding species may **condition** the environment for subsequent species. Many other factors such as grazing by invertebrates or shading by other algae may also influence succession. A predictable succession has been documented in benthic diatoms in the intertidal zone of the coast of Scotland. In this area five different species of diatoms dominate the benthic algal community from February to June; these species comprise 99% of the diatom population (Russell-Hunter, 1970). Not only is the population of each dominant alga adapted to different levels of light, temperature, and nutrients, but the preceding dominant diatom may also prepare (condition) the waters in some manner for the dominant species that follows. One mechanism of conditioning is the release of metabolites into the water, which may stimulate the growth of one alga or inhibit the growth of a competing alga.

In a subarctic lake a similar predictable succession occurs (Fig. 2-17). During the spring, diatoms, which grow well in cool, nutrient-rich waters, dominate, followed by green algae, which grow better at warmer temperatures and higher light intensities. In late July, as the nutrients become depleted, dinoflagellates (Pyrrhophyta) dominate. As autumn approaches and both temperature and light decrease, the lake

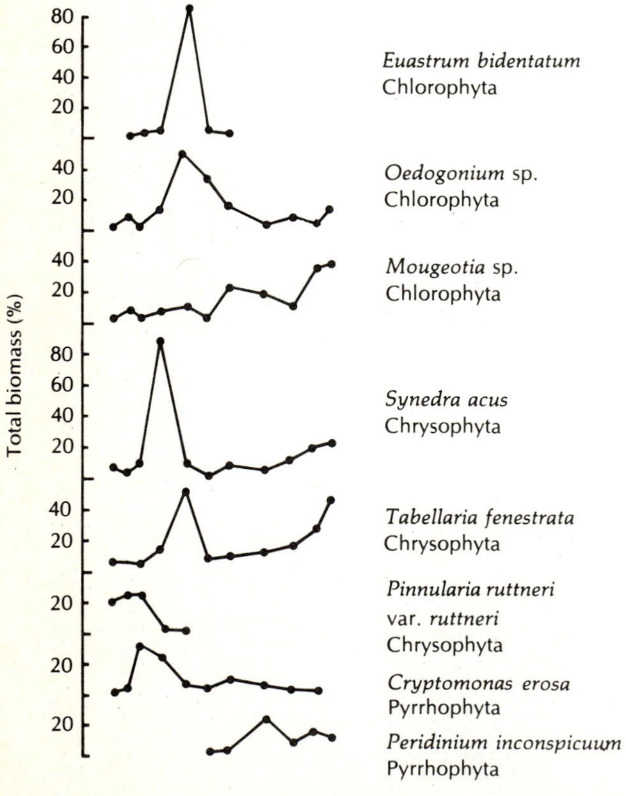

FIG. 2-17 Succession of major phytoplankton species in a subarctic lake.
In June green algae (Chlorophyta) and diatoms (Chrysophyta) dominate phytoplankton. In midsummer there are few phytoplankton—mainly Pyrrhophyta (dinoflagellates) and a filamentous green alga, *Mougeotia*. In late summer diatom populations start to increase and *Mougeotia* populations remain fairly high. (From Sheath, R.G., M. Munawar, and J.A. Hellebust. 1975. Can. J. Bot. **53**[19]:2240-2246.)

undergoes fall overturn, nutrients are again available in the photic zone, and there is a smaller peak of green algae, then diatoms (Sheath et al, 1975). In temperate freshwater lakes, the spring diatom "bloom" and the early summer green algal bloom may be followed by a blue-green algal (Cyanobacteria) bloom at summer's end. Some blue-green species can fix atmospheric nitrogen (N_2) into ammonia (NH_4^+), thereby enabling them to multiply rapidly when ammonia supplies are depleted in the photic zone (see Chapter 3). Following fall overturn nutrients are returned to the photic zone, and a smaller peak in diatoms may occur.

Not only does succession occur among algal species, but also among different algal divisions and different kingdoms. For example, on the rocky shores of temperate areas the bare rocks may initially be colonized by bacteria, followed by a diatom film, then brown algae, and finally by assorted benthic animals—barnacles, mussels, and sea stars (Sumich, 1980). Again, each group of organisms may condition the environment for the succeeding group. Chapman (1979) provides further discussion of succession, stimulation, and inhibition.

METHODS OF SAMPLING PHYTOPLANKTON

Phytoplankton are sampled by pulling a fine mesh silk net through the water (Fig. 2-18). If a flowmeter is attached to the net, quantitative measurements of number of phytoplankton per

FIG. 2-18 Wisconsin plankton net used to sample both phytoplankton and zooplankton.
Net mesh size is 60 μm. Net may be towed through water horizontally or vertically, or water samples may be poured through net. (From Lind, O.T. 1979. Handbook of common methods in limnology, ed. 2, St. Louis, The C.V. Mosby Co.)

TABLE 2-2. Size Grading of Plankton

Maximum cell size	Category
Larger than 1 mm	Macroplankton
Less than 1 mm but larger than 0.06 mm (60 μm)	Microplankton
5-60 μm	Nannoplankton
Less than 5 μm	Ultraplankton

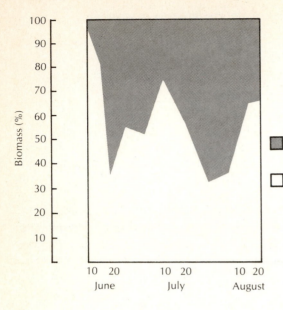

**FIG. 2-19 Percentage of nannoplankton in total phyto-
plankton biomass.**
In this subarctic lake nannoplankton may contribute
over 50% of phytoplanktonic biomass during
summer. (From Sheath, R.G., M. Munawar, and J.A.
Hellebust. 1975. Can. J. Bot., **53**[19]:2240-2246.)

m³ of water are possible. Most plankton nets
have a mesh opening of 60 μm so that phytoplankton cells less than 60 μm in size will pass
through the net and not be counted. Plankton
that are retained by 60 μm plankton nets are referred to as **net plankton,** while those that pass
through the net are called **nannoplankton.** Categories of plankton based on size are listed in
Table 2-2. Because of the difficulty in sampling,
little is known about the contribution of the nannoplankton to the primary productivity of various areas, but some estimates have been made
(Fig. 2-19). Many researchers think that the
nannoplankton contribute very significantly to
primary productivity. Further studies will define more clearly the role of the nannoplankton
in primary productivity.

UNDESIRABLE ALGAE

Although algae can be beneficial, they can
also cause many problems in fresh and salt
water. In freshwater lakes and rivers algae can
rapidly develop large populations that then decompose and deplete dissolved oxygen. When
certain algae build up large **algal blooms** in
water supply reservoirs, they may impart to the
drinking water an undesirable taste or odor that
is difficult to remove. Other algal species (mainly the Cyanobacteria) produce toxic metabolites
that, in large concentrations, can cause illness or
even death to fish and warmblooded vertebrates, including man (see Chapter 3).

In the marine environment fish, sea birds,
and mammals may be sickened or killed by toxins, usually produced by planktonic dinoflagellates (Pyrrhophyta) during rapid algal increases
called **red tides** (see Chapter 9).

Control of undesirable algae is often necessary. Herbicides such as 2,4-D (dichlorophen-
oxyacetic acid) or copper sulfate may be used to
temporarily control algal populations, especially
in lakes or reservoirs.

COMMERCIAL USES OF ALGAE

Brown and red macroalgae are harvested
commercially for the carbohydrates they contain
(see Chapters 7 and 11). The polysaccharides,
algin from brown algae and carrageenan and
agar from red algae, are used in a variety of
commercial applications such as stabilizers to

keep paints from separating, thickeners for cosmetics, media for bacterial growth, and in dairy products and other foods. More than 800,000 dry metric tons of macroalgae are harvested annually; the harvesting and processing of this algae comprises a billion dollar industry.

Algae also appear to have great medical potential. Extracts from various algae have improved neuromuscular disorders, stomach cancers, and mouse leukemia. Antibacterial activity has also been demonstrated for several algae. Volesky et al (1970) and Hellebust (1974) discuss antibiotic substances produced by algae.

ALGAE AS A SOURCE OF FOOD

As human populations rapidly increase and food supplies decrease, attention is being focused in the western hemisphere on the ubiquitous algae as a source of food. Macroalgae (multicellular marine algae, often called seaweed) have been used for food by residents of coastal areas for many centuries. Macroalgae have been a staple in the diet of the people of coastal Asia, the Pacific Islands, Australia, New Zealand, the west European coast, Iceland, the British Isles, and the coasts of North and South America. Japan has been the leader in both the use of locally abundant macroalgae and in the cultivation of macroalgae. In the nineteenth century the Polynesians made use of 75 different species of macroalgae as food; several of these species were cultivated in ponds in Hawaii and considered great delicacies. The Pacific Ocean countries of Japan, China, Hawaii, the Philippines, Malaysia, and Indonesia continue to use the largest amount of macroalgae for food at the present time.

Macroalgae, used as a marine vegetable, may be eaten raw or cooked with other foods. Some macroalgae are highly prized gourmet treats in the oriental countries. In Japan the red alga, *Porphyra* ("nori") is cultivated in shallow bays, and the supply never exceeds the demand. The culture of *Porphyra* in Tokyo Bay is the largest macroalgae cultivation project for food in the world (see Chapter 11). The Chinese have successfully cultured brown macroalgae (especially *Laminaria*) for both food and algin.

The nutritive value of macroalgae is not very high, because humans do not have the enzymes to break down many of the complex algal carbohydrates. Macroalgae, however, do contain significant levels of vitamins and minerals, and some algae (*Porphyra* and the green algae *Caulerpa* and *Enteromorpha*) do contain a significant proportion (12% to 30%) of protein.

Harvesting macroalgae from coastal areas is a time-consuming and labor-intensive process, making commercial harvesting uneconomical unless the algae are cultivated in confined areas as is practiced in Japan and other parts of the Orient.

In some coastal areas of Ireland, Scotland, France, Scandinavia, and Iceland, cattle and sheep graze the intertidal zone at low tide, using the copious macroalgae as fodder. Many macroalgae contain as much protein (10% dry weight) as dry oats or hay. Other sublittoral macroalgae are processed into a meal or powder and fed to poultry, hogs, or cattle as a food supplement that is high in minerals and vitamins.

Considerable research has been conducted on single-celled algae as potential food sources. The green algae *Chlorella* and *Scenedesmus* (see Fig. 4-15) have been cultured on a large scale using either sunlight, artificial light, or no light with a source of organic carbon. These algae have been considered as a potential food source for humans in outer space. Not only can these algae be used as food, but they also remove carbon dioxide from the atmosphere. The **microalgae** are easy to maintain in mass culture and have a more rapid growth rate than the macroalgae. The percentage of protein, fat, and carbohydrate in the cells may be controlled by the available nutrients, among other factors. The protein content of *Chlorella* and *Scenedesmus* may be quite high (up to 74%), but the algal cell walls may irritate the human digestive tract, and the flavor may be unpleasant and unappetizing. In the early 1970s the expense of producing a pound of *Chlorella* was approximately eight times the cost of producing a pound of soybeans. High levels of primary productivity were obtained in *Chlorella* cultures in a closed system, but these levels were maintained only over a short period and probably could not be maintained over a long period. Maximum net productivity of 28 g dry organic material/m²/day was obtained with a mean of 12.4 g/m²/day (Ryther, 1959). The major cost in harvesting *Chlorella* or *Scenedesmus* is extracting the cells from the medium, because very large quantities of water must be processed in order to concentrate the cells. Dubinsky et al (1978) discuss the large-scale production of algae for use as animal feed.

Many researchers believe that the culture of single-celled algae has great potential for increasing the world's food supply, but at this time the cost of production from microalgae greatly exceeds the cost of conventionally grown crops. In the future such cultures of **single-celled protein** (SCP) may become economically feasible as food prices continue to rise.

The cyanobacterium *Spirulina* (see Fig. 3-14) has also been cultivated on a large scale. *Spirulina* may contain up to 65% protein (dry weight) and has been harvested for food from alkaline lakes in Africa and Mexico for hundreds of years. *Spirulina* can tolerate a pH up to 9.5 and is easier to harvest than the green microalgae, because it forms clumps and can be skimmed off the surface without processing all the water medium. The cyanobacterium *Nostoc*, which has also been used as a food, may contain up to 20% raw protein (Chapman, 1970). The Cyanobacteria do well in warm tropical areas such as Israel, Mexico, Africa, and northern South America.

Algae may also be used to remove the nutrients, nitrate and orthophosphate, from sewage treatment ponds. Bacteria in the ponds must first decompose the organic material to inorganic material. Then the algae (green algae and Cyanobacteria, usually) rapidly take up the inorganic nutrients from the sewage and use them for growth and reproduction. If the algae are removed from the ponds and used as crop fertilizer or animal feed, the nutrients are removed from the aquatic system and not released into the receiving waters, where they might cause eutrophication problems. Instead the nutrients are recovered from the sewage and transferred to the land, where they nourish plant or animal crops. Wiedeman (1970) discusses the algae of waste-stabilization ponds, which include ponds for treatment of sewage and other organic wastes. The recycling of nutrients from sewage into food production is discussed by Ryther et al (1972).

Another use of algae is for food for such commercially important marine invertebrates as clams and oysters. Considerable research on this project has been done at the Woods Hole Oceanographic Institute in Massachusetts. Here algae are grown in a mixture of sewage effluent and sea water and fed to filter-feeding shellfish, which are then harvested. Many similar projects involving intense cultivation of algae and shellfish are located in estuaries and bays along coastal areas. The intense culture of marine algae and shellfish is called **mariculture.**

Wilcox (1976) proposed an "Ocean Farm Project" and completed pilot plant plans (Leeper, 1976). The ocean farm was to grow kelp (a brown macroalga), using nutrient-rich upwelling waters. The kelp "farms" would be highly productive, and the majority of kelp could be used for human or domestic animal food or for crop fertilizer. The remaining kelp would be processed anaerobically (without oxygen) to generate methane gas, a substitute for natural gas. At the present time the ocean farm is not economically feasible, since the cost of the kelp and the methane would not be competitive in price. As food, fertilizer, and energy costs continue to escalate, however, the ocean farm may become economically viable.

As the amount of arable land continues to decrease and food and energy costs continue to rise, researchers will look with increasing frequency to both fresh water and salt water for food and energy. Nutrients from upwelling areas or from sewage effluents can nourish large crops of algae, which in turn can feed invertebrates or vertebrates that contain high levels of protein. If the organism to be used for food occupies a low trophic level (producer, herbivore, or detritivore) more energy (calories) is available than if the food organism occupies a higher trophic level (primary or secondary carnivore). Energy (70% to 90%) is lost at each trophic level. In order to produce enough food for the world's rapidly growing population, emphasis must be placed on using food sources at the lower trophic levels. Eating secondary or even primary carnivores entails a large waste of potentially available energy.

In addition to eating food from lower trophic levels, nutrients must be recycled onto the land as much as possible. Many nutrients dumped into water are lost to the ocean floor, not to be recycled into the ecosystem for thousands of years. In attempting to produce more food from the sea it should be kept in mind that the primary productivity of both oceans and land is limited by such factors as available light, nutrients, temperature, and the efficiency of the photosynthetic process (only about 1% to 2% of visible light energy is converted into chemical energy). In the future emphasis must be placed on recycling resources, using renewable energy sources, and growing food low on the food pyramid with a minimum of environmental impact.

SELECTED REFERENCES

Chapman, A.R.O. 1979. Biology of seaweeds: levels of organization. University Park Press, Baltimore. 134 pp.

Falkowski, P.G. (ed.). 1980. Primary productivity in the sea. Vol. 19. Environmental Science Research Service. Plenum Publishing Corp., New York. 335 pp.

Fogg, G.E. 1980. Phytoplanktonic primary production. In R.S.K. Barnes and K.H. Mann (eds.). Fundamentals of aquatic ecosystems. Blackwell Scientific Publications, Oxford, England. pp. 24-45.

Lieth, H., and R.H. Whittaker. 1975. (eds.). The primary productivity of the biosphere. Springer, New York. 339 pp.

McConnaughey, B.H. 1978. Introduction to marine biology. Ed. 3. The C.V. Mosby Co., St. Louis. 624 pp.

Morris, I. (ed.). 1981. The physiological ecology of phytoplankton. Studies in ecology. Vol. 7. University of California Press, Berkeley. 625 pp.

Odum, E.P. 1971. Fundamentals of ecology. Ed. 3. W.B. Saunders Co., Philadelphia. 555 pp.

Reid, G.K., and R.D. Wood. 1976. Ecology of inland waters and estuaries. Ed. 2. D. Van Nostrand Co., New York. 485 pp.

Rosowski, J.R., and B.C. Parker (eds.). 1982. Selected papers in phycology II. Phycol. Soc. Am., Lawrence, Kansas.

Round, F.E. 1981. The ecology of algae. Cambridge University Press, Cambridge, England. 653 pp.

Russell-Hunter, W.D. 1970. Aquatic productivity. Macmillan, Inc., New York. 306 pp.

Sumich, J.L. 1980. An introduction to the biology of marine life. Ed. 2. Wm. C. Brown Co., Publishers, Dubuque, Iowa. 359 pp.

Wetzel, R.G. 1975. Limnology. W.B. Saunders Co., Philadelphia. 743 pp.

Whittaker, R.H. 1975. Communities and ecosystems. Ed. 2. Macmillan, Inc., New York. 385 pp.

3

CLASSIFICATION*

Division: Cyanobacteria (blue-green algae): 165 genera, 1500 species

Class: Cyanophyceae

ORDER 1. Chroococcales: cells solitary, loosely colonial or palmelloid (embedded in a gelatinous matrix); non-spore-forming; mostly fresh water; *Chroococcus, Coelosphaerium, Cyanidium, Gloeocapsa, Merismopedia, Microcystis*

ORDER 2. Chamaesiphonales: cells solitary or loosely colonial epiphytes; some filamentous and some forming crusts; some forming exospores and some forming endospores; mostly marine; not common; *Chamaesiphon, Dermocarpa*

ORDER 3. Oscillatoriales (Hormogonales): filamentous, lacking exospores; some branching; producing hormogones (a short, motile filament); some with heterocysts and some with akinetes; some with both; *Anabaena, Calothrix, Cylindrospermum, Fischerella, Gloeotrichia, Hyella, Lyngbya, Nostoc, Oscillatoria, Phormidium, Rivularia, Spirulina, Tolypothrix, Trichodesmium*

CYANOBACTERIA

*See Drouet (1968, 1973, 1978) and Drouet and Daily (1956) for an alternate classification used by some investigators.

The organisms classified in the division Cyanobacteria, sometimes known as blue-green algae, blue-green bacteria, or slime algae, are members of the kingdom Monera and are therefore composed of prokaryotic cells lacking a nuclear membrane and nucleoli. The cells have no flagella or any other type of locomotor organelle, although many move by an oscillating mechanism. In many ways the blue-green cells resemble the bacteria (division Schizophyta), but they differ from the bacteria by containing chlorophyll *a* and producing oxygen as a by-product to their photosynthetic activity (Table 3-1). Fossil evidence indicates that Cyanobacteria are approximately 3 billion years old, dating well into the Precambrian era; they were probably the first oxygen-evolving organisms to exist on earth.

Cyanobacteria are ubiquitous in their distribution and have invaded almost all ecological niches from hot springs (at 90° C) to glaciers (at 0° C). Not only are they important members of the flora in both salt and freshwater environments, but they are also capable of growing wherever some moisture exists, including damp soil, shaded terrestrial habitats, and rock crevices. They have even been reported airborne, both as spores and cells. Some Cyanobacteria occur as nitrogen-fixing symbiotic organisms and as such may be found in various unicellular flagellates, lichens, mosses, ferns, gymnosperms, and angiosperms. The color of the blue-greens ranges from the classic blue-green through all shades of blue and green to black and even red. The particular hue depends upon the relative abundance of specific photosynthetic pigments and mineral deposits associated with the cells.

The blue-greens may assume numerous forms, including single cells *(Chroococcus)*, aggregates of single cells within a gelatinous mass *(Gloeocapsa)*, flat plates of cells *(Merismopedia)*, filaments

TABLE 3-1. A Comparison of Bacteria and Cyanobacteria

Structure/function	Bacteria	Cyanobacteria
Membrane-bound organelles	None	None
Chromosomes	Not associated with basic proteins (histones)	Same
Ribosomes	Sedimentation coefficient: 70s	Same
Cell wall	Muramic acid, diaminopimelic acid and other organic acids	Muramic acid, diaminopimelic acid, glucosamines, alanine, glutamic acid
Lysozyme reaction	Dissolves cell wall	Same
Vegetative reproduction	Spores, fission, fragmentation	Spores, fission, fragmentation hormogones, akinetes
Sexual activity	Bacterial genetic recombination (transformation, transduction, or conjugation)	Recombination documented for a few mutant species
Nitrogen fixation	Some bacteria species	Some blue-green species, especially those having heterocysts
Light-receptive molecules	Bacteriochlorophylls	Chlorophyll a, carotenoids, phycobilins
Photosynthetic environment	Anaerobic	**Aerobic** or **anaerobic**
Photosynthetic by-products	Hydrogen, sulfur, organic compounds	Oxygen
Materials required for photosynthesis	H_2, H_2S, CO_2, organic compounds, light	H_2O, CO_2, minerals, light

(*Lyngbya*), or, rarely, as branching filaments (*Fischerella*), as shown in Fig. 3-1.

CELLULAR STRUCTURE

Fig. 3-2 shows a generalized cyanobacterial cell containing structures typical of photosynthetic prokaryotic cells. These cells are small, generally between 1 and 6 μm,* but they may measure less than 1 μm. In general the cells are larger than bacteria but smaller than most algae. One of the most striking things about these cells is the total absence of membrane-bound organelles such as the nucleus, mitochondrion,

_____
*One micrometer = 1 μm = 1/10,000 cm; formerly called 1 micron.

chloroplast, and endoplasmic reticulum that are so common to eukaryotic cells. It is clear that cyanobacteria are structurally more closely related to the bacteria than to any other group.

Slime Sheath

The slime sheath is a gelatinous mass comprised of **pectic acids** and acid **mucopolysaccharides.** The precise content of these sheaths is quite complex, including a number of uronic acids, many sugars, and a few enzymes. The sheaths are not always apparent with the light microscope but are made readily visible with the phase microscope or with staining procedures for mucopolysaccharides (e.g., Alcian blue). It is the presence of the gelatinous mucilage associated with the sheath that gives the

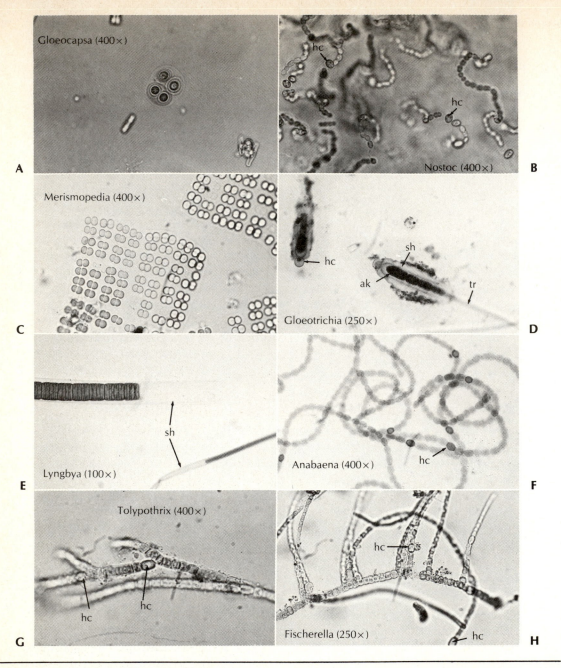

FIG. 3-1 Common forms of blue-green algae.

A, *Gloeocapsa* spp. (400×); note mucilaginous rings around individual cells. **B,** *Nostoc* spp. (400×); note that all cells in this filamentous genus are embedded in a gelatinous matrix. There are occasional heterocysts (*hc*) in the filaments. **C,** *Merismopedia* spp. (400×); note two-dimensional, flat sheet of cells. **D,** *Gloeotrichia* spp. (250×); note large basal akinete (*ak*) encased within a gelatinous sheath (*sh*). Also note tapering filament of trichome (*tr*) and basal heterocyst (*hc*). *Gloeotrichia* trichomes occur in hemispherical or globular colonies with their heterocysts in the center, thus forming a spiny ball. **E,** *Lyngbya* spp. (100×); note filament of disk-shaped cells encased within a gelatinous sheath (*sh*). **F,** *Anabaena* spp. (400×); note beadlike appearance of cells arranged in filaments containing occasional heterocysts (*hc*). **G,** *Tolypothrix* spp. (400×); note frequent "branching" arising from growth where heterocysts (*hc*) occur, thus producing "false branches." **H,** *Fischerella* spp. (250×); note true branching of filaments, appearance of occasional heterocysts (*hc*), and heterotrichous habit of this genus.

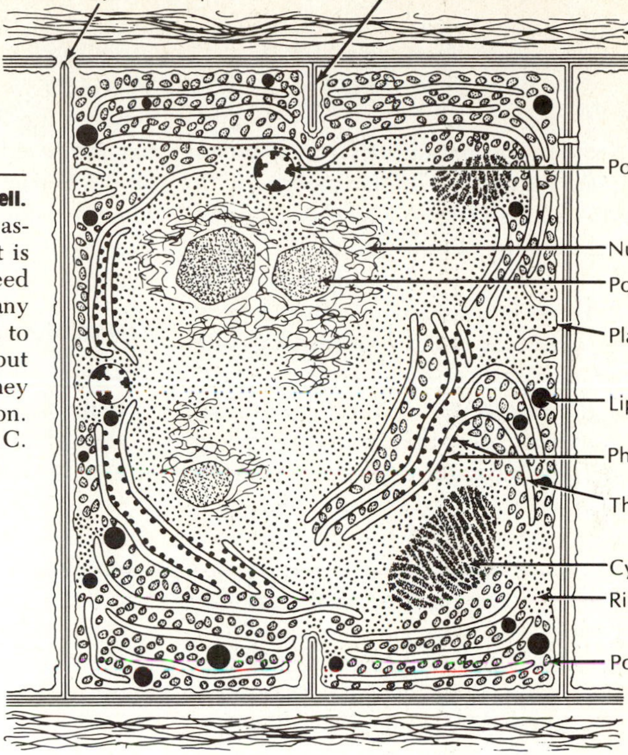
Junctional pore
Cell wall elaboration
Slime sheath
Cell wall
Polyphosphate body
Nucleoplasm with DNA
Polyhedral body
Plasma membrane
Lipid body
Phycobilisomes
Thylakoid lamella
Cyanophycin granule
Ribosomes
Polyglucan granule

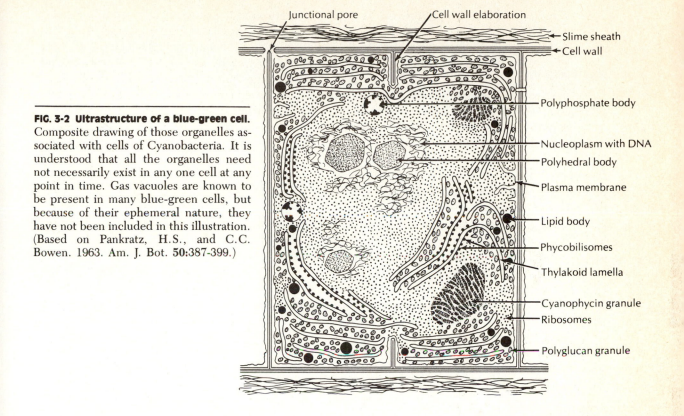

FIG. 3-2 Ultrastructure of a blue-green cell. Composite drawing of those organelles associated with cells of Cyanobacteria. It is understood that all the organelles need not necessarily exist in any one cell at any point in time. Gas vacuoles are known to be present in many blue-green cells, but because of their ephemeral nature, they have not been included in this illustration. (Based on Pankratz, H.S., and C.C. Bowen. 1963. Am. J. Bot. **50**:387-399.)

Cyanobacteria their common name, slime algae. Electron microscopic studies indicate the slime layer is structurally composed of microfibrils.

Cell Wall

The cell wall is a complex, four-layered structure containing many microplasmodesmata between adjacent vegetative cells. Other larger porelike structures—"junctional pores"—are narrow channels through which mucilage may be secreted. The walls also contain some "elaborations" that extend the walls into the protoplast, providing a certain degree of support for the cell contents.

Chemically the cell walls have affinities with the bacteria, which include compounds similar to those found in the bacterium *Escherichia coli*. Two amino sugars (muramic acid and N-acetylglucosamine) and three amino acids (alanine, glutamic acid, and diaminopimelic acid) have been identified in the cell walls of diverse cyanophytes. **Muramic acid** and **diaminopimelic acid** are characteristic of all prokaryotic organisms. Furthermore, a lipopolysaccharide containing mannose has been identified in blue-green cells that is chemically similar to one that has been isolated from some bacteria. Cellulose has not been definitely identified in blue-green cell walls, but many diverse sugars have been isolated. Finally, a distinctive feature of the cell walls is their vulnerability to digestion by lysozymes—enzymes from eukaryotic cells that also digest bacterial cell walls. Analysis has shown that the lysozyme-soluble part composes up to 50% of the cell wall; this appears to be the muramic acid fraction. All these findings indicate an evolutionary relationship between cyanophytes and bacteria. A good account of blue-green cell walls is given by Drews (1973).

Nucleoplasm

The nucleoplasmic area of the Cyanobacteria cell is composed mainly of DNA associated with a three-dimensional arrangement of protein fibrils. There are no chromosomes like those found in the nucleus of eukaryotic cells. The DNA is not found in association with the basic proteins (**histones**), as it is in most eukaryotic cells. The arrangement of the genetic materials in the blue-greens resembles more closely that found in bacterial cells than in the cells of eukaryotes. The exchange of genetic materials has been demonstrated in some blue-green species, but the process, which is quite distinct from meiotic segregation, remains poorly understood. One enigmatic genus of blue-greens, *Cyanidium*, possesses a membrane-bound nucleus. This organism is sometimes classified as a red alga (Rhodophyta) but is retained here as a blue-green because of its pigment complex.

Polyhedral Bodies

Polyhedral bodies are large (200 to 300 nm in diameter), polygon-shaped structures found in many vegetative cells and spores in association with the nucleoplasmic area. They are absent in heterocyst cells. They often extend longitudinally between two adjacent cells. Their association with DNA has lead to the theory that they play some role in the **transcription** process in which RNA is synthesized and stored in the form of nucleoprotein. Another theory is that they are the site for synthesis of ribulose 1,5-diphosphate carboxylase, an enzyme required for photosynthesis. Since polyhedral bodies are not found in heterocysts, and heterocysts lack Photosystem II, perhaps this theory is correct (Steward and Codd, 1975).

Polyphosphate Bodies

Polyphosphate bodies are large (100 to 500 nm), oval or round structures also known as **metachromatic granules** or **volutin granules** because of their staining properties. They are composed of linear polyphosphates of high molecular weight (Kessel, 1977). Their size and number increase with the age of the cell. They apparently function as a phosphate reservoir to accommodate rapid growth when environmental conditions are favorable. They are easily observed with the light microscope and may comprise a large volume of individual cells.

Thylakoid Lamellae

These structures are flattened, membrane systems that form the photosynthetic apparatus of the cell. They are prominent structures in all vegetative cells but may be altered in those differentiated cells known as heterocysts and akinetes. They contain, in association with their membranous structure, the pigments necessary for photosynthesis. These include chlorophyll *a*, various carotenoids, and assorted proteins including ferredoxins, cytochromes, and plastocyanin, all of which function in the photosynthetic process. The lamellae are often associated with phycobilisomes (see Fig. 1-7).

Unlike eukaryotic cells, where the photosynthetic apparatus is found in chloroplasts composed of stacks of grana formed by thylakoids, the blue green cells have the thylakoids distributed throughout the cytoplasm. The position and form of the thylakoids varies considerably in blue-green cells: they may occur peripherally scattered throughout the cells, or in a chainlike arrangement. They may also occur in groups approaching the structure of a granum.

Phycobilisomes

Phycobilisomes are small (10 to 15 nm), round, or rod-shaped structures that contain accessory photosynthetic pigments called **phycobilins**. Three types of phycobilins are found in these structures: the red-colored **phycoerythrin,** and the blue-green-colored **phycocyanin** and **allophycocyanin.** The phycobilisomes are always attached to the outer surface of the thylakoids. The phycobilins that are found in these structures are chemically bound to proteins, the **phycobiliproteins.** All three phycobilins act as accessory pigments to photosynthesis, probably the Photosystem II mechanism. These pigments, then, are capable of absorbing light in different wavelengths from chlorophyll *a* (see Fig. 2-7), thus enabling cyanophytes to live in environments where other photosynthetic organisms cannot. A review of phycobilisome structure and function may be found in Gantt (1980a).

Lipid Bodies

Scattered throughout the cytoplasm of most cyanophyte cells are small (30 to 90 nm) bodies that are **osmiophilic,** indicating that they are primarily lipid in nature. They are usually found near the plasma membrane. Their precise function has not been demonstrated, but it is hypothesized that they are reserve storage structures for high-energy-yielding metabolic products.

Cyanophycin Granules

Cyanophycin granules are structured granules larger than the lipid bodies, measuring up to 500 nm in diameter. Cytochemical tests indicate that these granules are composed of an unusual protein containing only two amino acids, arginine and aspartic acid. While no specific function has been demonstrated for these structures, it has been suggested that they act as reservoirs for nitrogen through the four nitrogen atoms associated with arginine. Supporting evidence for their role as nitrogen reservoirs is found in their absence in young cells, and in cells grown in nitrogen-rich media (Simon, 1973).

Polyglucan Granules

Polyglucan granules are numerous, small (25 nm) bodies sometimes occurring in stacks of elongate structures. They are composed of a polysaccharide material similar to animal starch (glycogen), and formed by branched polymers of α 1-4, 1-6–linked glucose molecules, called **myxophycean starch** by some investigators. These structures, probably the major storage point for sugars in blue-green bacterial cells, were formerly referred to as "cyanophycean starch granules."

Plasma Membrane

Just beneath the cell wall is found a structure composed of proteins and lipids called the plasma membrane. It is a typical fluid mosaic membrane (7 nm thick) found in all biological systems. Its function is to control the passage of chemical substances into and out of the cell. The membrane appears to have a complex structure, having many invaginations that protrude into the cytoplasm of the cells. It is thought that these invaginations give rise to new thylakoid membranes or other cellular inclusions.

Gas Vacuoles

Sometimes cyanobacterial cells produce elongate structures that are membrane bound and measure 70 nm in diameter. Their presence and size are variable, but these structures resemble the gas vesicles found in some bacteria (Walsby, 1972, 1977). These vacuoles can be viewed with the light microscope but are difficult to distinguish from other cellular inclusions. They often form in cells in response to environmental changes such as increasing light intensity and changing salinity or pH. Cyanobacteria involved in algal blooms often have many gas vacuoles in their cells. These ephemeral vacuoles play a dual protective role in cyanobacterial cells: not only do the vacuoles shield the photosensitive pigments from photooxidation by intense light, but they also protect the nucleic acids from ultraviolet rays. In addition, the vacuoles provide a means of buoyancy to otherwise dense cells that are unable to float. This latter function would certainly provide an ecological advantage, not only by enabling the alga to remain in the photic zone, but also by providing buoyancy to dispersal structures such as **hormogonia.**

Ribosomes

Ribosomes, which are 10 to 15 nm in diameter, are structures found in all living cells. In the cyanobacterial cell they are usually found in the area of the nucleoplasm but may be located anywhere within the plasma membrane. The sedimentation rates of ribosomes of the blue-green bacterial cell are similar to those found in bacteria. Both have 70s ribosomes (where s = Svedberg sedimentation coefficient), while eukaryotic cells have 80s ribosomes. The ribosomes of all cells are comprised mainly of RNA, and ribosomal RNA of cyanobacteria is similar in molecular weight to bacterial ribosomal RNA. Cyanobacterial ribosomes evidently function in protein synthesis in the same manner as RNA in all living cells, both prokaryotic and eukaryotic.

REPRODUCTION

Classic sexual reproduction in the Cyanobacteria has not been observed because no egg or sperm cells are produced, and meiosis does not occur. However, gene recombination through a parasexual mechanism does occur; it appears to be similar to the transformation process demonstrated in bacteria. Viral transduction may even cause genetic mixing, but this has not been demonstrated experimentally. Genetic exchange has been demonstrated in the laboratory for a few mutant cyanobacteria but has not been observed in naturally occurring populations. Most investigators think genetic recombination occurs and will eventually be definitively demonstrated in nature.

Asexual reproduction in the Cyanobacteria is accomplished in a number of ways. The most common is simple cell division. However, cell division of the type seen in eukaryotic cells is not found in the blue-greens, and therefore classic mitosis and meiosis do not occur. The cell division process closely resembles bacterial cell division, which is initiated through invaginations of the cell wall. The organelles distribute themselves equally on either side of the ingrowing wall. The manner by which the genetic material is equally distributed between two adjacent cells is not known. The entire process is often called binary fission or simply fission.

Hormogonia are short filaments of cells characteristic of filamentous cyanobacteria (the Oscillatoriales). They become separated from the parent **trichome** or filament and move away with a gliding motion typical of filamentous blue-green algae. Hormogonia are usually formed by separating a short filament of cells from the parent filament through **separation disks** (Fig. 3-3). These are nonliving cells that appear as biconcave disks. The filaments break at these points, liberating hormogonia, which may be formed at terminal or **intercalary** positions. Environmental alterations such as changes in humidity, temperature, pH, and salinity may induce hormogone formation. Hormogonia therefore represent not only a means of asexual reproduction but also a method of transporting the organism to a different environment. This becomes especially important when the organism is threatened by desiccation.

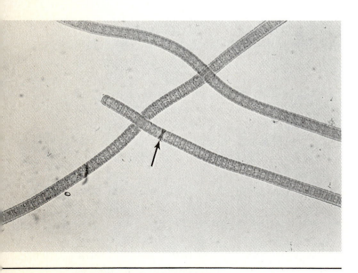

FIG. 3-3 *Oscillatoria* **spp. habit.**
Filaments of *Oscillatoria*, one of the most common Cyanobacteria in freshwater environments. Note separation disk indicated by arrow.

Some species of blue-greens also produce specialized reproductive cells called spores. **Exospores,** which function as a dispersal mechanism, are thick-walled spores produced at the end of a filament by cell division. They are formed basipetally one at a time and are especially common in the Chamaesiphonales (Fig. 3-4).

Endospores are formed within the parent cell protoplast, usually by multiple fission of the parent genome. They are usually thin-walled, and their major function appears to be dispersal (Fig. 3-5). Endospores are characteristic of some Chamaesiphonales (e.g., *Dermocarpa*).

Akinetes are thick-walled dormant spores formed within the original cell wall and are therefore a type of endospore. They are considerably larger than the ordinary cell and contain rich supplies of cyanophycin granules, representing a rich protein reservoir. They also contain 10 to 20 times the ordinary DNA content. These cells are resistant to adverse environmental conditions and may be viable for 70 to 80 years. The cell walls often produce many elaborations and are very thick, giving the cell a darker color than the vegetative cells (Figs. 3-1, *D,* and 3-6). Akinetes will germinate under favorable environmental conditions and form a vegetative trichome. They have been reported to occur in some unicellular cyanobacteria, but are much more common in the filamentous forms. In some filaments akinetes are formed randomly throughout the trichome, but in others they are formed only in apical or subapical positions.

A final mode of reproduction in the blue-greens is fragmentation. Most filamentous forms are rather fragile and are easily broken by various physical means. The resulting filaments merely float away and grow at some other favorable location. Unicellular forms are also dispersed by fragmentation.

Heterocysts (Fig. 3-1) are specialized cells that are differentiated from vegetative cells and are found only in the filamentous blue-greens, Oscillatoriales. The heterocysts and the akinetes are the most distinctive cells found in Cyanobacteria. Heterocysts, however, are not as large as akinetes, and their internal structure and contents are quite different. Heterocysts appear

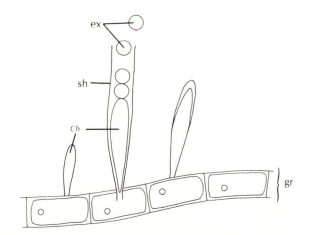

FIG. 3-4 *Chamaesiphon* **habit.**
Three filaments of *Chamaesiphon (Ch)* growing epiphytically on filament of green algae *(gr)*. Center *Chamaesiphon* is mature and producing exospores *(ex)*. Note sheath *(sh)* surrounding filaments.

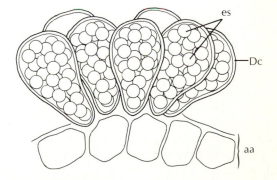

FIG. 3-5 *Dermocarpa* **habit.**
Seven *Dermocarpa* cells *(Dc)* growing epiphytically on aquatic angiosperm *(aa)*. Note that five of the *Dermocarpa* cells have produced multiple endospores *(es)*.

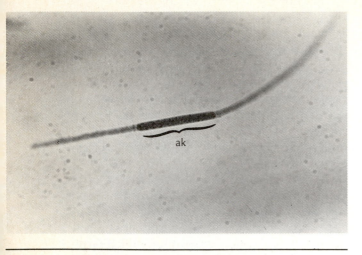

ak

FIG. 3-6 Aphanizomenon spp. habit.
Single filament of *Aphanizomenon* cells with centrally located akinete *(ak)*. In nature these filaments occur in short bundles, often in such abundance as to make ponds and lakes appear to be choked with pieces of chopped vegetation. Filaments are buoyant and are easily seen floating on water's surface (400×).

nearly empty when viewed with the light microscope. They are colorless except for two dark, distinctive polar nodules. The electron microscope reveals an internal series of folded membranes and some rather dense electron-opaque material. The polyhedral bodies, polyphosphate, and cyanophycin granules disappear as heterocysts develop. Pore channels are evident at either end of the cell through which cytoplasmic connections with adjacent cells are maintained.

The function of heterocysts has been debated and challenged many times. Their proposed role as reproductive spores, promoters of vegetative reproduction, points of attachment, and regulators of akinete and exospore formation has been hypothesized. However, none of these suggestions has been supported experimentally. In 1968 a new hypothesis was presented that promoted the idea that heterocysts were the site of nitrogen fixation (Fay et al, 1968). Since that time much interest has been generated, and ex-

perimental data have been gathered to support this theory. For instance, heterocysts are commonly found in those species of blue-greens capable of nitrogen fixation. Heterocysts are common in organisms grown on nitrogen-deficient media, but when grown on nitrogen-rich media, they are sparse or absent. With the establishment of cyanobacterial cells as fixers of atmospheric nitrogen, there has been renewed interest in the use of these organisms as instruments of soil revitalizations (Stewart, 1974b; Watanabe et al, 1977; Peters, 1978).

PHYSIOLOGY

Unlike the bacteria, blue-green cells produce molecular oxygen when they photosynthesize. The process follows this equation:

$$6\,CO_2 + 12\,H_2O \xrightarrow[\text{Light}]{\text{Chlorophyll}}$$
$$[C_6H_{12}O_6]_n + 6O_2 + 6H_2O$$

This equation is the same as the photosynthetic reactions in both eukaryotic algae and all higher plants. Also, the major photosynthetic pigment, chlorophyll *a*, is found in photosynthesizing eukaryotic cells. The major accessory photosynthetic pigments, however, are distinctive to the blue-green cells, although similar pigments can be found in some red algae (Rhodophyta) and some cryptomonads (Pyrrhophyta). These accessory pigments are composed of phycobilins, which are primarily found in the phycobilisomes. The blue-green pigment, phycocyanin, is always present and is usually the dominant phycobilin, but the red phycoerythrin and the blue allophycocyanin may also be present in significant quantities. The accessory (antenna) pigments enable these cells to absorb solar energy in the blue, green, and yellow wavelengths (450 to 700 nm), thus extending their effective light-absorbing range (Fig. 2-7). Chlorophyll *a* absorbs light energy primarily in the 400 to 440 nm and 640 to 680 nm ranges (Fig. 2-7). Notice that the absorption range for the cyanobacterial accessory pigments gives them a selective advantage in certain ecological niches such as deep water, where only blue-green light (450 to 550 nm) penetrates (Fig. 2-5).

A few species of blue-greens, such as *Oscillatoria limnetica*, are **chemoautotrophic** and capable of utilizing proton donors other than water in photosynthesis (Cohen et al, 1975). Some of these species are similar to hydrogen sulfide bacteria, because they obtain their electrons and protons by oxidizing hydrogen sulfide with the consequent production of elemental sulfur instead of oxygen. The photosynthetic equation for this is:

$$CO_2 + 2H_2S \xrightarrow[\text{Light}]{\text{Chlorophyll}} (CH_2O)_n + 2S\downarrow + H_2O$$

A few other chemoautotrophic reactions have been recorded for some Cyanobacteria species, thus giving further credence to the theory of their close relationship with the bacteria (Table 3-1).

Some blue-greens can also absorb and utilize preformed organic molecules as a source of energy, but most Cyanobacteria cannot live exclusively on organic materials. Saunders (1972) and Ellis and Stanford (1982) demonstrated that *Oscillatoria agardhii* living deep in the photic zone could transport organic nutrients into its cells. Most blue-green species, however, are autotrophs and utilize only inorganic molecules. The marine Cyanobacteria, however, evidently are auxotrophs, requiring at least vitamin B_{12} for growth.

Some Cyanobacteria have the ability to reduce atmospheric nitrogen (N_2) to ammonium salts (NH_4^+). This is an important biochemical reaction, because elemental nitrogen is not usable as a nutrient by other plants, but the ammonium salts are. Over 60% of the fixed nitrogen may be released into the environment and utilized by adjacent green plants. The **micronutrient** molybdenum is essential for nitrogen reduction; the entire process is called "nitrogen fixation." It has been reported most frequently in the filamentous Oscillatoriales, which produce heterocysts. However, nonheterocystous forms may also fix nitrogen under certain conditions (see below).

The heterocyst is thought to be the site of nitrogen fixation because it contains significant quantities of the oxygen-sensitive **nitrogenase** enzyme system. The thick cell walls of the heterocyst provide an anaerobic environment for this enzymatic activity. The enzyme nitrogenase is irreversibly damaged by oxygen and must be maintained in an anaerobic surrounding. Experiments have indicated that the frequency of heterocyst production in filamentous blue-greens is positively correlated with nitrogenase activity (Horne et al, 1972). When usable nitrogen sources are available in the form of ammonium compounds or nitrates, both the production of heterocysts and the amount of nitrogenase activity are greatly reduced (Fogg et al, 1973). Depletion of nitrogen sources results in increasing differentiation of heterocysts (Kulasooriya et al, 1972). The highest rate of nitrogen fixation occurs under conditions of low oxygen tensions. However, under artificially produced anaerobic conditions, ordinary vegetative cells of heterocyst-forming species may also show some nitrogenase activity. Evidently all blue-green bacterial cells can produce some nitrogenase, but the enzyme activity is inhibited by molecular oxygen. Some filamentous species that do not form heterocysts may fix nitrogen under conditions of low oxygen tension and reduced light intensity. One unicellular genus *(Gloeocapsa)* has been reported to be capable of nitrogen fixation (Stewart, 1977). Heterocysts apparently lack the oxygen-producing component of photosynthesis and thus have a low oxygen tension that protects the nitrogenase enzyme system required for nitrogen fixation (Tel-Or and Stewart, 1975). In addition, the heterocysts have active respiratory enzymes that can not only rapidly utilize any oxygen that might diffuse into the cell (Watley, 1976) but also produce the energy (ATP) required for reduction. Fogg (1974) and Peters (1978) discuss nitrogen fixation in greater detail. Nitrogen-fixing species have been isolated from such

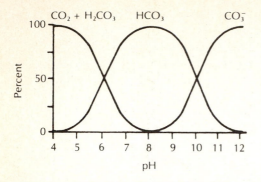

diverse environments as 50° C hot springs and from lichens growing at 0° C.

Most algae use free CO_2 as a carbon source for photosynthesis. The blue-greens are more efficient than other algae groups in obtaining carbon dioxide from waters low in free CO_2 (Shapiro, 1973). In neutral and alkaline waters most

FIG. 3-7 Theoretical relative proportions of different forms of carbon dioxide in water in relation to pH.
(From Cole, G.A. 1983. Textbook of limnology, ed. 3, St. Louis, The C.V. Mosby Co.)

FIG. 3-8 Stromatolites.
A, Stromatolite rock from the Cambrian dolostone of the Allentown formation; note the layered structure, presumably from cyanophyte algal mats. **B,** Section from columnar stromatolites taken from the Svanbergfjellet formation, Svalbard (Arctic Ocean) (800×). (**A** courtesy Donald Ryan and D. Spess, Department of Biology, Lehigh University, Bethlehem, Pa.; **B** courtesy Andrew H. Knoll, Biological Laboratories, Harvard University, Cambridge, Mass.)

A

B

of the available carbon is in the form of the bicarbonate ion (HCO_3^-), and most Cyanobacteria apparently can use this ion as a carbon source (Fig. 3-7). This ability to use the bicarbonate ion gives the Cyanobacteria a competitive advantage over other algae in neutral and alkaline waters. When the bicarbonate ion is removed via photosynthesis, the pH may rise to 9 or 10 and blue-greens will dominate the flora. Cyanobacteria generally dominate alkaline waters, while the green algae (Chlorophyta) may dominate slightly acid waters. Experimental work (Shapiro, 1973) has shown planktonic blue-green populations can be reduced by lowering the pH.

During photosynthesis by Cyanobacteria, calcium carbonate (calcite) often precipitates out of the surrounding waters onto the algal filaments. These carbonates may be deposited on the cell wall or on the mucilaginous sheath. Many species can also actively precipitate small amounts of $CaCO_3$. The deposition of calcium carbonate can alter the complex geochemical balance of carbon dioxide and calcium, resulting in an alkaline pH (Fig. 3-7). Marine Cyanobacteria may also precipitate some calcium carbonate, but most carbonate associated with marine blue-greens is physically trapped and bound in the mats that they form. Cyanobacteria therefore function as important geological agents in the formation of both coral reefs and stromatolites.

Geological phenomena known as **stromatolites,** rounded layers of calcium carbonate, are considered to be the result of ancient blue-green mats depositing carbonates in intertidal waters many millions of years ago (Fig. 3-8). Cell structure is not evident in most stromatolites, but a few well-preserved specimens indicate cellular organization.

Several species of Cyanobacteria produce an **endotoxin,** a cyclic polypeptide, which is retained in the cells and released upon cell death. The toxin affects the nervous system of vertebrates, causing paralysis, convulsions, or death. Drinking water with large cyanobacterial populations has been the reported cause of death of many cattle. Large blooms of some species of

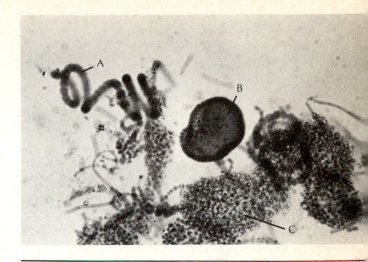

FIG. 3-9 Freshwater aquatic bloom.
Three genera often implicated in freshwater aquatic blooms are Cyanobacteria: A, *Anabaena* spp.; B, *Coelosphaerium* spp.; and C, *Microcystis* spp. (250×).

Aphanizomenon (Fig. 3-6), *Anabaena,* and *Microcystis* (Fig. 3-9) are known to produce toxic metabolites that can kill aquatic animals, farm animals, and birds. Most research on cyanobacterial toxins has been carried out using *Microcystis,* a common genus of summer aquatic blooms. Toxic blue-green blooms are a problem especially in areas around the fortieth lattitude (Canada, United States, and Australia), where environmental conditions favor the development of large populations. Blooms have also been implicated as causing human dermatitis. For example, a species of *Lyngbya* from Hawaii was shown to cause dermatitis in swimmers (Moikeha and Chu, 1971). Some people may be allergic to *Oscillatoria* or its metabolites and develop respiratory problems (hay fever, asthma) or eye irritations (conjunctivitis) after swimming near blooms.

EVOLUTION

Hetrotrophic bacteria are considered to have been the first living organisms on earth. These bacteria (prokaryotes) lived on the supply of organic molecules that were prevalent in the earth's waters at least 3.4 billion years ago. As the supply of organic molecules became smaller, natural selection favored evolving photosynthetic organisms. These organisms could make their own organic molecules from the readily available inorganic molecules (carbon dioxide, water, and minerals) using chlorophyll and solar energy. Autotrophic organisms prospered and multiplied because they could make their own food. Molecular oxygen (O_2) was a by-product of photosynthesis, and as autotrophic populations increased, molecular oxygen built up in the atmosphere. At the same time more organic molecules were added to the biosphere by the reduction of carbon dioxide during photosynthesis. Before the evolution of the process of photosynthesis, there was no molecular oxygen in the atmosphere. Another consequence of the evolution of molecular oxygen was the formation of the ozone (O_3) layer in the upper atmosphere, which reduced the amount of toxic ultraviolet radiation reaching the earth's surface.

Cyanobacteria probably were the first oxygen-evolving photosynthetic organisms. Their fossils, including unicellular, spherical cells resembling the Chroococcales, have been found with bacteria in the Precambrian Fig Tree chert from South Africa. These fossils are approximately 3.1 billion years old and are considered the oldest photosynthetic organisms. The geological time scale (see inside back cover of this book) shows the different geological eras and their time spans.

The bacteria and the blue-greens comprise the kingdom Monera. Blue-greens and bacteria share many characteristics (Table 3-1), including

FIG. 3-10 Fossil microorganisms.
Stromatolitic microorganisms in a section taken from the Gunflint chert formation, Ontario, Canada. Note the single-celled and filamentous forms, most of which are Cyanobacteria (1900×). Diameter of filaments approximately 1 μm. (Courtesy Andrew H. Knoll, Biological Laboratories, Harvard University, Cambridge, Mass.)

antiobiotic sensitivity and insensitivity to ultra-violet light, which was very prevalent 3 billion years ago. Some bacteria are photosynthetic and have the ability to convert inorganic molecules to organic molecules using solar energy. This process, however, occurs anaerobically using bacteriochlorophyll, which is chemically different from chlorophyll *a* and which absorbs light in the 700 to 750 nm range. Also the bacterial photosynthetic process is not linked to the photolysis of water, as is photosynthesis in the Cyanobacteria and in all other green plants.

Unicellular and filamentous Cyanobacteria are also reported from the Precambrian 2-billion-year-old Gunflint chert in Ontario (Fig. 3-10) and from the 1-billion-year-old Bitter Springs formation in Australia (Banks, 1970) (Fig. 3-11). The oldest known stromatolites are 2.7 billion years old and are thought to be of blue-green origin because of their resemblance to the laminated structures currently produced by Cyanobacteria. Large populations of blue-greens may build up in thick mats in calm estuaries. As these mats trap sand and carbonates precipitated out of the water, layered structures are produced that closely resemble Precambrian stromatolites. Such mats are currently formed in a bay on the Australian west coast (Playford, 1980) and in Antarctic lakes (Parker et al, 1981). Cyanobacteria have been extremely important in the deposition of carbonate sediments.

One unicellular, spherical form, *Cyanidium*, appears to be a transitional form between prokaryotic Monera and primitive eukaryotic organisms. *Cyanidium* is an enigmatic organism,

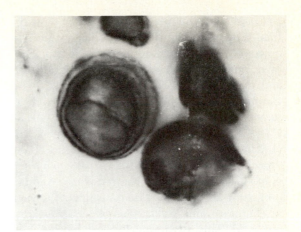

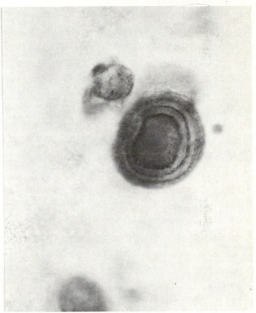

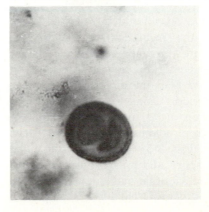

FIG. 3-11 Fossil microorganisms.
Three specimens of *Gloeodiniopsis lamellosa*, a fossil chroococcalean cyanobacterium from the Bitter Springs formation in Australia. Note concentric layers of the mucilaginous sheath and resemblance of this fossil to the modern genus *Gloeocapsa* seen in Fig. 3-1, *A*. For more data concerning this genus see Knoll and Golubic (1979). Maximum cell diameter approximately 25 μm (850×). (Courtesy Andrew H. Knoll, Biological Laboratories, Harvard University, Cambridge, Mass.)

because its pigment complex allies it with the blue-greens, while its morphology indicates eukaryotic affinities resulting from the presence of a membrane-bound nucleus, a chloroplast, and mitochondria. Klein and Cronquist (1967) claimed *Cyanidium* to be a transitional form between blue-greens and early green algae. However, Frederick (1979; 1980), upon biochemical analysis of food storage products and enzyme systems, demonstrates that *Cyanidium* is more closely allied to the red algae. Another enigmatic alga that may be transitional between prokaryotes and eukaryotes is *Synechocystis didemni*. This organism is prokaryotic, lacks phycobilins, and contains both chlorophyll *a* and *b* (Lewin and Withers, 1975). Chlorophyll *a* and *b* are generally found only in the green algae (Chlorophyta) and the higher green plants. *S. didemni* was found growing symbiotically in sessile sea squirts (tunicates) in the Marshall Islands and Hawaii. Lewin (1976) proposes the alga be reassigned to the genus *Prochloron* and placed in a new division, the Prochlorophyta, because it fits into neither the Cyanobacteria nor the Chlorophyta.

Since the evolution of blue-green algae over 3 billion years ago, they evidently have undergone few structural changes. Although the number of warm seas has decreased considerably, the aquatic environment is relatively stable compared to the terrestrial environment. Thus there has been little selection pressure to change from the highly successful ancient forms. A unicellular spheroid form such as *Chroococcus* is considered the most primitive form, while the palmelloid and filamentous forms are more advanced respectively. Fossil evidence suggests divergence between the unicellular and filamentous forms in the early Cambrian period. Cloud et al (1975) propose that the green algae may have arisen from the endospore-producing blue-greens (Chamaesiphonales, Fig. 3-5) because of the similarities of endospore and aplanospore formation. They also postulate that filamentous non-spore-forming blue-greens (Oscillatoriales) might have

given rise to the Rhodophyta, because spore formation in the red algae is quite different from that observed in the Cyanobacteria.

ECOLOGY

Cyanobacteria inhabit all corners of the world: they have been found in the atmosphere, as primary colonizers on newly exposed rock or soil, on the Greenland ice cap, in the highly saline waters of the Dead Sea, in Arctic and Antarctic lakes, in alkaline hot springs, in damp soils to a depth of one meter, in desert rocks, and in many other diverse freshwater and saltwater environments. In the desert environment they can survive extreme dessication. Aquatic species may be planktonic, benthic, **epilithic, endolithic,** or epiphytic and have been recovered from marine depths of 36 m. Blue-greens are capable of colonizing substrates that are uninhabitable to other primary producers.

Blue-greens may appear as patches of color—blue-green, brown, or black smears on rocks, soils, or trees. They are important members of the intertidal zone community (e.g., *Calothrix, Rivularia*, Fig. 3-12), often forming a blackish layer high on the rocks (see Fig. 2-16). The perceived color of the organism depends on which pigment or minerals predominate. The Red Sea is named for the color produced by the cyanobacterium *Trichodesmium erythraeum* (also called *Oscillatoria erythraeum**), which often dominates the algal community. The prevalent pigment, phycoerythrin, is water soluble and, when released from the cells, colors the water red. The Gulf of California is often called the Vermillion Sea because of similar *Trichodesmium* blooms. *Trichodesmium* is the most common marine planktonic Cyanobacteria; blooms have also been recorded frequently from the Indian Ocean and the Arabian Sea.

Trichodesmium can also fix nitrogen, not only in the tropics (Goering et al, 1966) but also in

*Humm and Wicks (1980) classify four species of *Trichodesmium* into one *Oscillatoria* species—*erythraea*.

the subtropical Caribbean Sea, where it may be responsible for up to 20% of the primary productivity (Carpenter and Price, 1977). *Trichodesmium* evidently can fix nitrogen in calm seas under aerobic conditions because the nitrogen-fixing cells are found near the center of the large floating algal mat where they are in a partially anaerobic environment (Carpenter and Price, 1976). When seas are rough, the mat is broken up, all cells are exposed to air, and nitrogen fixation ceases due to the sensitivity of the nitrogen reduction process to oxygen. Such mats form miniature ecosystems with associated bacteria, protozoa, and other organisms which become entangled in the mat (Andersen, 1977). There is some skepticism concerning the ability of *Trichodesmium* to fix nitrogen because the organism has not been grown in pure culture, and it may be the associated bacteria rather than *Trichodesmium* that is fixing nitrogen (Humm and Wicks, 1980).

Cyanobacteria are the major primary producers in mineral hot springs such as those found in Yellowstone National Park. Blue-greens have been reported growing at 90° C and successfully reproducing at temperatures only a few degrees lower. In these alkaline hot springs the blue-greens form extensive carbonate deposits called **marl** as they utilize the bicarbonate ion in the photosynthetic process and calcium carbonate precipitates. Deposition rates of 2 to 4 mm of carbonate per week have been recorded. This ability of the blue-greens to live and reproduce in alkaline, mineral hot springs gave rise to the theory that life evolved on earth approximately 3.5 billion years ago in hot, highly mineralized waters that covered the earth. Cyanobacteria thrive in very warm environments, including the tropics. The phycobilin pigments help shield against the high light intensity, and the mucilaginous sheaths insulate the protoplasm from high temperatures. Many present-day **thermophilic** Cyanobacteria grow readily in such heated waters (75° to 85° C) as the warm-water discharge canals from industrial and power plants. Extant **halophilic** (salt-loving)

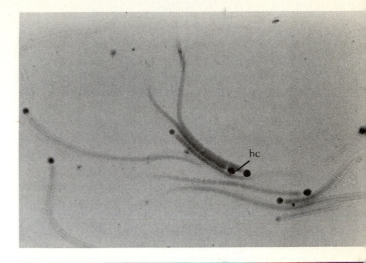

FIG. 3-12 *Rivularia* spp.
Filaments (trichomes) of *Rivularia* separated from their natural habit in which they occur packed within a mucilage, forming globular colonies, or produce a lumpy mucilaginous coating. Note appearance of basal heterocyst *(hc)*, tapering trichomes, and absence of akinetes (250×).

blue-green species can also be isolated from such highly saline environments as the Dead Sea (226 °/oo* salinity), the Great Salt Lake (200 °/oo salinity), and other areas with high evaporation rates such as isolated tidal pools.

Large populations (blooms) of Cyanobacteria can cause problems in ponds, lakes, reservoirs, and occasionally in flowing waters. The blue-greens rapidly build up a large biomass. Large populations persist because the Cyanobacteria are resistant to grazing and are not the preferred food of zooplankton or larger grazers. The crustaceans *Daphnia*, common zooplankton members, cannot extract enough food value from the blue-greens to support their population without additional food sources (Arnold, 1971). Certain species of herbivorous fish, *Tilapia*, however, can obtain adequate nutrition from Cyanobacteria and feed upon the dominant blue-greens in

*°/oo = parts per thousand.

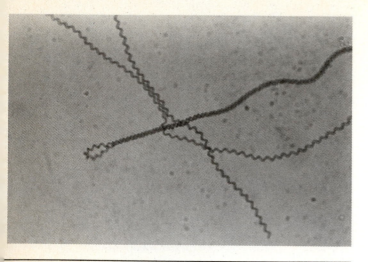

FIG. 3-13 *Spirulina* spp. habit.
Filaments of *Spirulina* cells growing in a pond. Note coiled nature of these filaments. This genus may grow as single coiled filaments or as bunches of coiled filaments, often intermingled with other genera of various algae. Since they have a high protein content, they are a primary food source for flamingoes in some African lakes (400×).

alkaline African lakes. Birds may also utilize the blue-greens. For example, it is estimated that a million flamingos of these African soda lakes may eat daily as much as 180 metric tons of blue-greens, primarily *Spirulina* (Fig. 3-13), (Johnson, 1976).

When a nutrient, light, or temperature becomes a limiting factor, large numbers of blue-greens in the **water bloom** start to die. Oxygen is essential for the breakdown of this large organic biomass. Dissolved oxygen levels in the lake may drop to near zero at night when oxygen-dependent bacterial decomposition continues, but no oxygen is generated in the water by photosynthesis. Oxygen depletion of this type results in fish kills and death of other oxygen-dependent organisms.

A bloom of Cyanobacteria is favored by high temperatures, long days, plentiful orthophosphate, available nitrogen, and an alkaline pH. Nitrogen-fixing species may dominate the phytoplankton when available nitrogen supplies are low. Obnoxious blue-green populations cause problems most often in late summer and early fall. Such blooms have produced a scum up to 20 cm thick on heavily fertilized fish ponds in Israel. The ability of many cyanobacterial genera to use the bicarbonate ion may interfere with the natural buffering system in the water and cause the pH to rise to 9 or 10. For example, a pH of 9.3 was recorded in a limestone stream during a September *Oscillatoria* (Fig. 3-3) bloom (Bradt, 1974).

Late summer blooms in Colorado have recorded populations of 62 million cells/liter of *Chroococcus* and 41 million cells/liter of *Schizothrix* (Reid and Wood, 1976). Such water blooms can render a lake undesirable for recreation as a result of scums and objectionable odors. Cyanobacteria are often the dominant flora of eutrophic lakes. Edmundson (1970) stated that the extensive growth of blue-greens signaled the onset of serious eutrophication problems in Lake Washington in Seattle. Such eutrophic lakes are highly productive, but Whittaker (1975) thinks this is a distorted and unbalanced increase in primary productivity. Genera that may build up large biomasses in lakes polluted with organic material are *Oscillatoria*, *Aphanizomenon*, and *Lyngbya*. Nitrogen fixation often occurs at high rates in eutrophic lakes, while it is less common in oligotrophic lakes (Tilson et al, 1977).

Approximately 20% of the known cyanobacterial genera occur in the marine environment. These marine species are important primary producers in intertidal zones (*Lyngbya*), or salt marsh muds (*Oscillatoria*), as phytoplankton (*Trichodesmium*), and on coral reefs (*Calothrix*, *Lyngbya*). Some blue-greens can cause erosion by secreting lime-dissolving substances, thereby enabling these organisms (*Hyella*, for example) to bore into mollusc shells, coral reefs, and limestone cliffs. Colman and Stephenson (1966) describe the boring Cyanobacteria of limestone cliffs along the Mediterranean shore. In estuaries about one third of the blue-green species are nitrogen fixers, while in the open ocean nitrogen fixation is negligible.

Blue-greens are found living symbiotically with other plants—e.g., diatoms, bryophytes, water ferns, gymnosperms, and angiosperms—and with some flagellated protozoans. Several nitrogen-fixing genera (*Nostoc, Gloeocapsa*) have been isolated from lichens, where they live in extremely close association with fungi (Chapter 20). *Anabaena azollae* is a nitrogen-fixing species (Fig. 3-14) found in the leaves of the water fern, *Azollae*. The *Azollae-Anabaena* complex has been used in the Far East to increase the fertility of rice fields (Peters, 1978). The complex is used in 1.3 million hectares of rice paddies in China and supplies up to 192 kg per hectare per year of useable nitrogen, the equivalent of approximately 90,000 metric tons of nitrogen fertilizer (Clark, 1980). Nitrogen-fixing cyanobacterial species have also been isolated from root nodules on the leguminous angiosperm *Trifolium* (clover).

Blue-greens rapidly colonize newly exposed soils; they are important soil stabilizers, since mats of blue-greens bind sandy or prairie soils, reduce erosion, and increase the organic content. These organisms are also used to reclaim soils that are depleted of nutrients and organic matter as the result of overuse. The lands are flooded, blue-green mats grow, and the land is subsequently drained. The mats dry on the soil surface, binding the soil, adding nutrients, and increasing the water-holding capacity. Such blue-green mats may be observed growing in the Florida Everglades during the rainy season. When the land drains, the blue-green mats add nutrients, organic material, and water-absorptive ability to the soils during the dry season.

Cyanobacteria appear to have an unusual function in some lakes in Antarctica: The blue-greens form thick mats on the lake bottom, and these mats break away and float to the top of the water, but under the ice. In the course of 5 to 10 years the mats work their way to the ice surface and are broken up and blown away by the intense winds. As the mats blow away from the

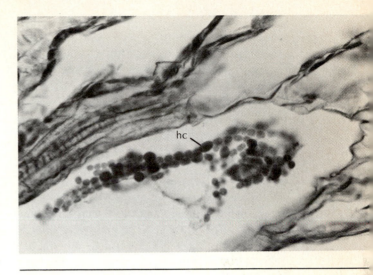

FIG. 3-14 *Anabaena azollae.*
Filaments of *A. azollae* growing within tissues of the aquatic fern *Azollae caroliniana*. Note presence of heterocysts (*hc*) in filaments of barrel-shaped cells.

lakes they take with them nutrients and other organic materials that are thereby removed from the lake. The lake, then, remains in an oligotrophic condition because the nutrients are being constantly removed (Parker et al, 1982). *Phormidium frigidum* dominates in the Antarctic lakes that have been studied to date.

ECONOMIC IMPORTANCE

Blue-greens are often the dominant forms in water blooms in ponds, lakes, and reservoirs. These blooms usually occur in late summer and early fall, forming large areas of gelatinous material that floats on the lake surface or covers the bottom of streams or rivers. A large Cyanobacteria bloom in a water supply reservoir may cause problems because they produce objectionable tastes and odors. The large mass of algae may also clog sand filters through which the water must pass. Genera most often causing water supply problems are *Anabaena, Aphanizomenon, Microcystis, Oscillatoria,* and *Coelosphaerium* (Figs. 3-6 and 3-9). Copper sulfate (1.0 ppm) may be used to reduce cyanobacterial populations, while activated carbon will effectively remove undesirable tastes and odors. The ability of some Cyanobacteria to remove carbonates may benefit water treatment, because the hardness of the water is reduced by such carbonate removal, thereby reducing the need for additional water treatment.

Toxic Cyanobacteria blooms have been responsible for illness and death in birds and domestic and wild animals in many countries of the world. Often it is not possible to establish definitely the causative agent in such toxic blooms. Schwimmer and Schwimmer (1968) discuss the incidences of known poisonings from Cyanobacteria toxins. Freshwater blue-greens are usually implicated as toxin-producing, but blooms of the marine *Trichodesmium* have caused toxicity to both fish and invertebrates. A species of *Lyngbya* that grows near tropical

Pacific coral reefs has been implicated in **ciguatera** poisoning of both fish and humans. Ciguatera in humans effects both the nervous and digestive systems. The herbivorous fish feed on the *Lyngbya,* and they in turn are eaten by carnivorous fish that are eaten by humans (Scheuer, 1977). Moore (1977) and Carmichael (1981) discuss both freshwater and marine toxin-producing Cyanobacteria.

The ability of blue-greens to fix nitrogen has led to their extensive use where available nitrogen may be limiting to growth. Use is made of this biochemical ability in some parts of the world in order to maintain high agricultural productivity. For example, in the Far East soils are inoculated with nitrogen-fixing Cyanobacteria to maintain high yields in rice paddies. Calculations show that blue-greens are capable of fixing 92 to 192 kg of nitrogen/hectare/year, thus substantially reducing the need for additional fertilizers (Fogg et al, 1973). Furthermore, rice paddies are ranked among the most highly productive managed areas of the biosphere, producing (net) 3000 g dry organic material/m²/yr (Whittaker, 1975). In Japan nitrogen-fixing blue-greens are cultured in tanks and spread on the rice fields annually. Introducing these Cyanobacteria onto the soil not only improves both organic content and water holding capacity, but also furnishes available nitrogen for the growing rice. Rice yields were increased by 20% by adding certain species of *Tolypothrix* (Fig. 3-1) to the soil, and the need for nitrogen fertilizer was greatly reduced (Tamiya, 1957).

Dried blue-greens are used as a food source in Mexico, South America, Africa, and China. *Nostoc, Spirulina, Phormidium,* and *Chroococcus* are most often used. *Nostoc* balls (Fig 3-15) are sold in South American and African markets. The *Spirulina* (Fig. 3-13) that grows in alkaline lakes in Mexico and Africa may have a protein composition as high as 65% to 70% of the dry weight and is used as a local food source. In Mexico 900 metric tons are produced per year in *Spirulina* ponds that measure 3 k in diameter.

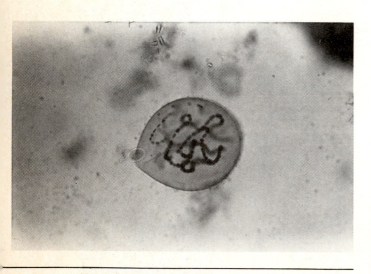

FIG. 3-15 *Nostoc* **spp. habit.**
Genus *Nostoc* consists of filaments of cells growing in a gelatinous matrix, forming balls of various sizes. The genus has been used as a human food source in some parts of the world (250×).

Spirulina is also being advocated as a nutritious supplement to diets in developed countries (Johnson, 1976).

Cyanobacteria are a potential source of both antibiotic and anticancer compounds. Several species have antibiotic activity (Starr et al, 1962) while extracts from several marine Oscillatoriales from the Marshall Islands demonstrated antileukemic activity against mouse leukemia (Mynderse et al, 1977). A compound isolated from a species of *Lyngbya* was not only antileukemic but it also irritated skin and was a potential cause of contact dermatitis ("swimmer's itch").

SUMMARY

The ubiquitous Cyanobacteria are important primary producers throughout the biosphere. They may be found converting solar energy to chemical energy in such inhospitable environments as hot springs, glaciers, and hypersaline lakes. The blue-greens (division: Cyanobacteria) and the bacteria (division: Schizophyta) are the only members of the kingdom Monera and share the following prokaryotic characteristics: (1) They have no membrane-bound organelles, (2) cell division is by fission, (3) there are similar chemical components in the cell wall, (4) there is an absence of typical sexual reproductive processes found in eukaryotic organisms, and (5) some species can fix nitrogen. Both biochemical and electron microscope studies contribute evidence to the close relationship of the blue-greens and the bacteria.

Heterotrophic bacteria are postulated to have been the first living organisms on earth (3.4 billion years ago), followed by the autotrophic blue-greens that could metabolize low-energy molecules (carbon dioxide and water) and produce high-energy organic molecules, using the sun as an energy source. In the photosynthetic process molecular oxygen (O_2) was produced as a by-product. As molecular oxygen built up in the atmosphere aerobic respiration became possible, and organisms evolved that could utilize organic food molecules very efficiently.

Cyanobacteria may occur as single cells, single cells in a gelatinous matrix, or as filaments. Evidence for genetic exchange is definitive for only a few mutant species. Vegetative reproduction occurs by akinetes, heterocysts, hormogones, or by fragmentation. Blue-greens can vegetatively multiply rapidly and built up a large biomass, occasionally causing problems in reservoirs, lakes, and estuaries.

Biochemically Cyanobacteria are extremely versatile, hence their capacity to exist under conditions where no other autotrophic organisms can. Their versatility may also cause problems as they exploit favorable environments. Morphologically the blue-greens apparently have not changed significantly since Precambrian times. They are highly successful organisms, having been able to live and multiply on the earth for 3 billion years in conservative, relatively unchanging aquatic environments.

SELECTED REFERENCES

Carr, N., and B. Whitton. 1973. The biology of the blue-green algae. University of California Press, Berkeley. 676 pp.

Desikachary, T. 1959. Cyanophyta. Indian Council of Agricultural Research, New Delhi, India. 686 pp.

Fogg, G.E., W.D.P. Stewart, P. Fay, and A.E. Walsby. 1973. The blue-green algae. Academic Press, Inc., New York. 459 pp.

Haselkorn, R. 1978. Heterocysts. Ann. Rev. Plant Physiol. **29**:319-344.

Humm, H.J., and S.R. Wicks. 1980. Introduction and guide to the marine blue-green algae. John Wiley & Sons, New York. 194 pp.

Orme-Johnson, W.H., and W.E. Newton (eds.). 1980. Symbiotic associations and blue-green algae. Vol. II. University Park Press, Baltimore. 352 pp.

Peters, G.A. 1978. Blue-green algae and algal associations. BioScience **28**(9):580-585.

Stewart, W.D.P. 1974. Blue-green algae. In The biology of nitrogen fixation. pp. 202-237. A. Quispel (ed.). American Elsevier Publishers, Inc., New York.

Stewart, W.D.P. 1977. Blue-green algae. In Hardy, R.W.F., and W.S. Silver (eds.). A treatise on dinitrogen fixation. pp. 36-121. Sect. III: Biology. John Wiley & Sons, Inc., New York.

4

CLASSIFICATION

Division: Chlorophyta (green algae): 450 genera, 8000 species; 87% freshwater; 13% marine

Class: Chlorophyceae

ORDER 1. Volvocales: unicellular to colonial (coenobia); motile, biflagellate cells; *Chlamydomonas, Dunallielia, Eudorina, Gonium, Pandorina, Polytoma, Volvox*

ORDER 2. Tetrasporales*: unicellular to colonial; mostly nonmotile; some in a gelatinous matrix; *Ankistrodesmus, Chlorella, Hydrodictyon, Pediastrum, Prototheca, Scenedesmus, Tetraspora*

ORDER 3. Ulotrichales: branched or unbranched filaments with a holdfast; uninucleate cells; *Chaetophora, Draparnaldia, Fritschiella, Microspora, Stigeoclonium, Trentepohlia, Ulothrix*

ORDER 4. Coleochaetales: branched filaments; uninucleate cells; heterotrichous habit; many epiphytes; *Coleochaete*

ORDER 5. Oedogoniales: branched or unbranched filaments; cell walls with terminal rings; large oogonia; fresh water only; *Oedogonium*

ORDER 6. Ulvales: sheetlike *(Ulva)*, tube-like *(Enteromorpha)*, or solid cylinders *(Schizomeris)* of cells; uninucleate cells

ORDER 7. Zygnematales†: uninucleate cells with distinctive chloroplasts; amoeboid gametes; no flagella; *Mougeotia, Spirogyra, Zygnema*, desmids such as *Closterium, Cosmarium, Micrasterias, Staurastrum*

ORDER 8. Cladophorales: branched or unbranched filaments; multinucleate (coenocytic) cells; *Chaetomorpha, Cladophora, Rhizoclonium*

ORDER 9. Caulerpales: siphonaceous or coenocytic structure; often depositing calcium carbonate; mostly marine; *Halimeda, Neomeris, Penicillus, Udotea*; also non-calcium-carbonate forms such as *Bryopsis, Caulerpa,* and *Codium*

ORDER 10. Siphonocladales: siphonaceous or coenocytic structure; becoming septate with maturity; marine; *Anadyomene, Valonia*

ORDER 11. Dasycladales: siphonaceous or coenocytic structure; becoming septate with maturity; marine; radially symmetrical or whorled branches; *Acetabularia, Cymopolia*

---

*The order Tetrasporales may be divided in three other orders, the Chlorococcales, the Chlorosarcinales, and the Chlorellales.

†The Zygnematales (Conjugales), because of their unique chloroplasts and lack of flagellate cells, are sometimes considered as a separate class, the Zygnemaphyceae.

CHLOROPHYTA

Division Chlorophyta belongs to one of the three main lines of algae that have evolved in the kingdom Protista and are therefore composed of eukaryotic (nucleate) cells. In general these algae appear to be grass-green and are photosynthetic, autotrophic organisms, but a few species can maintain a heterotrophic existence. Most of them are freshwater organisms, although some are found in the marine environment. They occur as unicellular forms or as multicellular organisms having colonial, filamentous, sheetlike or three-dimensional **pseudoparenchymatous** or parenchymatous organizations. Some organisms attain heights greater than 5 to 10 cm; however, most forms require a microscope for identification.

The green algae exhibit great diversity of form, ranging from microscopic motile unicells (*Chlamydomonas*, Fig. 4-1) and non-motile unicells (*Chlorella*) to the unicellular *Acetabularia*, which may be 8 cm high (Figs. 4-2 and 4-19). Some form microscopic, multicellular, netlike colonies (*Hydrodictyon*, Fig. 4-3), while others develop into spherical colonies (*Volvox*, Fig. 4-4). Some filamentous forms are only a few cells long, while others (*Cladophora*, Fig. 4-5) may reach lengths of 3 to 4 m. *Coleochaete* develops into small macroscopic disks. *Ulva* (Fig. 4-6) forms leaflike plants that may be only two cells thick. Some genera such as *Udotea* are capable of depositing calcium carbonate in their cell walls, thus producing hard, encrusting, coral-like forms. And finally, a pseudoparenchymatous species of *Codium* (Fig. 4-7) from the Gulf of Mexico may reach heights of 8 m.

Like all eukaryotes, the cells of the green algae contain membrane-bound nuclei plus all of the many organelles associated with eukaryotic cells, including chloroplasts and other plastids, endoplasmic reticula, Golgi bodies, microtubules, ribosomes, chromosomes, nucleoli, plasma membranes, nuclei, and cell walls. The

Chapter 4

FIG. 4-1 *Chlamydomonas* **structure.**

Internal anatomy of unicellular, biflagellate vegetative cell of *Chlamydomonas*, a common freshwater planktonic green alga. Note single cup-shaped chloroplast containing single large pyrenoid, thylakoids in bands of one to six, and eyespot. Note also presence of contractile vacuole as well as other usual organelles associated with eukaryotic cells. (After Walne, P.L., 1967. Am. J. Bot. **54:**564-577.)

FIG. 4-2 **Life cycle of** *Acetabularia.*

Acetabularia is a good example of the complexity that can be achieved in the life history of a single-celled organism. Life cycle is diplohaplontic type with isogamous reproduction. Young adult plant, which reaches heights of 3 to 7 cm, is especially interesting, since it is a single cell containing a single "primary" diploid nucleus within the rhizoidal area. Apex of cell differentiates into gametangial rays. When gametangial rays are completely formed, primary nucleus enlarges and divides meiotically, producing multiple haploid secondary nuclei. Haploid nuclei travel by cyclosis to gametangial rays, where they form resistant, multinuclear, walled cysts. Cysts undergo obligatory period of dormancy for 12 to 15 weeks before germination and release of biflagellate isogametes (see Koop, 1975a, 1975b).

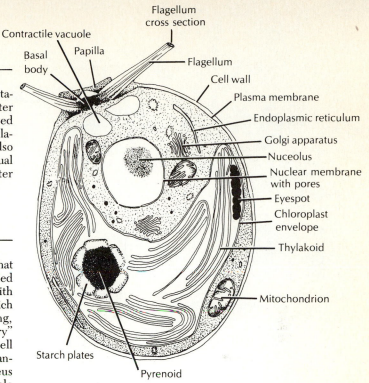

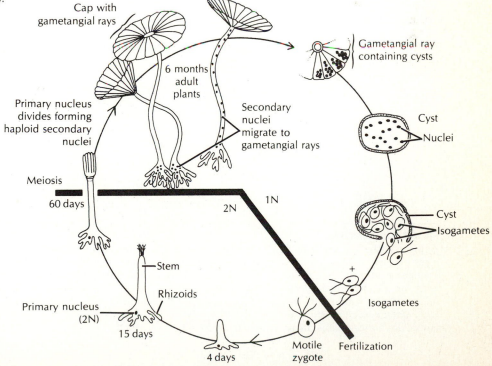

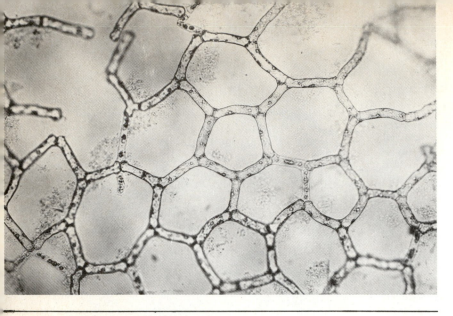

FIG. 4-3 Hydrodictyon habit.
Portion of a network of cells composing genus *Hydrodictyon*, or "water net"(125×). Such nets may reach sizes extending to 60 cm or more and may become an annoying weed in freshwater ponds and lakes.

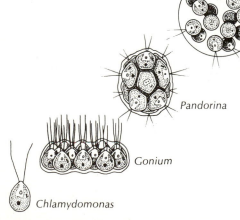

FIG. 4-4 Volvocine line.
Representative genera from Volvocine line of green algal evolution. Each group of cells is a coenobium containing a specified number and arrangement of cells for an individual species within the specified genus. Note gradual transition from unicellular *Chlamydomonas* to *Volvox* with its 1,000 to 40,000 cells. Also notice gradual increase in cell numbers per colony, increase in colony size, and increasing degree of cellular differentiation within each colony. *Chlamydomonas* is single celled. *Gonium* is a flat colony of 4 to 32 *Chlamydomonas*-like cells. Species illustrated has 16 cells. *Pandorina* is an eliptical colony of tightly packed, wedge-shaped cells. Anterior cells contain larger eyespots than posterior cells. *Eudorina* is a hollow colony of 16 to 64 cells. *Pleodorina* is a hollow sphere of 32 to 128 cells. Both *Eudorina* and *Pleodorina* have cells of different sizes with smaller cells posterior and larger cells anterior. The same is true of *Volvox*, a hollow sphere of cells with cytoplasmic connections (plasmodesmata) between each cell. *Volvox* represents the most highly advanced member of the Volvocine line. Sexual reproduction is oogamous. Polarity in colonies is easily determined by the presence of larger cells and larger eyespots in anteriorly located cells. Note formation of asexual reproductive "daughter colonies" within adult *Volvox* colony illustrated.

cells undergo regular mitosis and meiosis. The adult organisms are capable of reproducing both sexually and asexually. Motile biflagellate cells are common to all chlorophytes except the Zygnematales, whose only motile cells are amoeboid gametes. The two anterior whip-lash-type flagella are equal in length (isokont). These motile cells are called **zoospores** or **plano-spores.**

The biochemical nature of three major cellular components provides chemical characteristics that aid in the classification of most algae. The chemistry of the chlorophyll types, the molecular arrangement of the energy storage compounds, and the composition of the cell wall materials are used in this respect (see Table 1-1). In the green line of algal evolution chlorophyll types *a* and *b* are always present. The presence of chlorophyll *b*, which is unique to this line; indicates a relationship between green algae, bryophytes, and vascular plants.

The energy storage product is starch, a polymer of glucose molecules having an α 1-4 linkage. Actually this "plant" starch is a mixture of

FIG. 4-5 *Cladophora* habit.
Branching filaments of multinucleate *Cladophora* cells with netlike chloroplast (30.5×).

FIG. 4-6 Life cycle of *Ulva*.
Life cycle of *Ulva* represents an example of a diplohaplontic life cycle with isomorphic alternations of haploid and diploid generations and gametophytic generation that is dioecious. *Ulva* is a common floating or attached, shallow-water, marine, tropical or temperate, green alga known as "sea lettuce." Adult thalli are only two cells thick.

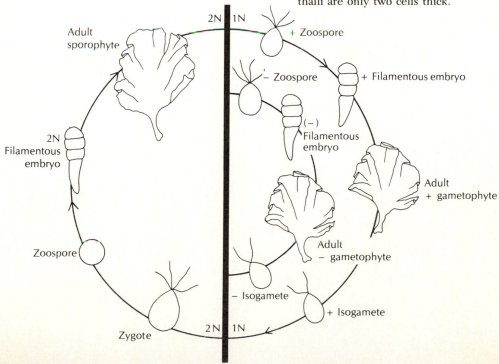

amylose, an unbranched polymer of glucose having an α 1-4 linkage, which is joined to amylopectin by an α 1-6 linkage. The amylopectin is a branched polymer of glucose that has the same structure as **glycogen** (animal starch), although it does not possess as many side branches. Starch is retained and stored by algal cells in various colorless plastids called leucoplasts or amyloplasts, or sometimes in **pyrenoids.**

The cell walls of most green algae are composed of cellulose and various mucilaginous substances, although the flagellate forms generally have walls composed of other substances. Biochemically and structurally, the wall materials of the green algae resemble those found in bryophytes and vascular plants.

CELL STRUCTURE

In most of the green algae the cell walls are composed of an inner layer of cellulose containing a linear polymer of glucose molecules and a thinner outer layer of nitrogen-containing pectins. The outer layer may become impregnated with calcium carbonate ($CaCO_3$) in some members of Caulerpales, Dasycladales, and Siphonocladales. The walls vary in chemical composition among chlorophytes. Most species contain

FIG. 4-7 Life cycle of *Codium*.
Life cycle of this monoecious species of *Codium* is shown as an example of a diplontic life cycle in which anisogamous gametes are produced. Adult vegetative plant is dark green, 25 to 40 cm high, and has a three-dimensional, spongy consistency that led to its common name, "Dead-man's fingers." It is a common alga in tropical and temperate shallow marine waters, where it grows attached to such solid substrates as rocks and shells. It is sometimes a serious "weed" in commercially important oyster beds.

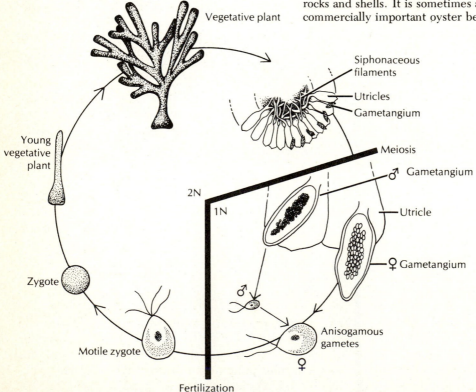

Vegetative plant

Siphonaceous filaments

Utricles

Gametangium

Meiosis

♂ Gametangium

Utricle

♀ Gametangium

2N

1N

Young vegetative plant

Zygote

Motile zygote

Fertilization

Anisogamous gametes

walls composed primarily of cellulose plus a number of other substances, including sugars and glycoproteins. True cell walls are rare in the flagellate forms, except for cellulosic cysts, zygotes, and other nonmotile forms (Stewart and Mattox, 1980). In flagellate green algae the cell covering is composed of polymers of mannose (mannans) or xylose (xylans). Amino acids, especially, hydroxyproline, are also present; they have been isolated from the cell covering of noncellulosic *Chlamydomonas* spp. (Lembi, 1980).

The presence of chlorophyll types *a* and *b*, starch, and cellulose cell walls supports a major argument seeking to establish an evolutionary link between the chlorophytes and the higher vascular plants. As will be seen later, these three characteristics are not the only evidence for an evolutionary relationship. Current biochemical, immunological, and morphological research is constantly adding to information in this area. For instance, the type of flagellation, the ultrastructure of the chloroplasts, and the mode of cell reproduction are all evidence for relating the chlorophytes to the higher plants.

Classically the green algae are viewed to have evolved from single-celled biflagellate organisms resembling the modern genus *Chlamydomonas*. It is not difficult to derive the entire green algae division from such hypothetical cells (Fig. 4-8). Fossil green algae have been found in the geological record dating from Precambrian deposits that are one billion years old. It is interesting to note that modern vascular plants did not begin inhabiting the earth until the Silurian period 65 million years later (See the geological time scale inside the back cover of this text.)

The structure of green algal cells may be considered as a series of variations on the structure of the *Chlamydomonas* cell (Fig. 4-1). All the cells of the green algae are eukaryotic; that is, they possess membrane-bound nuclei within which are located one or more nucleoli, numerous chromosomes, and fluid nucleoplasm. The nucleolus—the most clearly defined structure within the nucleus—is composed largely of granules that are rich in messenger RNA. It is not always present, and the size and number of nucleoli per nucleus depends upon the metabolic state of the cell. In cells actively growing or synthesizing protein, the nucleoli may become very large, because it is the messenger RNA in the nucleoli that is used to assemble ribosomes in the cytoplasm that in turn produce protein and subsequent growth.

Chromosomes contain the DNA necessary for the transmission of hereditary information. The number of chromosomes per nucleus varies depending upon the individual species and whether the cell is haploid or diploid. Within any given species the number of chromosomes remains constant. The chemical composition of the chromosomes is complex, with the chief component being DNA. However, significant quantities of histone protein and **nonhistone chromosomal protein** also occur. Histones, which are proteins containing a positive (basic) charge, function in conjunction with DNA to control gene expression and **morphogenesis.** In many cells the DNA/histone ratio appears to be 1:1. Besides the protein moiety, chromosomes also contain small quantities of lipids and RNA.

The structure of chromosomes varies with the stage of the nucleus in the cell cycle. Usually they are distended and dispersed throughout the nucleus and are referred to as "chromatin material." When a cell divides by mitosis or meiosis, the chromosomes condense and are visible through a light microscope. At this time chromosomes are composed of one or two chromatids joined at the centromere. These chromosomes are readily stainable by special techniques and are easily seen with a light microscope.

The nucleus is bound by a nuclear membrane that possesses many pores through which materials are exchanged with the cytoplasm (Fig. 4-1). The cytoplasm contains a great variety of inclusions and organelles, many of which are visible only with electron microscope preparations. The most prominent feature of the cytoplasm is the chloroplast; this large green structure is often

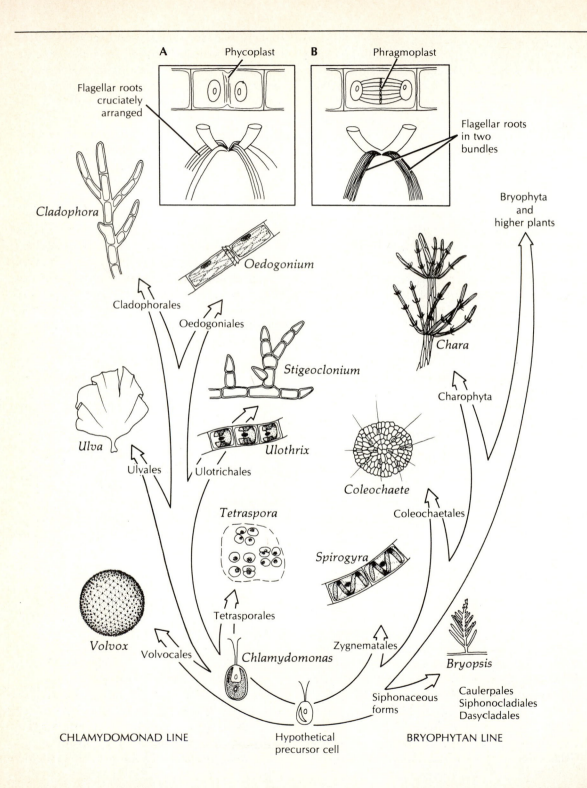

A   Phycoplast

B   Phragmoplast

Flagellar roots cruciately arranged

Flagellar roots in two bundles

Cladophora

Oedogonium

Bryophyta and higher plants

Cladophorales

Oedogoniales

Stigeoclonium

Chara

Ulva

Charophyta

Ulvales

Ulotrichales

Ulothrix

Coleochaete

Coleochaetales

Tetraspora

Spirogyra

Tetrasporales

Zygnematales

Bryopsis

Volvox

Volvocales

Chlamydomonas

Siphonaceous forms

Caulerpales
Siphonocladiales
Dasycladales

CHLAMYDOMONAD LINE

Hypothetical precursor cell

BRYOPHYTAN LINE

FIG. 4-8 Evolution in the Chlorophyta.
Evolutionary tree for green algae based on: (1) presence of phycoplast during cell division and cruciately arranged flagellar roots, or (2) presence of phragmoplast during cell division and flagellar roots arranged in two bundles. The two criteria are illustrated in boxes **A** and **B**. Those algal orders evolving to the left of the precursor cell have cellular characteristics seen in box **A** and are called the "chlamydomonad line." Those orders evolving to the right of the precursor cell have the cellular characteristics seen in box **B** and are called the "bryophytan line." Note that *Volvox* (Volvocales) and other volvocine algae are a branch of the chlamydomonad line. This branch is followed in sequence by *Tetraspora* (Tetrasporales: uninucleate cells in a gelatinous matrix); *Ulva* (Ulvales: uninucleate cells in two dimensions); *Ulothrix* and *Stigeoclonium* (Ulotrichales: uninucleate cells in filaments and branching filaments respectively); *Oedogonium* (Oedogoniales); and *Cladophora* (Cladophorales: multinucleate filaments).

The bryophytan line represents those green algal orders that give rise to vascular plants. *Bryopsis* (Caulerpales) represents all the siphonaceous orders, including Siphonocladiales and Dasycladales. *Spirogyra* (Zygenematales) represents those green algae with amoeboid gametes. *Coleochaete* (Coleochaetales) with its heterotrichous habit and sheathed gametes represents those green algae closest to bryophytes and vascular plants. *Chara* (Charophyta) is a separate division of closely related algae.

used to identify various algae because of its distinctive structure (Fig. 4-9). The ultrastructure of the chloroplast is also distinctive and provides morphological evidence for assigning some questionable organisms to specific taxonomic groups. The green algal chloroplast, for instance, has a double membrane-bound envelope, within which are located many disk-shaped thylakoids stacked in groups of two to six and suspended in a fluid matrix (see Fig. 1-7). The envelope lacks the CER associated with the plastids in most other algal groups. The membranes of the thylakoids contain the pigments chlorophyll *a* and *b* together with the many other pigments, enzymes, coenzymes, and various proteins associated with photosynthesis. The pigment complex is often distinctive. Specific **carotenes** (yellow-orange) and **xanthophylls**

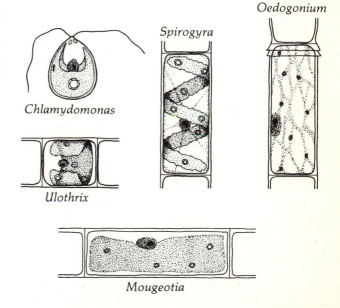

FIG. 4-9 Chloroplast shapes in green algae.
Note cup-shaped chloroplast in *Chlamydomonas;* band-shaped, braceletlike chloroplast in *Ulothrix;* spiral or helical chloroplast in *Spirogyra;* reticulate or netlike chloroplast in *Oedogonium;* flat, platelike chloroplast in *Mougeotia;* stellate (star-shaped) chloroplasts in *Zygnema;* and multiple oval or disk-shaped chloroplasts in this uninucleate cell of *Cladophora.*

(orange-red) may be present (see Table 1-1). These colored pigments, under certain conditions such as nitrogen deficiency, may be located throughout the cytoplasm, giving the entire cell an orange or red color. Some small quantities of starch, lipid, and DNA may be found within the chloroplast matrix.

Green algal chloroplasts often contain one or more pyrenoids, a proteinaceous body that is readily seen with the light microscope and that apparently functions· as the site of starch synthesis. Evidence for this is found in the presence of starch grains and starch plates that may be associated with the pyrenoids. However, because not all genera that synthesize starch contain these structures (e.g., *Microspora*), the precise manner in which pyrenoids function is still a matter for research.

Also associated with chloroplasts in many motile green algae is a structure called the **eyespot** or **stigma,** which is composed of tightly stacked droplets of carotene-lipid complexes. Its color is usually red but may be orange. Researchers dispute the function of this organelle. Some consider the eyespot to be the site of light reception, whereas others believe it functions as a shading device for the true photoreceptor located at the base of a flagellum. In either case the eyespot does function (directly or indirectly) in photoreception. Eyespots are found only in motile algae or the motile gametes of benthic or attached forms.

The cytoplasm of the chlorophyte cell also contains a well-developed network of membrane-bound channels called the endoplasmic reticulum (ER), which may or may not have ribosomes associated with it. The ribosomes are the site of protein synthesis, while the endoplasmic reticulum provides a transportation system throughout the cells for products synthesized at the ribosomes.

One or more Golgi bodies (**dictyosomes**) may be found in the cytoplasm of green algal cells. This structure consists of a series of stacked, flattened membranes. It functions in the formation of secretory vacuoles, which are composed of substances synthesized by ribosomes in association with the ER.

Contractile vacuoles also form a part of many freshwater, unicellular, motile green algae (e.g., *Chlamydomonas*). These membrane-bound structures function as **osmoregulators.** As such these organelles are very important to cells living in **hypotonic** environments, since they regulate water balance between the cell and its immediate milieu.

Mitochondria are double membrane bound structures that are sometimes visible with the light microscope (Fig. 1-2). They usually appear as oval bodies. The inner membrane is folded so that many membranous **cristae** protrude into the central part of the mitochondrion. The respiratory enzymes occur in association with the cristae, and hence it is here that oxygen is consumed, carbon dioxide is produced, and high-energy ATP molecules are formed. The mitochondrion provides all of the ATP to carry out the many metabolic activities of the cell, and this is accomplished through enzyme-mediated reactions of the tricarboxylic acid cycle (Krebs cycle) during the process of respiration.

Motile green algal cells generally possess two to four anterior flagella of equal length (isokont flagella). A notable exception to this is *Oedogonium*, which produces gametes with a ring of short flagella at one end of the cell (stephanokont flagella; see Fig. 1-9). The flagella are smooth, lacking any side hairs (mastigonemes), and are referred to as "whiplash flagella." The long axis of a flagellum is called the **axoneme,** and cross sections through it at various levels reveal a specific pattern of protein microtubules. This classic 9 + 2 pattern of microtubule doublets is the most common arrangement, but it may vary to some extent (see Fig. 1-8). The microtubules extend from the external flagellum to a basal body within the cell wall and sometimes extend beyond the basal body into the cytoplasm, forming microtubular bundles that are important phylogenetically.

The flagellar apparatus is a rather plastic, ephemeral structure in many species. Flagella are resorbed in all the sessile forms and are reassembled at the time of gamete or zoospore formation during sexual and asexual reproduction.

Recent work with the electron microscope has revealed two significant ultrastructural characteristics of the green algae that have caused a complete revision of the established theories of evolutionary relationships within Chlorophyta (see Pickett-Heaps, 1975; Stewart and Mattox, 1975). First, these investigators found differences in the microtubular arrangement of the flagellar roots. On the one hand there are green algal organisms that have cruciately arranged microtubules extending into the cytoplasm from the basal bodies at the base of each flagellum (Fig. 4-8). On the other hand there are green algal organisms in which the microtubules from each flagellum form a single broad bundle of nine microtubules extending into the cytoplasm toward the nucleus. Ultrathin sections of such cells often cut these bundles of microtubules in cross section, thus resembling a **centriole,** the organelle so common to animal cells.

The second significant ultrastructural feature occurs during cell division and can be seen in the manner by which cytoplasmic division (cytokinesis) is accomplished after nuclear division (mitosis). Some green algal organisms produce a **phycoplast,** a structure composed of microtubules arranged perpendicularly to an ephemeral spindle during telophase. The daughter nuclei of cells with a phycoplast appear to be very close together. The phycoplast initiates cytokinesis by furrowing (an inward growth of the parent cell wall) or by cell plate formation in which new cell wall materials are formed between the nearby daughter nuclei. However, some green algal cells produce a **phragmoplast,** a structure composed of microtubules arranged perpendicularly to a persistent spindle apparatus during telophase. The daughter nuclei are widely separated by a visible spindle. The phragmoplast forms a new cell wall between the newly formed daughter nuclei.

The presence of the cruciately arranged flagellar-root microtubules in conjunction with a phycoplast indicates one line of chlorophyte evolution designated the **chlamydomonad line,** since these characteristics are found in *Chlamydomonas*. The presence of two bundles of flagellar-root microtubules in conjunction with a phragmoplast indicates a second evolutionary line called the **bryophytan line,** since it is these characteristics that have led to the phylogenetic development of such forms as *Coleochaete*, Charophyta, Bryophyta, and the vascular plants.

Some green algae have developed **siphonaceous (coenocytic)** cells in which some or all of the cells have more than one nucleus. This condition has apparently developed twice during the course of evolution—once in the chlamydomonad line (Cladophorales), and once in the byrophytan line (Caulerpales, Siphonocladales, and Dasycladales). In some of those algae there are no **septa** dividing cells throughout the vegetative phase of the life cycle (Caulerpales). *Codium magnum*, the largest of all the chlorophytes, is a member of this order. This alga, which may reach a length of 8 m, is still a single multinucleate cell that develops into many-branched tubular filaments that interweave to give the plant support.

In other coenocytic greens, septa develop within the multinucleate cells only after vegetative maturity is attained (Siphonocladales and Dasycladales).

REPRODUCTION

Both sexual and asexual reproductive mechanisms are common to most chlorophytes. Sexual reproduction has been observed in all but a few members of the division. Among the more primitive members sexual reproduction is isogamous (e.g., *Chlamydomonas*), but anisogamy (e.g., *Codium, Enteromorpha*) and oogamy (e.g., *Volvox, Oedogonium, Coleochaete*) may be found in some genera. Actually different species of *Chlamydomonas* exhibit all three sexual reproductive modes—isogamy, anisogamy, and oogamy. A good account of sexuality in plants is given by Ende (1976).

Oogamous green alga exhibit chemotactic mechanisms whereby motile, flagellate sperm are attracted to large, nonmotile eggs; this phenomenon (**chemotaxis**) has been studied in a number of green algae, including *Volvox* (Starr, 1972; Starr and Jaenicke, 1974) and *Oedogonium* (Hoffman, 1973). It is discussed in detail by Bean (1979). Isogamous species apparently lack any chemotactic mechanism, and motile isogametes locate each other in a random fashion. In *Chlamydomonas* (Fig. 4-10), after isogametes have randomly found each other, they fuse or agglutinate in a so-called mating-type reaction brought about by specific proteins found in association with their flagella. This has

been interpreted as a typical antigen-antibody type of reaction (Goodenough, 1980). Soon after the flagella have joined, the isogametes unite and nuclear fusion (syngamy) takes place, producing a zygote. The act of fertilization in other isogamous and anisogamous species is believed to be similar to that observed in *Chlamydomonas*.

FIG. 4-10 Life cycle of *Chlamydomonas.*

Good example of haplontic life cycle with isogamous sexual reproduction. Cycle shows fusion of isogametes forming zygote that develops into thick-walled, dormant zygospore. Zygospores germinate by meiosis, producing four meiospores representing (+) female and (−) male vegetative organisms. Adult vegetative cells reproduce by mitosis, forming gametangia containing 4, 8, 16 or 32 isogametes. Released isogametes may fuse in a sexual process or may divide mitotically, extending vegetative life of organism. Note that gametangia are produced by single vegetative cell protoplast dividing within parent cell wall. Isogametes are thus somewhat smaller than vegetative cells.

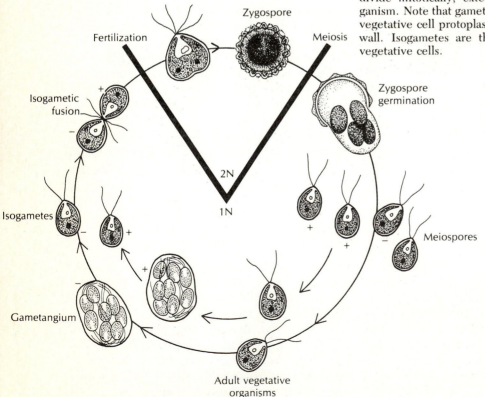

Volvox colonies may be **monoecious** (with both male and female characteristics) or **dioecious** (with male and female structures produced in separate colonies). In dioecious species, male colonies are known to produce an "inducing factor" composed of a glycoprotein (protein-polysaccharide mixture). This substance causes nearby asexual colonies of *Volvox*

Text continued on p. 90.

FIG. 4-11 Life cycles of *Oedogonium.*

Oedogonium is a good example of the degree of plasticity within a given genus. Life cycles of *Oedogonium* may be monoecious (homothallic) or dioecious (heterothallic). In addition, members of this genus may be macrandrous, with male and female vegetative filaments identical, or nannandrous, with small male filaments growing epiphytically on female filaments.

A through **E** show successive stages in sexual reproduction in macrandrous species. Note that both male and female reproductive structures are on same filament (monoecious). Dioecious macrandrous species are also known. Stephanokont sperm cells are attracted to oogonium by chemical substance secreted by the oogonium.

F through **K** show successive stages in development of sexual reproduction in dioecious nannandrous species. Note formation of androsporangia that produce single stephanokont androspores. Released androspores are attracted to developing female oospore mother cells. They attach to cell and induce it to divide, forming oogonium and support cell. Androspore develops into three-celled filament called a "dwarf male." Anterior two cells form two stephanokont sperm.

Note also *(ai)* immature antheridium; *(am)* mature antheridium; *(amc)* antheridial mother cell; *(anth)* antheridium; *(and)* androsporangium; *(asp)* androspore; *(dm)* dwarf male; *(oc)* oogonium; *(omc)* oogonial mother cell; *(op)* oogonial pore; *(sc)* sperm cell; *(sec)* gelatinous secretion; *(suc)* support cell; *(tomc)* oogonial mother cell in telophase; *(tr)* terminal rings; *(zyg)* zygospore.

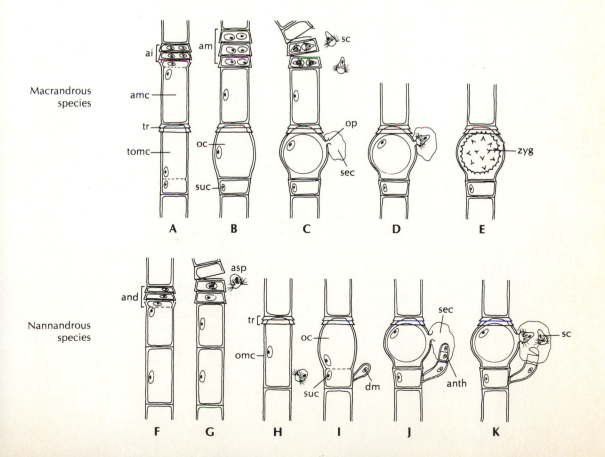

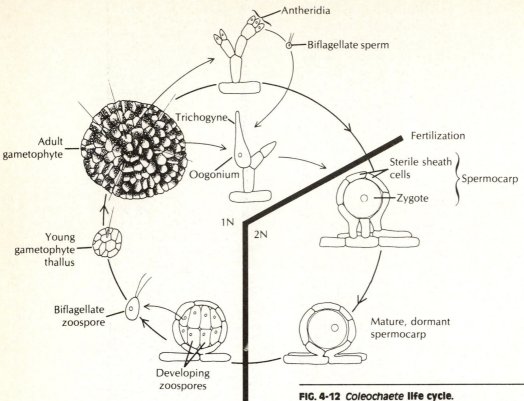

FIG. 4-12 *Coleochaete* **life cycle.**

Coleochaete is a fresh water alga that represents the epitome of the bryophytan line of green algal evolution. It is very close to the main evolutionary line of terrestrial plants based on its biochemistry, cellular ultrastructure, and reproductive mechanisms. Flagellate cells produce microtubules forming two broad bundles of flagellar roots, and dividing cells produce a phragmoplast (see also Fig. 4-8). Life cycle is haplontic, and zygote is enclosed within sterile sheath of gametophytic cells. Note egg cell remains attached to gametophyte both during and after fertilization. Antheridia and oogonia are produced on modified branches associated with surface of flat, disk-shaped gametophyte. After fertilization zygote is enclosed within thickened wall of sterile cells produced by modified branches of gametophyte and is called a spermocarp. Spermocarp develops into enlarged red sphere that remains dormant for extended periods (e.g., over winter). In spring, zygote divides by meiosis, producing biflagellate zoospores that disperse, settle onto suitable substrates, and develop into gametophytes. Suitable substrates are often found on submerged vegetation where gametophytes develop as epiphytes. Both monoecious and dioecious gametophytes have been reported.

to commence sexual differentiation (Darden, 1966; Ely and Darden, 1972). Production of the inducing factor is stimulated by decreasing concentrations of essential nutrients in the environment, which may include declining nitrates, decreasing light intensities, and other factors. These conditions are often experienced by the organisms during the fall of the year.

The filamentous, oogamous genus *Oedogonium* (Fig. 4-11) is interesting because it illustrates the various ways in which oogamy may be expressed. *Oedogonium* filaments may produce both male and female reproductive structures on one filament (monoecious species), or they may be produced in separate filaments (dioecious species). The male and female filaments may be the same size (**macandrous** species), or the male filament may be relatively small, con-

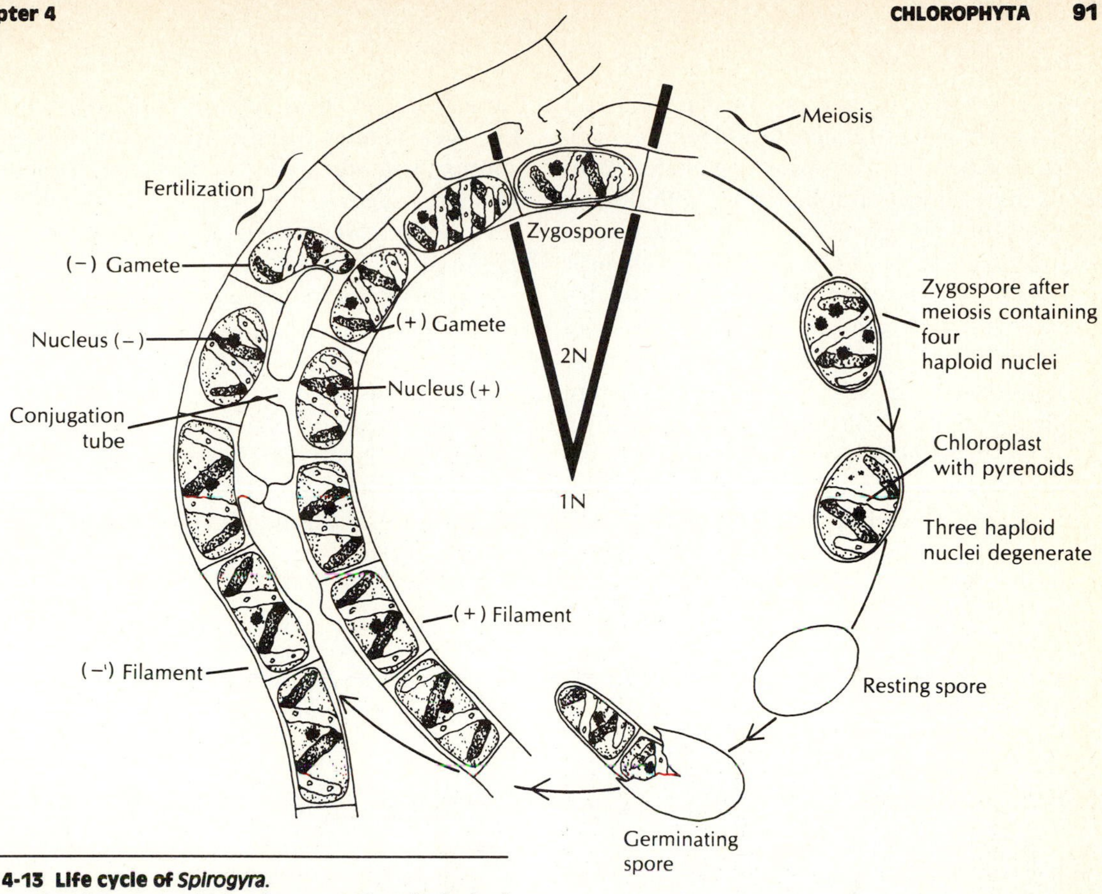

FIG. 4-13 Life cycle of *Spirogyra*.

Life cycle of *Spirogyra* is haplontic, with only diploid phase being the zygote. Note presence of two filaments of cells represented as (+) and (−) filaments. Process of fusion of two compatible filaments is called conjugation. Note also amoeboid nature of gamete produced by (−) filament as it moves across conjugation tube to unite with gamete from (+) filament.

sisting of only 2 to 3 cells that develop as an epiphyte on the larger female filaments (**nannandrous** species, Fig. 4-11). Both macandrous and nannandrous species produce stephanokont sperm in specially differentiated segments of the filament called **antheridia**. Futhermore, the larger filament in nannandrous species may produce stephanokont **androspores** in special **androsporangia**. Androspores are chemotactically attracted to immature oogonia, where they attach to the female filament and develop into a dwarf male (Fig. 4-11). Once growth of the dwarf male has begun, the oogonial mother

cell continues to grow and differentiate into a mature **oogonium.** The oogonium secretes a gelatinous sheath that entraps the stephanokont sperm released by the mature dwarf male, thus ensuring that fertilization will occur.

Coleochaete (Fig. 4-12), with its haplontic life cycle and oogamous sexual reproduction, is a green alga that closely resembles bryophytes in life style and cellular ultrastructure. Unlike most algae, the larger egg cell remains attached to the haploid gametophyte even after fertilization by the biflagellate sperm that are produced in antheridia. The zygote is encased within a cellular wall of gametophytic tissue forming a **spermocarp.**

In *Spirogyra* and other members of the Zygnematales the gametes are amoeboid, and sexual reproduction is accomplished by a special process called **conjugation** (Fig. 4-13). This process is referred to as "physiological anisogamy," since the behavior of the male and female isogametes is different. The male isogamete moves across the **conjugation tube** to fuse with the stationary female isogamete, thus forming a zygote, or **zygospore.** It secretes thick, cellulosic cell walls that are often colored and have distinctive surface ornamentations. Zygospore formation is stimulated by changing environmental conditions; this is especially apparent among members of this order in the fall. They then exist during the winter as zygospores in the muddy bottoms of ponds or streams. Meiosis occurs during the winter with formation of four haploid nuclei, three of which degenerate before germination of the spore the following spring. The life cycle is therefore haplontic. Desmids such as *Cosmarium* also reproduce sexually by conjugation and the formation of a zygospore (Fig. 4-14).

Asexual reproduction among the Chlorophyta is very common and is accomplished in a number of ways. Fragmentation of filaments and thalloid forms is one method used by all green algae except unicellular types. Regular mitosis followed by cytokinesis is common to all members of the division, including unicellular types. Most species also produce zoospores, which are

---

FIG. 4-14 Zygospore of *Cosmarium*.
Scanning electron photomicrograph of *Cosmarium* zygote, the result of conjugation between two desmids. Male and female cells secrete mucilage *(M),* which binds them together. Amoeboid gametes emerge from cell and fuse to form zygote, which then secretes a thick, spiny wall to form zygospore *(Z),* which may remain dormant for several months. Semicell *(SC)* is also called hemicell (1500×). (From Lott, J.N.A.: A scanning electron microscope study of green plants, St. Louis, 1976, The C.V. Mosby Co.)

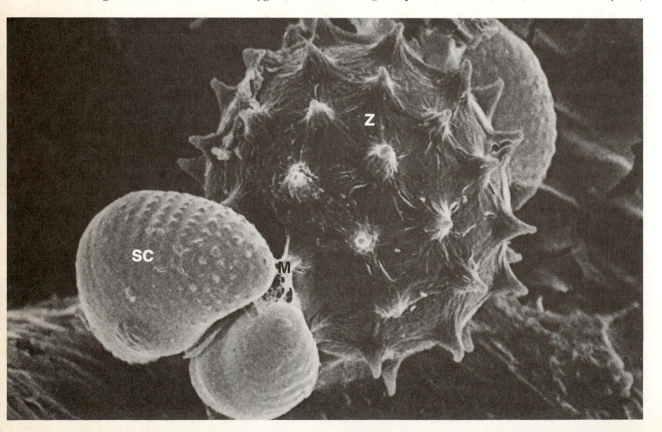

biflagellate or quadriflagellate motile spores. A few green algae also produce nonmotile spores called **aplanospores.**

In *Volvox* specialized asexual reproductive cells called **gonidia** are produced; these cells give rise to asexually produced colonies or **daughter colonies.** Five such daughter colonies can be seen in Fig. 4-4. Other members of the **volvocine line** (Volvocales) and some members of Tetrasporales (e.g., *Scenedesmus*, Fig. 4-15; *Pediastrum*) produce colonies of cells arranged in a definite pattern and composed of a definite number of cells. Such an arrangement is called a **coenobium.** Coenobia are capable of

asexually reproducing entire colonies by simultaneous division of the cells, thus producing daughter colonies with a definite number of cells in a particular arrangement. Daughter colonies grow in size, but the cell numbers remain constant throughout the existence of the coenobium.

PHYSIOLOGY

Most physiological research with green algae has been done using unialgal, **axenic** cultures in closed systems with a high cell density. Some genera commonly used for such experiments are *Chlorella*, *Scenedesmus*, *Chlamydomonas*, and *Caulerpa.* Much of the knowledge concerning photosynthesis and algal nutrition is derived from these studies. The extent to which results can be extrapolated from carefully controlled laboratory conditions to unpredictable field con-

FIG. 4-15 Some common green algae.
Some common planktonic algae, including: *A, Scenedesmus* sp.; *B, Ankistrodesmus* sp.; *C, Cosmarium* sp., a desmid; and *D, Staurastrum* sp., also a desmid (250×).

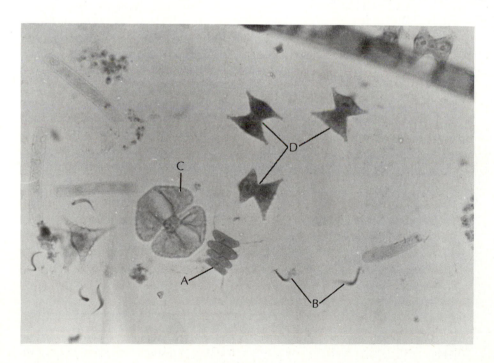

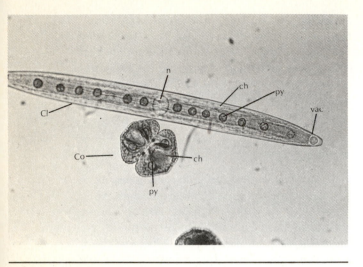

FIG. 4-16 Desmids.
Two types of desmids including genera *Closterium*
(*Cl*) and *Cosmarium* (*Co*). Note nucleus (*n*), chloroplasts (*ch*), pyrenoids (*py*) and vacuoles (*vac*) at the
tips of *Closterium* cell that contain crystals of calcium
sulfate that in living specimens may be observed
undergoing Brownian movement (250×).

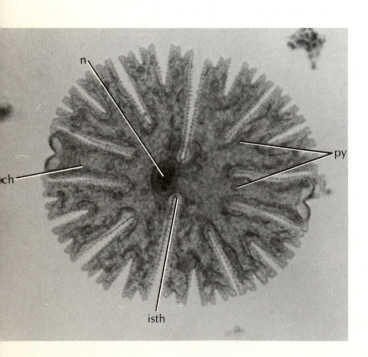

ditions is a major problem confronting all workers in algal physiology. Some phenomena observed in the laboratory may never occur in an
open environment, and some events may occur
under field conditions that have never been
seen in the laboratory.

The majority of Chlorophyta are capable of
photosynthesis. Many of them use only free carbon dioxide as their carbon source. Because of
this, most green algae are found in acid or neutral waters where free carbon dioxide abounds
(Fig. 3-7). Some specialized genera, especially
the desmids (*Closterium*, Fig. 4-16; *Micrasterias*, Fig. 4-17; *Cosmarium*, and others) are
quite tolerant of pHs as low as 3.5 to 4.5. Several species of *Chlamydomonas* also are tolerant
of acid conditions (pH 2.0 to 4.0) and may be
found growing in acid bogs. As with other acid-
tolerant algae, there appears to be a mechanism
by which the alga excludes hydrogen ions from
its protoplasm. The alga remains green even
when growing at pHs low enough to cause
chlorophyll to be degraded to **phaeophytin**
(Cassin, 1974). Other acid-tolerant green algae
have been found growing in very acid pools in
strip-mined areas.

Other green algae such as *Cladophora* and
Scenedesmus are common in alkaline waters
where the major source of carbon is the bicarbonate ion, HCO_3^-. Evidently these algae either
utilize the bicarbonate ion in photosynthesis or
efficiently extract the small amount of free carbon dioxide available at pHs above 7.0 (see Fig.
3-7). Nutrition among green algal cells is primarily **photoautotrophic**, using light as the
primary energy source. However, some genera

FIG. 4-17 *Micrasterias* habit.
Micrasterias is a desmid—a type of green alga whose
single-cell structure appears to be composed of two
cells joined at an isthmus (*isth*). These "semicells"
contain a single nucleus located in the isthmus. Note
nucleus (*n*), chloroplast (*ch*), and multiple pyrenoids
(*py*) (250×).

are capable of heterotrophic existence, utilizing preformed organic molecules as their energy source. For example, some members of the Volvocales such as *Gonium* and *Pandorina* (Fig. 4-4) are capable of metabolizing acetate molecules. The frequently used laboratory genera *Scenedesmus* and *Chlorella* each have strains that can grow heterotrophically in the dark on glucose or acetate (Droop, 1974).

In some cases autotrophic and heterotrophic chlorophyte genera appear to be closely related morphologically. The heterotrophic *Polytoma* closely resembles *Chlamydomonas*, except that it lacks chloroplasts. It does, however, have colorless, starch-containing **plastids** (Lang, 1963). The heterotrophic *Prototheca* closely resembles the autotrophic *Chlorella*. A species of *Prototheca* isolated from the oil-contaminated waters of Baltimore Harbor was capable of bio-degrading petroleum hydrocarbons (Walker et al, 1975). Other species of *Prototheca* may parasitize humans (**protothecosis**), causing a severe skin infection that may spread to internal organs. The first case of a *Prototheca* infection was described in 1964; this alga is the only one known to be pathogenic to humans. Sudman (1974) reviews the incidence of the disease. In nature *Prototheca* may occur in the sticky secretions (**slime flux**) covering wounds on trees.

Starch is the major energy storage compound in green algae, but some marine forms contain low-molecular-weight carbohydrates such as sucrose and glycerol (Craigie et al, 1967). These carbohydrate molecules are evidently used to regulate osmotic pressure in the marine environment. In the marine alga *Dunaliella* osmoregulation is accomplished by the synthesis and breakdown of glycerol in response to changing salinities (Ben-Amotz, 1975).

Most flagellate green algae are positively phototactic, but negative **phototaxis** may occur if the light intensity becomes too high. The filamentous and attached forms are not motile; therefore their continued existence depends on their being located in areas where the light is adequate for photosynthesis. Many filamentous freshwater algae such as *Spirogyra* and *Hydrodictyon* may form tangled mats that float on the water's surface where light is plentiful. The freshwater filamentous genera *Zygnema* (Fig. 4-9) and *Mougeotia* (Figs. 4-9 and 4-18) produce mucilage, which may aid flotation.

Green algae require the usual plant nutrients—nitrogen, phosphorous, potassium, sulfur, magnesium, and calcium. Additional trace elements may also be needed. Some Chlorophyta can absorb the essential nutrients nitrogen (ammonium or nitrate) and phosphorous (orthophosphate) in excess of their immediate needs. The ability of *Volvox* and *Spirogyra*, for example, to absorb and store phosphate in excess of their needs enables these algae to thrive in water that is deficient in this nutrient. The marine algae *Valonia* and *Derbesia* both concentrate nitrates in excess of their needs in a similar manner. Unicellular green

---

FIG. 4-18 *Mougeotia.*
Filamentous genus *Mougeotia;* cells possess platelike chloroplast *(ch)* with pyrenoids *(py)* (250×). Compare with Fig. 4-9.

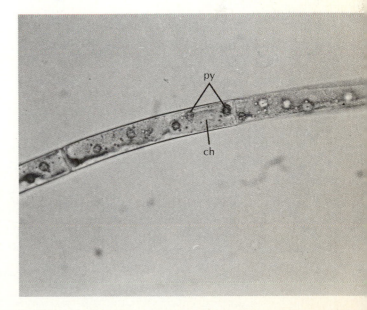

algae preferentially utilize ammonium nitrogen rather than nitrate, and many of these species may grow well in such ammonia-rich environments as sewage treatment ponds.

Copper is required by the green algae but in very low concentrations. At high concentrations copper is toxic to algae; in fact, copper sulfate at 1 part per million is an effective and widely used **algicide.** Copper concentrations in freshwaters may reach toxic proportions at the end of the summer stratification (see Chapter 2). This increase in copper concentration may be responsible for the population decrease in some species.

Certain green algae can grow at very low temperatures. Several unicellular chlorophytes regularly occur as **snow algae,** forming reddish patches on snow. A common snow alga is *Chlamydomonas nivalis.* Its red pigment has been identified as the xanthophyll astaxanthin (Viala, 1966). Another investigator (Czygan, 1970) reported that the red pigments in other snow algae were carotenoids that were formed as chlorophyll decomposed and nitrogen became depleted. The production of red pigments shields the chlorophyll from high light intensity and resulting ultraviolet damage. Light intensities in alpine or arctic and antarctic environments may be very high—up to 86,000 lux (Mosser et al, 1977). Optimum growth in these algae occurs at 4° C, and photosynthesis continues at temperatures ranging from 0° to 20° C. Additional discussion of snow algae is found in the section on ecology, p. 103.

Many marine Chlorophyta deposit calcium carbonate in or on their cell walls. *Acetabularia,*

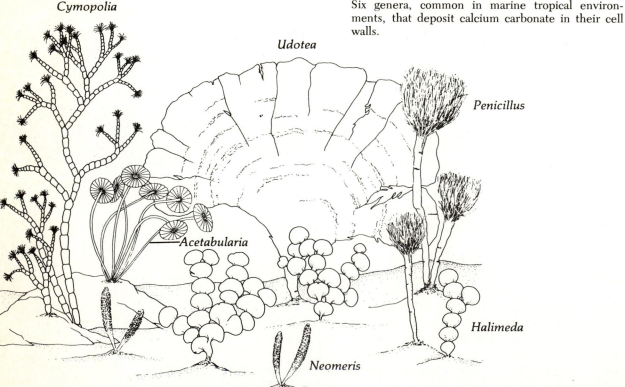

Cymopolia

Udotea

Penicillus

Acetabularia

Halimeda

Neomeris

FIG. 4-19 Calcareous green algae.
Six genera, common in marine tropical environments, that deposit calcium carbonate in their cell walls.

Cymopolia, Halimeda, Neomeris, Penicillus, and *Udotea* (Fig. 4-19) are common calcareous benthic algae found in shallow subtropical waters. These algae often appear whitish rather than green as a result of the precipitation of calcium carbonate. Several freshwater genera such as *Cladophora* may also deposit calcium or magnesium carbonate in alkaline waters (Wood, 1975). In cool waters the calcium carbonate is usually deposited in the form of calcite, while aragonite is the prevalent form in warm tropical waters. Carbon dioxide utilization in photosynthesis increases the bicarbonate concentration of the waters by reducing the amount of free carbon dioxide, thereby raising the pH (see Chapter 1). As the bicarbonate concentration increases it is utilized by some algae, the chemical equilibrium proceeds to the right, and calcium carbonate is deposited on or in the algal cell wall. Accumulated calcium carbonate deposits from genera like *Halimeda* have aided coral reef formation in tropical and subtropical areas by adding to the accumulated limestone.

EVOLUTION

Chlorophyta are classically viewed as having evolved along three divergent lines, the colonial (volvocine) line (Fig. 4-4), the uninucleate (**tetrasporine**) line, and the multinucleate (siphonaceous) line. This classic view of green algal evolution has recently been revised by Pickett-Heaps (1975) and by Stewart and Mattox (1975) on the basis of their ultrastructural studies with the electron microscope (Fig. 4-8). They divide Chlorophyta into two lines of evolutionary development—the chlamydomonads and the bryophytans. This revised evolutionary tree helps to explain the disputed positions of Oedogoniales with their stephanokont flagella and Zygnematales with their amoeboid gametes. Indications are that the new evolutionary ideas are correct, but much more ultrastructural evidence is needed to verify this, especially with the siphonaceous forms, where little work has been done.

The divergent evolutionary lines in the green algae apparently developed from a biflagellate, uninucleate cell similar to but not ultrastructurally identical to *Chlamydomonas.* This hypothetical primitive "stem" cell is often referred to as the archetypal cell. (For a review of flagellated organisms in algal evolution, see Lembi, 1980, and Stewart and Mattox, 1980.) The evolutionarily advanced organisms, including colonial, filamentous, parenchymetous, and siphonaceous forms, are all viewed as descendants of the primitive "stem" flagellate cell. Ultrastructural research reinforces this designation of "primitive" and "advanced" characteristics in members of the colonial and the uninucleate lines. However, to date, the ultrastructural data for the siphonaceous algae are incomplete and their position on the evolutionary tree is uncertain.

The morphological progression of algae through the unicellular, filamentous, thalloid, and parenchymatous sequence is well established. However, those members of the division having complex thalli must revert to the primitive, unicellular, biflagellate form for sexual reproduction. Once the organism has found a favorable environment it can conserve energy by becoming sessile and only forming motile cells for reproduction. With less energy used for motility, the alga may develop a complex, branched thallus. The morphology of the motile reproductive cells of these advanced forms often can be more helpful in determining evolutionary relationships than the morphology of the complex thalli.

Evidently the vegetative morphological organization of the green algae has evolved independently of the motile reproductive cells so that some simple single-celled forms are oogamous (some species of *Chlamydomonas*), while some more complex filamentous forms are isogamous *(Microspora).* These apparent evolutionary contradictions are most often found within the chlamydomonad line.

The earliest fossil evidence for Chlorophyta is found in the unmetamorphosed chert from Bitter Springs in central Australia (Fig. 4-20). This deposit has furnished the oldest (approximately 1 billion years old—Precambrian period) evidence for eukaryotic algal forms. These fossils reveal single-celled green algae in association with both unicellular and filamentous blue-green algae, aquatic fungi, and bacteria. Some evidence for cell division by mitosis can be found among these green algal cells (Schopf, 1968).

Unfortunately, the fossil record for Chlorophyta is quite incomplete. Most fossils found in marine sediments are multicellular forms that deposit calcium carbonate. The multicellular Dasycladales have been found in marine sediments from the late Cambrian period and represent the earliest record to date of Chlorophyta

with complex thalli. The soft-bodied algae, however, do not precipitate calcium carbonate and do not, therefore, leave many fossils. Recently investigators have found fossilized zygospores of some noncalcareous algae. Zygospores of some Zygnematales have been found in rocks of the Tertiary and Quaternary geological periods (van Geel and van der Hammen, 1978; Jarzens, 1979). As more fossilized zygospores are found, more information will become available concerning the evolution of the noncalcareous Chlorophyta.

Fossils closely resembling the modern marine alga *Halimeda* have been found in rocks from the Ordovician period. Devonian rocks have yielded fossils resembling the freshwater forms of *Oedogonium* and the desmid, *Closterium*,

FIG. 4-20 Sections of chert from Australian Bitter Springs Formation.

These thin sections of Precambrian chert show a unicellular green alga, *Caryosphaeroides*. This chert is approximately 1 billion years old. These fossils represent the oldest unicellular green algae found to date in the fossil record. (Reprinted with permission from J.W. Schopf, 1968. J. Paleontology **42**:657.)

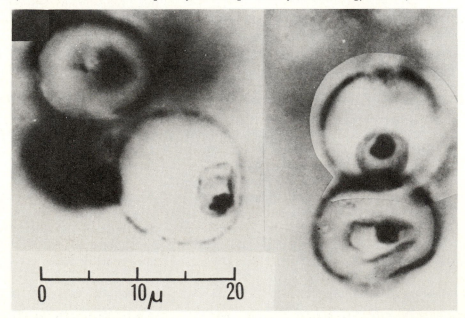

suggesting that freshwater environments had been successfully colonized by Chlorophyta by the Devonian period.

Most botanists agree that the bryophytan line of Chlorophyta is ancestral to land plants; the theory is reinforced by observation that the bryophytan line and land plants share these characteristics: (1) starch as an energy storage compound, (2) chlorophyll *a* and *b*, (3) similar chloroplast morphological organization and (4) a phragmoplast during cell division.

Further evidence for the relationship between the bryophytan line and vascular plants comes from Chlorophyta's well-developed alternation of generations, which is considered essential for land colonization. Gametophytes are restricted to water because they produce motile sperm that require an aquatic medium for reaching the egg. The sporophyte, however, has no such restriction. It may have a heterotrichous (i.e., upright) habit and produce nonmotile spores that are carried by air currents. Some biologists (Fritsch, 1945) believe that the first algae to invade land successfully had an isomorphic alternation of generations and a heterotrichous habit. Several amphibian genera partially meet these requirements—*Stigeoclonium* (Fig. 4-21) and *Fritschiella*—but they do not meet the ultrastructural requirements established by Pickett-Heaps and Marchant (1972) and they do not appear to be closely related to the Bryophyta. The green algae that were the progenitors of the land plants are very probably extinct, but their form, ultrastructure, and re-

FIG. 4-21 *Stigeoclonium.*
Alga found growing on trickling filter in sewage treatment plant. Note branching (250×).

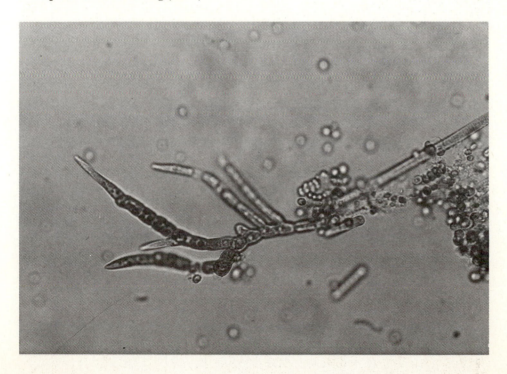

productive mechanisms remain a subject for a great deal of speculation. Only the oogamous and aquatic *Coleochaete* (Fig. 4-12) appears to have the advanced ultrastructural characteristics common to mosses and vascular plants.

If the first land alga did have a diplohaplontic life cycle and was heterotrichous, the sporophyte generation would have had to become dominant for continued successful colonization of the land. The development of an extensive diploid stage is considered essential for evolutionary advancement. Diploid organisms contain twice as much genetic material as haploid organisms and therefore have more genetic plasticity. Certain recessive characteristics may be repressed by dominant ones in diploid organisms, but these traits are not repressed in haploid individuals. Haploid organisms have no "diploid buffering of genetic changes" (Pickett-Heaps, 1975). Raper and Flexer (1970) state that diploid organisms have much greater potential for variability than haploid organisms, and therefore there is a "totally compelling correlation between somatic or vegetative diploidity and biological success."

As water levels receded as the result of geological uplift, some Chlorophyta were probably stranded on damp or drying shores. Those members of the division possessing advanced attributes such as a dominant sporophyte, oogamy, and resistance to desiccation were preadapted and had the best chance of surviving prolonged periods of dryness. Airborne spores and decreasing dependence on water for fertilization increased the chances of surviving on land. Further research may add to our knowledge of the important precursors of modern land plants.

The invasion of land by photosynthetic organisms was a very important step in the colonization of the terrestrial environment by all living organisms. When autotrophic organisms became established on land, they stabilized the soil and added organic matter to the substrate. They also formed the base of the terrestrial food chain that was later exploited by the animals as they too left the aquatic environment and invaded the many new ecological niches on land.

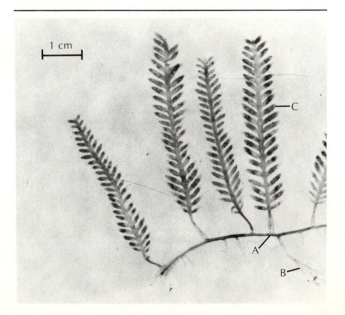

ECOLOGY

Chlorophyta have invaded many habitats—freshwater, saltwater, and terrestrial. They are widely distributed over the earth and may be found in a wide variety of ecological niches. The only niches they have not colonized are hot springs and deserts. The green algae inhabit relatively shallow waters because their chlorophyll absorbs light in the red and blue ranges of the visible spectrum, and these wavelengths (red: 650 nm, and blue: 450 nm) do not penetrate deeply into water (Fig. 2-5). Light in the green range of the spectrum is reflected by chlorophyll *a* and *b,* thus giving Chlorophyta their characteristic grass-green color. Most green algae are freshwater organisms, but some such as *Ulva, Enteromorpha,* and *Codium* live in marine or brackish waters, especially in rocky intertidal zones of temperate areas where holdfasts anchor them to the hard substrate. Other marine forms such as *Halimeda, Acetabularia,* and *Penicillus* (Fig. 4-19) are common in warm shallow lagoons where waters are often saturated with calcium carbonate. These subtropical

algae are often siphonaceous, deposit calcium carbonate in their cell walls, and appear whitish in color because of the encrusting lime. Certain noncalcareous greens such as *Caulerpa* (Fig. 4-22) and *Bryopsis* also occur in warm shallow waters. In tropical and subtropical marine waters where light penetrates deeply into the clear, less dense, warm water, many green algae may be found growing at depths of 100 m. One genus, *Anadyomene* (Fig. 4-23), has been recorded at 200 m in the Gulf of Mexico.

One biflagellate chlamydomonadlike green alga, *Dunaliella salinas,* occurs in hypersaline waters whose salt concentrations are as high as 50 to 100 %₀.* These conditions can be found in the Great Salt Lake, the Dead Sea, and marine tide pools where significant evaporation has occurred.

Other species of *Dunaliella* are common members of marine phytoplankton. This alga is

*Salinity is usually expressed as grams of salt/liter, and is represented by %₀ (parts per thousand). Normal sea water is usually 33 %₀, but may vary in different environments.

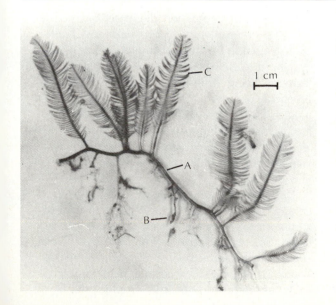

FIG. 4-22 *Caulerpa* **habit.**
Herbarium specimens of *Caulerpa mexicana* (far left) and *C. sertularioid* (near left), common to the shallows around Bermuda. Note division of this siphonaceous alga into horizontal, stolonlike portion *(A);* rhizoidal holdfasts *(B),* which grow into the sand or attach to rocks; and upright, feathery branches *(C).* Extensive colonies of these algae are associated with reefs and lagoons. In both photographs the upright portion is 4 to 5 cm tall. (Photographs courtesy D. Spess, Biology Department, Lehigh University, Bethlehem, Pa.)

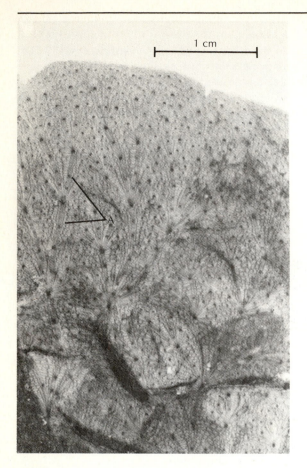

1 cm

FIG. 4-23 *Anadyomene* **habit.**
Portion of herbarium specimen of the foliaceous alga *Anadyomene stellata* collected in deeper waters around Bermuda. Note differentiation of siphonaceous cells into larger cells forming dichotomously branched ribs *(A)* within fan-shaped thallus and smaller cells that occur between ribs and that may give rise to biflagellate swarm cells. (Photograph courtesy D. Spess, Biology Department, Lehigh University, Bethlehem, Pa.)

freshwater ponds and lakes (*Chlamydomonas, Chlorella, Scenedesmus,* desmids, and others), but they are usually sparsely distributed in marine waters where diatoms (Chrysophyta) and dinoflagellates (Pyrrhophyta) generally predominate. In greenhouse ponds and aquaria the genus *Ankistrodesmus* (Fig. 4-15) often grows to the exclusion of all other algae. Such aquatic blooms are common in small bodies of confined water, in freshwater ponds and lakes, and, to a lesser extent, in brackish and marine environments. The blooms may indicate a natural or cultural eutrophication process (see Chapter 2) and high concentrations of phosphate or occasionally of nitrate or ammonia.

In temperate regions the planktonic Chlorophyta usually bloom in late spring following the spring diatom population decrease and before the blue-green blooms of midsummer (Chapter 2). Nitrogen in the form of ammonium or nitrate may be depleted by the green algal bloom; the nitrogen-fixing blue-green species follow the green algae because they can fix atmospheric nitrogen. The green flagellates are preferred food over the blue-greens by the herbivorous zooplankton, so the green algae are rapidly consumed, but the blue-greens are not as heavily grazed.

The desmids often comprise a large percentage of the algae in acid bogs and oligotrophic lakes. Certain desmids such as *Micrasterias* and *Staurastrum* (Fig. 4-15) are considered to be indicative of oligotrophic waters. Brook (1965) classifies certain planktonic desmids as indicators of the trophic status of

quite tolerant of organic pollutants such as the **chlorinated hydrocarbon** pesticides and petroleum hydrocarbons. The marine diatoms are more sensitive to these organic pollutants than *Dunaliella,* so the contamination of the marine environment with these pollutants could cause a change in the species composition of the phytoplankton. Such a shift in algal species could cause a resulting shift in herbivorous invertebrates and ultimately, perhaps, a shift in fish species. Many invertebrates and fish are highly selective in their choice of prey species.

Many other microscopic green algae are also important members of the plankton community. They are often the dominant phytoplankton in lakes—oligotrophic, mesotrophic, or eutrophic.

Filamentous floating green algae such as *Hydrodictyon*, *Spirogyra*, and *Zygnema* sometimes dominate the flora, creating green scums on small ponds or sheltered coves of lakes. Other filamentous forms such as *Cladophora*, *Draparnaldia*, *Stigeoclonium* (Fig. 4-21), and *Ulothrix* (Fig. 4-9) have holdfasts and grow attached to rocky substrates in streams and rivers, where they are well adapted to life in fast currents. *Cladophora* and *Stigeoclonium* are usually covered with epiphytic algae, especially diatoms (Figs. 8-14 and 8-15); these green algae may appear brown rather than green. Other filamentous genera such as *Draparnaldia* and *Zygnema* secrete a mucilaginous sheath, and epiphytic algae do not grow readily on these mucilage-covered forms.

Cladophora blooms have been a problem in the Great Lakes of North America. Lakes Erie and Ontario have had large blooms occurring in sheltered bays near sewage outfalls. Masses of the alga have been reported that were 15 m wide and 1 m deep on the north shore of Lake Erie (Neil and Owen, 1964). The alga, which grows from a perennial rhizoid, causes unpleasant blooms in late spring or summer in the littoral zones of the lakes. As large masses of the alga decompose, objectionable odors pervade the area. *Cladophora* is not grazed by herbivores; in fact, it appears to be toxic to many invertebrates (Hutchinson, 1981). Several studies on these nuisance blooms suggest that phosphate may be the limiting factor of the growth of *Cladophora* in the Great Lakes. *Cladophora* requires fast-moving water for its growth and may be found on wave-washed stony lake shores or in fast-flowing streams. Waves or currents may cause the filaments to form into lime-encrusted balls that may cover the substrate.

In some poorly buffered oligotrophic lakes in southern Scandinavia, eastern Canada, and the northeast United States, acid precipitation has been lowering the pH of the lake waters. The results of this decreasing pH on algae include a reduced number of Chlorophyta species (Hendrey and Wright, 1976), the dominance of the flora by acid-tolerant filamentous algae such as *Mougeotia* (Fig. 4-18), and a reduced efficiency in the assimilation of carbon by the algae (Raddum and Saether, 1981).

Some members of Chlorophyta have successfully invaded moist terrestrial habitats. They, along with the Cyanobacteria and the Chrysophyta (diatoms), are the only algae that occur on land to any great extent. They may be found wherever enough light exists to carry out photosynthesis and enough moisture is present to prevent desiccation. Such areas may be found on rocks, sheltered sides of buildings, or on the shaded sides of tree trunks (e.g., *Desmococcus*),* where they obtain most of their required nutrients from rainfall. Other genera such as *Chlorella*, *Fritschiella*, and some desmids are common among the algae found in soils, where they may aerate and stabilize the soil in addition to increasing the organic content. Green algae, along with the fungi and bacteria, may be among the first colonizers of newly exposed soils.

Another terrestrial green alga, *Chlamydomonas nivalis*, is one of a number of unicellular algae responsible for the phenomenon known as "red snow." Vast numbers of these algae growing on the surface of melting snows in the spring often produce a pink tinge to the snow because of reddish carotenoid pigments (which include the xanthophylls) within the algal cells. These species overwinter as zygospores in the soil beneath the snow's surface. In spring, as meltwater develops in the snowbank, the zygospores germinate at the soil-snow interface. The germinated zoospores "swim" to the surface of the snow, where they cause visible reddish blooms (Hoham, 1980).

---

*Unicellular green algae growing in moist terrestrial environments are difficult to identify. *Desmococcus* and *Apatococcus* (also known as *Protococcus* or *Pleurococcus*) grow on tree trunks, boards, rocks, etc. *Chlorococcum* is common on rocks. Other green unicellular genera also exist. Prescott (1978) may be used for the identification of terrestrial unicellular green algae.

The ubiquitous Chlorophyta have also invaded a number of other unusual habitats. *Trebouxia* and *Trentepohlia* live as symbiotic organisms growing in close association with certain fungi to produce lichens (see Chapter 20). Other green algae are epiphytes growing on aquatic plants or terrestrial bracket fungi (Chapter 18). Other Chlorophyta have evolved a symbiotic relationship with protozoans such as *Paramecium*, poriferans such as *Spongilla*, cnidarians such as *Hydra* and sea anemones, and various marine platyhelminths. This association usually results in the organism's being colored green by the intracellular algal cells. Members of Chlorophyta that are capable of this type of symbiosis are called **zoochlorellae** and are most often found to be species of the genus *Chlorella*. An unusual symbiotic relationship occurs between green algal chloroplasts and marine gastropods. Some marine sea slugs (order: Sacoglossa) have individual chloroplasts in the cells of their digestive system. These herbivorous slugs graze on the siphonaceous algae *Codium* and *Caulerpa;* the chloroplasts of these algae are retained by the digestive cells of the slug. The chloroplasts evidently continue to photosynthesize in the invertebrate cells, and **photosynthate** from the chloroplasts is passed on to the slugs (Trench et al, 1970, 1972).

ECONOMIC IMPORTANCE

Single-celled green algae have been explored as a potential source of food. In outdoor, carefully controlled cultures *Chlorella* has produced 12.3 g organic matter/m²/day (net productivity); this productivity exceeds that of most terrestrial crops. *Chlorella* may contain up to 50% protein and also contains all the essential amino acids. However, humans cannot digest *Chlorella* well, so its use as a human food supplement is not widespread. Desmids have also been used in mass culture in closed systems to achieve high rates of productivity. Algae grown in this type of culture could be used as food for humans or lower animals and could potentially increase food production in many areas of the earth.

They might even provide an energy and food source for closed systems in outer space. The Japanese have processed *Chlorella* into a white, tasteless powder that is high in protein and vitamins and have added it to regular wheat flour (Atlas and Bartha, 1981). At this time, however, the process of harvesting and processing algae from mass culture is expensive and has not received wide acceptance.

Single-celled and other small algae have been introduced successfully into oxidation ponds in sewage treatment plants. In these ponds bacteria break down organic matter to inorganic substances that are, in turn, used with sunlight by the algae to produce more algae that can be harvested and fed to livestock. The algae also add oxygen that can then be used by the bacteria for further decomposition of organic material.

Chlorella, Scenedesmus, and several members of the Volvocales have been identified as dominant members of these sewage treatment flora. These ponds are productive at warm temperatures but lose their effectiveness rapidly in cold weather and reduced hours of sunlight. Green algae such as *Stigeoclonium* (Fig. 4-21) are also important members of the trickling filter systems found in many sewage treatment plants. These filters function by dripping the sewage water over stones covered with algae, bacteria, and fungi, which oxidize and utilize the organic and inorganic nutrients in the sewage.

Spirogyra, Tetraspora (Fig. 4-24), and *Chaetophora* may cause problems of taste and odor when they develop large populations in drinking water supply reservoirs. Growth of *Cladophora* may also cause problems not only in large water supply reservoirs but also in rivers. *Cladophora* growth in British waterways may become so excessive that the rivers become clogged; then large quantities of the alga must be removed mechanically or chemically, because it interferes with boating and fishing.

In some countries green algae are eaten as food. Marine algae such as *Ulva, Enteromorpha, Monostroma,* and *Codium* may be used as salad supplements. *Caulerpa* is high in protein and may also be used in salads or soups, but, unfortunately, some species may be poisonous. The Japanese have been leaders in the use of algae for food.

One member of Tetrasporales *(Botryococcus)* has been implicated in the formation of oil shales and peaty coal deposits of the Tertiary period, 26 million years ago. The modern *Botryococcus* contains high concentrations of hydrocarbons—up to two thirds of its dry weight. Future investigations may show that these green algae have made important contributions to oil and coal deposits throughout the world (Knights et al, 1970).

Several substances having antibiotic activity have been identified from certain genera of green algae. *Halimeda* and *Chlorella* are known to produce products with antibiotic properties, but none of these substances have been produced on a commercial basis.

Medically the green algae have been of little interest. While some may cause dermatological problems because of specific allergic reactions to the proteins they secrete, very few cases have been reported. Spores from terrestrial forms have been identified in the air; these may be responsible for some respiratory allergies. Medically the most significant green alga is the genus *Prototheca*, a common inhabitant of the soil. This heterotrophic alga causes a dermatological disease in man known as **prototecosis**. It is also capable of causing systemic infections in livestock and other animals (Davies and Wilkinson, 1967; Mars et al, 1971).

A few of the green algae have become plant parasites in the leaves of *Magnolia*, *Thea* (tea), *Citrus*, and *Rhododendron*, in warmer climates. *Cephaleuros* damages leaves of tea crops by causing discoloration and necrosis in the infected areas. *Stomatochroon* may also infect ornamental plants in tropical areas.

The marine calcareous alga *Acetabularia* has been used frequently in the laboratory to study nuclear-cytoplasmic interactions. This alga is one large cell, and the nucleus-containing portion of the cell is readily separated from the rest

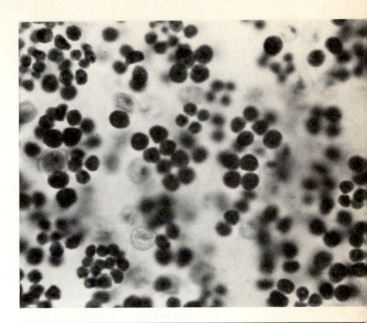

FIG. 4-24 *Tetraspora* **habit.**
Nature of the growth of *Tetraspora* spp. (Chlorophyta) as it might appear during an aquatic bloom in a freshwater pond. Note nucleate cells occurring in groups of four (thus the name *Tetraspora*), all embedded in a gelatinous matrix forming a palmella-type organism (400×).

of the cell. A single cell of *Acetabularia* may measure up to 30 mm. (See Gibor, 1966, for a discussion of some of the research conducted with this alga.)

Both marine and freshwater Chlorophyta are responsible for a large amount of primary productivity. Attached green algae such as *Codium*, *Enteromorpha*, and *Ulva* in shallow or intertidal salt water are highly productive. In freshwater lakes not only do the planktonic green algae contribute significantly to productivity, but the benthic macroalgae such as *Cladophora* may also be highly productive. In flowing waters the benthic algae contribute the most primary productivity to the ecosystem. Chlorophyta are generally considered to be more important in fresh water than in salt water. Eighty-seven percent of the Chlorophyta species are freshwater forms, while only 13% are saltwater forms, although the latter may build up a large biomass, thereby contributing significantly to primary productivity.

SUMMARY

Chlorophyta are the best known freshwater algae, but they have many representatives in the marine environment, especially tropical seas. They have a wide diversity of form and have invaded most habitats throughout the world. The specific adaptations to the wide variety of ecological niches occupied by the green algae has helped them to become the most diversified of all the algal divisions. Furthermore, this variation has led to the invasion of terrestrial environments, the eventual development of vascular plants, and the habitation of land.

Green algal diversity is reflected in their cellular structure, reproductive patterns, growth, physiology, and ecology. They have probably evolved along two separate avenues, the chlamydomonad and the bryophytan lines. These lines are based upon the ultrastructure of the flagellar roots and the presence of a phycoplast or a phragmoplast during mitotic cell divisions. The bryophytan line is considered ancestral to the land plants.

Reproduction among the green algae shows a progression from isogamy to anisogamy to oogamy during the course of evolution. This progression correlates well with the life cycle development from haplontic to diplohaplontic to diplontic cycles in highly specialized forms.

Nutrition among the chlorophytes is mostly autotrophic. However, some species are capable of a heterotrophic existence, and a few have even lost the ability to produce chlorophyll and are **obligate** heterotrophs. It is difficult to separate the latter forms from the protozoa, especially if the alga has also lost its cellulosic cell wall.

The green algae are very important to the overall ecology of both freshwater and terrestrial ecosystems, where they may comprise the major organisms responsible for primary productivity. Many streams, ponds, and lakes are almost entirely dependent upon the green algae to form the base of the food web. The phytoplankton and benthos in these habitats are often entirely composed of green algae. To a lesser extent they are also important in the marine environment.

These algae are potentially of economic importance to man. Some species are grown for food and some are harvested for chemical and medicinal properties. A few genera have been implicated as the cause of both plant and animal (including human) diseases.

SELECTED REFERENCES

Cole, G.A. 1979. Textbook of limnology. Ed. 2 The C.V. Mosby Co., St. Louis. 402 pp.

Hutchinson, G.E. 1967. A treatise on limnology. Vol. 2: Introduction to lake biology and the limnoplankton. John Wiley & Sons, Inc., New York. 1115 pp.

Lembi, C.A. 1980. Unicellular chlorophytes. In Phytoflagellates, E.R. Cox (ed.). Elsevier North-Holland, Inc., New York. pp. 5-59.

Palmer, C.M. 1977. Algae and water pollution. Office of Research and Development, U.S. Environmental Protection Agency, Cincinnati. 124 pp.

Pickett-Heaps, J.D., and H.J. Marchant. 1972. The phylogeny of the green algae: a new proposal. Cytobios **6:**255-264.

Pickett-Heaps, J.D. 1975. Green algae: structure, reproduction and evolution in selected genera. Sinauer Associates, Inc. Sunderland, Mass. 606 pp.

Prescott, G.W. 1978. How to know the freshwater algae. Ed. 3. Wm. C. Brown Co., Publishers, Dubuque, Iowa. 293 pp.

Smith, G.M. 1950. The freshwater algae of the United States. McGraw-Hill, Inc., New York. 719 pp.

Stewart, K.D., and K. Mattox. 1980. Phylogeny of phytoflagellates. In Phytoflagellates, E.R. Cox (ed.) Elsevier North-Holland, Inc, New York. pp. 433-462.

Zahl, P.A. 1974. Algae: the life givers. Nat. Geog. **145:**361-367.

5

CLASSIFICATION

Division: Charophyta (stoneworts, brittleworts, muskgrass): 6 genera, 100 species; mostly freshwater; nodal-internodal construction with a central axis, lateral appendages and rhizoids*; distinctive sex structures

Class: Charophyceae

ORDER. Charales: macroscopic (up to 40 cm tall), aquatic plants with whorls of lateral appendages occurring at nodes; multicellular reproductive organs; *Chara, Nitella, Tolypella.*

*These three characteristics resemble the stems, leaves, and roots of higher plants, even though their internal structure is quite distinct.

CHAROPHYTA

Charophyta are distinctive macroscopic plants of shallow freshwater and brackish water environments. They resemble small, rooted aquatic angiosperms, because their habit of **nodes** and **internodes** is reminiscent of roots, stems, and leaves (Fig. 5-1); however, they lack the vascular tissues and reproductive structures of higher plants.

The biochemistry of the charophytes is identical to that found in the higher plants and Chlorophyta: the presence of chlorophyll a and b, starch as an energy storage product, and cellulosic cell walls. Because of the peculiar combination of morphological and biochemical characteristics, Charophyta have historically represented a problem in classification to systematists. They have been grouped with the green algae on the one hand, and the bryophytes on the other hand.

Round (1971) and Bold and Wynne (1978) have assigned organisms from this group to a separate division, Charophyta, based upon their vegetative structure, sexual reproduction, and **protonemal** development of the germinating zygote. As such the charophytes are viewed as an evolutionary offshoot on the path from the green algae to the bryophytes. This approach is supported by the ultrastructural studies of Pickett-Heaps (1975), who finds significant differences between the cells of the bryophytan line of the green algae and the cells of the charophytes in their methods of cytokinesis and in the complexity of their motile sperm. Earlier authors considered the group to belong to the division Chlorophyta (green algae). (The reader is referred to the excellent monograph of Wood, 1965, and to Imahori, 1964.)

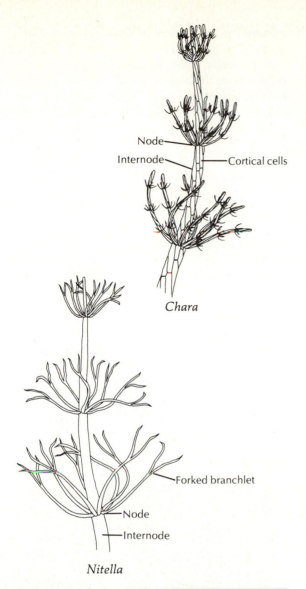

Chara

Nitella

FIG. 5-1 *Chara* and *Nitella* **vegetative habits.**
Two most common genera of Charophyta; note that both are divided into nodes and internodes. In *Chara* most species have internodes that are corticated by cells growing upward and downward from the nodes; in *Nitella* none of the species are corticated. Note also that branchlets in *Nitella* are forked, while those of *Chara* are not, even though they are divided in nodes and internodes. Not illustrated are lime ($CaCO_3$) encrustations that often occur on the surface of *Chara*— but not on *Nitella*.

CELL STRUCTURE

Charophyte cells resemble those of the green algae in their biochemistry, but structurally they are much more highly differentiated. The plants of the charophytes are divided into horizontal rhizoidal branches that function as holdfasts, upright central axes that are divided into nodal and internodal areas, and lateral appendages that occur as whorls of unicellular filaments at each node. At the tip of each central axis is a dome-shaped unicellular **apical meristematic** cell whose growth pattern is **indeterminate** (Fig. 5-2).

Internodal cells are elongate and extend from one node to another. These cells undergo mitotic divisions, forming multinucleate (coenocytic) cells, sometimes with as many as 1,500 or more pleomorphic nuclei per cell. They also have a very large central vacuole that occupies a major portion of the volume of the cell. In some species of *Chara* (Fig. 5-3) the internodal cells are ensheathed in a layer of cortical cells resembling a parenchymatous construction. Active cytoplasmic streaming (**cyclosis**) is readily observable in the elongate internodal cells.

The uninucleate nodal cells are much smaller than the internodal cells. These cells divide in a certain pattern, creating the whorls of small branches that superficially resemble leaves. The growth of these branchlets is **determinate.** This growth pattern, together with the indeterminate apical meristem, gives rise to a distinctive habit consisting of apical meristem–node–internode–node–internode, and so on. Such morphological patterns are not unusual among the higher plants but are rare among the algae.

Each cell contains hundreds of small stationary chloroplasts arranged in longitudinal rows. There are no pyrenoids. The chloroplasts have been seen to divide by simple fission in *Nitella* (Green, 1964). The ultrastructure of each plastid closely resembles those associated with green algal cells, with a double membrane envelope encasing multiple series of two to six thylakoids arranged in stacks or bands (see Fig. 1-7).

The cell walls of some charophytes may be encrusted with precipitated lime in the form of calcium or magnesium carbonate. These deposits are found in *Chara* but not in *Nitella* and may lead to extensive deposits of marl. It is the deposition of such carbonates that has lead to the common names "stonewort" or "brittlewort."

Other ultrastructural features of charophyte cells include the usual complement of endoplasmic reticula, ribosomes, Golgi bodies, and starch grains associated with eukaryotic plant cells.

REPRODUCTION

Asexual reproduction in the charophytes is accomplished by fragments of the plant body that have the capability of developing adventitious **rhizoids** and branches. Rhizoids may also give rise to new shoot systems forming colonies of the plant spreading over a wide area. Such colonies of *Chara* or *Nitella* are common in the quiet silt-bottomed waters of lakes and streams.

Uninucleate cells of charophytes undergo regular mitosis and meiosis. The cell cycle resembles that found in the bryophytan line of Chlorophyta, the bryophytes, and the vascular plants, all of which produce a prominent phragmoplast located between telophase nuclei. No zoospores are formed in Charophyta.

Sexual reproduction in Charophyta involves such dramatic differences from all other algae that special terminology is sometimes used to describe the structures involved. The distinctive egg-producing organ is the oogonium (often referred to as the **nucule**), while the sperm are formed in an **antheridium** (referred to by some writers as the **globule**). In monoecious forms, the oogonium is located at the node above the adjacent antheridium (Fig. 5-3). The most distinctive feature of these organs is their multicellular nature. Both structures contain many sterile support cells associated with the developing gametes. No other alga has a protective layer of vegetative cells associated with its gametes. In fact, one of the major characteristics of an alga is the nakedness of its developing gametic cells. The nonreproductive support cells of both the oogonium and the antheridium take on distinctive colors upon maturation: the shield cells of the antheridium turn reddish as a result of the accumulation of carotenoid pigments, and the oogonium becomes yellow-brown.

The oogonium is a multicellular structure

FIG. 5-2 Apical cell of *Chara* spp.
Median longitudinal section of vegetative thallus of *Chara*. Note single apical meristematic cell *(ams)* and cells making up internode *(in)* and node *(n)* (250×).

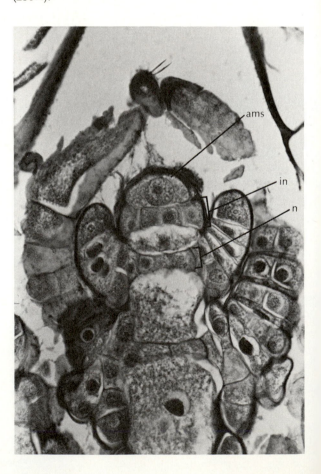

containing a single large egg cell laden with starch grains and other nutrients. A single layer of spirally arranged tube cells surrounds the egg cell (Fig. 5-3). The tip of the oogonium is capped with a series of crown cells forming a corona. There are five crown cells in *Chara* and ten crown cells in *Nitella* and *Tolypella*. Swimming sperm cells enter the egg cell through openings that occur between the crown cells and the tube cells. Syngamy is accomplished while the egg is still attached to the vegetative plant.

The antheridium is a complex multicellular structure unique in the plant world (Fig. 5-3). Sperm cells are formed within the walls of the cells that form the antheridial filaments. Mature sperm are liberated by the cells of the antheridial filament and emerge through degenerating **shield cells.** The sperm cells are also unique among the algae in their degree of cellular com-

FIG. 5-3 Life cycle of *Chara*.

Complete haplontic life cycle for a member of genus *Chara*. Notice that this species is monoecious. The oogonium is a multicellular structure containing an egg cell. In *Chara* the oogonium has 5 crown cells, while *Nitella* has 10 crown cells.

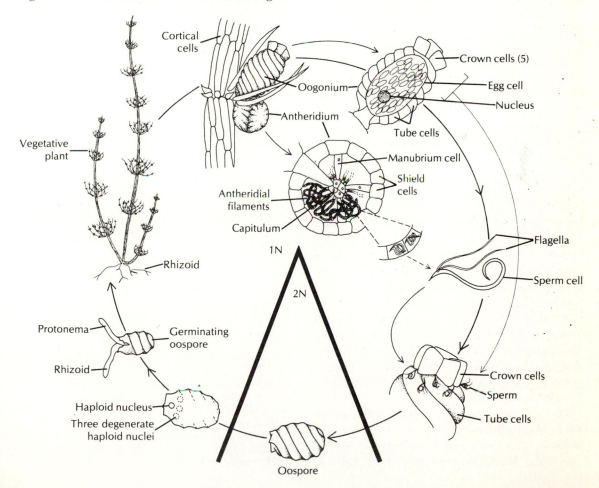

plexity. They are highly differentiated, elongate, subapically biflagellate cells that resemble the sperm found in some bryophytes. Among the algae, only the sperm of *Coleochaete* (Chlorophyta) come close to the degree of complexity in the sperm of *Chara* or *Nitella* (Fig. 5-3). Maturation of both oogonia and antheridia occurs during the fall of the year in temperate zones. Whether differentiation of these structures is controlled by photoperiod or by hormonal regulation is not known.

The zygote develops into a dark-colored, thick-walled oospore that is released into the environment when the vegetative tube cells degenerate. Such oospores form a resistant stage that is capable of living for months in the mud and silt before germination the following spring. When the zygote germinates it does so through meiotic type divisions (zygotic meiosis), giving rise to a filamentous, protonema-like stage of development. Cellular polarity is established at germination. Nonphotosynthetic rhizoids develop from the photosynthetic filament. Apical growth of the filament gives rise to a mature plant. Filamentous protonema stages are common among bryophytes and the lower vascular plants. Hutchinson (1975) gives a detailed account of reproduction and life cycles in individual species.

PHYSIOLOGY

Cells from members of the charophytes have been used for years to study cell sap streaming. The internodal cells are large, and the streaming cytoplasm is easily seen. Furthermore, these cells may be easily penetrated with electrodes, enabling physiologists to make accurate measurements of action potentials, turgor pressure, and ion transport. (The reader is referred to Green et al [1970] and Kamitsubo [1972] for further reviews of this work.)

The growth of the various parts of the charophyte plant exhibits positive and negative **geotropism.** The colorless rhizoids, composed of unicellular filaments, grow only at the tip, and

these growth areas are positively geotropic. Located within the growing filament is a series of **statoliths** (Fig. 5-4) composed of various membrane-bound, inorganic substances (Sievers and Schröter, 1971). The downward growth of the rhizoids results from the displacement by gravity of the heavy statoliths to one side of the growing cell; this displacement causes an asymmetrical distribution of cell wall materials at the growing tip, with the result that greater growth occurs on the upper side of the tip, turning it downward (positive geotropism).

At the other end of the charophytan plant is a negatively geotropic apical meristem composed of a single apical meristematic cell (Fig. 5-2). This cell evidently exhibits a type of apical dominance over the development of the alga in much the same manner as that observed in bryophytes and higher plants. It is not known

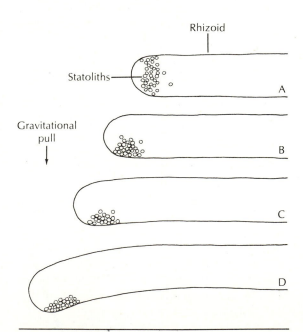

FIG. 5-4 Geotropism in *Chara* rhizoids.
Successive stages (*A–B–C–D*) in growth of a *Chara* rhizoid showing positive geotropism and its cause—downward displacement of heavy statoliths at growing rhizoid tip. (After Sievers, A., and K. Schroter. 1971. Planta **96**:339-353.)

whether plant hormones (**auxins**) are involved in the process.

Calcium carbonate deposition by some species of *Chara* is not the same as that observed in some members of Chlorophyta, Phaeophyta, and Rhodophyta. In *Chara*, calcite is apparently precipitated as an outer covering rather than being incorporated into the cell walls. This process occurs only in those plants found growing in waters having a high lime content.

Chara has been grown in the laboratory in unialgal culture and has been used experimentally to determine the minimum amount of light required for photosynthesis. *Chara* was chosen for these experiments because it had been found growing in Lake Tahoe, California, at a depth of 65.5 m, where light intensity is approximately 2% of full sunlight, and the habitat is well below the depth where angiosperms flourish (Frantz and Cordone, 1967).

All members of Charophyta are autotrophic. They carry out photosynthesis, with their only requirements being a regular light source, inorganic nutrients, and carbon dioxide. An excellent account of environmental conditions under which charophytes grow is given by Hutchinson (1975).

EVOLUTION

The biochemical similarities between Charophyta and Chlorophyta indicate an unmistakable relationship between the two. The geological record indicates that the charophytes are somewhat younger than the green algae, having evolved during the Silurian period. The distinctive morphological structure of Charophyta suggests that they diverged from the green algae early in the Silurian period and have since evolved along a separate evolutionary path. By the Devonian period they were highly specialized organisms with structures similar to those of the modern genera. The deposition of lime by their cells insured a good fossil record. By the late Jurassic period, shallow lakes contained a highly specialized charophyte flora (Harris,

1939). In the late Paleozoic and through the mid-Mesozoic eras this division probably contributed more to the primary productivity of freshwater ecosystems than it does now. There was little competition during these eras from the angiosperms, which did not evolve until the Cretaceous period.

Banks (1970) has proposed that the charophytes successfully invaded brackish and freshwater habitats during the Devonian period. If algal evolution began in the sea, then algae must have been moving toward the freshwater habitat at that time. The zygote, protected from desiccation by its thick, cellulose–calcium carbonate wall, would have been an adaptation to environments where periodic drought occurred. At this time the marine fossil record through the late Silurian period does not contain any of this type of thick-walled zygospore. Additional information from the fossil record should aid in determining more precisely the time at which algae started the colonization of fresh water. Peck (1953) reviews the history of this division.

ECOLOGY

Charophyta are benthic plants of clear, hard, fresh waters, where they often form extensive submerged meadows growing on sandy or muddy bottoms. They are well adapted to deep waters where the light intensity is low and where they have little competition from the angiosperms. A few species of *Chara* and *Tolypella* may occur among the littoral flora of brackish waters such as the Baltic Sea, but there are none that are strictly marine. Wood (1965) and Imahori (1964) recognize six genera. The three genera listed earlier are the most common. Their distribution is world-wide, from the arctic to the equator and from alpine habitats (up to 4600 m) to sea level. Indications are that Charophyta evolved in the New World and radiated from there to the rest of the world. This conclusion is based upon the numerous species that occur in temperate North America when compared with those occurring in the Old World.

The branched thallus of the charophytes usually does not exceed heights of 30 to 40 cm, but some may measure up to 1 m. However, the habit may be greatly affected by environmental conditions: intense light produces stunted bushy plants, while reduced light induces elongate diffuse plants. Quiet water, running water, temperatures, salinity, and available nutrients all affect the habit of individual plants, producing morphological structures that are environmentally rather than genetically controlled.

The genus *Chara* is the most common of the charophytes. It inhabits lakes, ponds and, slow-moving alkaline (pH 7.2 to 9.5) streams throughout the world. High levels of calcium carbonate are often found in waters where *Chara* abounds, and precipitated lime is often associated with the central axes and lateral branches of the plants. These lime deposits give the alga a rough feel and often give them a grey-white color. *Chara* may also be identified by its musky, skunklike or garlicky odor.

An extensive and characteristic epiphytic community may develop on *Chara*, and very large numbers of diatoms, especially, may be found. Allanson (1973) discusses the development of this community and its role in the aquatic ecosystem. Moss (1976) documents the effect of excess nutrients on such epiphytes.

Nitella (Fig. 5-1) is less-common and often found in acid waters such as acid lakes and bogs where pH ranges from 4.6 to 7.0. The genus can readily utilize the carbon dioxide available at acid pHs (Fig. 3-7) but cannot utilize carbon in the form of bicarbonate, which prevails at alkaline pHs (Smith, 1968). It is found in deep waters and is often collected only with a dredge. *Nitella* usually has a glassy, grass-green appearance, does not deposit lime, and commonly has an endophytic green alga, *Coleochaete nitellarum*, growing in its cell wall.

Tolypella is a solitary green plant usually found in alkaline waters and often in brackish estuaries. It is difficult to separate *Tolypella* from *Nitella* taxonomically, and in some instances it appears that the two intergrade. Such is the case with the *Tolypella*-like *Nitella* species from Japan, *N. imahorii* (Wood, 1965). Prescott (1978) differentiates the two on the basis of branching patterns, position of antheridia, and color (*Tolypella* is gray-green, while *Nitella* is dark green). *Tolypella* is also more common in the southern hemisphere.

Excess phosphate appears to inhibit the charophytes; this fact may explain the disappearance of this division from eutrophic lakes (Forsberg, 1965). The group is generally restricted to oligotrophic freshwater systems.

ECONOMIC IMPORTANCE

Economically Charophyta have not been of any great importance. Plantings of *Chara* have been used in aquatic insect control, as food for animals and waterfowl, and as stabilizing agents to settle silt. Organic compounds secreted by some species of *Chara* and *Nitella* kill mosquito larvae (Imahori, 1954) and repel planktonic arthropods such as *Daphnia* (Pennak, 1973). Hutchinson (1975) reviews the mosquito-larvicidal effects, concluding that the effects are inconsistent and appear dependent not only on the age and species of larva but also on the specific strain of the alga. The presence of certain charophytes in a body of water also discourages female mosquitoes from laying their eggs in that water. More research is needed before Charophyta can effectively be used in mosquito control, although they evidently have a potential for use in this important area.

Sometimes overgrowths of charophytes become a problem in municipal water supplies because they clog the streams and reservoirs, and some species of *Chara* and *Nitella* give the water a distinctive taste.

Because calcium carbonate may make up to 30% of the dry weight of some *Chara* species, extensive growths of this alga may form deposits of marl, a combination of clay, silt, and carbonates. These deposits have been used as fertilizers in lime-deficient soil and for making bricks.

SUMMARY

Charophyta are well-known freshwater plants throughout the world. Their distinctive habit, composed of apical growth, nodes, and internodes, separates them from all other green-colored algae. Their reproductive structures contain layers of sterile vegetative cells that afford some protection to the developing gametes. Meiosis is zygotic, leading to a haplontic life cycle. Germination of the zygospore produces a filamentous protonemalike stage of growth much like the early developmental stages found in Bryophyta. These factors have led to the assignment of Charophyta to a separate division within the green line of algal evolution.

Charophyta have never been very important to the economy of the world, but they are significant members in the ecosystems of alkaline freshwater habitats, where they contribute to the primary productivity of the environment. *Nitella* may also be an important producer in deep or acidic waters.

SELECTED REFERENCES

Gramblast, L.J. 1974. Phylogeny of the Charophyta. Taxon **23:**463-481 pp.

Harris, T.M. 1939. British Purbeck Charophyta. British Museum (N.H.) IX, London. 83 pp.

Imahori, K. 1954. Ecology, phytogeography, and taxonomy of the Japanese Charophyta. University of Kanazawa, Japan. 234 pp.

Peck, R.E. 1953. The fossil charophytes. Bot. Rev. **19:**209-227 pp.

Wood, R.D. 1965. Monograph of the Characeae. In R.D. Wood and K. Imahori, A revision of the Characeae, Vol. I. Cramer, Weinheim, Germany. 904 pp.

Wood, R.D., and K. Imahori. 1964. Iconograph of the Characeae. In R.D. Wood and K. Imahori, A revision of the Characeae, Vol. II. Cramer, Weinheim, Germany. 401 pp.

6

CLASSIFICATION

Division: Euglenophyta: (euglenoids) 35± genera, 800± species; mostly freshwater and brackish-water unicellular flagellates; some colonial; prominent eyespots (stigmata)

Class: Euglenophyceae

ORDER 1. Eutreptiales: primarily marine; two isokont anterior flagella; *Distigma* (a freshwater genus), *Eutreptia*

ORDER 2. Euglenales: primarily fresh water unicells with a single emergent anterior flagellum; *Astasia, Colacium, Euglena, Phacus, Trachelomonas*

ORDER 3. Heteronematales: colorless, phagotrophic unicells with one long anterior flagellum and one shorter trailing flagellum; no eyespots (stigmata); *Peranema*

EUGLENOPHYTA

Division Euglenophyta represents a primitive flagellate group possessing both plant and animal affinities and features. Two thirds of the genera are heterotrophic, and none of the species is purely autotrophic. Some possess chlorophyll types *a* and *b*, but most of them lack chlorophyll entirely, depending upon ingestion of organic materials for their energy. Most euglenoids are unicellular, but a few may be palmelloid or colonial. They are common inhabitants of small bodies of water where high concentrations of dissolved organic material are present. Most euglenoids are freshwater organisms, but a few inhabit marine or brackish waters, where they often form an important component of the **mud algae** in salt marshes and estuaries. The most common genus is *Euglena* (Fig. 6-1). Current investigations on euglenoid organisms are reviewed by Walne (1980).

Superficially euglenophytes appear to belong to Chlorophyta, but they differ from them in a number of significant respects. They resemble green algae and Charophyta in having the same chlorophyll types and a similar grass-green color; however, their energy storage product, method of cell division, nuclei, cell walls, and ultrastructural characteristics are quite different. Also, sexual processes have not been unequivocably observed in any of the euglenoids. And finally, they are not obligate autotrophs, because even the photosynthetic euglenoids require organic supplements to their diet.

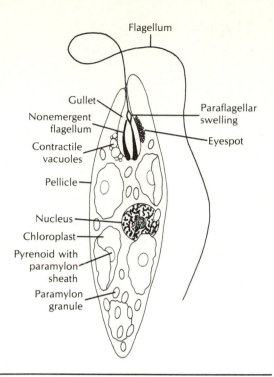

FIG. 6-1 Structure of *Euglena*.
Nearly 60 *Euglena* spp. are found in both marine and fresh waters of North America. They are a common organism from organically rich ponds and sloughs. Note presence of anteriorly located gullet and single emergent flagellum. Most species have a single conspicuous eyespot and multiple contractile vacuoles as well as numerous disk-shaped chloroplasts containing protruding pyrenoids ensheathed within a paramylon membrane.

CELL STRUCTURE

Euglenoid cells range in size from 35 μm wide to 400 μm long. They lack the rigid cellulosic cell wall found in many algal groups. In its place, euglenoid cells have a proteinaceous, flexible **pellicle.** This semirigid structure enables the cell to constantly change shape, allowing it to squeeze through small openings encountered in its environment. This change in shape is called **metaboly.** The pellicle is located inside the plasma membrane, which is the outer boundary of the cell (Fig. 6-2). It consists of a variably thickened, helically wound series of ridges and grooves whose composition is over 80% protein, with the rest being lipid and carbohydrate. Its thickness depends upon the individual species. The thicker it is, the more rigid the cell is, and the more likely it is for the cell to exhibit some sculpturing associated with the cell surface (e.g., *Phacus*, Fig. 6-3).

Just beneath the pellicle are located many microtubules and **muciferous bodies.** The microtubules appear to function in the formation of new pellicle materials, while the muciferous bodies produce a constant secretion of mucilage, which coats the outer surface of the cell. Mucilage may be observed as a small tail at the posterior end of the cell (e.g., *Phacus*, Fig. 6-3). Species having large muciferous bodies often secrete enough mucilage to form a stalk, as is the case in the sessile form of *Colacium* (Fig. 6-4). In other genera these bodies secrete a rigid

FIG. 6-2 Euglenoid pellicle.
Pellicle of euglenoid cell as seen through electron microscope. Note series of grooves and ridges that run spirally on the cell surface. Microtubules provide support and form new protein materials to maintain the pellicle. They vary in location and number. Muciferous bodies are located beneath ridges in the cytoplasm of cell. They are connected with cell surface through individual canals that carry the mucilaginous product of the muciferous bodies to the cell surface. (From Leedale, G.F. 1967. Euglenoid flagellates. Prentice-Hall, Inc., Englewood Cliffs, N.J.)

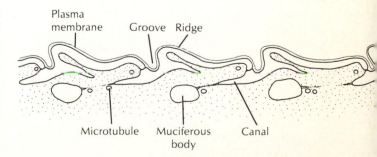

FIG. 6-3 Structure In *Phacus*.

Phacus cells are usually flattened and disklike with a pellicle containing numerous longitudinal ridges. Posterior of cell ends with tail-like structure secreted by mucilaginous bodies and composed of same substances as pellicle. At anterior end of cell note single emergent flagellum, gullet, eyespot, and numerous contractile vacuoles. Numerous small discoid chloroplasts are located in the cytoplasm. Ring-shaped paramylon granules are variable in size and sometimes get so large that they are the most conspicuous bodies in the cell.

Like *Euglena*, *Phacus* spp. are both marine and freshwater. They are most common in those organic waters in which *Euglena* thrive. (From Leedale, G. F. 1967. Euglenoid flagellates. Prentice-Hall, Inc., Englewood Cliffs, N.J.)

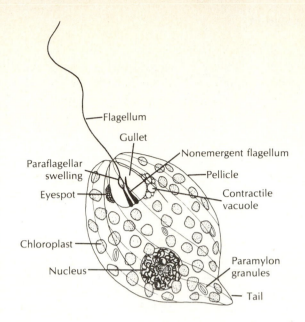

FIG. 6-4 Structure and habit of *Colacium mucronatum*.

Colacium sp. is representative of a group of sessile euglenoids. The free-swimming cells possess a single emergent flagellum that is resorbed when organism settles on a substrate and secretes a rigid mucilaginous stalk at its anterior end. Microscopic plants and animals (especially crustaceans) may appear green because of growths of *Colacium* spp., a type of symbiosis that is not well understood. In cell shown below, note projecting, cup-shaped pyrenoids encased within a paramylon sheath. (From Leedale, G.F. 1967. Euglenoid flagellates. Prentice-Hall, Inc., Englewood Cliffs, N.J.)

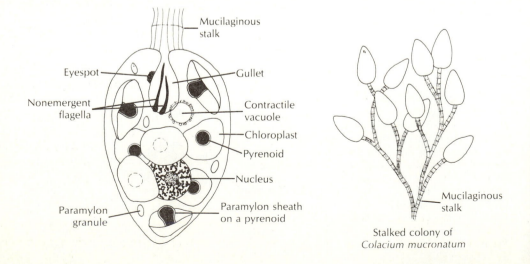

Stalked colony of
Colacium mucronatum

outer shell called a **lorica,** which is outside the plasma membrane and may be impregnated with ferric hydroxide or manganese compounds, such as in the free-swimming euglenoid *Trachelomonas* (Fig. 6-5).

In some genera, most notably *Euglena,* a heavy secretion of mucilage by the muciferous bodies forms a protective cyst. Encystment is a response to adverse environmental conditions such as desiccation, high light intensity, or changing temperatures. The cells slow down, become rounded, lose their flagella, and secrete an exterior coating of mucilage. Cysts may survive for months and emerge when favorable conditions return. Cell divisions sometimes occur within the mucilage before the flagella appear, forming a multicellular palmella stage.

FIG. 6-5 Structure of *Trachelomonas.*
One of several hundred species of *Trachelomonas* characteristic of freshwater ditches and bogs. Lorica is different in shape and ornamentation in each species. Its color ranges from yellow through red-brown to brown, depending on presence of iron and manganese. Note stalked pyrenoids encased within paramylon sheath and single emergent flagellum. (From Leedale, G.F. 1967 Euglenoid flagellates. Prentice-Hall, Inc., Englewood Cliffs, N.J.)

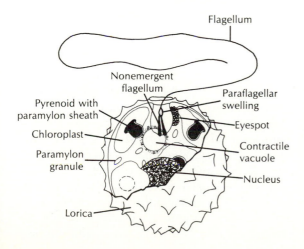

Germinating cysts are often responsible for the green euglenoid blooms that occur after heavy rains in puddles and ruts along dirt roads.

Chloroplast structure in euglenoids differs significantly from that observed in the green algae. With the light microscope the plastids, floating free in the cytoplasm, appear as numerous oval or disk-shaped bodies containing a central pyrenoid. The pyrenoid is apparently the formation site for the energy storage product **paramylon.** Ultrastructurally the chloroplasts are encased in a double unit membrane with an additional outer membrane of chloroplast endoplasmic reticulum (CER) containing ribosomes (see Fig. 1-7). The thylakoid arrangement is different from the green algae in that they are stacked in groups of three. Associated with the thylakoid membranes are chlorophyll *a* and *b* and various carotene and xanthophyll pigments (see Table 1-1).

Paramylon, the energy storage product in the euglenoids, is not found within the pyrenoids but is found in the cytoplasm in paramylon granules or as an encasing sheath around individual pyrenoids, as in *Euglena* (Fig. 6-1), *Colacium* (Fig. 6-4), and *Trachelomonas* (Fig. 6-5). These structures are composed of an insoluble polymer of glucose molecules having β 1-3 glucan linkages. Paramylon does not stain with iodine, the classic test for starch. The structure and properties of paramylon are very similar to the β 1-3 glucans found in diatoms (Chrysophyta) and the brown algae (Phaeophyta), but paramylon is not as highly branched.

The eyespot (stigma) of euglenoids differs from that of green algae in being independent of the chloroplasts. It is composed primarily of a number of spherical droplets of various pigments, among which the carotenoids (β-carotene, diatoxanthin, and diadinoxanthin) are present in greatest quantities (Heelis et al, 1979). Stigmata are located at the anterior end of the cell near the basal **paraflagellar swelling** associated with the longer of the two flagella (Fig. 6-1). It is believed that stigmata function as shading or filtering devices for the major photoreceptor organelle, the paraflagellar swelling (Diehn, 1969; Diehn and Kint, 1970). Creutz and Diehn (1976) have also shown that the paraflagellar swelling is sensitive to polarized light and that it may be affected by gravitational pull.

A contractile vacuole is located in the anterior end of the cell near the reservoir. This organelle functions as an osmoregulator, periodically discharging its contents into the external environment through the reservoir and its connecting canal.

The two flagella of the euglenoids have different shapes and positions, providing a feature for the identification and classification of the organisms (see classification on p. 117). All flagella have a classic 9 + 2 protein microtubular arrangement. The axonemic portion of the flagella have many fine hairlike projections and are therefore called tinsel-type flagella. They emerge through the anterior canal. The basal portion of the longer flagellum broadens into a paraflagellar swelling, which is the site of photoreception. It is shaded by the eyespot. The flagella extend into the cytoplasm beyond the basal bodies, and in cross sections of cells these extensions are sometimes mistaken for centrioles.

In phagotrophic genera such as *Peranema* (Heteronematales), there are two rigid **ingestion rods** adjacent to the anterior canal and reservoir (Fig. 6-6). These rods may be protruded and attached to prey before its ingestion. The prey is usually another algal cell such as *Euglena* or *Chlorella*.

Other cytoplasmic features of euglenophytes include the usual complement of endoplasmic reticulum, ribosomes, Golgi apparatus, and mitochondria. The mitochondria are more numerous in nonphotosynthetic heterotrophic euglenoids.

The nucleus is enclosed in a nuclear membrane containing numerous nuclear pores. Within the membrane is found the nucleolus, the nucleoplasm, and the chromosomes. The nuclear organization, however, is significantly different from most eukaryotic organisms and is

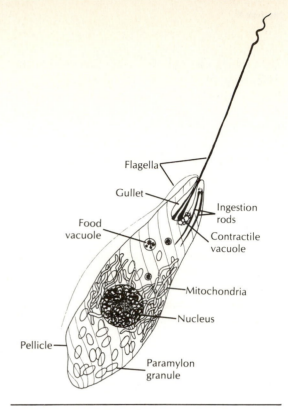

FIG. 6-6 Structure of *Peranema*.

Peranema spp. are colorless, heterotrophic euglenoids that stalk and capture prey using the two anterior ingestion rods that are protruded and attached to the victim before ingestion and formation of a food vacuole. Note presence of two anterior flagella, one directed forward and one directed backward. The larger, anterior flagellum is active in motility, especially near its tip. Cell interiors often contain numerous paramylon granules as well as elongate mitochondria that are visible with a light microscope. (From Leedale, G.F. 1967 Euglenoid flagellates. Prentice-Hall, Inc., Englewood Cliffs, N.J.)

called a **mesokaryotic** type by some investigators (Lee, 1977). The chromosomes are permanently condensed, and they lack the protein-histone complement of most eukaryotes. They resemble the chromosomes of some dinoflagellates (Pyrrhophyta) in this respect. The nucleolus is sometimes referred to as the **endosome,** because it is a permanent structure and does not become resorbed during mitosis. Some strains of *Euglena* and *Trachelomonas* contain reproducing bacterial populations within their nuclei (Leedale, 1969). The role of these symbiotic organisms is not known.

REPRODUCTION

Reproduction in the Euglenophyta is synonymous with cell division. No sexual union of gametes has been unequivocally reported in the literature, although evidence for meiosis was reported in two genera of euglenoids by Krichenbauer (1937) and Leedale (1962). The mitotic cycle in euglenoids is altered somewhat because of the permanently condensed chromosomes. During mitosis the nuclear membrane remains intact and no spindle microtubules appear. The chromosomes separate and sister chromatids migrate to opposite poles of the nucleus. The nuclear membrane pinches inward from opposite sides to accomplish nuclear division. Cytokinesis occurs by a longitudinal division of the cytoplasm.

PHYSIOLOGY

Euglena gracilis is the only species of this division that has been successfully grown in axenic culture; therefore much of the knowledge of euglenoid physiology is based upon this organism. Better than any other division, the euglenoids exemplify the problems involved in designating an organism as either a plant or an animal. Some euglenoids are green and photosynthetic (i.e., plantlike), but they require at least one vitamin, usually vitamin B_{12}. Such organisms are auxotrophic. Other euglenoids are heterotrophic and nonphotosynthetic (i.e., animal-like). Furthermore, when dividing cells of a photosynthetic *E. gracilis* are grown in the dark for extended periods, the chloroplast lamellations and chlorophyll disappear (Ben-Shaul et al, 1965). Subsequent generations can live heterotrophically in the dark for several years as long as organic nutrients, usually acetate, are supplied. These colorless *E. gracilis* closely resemble a species of the heterotrophic *Astasia*. Organelles called

proplastids remain and divide along with the cells. When the *E. gracilis* are returned to light the proplastids enlarge, become functional chloroplasts, and start to synthesize chlorophyll. The organisms again become photosynthetic (Leedale, 1967). Light is apparently essential for chlorophyll synthesis in *E. gracilis*, although some other algae can synthesize chlorophyll while growing in the dark. Schiff (1978) discusses further the control by light of chloroplast development in *Euglena*.

When dividing *E. gracillis* cultures are treated experimentally with ultraviolet light (Lyman et al, 1959), antibiotics such as streptomycin (Provasoli et al, 1948), or temperatures of 34° C (Pringsheim and Pringsheim, 1952), a permanent loss of chloroplasts takes place over time. In these "bleaching" experiments, evidently chloroplast replication is inhibited, no new chloroplasts are formed, the number of chloroplasts per cells decreases, and cells without chloroplasts result. Colorless and heterotrophic *E. gracilis* cells occur in the population in growing numbers, eventually resulting in a permanently colorless race. Nuclear replication is not inhibited by these treatments, which indicates that the chloroplast functions and replicates without control from the nucleus. Lyman and Traverse (1980) summarize the research on chloroplast bleaching and nucleus-organelle interactions in *Euglena*.

Paramylon is the energy storage compound seen in all photosynthetic euglenophytes. Paramylon is also synthesized by some heterotrophic euglenoids which use various organic compounds as carbon sources.

Nutrition in the Euglenophyta is never completely autotrophic, as previously discussed. Vitamin supplements, usually B_{12}, are required by all the green euglenoids. Vitamin B_{12} acts as a coenzyme for an enzyme that is essential to cell division (Gleason and Wood, 1977). The heterotrophic forms are capable of utilizing a number of organic substances; laboratory cultures may be maintained on substrates of acetates or alcohols at an alkaline pH. The nonphotosynthetic forms are obligate heterotrophs, while some of the photosynthetic types are facultative heterotrophs that are capable of utilizing organic molecules as an additional energy source. Both heterotrophs and auxotrophs can absorb dissolved nutrient substances directly from their environment, while the Heteronematales may also ingest particulate food phagotrophically.

Certain carotenoid pigments may be unique to the Euglenophyta. Antheraxanthin has been reported only from euglenoids, while **astaxanthin** (also called hematochrome) is found only in euglenoids and a few unicellular green algae. Astaxanthin, when present in high concentrations (as a response to high light intensity), is capable of imparting a red hue to the entire cell. *Euglena sanguinea* may produce enough astaxanthin to give a concentrated large population of cells a blood-red appearance.

Phototaxis is the movement of an organism in response to light. Positive phototaxis is movement toward a light source, while negative phototaxis is movement away. Both positive and negative phototaxis may be observed in euglenophytes. Efficient photosynthesis is accomplished only in those areas of light intensity where an optimal amount of radiant energy is absorbed by the photosynthetic pigments. Euglenophytes are positively phototactic in low light intensities and negatively phototactic in high light intensities. The receptor organ for light is apparently the paraflagellar swelling associated with the longer flagellum. When some euglenoids swim in the characteristic spiralling manner, the eyespot alternately shades and exposes the light-sensitive paraflagellar swelling. First filtered, then nonfiltered, light reaches the swelling, and this alternating light regimen directs the organism toward light intensities appropriate for efficient photosynthesis.

Tidal mud flats usually contain photosynthetic euglenoids that are positively phototactic at low tides and negatively phototactic at high tides. They also exhibit positive **thigmotaxis,** the behavioral tendency to cling to surfaces, during high tides. These behavioral responses to environmental conditions allow the euglenoid to burrow into the mud at high tide and to migrate back to the surface at low tide. Even when *Euglena* from tidal flats is removed to the laboratory it maintains an endogenous **circadian rhythm** of burrowing. The rhythm may persist for days, even in constant light (Pohl, 1949). Periodic vertical migrations of this sort are common among mud and sand algae.

Certain species of *Euglena* are able to concentrate dissolved phosphates in their cytoplasm up to 100,000 times the concentration in the surrounding water. The survival value of this phenomenon is clear, considering that in freshwater ponds and lakes phosphate is often the limiting nutrient in the growth of algal populations (see Chapter 2). Those organisms that can store phosphates for future use have a distinct advantage when phosphate supplies are depleted, as they may be during an aquatic bloom.

EVOLUTION

The origins of the euglenophytes are obscure, since these cellulose-free and carbonate-free cells have left no fossil record. Because of the various cellular characteristics discussed earlier, it appears that this division has certain affinities not only with the green line of algal evolution but also with flagellated protozoans. On the one hand, the presence of chlorophyll type *a* and *b* biochemically relates the euglenoids to Chlorophyta; on the other hand, the lack of cellulose and starch is not consistent with this relationship. The presence of paramylon and the mesokaryotic nucleus would seem to indicate a relationship to the brown line of algal evolution, especially to the Pyrrhophyta. However, the obligate heterotrophic nutrition of many euglenoids such as *Astasia* and *Peranema* appears to relate them to the protozoa. In fact, many colorless euglenoids are often classified as protozoans by zoologists.

For these reasons the precise position of Euglenophyta on the phylogenetic tree is not known. Some biologists (Leedale, 1967), consider the photosynthetic forms to be primitive, the heterotrophic forms to be derived from them, and the **phagotrophic** euglenoids to be the most highly advanced members of this division. These researchers postulate that colorless, heterotrophic euglenoids may have evolved from photosynthetic forms under natural conditions similar to those experimental conditions that induce chloroplast inactivation or loss. Leedale (1978) recently revised the phylogenetic scheme to take into consideration the theory that chloroplasts originated from **endosymbiotic** algae (Margulis, 1970; 1981). Klein et al (1972) proposes that the original symbiotic alga was a blue-green alga, while Stewart and Mattox (1975) and Gibbs (1978) think the symbiotic alga was a green alga because of similar carotenoids. Current research on the origin of phytoflagellates in general suggests that the heterotrophic flagellates may be primitive and that the photosynthetic flagellates are derived by the phagotrophic ingestion of algae, followed by the establishment of endosymbiosis (Stewart and Mattox, 1980). Only additional studies will establish the definite phylogenetic scheme by which Euglenophyta evolved.

ECOLOGY

Euglenophytes are primarily inhabitants of slow-moving or stagnant freshwater environments. A few species, however, inhabit salt marsh sediments, brackish water, and oceans. The photosynthetic euglenoids are important primary producers in small farm ponds, which are rich in organic matter. Species of *Euglena* may be the dominant alga in such waters, forming green or red floating scums. Bright red blooms in ponds—"blood lakes"—are caused by *E. sanguinea*, which produces the red astaxanthin. The biblical references in Exodus (Chapters 4 and 7) to rivers and ponds turning to "blood" may have been inspired by blooms of *E. sanguinea*. Revelations also contains references to waters turning to "blood."

Many euglenoids are tolerant of water that is polluted with organic material and low in dis-

solved oxygen. For instnce, *Phacus* and *Euglena* are often found in polluted conditions, even though *Phacus* appears somewhat more sensitive to pollution than *Euglena*. Some euglenoids are distributed worldwide (e.g., *E. gracilis*) and have been reported from waters with a range of pH from 3.2 to 9.9. Species of *Euglena* may be found in streams polluted with acid mine drainage where the pH is as low as 3.2. Such euglenoids as *Distigma* are often used as indicators of acid waters for industrial waste effluents or for naturally occurring acid peat bogs or acid lakes.

The heterotrophic euglenoids are often found where putrefaction is taking place and decaying organic material is being **mineralized.** In sewage oxidation ponds and in other waters with high concentrations of organic molecules, the heterotrophic euglenoids often dominate the plankton. The euglenophytes have a competitive advantage over other organisms, because not only can they utilize the small organic molecules, acetate and glucose, but they can also utilize ammonium, the prevalent form of nitrogen in sewage. The phagotrophic genus *Peranema* is also common in waters rich in organic debris and other particulate food materials.

The heterotrophic *Astasia* lives in soil or water where actinomycete bacteria are also found. Many free-living actinomycetes produce antibiotics such as streptomycin. Some investigators have proposed that *Astasia* evolved from photosynthetic euglenophytes that were "bleached" by antibiotic activity from soil organisms in a manner similar to the bleaching of *Euglena* by streptomycin.

The lorica-covered *Trachelomonas* (Fig. 6-5) is often found in warm waters in such places as shallow ponds or lakes or among aquatic plants along lake shores. Iron in the form of ferric hydroxide may be deposited in the lorica; the greater the amount of iron in the water, the darker brown the lorica.

Heterotrophic euglenophytes may parasitize amphibian respiratory tracts or live symbiotically in rotifers, nematodes, or copepods. Of special note is the colonial stalked species *Co-*

lacium libellae (a species similar to *Colacium mucronatum*, Fig. 6-4), which lives in the rectum of dragonfly nymphs in the winter, causing the posterior of the nymph to turn green (Roskowski and Willey, 1975). These organisms emerge from the rectum in the spring to live as free-swimming cells. Other species of *Colacium* may grow on microscopic zooplankton, especially crustaceans such as seed shrimp (Ostracoda), causing them to appear green.

Many members of this division can form thick-walled, reddish cysts in response to unfavorable environmental conditions. Cysts may remain dormant in muds and soils until spring rains or other such moisture provides suitable conditions for growth. Germination of cysts may suddenly cover a temporary pool or puddle with a green or reddish scum.

Most Euglenophyta live in organically enriched fresh waters. There are, however, several marine species. Many species of *Eutreptia* are members of the marine phytoplankton and may be very numerous in eutrophic temperate estuaries (Lackey, 1967). There are several freshwater *Eutreptia* species that may become numerous in acid lakes.

ECONOMIC IMPORTANCE

The euglenophytes contribute a significant amount of primary productivity to the biosphere. They are important phytoplankters in eutrophic fresh waters and estuaries, although they seldom outnumber the dinoflagellates or diatoms in lakes or oceans. Both heterotrophic and auxotrophic members may have large populations in sewage oxidation ponds. In some subtropical areas such sewage oxidation ponds are called "*Euglena* ponds" because *Euglena* spp. dominate the flora. Heterotrophic forms break down the organic materials, while the photosynthetic forms utilize the abundant inorganic nutrients. These ponds oxidize organic materials efficiently in warm climates where temperatures and sunlight are conducive to rapid growth rates of the euglenoids.

Large populations of euglenoids may have an adverse effect on municipal water supplies, causing fishy odors and tastes. *Trachelomonas,* with its brown-red-colored lorica, may cause the water to turn brown when large populations develop. It also may cause a problem by clogging water filter systems.

In the laboratory euglenoids have been used for research on nucleic acids, plastids, mitochondria, and enzyme analysis. One particular strain of *E. gracilis* is the organism of choice for vitamin B_{12} (cobalamin) assays. Other euglenoids have been used for nutritional research because of their specific individual requirements for certain vitamins, minerals, and amino acids. Students may also use *Euglena* populations for experiments on population growth rates, phototaxis, photosynthetic rates, and other physiological mechanisms.

SUMMARY

The Euglenophyta have historically occupied a unique position in science. They have been considered to have both plant and animal affinities. Even today they remain an evolutionary enigma because they have characteristics that appear to relate them with a number of algal, protozoan, and moneran organisms. Some evidence indicates that the photosynthetic euglenoids are the most primitive, while the heterotrophic forms are derived from their photosynthetic cousins. Other evidence suggests an endosymbiotic origin of the chloroplasts, by the ingestion of algae by primitive heterotrophic forms.

The life cycles of euglenoids appear to be relatively simple, since well-documented sexual reproduction is currently unknown. Apparently all euglenoid cells are haploid and lack sexuality.

Economically this division has been of little importance. Their role in the ecosystem, however, has been significant because of their remarkable ability to rapidly produce large numbers of cells, thereby taking advantage of ephemeral aquatic habitats. They can adapt to changing environmental conditions very quickly by expressing their auxotrophic and heterotrophic modes of nutrition (often within the same species).

SELECTED REFERENCES

Buetow, D.E. (ed.). 1968. The biology of *Euglena*, 2 vols. Academic Press, Inc., New York. Vol. I., 361 pp.; Vol. II., 417 pp.

Gibbs, S.P. 1978. The chloroplasts of *Euglena* may have evolved from symbiotic green algae. Can. J. Bot. **56**:2883-2889.

Gojdics, M. 1953. The genus *Euglena*. University of Wisconsin Press, Madison. 268 pp.

Heywood, P. 1980. Chloromonads. In Phytoflagellates, E.R. Cox (ed.) Elsevier North-Holland, Inc., New York, pp. 351-379.

Leedale, G.F. 1967. Euglenoid flagellates. Prentice-Hall, Inc., Englewood Cliffs, N.J. 242 pp.

Leedale, G., and P. Walne. 1971. Bibliography on Euglenophyta. In J. Rosowski and B. Parker (eds.), Selected papers in phycology. Department Botany, University of Nebraska, Lincoln, pp. 797-802.

Lyman, H., and K. Traverse. 1980. Euglena: mutations, chloroplast "bleaching" and differentiation. In Handbook of phycological methods, developmental and cytological, E. Gantt (ed.). Cambridge University Press, Cambridge, England, pp. 107-141.

Walne, P.L. 1980. Euglenoid flagellates. In Phytoflagellates, E.R. Cox (ed.), Elsevier North-Holland, Inc., New York, pp. 165-212.

7

CLASSIFICATION

Division: Phaeophyta (brown algae): 250 genera, 1500 species; mostly macroscopic marine benthic algae

  Class: Phaeophyceae; greenish-brown to black-colored algae

    ORDER 1. Ectocarpales: filamentous or pseudoparenchymatous growth; isomorphic alternation of generations; sessile; isogamous or anisogamous; *Bachelotia, Ectocarpus*

    ORDER 2. Ralfsiales: encrusting brown algae; *Ralfsia*

    ORDER 3. Chordariales: similar to Ectocarpales but with heteromorphic alternations of generations; isogamous; *Chordaria, Myrionema, Leathesia*

    ORDER 4. Sporochnales: uncommon in northern hemisphere, except *Sporochnus* and *Nereia* from southeastern United States

    ORDER 5. Desmarestiales: trichothallic growth; oogamous; cold sublittoral waters or Arctic and Antarctic Oceans; *Desmarestia*

    ORDER 6. Cutleriales: trichothallic growth; anisogamous; *Cutleria*

    ORDER 7. Sphacelariales: apical growth, isogamous to oogamous, mostly epiphytes; *Sphacelaria*

    ORDER 8. Tilopteridales: a few genera of *Ectocarpus*-like organisms

    ORDER 9. Dictyotales: isomorphic alternation of generations; dioecious gametophytes; oogamous; foliaceous habit; *Dictyota, Dictyopteris, Padina, Zonaria*

    ORDER 10. Dictyosiphonales: macroscopic sporophyte and microscopic gametophyte; isogamous; *Dictyosiphon*

    ORDER 11. Scytosiphonales: macroscopic sporophyte and microscopic gametophyte; only plurilocular sporangia in sporophyte; cells with a single chloroplast; oogamous; *Petalonia, Scytosiphon*

    ORDER 12. Laminariales: sessile, large, structurally complex sporophyte; microscopic gametophytes; oogamous; the "kelps"; *Alaria, Chorda, Egregia, Laminaria, Macrocystis, Nereocystis, Pelagophycus, Postelsia, Pterygophora*

    ORDER 13. Fucales: diplontic life cycles; oogamous; apical growth; structurally complex parenchymatous habit; *Ascophyllum, Durvillea, Fucus, Homosira, Pelvetia, Sargassum, Turbinaria*

PHAEOPHYTA

Phaeophyta are all multicellular, sessile plants—the most highly advanced division in the brown line of algal evolution (see Fig. 1-5). Most of them are large, brown-colored marine algae growing attached to rocky substrates in shallow water. They include the common cold water **kelps,** which form dense subtidal forests along the coastal areas of temperate, arctic, and antarctic regions. Some are very large, measuring 100 m with weights up to 250 kg. They are among the largest organisms in the world, often rivaling the giant *Sequoia* trees, which also grow to heights of 100 m.

Also included among Phaeophyta are the common **rockweeds** that grow on rocky outcroppings in the intertidal zone, where they form the most conspicuous of all the seaweeds associated with temperate rocky coastal areas. Many of the larger brown algal individuals are differentiated into plants with specialized organs—the holdfast, stipe, and blade. A few rare genera have also been reported from freshwater environments.

The brown algae are primarily cold-water marine plants. They are less common in warmer tropical waters, although there are a few notable exceptions to this rule (e.g., *Dictyota, Padina, Sargassum,* and *Zonaria*). In general the colder marine waters are inhabited by a higher percentage of benthic brown algae, while warmer seas are populated by a higher percentage of red algae (Rhodophyta).

CELL STRUCTURE

Brown algal cell structure follows the same basic pattern as in most eukaryotic cells. There are, however, a few distinctive variations on the basic features that will be discussed in some detail.

The cell walls of the brown algae are chemically distinctive. They are composed of two layers: (1) an inner rigid structural layer of cellulose and **alginic acid** that varies in thickness from cell to cell and species to species; and (2) an outer slimy layer composed of mostly (up to 97%) alginic acid that is mixed with lesser quantities of **fucoidin.** Alginic acid and fucoidin are complex colloidal substances called **phycocolloids.** Alginic acid (**algin**) is a polymer of two 5-carbon acids, D-mannuronic and L-guluronic acid. These polyuronic acids are economically and biologically important and chemically unique, since they have the ability to form viscous solutions and gels by absorbing up to 20 times their own weight in water. Fucoidin is formed from a complex mixture of various sulfated polysaccharides consisting mainly of L-fructose, D-xylose, D-galactose, and some uronic acids.

One genus of brown algae, *Padina*, deposits minute calcium carbonate crystals in the form of aragonite in the outer cell wall.

Structurally the walls of adjacent cells contain small channels called plasmodesmata through which the cytoplasms of the cells communicate. The stipes of some of the large kelps contain specialized, elongate, conducting cells shaped like a trumpet (and therefore called **trumpet cells**); they are also called **sieve cells,** since their end walls abut with adjacent sieve cells and contain many pores, producing a **sieve plate** (Fig. 7-1). These structures apparently function in the process of translocation in much the same manner as the sieve plates in phloem tissues of vascular plants (Sideman and Scheirer, 1977).

The chloroplasts of the brown algae, like those of the green algae, are diversified in their structure and may provide a useful clue to the taxonomy of any given organism. They range in

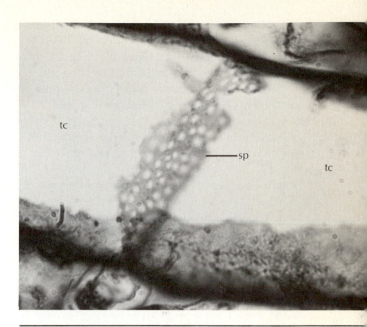

FIG. 7-1 Trumpet cells with sieve plate.
Longitudinal section through the stipe of a *Macrocystis* sp. Note sieve plate *(sp)* separating cytoplasm of two trumpet cells *(tc).* Sieve plate contains many perforations (900×).

form from the reticulate, net-shaped plastids of *Ectocarpus*, to star-shaped plastids in *Bachelotia*, to small oval disks in *Sphacelaria*, to laminar plates in *Scytosiphon*. The number of chloroplasts varies—but is constant within the cells of any given species—providing another taxonomic character.

The ultrastructure of the chloroplasts is quite distinctive, providing one means of determining the phylogenetic development of the Phaeophyta. The chloroplasts are surrounded by a double membrane envelope that is, in turn, covered by two outer membranes of chloroplast endoplasmic reticulum (CER), (see Fig. 1-7). The outer membrane of the CER is continuous with the nuclear envelope in some orders. This arrangement of CER membranes has become a taxonomic tool and is a major piece of morphological evidence relating Phaeophyta with Chrysophyta; Xanthophyta have a similar CER.

The interior of the chloroplasts is filled with a stroma in which are suspended bands of thylakoid membranes arranged in groups of three (see Fig. 1-7). The thylakoid membranes contain the photosynthetic pigments and accessory pigments, including chlorophyll a and c, β-carotene, and a distinctive xanthophyll called **fucoxanthin.** The brown-colored fucoxanthin is often present in such great quantities that it masks the green chlorophyll pigments, giving the alga a brown hue and therefore the common name "brown algae."

Also contained within the chloroplasts are significant quantities of lipid globules and DNA, which is often associated with fibrils.

Chloroplasts of brown algae usually resemble the structure illustrated in Fig. 1-7; however, in some species pyrenoids may be associated with the chloroplasts. When present, pyrenoids are attached to the chloroplast by a short stalk and are surrounded by the outer CER membrane. Pyrenoids are presumably the cite of synthesis for the major energy storage product, **laminarin,** a branched polymer of β 1-3, 1-6–linked glucose

molecules. Concentrations of laminarin are often located within a membrane-bound sac that surrounds the pyrenoid. Lesser quantities of D-mannitol and other alcohols and carbohydrates may also be isolated from this sac. Laminarin is chemically similar to the energy storage product found in both Chrysophyta and Euglenophyta.

Red eyespots (stigmata) are characteristic of all the motile cells (zoospores and male gametes). They are composed of a red carotenoid pigment arranged in numerous globules and located within the chloroplasts. The eyespot is always found in close proximity to the basal swelling of the shorter trailing flagellum (Dodge, 1973). The vegetative cells of the brown algae do not contain eyespots.

All brown algae release pyriform (pear-shaped) motile zoospores and male gametes at some point in their life. These cells move by means of two subapically inserted heterokont flagella. One flagellum is longer, directed anteriorly, and of the tinsel-type, bearing many hairlike projections (mastigonemes). The shorter trailing flagellum is of the whiplash type, being smooth and lacking appendages (see Fig.

FIG. 7-2 Interior of the blade of *Laminaria.*
Cross sections through a blade of *Laminaria* spp. Photo (near right) shows epidermal layer containing unilocular meiosporangia *(um)* containing meiospores. Beneath the epidermis is the cortex *(cor)* containing mucilaginous ducts *(md)*, which are partially lined with secretory cells *(sc)*. Note how these ducts open to the surface (arrow) (125×). Photo (far right) shows an individual mucilaginous duct (400×).

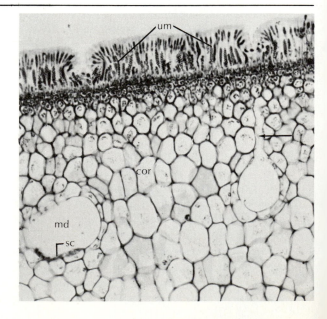

1-9). This flagellum bears a basal swelling located near the eyespot; it is presumably the site of light detection.

The interior structure of both flagella is identical to flagella of all plants and animals and has a typical 9 + 2 protein fibrillar construction.

In some brown algal cells there are conspicuous membrane-bound refractile vesicles called **fucosan** or phaeophyte **tannin** granules. They are generally clustered around the nucleus and contain a wide variety of organic acids that form a black pigment when exposed to air. Most herbarium specimens turn black because of this reaction.

The stipe, blade, and holdfast of the kelps (Laminariales) are the most structurally complex organs found among the algae. They are composed of three tissues: the **epidermis, cortex,** and **medulla.** Each tissue contains distinctive cell types. The epidermis is composed of epidermal cells, the meristematic cells of the **meristoderm, paraphyses,** and **unilocular sporangia.** Sporangia are located in dark brown spots on the thallus called **sori** (sing., sorus).

Haploid biflagellate meiospores are produced in unilocular sporangia within the sorus (Fig. 7-4).

The cortex is composed primarily of parenchyma cells similar to those seen in vascular plants. However, extending longitudinally and horizontally throughout the cortex is a series of mucilaginous ducts lined by distinctive **secretory cells** (Fig. 7-2). These cells secrete mucilage (mostly fucoidin, a complex sulfated polysaccharide) that is conducted to all parts of the plant through the ducts. The mucilage forms a slimy surface layer outside the epidermis.

The medullary tissue contains two cell types: parenchyma cells and sieve or trumpet cells. The elongate sieve cells resemble the sieve-tube elements of vascular plants. They contain within their cell walls a cytoplasm that has many mitochondria, and they function as the transportation system for the products of photosynthesis (i.e., translocation)—primarily in the form of the alcohol mannitol. Movement of these products is both up and down the stipe. The Laminariales are the only order of algae to have evolved conducting tissues of this type.

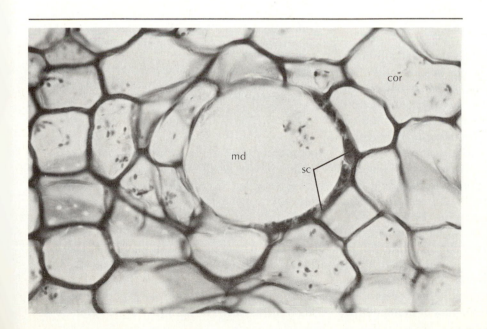

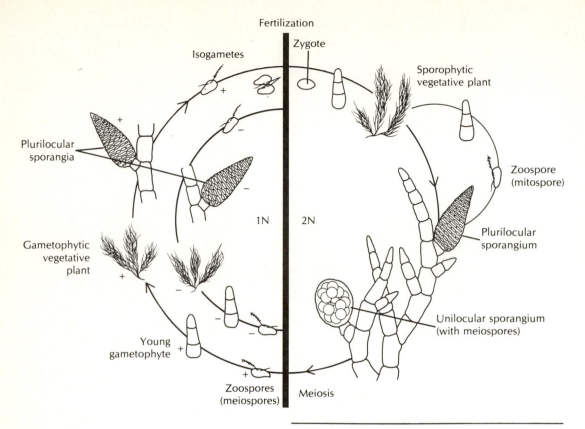

FIG. 7-3 Life cycle of *Ectocarpus*.
Ectocarpus is an example of a primitive brown alga with diplohaplontic life cycle containing isomorphic alternation of haploid and diploid generations. Gametophytes are dioecious and produce isogamous gametes. Note formation of plurilocular sporangia on the gametophyte and plurilocular and unilocular sporangia on the sporophyte.

REPRODUCTION

Asexual reproduction in the phaeophytes is not as well developed as it is in some of the less highly specialized divisions. Some browns are still capable of some degree of **fragmentation,** especially among the primitive orders, but more often their highly complex tissue-level organization and cellular differentiation prevents groups of cells from developing into a whole plant. A notable exception to this is free-floating *Sargassum,* whose only method of reproduction is through thallus fragmentation. However, benthic species of *Sargassum,* which grow anchored in the midlittoral zone in subtropical waters, reproduce sexually. Another exception is the vegetative reproduction in the Sphacelariales, which are mostly epiphytic and reproduce mainly through the production of multicellular,

nonmotile, distinctive structures called **propagules.** These small leaflike structures bear an apical cell that is capable of developing into a new adult organism.

Most brown algae have diplohaplontic life cycles with isomorphic or heteromorphic alternations of generations. Exceptions to this may be found in order Fucales, where specialized diplontic cycles occur (see *Fucus,* Fig. 7-5). Some phaeophytes possess complex, poorly understood life cycles. For example, *Scytosiphon* and

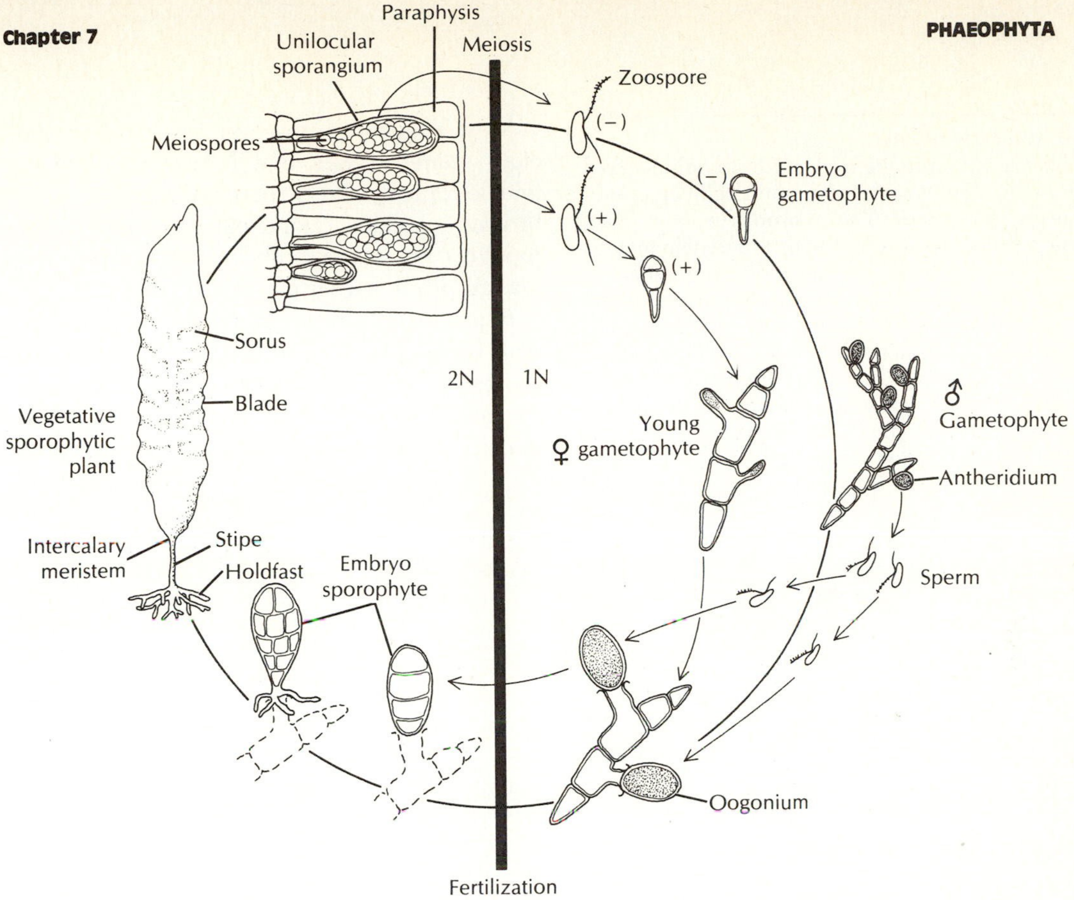

FIG. 7-4 Life cycle of _Laminaria_.

Laminaria has a diplohaplontic, heteromorphic life cycle with a large macroscopic sporophytic generation and a small, microscopic, gametophytic generation of few cells. The gametophytes are dioecious. Large sporophyte is differentiated into holdfast, stipe, and blade. The blade is the site of many sori (sing. sorus) or fruit dots, which may be seen with the naked eye. Sori are the site of unilocular sporangia where meiosis occurs and biflagellate zoospores (meiospores) are produced. _Laminaria_ species are common, shallow-water marine plants along rocky temperate coasts.

Petalonia are believed to have stages in their life cycles which closely resemble the crustose brown alga, _Ralfsia_. (For a thorough discussion of the life histories of brown algae see Wynn and Loiseau, 1976.)

Most brown algae have sporophytic generations that reproduce asexually through formation of zoospores produced in **plurilocular** (many-chambered) sporangia (Fig. 7-3). These zoospores, called mitospores, are produced by mitotic cell divisions. They may settle on suitable substrates and develop directly into other sporophytes.

Sporophytes also may produce unilocular (single-chambered) sporangia (Fig. 7-3). Zoospores produced in unilocular sporangia are formed by meiotic cell divisions and are called meiospores. There are usually 16 to 32 or 64 meiospores produced in each unilocular sporangium. Each of these zoospores may settle on a suitable substrate and develop directly into a gametophyte (Figs. 7-3 and 7-4).

FIG. 7-5 Life cycle of *Fucus*.

Fucus has a diplontic life cycle with its only haploid phase being the gamete. No multicellular gametophyte is produced. The sporophyte is a macroscopic, multicellular plant with a parenchymatous thallus and considerable cellular differentiation. It grows attached to solid substrates in the marine intertidal and shallow-water habitats. It is common in salt marshes. Because it is often found growing attached to rock in the intertidal zone, it is often referred to as "rockweed."*Fucus* plants are common in the colder temperate waters of the Northern Hemisphere. In the Southern Hemisphere, identical ecological niches are filled by the genus *Homosira*.

Most species of phaeophytes produce dioecious gametophytes that form either male or female reproductive structures. However, some browns produce monoecious haploid gametophytes on which both antheridia and oogonia are produced on the same thallus.

Reproduction in the gametophytic generation may be asexual by production of haploid zoospores or sexual by production of eggs and sperm. In some (*Ectocarpus*) sexual reproduction is isogamous (Fig. 7-3), whereas in others

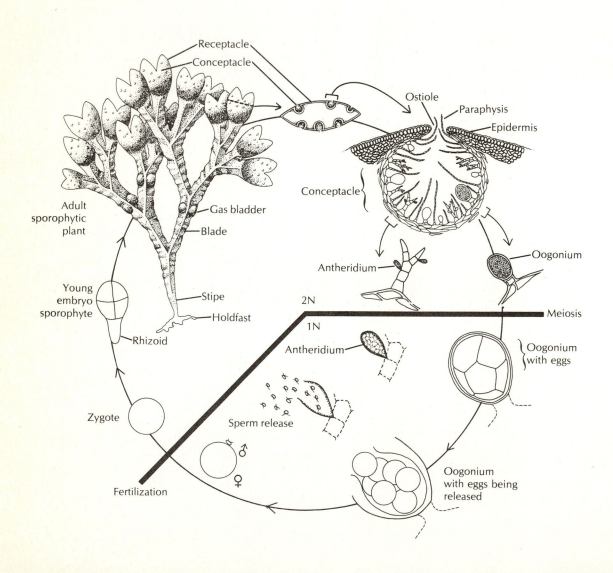

(Laminaria) it is oogamous (Fig. 7-4). In those browns that have isomorphic alternations of generations (e.g., the Ectocarpales), the haploid generation produces only plurilocular sporangia, some of which act as gametangia and produce isogametes.* Both plurilocular and unilocular sporangia can be seen in the life cycle of *Ectocarpus* (Fig. 7-3).

In those brown algae that have heteromorphic alternations of generations, the free-living, microscopic gametophytes are reduced to a small number of cells and are easily confused with small members of the phytoplankton community. Gametes of these species are produced in specialized areas known as antheridia (sperm) and oogonia (eggs). Fertilization takes place while the egg cell is still attached to the female gametophyte in *Laminaria* (Fig. 7-4).

Rockweeds (order Fucales) are a special case among brown algae, since they have diplontic life cycles (see *Fucus*, Fig. 7-5). Thalli of fucalean algae differentiate into a number of distinctive structures including apical meristems, gas vesicles, and receptacles. Receptacles are enlarged or swollen tips of thalli which contain special cavities called **conceptacles.** Within conceptacles three structures may be differentiated: oogonia, antheridia, and sterile hairs called paraphyses (Figs. 7-5 and 7-6). Individual species may be monoecious, as is the case with most species of *Fucus*, or dioecious, as is the case with *Fucus vesiculosus*, where antheridial conceptales are orange when opened, while oogonial conceptacles are olive-green (Taylor, 1957). *F. vesiculosus* may be collected from the ocean (keeping male and female thalli separate), transported to the laboratory, and fertilized in dishes beneath dissecting microscopes so that zygotes and young embryos may be easily observed. Sterile conceptacles containing many paraphyses are usually scattered throughout the

thalli. In the sea both male and female gametes are released from the thallus before fertilization.

Oogamous species of *Fucus*, isogamous species of *Cutleria*, and isogamous species of Ectocarpus have been shown to be under the influence of chemotactic hormones released by the female gamete (Ende, 1976). These organic chemotactic molecules attract the male gamete to the female gamete.

Patterns of growth and development among the brown algae are complex because of the tissue-level organization that forms the holdfast, stipe, and blade. Kelps and rockweeds have a true parenchymatous construction with epidermis, cortex, and medulla tissues forming these organs. More cellular differentiation exists in this division than in any other group of algae.

The structurally simpler, more primitive orders, however, have a pseudoparenchymatous construction produced by many individual algal filaments growing in close proximity. This growth pattern produces a **multiseriate (multiaxial)** structure that superficially resembles a true parenchyma. When filaments grow singly they form a **uniseriate (uniaxial)** construction. *Desmarestia* has both uniseriate (filamentous) and multiseriate (pseudoparenchymatous) growth patterns (Fig. 7-7).

The type of construction in any given brown alga is a reflection of the different types of meristematic activity exhibited by the cells of the various species. **Meristem** is used to describe areas of any plant where mitotic activity is initiated. Among the algae these areas are often poorly defined because all of the cells of many algae are totipotent.* Therefore, terms such as "diffuse meristem," "meristoderm," and "regenerative meristem" are often employed to describe mitotic activity over wide areas of an individual plant. Apical, marginal, trichothallic, and intercalary meristems are terms used to de-

---

*Use of the term "sporangium" to indicate a structure that forms gametes is confusing. Those plurilocular sporangia that produce isogametes are actually functioning as plurilocular gametangia.

*Totipotent cells are capable of differentiating into an entirely new organism.

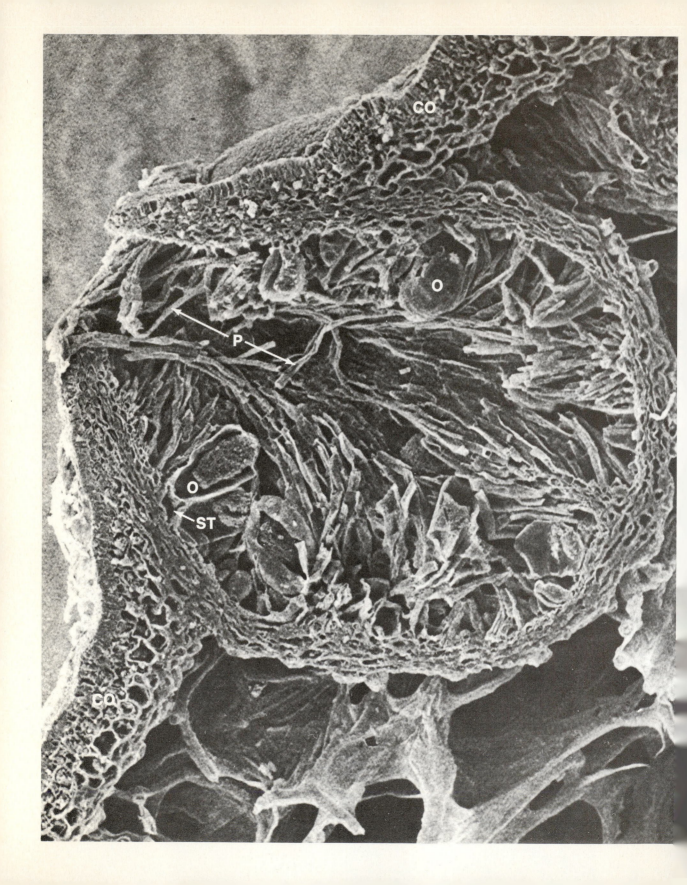

FIG. 7-6 Conceptacle of *Fucus*.
Scanning electron photomicrograph showing inside of
a conceptacle from a female monoecious species of
Fucus. Note single chambered oogonia *(O)* and sterile
hairs called paraphyses *(P)*. Also seen are stalks *(ST)* of
oogonia and cell layers of the thallus cortex *(CO)*.
(From Lott, J.N.A. 1976. A scanning electron microscope study of green plants. St. Louis, The C.V.
Mosby Co.)

scribe mitosis in more restricted parts of the
plant.

Diffuse meristems consist of actively dividing
cells throughout the plant body. A simple
example would be a filamentous alga in which
all the cells of the filament have the ability to
divide. This type of meristem can be found in
Ectocarpus, the green alga *Ulothrix*, and the
red alga *Porphyra*.

A unique meristem, the meristoderm, is associated with species of the Laminariales. It
consists of an ill-defined group of embryonic
cells associated with the epidermis of young
plants and with cells immediately below the epidermis in older plants. Cell divisions by the
meristoderm add girth to the stipe, producing
concentric seasonal growth rings that resemble
the pattern of annual rings seen in cross sections
of trees. The meristoderm is a special diffuse
meristem found in the large kelps. These meristems account for many of the growth forms
found among the brown algae.

When kelps or rockweeds are injured or
grazed by feeding animals, the cells at the surface of the injury form a regenerative meristem,
a special type of diffuse meristem that is active
following injury to the plant. These are common
among the intertidal and benthic brown and red
algae where animals often eat pieces of blades or
stipes or injury occurs from the constant beating
of the surf. Growth from these meristems results in many bizarre forms for any given species.

Intercalary meristems consist of embryonic
cells located between the two extremes of any
plant, i.e., somewhere distant from the tips of
the thallus. They may occur as single cells at the
base of delicate unicellular filaments, resulting
in **trichothallic growth** (e.g., *Desmarestia*, Fig.
7-7). They may also occur on a larger scale in
areas between the stipe and blade or the stipe
and holdfast in the larger brown algae, resulting
in a massive intercalary meristem of the sort
found in *Laminaria* or *Macrocystis* (Fig. 7-4).

FIG. 7-7 *Desmarestia*, growing tip.
Growing tips of *Desmarestia* show both uniseriate
(uniaxial) construction, composed of a single filament
of cells, and multiseriate (multiaxial) construction,
composed of many filaments growing close together.
Multiseriate growth creates a pattern that appears to
be parenchymatous but is not formed in the same
manner as a true parenchyma. it is therefore referred
to as "pseudoparenchymatous" growth. Intercalary
meristem at the base of the uniaxial filament produces
uniaxial growth in one direction and multiseriate
growth in the other. This type of growth is called
"trichothallic."

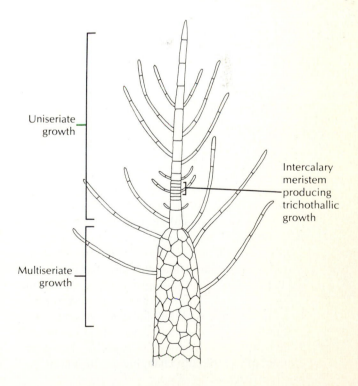

Uniseriate
growth

Intercalary
meristem
producing
trichothallic
growth

Multiseriate
growth

Apical meristems are not a common feature among most algal divisions except for the Phaeophyta, Rhodophyta, and Charophyta. These divisions have unicellular apical meristematic cells whose mitotic activity directs the growth pattern for the entire thallus. The apical cells are easily identified as the very large cell at the growing tip of the thallus. In *Dictyota* this cell protrudes from the surface of the thallus (Fig. 7-8). In the Fucales the apical cell is located in a deep depression at the tip of the thallus (Fig. 7-9). The fucalean meristematic cell is evolutionarily derived from the trichothallic growth pattern of the more primitive brown algae. During the ontogenic development of *Fucus, Sargassum, Ascophyllum,* and other Fucales, a terminal hairlike filament of cells is lost, leaving only a basal cell that becomes the apical meristematic cell (Fig. 7-9).

Marginal meristems may be found in foliose thalli such as *Padina* or *Zonaria*. In these organisms growth is accomplished through mitoses by a single row of cells along the margin of the fan-shaped thallus. This type of meristem resembles an apical meristem that has been spread along the edge of the leaflike plant.

FIG. 7-8 *Dictyota*, **apical meristematic cell.**
Growing region of *Dictyota* thallus with a single apical meristematic cell *(ac)*, responsible for the growth of the alga (125×).

PHYSIOLOGY

Biochemically the brown algae are characterized by the photosensitive pigments chlorophyll a, c_1, and c_2 and various carotenoids, including the xanthophyll fucoxanthin. The abundant fucoxanthin masks the chlorophylls giving the algae a brown or olive-green color. The cell walls and the interstices between the cells (e.g., the middle lamella) contain the phycocolloids alginic acid and fucoidin. These phycocolloids can absorb large amounts of water and therefore help protect intertidal algae against desiccation at low tide. They also provide some structural support and buffering action against extreme salinity changes. Intertidal algae generally contain more phycocolloids than algae of the sublittoral zone, and they are thereby able to survive in this rapidly changing environment.

Laminarin is the most commonly found carbohydrate storage product; it may comprise up to 34% of the dry weight of some brown algae (Powell and Meeuse, 1964). The amount of laminarin varies seasonally, with the highest levels being found in the spring and summer and the lowest levels in the winter (Meeuse, 1962). Other storage products include fats, various carbohydrates, and alcohols, with mannitol being the prevalent alcohol.

Chemical energy from photosynthesis is translocated in the form of mannitol in *Macrocystis* and *Alaria* and probably in other kelps (Schmitz and Srivastava, 1975). Translocation via mannitol has been demonstrated by the use of the radioactive tracer carbon-14 (Schmitz et al, 1972). Carbohydrates are synthesized via

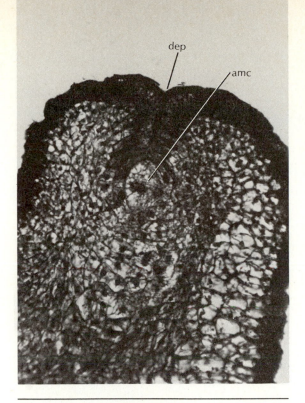

FIG. 7-9 *Sargassum*, apical region.
Medial longitudinal section of apical region of the common fucoid *Sargassum*. Note deep depression (*dep*) at the growing tip with a single apical meristematic cell (*amc*) at the base of the depression (250×).

photosynthesis by the upper parts of the blade in the photic zone and transported to other parts of the kelps below the photic zone (Schmitz and Lobban, 1976). The intercalary meristem especially needs energy for growth and is often located below the photic zone in giant kelps. Translocation rates in kelps have been recorded from 50 to 780 cm/hour (Parker, 1965; 1966); apparently this process takes place in the central portion of the stipe where the sieve cells are situated. By the use of microautoradiography, researchers have shown sieve cells to be the conducting elements in one species of *Laminaria* (Steinbiss and Schmitz, 1973); most likely they serve a similar function in other Laminariales. Translocation of photosynthate also takes place laterally through pores in the sieve cell side walls. Lateral translocation is usually from the photosynthetic outer portions of the blade or stipe to the inner portion where vertical translocation takes place.

The kelps may be either annual (*Chorda, Nereocystis,* and *Postelsia*) or perennial (*Laminaria, Macrocystis,* and *Pelagophycus*). Growth rates in *Nereocystis* can be very rapid and have been recorded at 3 to 6 cm/day. The holdfast overwinters and stores enough carbohydrate to produce very rapid growth in the spring. A 36 m sporophyte weighing 120 to 125 kg can grow from the holdfast in 12 to 16 weeks. Several North Atlantic kelps grow so rapidly that they replace all their blade tissue up to five times a year; these kelps also keep growing during the winter (Mann, 1973).

A very fast growth rate of up to 45 cm/day was also recorded in the perennial *Macrocystis* (Fig. 7-10) (Clendenning, 1964). Growth rates in the kelps can be difficult to measure because the blades are eroded away at the top by wind and wave action but new tissue is added from the basal or intercalary meristem. Growth may be monitored by punching two holes in the blade and measuring their distance from the substrate over time (Field et al, 1977). An individual *Macrocystis* plant may weigh 250 kg and have a holdfast up to 3 m in diameter. Growth in width of the stipe also occurs in some perennial species; concentric rings similar to those in trees may help in determining the age of individual plants.

Kelps grow so rapidly because they live in areas of the ocean that receive adequate light and ample nutrients from upwelling and land runoff. One *Laminaria* species can accumulate nitrate ions from the water in winter up to 28,000 times the concentration in the surrounding water (Chapman and Craigie, 1977). The kelp then has a store of nitrate that enables it to continue rapid growth in the spring, even after nitrate has been depleted in the water column.

Several perennial kelps live many years. The longest-lived kelp is *Pterygophora*, found off the west coast of North America. Individual plants

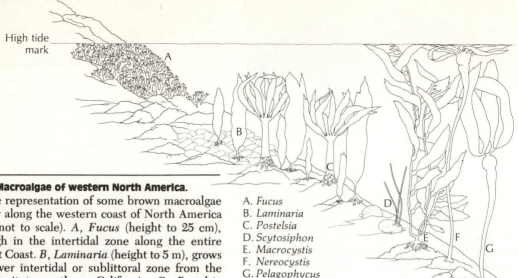

FIG. 7-10 Macroalgae of western North America.

Schematic representation of some brown macroalgae that occur along the western coast of North America (drawing not to scale). *A, Fucus* (height to 25 cm), grows high in the intertidal zone along the entire U.S. West Coast. *B, Laminaria* (height to 5 m), grows in the lower intertidal or sublittoral zone from the Bering Strait to southern California. *C, Postelsia* (height to 60 cm), is also an intertidal alga; its habitat is the rocky coasts of Vancouver, Washington, and Oregon. *D, Scytosiphon* (height to 70 cm), grows low in the intertidal zone and is widely distributed along the entire coast. *E, Macrocystis, F, Nereocystis,* and *G, Pelagophycus* are giant kelps that grow in the sublittoral zone to depths of 90 m. *Macrocystis pyrifera* (height to 50 m), widely distributed from Alaska to southern California, is the dominant kelp in the underwater "forests." *Nereocystis* (height to 40 m) grows on the coast from Alaska to central California. *Pelagophycus* (height to 47 m) is found along the southern California coast.

A. *Fucus*
B. *Laminaria*
C. *Postelsia*
D. *Scytosiphon*
E. *Macrocystis*
F. *Nereocystis*
G. *Pelagophycus*

have been found that are 17 years old. *Macrocystis* usually lives about 5 years, and *Laminaria* sporophytes may live several years. The fucalean genus *Ascophylum*, however, may live for several decades (Lewis, 1977). The age of *Ascophyllum* is determined readily because it produces one gas-filled float on each shoot each year.

Many Phaeophyta have gas-filled floats or bladders that keep the photosynthetic portions of the plant in the photic zone. These floats are especially prominent on *Fucus* (Fig. 7-5), *Sargassum* (Fig. 7-11), and on many kelps (Fig. 7-10). The floats, also called **pneumatocysts,**

usually contain the same gases prevalent in air (i.e., nitrogen, oxygen, and carbon dioxide). In *Nereocystis* and *Pelagophycus*, however, carbon monoxide (up to 2%) has been found in the floats. The source of the toxic carbon monoxide is unknown, but it is thought to be metabolic in origin. In *Pelagophycus*, the elkhorn kelp, the single float is very large, often reaching the size of a basketball (Fig. 7-10).

Vacuoles in the cells of the cold-water brown alga *Desmarestia* contain sulfuric acid (Meeuse, 1956). The pH of vacuole liquid may be as low as 0.8 (Schiff, 1962). The acid is released from the vacuoles soon after the alga has been removed from its habitat. The acid can rapidly discolor and destroy other algae in the collecting vessel, so *Desmarestia* must be kept separately.

The release of gametes in *Fucus* and the maturation of oogonia in *Dictyota* are timed to correspond with tidal cycles. Such endogenous rhythms of gamete production are independent of environmental change and persist if the organism is brought to the laboratory (Sweeney and Hastings, 1962). The female gametes of

Fucus and *Ectocarpus* secrete a chemical attractant that is not species specific (Müller et al, 1971; Müller and Jaenicke, 1973). A simple aromatic compound such as *n*-hexane will attract male gametes of *Fucus, Ascophyllum,* and *Ectocarpus* (Cook and Elvidge, 1951; Müller, 1968).

The Phaeophyta are predominantly photoautotrophic; however, a few species are auxotrophic, requiring vitamins for growth. No parasitic forms are known and no browns have been found that can grow exclusively in the dark and utilize organic carbon (Darley, 1974). However, Wilce's (1959) study of Arctic and Antarctic phaeophytes indicated that these algae can survive during the long periods of darkness or reduced illumination. The algae often live below the compensation depth for many months of the year. Wilce suggests that these algae may be living heterotrophically on dissolved organic matter or on stored food reserves.

Interaction of environmental factors such as length of **photoperiod,** quality of light, temperature, and nutrient levels may influence morphogenesis. In *Petalonia* and *Scytosiphon,* a **crustose,** inconspicuous microthallus is produced during long photoperiods and high temperatures, while a macrothallus (blade) is produced during short photoperiods and cooler temperatures (Wynne, 1969). The crustose thallus closely resembles *Ralfsia,* which forms a crust on intertidal rocks. Other studies indicate that the quality of light also is influential in determining whether a microthallus or macrothallus is formed in certain Phaeophyta species (Lüning and Dring, 1973). Quality of light can also induce differences in the ratio of chlorophyll *a* to chlorophyll *c*. In the dim blue-green light of deep waters, chlorophyll *c* predominates, thereby enabling the alga to utilize those light waves that penetrate deeply into the water column.

Temperature, light, and nutrients can also influence reproduction. In *Desmarestia* the gametophytes only reproduce sexually when the water is cold (3° to 5° C) and when light quality

Bladder —

FIG. 7-11 Habit of *Sargassum fluitans* (gulfweed).
S. fluitans is a macroscopic member of the phytoplankton community. Algae like this one and *S. natans* make up more than 90% of the biological bulk of the Sargasso Sea, an extensive area of the mid-Atlantic Ocean. In vast expanses of the Sargasso Sea the only visible vegetation is the floating windrows of the two species of *Sargassum,* thus producing a unique habitat for mid-ocean invertebrates and vertebrates. Planktonic *Sargassum* species are sterile; reproduction is by fragmentation only. Benthic species of *Sargassum,* however, reproduce sexually.

and quantity are adequate (Chapman and Burrows, 1970.) The sporophytes grow large (up to 5 m) in the warm waters of the Caribbean Sea, but they are only found occasionally because they cannot reproduce sexually in that environment; the usual habitat of *Desmarestia* is the cold waters of the temperate zones.

Both temperature and light evidently determine the length of the gametophyte and sporophyte state in *Ectocarpus*. In the cold waters of northern Europe the sporophyte is usually found, while the gametophyte occurs more frequently in the warm Mediterranean waters (Müller, 1964). Both sporophytes and gametophytes are found off the New England coast. *Ectocarpus siliculosus* has been studied extensively and has a complex life cycle that is influenced by several environmental factors. Edwards (1969) and Müller (1972; 1975) discuss the life cycle of *E. siliculosus*. Dring (1974) gives an account of the role of environmental factors in controlling algal reproduction.

Several genera of the Fucales (e.g., *Pelvetia*, *Fucus*, and *Ascophyllum*) are able to concentrate such trace metals as titanium, zinc, nickel, and strontium several hundred times over the ambient concentration in the seawater. *Ascophyllum* can also concentrate iodine up to 220 times the level found in seawater.

EVOLUTION

Phaeophyta have evolved as the most complex structures of all the algae. They have many characteristics found in vascular plants. Some of the features common to the large browns (Laminariales and Fucales) and vascular plants are: (1) dominant large sporophyte; (2) gametophytes reduced to a few cells; (3) sieve cells for internal photosynthate transport; (4) differentiation of plant body into rootlike holdfast, stemlike stipe, and leaflike blade; (5) motility reduced to gametes and zoospores; and (6) growth in width. Similarity between the advanced brown algae and vascular plants has led some investigators to speculate that the browns were ancestors of vascular plants. However, biochemically Phaeophyta differ significantly from the vascular plants (e.g., in chlorophyll types, carbohydrate storage product, and cell wall composition), and most investigators conclude that the two groups exhibit convergent evolution—two distantly related groups evolving similar characteristics

to cope with similar environmental problems.

No single-celled vegetative brown algae are known, but it is postulated that the browns evolved from such a biflagellate single-celled ancestor. The present-day motile gametes and zoospores are probably the closest existing structures to the hypothetical ancestor. Perhaps unicellular browns may be found among little known nannoplankton (less than 70 μm). No known forms appear to be transitional between the proposed ancestor and the differentiated multicellular algae.

The evolutionary trends within the Phaeophyta are similar to the trends observed within the Chlorophyta: single-celled to branched filament, prostrate to heterotrichous habit, isogamy to oogamy, isomorphic to heteromorphic alternation of generations, haplontic to diplontic life cycles, and gametophyte dominance to sporophyte dominance. The browns, however, have evolved larger, more complex thalli.

In Phaeophyta, evolution culminates in the Fucales, where the haploid generation is reduced to from one to several cells (Fig. 7-5). The diplontic life cycle observed in *Fucus* apparently evolved independently and did not lead to land plants. The Fucales have evidently been separate from the other orders for a long time, because their mode of reproduction and sperm flagellation are quite distinct from the other browns. However, they retain biochemical affinities to the rest of the orders.

The fossil records produce little evidence to help determine the evolutionary relationships of the brown algae. The soft-bodied browns do not leave good fossils. The first definite fossil record is found in rocks of the Triassic period. Some fossils dating back to the early Paleozoic era (Silurian and Devonian periods) resemble members of the Fucales. At present these fossils are not definitely assigned to Phaeophyta (Taggart and Parker, 1976). Several fossils that are prominent in North American petroleum-bearing, precarboniferous shales appear similar to present-day browns, but affinities are not certain (Arnold, 1947). Some evidence suggests that browns were well diversified by the late Devonian, but, again, this evidence is not yet definitive. Other fossil evidence suggests that intertidal browns may have colonized shores in the early Paleozoic. Perhaps several Phaeophytes, inhabitants of the intertidal zone and able to resist desiccation, invaded the land temporarily, then died out.

Phaeophyta are biochemically related to both Chrysophyta and Pyrrhophyta through mutual presence of chlorophyll *a* and *c*, fucoxanthin, and similar energy storage compounds. Some researchers propose a common unicellular ancestor for the Chrysophyta and the Phaeophyta. Structurally, however, the brown algae display much more complexity.

ECOLOGY

Division Phaeophyta contains plants ranging in size from microscopic epiphytes *(Myrionema)* to some of the largest plants on earth *(Macrocystis*—100 m long, 250 kg). The division is primarily marine, but a few freshwater genera have been recorded. Members of this division dominate rocky intertidal and littoral zones in temperate and polar regions but are less prominent in subtropical and tropical seas. Some plants grow totally submerged in the sublittoral zone, while others are important members of the intertidal zone community and must endure regular periodic tidal wetting and drying. Intertidal and sublittoral species occupy a narrow band around the edges of continents and are therefore vulnerable to pollution from land drainage. The phaeophytes are important primary producers in coastal upwelling regions and polar areas.

In the intertidal zone those brown algae that contain large amounts of algin or fucoidin can withstand the periods of desiccation when they are exposed to air at low tide. Phycocolloids, like algin, not only act as shock absorbers to cushion the plants from pounding surf but also as protection against freezing temperatures. An important intertidal genus is the perennial rock-

weed, *Fucus* (Fig. 7-5), which is prevalent on the rocky eastern and western coasts of North America, Europe, and Asia. *Ascophyllum* (Fig. 7-12) is often found in association with or competing with *Fucus* for intertidal space. Several environmental factors such as grazing pressure

FIG. 7-12 *Ascophyllum nodosum* **habit.**
Herbarium specimen of *Ascophyllum nodosum*, a rockweed common to intertidal zones of rocky coasts in eastern North America. Bladder contains gasses that assist in flotation during high tide. The swollen receptacles appear in late winter or early spring and contain gamete-forming conceptacles. *A. nodosum* grows profusely over rocky coastal areas, providing shelter for many invertebrates. The red alga *Polysiphonia lanosa* is a common epiphyte on this species. (Courtesy D. Spess, Biology Department, Lehigh University, Bethlehem, Pa.)

from invertebrates and differential harvesting may determine which genus dominates. In the southern hemisphere the ecological niche of *Fucus* is occupied by a similar genus, *Homosira*. *Postelsia palmaeformis*, the sea palm (Fig. 7-10), populates the rugged rocky coasts of Oregon and Washington, forming dense stands and smothering competing algae. All these important brown algae of the intertidal zone have exceptionally strong holdfasts that anchor them firmly to the rocks. Intertidal species may have rates of photosynthesis in air that exceed the rates in water. Intertidal algae are subjected to extreme daily changes in light intensity, temperature, salinity, and moisture, but this group has evolved physiological and structural mechanisms for successfully coping with this harsh environment. Some intertidal species can withstand days of exposure to sun and air and still survive (e.g., *Pelvetia*).

In warmer waters, a fucalean genus, *Sargassum*, may occupy the mid and sublittoral zones. These species of *Sargassum* which are attached to the ocean floor reproduce sexually. Two species of *Sargassum* (*natans* and *fluitans*) are free floating and only reproduce asexually by fragmentation. Free-floating, long-lived *Sargassum* plants (Fig. 7-11) form enormous masses in the Sargasso Sea, an area of 6.5 million km^2 in a large oceanic eddy in the mid-Atlantic east of Bermuda. In no other area of the earth is there such a large area dominated by a single plant species. Total mass of the floating *Sargassum* is estimated at 3.5 to 9 million metric tons. In 1492 Columbus supposedly encouraged his sailors by convincing them that they were near land when they encountered thousands of square kilometers of this floating gulf weed, *Sargassum* spp. Peculiar weather and current patterns cause the Sargasso Sea to be quite calm, allowing floating *Sargassum* to accumulate in the area, up to 1.5 metric tons per square kilometer.

An attached species of *Sargassum* from the Sea of Japan was inadvertently introduced into England and northwest America with Japanese oysters. It is rapidly becoming a nuisance, outcompeting native attached brown algae for nutrients and space in the intertidal zone (Fletcher and Fletcher, 1975).

Phaeophyta may be distributed vertically into zones along sea coasts in temperate regions (Fig. 7-10) in a manner similar to the zonation observed in other algae (see Fig. 2-16). Dayton (1975) documents such vertical zonation in the kelp beds of Amchitka Island in Alaska. The vertical zonation observed generally in coastal algae is partially the result of the differing ability of individual species to withstand the various stresses of the intertidal or sublittoral areas. Biological interactions between different algal species may also determine in which vertical zone the alga will be able to photosynthesize, grow, and reproduce. Chapman (1979) discusses the structure of benthic algae communities and their interactions and species succession.

The sublittoral zones of many polar and temperate regions are populated by kelps, the Laminariales. Extensive kelp beds are found north of latitude 30 degrees N and south of latitude 30 degrees S. The giant kelps, *Macrocystis* (giant or vine kelp), *Nereocystis* (bladder kelp), and *Pelagophycus* (elkhorn kelp), are some of the largest plants on earth (Fig. 7-10). Other large kelps include the east coast genus *Chorda* (up to 10 m) and the widely distributed *Laminaria* (up to 18 m). *Durvillea*, the bull kelp of the southern hemisphere, is a member of the Fucales, not the Laminariales.

The giant kelps form dense underwater "forests" that are highly productive, not only in the amount of organic material produced from sunlight, but also in the number of invertebrates and vertebrates for which they are a source of food. The average primary production of a kelp forest has been measured at 3,400 g dry organic matter/m^2/yr with a maximum productivity of 4,250 g/m^2/yr (Teal, 1980). In Nova Scotia, Mann (1973) reported primary productivity of 3,850 g organic matter/m^2/yr. (See Field et al, 1977, for a description of the energy flow through a kelp forest.) These forests are breeding and grazing grounds for a myriad of marine animals (up to 750 different kinds) including many sport fishes. The enormous kelp blades are covered with epiphytic animals and maybe even a miniscule gametophyte of their own species. The minute gametophytes of these heteromorphic plants (Fig. 7-4) may restrict the distribution of the kelps because they often grown on the ocean bottom in deep water and must obtain enough light in the "forest" to photosynthesize and to produce viable gametes. In addition some gametophytes, *Laminaria* for example, must have cool water (less than 16° C) for successful gamete production, while other gametophytes require blue light (460 to 500 nm) for gamete development. Blue light abounds in the sublittoral zone because, unlike other light waves, it penetrates deeply into the water column (see Fig. 2-5).

The highly productive kelp beds off the California coast suffered severe destruction from large populations of grazing sea urchins during the 1960s. The kelp beds were reduced to one tenth of their pre-1960 size. Increase in the sea urchin population was caused by the reduction in numbers of their major predator, the sea otter, which was endangered by hunters seeking pelts. The sea urchins, without a check on their population, ate not only the usual detritus in the kelp forests but also the live kelps—and the kelp population declined rapidly. At the same time unusual warm water currents bathed the kelps, causing reproductive problems because many kelp gametophytes need cool water for gamete production. Under protection, the sea otter population increased, the sea urchins were kept in check, and the kelps are recovering (Branning, 1976).

Sea urchins and their predators and prey are important in determining the structure of the kelp communities (Simenstad et al, 1978). A study in Alaska demonstrated that when sea urchins fed upon extremely plentiful diatoms the young kelps (*Alaria, Nereocystis*) were left alone and grew large enough to resist grazing by the sea urchins (Duggins, 1981).

Large kelp forests support very large numbers of invertebrates and vertebrates. During their reproductive seasons gamete and zoospore concentrations can reach 3 million cells per liter in the plankton. In addition, dead and decaying plant parts contribute significantly to the energy in the detritus food chain. Large populations of kelp are a food source for marine animals and also provide shelter and protection for spawning fishes and their young. A community of algae and invertebrates also develops around the kelp holdfasts. Kelp forests contribute significantly to both primary and secondary productivity in temperate littoral areas, forming a complicated series of algal and animal interrelationships. (See Earl, 1980, for a discussion of life in a kelp forest off southern California.)

The Phaeophytes from tropical and subtropical regions do not develop the large size of the kelps found in temperate and arctic waters. The numbers of brown algal species in the warmer waters are also notably smaller, although the calcium carbonate–depositing *Padina*, the reef inhabitants *Dictyota* and *Turbinaria* (Fig. 7-13), and the lagoon *Sargassum* are often quite common.

The photosynthetic pigment fucoxanthin enables brown algae to convert light energy to chemical energy at depths where chlorophyll *a* and *c* are not very effective. Fucoxanthin absorbs energy in those green light wavelengths (500 to 550 nm) that penetrate deeply into water and then transfers this energy to chlorophyll for the manufacture of carbohydrates. In turbid, temperate waters the brown algae are found to depths of 35 m, in the clear Mediterranean to 100 m, and in the very clear tropical waters to 200 m.

Freshwater genera include a member of the Ectocarpales, *Heribaudiella*, which grows on rocks in fast-flowing streams. To date, this genus has been reported growing in several European streams and there is a disputed report of *Herbaudiella* in a Connecticut stream in the United States. Four other freshwater phaeophyte genera have been reported.

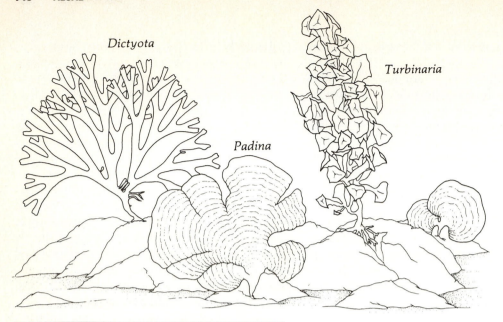

Dictyota

Turbinaria

Padina

FIG. 7-13 Three common tropical brown algae.
Habits of three common marine, shallow-water, benthic brown algae, *Dictyota*, *Padina*, and *Turbinaria*. Note dichotomous branching in the *Dictyota* sp., triangular arrangement of branches in *Turbinaria*, and fan-shaped thallus of *Padina*. *Padina* is the only genus of brown algae that deposits calcium carbonate on the surface of the thallus; it therefore appears white when it is growing.

ECONOMIC IMPORTANCE

Phaeophyta have been used for centuries for food, fertilizer, animal fodder, and as sources of calcium, iodine, and potassium. More recently the giant kelps have been harvested for their algin content. The term "algin" refers to derivatives of alginic acid—especially the calcium, sodium, potassium, or ammonium salts (alginates). Sodium alginate is the best known algin, but the other salts are also used commercially. The alginate industry started growing in the 1920s; the economic importance of the brown algae has increased substantially since then as the demand for algin exceeded the supply. The kelps are harvested in Europe, North America, and the Orient. Algin content of the giant kelps can range from 14% to 40% of the wet weight and varies seasonally, with more algin present in the winter.

The term "kelp" originally meant the burned ash of seaweeds (primarily brown algae), but the term now is applied to the actual plant—usually a member of the Laminariales. In Europe in the seventeenth century kelp was used to make soda (sodium carbonate) for glass manufacturing and pottery glazes. In the eighteenth century Scotland produced 18,000 metric tons of soda annually from 360 metric tons of wet kelps (Chapman, 1970). The seaweeds (mostly Laminariales and Fucales) were harvested by hand from the shores where they were washed up or exposed at low tide. "Algae or seaweed rights" to harvest seaweed along the beaches were frequently disputed.

In the mid-nineteenth century kelps were harvested primarily for their iodine and potassium content. Kelps may contain 1.4% to 1.8% iodine, concentrating these minerals from seawater. As terrestrial mineral deposits of these essential elements became available, the harvesting of kelps for iodine and potassium decreased. Today only the U.S.S.R. and Japan produce iodine from seaweed.

Giant kelps that are commercially harvested are *Macrocystis pyrifera* on the U.S. west coast and *Alaria* and *Laminaria* in the United Kingdom. Limited commercial harvesting occurs in the Canadian Maritime Provinces and Australia. These kelps grow rapidly, up to 0.5 m a day, and can be repeatedly harvested if only their tops are removed. Large floating barges equipped with mowers cut off the blades about 1 m deep. A crew of five men can harvest 275 metric tons of kelp a day. Approximately 130,000 metric tons (wet weight) of *Macrocystis* is harvested per year in southern California (Abbott and Hollenberg, 1976). (See Kuwabara, 1982, for recent work with micronutrients and kelp cultivation in California.)

Algin extracted from the giant kelps is a gummy, hydrophilic substance used commercially as a stabilizer and emulsifier in such diverse products as ice cream, soup, salad dressing, puddings, paints, cosmetics, shaving cream, soap, medicines, buttons, dental impressions, and photography chemicals. Ammonium alginate is used as a fireproofing substance, calcium alginate is used in plastics and laundry starch, and sodium alginate is widely used in the dairy industry. Nontoxic algins help to stabilize consistency and to retain moisture in candy, marshmallows, mayonnaise, and cheese. Commercial harvesting of kelps is economically feasible only where high concentrations of the seaweed occur. California is the major algin-producing area in the world. Over 17,000 metric tons of alginate are sold annually throughout the world, a $130 million a year industry. Percival and McDowell (1967) review the many different uses of algin.

Many thousands of dollars have been spent on kelp bed research in California. Recent research has focused on management techniques for the kelp forests (North, 1971). Based on fish, shellfish (mainly abalone), and algin production, the value of 2.6 km² (one square mile) of *Macrocystis* forest is estimated $1 million per year

(Earl, 1980). Kelp "farming" has been proposed, with the kelp biomass used to generate methane (North, 1979; Bungay, 1981).

In Japan members of the Laminarilaes are harvested, dried, pressed into blocks, and used as food called "kombu." *Laminaria* and *Alaria* are the principal kelps used for kombu, which is eaten as a cooked vegetable, boiled in soup, or served with fish or meat. Before World War I, 225,000 metric tons of kelp per year were harvested in Japan. Through cultural techniques both the Japanese and Chinese have been able to double the growth rate of *Laminaria*. *Laminaria japonica*, called "haidai" ("seabelt") has been used in China for hundreds of years as both a food and a medicine. The development of the mass culture of this kelp has been a major achievement of the Chinese government programs in aquaculture (Cheng, 1969).

Algins, which have a high percentage of guluronic acid, have a specificity for divalent ions such as calcium, barium, and strontium and have been used medically to remove toxic cations such as strontium from the bodies of exposed individuals (Sutton et al, 1971).

Farmers who live near seacoasts often harvest brown algae such as *Fucus* and *Ascophyllum* from intertidal areas for use as fertilizers and animal feed. Fresh seaweed is similar to barnyard manure in its nutrient content, although it contains more potassium and less phosphorous.

SUMMARY

Phaeophyta are the top of the brown line of algal evolution. Their growth patterns are directed by a number of cellular meristematic regions that enable these plants to develop highly differentiated organ level structures—the holdfast, stipe, and blade. Cellular differentiation among the brown algae is much greater than in any other group of nonvascular plants, and tissue-level organization is common to all members of the division.

The brown algae include some of the largest plants known to man—the giant kelps. They also include organisms that grow in one of the earth's harshest ecological niches, the intertidal zone. Here the algae must cope with twice-daily tidal variations that include alternation of wet-dry, cold-hot, light-dark, and mechanical agitation caused by the pounding surf. The browns are primarily shallow-water marine organisms common to the cooler waters of temperate and arctic regions. They occupy a narrow band of the earth's surface adjacent to land masses. They sometimes form dense underwater "forests" that ecologically resemble terrestrial tropical rain forests.

Reproduction among the brown algae is generally sexual, and most have diplohaplontic life cycles with isomorphic or heteromorphic alternations of generations. The life cycles of the giant kelps are similar to those of the ferns with microscopic haploid gametophytes. One order, Fucales, has evolved a diplontic life cycle.

Most phaeophytes are photoautotrophic. A few are auxotrophic, but no exclusively heterotrophic forms are known. They are all multicellular and are related to other divisions in the brown line of evolution through the presence of chlorophyll *a* and *c*, fucoxanthin, and similar energy storage compounds.

Economically the brown algae are an important product of the oceans. Entire industries have developed around the production and harvesting of these algae. Their most important product is algin, an emulsifying agent and stabilizer used in packaging many consumable foods and commercial products.

SELECTED REFERENCES

Abbot, I.A., and G.J. Hollenberg. 1976. Marine algae of California. Stanford University Press, Stanford, Calif. 827 pp.

Branning, T.G. 1976. Giant kelp: its comeback against urchins, sewage. Smithsonian 7(6):102-109.

Chapman, V.J., and D.J. Chapman. 1980. Seaweeds and their uses. Ed. 3. Methuen & Co., London. 334 pp.

Cheng, T.H. 1969. Production of kelp—a major aspect of China's exploitation of the sea. Econ. Botany 23:215-236.

Earle, S.A. 1980. Undersea world of a kelp forest. Nat. Geog. 158(3):411-426.

Field, J.G., N.G. Jarman, G.S. Dieckmann, C.L. Griffiths, B. Velimirov, and P. Zoutendyk. 1977. Sun, waves, seaweed, and lobsters: the dynamics of a west coast kelp bed. S. Afr. J. Sci. 73:7-10.

Lobban, C.S., and M.J. Wynne (eds.). 1982. The biology of seaweeds. Bot. Monogr. 17. University of California, Berkley. 798 pp.

Teal, J.M. 1980. Primary production of benthic and fringing plant communities. In R.S.K. Barnes and K.H. Mann (eds.), Fundamentals of aquatic ecosystems, pp. 67-83, Blackwell Scientific Publications, Oxford, England.

Taylor, W.R. 1957. Marine algae of the northeastern coast of North America. University of Michigan Press, Ann Arbor. 509 pp.

Taylor, W.R. 1960. Marine algae of the eastern tropical and subtropical coasts of the Americas. University of Michigan Press, Ann Arbor. 870 pp.

8

CLASSIFICATION

Division: Chrysophyta (golden-brown algae): 425 genera, 7000 species; a diverse assemblage of organisms containing chlorophyll *a* and in many species *c*, chrysolaminarin, and fucoxanthin

Class 1: Chrysophyceae: cell walls absent or composed of cellulose or silicates; isogamous; haplontic life cycles; statospore formation; 75 genera, 300 species

ORDER 1. Ochromonadales: naked biflagellate cells; some colonial; *Dinobryon, Ochromonas, Paraphysomonas, Spumella, Synura*

ORDER 2. Dictyochales: mostly marine fossils; delicate siliceous skeletons; silicoflagellates; *Dictyocha, Distephanus*

Class 2: Prymnesiophyceae (= Haptophyceae): cell walls not cellulosic; biflagellate plus a haptonema; 150 genera, 600 species

ORDER 3. Prymnesiales: some with cell walls containing calcium carbonate disks (coccoliths); many fossil forms; coccolithophorids; *Apistonema, Coccolithus, Discophaera, Florisphaera, Hymenomonas, Phaeocystis, Prymnesium, Rhabdosphaera, Syracosphaera, Umbellosphaera*

Class 3: Bacillariophyceae: siliceous cell walls in two overlapping halves called frustules; unicellular or colonial; auxospore formation; diatoms; 200 genera, 6000 species

ORDER 4. Centrales: radial symmetry; *Biddulphia, Chaetoceros, Coscinodiscus, Cyclotella, Melosira, Rhizosolenia, Stictodiscus*

ORDER 5. Pennales: bilateral symmetry; *Ditylum, Eunotia, Fragilaria, Frustulia, Gomphonema, Hantzschia, Licmophora, Navicula, Nitzschia, Pinnularia*

CHRYSOPHYTA

Chrysophyta are often called the golden-brown algae because their cells contain yellow carotenoid pigments, including the brown pigment fucoxanthin. These pigments usually mask the green color of the chlorophyll that is also present. The division is primarily composed of single-celled organisms, including the groups known as **diatoms, silicoflagellates,** and **coccolithophorids.** However, they also occur in colonial and filamentous forms. These organisms are not well known to the general public and beginning students, although they play an extremely important role in our biosphere. The diatoms, for instance, are the major primary producers in temperate and arctic oceans and in freshwater lakes and may be the world's most important fixers of carbon dioxide via photosynthesis. They are the grasses of the sea, the jewels of the plant world. They are also important economically, especially in the petroleum industry.

The division contains three rather diverse classes of organisms: Chrysophyceae, which includes the silicoflagellates; Prymnesiophyceae (= Haptophyceae), which includes the coccolithophorids; and Bacillariophyceae, the diatoms. The three classes are related on the basis of their chlorophyll content (a, c_1, and c_2), their energy storage product (**chrysolaminarin**), their photosynthetic accessory pigments, including fucoxanthin, and the ultrastructure of their chloroplasts. These chloroplasts are enclosed within two chloroplast membranes plus two endoplasmic reticulum membranes, the outer one of which is continuous with the nuclear membrane (see Fig. 1-7). Both the chloroplast structure and the presence of fucoxanthin are characteristic of the brown line of algal evolution.

CELL STRUCTURE

The morphological organization of individuals in the three classes of chrysophytes varies because of the diversity in the cell coverings. Within Chrysophyceae, cell coverings range from completely naked plasma membranes (*Ochromonas*), to thickened periplasts, to delicate siliceous scales (*Synura*), to gelatinous loricas that are often impregnated with minerals (e.g., iron in *Dinobryon*). Traces of silicon, cellulose, protein, calcium carbonate, and various mineral substances are reported in connection with these cell coverings. The lorica is the most distinctive structure; it is an incomplete cell wall that does not touch the plasma membrane and is characteristic of some genera (e.g., *Dinobryon*, Fig. 8-1). Its chemical composition may vary, but in all cases it functions as a protective layer for the contained algal cell.

Chrysophyceae also contain genera that secrete siliceous, tubular, star-shaped skeletons (the silicoflagellates, Fig. 8-2). Evidence indicates that the siliceous skeletons are secreted internally by the Golgi apparatus and are extruded by some unknown mechanism to the exterior of the cell (Van Valkenburg, 1980).

The most characteristic structure in Chrysophyceae is the **statospore,** a type of cyst common to all species (Fig. 8-3); it is formed internally through the deposition of a distinctively sculptured siliceous cell wall that contains an unsilicified plug. The urn-shaped statospore functions as a resting spore formed in response to adverse environmental conditions. Sculpturing of the cell wall is species specific. Germination of the spore takes place when environmental conditions are favorable. The protoplast contained within the spore divides internally, forming numerous zoospores that emerge by displacing the plug in the cell wall.

Chrysophyceae have two heterokont flagella, one of which is long and of the tinsel type, while the other is of a shorter whiplash type. The shorter flagellum possesses a basal swelling that may be associated with a red eyespot (stigma).

Inside the plasma membrane the usual complement of eukaryotic organelles can be identified, including Golgi bodies, endoplasmic reticulum, ribosomes, mitochondria, chloroplasts, and contractile vacuoles. When present, chloroplasts occur as one or two parietal structures in each cell. Each chloroplast is characteristic of the brown line of algal evolution (see Fig. 1-7). The thylakoid membranes occur in bands of three. Chlorophyll *a* and *c*, large quan-

FIG. 8-1 Representative freshwater Chrysophyta. Scanning electron photomicrograph of some chrysophytes from Lake Michigan. *A, Dinobryon* sp., showing the lorica. *B, Fragilaria* sp., showing attachment of the cells by their sides, forming ribbons. *C, Tabellaria* sp., showing how the cells adhere to each other at the corners. *D, Asterionella* sp., showing how cells adhere, forming a star-shaped colony. (Courtesy Jon I. Parker, Department of Biology, Lehigh University, Bethlehem, Pa.)

FIG. 8-2 Silicoflagellates.

Tubular, siliceous skeletons of two silicoflagellate species from the Sargasso Sea as seen through the scanning electron microscope. The protoplasm—with its chloroplasts and other organelles, including the single flagellum—is gone. **A,** Three specimens of *Distephanus* sp. (1160×). **B,** Single skeleton of *Dictyocha* sp. (2770×). (Courtesy Margaret Goreau, Rosenteil School of Marine and Atmospheric Science, University of Miami, Fla.)

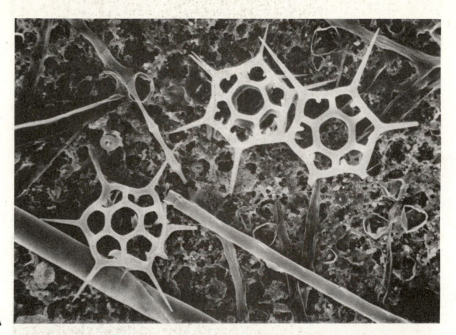

A

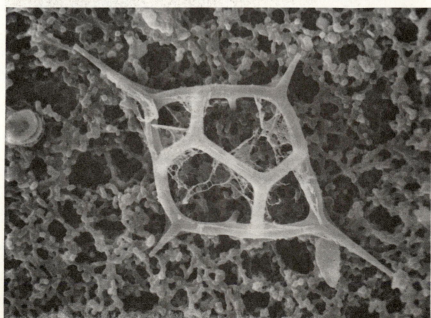

B

FIG. 8-3　A statospore.
Mature statospore of a species from class Chrysophyceae. Note presence of a series of spines *(sp)*
emanating from the siliceous cell wall *(cw)* and the
nonsilicified plug *(pl)* located in the pore and surrounded by the collar *(co)*. Within the statospore is
the live protoplast with its plasma membrane *(pm)*
and nucleus *(n)*.

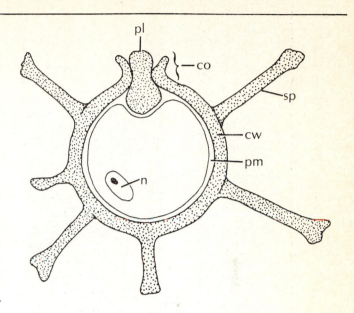

tities of the brown pigment fucoxanthin, and
other carotenoids are located within the chloroplasts. Pyrenoids are common in some genera.
Flagellate cells contain a red eyespot located
within the chloroplast membrane and situated
in close association with the basal swelling of the
shorter flagellum. Some genera contain rather
large membrane-bound vesicles composed of
chrysolaminarin, the energy storage compound.
Contractile vacuoles are also common in the
freshwater species.

Prymnesiophyceae (= Haptophyceae) differ in
some respects from Chrysophyceae. The cells
have two isokont whiplash-type flagella. However, they also possess an additional appendage,
the **haptonema,** which emerges between the
two flagella (see Fig. 1-9). The haptonema is a
unique structure of variable length containing
six internal protein fibrils (Fig. 1-8). The function of the haptonema has not been established,
although it seems to act as an attachment organelle in some genera, enabling them to fasten
themselves to some substrate while continuing
flagellar movement in search of food. However,
it is known that the haptonema can undergo
rapid uncoiling movements (Hibberd, 1980a); it
has been hypothesized that these movements
may play a role in changing turgor pressure
within the cell.

Another unique feature of Prymnesiophyceae
is the nature of the cell covering. In some genera the plasma membrane is covered with a layer of proteinaceous scales (e.g., *Prymnesium*).
Other genera produce many small calcified
disks embedded in a thick organic matrix (e.g.,
Syracosphaera, Fig. 8-4). Such calcite or arago-

FIG. 8-4　Coccoliths.
Scanning electron photomicrograph of the surface of
Syracosphaera mediterranea, a marine planktonic
coccolithophorid from the Sargasso Sea, showing
appearance of calcified plates called coccoliths
(10200×). (Courtesy Margaret Goreau, Rosenteil
School of Marine and Atmospheric Science, University of Miami, Fla.)

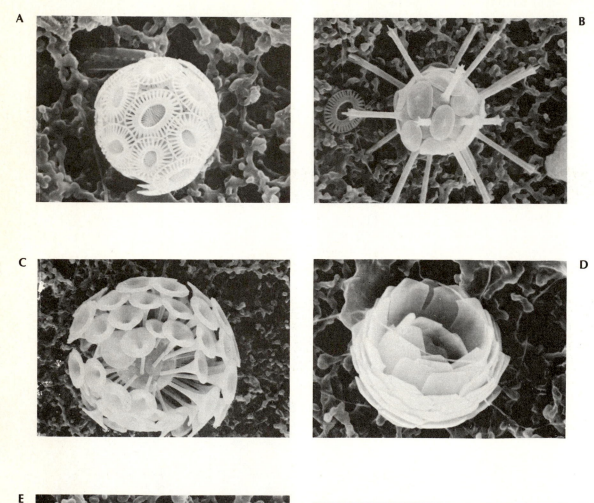

FIG. 8-5 Coccolithophorids.
Scanning electron photomicrographs of five species of planktonic coccolithophorids from the Sargasso Sea, showing the nature of the calcium carbonate disks that cover their surface. **A,** *Coccolithus huxleyi* (also called *Emiliania huxleyi* and *Gephyrocapsa huxleyi*), (8440×); **B,** *Rhabdosphaera stylifera* (with a coccolith from *C. huxleyi* in the background), (5730×); **C,** *Discosphaera tubifera* (4300×); **D,** *Florisphaera profunda* (7680×); **E,** *Umbellosphaera irregularis* (5120×). (Courtesy Margaret Goreau, Rosenteil School of Marine and Atmospheric Science, University of Miami, Fla.)

nite disks are called **coccoliths;** organisms that produce them are known as coccolithophorids (Fig. 8-5). They are primarily marine and may form a major part of the nannoplankton from tropical waters saturated with calcium carbonate. In these areas the coccolithophorids rival the diatoms and dinoflagellates as the major source of primary productivity.

Bacillariophyceae differ from the other two classes of Chrysophyta in three major respects. First, their cell walls are composed almost entirely of silicates, are distinctively sculptured, and they contain no cellulose. Second, they have diplontic life cycles with the vegetative cell always being diploid and the only haploid phase being the gamete. Third, in those species that produce flagellate sperm cells, the flagella are single whiplash-type flagella that in cross section lack the two central protein microtubules. The Bacillariophyceae may be artificially divided into two orders—the radially symmetrical **Centrales** and the bilaterally symmetrical **Pennales.**

The siliceous cell walls of diatoms are composed of more than 96% silica dioxide (SiO_2), with the remainder being various mineral substances, amino acids, uronic acids, and sugars. These walls are so unique structurally that some special terminology is used to describe their construction. The walls are called **frustules,** and they occur in two overlapping halves called **valves.** The valves form a pillboxlike structure with an outer valve, the **epitheca,** and an inner valve, the **hypotheca.** When viewed from above or below, the frustule presents a valve view. When viewed from the side, a girdle view is observed (Fig. 8-6). The architecture of the frustule provides taxonomists with an excellent tool for the identification of different diatom species. The sculpturing of the surface of the frustules is very complex, and an entire diatom vocabulary has been developed to describe it. Each species contains a specific set of surface configurations composed of various spines (processes), pores (**punctae**), rows of contiguous pores (**striae**), and ridges (**costae**) (Figs. 8-6 and 8-7). **Pennate** diatoms also contain structures associated with

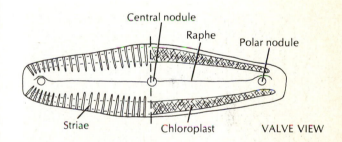

VALVE VIEW

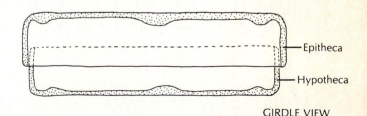

GIRDLE VIEW

FIG. 8-6 Diatom structure.
Valve, girdle, and end views of a pennate diatom as they would appear if seen under a light and electron microscope. In the valve view, note that the two chloroplasts are clearly visible. In the girdle view, the frustules (cell walls) are clearly seen as two overlapping valves, the epitheca and the hypotheca. Sections through the frustules, seen in end view, show the raphe, which completely traverses both epitheca and hypotheca. This is not the case as indicated in the frustule section, showing a pseudoraphe.

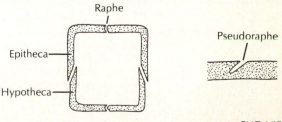

END VIEW

their main axis, the **raphe, pseudoraphe, polar nodules,** and a **central nodule** (Fig. 8-6). Furthermore, the inorganic chemical nature of the cell wall insures that the frustules persist long after the protoplast within has ceased to live. Excellent diatomaceous fossils, millions of years old, can be found in most geological strata. These frustules are often used as time markers in petrographic and oil research.

The raphe is a dominant feature in motile pennate diatoms. It forms a slitlike opening in the frustule that extends between the polar nodules, providing a channel through which cytoplasm flows and allowing for movement of

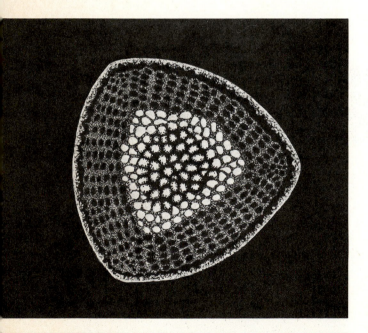

FIG. 8-7 *Stictodiscus* **habit.**
Valve view of centric (Centrales) diatom showing numerous pores (punctae) through the frustule. Not all centrate diatoms are triangular like this one; they may be round or oval. (Adapted from photograph by Ruth F. Patrick and Jay Sacks, Academy of Natural Sciences of Philadelphia.)

the cell. A pseudoraphe, found in some nonmotile pennate diatoms, is a slitlike opening in the frustule that does not pass all the way through the wall (Fig. 8-6). It may also be an area of the frustule that has no sculpturing and is therefore just a flat space between adjacent costae, striae, or puncta.

The protoplast, which lives within the siliceous frustule, possesses all of the usual organelles associated with eukaryotic cells. The chloroplasts are the most prominent features within the protoplast. Their ultrastructural characteristics indicate that diatoms are members of the brown line of algal evolution, since they have a double membrane CER and thylakoids in bands of three (see Fig. 1-7). In pennate diatoms the chloroplasts occur as two large golden-brown parietal structures adhering closely to the sides of the cell. The green color of their chlorophyll content is masked by the presence of fucoxanthin. In valve view under the light microscope, the chloroplasts appear as two broad brownish stripes on either side of the cell (Fig. 8-6). In centric diatoms, chloroplasts exist as numerous, small, disk-shaped, greenish-brown structures floating free within the cytoplasm.

All chloroplasts contain chlorophyll a and c (c_1 and c_2) plus large quantities of photosynthetically active accessory pigments, including the carotenoid fucoxanthin. The exact color of the plastid depends upon the relative amounts of green, brown, and yellow pigments present. Some chloroplasts possess pyrenoids, which apparently function as sites for the synthesis of the energy storage product, chrysolaminarin. This substance resembles the energy storage product of the brown algae, laminarin, which is a β-1,3–linked polymer of D-glucose molecules. However, unlike laminarin, chrysolaminarin contains two β-1,6 linkages of D-glucose in each molecule. This chemical characteristic appears trivial, but it alters the physical-chemical properties of the energy storage product and therefore has both ecological and phylogenetic significance.

REPRODUCTION

The three classes of Chrysophyta have each developed different modes of sexual and asexual reproduction. Asexual reproduction by fragmentation occurs only in those organisms that form colonies or filaments. Most Chrysophyta are unicellular; therefore, increases in cell numbers are usually accomplished by mitosis followed by cytokinesis. Motile zoospores and nonmotile aplanospores may be formed by members of the division.

Genetic recombination through sexual reproduction has been observed in all three classes, although it is not especially common in any of them. Chrysophyceae produce motile isogametes as a response to changing environmental conditions. Statospore formation is often a result of the union of two isogametes (Fott, 1964); therefore, statospores function as zygotes. They germinate by meiotic cell divisions, producing motile zoospores. Zygotic meiosis of this type is characteristic of haplontic life cycles.

The class Prymnesiophyceae has a considerable degree of **polymorphism** among its members, with some species taking on different shapes under different conditions. Sexual reproduction is isogamous but their life cycles are diplohaplontic. The haploid and diploid generations are often difficult to distinguish from each other and from similar organisms. In some genera, as with *Apistonema* (1N) and *Hymenomonas* (2N), different generic names have been unwittingly assigned to the haploid and diploid generations of a single species. The *Apistonema* genus is now known as the apistonema stage of *Hymenomonas*. The apistonema (gametophytic) stage produces isogametes that fuse, forming the diploid (sporophytic) phase of *Hymenomonas* (Leadbeater, 1970). The diploid phase produces the calcareous cell coverings called coccoliths.

Because of their small size, polymorphic nature, and warm, open-ocean habitat, Prymnesiophyceae have been difficult to study. Cell cultures are difficult to maintain because most species are auxotrophic, requiring certain vitamin additives. Furthermore, the appearance of cultured cells may vary with such conditions as

---

FIG. 8-8 Reproduction in diatoms.
Diagrammatic representation of progressive diminution in the average size of cell frustules in any diatom population. Notice that each epitheca synthesizes a new hypotheca, and each hypotheca also synthesizes a new hypotheca (and thus becomes the epitheca of the new cell). Over extended periods, this type of growth pattern has an effect on the average cell size in any population of diatoms.

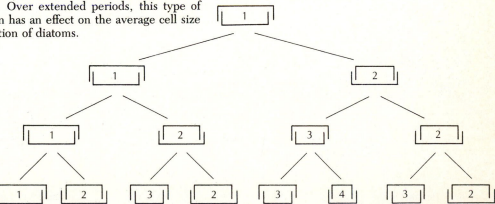

crowding, age, and available nutrients (Green and Parke, 1975), adding to the confusion and difficulty.

Mitosis in Bacillariophyceae presents a special problem, unique in the biological world, because of the cell wall architecture. The two valves of a cell are not exactly the same size. When mitosis occurs, the overlapping valves come apart and each half produces a new bottom (hypotheca). The hypotheca of the parent cell thus becomes the epitheca of one of the daughter cells (Fig. 8-8). Successive cell divisions over long periods of time will result in a population of cells with an average diameter that is considerably smaller than the original parent cell. The progressive decrease in average cell diameter in diatom populations was first observed by Pfitzer (1871). He discovered that when a certain minimal average cell size was reached (about three-fifths the size of the original cell), a sudden increase in size took place and the cells were the same size as the original parent cell. This dramatic size increase is caused by the formation of special growth spores called **auxospores** (Fig. 8-9). They are usually formed by a sexual process through the union of two gametes (and are therefore zygospores), but they may also be formed by asexual processes in some genera (e.g., *Cyclotella* and *Ditylum*).

Formation of auxospores under natural conditions is not a commonly observed characteristic except in laboratory cultures. Haploid nuclei that fuse to produce auxospores may be formed by two different methods. First, haploid gametes may be the result of meiotic divisions in separate individuals producing a true sexual mixing of genetic information. A second method is produced through **autogamy**, whereby two haploid nuclei produced by a single diploid nucleus fuse to form the diploid zygote. Autogamy accomplishes a reshuffling of the genetic composition of a single organism through crossing over and recombinations within the single genome. It is also common among some protozoans and may be more important among the algae than was previously supposed.

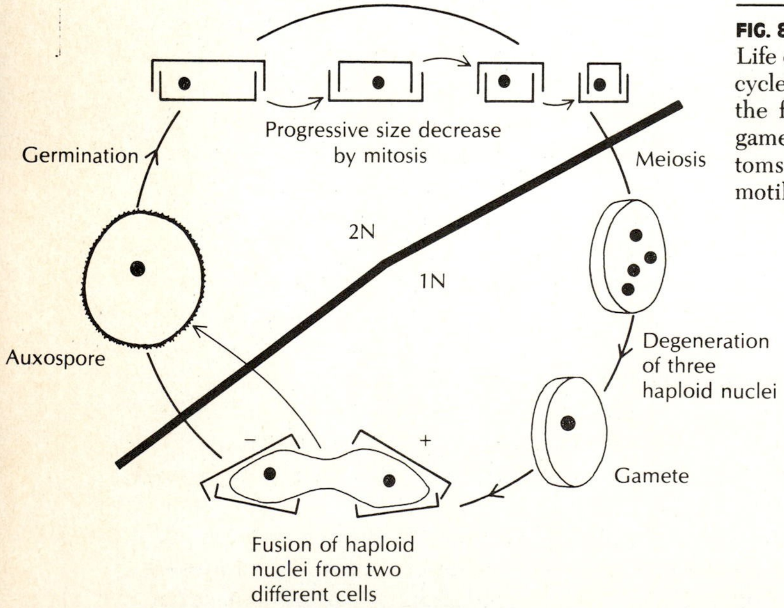

FIG. 8-9 Diatom life history.
Life cycle of a typical pennate diatom. Notice that life cycle is diplontic, with the only haploid phase being the formation of gametes. In pennate diatoms, the gametes are amoeboid, as illustrated. In centric diatoms sexual reproduction may be oogamous with motile flagellate sperm.

Germination

Progressive size decrease by mitosis

Meiosis

2N

1N

Auxospore

Degeneration of three haploid nuclei

−   +

Gamete

Fusion of haploid nuclei from two different cells

In the Pennales auxospore formation is a result of the union of isogamous naked gametes that are usually amoeboid (Fig. 8-9). In the Centrales oogamy is common, and flagellate sperm cells are produced. Gamete formation results from meiotic cell divisions within certain diploid vegetative cells. The life cycles are therefore diplontic, with the gametes being the only haploid phase.

Auxospore formation may occur within a vegetative filament, at the tip of a vegetative filament, or as a free-floating cell. The spores are invariably much larger than the parental cells that produced them. After the naked gametes fuse, a siliceous, ornamented, thick cell wall is produced around the zygote. The spore may then undergo a rest period before germination. When an auxospore germinates, it forms a new diatom cell with a frustule whose dimensions are the largest within that species (Fig. 8-9); thus the original cell size for that species is restored.

Indications are that auxospores are formed by the smaller cells of a population as a result of environmental changes in temperature, light, and available nutrients (Holmes, 1966; Neuville and Daste, 1975). When the seasons change, phytoplankton members form auxospores that sink to the bottom of lakes or neritic seas. Other genera may form resting spores under conditions of decreasing light, temperatures, and nutrients. Such phenomena could account for the waxing and waning of certain diatom species in various freshwater and saltwater systems throughout the year.

PHYSIOLOGY

Much of the knowledge concerning the physiology of Chrysophyta has been obtained through laboratory cultures. Autotrophic, auxotrophic, and heterotrophic species are known in all three classes. Among the Chrysophyceae some species of *Ochromonas* actively photosynthesize, but they do not contain enough chlorophyll for photosynthesis to compensate for respiration. Therefore, organic substances (vitamin B_{12}, sugars, fats, and amino acids) must be supplied. Other species of *Ochromonas* may respire anaerobically, producing ethanol, lactic acid, and carbon dioxide. Certain species of *Ochromonas* may be phagotrophic, actively ingesting such particulate food as bacteria. Heterotrophic species have lost the genetic ability to synthesize chlorophyll and are colorless but may resemble photosynthetic species. For example, *Spumella* closely resembles *Ochromonas* but lacks chlorophyll, while *Paraphysomonas* is very similar to a colorless form of *Synura*.

Most members of Prymnesiophyceae are auxotrophic and require vitamin B_{12} or thiamine for growth and reproduction. Some members of this class, the coccolithophorids, have the highest reproductive rate of any alga. They are capable of dividing more than twice each day under optimum conditions. The coccolithophorids also have the ability to deposit calcium carbonate in the form of calcite or aragonite in the small scalelike coccoliths attached to their cell surface (Fig. 8-4). As many as 30 coccoliths per cell may be formed. One species, *Coccolithus pelagicus*, has both motile and nonmotile stages, and each stage has a different coccolith type (Parke and Adams, 1960). However, coccolith deposition is not essential for the cell and the organisms exhibit considerable morphological polymorphism. Organisms that do not live in calcium carbonate–saturated warm tropical waters do not form calcified coccoliths but form cellulose scales instead. In culture the coccolithophorids require not only water saturated with calcium carbonate but also light for photosynthesis and a neutral or alkaline pH to form coccoliths. Some coccolithophorids, however, do not require light: some have been found in very deep open ocean water, where they may be either living heterotrophically or in a resting stage.

Bacillariophyceae differ from the other two classes in their nutrient requirements. Most are entirely autotrophic, though some genera are auxotrophic. A few are entirely heterotrophic and completely lack chlorophyll. For example, *Nitzschia alba* is a marine epiphyte that has lost the ability to produce chlorophyll but very closely resembles the autotrophic *Nitzschia laevis*. Several marine species have been grown heterotrophically in the dark with glucose as a carbon source (Lewin and Lewin, 1960). Diatoms have an absolute requirement for soluble silica (SiO_2). Dissolved silica is then used by the cell to produce polymerized opaline silica ($SiO_2 \cdot nH_2O$). This siliceous compound may comprise up to 96% of the cell wall, depending upon the species. Diatom cells usually cannot divide unless silica is available, although there is

a wide range (0.6 mg/l to 13.4 mg/l) of silica required for growth, depending on the species (Kilham, 1971). Some diatoms, however, may be induced experimentally to divide without silica, producing naked protoplasts. These cells will form cell walls when transferred to media containing silica. Silica concentrations in salt water are generally adequate for diatom growth, but silica may often be a limiting factor to diatom populations in fresh water. As diatom numbers increase, silica levels decrease (Fig. 8-10). Freshwater and estuarine diatom blooms occasionally may completely deplete silica so that it is undetectable in the water column (Munawar and Munawar, 1975; Paasche and Østergren,

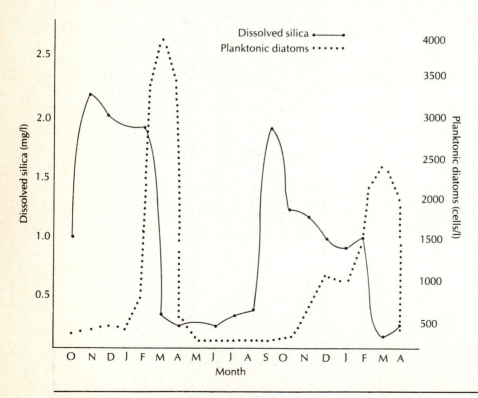

FIG. 8-10 Diatom growth and silica content.
Graph of silica content and planktonic diatom growth in a freshwater habitat over a 19-month period. Notice that when dissolved silica decreases, diatom numbers increase shortly afterward, indicating that the silica is being used by the diatoms for frustule synthesis. (Data from Lund, 1964.)

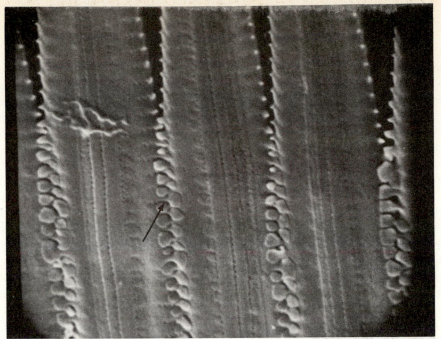

A

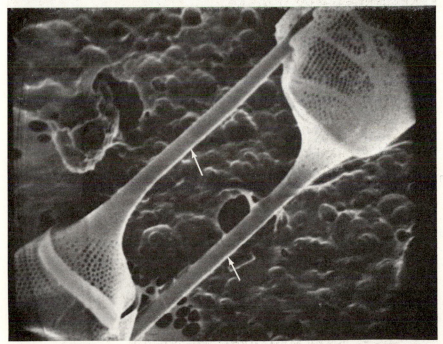

B

FIG. 8-11 Interlocking spines of diatoms.

Scanning electron photomicrographs of two species of diatoms, both of which produce siliceous spines associated with their cell walls. **A,** Series of cells of *Fragilaria crotonensis* joined by interlocking spines (arrow), (7500×); **B,** Tips of two cells of *Rhizosolenia eriensis*, each of which produces a long spine (arrows) that attaches to the adjacent cell (6500×). (Courtesy Jon I. Parker, Department of Biology, Lehigh University, Bethlehem, Pa.)

1980). Wetzel (1975) discusses in greater detail the cycle of silica in lakes.

Diatoms are a major component of the phytoplankton in freshwater and marine environments. They have a series of mechanisms that enables them to maintain their position in the photic zone. They have evolved a number of flotation strategies, including storage of oil droplets in their cytoplasm, production of gas, mucus secretions, reduction in cell density by reducing cytoplasmic ionic concentrations, and increase in the surface-to-volume ratio of the cells. The latter is reflected in the cell morphology. Some diatoms are flattened, some form filaments that are loosely joined through mucus secretions or interlocking spines, and some produce long siliceous spines (Fig. 8-11).

Waters of the open ocean and the tropics are less dense than the waters of shallow seas and temperate and arctic areas (see Chapter 2).

Open-ocean (pelagic) and warm-water species produce long spines that enable them to stay in the photic zone in the less-dense waters. Similar diatom species in the denser cool or neritic waters have shorter spines (Fig. 8-12). The life of a planktonic alga depends upon its ability to stay in the photic zone and to photosynthesize. Should the alga sink below the photic zone for any appreciable length of time, the chemical energy produced by photosynthesis will not exceed the energy expended in respiration and the cell may die and sink to the bottom. Raymont (1980) provides additional information about diatom flotation.

Motility in Chrysophyceae and Prymnesiophyceae depends upon the presence of flagella in the vegetative cells and gametes. Diatoms, however, do not possess flagella in their vegetative cells, but many benthic and epiphytic forms exhibit a peculiar gliding motion. Diatom motil-

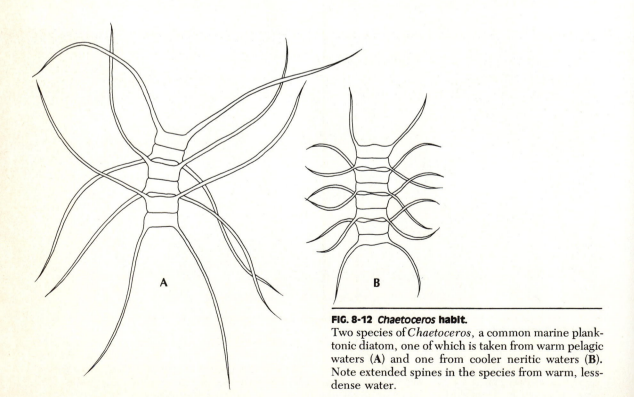

FIG. 8-12 *Chaetoceros* **habit.**
Two species of *Chaetoceros*, a common marine planktonic diatom, one of which is taken from warm pelagic waters (**A**) and one from cooler neritic waters (**B**). Note extended spines in the species from warm, less-dense water.

A

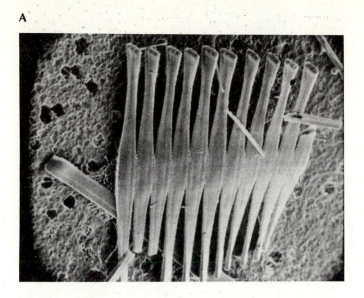

B

C

D

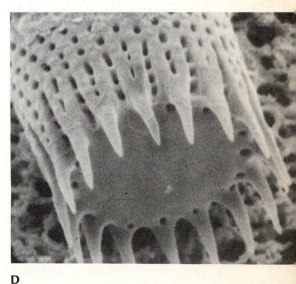

FIG. 8-13 Colony types in diatoms.

Scanning electron photomicrographs of various diatoms. **A,** Ribbon of *Fragilaria* cells (3500×); **B,** star-shaped colony of *Asterionella* cells (3000×); **C,** zig-zag chain of *Tabellaria* cells (4000×); **D,** spikes of a *Melosira* cell (4500×) that became joined to another set of spines from another *Melosira* cell, producing a filament of cells. (Courtesy Jon I. Parker, Department of Biology, Lehigh University, Bethlehem, Pa.)

ity is restricted to gametes and pennate forms possessing a raphe. One theory of motility is that mucus substances produced in the cytoplasm are carried by cytoplasmic streaming through the central or polar nodules and along the adjoining raphe. The mucus flows along the raphe (Fig. 8-6), creating friction between the cell and its substrate (Drum and Hopkins, 1966). A slime trail is left behind the advancing diatom. The raphe must be in contact with a substrate for movement to occur (Nultsch, 1974). Additional theories of motility are discussed by Raymont (1980). **Centric** diatoms and pennate diatoms with a pseudoraphe are not motile. Motility is usually a response to chemical, physical, or mechanical stimuli such as toxic substances, light changes, or receding tides. Motility has three environmental requirements: oxygen, light, and a suitable substrate. Different diatom species move at different rates. For example, *Hantzschia*, a genus that forms brownish-green patches on sandy beaches at low tide, can move at speeds of 2.7 μm/sec (Palmer, 1973). At high tide *Hantzschia* migrates down among the sand grains to prevent its being swept away (Palmer, 1975). Other genera such as *Navicula* and *Nitzschia* occupy similar niches among the muds of tidal flats. The range of speeds for these genera is 0.2 to 25 μm/sec (Harper, 1977). Such salt marsh and estuarine diatoms may form tubes of sediment and migrate up and down with the tides in an endogenous rhythm. Diatoms inhabiting fresh water also exhibit a vertical migration in the sediment; this endogenous rhythm may persist under constant conditions of dim light or darkness in the laboratory (Harper, 1969; see Harper, 1977, for a thorough discussion of diatom motility).

Mucus secretion is essential for diatom movement, but it is not restricted to motile forms. Mucilage secretions may bind planktonic cells into ribbons (*Fragilaria*), stars (*Asterionella*), or zigzag chains (*Tabellaria*), as shown in Figs. 8-1 and 8-13, thereby increasing both the surface-to-volume ratio and the area for absorption of nu-

FIG. 8-14 Stalked diatoms.
Gomphonema sp. growing as an epiphyte on a filament of *Cladophora* sp. (Chlorophyta) taken from a freshwater limestone stream in Pennsylvania. Note mucilaginous stalk *(st)* that attaches diatom to the green algal filament (250×).

trients. Many sessile diatoms secrete a mucus stalk that fastens them to a hard substrate such as a rock or another organism. Aquatic angiosperms and multicellular algae are often covered with stalked diatoms such as *Gomphonema* (Fig. 8-14). Freshwater species of *Vaucheria*, *Rhizoclonium*, and *Cladophora* also may be completely covered with epiphytic diatoms. *Cocconeis* (Fig. 8-15) often grows on both freshwater and marine organisms and may even cover the undersides of giant blue whales. The pennate diatom *Licmophora* lives symbiotically inside a marine flatworm (Turbellaria).

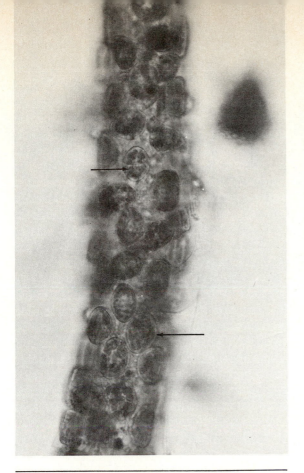

FIG. 8-15 Epiphytic diatoms.
Valve view of *Cocconeis pediculus* cells (arrows) growing as an epiphyte on filaments of the green alga *Cladophora* in a freshwater limestone stream in Pennsylvania. Note how diatom forms a coating over entire surface of the *Cladophora* filament. This species of diatom has a pseudoraphe on its epivalve and a true raphe with central and polar nodules on its hypovalve (250×).

EVOLUTION

The geological record indicates that the Chrysophyta are a relatively new group that evolved during the Mesozoic era. Among the Chrysophyceae only the silicoflagellates have left a good fossil record. They originated during the Cretaceous period when they were prominent members of marine cold-water plankton. Their delicate siliceous skeletons are common fossils in both chalky marine limestone deposits and **diatomaceous earth.** Evidently the silicoflagellates were more numerous in earlier geo-logical times than they are today. Their greatest diversity was during the Tertiary period.

The history of Prymnesiophyceae is best traced through the deposition of calcareous coccoliths in sediments. They are unknown before the Jurassic period, when they were a dominant organism in the warmer marine waters, but they are not as abundant today. Coccoliths occur in quantities in the limestone deposits of the Jurassic and Cretaceous periods and may be found in common blackboard chalk (Lee, 1980).

Diatoms (Bacillariophyceae) are also unknown before the Jurassic period. The fossil record indicates that they evolved rapidly and became the dominant organism in the seas of the Cretaceous period. The earliest genera were centric forms. The motile pennate diatoms evidently evolved later in freshwater environments of the Cretaceous period. Many fossil diatoms associated with diatomaceous earth deposits have been described. A large number of the fossil forms are similar to present-day genera with little or no change, indicating that these organisms can adapt to a wide range of environmental conditions. Fryxell (1983) discusses recent research in diatom evolution.

Well-preserved siliceous fossils of diatoms provide excellent clues to the time at which sediments were deposited. Diatom fossils are used to identify oil-bearing rocks and soils and to give indications of past ecological conditions. An example of this may be found in the study of freshwater sediments (Almer et al, 1978). Since diatom species have apparently changed very little over geological time and since the present environmental conditions required for current species are known, inferences may be made concerning past ecological conditions. Such inferences have been made about temperature, pH, primary productivity, and drainage basin history. At present lake sediment cores are being analyzed for diatom frustules of pH sensitive species in an effort to determine the pH changes that have occurred over time as the precipitation in northeast North America has become increasingly acidic (Whitehead et al,

1981). Analyses of this type are not as easy in the marine environment because oceanic diatoms not only may be transported thousands of miles by ocean currents but also may take many years to settle out of the water column.

Chrysophyta are considered to be related to the brown line of algal evolution because they contain chlorophyll *a* and in most cases *c* (c_1 and c_2), and they lack chlorophyll *b*. The three classes are biochemically related because of similar chlorophyll types, energy storage products, and accessory photosensitive pigments. Chrysophyceae are related to other flagellated organisms, and some investigators consider them related to the protozoa. *Ochromonas* is proposed as a common ancestor of the class (Hibberd, 1976). The relationship of the haptonema-bearing Prymnesiophyceae to the other two classes of this division is not clear, but they are first found in fossil records of the Jurassic period along with the centric diatoms.

ECOLOGY

Many chrysophytes are members of the phytoplankton at some point in their life cycle. They also occur in significant numbers as part of the benthic flora in both marine and freshwater habitats. Some grow attached to rocks, sand grains, mud particles in estuaries, coastal areas, streams, and lakes, or even as epiphytes on plants and epizoics on animals. These algae may be heavily grazed by assorted arthropods and molluscs.

Chrysophyceae are mostly freshwater algae from cold mountain streams, springs, and lakes. They are sensitive to temperature and pH changes and are therefore more common in the cold streams and lakes during spring and fall. They often occur as brown gelatinous or encrusting growths on rocks, twigs, or coarse gravel and may dominate the phytoplankton of acid lakes (Kelso et al, 1982). *Dinobryon* and *Synura* may be prevalent in aquatic blooms. Their metabolic by-products give waters a foul, fishy taste, and thus these organisms may cause

problems in municipal water supplies. Their sensitivity to environmental changes causes them to degenerate shortly after they are removed from their natural habitat. This means that diagnosis of *Dinobryon* or *Synura* populations must be made soon after collection because they do not preserve well.

The silicoflagellates were important primary producers in the Cretaceous seas. They are less common today but still form a significant part of cold-water marine phytoplankton. They are extremely temperature sensitive and cannot tolerate temperature variations greater than a few degrees (Centigrade) from the ambient temperature (Van Valkenburg, 1980).

Many chrysophytes belong to the plankton group known as nannoplankton, with diameters from 10 to 60 μm (see Table 2-2). Small diatoms, coccolithophorids, and silicoflagellates often comprise the majority of nannoplankton members. The small size of these organisms makes both sampling and intensive research difficult. Recent investigations have shown that the nannoplankton are the major primary producers in certain habitats (Malone, 1971; Sheath et al, 1975; Pingree et al, 1976). The nannoplankton cells are not captured by the standard plankton nets with mesh sizes of 0.06 mm (60 μm) that are routinely used (see Table 2-2). Hardy (1965) tabulated 727,000 diatoms and dinoflagellates per cubic meter in the North Atlantic Ocean (and 4,500 zooplankters per cubic meter). These phytoplankton figures, however, are underestimated, because it is now postulated that 90% of the phytoplankton was nannoplankton and therefore missed in Hardy's study. More realistic figures would be 6.5 million phytoplankters per cubic meter.

In the Coral Sea in the southern Pacific Ocean the nannoplankton contributes 90% of the primary productivity. It is clear that the nannoplankton and ultraplankton (0.5 to 10 μm) constitute relatively unknown but very important members of marine ecosystems. Frequently these organisms account for the great majority of total primary productivity.

Prymnesiophyceae are a little-studied group of rather small members of the nannoplankton. The best-known organisms in this class are the coccolithophorids, which are much more common than was once supposed. In warm tropical marine waters where high calcium carbonate levels are found, the coccolithophorids are often the major primary producers. Coccolithophorids make up 45% of the total phytoplankton in the pelagic areas of the middle latitudes of the Atlantic Ocean and from 75% to 95% of the phytoplankton in the Mediterranean Sea at certain times of the year (Dawson, 1966).

It appears that coccolithophorids are more important than diatoms as primary producers in nitrogen-deficient waters. This is usually the case in warmer tropical water. Corner and Davies (1971) demonstrated that coccolithophorids grown at low nitrogen concentrations have faster growth rates than diatoms. At the higher nitrogen levels the coccolithophorids are replaced by diatoms. In agreement with these data, documented changes in phytoplankton species have been observed off the coast of southern California when upwelling increases nitrate levels in the surface waters. One known freshwater coccolithophorid was reported from Ohio (Lackey, 1938).

Coccolithophorids have an interesting history. They were not recognized as algae before the turn of the century. They were originally described in the early 1800s as minute carbonate disks (coccoliths) in Cretaceous carbonate deposits, and they were thought to be of inorganic origin. Later they were discovered in the ocean bottom sediments that were examined during the laying of the first transatlantic cable in 1858. Their true nature as an algal component was not recognized until 40 years later in 1898 (Dawson, 1966). Their importance had been highly underestimated, since the role of these organisms in the energy economy of the oceans has now been well established. For example, the widely distributed *Coccolithus huxleyi* (Fig. 8-5) is an important primary producer in the Sargasso Sea. One of the reasons that the coc-

colithophorids flourish in warm waters may be the ability of the coccoliths to reflect high intensity light, thereby preventing damage to the photosynthetic pigments.

Other Prymnesiales occur as phytoplankton members in marine and brackish waters. *Prymnesium* blooms have caused problems in fish ponds in Israel because of the release of an extracellular toxin that may poison fish. Allen (1969) reviews fish poisoning from *Prymnesium* populations that may reach concentrations of 10^6 to 10^7 cells/l. The toxin forms only in light, and it alters the selective permeability of the fish gills (Hibberd, 1980a).

Phaeocystis may occur in large numbers in the ocean, causing the sea to appear oily and yellow-brown. Herring avoid such large populations during their migrations in the North Atlantic and North Sea. When the herring deviate from their usual migratory paths, catches of these fish are greatly reduced. (Boney, 1970, discusses these blooms in greater detail.)

Bacillariophyceae are the most important group of organisms in the oceans because they rank as the number one primary producers, especially in temperate seas. They fix more carbon dioxide into organic carbon than any other major plant group in the world. The diatom's ability to withstand oceanic pollution has recently been a major area of research. For example, workers are studying what effect the persistent chlorinated hydrocarbon pesticide DDT (1,1,1-trichloro-2,2-bis [p-chlorophenyl] ethane) has on the primary productivity of diatoms, considering that DDT is lipid soluble and that diatoms store considerable quantities of oil. Wurster (1968) demonstrated that DDT in parts per billion reduced photosynthesis rates in both the marine diatom *Skeletonema costatum* and the coccolithophorid *Coccolithus huxleyi*.

Marine and freshwater diatoms undergo regular seasonal increases in numbers called diatom pulses or blooms (see Fig. 2-7). These pulses have been studied for years, and indications are that they depend upon a number of factors: (1) availability of light in wavelengths effective for photosynthesis; (2) quantity and quality of dissolved nutrients, especially phosphates, nitrates, and silica; (3) availability of vitamin B_{12}; and (4) stability of the water column, which is influenced by temperature. Any of these factors may limit the growth of diatoms. Silica is frequently a limiting factor in freshwater, and nitrate is often a limiting factor in warm seas. Rising temperatures in spring cause water column stability (thermal stratification), while declining temperatures in autumn cause upwelling of the nutrient-rich underlying waters, bringing about water column instability, consequent turbulence, and nutrient redistribution (see Chapter 2). These factors all influence the pulses in the spring and fall and cause aquatic blooms in both fresh and salt water.

Densities of diatoms during a pulse have been reported to reach very high numbers—up to 1.2×10^7 cells/l in an English lake (Hutchinson, 1967) and up to 10^6 cells/l in estuaries (Guillard and Kilham, 1977). During favorable conditions cells may divide more than once per day. Diatom blooms may form an oily slick on the water's surface. The spring diatom increase (SDI) in the North Atlantic occurs at the same time as many invertebrate larvae are hatching. For example, *Skeletonema costatum* releases a substance that induces the hatching of the barnacle *Balanus balcinoides* (Russell-Hunter, 1970). Similar SDIs occur in lakes, streams, and rivers. For example, *Navicula, Melosira* (Fig. 8-16), and *Meridion* covered rocks with a brown scum in a Pennsylvania limestone stream in late January, responding to increasing light and adequate nutrients (Bradt, 1974; Bradt and Weiland, 1978). Patrick (1977) discusses various factors associated with blooms of diatoms.

The diversity of Bacillariophyceae species is generally high in pelagic waters where the density of cells is low, but in neritic and coastal waters the diversity is generally low and the density of cells generally high. The oligotrophic waters of the open ocean and infertile lakes usually have high diversity (i.e., numbers of species) and low total numbers, while the eutrophic waters of coast lines and fertile lakes have low diversity and high numbers of individuals. Conditions of stress such as pollution with organic or toxic materials tend to lower the diversity of organisms in an ecosystem. As stressful conditions become intolerable to sensitive species they disappear, and the more tolerant species survive and reproduce, so eventually only a few tolerant species will dominate in a severely stressed ecosystem. (See Pielou, 1975, for an

FIG. 8-16 *Melosira granulata* **cell wall skeleton.** Scanning electron photomicrograph of two cells of the filamentous, freshwater, centric planktonic diatom *M. granulata*. The two cells are joined at the center (arrow). Sphere to the left (double arrow) is a glassy particulate residue from the stack emission of a factory near Chicago. Note spines and pores associated with the cell wall. *Melosira* spp. may be either planktonic or benthic (2000×). (Courtesy Dr. Jon I. Parker, Department of Biology, Lehigh University, Bethlehem, Pa.)

excellent and thorough discussion of the diversity concept.)

Diatoms, because of their individual species requirements, make good water-quality indicators. Some are indicative of specific pH ranges, specific alkalinity ranges, specific temperatures, and eutrophic or oligotrophic conditions. For example, the recovery of Lake Washington, Seattle, from large influxes of sewage was followed by a documented reduction in the number of diatom species associated with eutrophic waters in addition to the reduction in nutrient input (Edmundson, 1970). Genera associated with mild organic enrichment are *Melosira*, *Rhizosolenia*, and *Fragilaria* (Figs. 8-1, 8-11, and 8-13). The diatoms *Frustulia*, *Eunotia*, and *Pinnularia* are often found in acid bogs in association with some Chrysophyceae—*Dinobryon* and *Synura*. Patrick (1977) notes a reduction in the number of species in polluted waters and has used diatom communities as indicators of pollution or stress, clean water, acid water, and other conditions. Cairns et al (1982) discuss a method for automated diatom identification for continuous monitoring of water quality. Such continuous long-term monitoring can detect changes in diatom communities that may not be discerned by short-term studies (Gale et al, 1979).

Diatoms have such specific requirements for growth, and their blooms occur at such regular intervals, that the path of water masses may be traced by diatom species as the water currents move from one part of the ocean to another. Species usually found in the English Channel occurred in the North Sea, leading to speculations about changing patterns of water flow (Reid, 1975). Different diatom species are indicative of various environmental parameters such as salinity, light, temperature, and so on. An expert in phytoplankton can study a plankton sample and obtain much information about the source of the sample. For example, diatoms from subtropical waters have longer spines than those from cooler waters (Fig. 8-12).

Diatoms can form mats on such man-made objects as docks, pilings, and boats. They are the second group in a succession of fouling organisms that may start with bacteria and end with such sessile invertebrates as barnacles and oysters in the marine environment.

Bacillariophyceae are also important members of the soil microbial flora. They often have large populations in salt marsh, forest, and agricultural soils. Cook and Whipple (1982) recount diatoms' role and distribution in salt marsh sediments. Such pennate genera as *Navicula*, *Pinnularia*, and *Nitzschia* are common soil inhabitants and usually produce a mucilaginous secretion that helps the cell retain water. A behavioral mechanism, vertical migration, also protects these organisms from desiccation. They migrate vertically in the soil, depending upon the amount of moisture available. Diatoms have been found at soil depths of 1.5 m (Moore and Carter, 1926). Some soil diatoms are capable of resuming growth after 48 years of dry storage, reinforcing their extraordinary and necessary ability to withstand desiccation (Bristol-Roach, 1919).

ECONOMIC IMPORTANCE

Millions of diatoms from thousands of years of primary production have formed extensive silicious deposits known as diatomaceous earth, **diatomite,** or **kieselgühr.** These deposits were formed in the sediments of shallow saltwater seas or freshwater lakes. The number of diatom frustules originally in the Lompoc, California, deposit is estimated at 10^{24}. It is the largest deposit in the world, 31 k² in area and 300 m deep. Such a large concentration of diatoms $(6 \times 10^6/mm^3)$ indicates a tremendous-amount of primary productivity in the shallow seas where these deposits occurred. Most of this production took place in the Tertiary and Quaternary periods. Geological change has raised these deposits above sea level where the diatomaceous earth may be quarried. When the diatoms died they sank to the bottom, and many silica-containing frustules persisted for millions of years because they are extremely resistant to decay (Werner, 1977a). Large deposits of diatom skeletons built up as the organic material of the cells was oxidized, leaving only the insoluble silica frustules. Not all extensive diatomaceous earth deposits are of marine origin; important freshwater deposits are quarried in the United Kingdom, Maryland, and Nevada.

The majority (66% in 1980) of diatomaceous earth production is used for industrial filtration of such liquids as wine, beer, and antibiotics, and sugar refinery wastes. Other uses for the chemically inert diatomite include boiler insulation, fillers in various products, and mild abrasives in such products as toothpaste and metal polishes. The United States is the major world supplier of diatomaceous earth, having mined over 227,000 metric tons in 1951 and 624,000 metric tons in 1980 (Meisinger, 1981).

Individual species of diatoms can be identified from diatomite samples (Fig. 8-17). Because many diatoms have not changed since Tertiary times, identification of genera is not difficult. Silicoflagellate fossils are also often found in the diatomite deposits. The diatomite deposits may contain up to 88% silicon dioxide and are entirely biogenic in origin.

Certain genera of diatoms are used as markers for geological deposits to identify potential oil- and gas-bearing strata. Diatoms have also been involved in the production of some oil deposits, because they produce large amounts of intracellular oils. Therefore, large populations of diatoms have contributed to some major oil deposits.

The delicate markings on the diatom frustules were formerly used to test the optical resolving power of microscope lenses. The distances between the pores and ridges are constant from species to species.

Several genera of diatoms can cause problems in water supply plants because they clog the sand filters or cause unpleasant "fishy" tastes. Genera that may cause problems include *Stephanodiscus*, *Asterionella*, *Fragilaria*, *Melosira*, and *Cyclotella*. Objectionable tastes may be removed with activated carbon filtration, which is used in many municipal water supply systems.

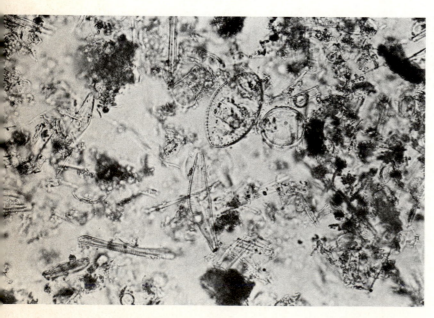

FIG. 8-17 Diatomaceous earth.
Light microscope view of diatomaceous earth mounted in water. Individual centric and pennate fossil diatoms are clearly visible in the strew (250×).

Diatom secretions may cause various problems. For example, mucilagenous secretions from the marine diatom *Coscinodiscus nobilis* clogged fishermen's nets in the English Channel in 1977. The gray slime was so abundant that trawling was restricted in some areas. Previously this organism had been reported only from the Java Sea and the Indian and Pacific Oceans; its presence in bloom proportions in the English Channel remains unexplained (Boalch and Harbour, 1977).

When water containing the coliform bacterium *Escherichia coli* is passed over a layer of *Nitzschia palea* that is actively photosynthesizing, the *E. coli* numbers are substantially reduced, indicating a type of antibacterial activity. Evidently this antibiotic activity is effective only when the alga is photosynthesizing. *N. palea* is a common inhabitant of sand filters in water supply plants, and its antibiotic action may help to reduce bacterial populations.

The most important economic contribution of diatoms is the very large amount of primary production, both freshwater and marine, for which they are directly responsible. They convert solar energy to chemical food energy, which is then utilized by animals. Diatoms are the major primary producers in both the fresh and salt waters of northern temperate zones, and the coccolithophorids may be the dominant phytoplankters in subtropical or tropical waters. Chrysophyta, therefore, contribute a very significant amount of primary productivity to the biosphere.

SUMMARY

The Chrysophyta are composed mostly of unicellular organisms whose nutrition is autotrophic, auxotrophic, or heterotrophic. The division contains three rather diverse classes that are unified on the basis of the presence of chlorophyll *a* and *c*, fucoxanthin, chrysolaminarin, and, to a lesser extent, the presence of silicates in their cell walls. The classes are: (1) Chrysophyceae, including organisms known as silicoflagellates; (2) Prymnesiophyceae (= Haptophyceae) composed largely of algae containing a haptonema, many of which are known as coccolithophorids; and (3) Bacillariophyceae, which are the diatoms.

The structure of individual cells varies from one class to another. Many Chrysophyceae produce special cysts called statospores that deposit silicon in their thick cell walls. Some members of this class, the silicoflagellates, produce delicate siliceous skeletons. The crysophycean cells are heterokont biflagellate types with one longer tinsel-type flagellum and one shorter whiplash-type flagellum. Loricas are formed in some genera.

The cells of Prymnesiophyceae all possess two isokont whiplash-type flagella plus an additional appendage, the haptonema. Many genera produce calcified disks called coccoliths in association with their cell walls.

Bacillariophyceae (diatoms) are not vegetatively flagellated. Some pennate forms are nevertheless motile. Their cell walls are composed of silicates and contain no cellulose. The surface of the walls is distinctively ornamented, providing an excellent tool for identification.

Reproduction in Chrysophyta is usually asexual by mitotic cell division. However, sexual processes involving genetic mixing are known in all three classes, although it is best known in the diatoms. Diatoms have diplontic life cycles, and union of isogametes or anisogametes results in formation of a special zygote known as an auxospore (growth spore).

The three classes in this division are all important members of the phytoplankton, especially the **microplankton** and nannoplankton. The diatoms are the most important primary producers on earth. In some of the warmer seas the diatoms are replaced by the coccolithophorids as the most important primary producers. Humans must recognize and carefully guard the fragile existence of these unicellular organisms.

Chrysophytes are a group of organisms that have evolved relatively recently. The earliest fossils date back only to the Jurassic period (191 to 205 million years ago). They have changed very little since their appearance, indicating that they are well adapted to their environment.

Economically the diatoms have provided an entire industry built around their extensive deposits in certain regions of the world. Diatomaceous earth has been found to have many commercial and industrial uses. Perhaps the most important economic service of this division is the role its members play as primary producers in aquatic habitats, for it is there that they form the base of many food chains.

SELECTED REFERENCES

Cairns, J., Jr., S.P. Almeida, and H. Fujii. 1982. Automated identification of diatoms. BioScience 32(2):98-102.

Drum, R.W. and J.T. Hopkins. 1966. Diatom locomotion: an explanation. Protoplasma 62:1-33.

Edmundson, W.T. 1970. Phosphorous, nitrogen and algae in Lake Washington after diversion of sewage. Science 169:690-691.

Fryxell, G.A. 1983. New evolutionary patterns in diatoms. BioScience 33(2):92-98.

Hibberd, D.J. 1980a. Prymnesiophytes (= Haptophytes). In Phytoflagellates, E.R. Cox (ed.), pp. 273-317. Elsevier North-Holland, New York.

Paasche, E., and I. Østergren. 1980. The annual cycle of plankton diatom growth and silica production in the inner Oslofjord. Limnol. Oceanogr. 25(3):481-494.

Patrick, R. 1977. Ecology of freshwater diatoms—diatom communities. In The biology of diatoms. D. Werner (ed.), p. 284-332. University of California Press, Berkeley.

Patrick, R., and C.W. Reimer. 1966. The diatoms of the United States exclusive of Alaska and Hawaii. Vol. 1. Monogr. of The Acad. of Natural Sci. of Philadelphia. No. 13. 688 pp.

Patrick, R. and C.W. Reimer. 1975. The diatoms of the United States exclusive of Alaska and Hawaii. Vol. 2. Part 1. Monogr. of The Acad. of Natural Sci. of Philadelphia. No. 13. 213 pp.

Van Valkenburg, S.D. 1980. Silicoflagellates. In Phytoflagellates. E.R. Cox (ed.), pp. 335-350. Elsevier North-Holland, New York.

Werner, D. (ed.) 1977b. The biology of diatoms. University of California Press, Berkeley. 498 pp.

9

CLASSIFICATION

Division: Pyrrhophyta (fire algae): 125 genera, 1,100 species; mostly bi-flagellate, motile unicells; important marine and freshwater phytoplankton; autotrophic and heterotrophic

Class 1*: Dinophyceae (dinoflagellates): motile biflagellate unicells with one circumferential flagellum and one trailing flagellum; mesokaryotic nuclei; trichocysts; the cause of some "red tides"

Selected Orders:

ORDER 1. Gymnodiniales: free-living, motile, unicells covered with delicate thecal plates; *Gymnodinium, Woloszynskia, Ptychodiscus*

ORDER 2. Noctilucales: naked free-living, motile unicells without chlorophyll; heterotrophic; cells up to 2 mm in diameter; frequent bioluminescence; *Noctiluca*

ORDER 3. Peridiniales: the armored dinoflagellates; free-living, motile unicells, covered with many thecal plates; some without chlorophyll; *Ceratium, Glenodinium, Gonyaulax, Nematodinium, Peridinium*

Class 2: Cryptophyceae (cryptomonads): asymmetric biflagellate, flattened unicells with a periplast; marine and freshwater, especially in mud; some with ejectosomes; *Chroomonas, Cryptomonas*

PYRRHOPHYTA

*At least four other smaller classes have been described, with as many as twelve orders. Only the most common classes and orders are listed here.

175

Pyrrhophyta are primarily unicellular marine phytoplankton, although freshwater forms are common in lakes and ponds. They are commonly referred to as the "fire algae" because of the frequent bioluminescence of some species. The division includes those unicellular algae known as the **dinoflagellates** that are often encountered as members of the phytoplankton community in warm, tropical marine waters. Also included are the **cryptomonads,** a small, poorly understood, little-studied group of marine and freshwater unicellular organisms that are common to muddy environments. Pyrrhophyta, together with Chrysophyta, comprise most of the phytoplanktonic biomass in the oceans. The dinoflagellates are second only to the diatoms as primary producers in the marine habitat. As a general rule the diatoms dominate the cooler waters, while the dinoflagellates prevail in warmer waters.

The cells of dinoflagellates have complex cell coverings that may include a cellulosic component forming distinctive plates. Some cells lack specialized cell coverings and appear as naked protoplasts lacking a cell wall. Most species are solitary unicellular forms, but palmelloid, colonial, and filamentous species have been described. Each dinoflagellate cell possesses two heterokont tinsel-type flagella, one fitting in the transverse circumferential groove, the **girdle,** while the other shorter flagellum trails posteriorly through the longitudinal groove, the **sulcus** (Fig. 9-1). Like other members of the brown line of algal evolution, the cells contain chlorophyll a and c (c_1 and c_2). Various carotenes and xanthophylls, some of which are unique to pyrrhophytes, are also present (e.g., **peridinin** and **dinoxanthin**).

Energy storage products in the pyrrhophytes include some complex oils and starch. The starch is the same chemical form of carbohydrate associated with Chlorophyta and vascular plants, although

these groups are not closely related. Most pyrrhophytes are autotrophic, although colorless
heterotrophic forms are known. There is a high
incidence of auxotrophy among the various species, many of them requiring exogenous vitamin
B_{12}. Some pyrrhophytes, referred to as **zooxanthellae,*** have developed a symbiotic existence with various animals.

---

*The term zooxanthellae is used to describe those symbiotic
algae having a red, rust, yellow, or brown color. They are
usually members of Pyrrhophyta or Chrysophyta. The term
zoochlorellae is used to describe green symbiotic cells usually belonging to the Chlorophyta (see Chapter 4).

CELL STRUCTURE

The cells of Pyrrhophyta are difficult to observe because they are not easily preserved;
some have a tendency to disintegrate or become
greatly distorted when they are removed from
their natural habitat. Some genera lack cell
coverings and are referred to as the naked or
unarmored dinoflagellates (*Noctiluca*, Fig. 9-1),
while others possess distinctive cell coverings
(**thecae**) composed of species-specific cellulosic
thecal plates (*Peridinium* and *Gonyaulax*, Fig.
9-1). Some genera produce diagnostic spines associated with the thecal plates (*Ceratium*, Figs.
9-7 and 9-8). When the thecal plates are large,

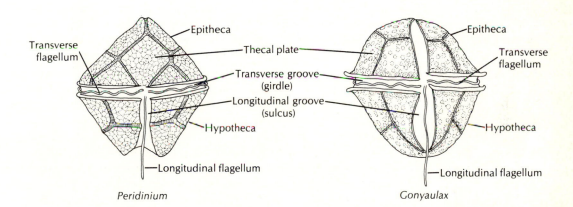

Peridinium   *Gonyaulax*

FIG. 9-1 Dinoflagellates.
Peridinium and *Gonyaulax* are armored dinoflagellates with characteristic thecal plates forming both
the epitheca and hypotheca. Their order, Peridiniales,
is the most common of the fire algae. *Noctiluca* is a
naked, colorless, heterotrophic organism. Its cells are
very large, with most of the volume composed of vacuoles. The unique tentacle is a food-procuring device. Two flagella develop in gametic forms of *Noctiluca* (Zingmark, 1970), and a single emergent flagellum may sometimes be seen near the base of the
tentacle.

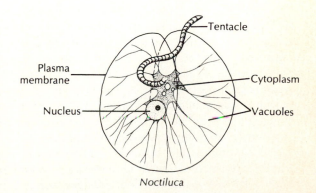

Noctiluca

as in members of Peridiniales, the cells are described as being armored because of the resemblance of the plates to the mail of knights' garments or to armored vehicles (note the parallel between the term dinosaur, or "armored reptile," and dinoflagellate, or "armored flagellate"). All dinoflagellates produce thecae to some extent, although these may not be readily visible in the light microscope. Thecae are formed through a combination of two thecal membranes between which a series of thecal vesicles is located. These thecal vesicles are the site of cellulosic plate formation (Steidinger and Cox, 1980). In some dinoflagellates the thecal plates are arranged into an epitheca and a hypotheca (the terms **epicone** and **hypocone** are used by some investigators) much in the same manner as that found in diatoms. The extent of cellulose deposition in thecal vesicles determines the size of the organism and provides the basis for most dinoflagellate taxonomy.

Each pyrrhophyte cell contains two tinsel-type flagella that emerge through an **equatorial pore** in the thecae. One flagellum extends horizontally around the cell in a thecal groove, the girdle. The shorter flagellum extends posteriorly in another thecal groove, the sulcus (Fig. 9-1). The girdle flagellum is flattened and is helically arranged. Its undulating movements propel the cell forward. The sulcus flagellum acts as a rudder, steering the cell and changing its direction.

The most striking cytological structure of a dinoflagellate cell is its mesokaryotic nucleus, the organization of which differs significantly from the eukaryotic nucleus of other protistan cells. The nucleus is membrane bound like eukaryotes, but there the similarity ceases. The mesokaryotic nucleus is relatively large in proportion to the cytoplasm, and it takes a polymorphic shape in many dinoflagellate species (Fig. 9-2). Within the nucleus the chromosomes apparently exist in a permanently condensed form not unlike that observed in many prokaryotic organisms. Such chromosomes have been observed resembling letter shapes such as

FIG. 9-2 Ultrastructure of *Ptychodiscus brevis* (= *Gymnodinium breve*).
Near-median longitudinal section of the dinoflagellate *P. brevis,* showing thickness of the epitheca and hypotheca and some of the internal organization, including the chloroplast *(c)*, pyrenoid *(py)* vesicles, and lipid globules. Note appearance of the mesokaryotic nucleus, containing many permanently condensed chromosomes *(ch)*. Also note position of the transverse groove or girdle *(gd)*, containing the transverse flagellum *(tf)*. (Courtesy K.A. Steidinger, Marine Research Laboratory, Florida Department of Natural Resources, St. Petersburg, Fla.)

O, U, V, or Y. They lack **centromeres** (which are the point of spindle attachment in eukaryotic organisms). In fact, dinoflagellates lack the spindle apparatus and microtubules associated with spindle formation. The chromosomes move apart during mitosis by attaching to the persistent nuclear membranes. The dimensions of individual chromosomes are 4 to 10 times smaller than those of most eukaryotes. Even the chemical composition is different, lacking the usual complement of histone (basic protein) associated with chromosomes of most eukaryotic cells (Rizzo and Nooden, 1974; Rizzo and Cox, 1977). A nonhistone chromosomal protein portion has been isolated from some dinoflagellates, and in this respect the chromosomes are more like those of other eukaryotes.

Dinoflagellate cells contain a small number of oval chloroplasts, each with a double membrane of chloroplast envelope and a single membrane of chloroplast endoplasmic reticulum (CER), (see Fig. 1-7). Unlike other plastids in the brown line of evolution, the CER is not continuous with the nuclear membrane. Within each chloroplast the thylakoids are arranged in stacks of three, and scattered among packets of thylakoids are lipid deposits and some strands of DNA. Some chloroplasts contain a central pyrenoid, but most lack this structure. If pyrenoids are present, they appear as granular, proteinaceous structures of variable form. They

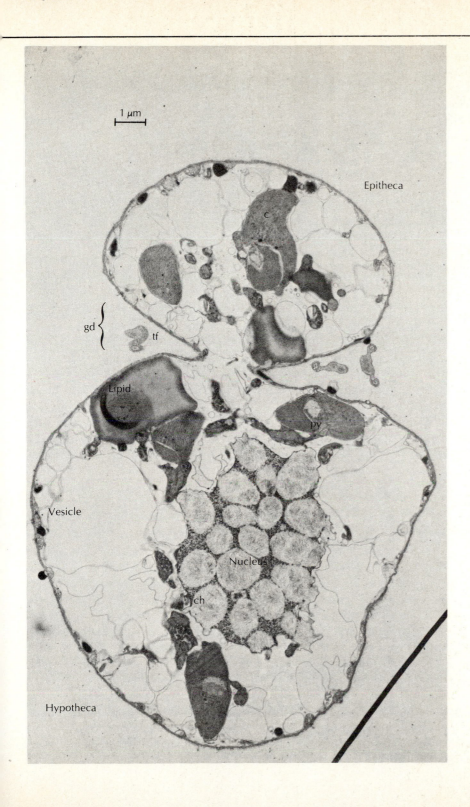

apparently produce the energy storage products, starch and oil products.

Pigments contained within chloroplasts of dinoflagellates include chlorophyll *a* and two types of chlorophyll *c* (*c*₁ and *c*₂). Chlorophyll c_2 is found in all species but chlorophyll c_1 in only a few species. Other pigments characteristic of Pyrrhophyta include peridinin and dinoxanthin, both of which are xanthophylls unique to dinoflagellates. They are, however, not found in all species. Four other carotenoid pigments have been isolated from dinoflagellates: β-carotene, diatoxanthin, diadinoxanthin, and fucoxanthin. The latter pigment is characteristic of algae in the brown line of algal evolution. Steidinger and Cox (1980) have developed an evolutionary scheme based upon the presence or absence of peridinin and fucoxanthin in the cells of dinoflagellates. Peridinin has attracted some attention because of its close chemical association with chlorophyll and protein, forming a peridinin-chlorophyll-protein complex within chloroplasts. Prezelin (1976) has shown that as light intensity decreases, the ratio of peridinin to chlorophyll *a* increases, indicating that peridinin is active in light absorption in dim light. Such alteration in light-absorbing pigments is known as chromatic or intensity adaptation, and it has significant adaptive value to organisms existing at various water depths where different wavelengths of light are available for photosynthesis (see Fig. 2-5 and Chapter 11, p. 213).

Although most dinoflagellates lack an eyespot (stigma), it does occur in some genera, located beneath the sulcus at a point near the emergence of the two flagella. The ultrastructure of the eyespot is variable within the division Pyrrhophyta. In some genera (*Woloszynskia*) it is not part of a chloroplast but is simply a cluster of carotenoid granules lying free in the cytoplasm. In *Glenodinium* the carotenoid pigments are located within a modified chloroplast; in *Peridinium* the eyespot forms a part of a chloroplast; and in *Nematodinium* the structure has evolved into an architectural complex involving a light-gathering lens (Dodge, 1973) that resembles the ocelli (eyes) found in some invertebrates.

Pyrrhophyte cells also contain organelles that may be ejected from the cell upon appropriate stimulation such as mechanical agitation, heat, or a change in the chemical environment (Dodge, 1973). In dinoflagellates these projectiles are membrane-bound, proteinaceous, threadlike structures called **trichocysts.** They resemble the trichocysts found in the chloromonads (Xanthophyta) and many protozoans and act like an attached rocket when they are ejected from the cell surface. However, they differ morphologically from these trichocysts. Trichocysts are formed by the Golgi apparatus, and they project through small pores in the surface of thecal plates (Fig. 9-1). Some cells contain 100 or more trichocysts. Their precise function is not fully understood; it has been hypothesized that they act to quickly move the cell away from areas of danger through their jetlike reaction. They may also be a predatory device or even an attachment organelle. Good scientific evidence for their function is still lacking. A few dinoflagellates contain more complex projectiles called **nematocysts,** which resemble the much larger structures of the same name found in animals (Coelenterata).

Some investigators group Cryptophyceae (cryptomonads) in a separate division, Cryptophyta, because they are significantly different cytologically from the dinoflagellates. They are included here for convenience as a class of Pyrrhophyta. The cryptomonads are a relatively small group of poorly studied organisms. (For a more comprehensive review, the reader is referred to Gantt, 1980b.) Their cells are dorsoventrally flattened with two subapically inserted heterokont flagella of the tinsel type (Fig. 9-3). The cell covering consists of a thickened **periplast** covered by a plasma membrane. It lacks cellulose and the thecal plates found in some dinoflagellates. Contained within the periplast are a series of structurally complex defensive structures, the **ejectosomes,** which are needlelike projectiles similar to but structurally distinct

FIG. 9-3 Ultrastructure of *Cryptomonas* **sp.**
Internal structure of *Cryptomonas* shows single, two-lobed chloroplast encased within a double membrane chloroplast endoplasmic reticulum (CER) that also encases the nucleus. Within the CER is another double membrane, the chloroplast envelope, which contains a single, large, centrally located pyrenoid and many thylakoids arranged in stacks of two. Starch grains are located between the CER and the chloroplast envelope. The cytoplasm also contains lipid bodies, mitochondria, ribosomes, and specialized defense structures, the ejectosomes. (Adapted with permission from Lee, Robert E. 1980. Phycology. Copyright, Cambridge University Press, New York.)

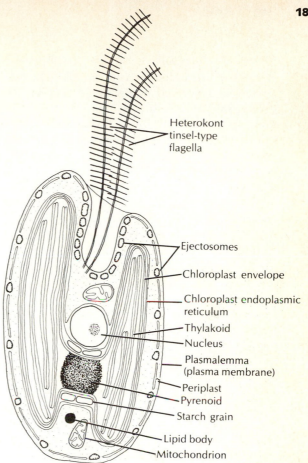

Heterokont tinsel-type flagella

Ejectosomes
Chloroplast envelope
Chloroplast endoplasmic reticulum
Thylakoid
Nucleus
Plasmalemma (plasma membrane)
Periplast
Pyrenoid
Starch grain
Lipid body
Mitochondrion

from the trichocysts of dinoflagellates (Hovasse et al, 1967).

Within the cells of cryptomonads is a single bilobed chloroplast that has an ultrastructure different in many respects from those in other algae (compare Fig. 9-3 with Fig. 1-7). The thylakoids are arranged in groups of two and appear to represent a transition between Cyanobacteria and the algae from the brown line of evolution. The CER lacks attached ribosomes and its outer membrane surrounds the nucleus. Each chloroplast contains a single large pyrenoid located in the isthmus connecting the two lobes of the plastid. Associated with the chloroplasts are chlorophyll *a* and *c*, various xanthophylls and carotenes, and, in some genera, unique photoreceptive pigments, the phycobilins, which include phycoerythrin (red) and phycocyanin (blue-green). These pigments are similar to the phycobiliproteins found in Cyanobacteria and Rhodophyta, although their absorption spectra and molecular size are different. With a complement of pigments like these, the color of cryptomonad cells may vary from red to blue to brown to yellow or green, depending upon the dominant pigment. Some are even colorless and heterotrophic. The ultrastructure of *Cryptomonas* is discussed more fully by Lucas (1970).

REPRODUCTION AND GROWTH

Reproduction among Pyrrhophyta is largely a matter of vegetative cell division, since the organisms are mostly unicellular. Perhaps because the structure and chemistry of the chromosomes in mesokaryotic nuclei varies so greatly from those of eukaryotic nuclei, the manner by which mitosis is accomplished differs significantly (Kubai and Ris, 1969). The usual textbook description of the mitotic cycle does not apply to the mesokaryotic nucleus: There is no nuclear membrane breakdown and no spindle apparatus is formed; there are no centrioles; the nucleolus and nuclear membrane are persistent throughout nuclear division. Individual chromosomes split longitudinally and attach to the nuclear membrane, which divides by a process called "furrowing" in a manner similar to cytoplasmic division in many eukaryotic cells. Sister chromosomes are separated by division of the nuclear membrane (to which they are attached) in a manner similar to bacterial fission where DNA attaches directly to the plasma membrane. The product of such nuclear division is a binucleate cell with both daughter nuclei located within a common cytoplasm. Cytoplasmic division (cytokinesis) usually follows shortly after nuclear division (mitosis) is accomplished, and two new daughter cells are thereby produced. In some species the cytoplasm fails to divide, forming binucleate cells (e.g., *Peridinium* spp.).

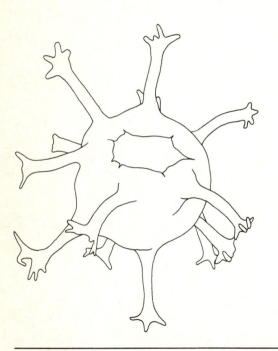

FIG. 9-4 Fossil dinoflagellate cyst.
Fossil cyst of a dinoflagellate *(Oligosphaeridium complex)* from the Cretaceous period. Projections are tubular in nature, with one projection for each thecal plate in the adult vegetative cell. Fossil cysts like this one are called "hystricospheres." (Adapted with permission from Sarjeant, W.A.S. 1974. Fossil and living dinoflagellates. Copyright Academic Press, Inc. [London] Ltd.)

Sexual reproduction has been observed in some genera (Pfeister, 1975; Walker and Steidinger, 1979; Spector et al, 1981). Both isogamous and anisogamous species are known. The life cycles are haplontic (with the exception of *Noctiluca*), with meiosis taking place upon germination of thick-walled resting zygotic cysts called **hypnospores** (Fig. 9-6). Fossil hypnospores often have elongate spines and are called **hystrichospheres** (Fig. 9-4). Sexuality may be induced in dinoflagellates by nitrogen deficiency in the environment, causing isogametes to develop within the parental theca. Fusion of isogametes produces a binucleate motile zygote (**planozygote**). Nuclear fusion occurs at a later time after the motile zygote has settled, lost its motility, and secreted a thick cell wall forming a **resting cyst** (Walker and Steidinger, 1979).

As indicated previously, an exceptional life cycle is found in the genus *Noctiluca* (Fig. 9-1), the only known dinoflagellate to have a diplontic life cycle. *Noctiluca* is peculiar in other respects also. Its nuclear structure more closely resembles other eukaryotes than other dinoflagellates, which have mesokaryotic nuclei. Its chromosomes are not permanently condensed as in other dinoflagellates. The cells of *Noctiluca* are very large (up to 2 mm in diameter) when compared to other dinoflagellates. Finally, bioluminescence is more frequent in *Noctiluca* than in other dinoflagellates.

No sexual reproduction has been reported among the cryptomonads.

PHYSIOLOGY

Nutrition in the dinoflagellates may be autotrophic or heterotrophic; some species combine both nutritional modes. Many dinoflagellates are facultative heterotrophs capable of utilizing available organic molecules. For example, *Ceratium hirundinella* is primarily photosynthetic but may supplement its feeding phagotrophically when necessary. Most photosynthetic species require B_{12} (cobalamin), while heterotrophic species have a wide range of requirements for organic nutrients. Heterotrophic species have been found living deep in the oceanic water column, where they exist saprophytically on organic debris without the need of solar energy. In this division food reserves are stored primarily as starch and occasionally as oils.

Dinoflagellates may parasitize annelids, copepods, and fish. When living as **ectoparasites** or **endoparasites** these organisms do not resemble the free-living dinoflagellates. They only reveal their affinities to Pyrrhophyta when they produce motile cells which closely resemble free-living dinoflagellates.

Symbiosis is a specialized life-style found among many algae, including the dinoflagellates and cryptomonads. Symbiotic cells may grow in and among the cells of many invertebrates. These cells, zooxanthellae, may provide up to 69% of the carbon needs of the invertebrates (Muscatine et al, 1981). Invertebrates that have symbiotic zooxanthellae include various proto-

zoans, sponges, jellyfish, sea anemones, corals, flatworms, snails, clams, and tunicates. Zooxanthellae are nonmotile in the vegetative state, but the reproductive cells assume a typical motile dinoflagellate form that morphologically resembles *Ptychodiscus* (Figs. 9-2 and 9-5).

In a classic example of symbiosis, zooxanthellae (up to 30,000 algae/mm^3 of coral) are found lining the internal cavity of reef-building corals (Coelenterata). The coral provides the alga with carbon dioxide, shade, protection, and waste products containing the vital plant nutrients nitrogen and phosphorus. The alga in turn provides the coral with oxygen, waste removal, and carbohydrates such as the alcohol glycerol. Studies with labeled carbon indicate that up to 60% of the carbon fixed by the alga is released in the surrounding medium and utilized by the coral (Muscatine, 1967). The release of high levels of glycerol by the zooxanthellae takes place only in the presence of

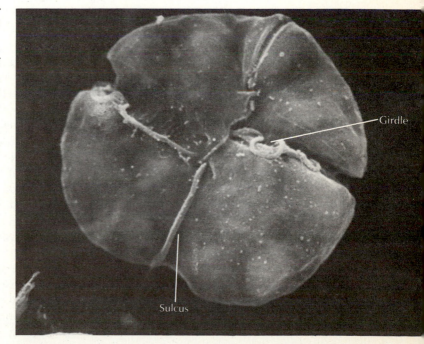

FIG. 9-5 Scanning electron photomicrograph of *Ptychodiscus brevis* (= *Gymnodinum breve*). This is the dinoflagellate that has been responsible for red tides on the west coast of Florida and other areas in southeastern United States. Note transverse groove (*girdle*) with its flagellum and longitudinal groove (*sulcus*) with its flagellum (2700×). (Courtesy L.S. Tester, Marine Research Laboratory, Florida Department of Natural Resources, St. Petersburg, Fla.)

Girdle

Sulcus

the host coral tissues (Von Holt and Von Holt, 1968). There has been considerable debate about the extent of interdependence of coral and zooxanthellae. There is, however, agreement that coral grows up to 10 times faster when zooxanthellae are present. The carbohydrates provided by the zooxanthellae are extremely important to the nutrition of coral reefs because they provide essential nutrients in the nutrient-poor tropical seas where coral reefs abound. The nutrient recycling that occurs between corals and the symbiotic algae is an important factor in the extremely high primary productivity of coral reefs (see Chapter 2). The animal coral is heterotrophic, but the reef as a whole appears to be autotrophic. The reef's growth depends on photosynthesis for the deposition of calcium carbonate, so coral reefs are restricted to shallow waters where adequate light can penetrate.

Twenty species of Pyrrhophyta can produce powerful toxins that may immobilize or kill invertebrates and vertebrates. Large populations (up to 8×10^6 cells/l) of dinoflagellates are called "red tides" because the water may appear reddish as a result of dominant carotenoid pigments. The toxins produced in these blooms not only cause massive fish kills and shore bird and mammal mortality but also can cause shellfish contamination. One toxin, **saxitoxin (STX),** has been isolated from both the Alaskan butter clam, *Saxidomus giganteus,* and California mussels, *Mytilus californianus.* These shellfish concentrate the toxin in their tissues while feeding on *Gonyaulax catenella* off the North American west coast. Saxitoxin is an endotoxin because it is produced within the cell and not released into the environment until the cell is crushed or destroyed. Saxitoxin is 50 times as potent as curare and is nontoxic to the shellfish but causes **paralytic shellfish poisoning (PSP)** in birds and mammals. A 20 g mouse given 0.18 mg of saxitoxin will die in 20 minutes (Trainor, 1978).

Another species of dinoflagellate, *Gonyaulax excavata* (= *G. tamarensis*) (Fig. 9-6), causes PSP off the New England and eastern Canadian coasts. Evidently *G. excavata* produces several toxins, one of which is saxitoxin (Schantz et al, 1975). Transformation of the toxins in *G. catenella* may take place with saxitoxin being the final product (Shimizu and Yoshioka, 1981). *Ptychodiscus brevis* (= *Gymnodinium breve;* see Fig. 9-5) is prevalent off the west coast of Florida and produces an endotoxin that has caused massive fish kills. The toxin has been analyzed and found to be less potent than saxitoxin, although it contains five different toxic components (Trieff et al, 1975).

The toxin metabolically produced by *Ptychodiscus brevis* (= *Gymnodium breve*) is readily liberated into the surrounding environment when the fragile cell is broken. The toxin can be isolated from seawater even when intact organisms are not present. In contrast, *Gonyaulax* cells are heavily armored and more robust than *Ptychodiscus* cells and the toxin is not readily released into the environment. Toxins from *Gonyaulax* have been found associated only with organisms or particulate matter and not in the surrounding medium (Twarog and Gilfillan, 1975). The toxins produced by these dinoflagellates are primarily **neurotoxins;** they prevent normal nerve impulse transmission by the inhibition of the passage of sodium ions into a nerve or muscle cell (Schantz et al, 1975). Some investigators have tried to find a beneficial use for the toxin. Clemons et al (1980) ground up the toxin-containing *Gonyaulax* and tested its insecticidal activity against the German cockroach.

Several genera of marine dinoflagellates exhibit **bioluminescence,** the biological generation of light without the production of heat. Two types of bioluminescence can be seen at night: sudden flashes when water is disturbed, as in boat wakes and breaking waves, or a persistent glow from patches of water. *Noctiluca* ("night light"), *Gonyaulax, Ptychodiscus,* and *Peridinium* (Figs. 9-1 and 9-2) all have bioluminescent species. Not all individuals in a species population in the same area will bioluminesce. Bioluminescence is prevalent in coastal and upwelling areas with high primary productivity

(Tett and Kelly, 1973). The generation of light requires energy (ATP), oxygen, the substrate **luciferin,** and the enzyme **luciferase.** Energy requirements for producing light are not high for dinoflagellates, because light production requires only 0.1% to 1.0% of the total energy budget (Schmidt et al, 1978). According to evolutionary theory, a mechanism persists only if it conveys a selective advantage to its possessor, and many investigators wonder what advantage bioluminescence conveys. Its function in dinoflagellates remains an enigma. Nealson (1981) discusses bioluminescence in detail.

As in many algae previously discussed, the Pyrrhophyta exhibit a circadian rhythm for several physiological functions. Such **endogenous rhythms** are seen in bioluminescence, vertical migration in the water column, photosynthesis rates, and cell division frequency. These rhythms will persist in the laboratory for several days when dinoflagellates are placed under constant environmental conditions. Bioluminescence peaks about midnight in those organisms that glow constantly at night. Vertical migration

during the day enables the organism to stay in that region of the water column where environmental conditions are optimal for photosynthesis. Photosynthetic rates in a *Gonyaulax* species peak at noon, while maximum cell division rates occur at dawn (McMurray and Hastings, 1972).

Under optimal conditions in the environment, dinoflagellate cell division can exceed one division per day and large populations can build up rapidly. The dinoflagellates, however, cannot reproduce as rapidly as the diatoms or coccolithophorids, which are usually their major competitors. Some dinoflagellates can assimilate nitrate in the dark giving them an advantage over the diatoms when competing for nutrients (Harrison, 1976).

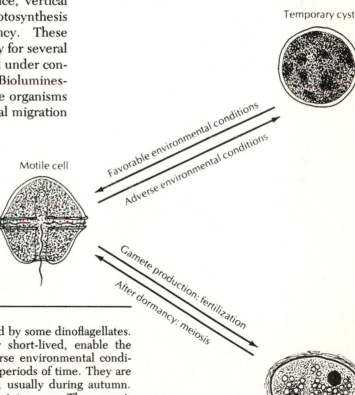

FIG. 9-6 *Gonyaulax excavata* **life history.**
Two types of cysts that may be produced by some dinoflagellates. Temporary cysts, which are relatively short-lived, enable the organism to withstand temporary adverse environmental conditions. Resting cysts may last for longer periods of time. They are formed by sexual fusion of isogametes, usually during autumn. Resting cysts provide a mechanism to winter over. They germinate in the spring by meiotic cell divisions, producing haploid, motile, vegetative cells. (Adapted with permission from Yentsch, C.M., C.M. Lewis, and C.S. Yentsch. 1980. Biological resting in the dinoflagellate *Gonyaulax excavata*. BioScience **30**:251-254. Copyright © 1980, American Institute of Biological Sciences.)

Many marine and freshwater dinoflagellate genera can survive unfavorable environmental conditions by producing cysts. Two types of cysts (Fig. 9-6) are produced in response to stress: **temporary cysts** and **resting cysts.** *Gonyaulax excavata,* for example, responds to stressful conditions by shedding its thecae, becoming nonmotile, and forming a temporary cyst. Such temporary cysts occur regularly in laboratory cultures; they can be induced by unfavorable conditions of light, temperature, or nutrient concentrations. When environmental conditions become more favorable, a temporary cyst will rapidly regenerate its flagella and thecae and resume its vegetative form (Schmitter, 1979).

Resting cysts are formed in response to changing environmental conditions such as decreasing light or nutrients that may accompany changing seasons. In temperate areas in autumn when temperatures drop below 5° C, no motile dinoflagellates can be found. Reorganization of the contents of the motile cell usually precedes cyst formation. The cyst is formed and drops to the bottom. In many dinoflagellates resting cysts have been found to be zygotes, the result of the fusion of two isogametes. The resting cysts of *Gonyaulax* remain dormant from 1 to 6 months (Steidinger and Haddad, 1981). Encysting dinoflagellates secrete heavy cyst walls that may be calcareous or siliceous.

In toxin-producing dinoflagellates the toxin content of the cysts is higher in the autumn than in the spring, indicating that a significant amount of metabolic activity occurs in resting cysts.

Most fossil forms of dinoflagellates that have been found are in the encysted state. Two types of cysts are found: a cyst that resembles the shape and form of the vegetative cell and a spherical cyst that is covered with tubular projections and is morphologically distinct from the vegetative cell (Fig. 9-4). Originally these latter types of cysts were thought to be a separate group of organisms, but their affinity to the dinoflagellates was revealed when cysts from modern sediments excysted (hatched) to become motile dinoflagellates (Yentsch et al, 1980). Fossil cysts with projections, referred to as "hystricospheres" by paleontologists, are reasonably abundant in the fossil record.

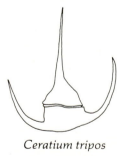

Ceratium tripos

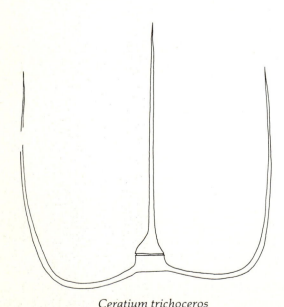

Ceratium trichoceros

FIG. 9-7 *Ceratium trichoceros* **and** *C. tripos.*
Two species of the genus *Ceratium* that are common in the phytoplankton of certain areas of the ocean. *C. trichoceros,* with its long, attenuated horns, usually lives in less-dense, warmer open ocean waters. *C. tripos,* with its shorter horns, occurs mostly in more-dense, cooler, shallow waters.

Like all planktonic algae, the dinoflagellates must be able to maintain themselves in the photic zone in order to photosynthesize. Species living in warm or low-salinity waters (i.e., waters of low density) have devices to increase their flotation capability in order to remain in the photic zone. *Noctiluca*, for example, increases its buoyance by selectively excluding heavy divalent ions (for example, Ca^{++} and $SO_4^=$) and by maintaining a higher relative concentration of lighter monovalent ions (Na^+ and NH_4^+) in the cell (Kesseler, 1966). Pyrrhophyta living in subtropical and tropical seas have longer spines and protuberances that increase their surface-to-volume ratio, enabling them to maintain themselves in the photic zone of the less-dense warmer waters. Different species of the same genus will vary morphologically according to their occurrence in warm or cold water (Fig. 9-7). A similar phenomenon was noted with the diatoms (see Fig. 8-12). Dinoflagellates are much more active swimmers than diatoms because dinoflagellates have two flagella and can maintain themselves in the photic zone in less-dense warm waters. Perhaps in consequence, the dinoflagellates are more prevalent than diatoms in warm seas. Some dinoflagellates can swim at 1 to 2 m/hour, which enables them to migrate vertically for considerable distances.

Blooms of cryptomonads may produce various effects on the color of the water in which they grow. In addition, the color may vary with the condition of the algae. For example, *Chromonas* cultures appear red when young but become greenish with age (Butcher, 1967). Cryptomonads are primarily autotrophic, but some species are heterotrophic.

EVOLUTION

The dinoflagellates are among the most ancient of the algae, since they probably existed in the Precambrian period more than 600 million years ago (Sarjeant, 1974). Most of our knowledge of this class is based upon fossilized cysts from the Silurian period when the dinoflagellates were dominant members of the marine phytoplankton (Yentsch et al, 1980). Originally the fossilized cysts were called "hystricospheres" (Gr. *hystrix* = porcupine) because of their characteristic spiny cell walls (Fig. 9-4). They were later found to be cysts of dinoflagellates. These cysts have been found in phosphate rocks, crude petroleum deposits, and even in rock salt.

Dinoflagellates have been important planktonic organisms in aquatic environments since the Silurian period. Undoubtedly they have played an extremely important role as primary producers in the oceans during the course of evolution. Diatoms date back only to the Jurassic period, so the dinoflagellates predate these other predominant phytoplankters. The fossil record indicates that the dinoflagellates were among the earliest eukaryotic forms to evolve. Their mesokaryotic nucleus would appear to place them between the prokaryotic and eukaryotic organisms (Sarjeant, 1974). Future research will have to establish definitely their position in algal evolution. Unfortunately, the fossil record is incomplete for many reasons, one of which is that only those dinoflagellates that encyst have left a significant number of fossils in the sediments. It is not possible to positively identify some cysts as dinoflagellate. The earliest fossils resembling cysts of present-day dinoflagellates do not occur until the Triassic period.

Recent investigations with dinoflagellates have increased the knowledge concerning the evolution of this class. Ultrastructural studies of two species of *Peridinium* (*P. foliaceum* and *P. balticum*) have revealed eukaryotic nuclei in addition to mesokaryotic nuclei (Dodge, 1971; Tomas et al, 1973). Both of these species also contain fucoxanthin but no peridinin. Tomas and Cox (1973a) propose that the lack of a chloroplast and heterotrophy may be primitive in the dinoflagellates and that subsequently symbiosis with autotrophic eukaryotic algae became established in some species. The presence of fucoxanthin and chlorophylls a, c_1 and c_2 in the

Peridinium species (Tomas and Cox, 1973a, b) studies suggests that the endosymbiont was a chrysophyte (Gibbs, 1978). Certain species of *Gymnodinium* have fucoxanthin and a distinctive chrysophytelike pyrenoid but do not have a eukaryotic nucleus. These species may have lost the eukaryotic nucleus over evolutionary time but retained the characteristic chrysophyte pyrenoid, fucoxanthin, and chlorophylls.

Loeblich (1976) and Dodge (1979) think that **endosymbiosis** took place several times during dinoflagellate evolution and that the variation in carotenoids within the class is the result of the photosynthetic endosymbionts being from different algal divisions. Steidinger and Cox (1980) propose three lines of dinoflagellate evolution from the primitive heterotrophic form: The first line is uninucleate, with neither fucoxanthin nor peridinin (*Ptyochodiscus brevis*); the second line is binucleate, with fucoxanthin and both chlorophyll c_1 and c_2 (*Peridinium foliaceum* and *P. balticum*) that eventually led to a uninucleate line when the eukaryotic nucleus was lost (*Gymnodinium* spp.); and the third line is the main line of dinoflagellate evolution in which the algae are uninucleate and contain peridinin and chlorophyll c_2. This theory that heterotrophic dinoflagellates acquired endosymbiotic eukaryotic autotrophs would agree with the proposal of Margulis (1970) that eukaryotic cells originated by the acquisition of endosymbiotic prokaryotic cells (see Chapter 1). For dinoflagellates, however, the proposed photosynthetic endosymbiont was a eukaryotic cell rather than a prokaryote.

Little is known of the evolutionary history of the cryptomonads. The possession of phycobilins by several genera would imply a relationship with the blue-greens and the red algae, but the cryptomonad phycobilins differ in several ways from the phycobilins of the blue-greens and reds. Several investigators (Scagel et al, 1965, 1982; Gantt, 1980b) think that the photosensitive phycobilins probably evolved more than once (**parallel evolution**) and that their presence in the cryptomonads does not imply a relationship to the blue-green or red algae. The occurrence of phycobilins also suggests that cryptomonads, like the dinoflagellates, may be a potential intermediary between prokaryotic and eukaryotic organisms. The structure of the thylakoids also appears to be intermediary between the prokaryotes and the brown line of evolution.

Several schemes for the evolution of the cryptomonads have been proposed (Ragan and Chapman, 1978). Before phylogenetic relationships between phycobilin-containing groups can be discerned, more information is needed concerning the biosynthetic pathways of these accessory photosynthetic pigments (Gantt, 1980b).

ECOLOGY

The largest concentrations of Pyrrhophyta are found in tropical and subtropical seas near the coasts at depths of 18 to 90 m. Dinoflagellates are responsible for approximately 30% of the primary productivity of the oceans and are second only to the diatoms in their importance as energy producers in the seas. The armored dinoflagellates are prevalent in coastal areas of both temperate and polar seas and in fresh water, while the naked forms are more common in the less-dense open ocean and warmer seas. The more active swimmers (e.g., *Peridinium* spp., Fig. 9-1) are usually found in the warm seas where they maintain their position in the photic zone by motion of the flagella. They can also migrate vertically to take advantage of optimum light and nutrient conditions. In addition to being planktonic, dinoflagellates may also occur in the benthos, as symbionts or as ectoparasites or endoparasites with other organisms.

Pyrrhophyta are also important members of the freshwater plankton, especially in small lakes and ponds with substantial emergent vegetation. They are less abundant—but usually present—in the plankton of larger lakes. The genera that occur most often in fresh water are *Ceratium*, *Peridinium*, and *Gymnodinium*. Large numbers of *Ceratium* (Fig. 9-8) may color the water of small lakes gray or brown (Prescott, 1978). One species, *Peridinium polaricum*, pro-

FIG. 9-8 Scanning electron photomicrograph of some freshwater phytoplankton.
A, Freshwater dinoflagellate *Ceratium hirundinella*. Note sculpturing on the thecae of the cell wall and characteristic spines associated with this genus, together with the transverse groove (sulcus). The other genera seen are all members of Chrysophyta: *B, Dinobryon,* showing its characteristic lorica and two diatoms; *C, Asterionella,* and *D, Fragilaria,* showing characteristic ribbons of individual cells. (600×). (Courtesy Jon I. Parker, Department of Biology, Lehigh University, Bethlehem, Pa.)

duces an extracellular toxin that kills fish (Nozawa, 1970). Other species dominate in acid lakes in Canada (NRCC, 1981). The common freshwater species *Ceratium hirundinella* (Fig. 9-8) exhibits seasonal polymorphism, with a reduction in cell size and the development of a fourth horn during the summer months, thereby increasing its surface-to-volume ratio as the water becomes warmer and less dense (Hutchinson, 1967). *C. hirundinella* may also produce planktonic cysts (Chapman et al, 1982).

Blooms of marine dinoflagellates may color waters red, green, yellow, or brown and may produce bioluminescence at night. Concentrations of cells of 200,000 to 500,000/l will usually color the water, depending on the species. However, cell concentrations of 2 to 8 million/l may be reached during a "red tide" phenomenon. If these cells release toxins into the water column, the effects on fish in the area may be quite dramatic, causing sizeable fish kills. Not all blooms of dinoflagellates produce toxins.

There are three categories of toxic red tides, each of which has a different effect on marine ecosystems. The first type produces a toxin that affects fish and a limited number of invertebrates. The tides that occur off the west coast of Florida in the Gulf of Mexico are a good example. They are generally produced by *Ptychodiscus brevis* (= *Gymnodinium breve,* Fig. 9-5). When these blooms occur near shore, millions of dead fish may be washed up on the beaches; this occurred in Florida in 1971 and again in 1974. Sometimes winds blowing over waters containing a red tide bloom will carry toxin aerosols to the land, causing irritated eyes and respiratory systems. Fish-killing red tides have been known in Florida since 1844. During a *P. brevis* bloom shellfish may also become contaminated with toxin, but not killed (Keys, 1975).

A second type of red tide, lethal to a limited number of invertebrates, is usually caused by members of the genus *Gonyaulax* (*G. monilata* or *G. veneficum*). A *G. monilata* bloom near Galveston, Texas, in 1971 killed primarily molluscs, arthropods, echinoderms, and a few fish (Wardle et al, 1975).

The third type of red tide does little damage to organisms living in the ocean. The dinoflagellate cells are eaten by filter-feeding invertebrates, and the toxin is concentrated and retained by invertebrates such as clams, mussels, scallops, and oysters. The high concentrations of toxins do not affect the molluscs, but the toxins produce extreme symptoms in those warm-blooded organisms that feed upon the molluscs. Thus birds, rodents, raccoons, and humans may suffer from paralytic shellfish poisoning (PSP) after feeding on shellfish that fed upon toxin-producing dinoflagellates. PSP is usually caused by *Gonyaulax excavata* in New England and the Canadian Maritime Provinces and by *G. catenella* in the Pacific northwest. Records of red tide deaths go back to 1799, when 100 members of a Russian expedition in Alaska reportedly died from eating affected shellfish. Red tide blooms in 1974 in New England caused a shutdown of the clam and mussel industry.

Toxins produced by the dinoflagellates are neurotoxins, which interfere with nerve impulse transmission in poikilothermic (cold-blooded) or homeothermic (warm-blooded) vertebrates. Only healthy, growing populations of dinoflagellates produce toxins (Hellebust, 1974). The dinoflagellates that have the metabolic ability to produce toxins are all common members of indigenous phytoplankton communities. Evidently the toxins are always produced by the algae, but as long as the algae are not too numerous, the toxins occur in low, harmless concentrations. An environmental problem is created when algal populations increase exponentially, causing large quantities of toxins to be produced (Sarjeant, 1974). Cell concentrations of *G. catenella* of 100,000 to 200,000/l can cause toxic concentrations in shellfish. Blooms, however, may not be visible until the cell concentrations exceed 1 million/l when large patches of reddish water or nighttime bioluminescence signal the presence of a red tide. Shellfish can therefore become toxic before a bloom is visible. Toxicity in shellfish usually disappears in 2 to 3 weeks, but longer retention of the toxin has been reported. Some Alaskan clams can retain up to 80% of the toxin and store it for up to a year. Copepod and barnacle larvae can accumulate the toxin of *G. excavata*, retain it up to 3 weeks, and pass it on to their fish predators (White, 1981).

The most intense dinoflagellate blooms (red tides) occur in protected areas and may cover many thousands of square kilometers. Extensive research has been conducted to try to identify those conditions that contribute to massive dinoflagellate blooms. Significant economic impact occurs from red tide outbreaks; therefore research has been focused on predicting, if not preventing, these blooms. Most red tides occur in conditions of high water temperatures (18 to 25° C), many hours of sunlight, ample supplies of nutrients from either upwelling or land runoff, increased stability of the water column, calm seas, and lowered salinities. Chelated iron, humic materials, and other substances from land runoff may also play a role in blooms. Previous phytoplankton populations of such algae as diatoms may "condition" the water for dinoflagellate blooms by providing vitamin B_{12}, thiamin, biotin, and other metabolites or organic nutrients (Chapter 2). The humic substances and organic chelators from land runoff may help to reduce copper concentrations that can be toxic to dinoflagellates. *G. excavata*, for example, is especially sensitive to copper (Martin and Martin, 1973). Water column stability and lack of winds can also lead to large populations concentrating in sheltered areas and to nutrient build-up above the thermocline. Under such ideal growth conditions dinoflagellate populations build up rapidly.

According to Steidinger (1975), dinoflagellate blooms begin as gradual increases in motile cells rather than explosive increases in rates of cell division. The motile cells may come from seed populations of temporary or resting cysts from the sea floor. These cysts may be brought to the surface in areas of upwelling. Over 70 extant genera may produce cysts as part of their life cycles (Steidinger and Cox, 1980). Red tides usually start 16 to 60 km offshore and may be carried to coastal areas by onshore winds and currents.

During large dinoflagellate blooms, death of fish may occur from the depletion of oxygen rather than from the production of toxin. Large algal populations photosynthesize and produce oxygen during the day, but at night these algae continue to respire and use up the dissolved oxygen. The oxygen cannot be replenished until photosynthesis resumes at daylight. Bacteria that decompose dead algae also require oxygen for their respiration so that oxygen concentrations may become very low in the early morning hours before sunrise, resulting in **dawn fish kills.** Such a situation occurred off the New Jersey coast in the late 1970s when a bloom of *Ceratium tripos* (Fig. 9-7) colored the water dark brown **(black tide),** causing fish kills from oxygen depletion rather than from toxic compounds. Bacterial populations may also build up in response to decaying algae and cause fish and shellfish mortality. Large concentrations of pathogenic bacteria apparently have killed innumerable fish during **mahogany tides** in the Chesapeake Bay (Seliger et al, 1975).

Red tides are usually self-limiting: nutrients become exhausted and the algae begin to perish because they are shaded from the sun by their own vast numbers. Predators such as the heterotrophic *Noctiluca* and various crustacea move in, populations are reduced, and the stage is set for the domination of the phytoplankton by another alga. Such a succession of different algae in the plankton is very well documented and is a regular seasonal occurrence (see Fig. 2-17).

Dinoflagellate blooms are yearly phenomena in such areas as the Gulf of Mexico, the east and west coasts of North America, in the vicinity of the 30th parallel of latitude, and in other highly productive coastal areas in Peru, Japan, Africa, and India. However, it is only when all conditions are favorable that these blooms reach toxic or harmful proportions. The evidence indicates that red tides have been occurring regularly for centuries—the biblical references in Exodus 7:20-21 to the waters of the Nile being changed to blood and fish dying may have been inspired by a red tide.* Euglenophyte blooms may also

*"And Moses and Aaron . . . smote the waters that were in the river; and all the waters that were in the river were turned to blood. And the fish that were in the river died; and the river became foul, and the Egyptians could not drink water from the river."

cause similar phenomena. Massive fish kills in upwelling areas millions of years ago may have been caused by red tides and may have contributed organic material for petroleum formation (Brongersma-Saunders, 1948).

Cryptomonads are common in fresh or salt waters having a high organic content. Some species, however, occur in waters low in nutrients. Some genera form part of the mud algal flora in salt marshes. *Cryptomonas* populations may occur fairly deep in the water column, demonstrating a predilection for low light intensity (Soeder and Stengel, 1974). Cryptomonads also form portions of the symbiotic zooxanthellae of marine coelenterates and protozoa.

ECONOMIC IMPORTANCE

Pyrrhophyta play an important economic role in coastal areas not only because of their contribution to primary production but also because of their propensity for causing red tides. As previously discussed, red waters have been recorded since the exodus of the Jews from Egypt. Records of red tide outbreaks in North America date back to 1799. Charles Darwin described dinoflagellate blooms during his voyage on the *Beagle* in 1832. The *Annals and Magazine of Natural History* carried an article in 1881 that said, "For the last two years there has been a very serious mortality among the fish in the Gulf of Mexico near Florida arising apparently from some peculiar condition of a belt of water at some small distance from the shore." A red tide threatened the pearl fishing industry in Japan in 1934, and India experienced a massive fish kill in 1948. In the United States 0.5 billion fish were killed during a bloom of *Ptychodiscus brevis* (= *Gymnodinium breve*) on the gulf coast of Florida in 1947. During this environmental disaster it was discovered that the dead fish were not the only hazard—airbone toxins produced by the algal bloom caused human respiratory distress, a burning sensation of the mucous membranes around the eyes, and irritation of exposed skin. The stench of decaying fish on the beaches coupled with the irritating aerosols reduced tourist trade along the Gulf Coast and had a detrimental effect on the fishing industry. Extensive red tides on the Florida west coast occurred again in 1971 and 1973, causing massive fish kills. Financial losses to the tourism and fishing industries were estimated at $20 million in 1973 (Habas and Gilbert, 1975). This was the twenty-fifth major red tide recorded in this area since 1844 when records were first kept.

At higher latitudes and even as far north as Iceland and Norway, red tides occur but are seldom lethal to the fish population. However, some armored dinoflagellates of northern temperate waters still produce toxic metabolic products that may concentrate in and contaminate shellfish, especially clams and mussels. In 1972 during a red tide off the New England coast, hundreds of gulls, ducks, and shore birds died after feeding on contaminated mussels and clams. The dead birds signaled the presence of shellfish contaminated with toxin from *G. excavata*. The toxin from *G. excavata* acts in a manner similar to curare, causing distal paralysis and respiratory failure in warm blooded organisms. The amount of paralysis is related to the concentration of toxin in the invertebrate tissue, which, in turn, is related to the amount of algae consumed by the shellfish.

During the 1972 episode in New England 26 people were striken with PSP and two became severely ill (Bicknell and Walsh, 1975). Innumerable birds and mammals also became ill. Cell concentration reached $2.6 \times 10^6/l$, with *G. excavata* dominating the phytoplankton population to the exclusion of other algae (Hurst, 1975). A health emergency was declared and the shipment and sale of shellfish from the affected coast was banned. The shellfish industry was closed; a *New York Times* headline read, "Infestation of Poisonous Algae Brings 3-State Ban on Mussels and Clams at a Loss of $1 Million a Week." The New England coast was declared a disaster area so that people dependent on the shellfish industry could collect unemployment

insurance and government assistance (Sarjeant, 1974). Another *G. excavata* bloom occurred in 1974 and clam flats were closed to harvesting (Mulligan, 1975). The shellfish industry in New England took a long time to recover; sales had not returned to their pre-1972 level by 1978. Closings of the shellfish industry in Maine have occurred almost every year since 1958. The cost of the red tide contamination of Maine shellfish has been estimated at $8-million (Yentsch and Incze, 1980).

Clearly red tides cause economic havoc when they occur. Paralytic shellfish poisoning cases total only 1,600 worldwide since records have been kept, so it is not a major public health problem, although 300 deaths have been reported. Death of fish at lower latitudes and death of birds and mammals at higher latitudes present economic and ecological problems. There is growing concern that PSP outbreaks are increasing in intensity and spreading to new areas (Prakash, 1975). It is known that red tides may occur naturally under certain environmental conditions. It is not known, however, to what extent man's manipulation of the environment induces red tides. Some investigators (Prakash, 1975) think that the increased frequency of dinoflagellate blooms is related to the increased enrichment (eutrophication) of coastal waters.

Shellfish in red tide–prone areas are regularly tested by the Federal Government using standardized bioassay methods to measure toxicity. Clem (1975) discusses surveillance methods in the United States. Often bioassay is the only way to determine the presence of the toxin. The toxin does not alter the shellfish in any noticeable way—they look and smell harmless. The toxins are only slightly denatured by cooking.

Considerable research has been focused on the environmental factors that trigger and aid the continuance of a red tide. For a population of dinoflagellates to explode and form a red tide, a particular series of events must occur. These include a seed population being brought to the photic zone by currents, a lack of predation and competition, and adequate levels of temperature, salinity, nutrients, and light (Steidinger and Haddad, 1981). Remotely sensed data from satellites is now being used in an attempt to predict and identify developing red tides in the Gulf of Mexico (Gordon et al, 1980). The use of remote sensing for red tide prediction may eventually be extended to all red tide–prone areas of the world.

SUMMARY

Pyrrhophyta are found in both salt and fresh waters. They are most often found as dominant members of the phytoplankton in subtropical and tropical coastal areas. The division is composed of two classes, Dinophyceae (dinoflagellates) and Cryptophyceae (cryptomonads). The dinoflagellates are responsible for approximately 30% of the primary productivity in the world, second only to the diatoms.

There are naked and armored pyrrhophytes; the heavier, armored species are found more frequently in the denser and cooler temperate waters. Photosynthetic pigments in this division include chlorophyll a, c_1 and c_2, β-carotene, and two xanthophylls. Some cryptomonads contain phycobilins. Food reserves are stored as starch or oils. The starch is biochemically similar to the starch found in Chlorophyta and vascular plants. The nucleus is mesokaryotic, and it undergoes mitosis in a different manner from eukaryotic nuclei. Some Dinophyceae also contain eukaryotic nuclei. Nutrition methods range from photoautotrophic to heterotrophic. Most reproduction is via vegetative cell division, but sexual reproduction has been demonstrated in some species. Bioluminescence is common among the dinoflagellates.

Many pyrrhophytes live as symbionts in various marine invertebrates. As endosymbionts in corals, the pyrrhophyte zooxanthellae contribute significantly to the high primary productivity of tropical coral reefs.

Dinoflagellates produce temporary and resting cysts in response to unfavorable environmental conditions. Cysts tentatively identified as dinoflagellates have been found in Precambrian rocks, indicating that the dinoflagellates may have evolved before the Cambrian period. They have been important members of the marine plankton since the Silurian period.

Large populations (blooms) of dinoflagellates are called "red tides" and may color hundreds of square kilometers of ocean water yellow, brown, or red. Some dinoflagellates produce toxins, and when blooms of toxin-producing species occur, poisoning of fish, shellfish, birds and mammals can result. The type of poisoning that occurs depends upon which species is causing the bloom. These blooms are especially prevalent in upwelling areas and in coastal areas with high nutrient concentrations. When large numbers of fish are killed or shellfish are contaminated by red tide toxins, the economic impact may exceed millions of dollars.

SELECTED REFERENCES

Gantt, E. 1980b. Phytosynthetic cryptophytes. In Phytoflagellates. E.R. Cox (ed.), pp. 381-405. Elsevier North-Holland, New York.

LoCicero, V.R. (ed.). 1975. Proceedings First International Conference on Toxic Dinoflagellate Blooms. Mass. Sci. and Tech. Found., Wakefield, Mass. 541 pp.

Nealson, K.H. 1981. Bioluminescence: current perspectives. IHRDC Publications, Boston. 165 pp.

Sarjeant, W.A.S. 1974. Fossil and living dinoflagellates. Academic Press, London. 182 pp.

Steidinger, K.A., and E.R. Cox. 1980. Free-living dinoflagellates. In Phytoflagellates. E.R. Cox (ed.), pp. 407-432. Elsevier North-Holland, New York.

Steindinger, K.A., and K. Haddad. 1981. Biological and hydrographic aspects of red tides. BioScience 31(11):814-819.

Taylor, D.L., and H.H. Seliger (eds.). 1979. Toxic dinoflagellate blooms. Elsevier North-Holand, New York. 505 pp.

Tett, P.B., and M.G. Kelley. 1973. Marine bioluminescence. Oceanogr. Mar. Biol. Annu. Rev. 11:89-173.

Yentsch, C.M., C.M. Lewis, and C.S. Yentsch. 1980. Biological resting in the dinoflagellate Gonyaulax excavata. BioScience 30(4):251-254.

10

CLASSIFICATION

Division: Xanthophyta (yellow-green algae): 90 genera, 600 species; some with cellulose cell walls occurring in two overlapping halves; mostly freshwater; chlorophyll *a;* no fucoxanthin

Class 1: Xanthophyceae; superficially resembling green algae, but lacking chlorophyll *b* and starch; chlorophyll c_1 and c_2 in some

ORDER 1. Heterochloridales: naked unicells with two subapically inserted heterokont flagella; *Olisthodiscus*

ORDER 2. Tribonematales: filaments of cells containing H-pieces within its cell walls that overlap, enclosing the protoplast; *Tribonema*

ORDER 3. Vaucheriales: multinucleate (coenocytic) filaments or spheres; some with chlorophyll c_1 or c_2; *Botrydium, Vaucheria*

Three other orders are listed by Bold and Wynne (1978)

Class 2: Eustigmatophyceae: Superficially resembling euglenoids; a few genera of unicellular freshwater and marine algae with a single emergent anterior flagellum; large eyespots; chlorophyll *a* only; *Chlorobotrys*

Class 3: Chloromonadophyceae (chloromonads): A few genera of bright green naked biflagellate cells usually associated with mud bottoms in bogs and ponds; chlorophyll *a* and *c;* some with trichocysts; one anterior tinsel-type flagellum and one trailing whiplash type flagellum; *Chattonella, Gonyostomum*

XANTHOPHYTA

Xanthophyta are a relatively small group of yellow-green algae that historically have presented problems in classification. They have been variously listed as members of Chlorophyta, Chrysophyta, and Pyrrhophyta. They are most often confused with the green algae because of their structure and color. Morphologically they are often similar to genera from other divisions: *Vaucheria* (Fig. 10-1) closely resembles a siphonaceous green alga; *Tribonema* (Fig. 10-2) looks like a filamentous diatom; the Eustigmatophyceae are easily mistaken for small euglenoids bearing relatively large eyespots (stigmata); the **chloromonads** are easily confused with the cryptomonads (Pyrrhophyta); and many of the rarely seen unicellular types are unfamiliar to even the best phycologists and are only poorly described in the literature. Hibberd (1980b, 1980c) and Heywood (1980) have attempted to condense what is known about this division.

Xanthophyta are separated from other algal divisions largely on the basis of the absence of certain biochemical substances. They lack chlorophyll *b* and starch, both major characteristics of Chlorophyta. They also lack the pigment fucoxanthin and are therefore not considered to be members of the brown line of algal evolution, even though some xanthophytes possess chlorophyll *c*. All xanthophytes contain chlorophyll *a*. Their taxonomic position is enigmatic; this is reflected in the position assigned to them, as shown in Fig. 1-5, at the base of the brown line of algal evolution.

Xanthophyta are mainly freshwater or damp soil organisms, although a few are marine and some are among the most obvious algae in salt marshes of various salinities. Some occur as free-floating members of the phytoplankton, while others are benthic. Autotrophic, auxotrophic, and heterotrophic nutritional strategies are known. The unicellular, biflagellate forms exhibit morphological and

196

FIG. 10-1 *Vaucheria* **habit.**
During sexual reproduction, the coenocytic vegetative filaments *(veg)* of monoecious species of *Vaucheria* produce sexual structures called oogonia *(oog)* and antheridia *(anth)*. After fertilization takes place, the egg cell within the oogonium is called an oospore (125×).

nutritional affinities to certain protozoan flagellates. The multinucleate (coenocytic) forms bear obvious morphological and biochemical similarities to a group of aquatic, coenocytic fungi, the Oomycetes. (Compare the sexual reproductive structures of *Saprolegnia*, Fig. 15-3, with those of *Vaucheria*, Fig. 10-1.)

Considerable morphological diversity exists within Xanthophyta, including motile flagellate types, nonmotile forms, and branched and unbranched filaments, some with coenocytic organization. Species are assigned to this division chiefly because the majority produce asexual zoospores at some time during their life history.

CELL STRUCTURE

The cells of Xanthophyta are similar in many respects to the cells of most algal divisions. Each contains the usual complement of mitochondria, Golgi apparatus, endoplasmic reticulum, ribosomes, nucleus, nucleolus, and chromosomes. However, they also contain some distinctive structures and biochemical characteristics. The

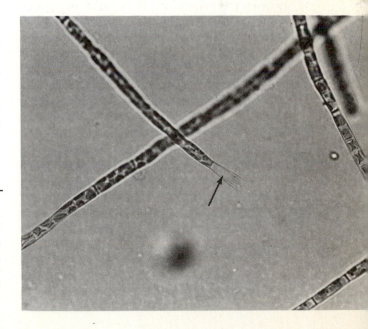

FIG. 10-2 *Tribonema* **habit.**
Characteristic elongate barrel-shaped cells of the filamentous *Tribonema* species, which is a common pioneer organism of newly formed bodies of water, including aquaria and pools formed by melting ice in temperate and northern latitudes. Note the characteristic H-piece (arrow) formed by the cell walls when a filament is broken. Also note the presence of disk-shaped parietal chloroplasts (400×).

flagellar apparatus of motile cells has been used by taxonomists to differentiate Xanthophyta from other algal divisions, since xanthophytes possess two heterokont flagella, thus leading to their older common name, the heterokonts.* Motile zoospores and sperm cells possess two anteriorly inserted flagella of different lengths. The longer, anteriorly directed flagellum bears many lateral hairlike projections called mastigonemes (Bouck, 1972). The shorter, posteriorly directed flagellum is a whiplash-type flagellum, lacking mastigonemes. An exception to this flagellar arrangement is found in the multiflagellate, multinucleate special zoospores produced by *Vaucheria* under certain environmental conditions, even though the same species is capable of producing typical heterokont sperm cells.

The chloroplast structure is also of some taxonomic value. Chloroplast numbers vary from a single plastid in some cells, to a few in others (e.g., *Tribonema*), to many in the coenocytic genera (e.g., *Vaucheria*, *Botrydium*). Each plastid has an ultrastructure like those of Chrysophyta and Phaeophyta with two membranes of CER enclosing each plastid and the outer membrane being continuous with the nuclear envelope (see Fig. 1-7). The double membrane-bound thylakoids occur in bundles of three. Pyrenoids are present in some—but not all—chloroplasts.

The motile flagellate cells generally possess eyespots (stigmata) that occur within the chloroplast envelope directly beneath a basal swelling on the shorter whiplash flagellum. However, there are exceptions to this: In *Vaucheria*, motile cells lack eyespots (Ott and Brown, 1974), while in the Eustigmatophyceae, the relatively large eyespots occur outside the chloroplast envelope. The latter arrangement is distinctive for the Eustigmatophyceae.

The photochemical pigments found within the chloroplasts include chlorophyll a, c_1, and c_2, plus a number of accessory pigments (Table 10-1). The xanthophytes may appear yellow-green because the predominating carotenes and xanthophylls mask the chlorophyll.

The products of photosynthesis are stored in the cytoplasm in membrane-bound vesicles called **leucosin** vesicles. Their content is a complex polysaccharide, leucosin, which is a β-1,3–linked polymer of glucose molecules. Leucosin is identical with the energy storage product chrysolaminarin found in Chrysophyta. Besides leucosin, various other sugars, fats, oils, and alcohols have been reported as reserves in some species. Tests for starch are always negative.

The cell walls of xanthophytes are distinctive in a number of respects. In some genera they occur as two overlapping halves much like the

*The designation "heterokonts" has been abandoned, since this is not the only division to exhibit heterokont flagella. Heterokonts may be found in both Chrysophyceae (Chrysophyta) and Cryptophyceae (Pyrrhophyta).

TABLE 10-1. Photosynthetic Pigments and Storage Products in Xanthophyta

Class	Chlorophylls	Carotenes	Xanthophylls	Storage product
Xanthophyceae (Hibberd, 1980b)	a, c_1 and c_2	β-carotene	Diatoxanthin Diadinoxanthin	Leucosin Alcohols
Eustigmatophyceae (Hibberd, 1980c)	a only	β-carotene	**Violaxanthin Vaucheriaxanthin** esther Others	Layered compound of unknown composition
Chloromonadophyceae (Heywood, 1980)	a and c	β-carotene	Antheraxanthin Others	Oils

walls described for diatoms (Chrysophyta). However, unlike diatoms, the walls do not contain silicates. In filamentous species (e.g., *Tribonema*), the walls are joined end to end by tapering, interlocking H-pieces (Fig. 10-2). This feature is diagnostic for *Tribonema*. H-pieces are easily discernible with the light microscope when one views the tips of broken filaments. In *Vaucheria* and other coenocytic forms, the cell walls lack cross walls (septa), thus producing a multinucleate (coenocytic) organism. Septa do form, however, when reproductive structures differentiate within the filament.

The cell walls are composed primarily of cellulose (at least in *Vaucheria:* Parker et al, 1963; Cleare and Percival, 1973). Smaller amounts of glucose, various polysaccharides, and uronic and pectic acids have also been reported. Curiously, the chemical nature of the walls are similar to that found in some aquatic fungi.

REPRODUCTION

Asexual reproduction by simple mitosis is the most common method of duplication in this division, since most species are unicellular. Production of motile zoospores and nonmotile aplanospores is also common. Spore motility seems to depend upon the environmental conditions. When plenty of water is available, motile zoospores are produced, but when conditions become dry, as in terrestrial soil forms, nonmotile aplanospores are produced. In *Vaucheria* a special motile, multinucleate, multiflagellate zoospore is produced in terminal club-shaped zoosporangia. In uninucleate filamentous forms, fragmentation of the filament often provides the species with an effective mode of reproduction (e.g., *Tribonema*). It is interesting that fragmentation, or the tearing apart, of coenocytic filaments (e.g., *Vaucheria*) is seldom a useful method of reproduction, even though septae do develop in response to this type of injury, thus preventing the cytoplasm from flowing out into the environment.

Sexual reproduction has been described in only three genera, *Tribonema*, *Botrydium*, and *Vaucheria*. *Tribonema* produces motile isogametes, while *Botrydium* may be isogamous or anisogamous, depending upon the individual species. *Vaucheria* is oogamous, with many small, colorless, heterokont, biflagellate sperm produced in specially differentiated antheridia and a single large nonmotile egg cell produced in special oogonia (Fig. 10-1). After fertilization the egg cell is called an **oospore.** In most species antheridia and oogonia are located near each other on the same coenocytic filament. They are, therefore, monoecious. However, dioecious species have been reported. Both antheridia and oogonia are separated from the coenocytic vegetative filament by a cellulose septum. The oogonium is filled with energy storage products. After fertilization the zygote develops into a thick-walled resting stage, the oospore. Oospores remain dormant for several months— usually during the winter—after which they germinate by meiotic cell divisions, producing a new haploid coenocytic filament. The life cycles of xanthophytes investigated to date are haplontic.

PHYSIOLOGY

The photosynthetic pigments in Xanthophyta are chlorophyll *a* and *c* (c_1 and c_2), β-carotene, and several xanthophylls, including the distinctive **diatoxanthin** and **diadinoxanthin** (Table 10-1). Chlorophyll c_1 and c_2 are found in very low concentrations in the Xanthophyceae (Guillard and Lorenzen, 1972; Jeffrey, 1976). Chlorophyll *c* is also found in the chloromonads, but none has been found in the eustigmatophytes. Each class in this division has unique and distinctive xanthophylls, while all three classes contain β-carotene (Table 10-1). Excess photosynthetic products are stored as leucosins, oils, lipids, alcohols, or other polysaccharides of undetermined composition. No starch has been found in this division; the absence of starch was one of the major reasons for the removal of these organisms from division Chlorophyta. Cellulose has been definitely identified in the cell wall of only one genus, *Vaucheria* (Cleare and Percival, 1973).

Nutrition among members of this division may be autotrophic, auxotrophic, or heterotrophic. Colorless saprophytic and phagotrophic forms have also been described. For example, some *Tribonema* species can metabolize glucose and acetate molecules when maintained in darkness (Droop, 1974).

The genus *Vaucheria* has been most studied by investigators. It may produce multinucleate, multiflagellate zoospores, the production of which can be induced by certain chemical substances, including **indole acetic acid (IAA)** and tryptophan. Zoospore production in nature is probably controlled by endogenous circadian rhythms (Rieth, 1959). Some genera (e.g., *Botrydium*) found growing on damp soils release motile zoospores or gametes only when the thallus has been flooded by heavy rains.

Chloroplast movement in the coenocytic cells of *Vaucheria* has been the subject of numerous investigations (Haupt and Schönbohm, 1970; Nultsch, 1974). In the dark, chloroplasts remain randomly distributed throughout the multinucleate filaments. However, when the filament is exposed to light the chloroplasts become oriented toward the light source. At low light intensities the chloroplasts move to positions where they receive maximum illumination, thus indicating the chloroplasts can alter position to receive the proper light intensity for effective photosynthesis.

The heterokont biflagellate marine and brackish-water genus *Olisthodiscus* is an important organism in a number of respects. First, it illustrates the phenomenon of organic inhibition, or the effect of one organism upon the subsequent growth and development of another. Pratt (1966) has shown that in Narragansett Bay (Rhode Island) when *Olisthodiscus* is present, the diatom *Skeletonema costatum* is absent. Evidently an extracellular product (**ectocrine**) occurs in concentrations high enough to inhibit *Skeletonema* growth. The ectocrine produced by *Olisthodiscus* appears to be a tannoid substance (Pratt, 1966). Mutual inhibition or stimulation by metabolic products is a process that affects populations in the environment but is difficult to demonstrate because the concentration of the inhibitors or stimulators is so small (see Chapter 2). Tomas (1980a, 1980b) discusses the ecology and the occurrence of *Olisthodiscus*.

EVOLUTION

The discovery of chlorophyll *c* in some species of *Vaucheria* and in the chloromonads indicates that these organisms are related to the brown line of algal evolution. Morphologically they have characteristics that resemble both Chlorophyta and Chrysophyta.

Xanthophyta are composed largely of naked cells or cells with biodegradable cell walls and therefore have left few evolutionary clues in the fossil record; little is known about their origins. Observation of the present members of the division yields little evidence to indicate where or when these organisms evolved, other than the implication that they developed somewhere near the origin of the brown line of evolution (see Fig. 1-5). Some investigators (Parker et al, 1963) propose a relationship between *Vaucheria* and the aquatic fungi (Oomycetes—see Chapter 15) because of the similarity of cell wall composition (cellulose and glucans) and methods of sexual reproduction (oogamy). The aquatic fungi might have arisen from Xanthophyceae by the loss of the ability to synthesize chlorophyll.

ECOLOGY

Xanthophyta are widespread in the environment but seldom become dominant or conspicuous. They occur in fresh water, brackish water, and salt water, with the majority found in temperate fresh waters. Several genera occur on damp soils or along the flood plains of streams and rivers that are periodically inundated. Still others may be found on tree trunks, damp stone walls, or among mosses.

Unicellular xanthophytes are found among the phytoplankton in lakes, ponds, and oceans.

Tarapchak (1972) has described 56 xanthophyte species in Minnesota lakes. The Eustigmatophyceae and the chloromonads are not very common and are often found under acid conditions in peat swamps and ponds. The genera most often encountered are the filamentous diatomlike *Tribonema*, the siphonaceous green *Vaucheria*, and the small green balloonlike *Botrydium*. *Vaucheria* is sometimes called "water felt" because it forms dark to light grass-green velvety mats growing on rocks beneath flowing water or covering damp soils. These mats provide a habitat for many macroscopic and microscopic invertebrates. Often *Vaucheria* is covered with calcium carbonate. In several small German lakes species of *Vaucheria* grow in close association with the charophytes (see Chapter 5) *Chara* and *Nitella* (Jeschke, 1963).

Some *Vaucheria* species frequently grow on intertidal mud flats or on quiet sandy bottoms, as in some Bermuda bays (Dawson, 1966; Abbott and Hollenberg, 1976). These species have colorless rhizoids that penetrate the substrate and act as holdfasts and the green photosynthetic parts of the filament extend above the sand surface. (Additional information on the ecology of *Vaucheria* in salt marshes may be found in Nienhuis and Simons, 1971.)

The unbranched filaments of *Tribonema* occasionally reach bloom proportions in ponds and ditches or in freshly prepared aquaria. The filaments may be attached or free floating. The genus is most often encountered in cold-water springs or waters rich in humic acids. In northern latitudes *Tribonema* is often one of the first algae to appear following the melting of snow and ice in the spring.

Botrydium is a frequent inhabitant of damp soils, along streams, or even in greenhouses. Colorless rhizoids anchor the green balloonlike thallus to the soil. These algae may be covered with deposits of calcium carbonate in certain environments. The thalli may measure up to 1 mm in diameter and are hence visible to the unaided eye. Cool temperatures favor the growth of *Botrydium*, so it is most often encountered in spring and autumn.

Eustigmatophyceae and Chloromonadophyceae are not frequently encountered in any abundance in the environment. The role they play in most ecosystems appears to be relatively insignificant. However, eustigmatophytes have been found in fresh water, salt water, and soils. They also occur regularly both in marine phytoplankton (Antia et al, 1975) and in the freshwater plankton (Prescott, 1978). *Chlorobotrys* is a freshwater genus that is surrounded by a gelatinous sheath and that may be found in soft-water bogs (Smith, 1950). As more investigators become familiar with this class, their occurrence may be reported more frequently (Hibberd, 1980c). The chloromonads are also found in both fresh water and marine waters and seem to prefer waters rich in organic materials. Several species (e.g., *Gonyostomum*) are found in acid bogs and lakes where the pH may be below 4.0 and where *Sphagnum* moss abounds (Fott, 1952). Other species may be found living over a wide range of pH. Blooms of marine chloromonads have been reported in Europe and India and concentrations of cells have reached 38,000/ml during these blooms (Margalef, 1968). (See Heywood, 1980, for further discussion of the habitats in which chloromonads have been found.)

ECONOMIC IMPORTANCE

The members of Xanthophyta have very little influence on any aspect of our economic system. One marine chloromonad (*Chattonella* = *Hornellia*) has been reported to cause fish kills in the Indian Ocean (Subrahmanyan, 1954), but the total effect has been negligible. As a group they certainly have a role as primary producers under specialized conditions, but their total impact is not pronounced.

One genus, *Olisthodiscus*, has been cultivated commercially. It is easily grown in culture and has been used as a food organism for the farming of shellfish such as clams and oysters. It releases carbohydrates (mannitol and glycolic acid) into the medium and thereby provides nutrients for the shellfish (Hellebust, 1965).

SUMMARY

Xanthophyta are sometimes called the yellow-green algae because of their color and are sometimes referred to as the heterokont algae because of the unequal length of their flagella. They constitute a relatively small group of mostly freshwater algae with only about 90 described genera. There is considerable morphological and biochemical diversity within the division, since it includes single cells as well as filaments of cells and siphonaceous types. All cells contain chlorophyll *a*, and two classes contain chlorophyll *c*. The chloroplast ultrastructure resembles that associated with the brown line of evolution. Sexual reproduction in the division ranges from isogamous to oogamous.

The evolutionary history of this division is poorly understood. They have left a sparse fossil record. They are classified largely on the basis of their biochemical and ultrastructural features, both of which are difficult to determine. However, members of this division are not frequently encountered in the environment in significant numbers.

Xanthophytes are largely freshwater or damp-soil organisms, although some marine species are known. Chloromonadophyceae are especially common in acid environmental conditions.

SELECTED REFERENCES

Ettl, H. 1978. Xanthophyceae. In Gerloff, H.J., and H. Heynig (eds.). Susswasserflora von Mitteleuropa, Bd. 3.1. Teil. Gustav. Fischer, Stuttgart, Germany.

Guillard, K., and C. Lorenzen. 1972. Yellow-green algae with chlorophyllide *c*. J. Phycol. **8**:10-14.

Heywood, P. 1980. Chloromonads. In Cox, E.R. (ed.). Phytoflagellates, pp. 351-379. Elsevier North-Holland, Inc., New York.

Hibberd, D.J. 1980b. Xanthophytes. In Cox, E.R. (ed.). Phytoflagellates, pp. 243-271. Elsevier North-Holland, Inc. New York.

Hibberd, D.J. 1980c. Eustigmatophytes. In: Cox, E.R. (ed.). Phytoflagellates, pp. 319-334. Elsevier North-Holland, Inc., New York.

Nienhuis, P.H., and J. Simons. 1971. *Vaucheria* species and some other algae on a Dutch salt marsh, with ecological notes on their periodicity. Acta Bot. Nederl. **20**:107-118.

Ott, D.W., and R.M. Brown. 1974. Devleopmental cytology of the genus *Vaucheria*. I. Organization of the vegetative filament. Br. Phycol. J. **9**:111-126.

11

CLASSIFICATION

Division: Rhodophyta (red algae): 800 genera, 6000 species; mostly macroscropic, marine benthic algae

Subclass: Bangiophycidae*: uninucleate cells; sexual reproduction is rare; intercalary meristems; some unicellular species; pit connections rare

ORDER 1. Porphyridiales: single cells with a stellate chloroplast; freshwater, marine, and terrestrial; *Porphyridium*

ORDER 2. Goniotrichales: small filamentous marine epiphytes; spores develop from transformed vegetative cells; *Asterocystis*

ORDER 3. Compsopogonales: freshwater, tropical, or subtropical; uniseriate growth that becomes corticated with smaller cells; *Compsopogon*

ORDER 4. Bangiales: horizontal disks, erect filaments and leafy thalli; uniseriate to multiseriate; holdfasts; some epiphytic; some with "conchocelis" stages; *Bangia, Porphyra, Smithora*

Subclass: Florideophycidae†: macroscopic multicelled organisms; multinucleate vegetative cells; prominent apical meristematic regions; prominent pit connections between cells; sexual reproduction with complicated life cycles (often triphasic)

ORDER 5. Nemaliales: heterogeneous group of algae with uniseriate or multiseriate growth; mostly marine; some freshwater; some calcium carbonate deposition; *Acrochaetium, Audouinella, Batrachospermum, Bonnemaisonia, Galaxaura, Gelidium, Lemanea, Liagora, Nemalion, Pterocladia*

ORDER 6. Cryptonemiales: filamentous or pseudoparenchymatous blades or crusts of red algae; deposit calcium carbonate; includes the crustose and articulated "coralline reds" (family: Corallinaceae); *Amphiroa, Constantinea, Corallina, Goniolithon, Hildenbrandia, Lithothamnium, Melobesia, Porolithon, Schmitziella, Solenopora*

ORDER 7. Gigartinales: moderate to large foliaceous or fleshy algae; taxonomy based upon reproductive organ cytology; *Chondrus crispus* (Irish moss), *Eucheuma, Gigartina, Gracilaria, Iridaea, Schizymenia*

ORDER 8. Rhodymeniales: moderately large plants with multiseriate growth; taxonomy based upon reproductive organ cytology; *Palmaria, Rhodymenia*

ORDER 9. Ceramiales: 250 genera of mostly filamentous or foliaceous forms; taxonomy based upon reproductive organ cytology; *Centroceras, Ceramium, Chondria, Laurencia, Polysiphonia*

---

*This subclass has life cycles that are generally characterized as haplontic or composed of two successive haploid generations.

†This subclass has life cycles that are mostly characterized by a progressive postponement of meiosis resulting in an alternation of generations, with two and frequently three separate generations.

RHODOPHYTA

Rhodophyta are distinct from other algal divisions in a number of respects, including their cellular structure, biochemistry, ecology, and life history. They are commonly known as the "red algae" because most of them have a red or purple color. Some, however, are black or dark green. They are the only division in the red line of algal evolution (see Fig. 1-5). Many of the macroscopic "seaweeds" found in shallow-water marine environments of tropical and subtropical areas are members of this division. They are not so apparent in the cold arctic waters, even though many species grow there. Some species have succeeded in invading the freshwater environments, where they grow on rocks in flowing or still cold mountain waters. Other red algae have adapted to growth in deep-water marine environments, where they may be found at depths of 200 m. Many of the calcareous algae belong to Rhodophyta, and some even appear more like rocks than algae because of their thick calcareous cell walls. These types are important in the formation of atolls and tropical reefs.

Most of the red algae are relatively small plants, seldom attaining more than 1 m in height. A few are unicellular, some are filamentous, but most tend to be macroscopic three-dimensional forms, with the adult organism usually a result of the aggregation of numerous filaments forming a pseudoparenchymatous, multiseriate type of construction that is reminiscent of that seen in the brown alga, *Desmarestia* (see Fig. 7-7).

Classification of the red algae has historically been based upon the life cycles of the particular species. Research within the past few decades has indicated that the life histories of many of the described species are much more complicated than was originally thought. Therefore the sharp lines drawn between various groups (especially the subclasses) should be considered more of a taxonomic aid than an evolutionary relationship. Much of the recent taxonomic work is based upon the cytology of the development of a third generation, the carposporophyte. The red algae have introduced a new generation into their life histories so that they have **triphasic life cycles** rather than the diphasic life cycles of the other algal divisions. Because of this complication phycologists employ a separate set of terms to describe the diploid generations—the **carposporophyte** and the **tetrasporophyte.**

CELL STRUCTURE

Rhodophyte cells are eukaryotic as in most other algae. Each cell has the usual complement of endoplasmic reticulum, ribosomes, Golgi apparatus, and mitochondria. The nuclei contain nucleoli and chromosomes; however, some differences exist between red algal cells and those of other algae. These differences provide the basis for segregation of the red algae into a separate division.

The cell walls of the rhodophytes are unique in many respects. They are constructed of an inner layer of carbohydrate material and an outer cuticle formed from various mucilaginous substances. The Bangiophycidae have an inner wall of microfibrils composed of a polymer of a special four-carbon sugar, xylose. The Florideophycidae contain inner walls formed by randomly arranged fibrils of cellulose.

The outer cuticular layer of the cell walls is of special interest because it is composed of amorphous, mucilaginous substances called phycocolloids, which have important commercial applications (see Economic Importance, pp. 226-229). Chemically the cuticle possesses a number of nitrogen-containing organic compounds. In the Bangiophycidae the cuticle is apparently formed by proteins, the specific character of which has not yet been determined. In the Florideophycidae it is composed of water-soluble mucilaginous compounds made up of **sulfated polysaccharides** (usually sulfated **galactans**) including a number of commercially important products such as **carrageenan** and **agar.** Up to 70% of the dry weight of some cells may be formed by the sulphated compounds. (An excellent review of the subject is given by Percival and McDowell, 1967.)

Calcium carbonate in the form of calcite or aragonite may also form a major part of the cell wall in some red algae (see Table 1-2). Mixtures of calcite and aragonite are *not* found among the algae as they are in some invertebrate organisms. The amount and form of the carbonate are apparently environmentally controlled, since members of the Florideophycidae in culture often fail to produce as much as they do under natural conditions.

The cell walls of the more advanced rhodophytes contain unique **pit connections** whose formation is a function of specific cellular activity and whose structure resembles the septal plugs found in fungal ascomycetes (Ramus, 1969). The pit connections, readily visible with the light microscope, often provide positive diagnostic features for identification of red algae. They occur commonly in filaments of the higher reds, but in the Bangiophycidae they occur only in **conchocelis stages.**

Pit connections consist of an opening in the cell wall that is usually plugged with an acid polysaccharide-protein complex (Ramus, 1971). Two types of pit connections may be distinguished based upon their ontogenetic development: **primary pit connections** are the result of equal division by cells growing in a filament; **secondary pit connections** are the result of an unequal cell division by a cell in one filament followed by cell fusion of the small cell with an adjacent cell or filament (Fig. 11-1). Such structures create a three-dimensional thallus that has been interpreted as a pseudoparenchyma because the secondary pit connections form a false parenchyma.* Many of the three-dimensional red algae grow in this manner.

*A true parenchymatous construction would involve a three-dimensional growth pattern from the apical meristem forming a cortex bound within the epidermis.

The chloroplasts of the red algae form a distinctive structure that may be taxonomically significant. Some of the more primitive unicellular or filamentous rhodophytes possess cells in which the chloroplast is a single star-shaped (stellate) structure containing one central pyrenoid. In the advanced, more complex species, chloroplasts usually appear as numerous disk-shaped cytoplasmic inclusions. The ultrastructure of the plastid is quite different from that observed in other eukaryotic algae. It is bounded by a double membrane but lacks the CER found in other algae divisions (see Fig. 1-7). Within the chloroplasts, thylakoids occur as single-membrane structures that are reminiscent of both the structure and biochemistry of the photosensitive apparatus associated with Cyanobacteria. This thylakoid structure is unique among the algae. Each thylakoid contains chlorophyll *a* and in some cases chlorophyll *d*. Attached to the thylakoids is a series of globules called phycobilisomes (see Fig. 1-7). Each phycobilisome is composed of phycobiliproteins containing various photoactive pigments, including phycocyanin, allophycocyanin, and phycoerythrin. These three pigments, collectively known as phycobilins, are generally responsible for the various colors expressed by rhodophyte thalli in natural environments. Large quantities of the red pigment phycoerythrin are found in most cells, accounting for the red color of most rhodophytes; however, when the ratio changes among phycoerythrin, blue-green-colored phycocyanin, and the green-colored chlorophyll, the alga's color will range in a manner reminiscent of the Cyanobacteria. Changes in pigment ratios often occur in response to altered environmental conditions; these are referred to as chromatic adaptations or intensity adaptations (Dring, 1981). Such accommodation is very important to algae living in deep water where only low-intensity blue-green light is available for absorption and photosynthesis.

Chloroplasts of red algae also contain relatively small quantities of carotenoid pigments, including the yellow-orange carotenes and the orange-red xanthophylls. Chloroplast DNA is also present.

The energy storage compound in the red algae is composed of a branched polymer of glucose and is similar in many respects to both the

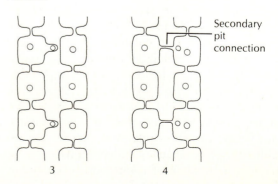

FIG. 11-1 Primary and secondary pit connections. Four-step sequence illustrating developments of secondary pit connections in multiseriate construction found in some red algae. Secondary pit connections form a structurally stable pseudoparenchymatous thallus.

Two adjacent cellular filaments

Primary pit connection

Mitosis

Secondary pit connection

1   2   3   4

amylopectin "starch" found in higher plants and green algae and the "myxophycean starch" of Cyanobacteria. It differs from these two starches in degree of branching associated with the glucose polymers and is therefore called **floridean starch** (Table 1-3). Chemically it is an α 1-4, 1-6 glucose polymer much like animal starch (glycogen). In many red algal cells the cytoplasm contains membrane-bound granules filled with floridean starch.

FIG. 11-2 Life cycles of red algae.
Three types of life cycles are represented in this outline: haplontic, diplohaplontic, and the triphasic diplo-diplohaplontic.

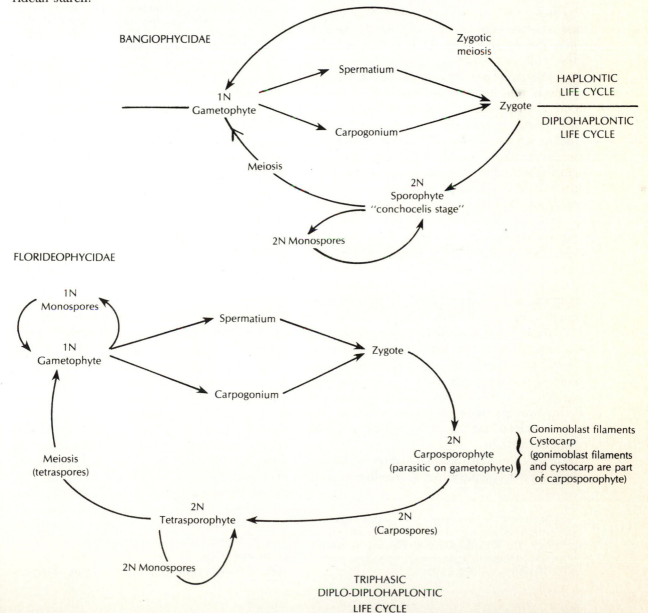

REPRODUCTION AND GROWTH

Reproduction in the Rhodophyta is unique; consequently, life histories in this division do not fit neatly into the classic diphasic haplontic, diplohaplontic, and diplontic life cycles presented in Chapter 1. The red algae have apparently evolved along a separate branch and have produced a triphasic type of life cycle that follows a 1N–2N–2N pattern,* thus introducing a second diploid generation (Fig. 11-2). Triphasic life cycles are common in the Florideophycidae and are best studied in characteristic genera such as *Nemalion* (Fig. 11-4) or *Polysiphonia* (Fig. 11-5).

In the more primitive subclass, the Bangiophycidae, life histories are less studied and in most species poorly understood, in part because most species apparently lack sexual processes. The first conclusive documentation of sexual fusion in *Porphyra* was given by Hawkes (1978). Evidently the life cycles in this group are haplontic in most organisms, although they are diplohaplontic in *Porphyra* (Fig. 11-3) and *Bangia*. In both these genera a microscopic, encrusting, filamentous diploid phase, the conchocelis stage, has been identified (Drew, 1949; Richardson and Dixon, 1958). The conchocelis stage of *Porphyra* exists as a reddish encrustation often found growing on shells and rocks or as an epiphyte on larger algae. In the past, conchocelis stages were identified as a separate genus, *Conchocelis*. It is illustrated in Fig. 11-3 as the diploid phase in the life cycle of *Porphyra*. As more of the Bangiophycidae are carefully cultured and studied, the diploid stages in their life cycles are classified, although some species appear to have lost the ability to form a diploid phase altogether.

Perhaps the most unusual feature of the red algae is their complete lack of flagellate cells at

---

*An independent 1N gametophyte produces a 2N epiphytic carposporophyte, which produces a 2N independent tetrasporophyte. The 2N tetrasporophyte then regenerates the independent 1N gametophyte through the meiotic production of 1N tetraspores (Fig. 11-2).

FIG. 11-3 Life cycle of *Porphyra* **(Bangiophycidae).**
Porphyra is a rather common, sheetlike, red-purple alga found growing attached or free floating in warm temperate and tropical marine shallow waters. It resembles *Ulva* in its habit. As seen here, *Porphyra* has a variable type of life cycle, depending on the particular species. Some are haplontic, producing carpospores that divide by meiosis, directly producing four haploid *Porphyra* plants. Others form microscopic diploid phases (conchocelis stages), thus following a diplohaplontic life cycle. A third type (not illustrated) produces a diploid, diminutive *Porphyra* vegetative thallus that resembles the gametophyte and that forms haploid monospores. These in turn develop into haploid gametophytic plants.

any time during their life histories. Because of this special quality the nonmotile sperm and stationary female reproductive structures are assigned special designations. Sperm cells are called **spermatia.** They are passively transferred to the female by water currents or through strands of secreted slime that stick to the hairlike **trichogynes** associated with the female reproductive cells (Neushul, 1972).

The female reproductive structure is composed of an egg cell (oogonium) to which is attached a long hairlike process, the trichogyne (Figs. 11-4 and 11-5). The oogonium with its trichogyne is called a **carpogonium.** In the more advanced orders, following fertilization there occurs a rather complicated series of cytological events involving nuclear transfer of the diploid zygote nucleus to other nearby cells called **auxiliary cells.** Cell fusion is followed by mitosis and cell wall formation, producing a new generation known as the carposporophyte. The epiphytic carposporophyte is composed of a series of diploid cells called the **gonimoblast filaments** (Fig. 11-4). All the gonimoblast filaments taken together form the diploid carposporophyte. If, as in some genera, the gonimoblast filaments are enclosed within an envelope of haploid cells (the **pericarp**), they are called a **cystocarp** (Figs. 11-4 and 11-5). A complete review of these events and the problems of repro-

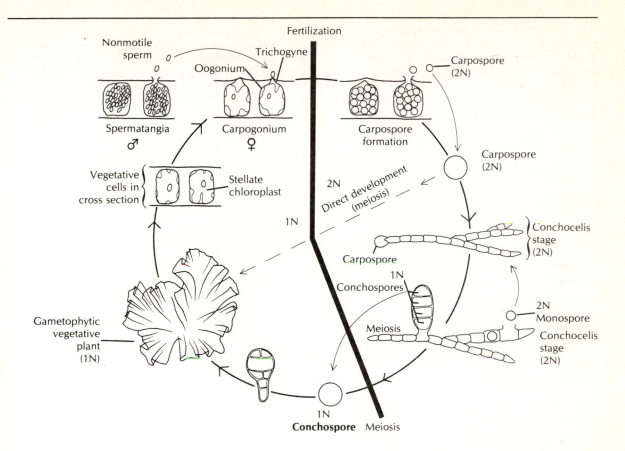

duction and life history development can be found in Dixon (1973).

The mature carposporophyte produces diploid **carpospores,** which are released and germinate to form an independent, free-living diploid generation, the tetrasporophyte. The tetrasporophyte may be morphologically identical (isomorphic) to the haploid gametophyte (e.g., *Polysiphonia,* Fig. 11-5) or it may have a different (heteromorphic) form, as in some *Nemalion* species (Fig. 11-4). The mature tetrasporophyte produces **tetrasporangia** in which cells undergo meiosis, producing haploid **tetraspores.** These haploid spores are released from the plant and germinate to produce the haploid gametophyte, thus completing the life cycle.

It would appear that during the course of evolution the red algae have developed a delay in the timing of meiotic cell divisions, probably as a device for ensuring some genetic variation (Searles, 1980). The more primitive early types are characterized by zygotic meiosis and therefore have haplontic life cycles. The more advanced (derived) forms have delayed meiosis until after the first diploid generation (as in some species of *Liagora*). Further delay in meiosis created the second diploid generation, producing a triphasic **diplo-diplohaplontic life cycle** (as in *Polysiphonia* and many other Florideophycidae).

A further complexity in the red algae is found when asexual reproduction is considered. Many

FIG. 11-4 Life cycle of *Nemalion* (Nemalionales).

Nemalion has a triphasic diplo-diplohaplontic life cycle with two successive diploid generations (carposporophyte and tetrasporophyte), followed by the haploid gametophyte. Illustration shows dioecious gametophytes; however, monoecious forms are also known. Note that the carposporophyte is parasitic on the female gametophyte; it consists of diploid gonimoblast filaments enclosed within a pericarp covering, thus forming a cystocarp. Carpospores, produced by the carposporophyte filaments, give rise to a microscopic, independent, diploid tetrasporophyte. It produces tetraspores by the process of meiosis. Tetraspores germinate to form independent gametophytes, thus completing the cycle.

Nemalion is a small (up to 25 cm) common marine alga in rock-bottomed, shallow-water habitats in the cooler temperate coastal regions. It is often found within the intertidal zones growing on rocks. The reddish-brown plants have a gelatinous consistency. (Note: The microscopic tetrasporophyte of *Nemalion*, which closely resembles the genus *Acrochaetium*, is sometimes called the "acrochaetium stage" in a manner similar to the "conchocelis stages" of other red algae (e.g., *Porphyra*).

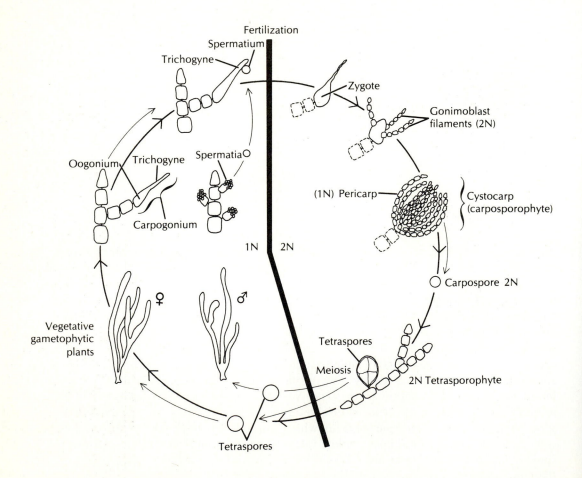

FIG. 11-5 Life cycle of *Polysiphonia* (Ceramiales).

Polysiphonia has a triphasic life cycle similar to that of *Nemalion*, except that the tetrasporophyte is macroscopic and is isomorphic with the dioecious gametophytic generation. The carposporophyte develops within a cystocarp and is parasitic upon the female gametophyte. Vegetative and reproductive structures of *Polysiphonia* and other red algae are complex and difficult to study because of the siphonaceous nature of the thalli and the degree of cellular differentiation associated with the reproductive structures.

Polysiphonia is a cosmopolitan genus of small size (5 to 40 cm in height). Its color ranges from bright red to purple to black. Its 150 species may be found in all marine habitats in all latitudes. One species, *P. nodosum*, is an epiphyte on rockweeds only (especially *Ascophyllum*).

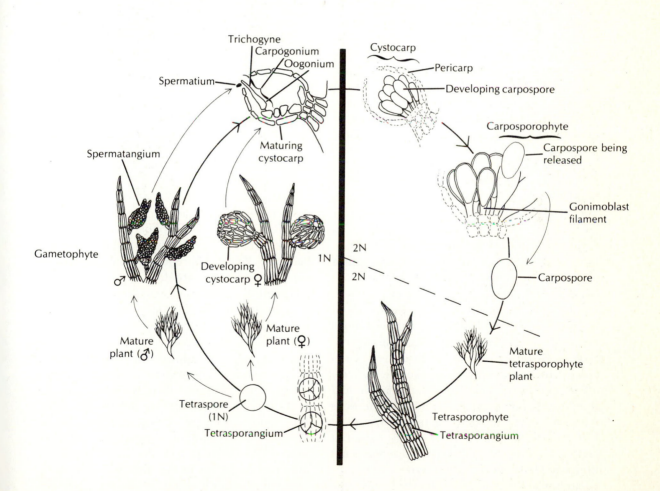

species have the capability of reproducing by fragmentation. Pieces of a filament of red algae may regenerate an entire new plant through this method. Well-developed fragmentation is seen in the marine epiphyte, *Centroceras*, which produces missile-shaped, multicellular, asexual structures. These specialized structures, called propagules, are carried by ocean currents to other host plants. When the propagules land on a suitable host, evidently they can put down rhizoids and elongate rapidly. Within 4 days some individuals are well enough established on their new hosts to launch propagules of their own (Lipkin, 1977). Propagules have also been reported in *Polysiphonia* (Kapraun, 1977).

Asexual spores are also produced by all genera. Several types of these spores may be identified, the most common of which is the **monospore** (Fig. 11-3). Indeed, in many of the Bangiophycidae, monospores are the only means of reproduction. They are produced in single-celled structures (**monosporangia**) or in sporangia containing a small number of spores (**parasporangia**), and they may be either haploid or diploid.

Growth in the Rhodophyta is generally accomplished through cell division in apical meristematic cells. In a few species of unicellular reds and filamentous forms, growth is a function of mitotic cell divisions throughout the organism; therefore, those species possess a **diffuse meristem.** The more advanced bladelike (**foliaceous**) and pseudoparenchymatous genera grow by cell divisions in specialized apical areas. In filamentous and pseudoparenchymatous forms there is a single dome-shaped apical cell; in foliaceous algae there is a marginal meristem— a single row of cells at the margin of the thallus.

The degree of cellular differentiation and therefore thallus complexity in red algae does not approach that found in the brown algae. The thalli of red algae have cells of various sizes, but the degree of morphological specialization is not pronounced. However, three types of cells associated with the surface of the thallus deserve special recognition. First, the secretory cells that are found on some Florideophycidae appear to function in halogen metabolism. These cells contain high concentrations of bromine or iodine (Fenical, 1975). Second, many of the foliaceous red algae produce colorless sterile hairs that have no apparent function. Could they possibly play a role in absorption of nutrients from the surrounding environment? These hairs are especially confusing because they resemble hairlike trichogynes found on carpogonia. Third, many red algae produce unicellular or multicellular rhizoidal holdfasts that serve to attach the thalli to hard substrates.

PHYSIOLOGY
Pigments and Storage Products

The rhodophytes contain the photosensitive pigments chlorophyll *a*, phycoerythrin, phycocyanin, α- and β-carotene, and the xanthophyll **lutein.** Several advanced members of the Florideophyceae also contain chlorophyll *d*, which appears to be a slightly altered form of chlorophyll *a* and absorbs light at 675 nm (Dixon, 1973; Meeks, 1974). Phycoerythrin and phycocyanin are water soluble and are found in Cyanobacteria and the cryptomonads (Chapter 9) in a slightly different form. Various phycobilins with different light-absorption peaks have been isolated from the rhodophytes and from Cyanobacteria. These different types of phycobilins are differentiated by prefixes (e.g., R, B, C) to designate their different absorption spectra.

Red algae of the sublittoral zone usually appear red due to the predominance of phycoerythrin. In the intertidal zone, however, the rhodophytes may appear blue-green, green, purple or black depending on the ratios of pigments. In addition phycoerythrin may be broken down by the intense light of the intertidal zone causing any red color in the algae to disappear. Two additional pigments—carmine-purple **floridorubin** and the blue allophycocyanin—may occur in the red algae. It is questionable whether floridorubin has any photosynthetic role. Allophycocyanin may be involved in the photoperiodism observed in some red algae because its chemical structure is similar to phytochrome, which is a nonphotosynthetic pigment involved in photoperiodism in the flowering plants (angiosperms). In the red algae and Cyanobacteria allophycocyanin absorbs light energy (Goodwin, 1974) and can transfer the energy to chlorophyll (Ley et al, 1977).

The major energy storage product, floridean starch, is a highly branched polysaccharide like the animal storage product glycogen, but chemical tests place it in the amylopectin category (Craigie, 1974). Red algae living deep in the ocean often have exceptionally large amounts of floridean starch that provide them with a reserve energy supply when adequate light energy is not available for photosynthesis. For example, the deep-water marine form *Constantinea* contains almost as much starch as a potato tuber, albeit floridean starch rather than regular starch (Dawson, 1966).

Other energy storage substances found in the red algae are galactosides (floridoside and isofloridoside), mannoglyceric acid, **trehalose** (a disaccharide), mannitol, and several other alcohols. Trehalose is found most often in the freshwater red algae.

The occurrence of accessory pigments (carotenes, xanthophylls, and phycobilins) in the rhodophytes enables them to utilize light energy in wavelengths that are not efficiently absorbed by chlorophyll *a*. The blue and green wavelengths penetrate deepest into water and it is these blue and green wavelengths that the phycobilins efficiently absorb (see Fig. 2-7). Light energy absorbed by accessory pigments is transferred to chlorophyll for the completion of the photosynthetic process. Studies of ultrastructure indicate that the accessory (antenna) pigments are physically located very close to chlorophyll in the cell, and the transfer of light energy, therefore, can take place very rapidly, in 0.5×10^{-9} seconds (Dixon, 1973).

When one or more accessory (antenna) pigments occur in combination with chlorophyll *a*, the photosynthetic efficiency of the alga is often increased. This increase in efficiency is known as the **Emerson enhancement effect;** this effect has been observed in many marine algae, including Rhodophyta (Govindjee and Braun, 1974).

Chromatic and Intensity Adaptation

Chromatic adaptation in the algae in general and in the Rhodophyta in particular has been widely debated. The theory of complementary chromatic adaptation maintains that the color of algae is complementary to the prevailing light that illuminates them (Engelmann, 1883). This theory has been used to explain the vertical distribution of algae in the coastal areas—greens in shallow water where all visible wavelengths abound, the browns in middepths where blue, green, and yellow light penetrates, and the reds in the deepest water where only green light is found (see Fig. 2-5). Some modern evidence has tended to support Englemann's theory. Levring (1947) demonstrates that photosynthesis in green algae decreased with depth in coastal waters more rapidly than photosynthesis in the red algae, indicating an ability in the reds to photosynthesize efficiently at greater depths. Further investigations determined that individual algae could alter the various pigment concentrations in an apparent response to different wavelengths of light in their environment (Govindjee and Braun, 1974). Pigment composition in algal cells is not static, and algae have some biochemical flexibility in responding to such environmental factors as light (Halldahl, 1970).

Other investigators (Oltmanns, 1892) maintained that changes in pigment concentrations took place in response to light quantity (intensity) rather than to light quality (wavelength). This theory is called **intensity adaptation.** Still other researchers thought that pigment changes were in response to both factors—light quality and quantity (Biebl, 1962).

Changes in pigment concentrations or ratios have been observed in most algal divisions, but the best example is seen in the chlorophyll *a*–

phycobilin complex of Rhodophyta and Cyanobacteria. The rhodophytes contain photosensitive pigments that enable them to absorb light energy from a wide spectrum of wavelengths. It is the presence of the accessory photosensitive pigments, the phycobilins and especially phycoerythrin, that enable these algae to efficiently photosynthesize in the green, low-intensity light of deep coastal waters (see Figs. 2-4 and 2-6). Green light is readily absorbed by phycoerythrin and to a lesser degree by phycocyanin (see Fig. 2-7). Red algae with high concentrations of phycoerythrin often grow deep in the sublittoral zone.

Some red algal genera can respond to variations in both light quality (wavelength or spectral composition) and light quantity (intensity) by altering the total pigment concentrations or the ratio of one pigment to another. Such differences in relative concentrations of pigments and proportions of the pigments can be observed in members of the same species from habitats with different light quality and quantity (Ramus et al, 1976a). Thus red algae located near the surface of the ocean where light intensities are high will often appear green or purple because of a high percentage of chlorophyll *a* and a low percentage of phycoerythrin. However, the same species from deeper waters will appear deep red or pink as a result of the high content of phycoerythrin. For example, *Chondrus crispus* growing in the intertidal zone will be bleached and greenish, while shaded or deep-water plants will appear dark red (Taylor, 1957). Deep-water rhodophytes may be killed by the high light intensities that illuminate the intertidal zone (Dawson, 1966).

It is well known that both light wavelength (quality) and light intensity (quantity) influence photosynthetic rates. The theory concerning chromatic adaptation implies that only light wavelength is important to photosynthesis. Dring (1981) and Ramus et al (1976a, 1976b) have demonstrated that the ratio of phycoerythrin to chlorophyll *a* increases with decreasing light intensity and increasing water depth. This was demonstrated in both laboratory and field studies involving a number of the Bangiophycidae and the Florideophycidae. Phycoerythrin content increased in response to low light intensities and not in response to green wavelengths (Dring, 1981). It would appear that chromatic adaptation is a by-product of the alga's adaptation to low light intensities and not a response to increasing green light.

As light travels through water, not only are the wavelengths of light altered but the intensity of light also diminishes (see Fig. 2-4). Algae living at significant aquatic depths must therefore adapt to photosynthesizing with green light of low intensity. The quality and quantity of light are separate variables, and it is difficult to isolate the effects of one from the other. Dring (1981) has experimentally accomplished their separation by keeping spectral composition (wavelength) constant while varying light intensity. The results indicate that the "chromatic adaptation" concept would be more accurate if it were renamed "intensity adaptation." However, it is clear that both the quality and the quantity of light are important in determining the distribution of algae in the environment. The relative roles of both will continue to be investigated along with the many other physical factors which determine algal distribution.

Another factor that must be considered in evaluating algal response to light is the morphological structure of the thallus. Ramus et al (1976a, 1976b) found that the membranous *Porphyra* (Fig. 11-3) and *Ulva* (Chlorophyta, Fig. 4-6) did not absorb incident light as well as the fleshy thalli of *Chondrus* (Fig. 11-9) and *Codium* (Chlorophyta, Fig. 4-7). If an alga can absorb most of the incident light it receives, it will be able to photosynthesize efficiently deep in the sublittoral zone—and this is one of the habitats in which both *Chondrus* and *Codium* are found. These algae appear black to divers observing them on the ocean floor, indicating that a maximum amount of light is being absorbed and a minimum amount is being reflected.

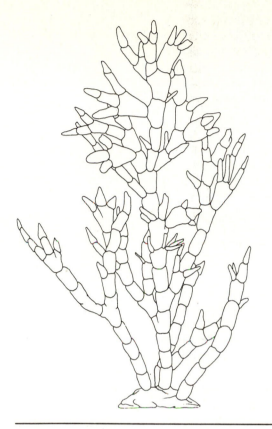

FIG. 11-6 Habit of *Corallina.*

Corallina is an example of one type of calcareous algae, the jointed or articulated type, found in the order Cryptonemiales. *Corallina* is a common warm- and temperate-water genus found growing in shallow marine areas (to depths of 30 m). *Corallina* grows in the intertidal and sublittoral zones.

Calcium Carbonate Deposition

Many red algae deposit calcium carbonate on their cell walls, especially in warm ocean waters. Since tropical and subtropical waters are usually saturated or supersaturated with calcium carbonate, it requires minimal energy for certain algae to deposit the material. The solubility of calcium carbonate is temperature dependent, i.e., solubility is decreased in warm water. Calcium carbonate deposition, therefore, is less prevalent in cold regions, but it does occur in many red algae of temperate regions and even in the polar regions. Deposition of calcium carbonate occurs in one of two crystalline forms—the rhombohedral calcite or the orthorhombic aragonite. These two forms differ in such physical properties as solubility, specific gravity, and hardness. Coralline algae deposit calcite and certain members of the Nemalionales contain aragonite. Magnesium and strontium carbonate are also deposited by some red algae. The mechanisms by which calcareous algae deposit calcium carbonate are complex—in the laboratory some red algae will not deposit calcium carbonate or will deposit much less than in the natural environment (Dixon, 1973; Digby, 1977).

The coralline algae (Cryptonemiales) contain two types of calcareous algae—**articulated** (jointed) and **nonarticulated** (**nullipore** or crustose). *Corallina* (Fig. 11-6) is an example of an articulated alga; *Porolithon* and *Lithothamnium* (Fig. 11-7) are crustose forms. The crustose red

FIG. 11-7 Habit of *Lithothamnium pacificum.*

Lithothamnium pacificum is a crustose rhodophyte that grows on the western coast of North America. The alga may be found on rocks and shells in the intertidal zone. Similar species of *Lithothamnium* are important in the formation of coral reefs. *Lithothamnium* spp. and *Porolithon* spp. are the dominant algae in the formation of the algal ridge on the windward side of reefs.

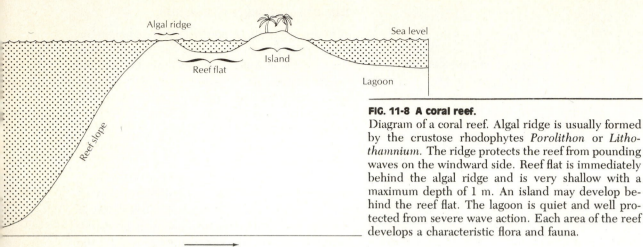

FIG. 11-8 A coral reef.
Diagram of a coral reef. Algal ridge is usually formed by the crustose rhodophytes *Porolithon* or *Lithothamnium*. The ridge protects the reef from pounding waves on the windward side. Reef flat is immediately behind the algal ridge and is very shallow with a maximum depth of 1 m. An island may develop behind the reef flat. The lagoon is quiet and well protected from severe wave action. Each area of the reef develops a characteristic flora and fauna.

algae are extremely important in the formation of coral reefs in shallow waters. These algae form an elevated algal ridge on the ocean side of the reef and help to hold the reef together (Fig. 11-8). Sometimes the red algae may totally exclude the coral animals from the reef. *Porolithon* and *Lithothamnium* are the most common algae of coral reefs and atolls.

Utilization of Organic Molecules

Several red algae can utilize organic molecules (heterotrophy). *Corallina* and *Gelidium*, for example, can assimilate glycine from dilute solutions. The ability to absorb organic compounds is advantageous to species living near sewage outfalls that are laden with organic material or in the arctic where light is scarce but where organic materials may be abundant from highly productive summer algal blooms. The coralline red algae, especially, appear well adapted to utilizing the organic nutrients found in sewage.

Halogen Metabolism

The metabolism of halogens is well developed in some genera of the Florideophycidae (e.g., *Bonnemaisonia*), therefore these genera produce halogen-containing metabolites. These metabolites are found not only in the secretory cells but also in some red algae without secretory cells (Dixon, 1973). Studies of the secretory cells demonstrate that they contain unusual chlorine, bromine, or iodine compounds and that they appear to be the sites of halogen metabolism. When iodine-containing genera are placed on herbarium paper, the concentration of iodine may be sufficient to leave a blue imprint on the starch-containing paper, since iodine turns starch blue. The function of the secretory cells is not certain at this time, but several beneficial functions have been proposed, including deterrent activity against bacteria and invertebrate grazers. (See Fenical, 1975, for a detailed review of halogen metabolism.)

Epiphytes, Endophytes, and Parasites

Many red algae grow epiphytically or parasitically on other algae, frequently on closely related taxa. The range of epiphytic reds includes: (1) superficial epiphytes that use the host only as a base on which to grow, (2) obligate epiphytes that grow only on specific hosts, (3) partial parasites that have some photosynthetic activity, and (4) parasites without chlorophyll that have no photosynthetic ability. Using radioactive labeled phosphorous (P[32]), researchers have demonstrated transfer of nutrients from the host alga to the epiphyte even in superficial epiphytes (Linskens, 1963). Some nutrient transfer from epiphyte to host also occurs, but not in large quantities (Dixon, 1973). Fan (1961) proposes that parasites develop easily from epiphytes in Florideophyceae, because members of this family readily form secondary pit connections and can therefore penetrate host tissue and absorb nutrients.

Forty genera of the Florideophycidae are parasitic and 90% of these parasitic forms are very closely related (Dawson, 1966). The extent of parasitism is questionable. Dixon (1973) thinks that the evidence is weak for the hypothesis that many epiphytes are actually parasitic on their hosts. Epiphytic or parasitic species often form **galls** or **tubercules** (small growths) on their very closely related (e.g., same family) host plant. Secondary pit connections (Fig. 11-1) are more easily formed with species that are closely related biochemically. For example, *Gracilaria* is "parasitized" by its minute and colorless close relative *Gracilariophila*, which forms tubercules on its host (Abbott and Hollenberg, 1976). The evolutionary origin of the "parasitic" reds is thought to have taken place in one of two ways—either from a spore that remained on the plant and mutated to become parasitic (Setchell, 1918) or from an epiphyte that lost its photosynthetic ability, formed secondary pit connections, and became parasitic (Sturch, 1926). Both epiphytic and parasitic red algae exhibit reduced morphological complexity and pigmentation, with such reduction being most pronounced in the parasites. Many of the epiphytes are obligate; for example, *Smithora* grows on the angiosperm eelgrass *Zostera* or on the surf grass *Phyllospadix*. Some rhodophytes are endophytic in other rhodophytes, in algae, or in vascular plants. For example, *Schmitziella* grows inside the green alga *Cladophora*.

Motility

Motility is not usually observed among the red algae, since the absence of flagella is a char-

acteristic of the division. However, some cells exhibit amoeboid movement. Monospores may move several cell diameters (Nichols and Lissant, 1967). Members of the unicellular order Porphyridiales show a gliding or amoeboid motion that leaves a mucilaginous trail. Some spermatia and tetraspores also exhibit amoeboid movement (West, 1968).

Much speculation has taken place about how the sperm reaches the egg in the red algae. The spermatia must rely on water currents to convey them to the carpogonium. Some mechanism of attraction, probably chemical, has been postulated, but this has not been demonstrated experimentally. Some attraction process must be operating, because the fertilization process is much more efficient than it would be if the sperm reached the egg only by chance. Some Rhodophyta produce mucilaginous strands that entrap the spermatia and facilitate their transfer to the carpogonium. Neushall (1972) proposes that the probability of fertilization is increased by the transfer of such spermatia-containing strands of slime to the carpogonium.

Influence of Environment on Life Cycles and Morphology

The life cycle of a red alga may be influenced by various environmental factors. The type of spore (carpospore or monospore) produced and the stage of the life cycle that develops may be regulated by either the length of daylight (photoperiod) or temperature. Nutrient levels also may regulate the stage of the life cycle that develops in some genera. *Bangia* provides a good example of the influence of the length of daylight. With less than 12 hours of sunlight some *Bangia* plants will produce monospores that will grow into a blade genetically identical to the plant that produced them. With more than 12 hours of sunlight carpospores will be produced and they will develop into the minute "conchocelis" stage (Richardson, 1970). The conchocelis stage can be maintained in the laboratory for up to 3 years as long as 16 hours of sunlight are provided. The life cycle of *Bangia* also apparently is influenced by temperature because only monospores are produced at high temperatures (20° to 22° C) while both carpospores and monospores may be produced at lower temperatures (8° to 10° C). Photoperiod evidently interacts with temperature in some manner to determine what type of spore is produced and which phase of the life cycle develops.

The morphological structure of the red algal thallus varies considerably under different chemical, geographical, physical, or mechanical conditions. These variations can cause difficulties in identifying red algae. Certain members of the Bangiophycidae especially seem to have considerable flexibility in the form of their thallus. For example, *Asterocystis* forms branched filaments in environments with adequate salinity, but in less favorable environments with reduced salinity it assumes a unicellular form (Lewin and Robertson, 1971).

Intertidal Red Algae

Rhodophytes living in the intertidal zone are subjected to wide environmental fluctations in temperature, light, salinity, and moisture and have evolved mechanisms for coping with these extreme conditions. A high intertidal zone alga such as *Bangia* can remain viable after 21 days of air drying at room temperature. Sublittoral zone algae, however, usually do not survive more than an hour in such a dry environment (Dawson, 1966). The intertidal algae must also withstand changes in pH. The pH of the ocean varies from 7.5 to 8.4, but values of 9 to 10 have been recorded in tidal pools where large amounts of photosynthesizing algae may use the bicarbonate ion in photosynthesis, thereby raising the pH (see Fig. 3-7). Intertidal *Porphyra* (Figs. 11-3 and 11-9) species can adjust their internal osmotic potential to the increased salinity found in such environments as tidal pools where evaporation has removed the water and concentrated the salts. The algae accomplish this by increasing the potassium content or the floridoside or isofloridoside content of the cells. By

FIG. 11-9 Diagram of rhodophytes along a rocky coast of New England.
All seven genera may grow in the intertidal zone. *Chondrus, Corallina, Polysiphonia,* and *Palmaria* may also grow deep into the sublittoral zone. Some species of *Ceramium, Porphyra,* and *Polysiphonia* may grow epiphytically. Note the pincerlike tips on *Ceramium* and the foliaceous thallus of *Porphyra.* It is often difficult to distinguish between the dichotomously branched *Gigartina* and *Chondrus. Corallina* is an articulated coralline alga. *Palmaria* may grow to be the largest of these genera, with a maximum height of 50 cm.

increasing the internal osmotic pressure, the organism avoids cell **plasmolysis** (i.e., water loss).

Iridescence

Some red algae appear blue or green in reflected light; this differs from the red color that characterizes these algae in transmitted light. This phenomenon, known as iridescence, is produced by physical interference of light as it passes through parts of the organism. In some genera iridescence has been attributed to intracellular **iridescent bodies** associated with vacuoles, as in *Chondria* (Feldman, 1970). In other genera such as *Iridea* the phenomenon is caused by the multilayered nature of the cell wall and its accompanying cuticle (Gerwick and Lang, 1977). Iridescence is frequently found in members of the Gigartinales and Rhodymeniales.

Galls

Irregular masses of cells known as galls are found on some rhodophytes. The galls vary in size, with many being macroscopic. Some galls may be caused by an epiphytic or endophytic red alga that may be closely related to the host. Potentially parasitic harpacticoid copepods, nematodes, and fungi have also been isolated from some galls. The incidence of galls appears to increase near sources of industrial waste. *Porphyra* species growing near chemical waste

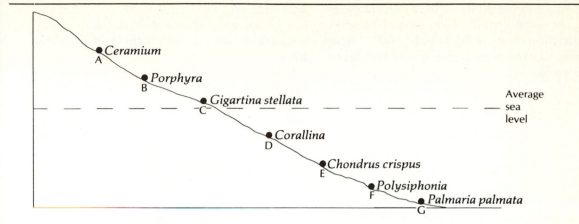

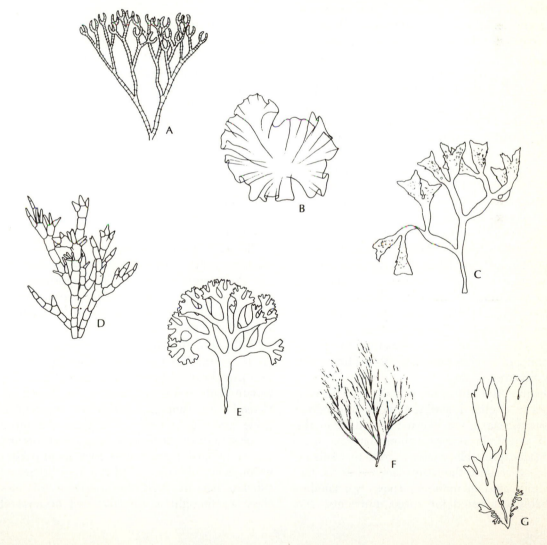

outflows in Japan exhibit a high percentage of such abnormal growths composed of binucleate and trinucleate cells (Dixon, 1973). These growths may represent a response by the alga to toxic wastes.

EVOLUTION

The calcareous rhodophytes have left an excellent fossil record. The noncalcareous reds, unfortunately, have left very few fossils because of the delicate nature of their thalli.

Fossils of red algae have been found in rock strata of the Cambrian period. Some investigators think the red algae may have originated in the Precambrian period. The fossils from the upper Cambrian period belong to an extinct red algal family, the calcareous Solenoporaceae. Members of this family were important rock builders from the Cambrian through the Cretaceous periods—a span of 470 million years. *Solenopora* was an important genus during these periods. During the Silurian period the Solenoporacea contributed significantly to the building of limestone coral reefs that are prevalent in the strata of this period. The fossil evidence indicates, however, that the coral animals were more important in the reef building process during this time than were the red algae. This is not always the case in modern times.

Another extinct family, the Gymnodiaceae, contributed to rock formation in Asian and European waters during the Permian period (Dixon, 1973). Members of this family resemble the modern genus *Galaxaura*. The Gymnodiaceae span the 150 million years from the Permian through the Cretaceous periods, after which they became extinct.

During the Jurassic period the Solenoporaceae declined and the modern coralline (Corallinaceae) algae became abundant in the fossil record. Researchers think that the coralline algae probably evolved from the Solenoporaceae because the latter declined as the former increased in number. These two families probably competed for space, nutrients, and

light and the Corallinaceae were successful. The two groups were classified in the same family until taxonomists separated them in the early 1900s.

Both encrusting (crustose) and erect articulated fossils of the coralline algae have been found in Jurassic rocks. By the Cretaceous period the coralline algae were widely distributed, and a continuous fossil record of these algae is available from the end of the Triassic period to the present. Since the Cretaceous period, it appears that the red algae have become more important than corals in the establishment and maintenance of coral reefs. Many modern algal genera are recognizable in the late Mesozoic and early Cenozoic fossils of coral reefs—for example, *Lithothamnium* (Fig. 11-7).

In the Tertiary period many coral reefs were formed, and these exist as active reefs or fossil reefs today. The Eniwetok Atoll, an active reef in the Marshall Islands, western Pacific Ocean, is about 1500 m deep and contains fossils dating back to the early Tertiary period. Other coral reefs may exceed 350 m in depth and contain many coralline algal fossils. In recent times oil deposits have been found associated with ancient coral reefs. This discovery has increased interest in the location and exploration of these fossil coral reefs.

The only fossil records extant of the noncalcareous algae are fossils of *Rhodymenia* and the Nemalionales, which are approximately 10,000 years old (Recent epoch). Another fossil, identified as *Gigartina*, was found in relatively recent rocks from Peru, about 600 years old.

Within the Rhodophyta, evolutionary trends are from primitive unicellular forms to the complex pseudoparenchymatous thallus. The multiseriate thallus seen in *Bangia* is considered advanced even though the subclass Bangiophycidae has been considered the less advanced subclass. Advanced reproductive structures and life cycles are seen in the Florideophycidae, which is usually considered the more advanced subclass. The thalli of some species in this subclass are thought to be advanced because of their heterotrichous habit and extensive branching. Oogamy is also very common in Florideophycidae and less common in Bangiophycidae.

Recent evidence reveals that the distinctions between the two subclasses are not clear cut. Pit connections in the conchocelis stage, apical growth, and sexual reproduction have all been documented in Bangiophycidae. These features were formerly considered "advanced" and characteristic of Florideophycidae. The Florideophycidae might have arisen from a pit connection–containing conchocelis form of a *Porphyra-* or *Bangia*-like extinct alga. Some investigators (Lee, 1980) have abandoned the subclass distinctions.

Skuja (1938) proposed an evolutionary scheme for the rhodophytes—that the reds originated in shallow marine waters and that the primitive reds contained phycocyanin primarily. These primitive reds then colonized shallow fresh water; many extant freshwater reds are considered primitive members of the Nemaliales. Over evolutionary time the rhodophytes developed increasing amounts of phycoerythrin and could therefore grow successfully in waters of greater depths.

The phylogeny of Rhodophyta remains clouded. Many researchers think that the red algae probably evolved from the Cyanobacteria. Evidence for this comes from the similarity of the energy storage compounds, the cell wall constituents, and the photosynthetic pigments, especially the phycobilins. *Cyanidium caldarium*, a thermophilic unicellular alga with a eukaryotic nucleus and phycobilins (see Chapter 3), may resemble an extinct intermediary form between the two groups. Similarities in chloroplast (thylakoid) ultrastructure and the lack of flagellated cells in both divisions provide further evidence for this relationship. Dixon (1973) thinks that because the biochemical similarities exceed the differences, the relationship between the reds and the blue-greens is strong. However, a major problem exists in establishing a direct connection between the two divisions because the blue-greens are prokaryotic while the reds are eukaryotic. Did membrane-bound organelles arise independently—once from the blue-greens to the reds and once from blue-greens to greens and browns? Or did membrane-bound organelles arise only once, giving rise to a single eukaryotic prototype that in turn gave rise to the red, green, and brown lines of algal evolution? Current scientific thinking favors the latter evolutionary scheme. Are the Rhodophyta an evolutionary dead end or did they give rise to a fungal group, the ascomycetes (Chapter 17)? Some investigators have proposed a relationship between the colorless parasitic red algae and the marine ascomycetous fungi (Denison and Carrol, 1966; Kohlmeyer, 1975). Other researchers, however, think that the apparent similarity between certain reproductive and other structures in the two groups occurred via parallel evolution rather than through any close phylogenetic relationship (Fritsch, 1945; Moore, 1971).

ECOLOGY

The red algae are found primarily in coastal marine environments, but freshwater and terrestrial forms also occur. Their greatest diversity is found in the warm tropical waters in both the intertidal and sublittoral zones. Algae of the coastal areas grow on hard substrates such as rocks or epiphytically on other algae or higher plants. They are found less frequently on the sandy bottoms of bays and lagoons. The sizes of rhodophytes range from microscopic unicells to some species of *Gigartina* measuring 1×2 m (Fig. 11-9). Most frequently the size range is 5 to 25 cm. Growth forms vary from filamentous to foliose to pseudoparenchymatous. A few species of red algae occasionally may be found in the phytoplankton (*Porphyridium*—Fig. 11-10), but the majority are benthic organisms. The rhodophytes are vertically distributed over a larger area than are the brown algae. Their range extends from the high intertidal zone to the sublittoral zone, which may be as deep as 200 m in clear tropical waters. The freshwater

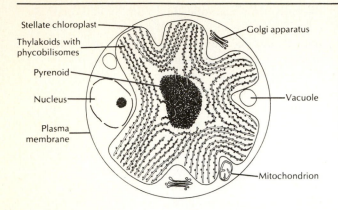

Stellate chloroplast

Thylakoids with
phycobilisomes

Pyrenoid

Nucleus

Plasma
membrane

Golgi apparatus

Vacuole

Mitochondrion

FIG. 11-10 *Porphyridium* **ultrastructure.**
Unicellular red alga *P. cruentum* as it appears in high-magnification electron microscopy. Note single star-shaped chloroplast, which contains a single centrally located pyrenoid and many single thylakoids with attached phycobilisomes (compare with Fig. 1-7). *Porphyridium* is an alga of damp soils, walls, and greenhouse areas. It grows in a gelatinous, red to purple mass. (Adapted with permission from Gantt, E., and S.F. Conti. 1965. Ultrastructure of *Porphyridium cruentum*. J. Cell Biol. **26**:365-381, Copyright 1965, The Rockefeller University Press, New York.)

reds are more often found in temperate regions in cool, fast-flowing water.

Porphyridium (Fig. 11-10) is a unicellular red alga that may be found growing terrestrially on damp soils. Sometimes it forms a blood-red, thin, gelatinous coating on greenhouse soils, pots, or walls and may form mucilaginous coatings on wet soil surfaces.

Coral Reefs

Several genera of Rhodophyta deposit calcium carbonate. Of the calcareous species, the crustose forms are especially important in providing primary production, cementing action, and strength for coral reefs and atolls. These coralline algae grow over the coral rubble, cementing and binding the plant and animal remains into a reef that is highly resistant to the pounding of the surf. The crustose algae *Lithothamnium* (Fig. 11-7) or *Porolithon*, which form an elevated ridge on the seaward margin of the reef (Fig. 11-8), are the principal cementing agents. This ridge is known as the *Porolithon* or *Lithothamnium* ridge, depending on the dominant genus. Coral reefs and atolls actually grow from the deposition of calcium by both the algae and the corals; the contribution of each organism may vary according to environmental conditions. The reefs (Fig. 11-8), which may be 45 to

60 km long, are most frequently found in tropical areas. Occasionally the red algae of coral reefs may grow so well that they exclude the corals, thereby forming a calcareous reef on top of the dead coral. Such calcareous reefs with no associated living corals are found in the Indian Ocean and occasionally in temperate and even arctic waters, even though calcium carbonate deposition occurs less readily in cool waters. Such "nullipore" reefs (also called "pavements") without corals, are found in New England and in the glacial fjords of both Norway and Greenland.

Some calcareous algae are easily confused with corals or even with inorganic rock formations. Linnaeus (1707-1778) treated calcareous reds as "zoophytes" because they were not like any other alga. Lamarck (1744-1829) used the term "nullipore" to distinguish them from the porous corals. It was not until the mid-1800s that these stony crusts (crustose) and branched nodules (articulated) became recognized as true photosynthetic organisms belonging to the red algae. These algae are now classified in the family Corallinaceae. The articulated and the crustose rhodophytes, the most prevalent calcified algae, are a part of the marine flora in all parts of the world. Their important role in the formation and maintenance of coral reefs were not recognized by ecologists until the beginning of the twentieth century. (See Goreau et al, 1979, for further information on coral reefs.)

The "nullipore" *Porolithon* thrives on the outer surface of the coral reefs. Its growth is not affected by the pounding of the surf. In fact, the stronger the surf, the better *Porolithon* grows. Evidently the limiting factor for *Porolithon* is the amount of oxygen available in the dark for respiration, and the pounding surf provides an abundant supply of dissolved oxygen even at night (Dawson, 1966). Extensive growth of the reef also occurs below the water's surface, and calcareous fragments of both algae and coral collect on the seafloor and are bound together by crustose red algae. The coralline algae often survive in areas where herbivorous sea urchins reduce the noncalcareous algae by grazing.

The coral reef as a whole is autotrophic, producing its own energy, whereas the coral is heterotrophic. The reef as an ecosystem is an extremely complex assemblage of organisms that live together in a delicate balance maintained by an equilibrium between construction and destruction. Coral reefs are some of the most productive areas in the world, and this delicate balance is essential for the maintenance of this high primary productivity. Much of the primary productivity is provided by the coralline algae and the zooxanthellae (see Chapter 9). Phytoplankton contribute little productivity to the coral reef areas, as the majority of photosynthesis takes place in the attached and endozoic algae.

Impoverished Flora

In some areas of the tropics and subtropics the algal population in general and the red algal population in particular is impoverished. Such an area is the Pacific coast of Costa Rica. On this coast a hostile environment exists for intertidal and shallow-water benthic algae. The combinations of high temperatures, high tide amplitudes, intense solar radiation, turbidity, and grazing pressure create extremely stressful conditions for attached algae. The only algae that survive in this unfavorable environment are the calcareous, crustose red algae (e.g., *Lithothamnium*), which are well adapted to low light intensities resulting from turbidity and which are resistant to heavy grazing pressure.

Coastal Benthic Algae

The benthic rhodophytes contribute significantly to primary productivity in tropical, temperate, and polar areas. The shallow-water benthos has often been overlooked as an important source of productivity. Actually, benthic productivity may exceed that of the phytoplankton in sublittoral areas.

The larger, fleshy rhodophytes are usually found in the cool waters of the temperate zone. Such algae as *Gigartina*, *Iridaea*, and *Schizymenia* grow on the Pacific coast of North America, where dissolved oxygen is plentiful because of lower temperatures (cold waters can hold more oxygen), abundant turbulance, and adequate nutrients. These plants may reach heights of 2 m. The growth of the larger benthic reds is usually physically controlled by such factors as storms, turbulence, and dissolved-oxygen levels. In the tropics the reds often dominate the aquatic flora, but they are generally small and inconspicuous. They grow in the intertidal zone and down to the depths of light penetration, which may be as deep as 200 m.

Along temperate zone rocky coasts, benthic rhodophytes can be so numerous that they can be commercially exploited for their phycocolloids. Extensive beds of *Chondrus crispus*, "Irish moss" (Fig. 11-9), are found on both coasts of the northern Atlantic. The alga usually grows on rocks (epilithic) from low in the intertidal zone down to 24 m (Dixon and Irving, 1977). Plants may live up to 6 years and may regenerate from the holdfast. Off the Canadian east coast *C. crispus* may grow underneath *Laminaria* "forests." Pure stands of *C. crispus* are found off the coast of Nova Scotia, and below these stands are large populations of *Corallina* (Fig. 11-6).

Gigartina stellata, "false Irish moss" (Fig. 11-9), may grow in the same areas as *C. crispus* along the North Atlantic coast. It is difficult to distinguish *G. stellata* from *C. crispus* because both exhibit considerable morphological flexibility and their environmental requirements are similar. *G. stellata*, however, may be found at lower latitudes than *C. crispus*. Burns and Mathieson (1972) have studied the growth and reproduction of *G. stellata* and *C. crispus* (Mathieson and Burns, 1975).

Most benthic rhodophytes populate the sublittoral zone, but several genera are widely distributed in the intertidal zone. These intertidal forms exhibit a tolerance for the extreme conditions of that zone that is not seen in most red algae (Hillson, 1977). *Bangia*, *Polysiphonia*, *Nemalion* (Fig. 11-4), and *Porphyra* are widely distributed on the temperate rocky coasts of the world. Both *Porphyra* (Fig. 11-3) and *Polysiphonia* (Fig. 11-5) also may be epiphytic. For example, *Polysiphonia lanosa* is a blackish epiphyte on the brown intertidal alga *Ascophyllum nodosum* (Fig. 7-12) or occasionally on other fucoids. However, no transfer of metabolites has been demonstrated between the fucoid host and the epiphyte (Turner and Evans, 1978).

The **agarophytes** *Gracilaria*, *Gelidium* (Fig. 11-11), and *Pterocladia* are found in coastal waters of the subtropics and tropics, whereas the carrageenan-containing *Chondrus* and *Gigartina* are found in cooler, temperate areas. *Gracilaria* thrives in shallow temperate and tropical estuaries with high nutrient contents. Most plants are small (to 20 cm) but one species may grow 2 m tall. The growth of *Gigartina verrucosa* off the North Carolina coast has been studied extensively because the alga was grown as a source of agar during World War II. The species has two growth forms—a benthic form that never becomes abundant and a planktonic, drifting form that may be very abundant. The drifting forms are concentrated by winds and currents in back bays and may be readily harvested (Michanek, 1975). Santelices (1974) provides further information on the ecology and uses of the alga. *Gelidium* grows well in south-

ern Europe, Japan, Australia, and on the eastern Pacific coast (California and Mexico), whereas *Pterocladia* is abundant in New Zealand and on the southern European coast. It may be difficult to separate taxonomically *Gelidium* and *Pterocladia*.

The calcareous reds also contribute to benthic productivity in temperate areas. The articulated coralline *Corallina officinalis* (Figs. 11-6 and 11-9) is abundant in deep northern waters, though it may also be found in tide pools or growing on

FIG. 11-11 *Gelidium* **habit.**
Gelidium spp. are indigenous to the warmer tropical and subtropical marine shallow and intertidal waters of both coasts of North America, although they are more common on the West Coast. The sulfated polysaccharides isolated from the cell walls of this genus make it a commercially important organism in some parts of the world. *G. cartilagineum* is harvested in Japan for the manufacture of agar.

mussels. Extensive "meadows" of *C. officinalis* may be found off the coast of Iceland. The articulated corallines may also grow either epiphytically in kelp "forests" (see Chapter 7) or benthically in the deep shade of these forests. Some coralline algae are colorless parasites, although others may be parasitic only during their juvenile stages. Johansen (1981) discusses coralline algae and Abbott and Hollenberg (1976) describe the many species of crustose and articulated rhodophytes along the California coast.

Rhodophytes also grow in the polar regions, where some species can survive frozen in the ice (Dixon and Irving, 1977). Generally, though, the brown (Phaeophyta) and green (Chlorophyta) algae are more numerous in polar regions.

Rhodophyte Genera with Freshwater and Marine Species

Several rhodophyte genera have freshwater and marine species. For example, the crustose but uncalcified *Hildenbrandia* (Cryptomoniales) is a reddish-brown crust that grows in the intertidal zone of both coasts of North America (Taylor, 1957; Abbott and Hollenberg, 1976). Other species grow deep in freshwater lakes at depths greater than the angiosperms grow or under stones in streams. *Hildenbrandia* grows very slowly and is usually found in soft water; it cannot grow on limestone (Hynes, 1970). Another genus, *Audouinella*, has many marine species that are epiphytic, epizoic, or endophytic and that grow on both sides of the North Atlantic. One species *(A. efflorescens)* is widely distributed in the Arctic Ocean. Dixon and Irving (1977) describe the many marine *Audouinella* species. In fresh water, *Audouinella* is free living and grows in the winter. It is replaced by *Batrachospermum* (Fig. 11-12) in the summer (Dillard, 1966). The limiting factor in the seasonal periodicity of these two algae appears to be water temperature, but light is also important.

One species of *Bangia* frequently found in the Great Lakes and in European hardwater lakes also grows in the intertidal zone. Sheath and Cole (1980) discuss the occurrence of this alga in the Great Lakes. Geesink (1973) demonstrated that the freshwater form could be adapted gradually to live in seawater—an ability not seen often in algae.

Freshwater Rhodophyta

Two percent of the described rhodophyte genera are exclusively freshwater. These algae are often not red but may appear green or gray.

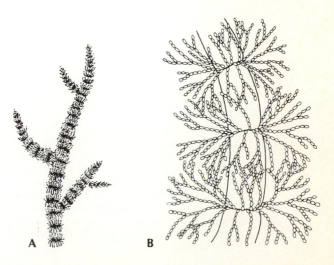

FIG. 11-12 *Batrachospermum* **habit.**
A, Freshwater species of *Batrachospermum* as it appears in cool, running mountain streams. **B,** Enlargement of the main filament with many side branches occurring at nodes. The side branch tips are the site of both male (spermatia) and female (carpogonium with attached trichogyne) reproductive structures. *Batrachospermum* appears as a green- or gray-colored plant with a mucilaginous consistency and a beadlike appearance to the unaided eye.

The most common genus *Batrachospermum* (Fig. 11-12) is found throughout the world in cold, fast-flowing waters or in cold ponds or springs. Some species may grow in *Sphagnum* (Bryophyta) bogs. Many species of *Batracho-spermum* are intolerant of high light intensities and grow in densely shaded waters or deep in lakes, down to 40 m (Hutchinson, 1975). Woelkering (1975) describes the environmental requirements of *Batrachospermum*, and Mori (1975) discusses the habitats of the alga in Japan.

Lemanea is a frequent inhabitant of cold swift water in such places as waterfalls and rapids and is intolerant of high temperatures. This alga is characterized by a 5 to 20 cm-long slender nodal thallus (Fig. 11-13) and often grows in hard-water streams where it may be covered with calcium carbonate. Magne (1969) documents the life cycle of this alga. Zonation in freshwater rhodophytes is documented by Minckley (1963). *Hildenbrandia*, *Audouinella*, and *Batracho-*

spermum are usually found upstream of areas where *Lemanea* grows. Other freshwater rhodophytes are *Compsopogon*, a tropical stream alga that may also grow epiphytically on mangrove roots in estuaries, and *Asterocystis*, which may be free-floating or epiphytic on *Cladophora* (Chlorophyta) (Prescott, 1978).

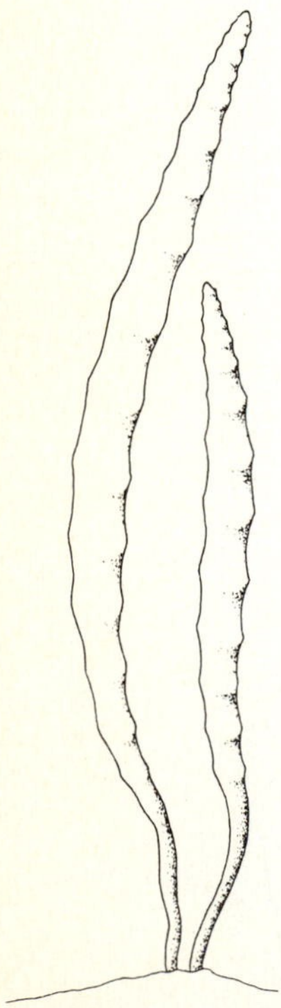

ECONOMIC IMPORTANCE

The major economic uses of the Rhodophyta are as raw material for the extraction of the commercially useful products carrageenan and agar and as human food, primarily in the Orient. These phycocolloids have the ability to form gels of varying degrees of strength when combined with water. The demand for phycocolloids, including algin from the brown algae, is increasing, and natural populations of algae are rapidly being depleted as yearly harvests increase. Percival (1979) discusses the structure and biosynthesis of these important polysaccharides. Harvesting of these algae in the wild has become very expensive because of labor costs, so efforts are being made to cultivate the algae in confined areas to reduce collecting time.

Carrageenan

The phycocolloid carrageenan,* a cellular component found in some members of the Gigartinales, is a sulphated galactan that may be extracted from *Chondrus crispus* (North Atlantic), *Iridaea* and *Gigartina stellata* (North America), and *Eucheuma* (Florida, East Indies, and the Philippines). The water-soluble carrageenan is extracted by placing the alga in boiling water and then concentrating and purifying the aqueous supernatant.

In Ireland several hundred years ago *Chondrus crispus,* "Irish moss," was gathered from the rocky shores, dried, and boiled with milk. The milk containing the extract was flavored with vanilla or fruit, allowed to cool and gel, then eaten as a pudding dessert, "blancmange."

FIG. 11-13 *Lemanea* habit.
Lemanea represents one of the few freshwater red algae. The gray-green, nonbranching, cartilaginous, spinelike growth may reach a length of 20 cm. It is characteristic of stream rapids and waterfalls, where the thalli become bent over and flattened against rocks and dam walls. Carpogonia occur internally at the swollen nodal areas, and spermatia are borne externally in clusters at the nodes.

*Various spellings are used—carrageen, carrageenin.

When the colonists arrived in the United States they missed their "blancmange" and imported *Chondrus crispus* (Fig. 11-9) so that they could make it. Eventually, to their delight, they discovered *Chondrus crispus* growing extensively on the New England rocky coast and importation was no longer necessary. "Irish moss" is still hand-harvested for personal and commercial use in New England and the Canadian Maritime Provinces.

In addition to "blancmange," carrageenan has a rapidly expanding number of commercial uses (Levring et al, 1969). *Gigartina stellata* (Fig. 11-9) has also been commercially harvested on the east and west North American coasts for its high carrageenan content. Mariculture of *G. stellata* in the sublittoral zone was attempted on the North American west coast in Puget Sound, but culture was hampered by adverse salinities and the lack of a suitable substrate. *Eucheuma* is being grown by mariculture in the Florida Keys, the Philippines, and Hawaii. Individual plants are capable of doubling their weight in 35 days. A density of 100,000 algal thalli per hectare can be maintained, but grazing sea urchin populations must be controlled to maintain this high density and productivity. Dawes et al (1974a, 1974b) discuss the ecology of *Eucheuma*. *Eucheuma* is harvested by hand in the Philippine Islands and the East Indies; these two areas supply about 20% of the world's carrageenan. North America, however, is the world's major carrageenan producer, using mostly extracts from *Chondrus crispus* and *Gigartina stellata*. France, Ireland, and Norway also harvest and use carrageenan extracted from *Chondrus*. The Canadian regions of Nova Scotia and Prince Edward Island are presently the largest producers of carrageenan. Eastern Canada harvested 43,150 metric tons of *Chondrus crispus* in 1969 with a value of 2.7 million U.S. dollars (French, 1970). For many years the center for carrageenan extraction and purification was at Scituate, Massachusetts, but the current center of production is Rockland, Maine. There are several other car-

rageenan-producing plants in Maine. Most raw material is imported from Canada.

Many investigations have been directed toward improving the carrageenan content of commercially important species. The carrageenan content of *C. crispus* varies with the season, the stage of the life cycle, and the habitat of the organism. *C. crispus* is tolerant of a wide range of temperature, light, and salinity and therefore can grow in many different habitats. Mathieson and Tveter (1975) discuss the variation in carrageenan content in relation to nutrients and other factors.

Carrageenan is used commercially as a stabilizer and thickener in dairy products, creamed soups, whipped cream, and other foods. It is also used as a stabilizer for paint emulsions and cosmetics, because it keeps substances homogeneous, and as a finishing compound for textiles and paper.

Agar

Agar, another complex mixture of sulfated polysaccharides, is used commercially for a variety of purposes. Rhodophyta that contain agar in their cell walls are sometimes referred to as agarophytes. These organisms include *Gelidium* (Fig. 11-11), which is harvested in Japan and California, and *Gracilaria* from Florida and Italy. Agar has less sulfate than carrageenan and forms a stronger gel. It melts in water at temperatures of 90° to 100° C but gels firmly at temperatures of 35° to 37° C. A solution (1% to 2%) of agar in water to which various nutrients have been added is an exceptionally good medium for growing both bacteria and fungi in the laboratory; in fact, agar was first used for laboratory culture of microorganisms. Its value for this was first recognized by the microbiologist Robert Koch in 1882.

Japan has been the major agar producer for many years. In Japanese waters the Pacific agar weed *Gelidium cartilagineum* grows very well, up to 1 m tall. During World War II the supply of agar to North American was terminated, so the culture of *Gelidium cartilagineum* was attempted in California and the culture of *Gracilaria verrucosa* (= *Gracilaria confervoides*) was pursued in North Carolina. In California in 1943, divers hand-harvested 70,000 kg, but the culture was abandoned after the war because it was too expensive. When the war was over, the Japanese regained their dominance of the agar market and continue to hold this dominance, although Spain and the Azores harvest a considerable amount. Japan exported 640 metric tons of agar in 1967 at a value of 4 million U.S. dollars (Michanek, 1975). In the Azores off the coast of Portugal, harvest of the agarophyte *Pterocladia* netted about $3 million in 1980 (Chernush, 1980). The growing agar industry may prove economically beneficial to coastal Third World countries. Italy cultures *Gracilaria* in shallow bays near nutrient-bearing freshwater outfalls. *Gracilaria* has also been cultured on the east and west coasts of Florida. India, South Africa, Denmark, Chile, Argentina, and Mexico also harvest and process the agarophytes. Chiang (1981) discusses the cultivation of *Gracilaria* in Taiwan.

When agarophytes are harvested they are first dried, then boiled, and the agar-containing extract is filtered and frozen to remove impurities. As with carrageenan, agar content in algal cell walls may vary seasonally, from population to population, and in different stages of the life cycle.

In addition to bacterial and fungal culture media, agar is sometimes used like carrageenan as a stabilizer for foods. It is used to keep cocoa in chocolate milk from separating, bakery products from drying out, and canned meats and fish from being fragmented in the can. When agar is added to jellies it aids rapid setting; when it is added to cheeses or puddings it promotes thickening. Agar is a good inert carrier for encapsulating such drugs as antibiotics and vitamins when slow release into the system is desired (Dixon, 1973). Frequently carrageenan may be substituted for agar, but the carrageenan gel is not quite as strong as that of agar.

Porphyra

The most extensive culture of rhodophytes is the cultivation of *Porphyra* (primarily *Porphyra tenera*) in Japan. *Porphyra*, which has been eaten for thousands of years in Japan, has been cultivated since the seventeenth century and its use continues to increase. The alga is sold as dried sheets, "nori," and eaten raw or cooked with other foods. *Porphyra* may also be eaten as "sushi," where a sheet of the dried alga is rolled around rice, vegetables, fish, or eggs.

Porphyra (Figs. 11-3 and 11-9) is cultivated on bamboo bundles or nets in shallow-bay plantations such as in Tokyo Bay, where there are about 500 km of nets covered with growing *Porphyra* plants. Harvesting of both cultivated and wild populations contributes to the overall yield. Currently the *Porphyra* industry is a multimillion dollar business. In the early 1960s, 120,000 metric tons (wet weight) were harvested per year and 500,000 workers were employed (Dawson, 1966). Oohusa (1971) estimated annual production at 123,000 metric tons with a value of 60 billion yen (260 million U.S. dollars in 1983). Cultivation of *Porphyra* employed over 60,000 families in 1973 (Michanek, 1975).

Extensive research has been conducted on the life cycle and growth of *Porphyra*. In 1949 Kathleen Drew, an English woman, discovered the heteromorphic alternate phase of *Porphyra*, called the "conchocelis" phase (Drew, 1949). Before 1950 the artificial cultivation of *Porphyra* had been hampered by a lack of understanding of the life cycle. The conchocelis phase is a small filamentous form that bores into shells. The Japanese were so grateful for Drew's breakthrough on the *Porphyra* life history that they erected a monument to her on Tokyo Bay. Drew (1956) reviews the life cycle of *Porphyra*.

The culture of *Porphyra* begins with the cultivation of the conchocelis form on oyster shells in tanks. The shells are subsequently used to "seed" the *Porphyra* beds in the bay. The spores settle on poles or nets in shallow water of relative high salinity. The higher saline environments favor spore settling (Dixon, 1973). When the *Porphyra* germlings have gotten a good start, they are transferred to less-saline environments near freshwater sewage outfalls. The lower saline waters and the nutrients from the outfalls stimulate growth of the sheetlike thalli, which are then harvested, dried, and sold. *Porphyra* cultivation has not been increasing as rapidly as the Japanese would like because the freshwater outfalls also bring industrial pollutants to the *Porphyra* beds and growth is inhibited or plants are killed. Oil pollution is a serious problem in Tokyo Bay, so *Porphyra* cultivation may have to be moved away from coastal areas to protect the plants from industrial water pollution.

Porphyra is widely distributed on coasts in northern latitudes. After their migration from Asia, the American Indians used *Porphyra* as a source of salt. The Chinese also cultivate considerable amounts of *Porphyra*. The Irish and Welsh eat *Porphyra* in a form called "laver bread," and the Welsh have produced up to 175 metric tons (wet weight) in a year (Dixon, 1973).

Additional Uses of Red Algae

Coralline algae may form loose calcareous beds in sublittoral areas of western Europe and the Mediterranean. *Lithothamnium* (Fig. 11-7) thalli, for example, get crushed in storms and the calcareous fragments are cast up on the shore, forming pink beaches that are later bleached brilliant white by the sun. The "maerl" (marl) on the southern England and western Ireland coasts was formed in such a way. These calcareous algal remains are used as a soil dressing in Spain, Brittany, and other western European coastal areas to neutralize acidic soils. With the availability of commercial limestone the use of algal "sand" for neutralization is decreasing (Dixon, 1973).

The Japanese also cultivate the rhodophyte *Gloiopeltis furcata* ("funori") from which they extract a glue and sizing agent that they call "funoran." Several other genera are also grown for "funori" content.

For centuries algae have been used as food, especially in Japan, China, Indo-Malaya, Hawaii, and the Philippines. The Chinese have been eating species of *Gracilaria* for thousands of years. About thirty algal species were eaten in Polynesia a century ago, but this use is declining. For centuries "dulse," or *Palmaria palmata* (= *Rhodymenia palmata*) (Fig. 11-9), has been commercially harvested and eaten raw or cooked in the British Isles and in Iceland. Presently no commercial harvesting is done in Scotland, but collecting for personal use continues. In Iceland the alga is now used only as a supplement to cattle feed in the winter. The Canadian Maritime Provinces, however, continue to harvest about 40 to 50 metric tons/year. Dried *Palmaria* may be found in bowls on tavern counters because it, like salty peanuts or pretzels, induces thirst in patrons. Is there any food value in these "seaweeds"? They are similar to lettuce and celery in that they provide fiber and vitamins, but humans lack enzymes to digest some of the complex polysaccharides found primarily in the cell walls. The seaweeds also stimulate the appetite and provide such trace elements as iodine and bromine.

Several genera of rhodophytes have medical applications. *Gelidium* and *Chondrus* extracts are used to treat stomach and intestinal disorders. In the Orient various red algae are used to treat hay fever. These marine rhodophytes contain substantial amounts of potassium chloride, which is known to moderate hay fever symptoms (Dawson, 1966). *Corallina* (Fig. 11-6) and *Alsidium*, "Corsican moss," are used as vermifuges to kill parasitic worms.

Several rhodophytes are used as fodder for domestic animals. In Iceland, Scotland, and Scandinavia, coastal farmers permit their cows, goats, and sheep to graze the shores at low tide. The stock usually graze selectively, preferring *Rhodymenia* and the brown alga *Alaria* to other intertidal seaweeds.

SUMMARY

Rhodophyta are widely distributed in benthic habitats of coastal areas of all the continents. Their greatest diversity occurs in the tropics, where they may inhabit the intertidal zone and the sublittoral zone to a depth of 200 m in very clear water. Most genera are marine, but there are a few freshwater genera and even a few terrestrial forms.

The division is composed of two subclasses—Bangiophycidae and Florideophycidae. The Bangiophycidae have poorly understood life cycles, but the life cycles of several genera (e.g., *Bangia* and *Porphyra*) are known and include a minute filamentous conchocelis stage that was originally thought to be a separate genus. The life cycles in the Florideophycidae are complex, often with a postponement of meiosis resulting in two diploid generations—the carposporophyte and the tetrasporophyte. No flagellated cells are found during any phase of red algal life cycles.

Members of the Florideophycidae contain pit connections, both primary and secondary. Pit connections are seen in both free-living algae and in parasitic forms, where they occur between the parasite and the host alga. The chloroplasts of Rhodophyta lack the CER seen in other algal divisions. The thylakoids have a single-unit membrane similar to the photosensitive apparatus of Cyanobacteria, and they contain chlorophyll *a*. Phycobilisomes containing the photosensitive pigments phycoerythrin, phycocyanin, and allophycocyanin are attached to the thylakoids. The phycobilin photosensitive pigments enable Rhodophyta to absorb light energy from green light, which penetrates most deeply into coastal waters, and therefore enables them to photosynthesize deep in the sublittoral zone.

The main energy storage compound in this division is a starch molecule called "floridean starch." It is chemically similar to the starch of Chlorophyta and vascular plants and to the myxophycean starch of Cyanobacteria but is not identical to either.

The forms of rhodophytan thalli are usually delicately branched and pseudoparenchymatous. Others forms are filamentous, foliose, or crustose. Many genera, especially intertidal forms, appear black or green, depending on the predominating pigment, but sublittoral genera usually appear red from the predominance of the red phycoerythrin pigment.

Florideophycidean cells contain phycocolloids; two of these phycocolloids, carrageenan and agar, are harvested and used commercially. Carrageenan is harvested from *Chondrus crispus*, *Iridaea*, *Gigartina*, and *Eucheuma*, and most production is in the Canadian Maritime Provinces. Japan leads the world in agar production, using the agarophyte *Gelidium cartilagineum*.

Many rhodophytes, especially in the tropics, have calcium carbonate in their cell walls. The calcified species have left an excellent fossil record that indicates that the division was present during the Cambrian period and perhaps in the Precambrian period. The calcareous "coralline" algae are important members of highly productive coral reefs and atolls, because they not only form an algal ridge that shelters the reef from the pounding surf but they also help to cement the reef together.

Rhodophyta form a separate line of algal evolution, the red line, and appear to be most closely related in biochemistry and ultrastructure to the prokaryotic Cyanobacteria.

Several red algal species are used for food, especially in the Orient. *Porphyra* has been cultivated in Japan since the seventeenth century and has been used as a food source for thousands of years. *Porphyra* culture is presently being threatened by water pollutants.

SELECTED REFERENCES

Abbott, I.A., and G.H. Hollenberg. 1976. Marine algae of California. Stanford University Press, Stanford, Calif. 827 pp.

Dawes, C.J. 1981. Marine botany. John Wiley & Sons, Inc., New York. 632 pp.

Dawson, E.Y. 1966. Marine botany. Holt, Rinehart & Winston, New York. 371 pp.

Dixon, P.S. 1973. The biology of the Rhodophyta. Hafner Press, New York. 285 pp.

Dixon, P.S., and L.M. Irving. 1977. Seaweeds of the British Isles. Vol. I. Rhodophyta. Part 1. Introduction: Nemaliales, Gigartinales. British Museum (Natural History), London. 252 pp.

Dring, M.J. 1983. The biology of marine plants. Edward Arnold, Baltimore. 208 pp.

Goreau, T.F., N.I. Goreau, and T.J. Goreau. 1979. Corals and coral reefs. Sci. Am. 241(2):124-136.

Hillson, C.J. 1977. Seaweeds. The Pennsylvania State University Press, University Park. 194 pp.

Johansen, H.W. 1981. Coralline algae, a first synthesis. CRC Press, Inc. Boca Raton, Fla. 256 pp.

Jones, O.A., and R. Endean. 1973. Biology and geology of coral reefs. Vol. II. Biology I. Academic Press, Inc., New York. 502 pp.

Jones, O.A., and R. Endean. 1976. Biology and geology of coral reefs. Vol. III. Biology II. Academic Press, Inc., New York. 457 pp.

Levring, T., H.A. Hoppe, and O.J. Schmid. 1969. Marine algae. A survey of research and utilization. Hamburg, Germany. 421 pp.

Michanek, G. 1975. Seaweed resources of the ocean. Food and Agriculture Organization of the United Nations. Fish. Tech. Paper 138. 127 pp.

van der Meer, J.P. 1983. The domestication of seaweeds. BioScience 33(3):172-176.

FUNGI

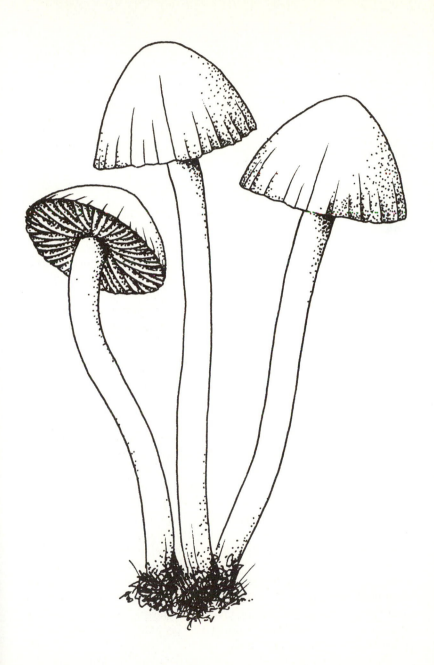

12

Introduction to the FUNGI

Under the classification system used in this text, fungi (sing. fungus) are members of a separate kingdom. The study of fungi is called mycology (fr. Greek *mykes* = mushroom + *logos* = discourse). The unifying factor for all fungi is their lack of plastids and chlorophyll at all times during their life cycles. They are, therefore, heterotrophic. They also possess discrete membrane-bound nuclei and are thus eukaryotic organisms (although the nuclei are minute and difficult to observe without special techniques). Some are unicellular but most are coenocytic (multinucleate). More than 8,000 genera and 150,000 species have been described, although many of these are not taxonomically valid, since most designations represent only stages in the life cycles of other fungi. With so many descriptions and so much error and duplication, fungal classification is obviously difficult, frustrating, and confusing. Even common mushrooms often present taxonomic problems that are difficult for trained mycologists. The classification system used in this text is that followed by Ainsworth, Sparrow, and Sussman (1973a, 1973b).

Besides being eukaryotic and nonphotosynthetic, most fungi have other characteristics in common. All reproduce by forming spores, most of which develop in special structures called **sporangia** (sing. **sporangium**), which are, in turn, often borne on special stalks called **sporangiophores.** Some fungi produce different kinds of spores; for example, many ascomycetes produce **conidia** (spores) borne on **conidiophores,** as well as **ascospores** borne within special sacs or **asci** (sing. **ascus**). The names given to different spores depend on the manner by which the spores are formed. These will be explained as the names are used.

Most fungi are filamentous organisms containing multinucleate, binucleate, or uninucleate cells. The number of nuclei per cell is related to the evolutionary development of the individual fungus.

Yeasts, however, have never progressed beyond the unicellular, uninucleate stage. Some fungi are **dimorphic,** which means that they may exist in a unicellular, yeastlike form or in a filamentous, multinucleate form. Dimorphic fungi are especially common among plant and animal pathogenic forms.

Fungal filaments typically possess rigid cell walls whose structure resembles that found in algae and higher plants but whose chemical composition is quite different from that of any other group of organisms. Mature cell walls are multilayered, with each layer being distinguishable by its distinctive structure or chemical nature. Different fungi have varying numbers of cell wall layers containing various chemical compositions. Fig. 12-1 shows a cell wall from *Neurospora crassa* as interpreted by Cole

(1981). The cell wall alters its structure and composition as it matures, becoming more complex in both form and chemical makeup. The major organic component is almost always **chitin** (see Fig. 1-3), which forms a complex with varying amounts of proteins, glycoproteins, lipids, and an outer layer of glucose polymers made up primarily of β-1,3 **glucans.** The glucose polymers may be cellulose, as in some members of Myxomycota and Mastigomycotina, or the polymers may be of mannose, as in some yeasts. The chemical nature of the fungal cell walls may be used as a taxonomic tool (Table 12-1).

The protein moiety of fungal cell walls represents both a structural and enzymatic component. The enzymes associated with individual walls are important to the overall metabolism of the fungus and to the environmental effects

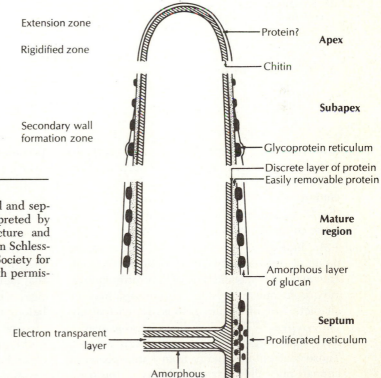

FIG. 12-1 Organization of fungal cell walls.
Diagrammatic representation of the hyphal and septal walls in *Neurospora crassa,* as interpreted by Cole. (From Cole, G.T. 1981. Architecture and chemistry of the cell walls of higher fungi. In Schlessinger, D. [ed.]. Microbiology. American Society for Microbiology, Washington, D.C. Used with permission.)

TABLE 12-1. Chemical Composition of Fungal Cell Walls

Carbohydrate polymer	Taxonomic group	Common name or characteristic
Cellulose	Acrasiomycetes	Only in spores of some cellular slime molds
Cellulose-glucans (Various-length polymers of glucose)	Oomycetes	Water molds and downy mildews
Cellulose-chitin	Hyphochytridiomycetes	Aquatic, posteriorly uniflagellate zoospores
Chitosan-chitin (Various-length polymers of N-acetylglucosamine)	Zygomycetes	Bread molds and black molds; no motility; zygospores
Chitin-glucan (Chitin with various-length polymers of glucose)	Chytridiomycetes	Aquatic, anteriorly uniflagellate zoospores
	Ascomycetes	"Sac fungi", including yeasts, *Penicillium, Neurospora*, etc.
	Basidiomycetes	"Club fungi", including mushrooms, smuts, and rusts
	Deuteromycetes	"Imperfect fungi"; no sexual stages
Mannan-glucan (Various-length polymers of mannose and glucose, mixed)	Hemiascomycetes	Yeasts
	Deuteromycetes	Yeastlike "imperfect fungi"; order Cryptococcales
Mannan-chitin (Various-length polymers of mannose attached to chitin)	Deuteromycetes	Yeastlike "imperfect fungi"; order Sporobolomycetales

caused by that fungus. The cell wall enzymes probably represent many of the enzymes required to break down foreign organic matter including wood. They are therefore important in the decay and recycling of both mineral and organic components.

As indicated previously, the heterotrophic mode of nutrition is characteristic of all fungi. However, the specific type of heterotrophy may vary from one fungus to another and may be characteristic of certain groups. Fungi that obtain energy by breaking down organic matter from dead organisms are called **saprophytic.** Those that obtain energy from living plants or animals are called **parasitic.** Many of the parasitic fungi are medically or economically important organisms, since they cause either plant or animal diseases. Saprophytic fungi are indispensable to the overall energy economy of our environment. Some fungi have evolved **symbiotic** relationships with algae (**lichens**) and vascular plants (**mycorrhizae**).

A few fungi are capable of actively engulfing their food by **phagocytosis,** a process similar to that found in many protozoans. These fungi lack rigid cell walls (except for their spore stages) and belong to a separate fungal division, the Myxomycota (slime molds). They present a special problem for taxonomists, since they differ from other fungi in their mode of nutrient procurement and their lack of cell walls. Most investigators have placed them with the protozoans in kingdom Protista. However, most fungi are unable to engulf their food. Digestion must therefore be accomplished outside of the cell through the production and secretion of **exoenzymes.** Digested materials are then absorbed into fungal hyphae by active transport of the molecules across the plasma membrane. Digestion, then, is similar to that which occurs in bacteria.

It seems clear that the fungi as a group do not have close affinities with the algae or the higher plants; thus it is logical to place them in a separate kingdom, the Fungi. Some mycologists, however, have proposed that the ascomycetes may be phylogenetically derived from parasitic members of the Rhodophyta, or red algae (Kohlmeyer, 1973). Some investigators prefer to place the fungi as derived from protistan flagellate ancestors lacking cellulosic cell walls. The precise direct-line relationship between the protozoans and the fungi is, however, difficult to establish (see Fig. 1-1).

The fungi as a group are universal in their distribution. Some are found throughout the world, whereas others are restricted to specific environmental niches controlled by chemical, physical, and biological factors. Some are found only in temperate or tropical environments. Others are strictly aquatic. Symbiosis with both plants and animals is common. Wherever they are found, the fungi play an important role in the flow and recycling of nutrients in the world's environment, and their importance to the economy of any ecosystem is often understated.

The general structure of a fungal thallus is somewhat different from that found among the algae; therefore, different terms are used in its description. The vegetative thallus of fungi is composed of filaments or threads of living material called **hyphae** (sing. **hypha;** fr. Greek *hyphae* = web). A mass of hyphae is collectively referred to as the **mycelium** (pl. **mycelia**). As with the algae, the term thallus can be used to describe the entire organism. When the vegetative body is composed of many loosely arranged hyphae, its nature may be delicate and

cottony as, for example, in bread molds. When the hyphae are arranged in a compact mass, the resultant structure may be hard or leathery, as in shelf fungi or mushrooms. The different classes of fungi have distinctive hyphae and may sometimes be identified by the use of hyphal morphology. When a thallus enters into sexual reproduction, it is said to be in a "perfect" stage of development. The asexual stages are called the "imperfect" stages and are usually accompanied by some type of spore production.

At certain stages in the life history of many fungi the mycelium organizes into a tissuelike complex that resembles the pseudoparenchymatous construction seen in many algae. Hyphal filaments become compact masses in which individual hyphae lose their identity. These structures are called **stromata** (sing. **stroma**) or **sclerotia** (sing. **sclerotium**). Stromata are pliable and mattresslike, whereas sclerotia are hard and woody. Stromata bear reproductive structures either internally or externally, depending upon the species. Sclerotia are resting bodies that are resistant to adverse environmental conditions; they may remain viable for years.

Fungal hyphae possess some interesting variations in their internal structure. The more primitive forms, belonging to Chytridiomycetes, Oomycetes, and Zygomycetes, produce multinucleate (coenocytic) hyphae that usually lack cross-walls or septa (sing. septum). If the hypha contains nuclei that are all of one genetic constitution, it is called a **homokaryon.** If the hypha contains nuclei that are genetically different (e.g., haploid and diploid nuclei or haploid nuclei from male [−] and female [+] hyphal strains), it is called a **heterokaryon. Heterokaryosis** (p. 243) is very important to fungi that do not undergo regular sexual cycles (e.g., the Deuteromycetes—Fungi Imperfecti).

The higher fungi belonging to the ascomycetes and some deuteromycetes produce hyphae that are partially septate. The septa contain pores through which cellular organelles, cytoplasm, and nuclei freely pass. Associated with the pores are crystalline organelles,

woronin bodies, which may plug the pores between adjacent parts of the hyphae, thus isolating nuclei (Gull, 1978). Nuclei are often isolated so that there are two in each hyphal segment (cell), thus forming a **dikaryotic hypha.** If there is only one nucleus per hyphal segment, it is called a **monokaryotic hypha.**

The higher fungi belonging to the basidiomycetes and some Deuteromycetes possess hyphae with septa containing a small central pore that is covered with a membranous cap. The cytoplasm is free to migrate through the pore, but nuclei and other organelles normally do not. Thus the monokaryotic or dikaryotic condition of the hyphal cells is maintained, unlike the condition in the more primitive fungi that lack septa altogether. The hyphal condition in the Ascomycetes, Basidiomycetes, and Deuteromycetes becomes especially significant when mitotic cell divisions occur (see discussion of crozier formation [p. 331; Fig. 17-4] and clamp connections [p. 356; Fig. 18-2]).

Variations in hyphal structure are common among all fungi. Striking examples of this are found in the formation of asexual and sexual reproductive structures (see descriptions for individual groups). A good asexual example is the formation of specialized hyphal segments called **haustoria** (sing. **haustorium**). They are most often found among parasitic fungi where intercellular hyphae produce specialized branches that penetrate the host cells (see Fig. 13-2). Haustoria function as absorptive structures, extracting nutrients from the cells of the host organism. Some investigations have shown haustoria to exchange materials both ways: from host to fungus, and from fungus to host. Host-to-fungus transfers are common in parasitic fungi such as the Peronosporales (downy mildews of the Oomycetes), the Erysiphales (powdery mildews of the ascomycetes), and the Uredinales (rusts of the basidiomycetes). Fungus-to-host and host-to-fungus transfers are common among the mycorrhizae found in many of the basidiomycetes (see Chapters 13 and 18).

The more primitive fungi are aquatic and produce flagellate reproductive cells at some time during their life history. The derived fungal forms tend to be terrestrial and have lost the ability to produce flagellate cells. Gametes of terrestrial forms move by amoeboid motion, whereas spores are dispersed by wind, water, or animals. Their motility, or lack thereof, is reflected in the classification of fungi used in this text.

CELL STRUCTURE

Fungal cells are similar in some respects to algal cells and the cells of higher plants. They contain all of the usual cellular organelles associated with other eukaryotic cells. However, some important differences help to separate the fungi from algae, vascular plants, and protistan cells.

First, fungi are eukaryotes possessing a well-defined, membrane-bound nucleus. However, the nucleus is usually much smaller than that associated with most algae. It is difficult to see the unstained fungal nucleus because its size is so small (1 to 2 μm). The nucleus is even difficult to identify in some electron photomicrographs.

The endoplasmic reticulum of fungi differs from that found in most eukaryotes because it produces enlarged sacs or vesicles at its ends, and it is often sparsely represented. Some electron photomicrographs show almost no endoplasmic reticulum.

The ribosomes in fungal cells are found floating free in the cytoplasm or are located along the endoplasmic reticulum as in other eukaryotes forming rough endoplasmic reticulum.

The mitochondria of fungi are similar to the mitochondria of other eukaryotic cells. The Golgi apparatus is not always present, but when it is, its structure is similar to that of other eukaryotes. Vacuoles are common in fungi, but they do not reach the great size found in higher plants. The number and size of vacuoles seem to increase with the age of the fungal cell. Energy storage products may also be found in the cytoplasm; these are usually found in a series of granules of glycogen clusters floating free in the cytoplasm. Glycogen is the major storage polysaccharide associated with fungi: it may comprise as much as 5% of the dry weight of an older thallus. It consists of a profusely branched polymer of α-1,4–linked glucose molecules containing glucose branches having an α-1,6 linkage. It is identical with the glycogen formed by animals as an energy storage compound (see Fig. 1-3).

Typically, in those primitive fungi that produce flagellate cells, paired cylindrical centrioles are found in close association with the nuclear membranes. Each centriole contains nine triplets of protein fibrils that are arranged in a ring, resembling the microfibrillar arrangement found in all flagella. Nonflagellate cells lack centrioles, and their specific location just outside the nuclear membrane is replaced by an electron-dense structure known as the **nucleus-associated organelle,** or **NAO.** It is also known as the **spindle pole body** or **SPB.** Heath (1981) gives compelling arguments for the preference of the term NAO, since the specific function of the structure has not been established and does not appear to be associated with spindle formation.

Outside the plasma membrane of single fungal cells or multinucleate hyphae are two distinctive structures. The first is a series of vesicles that lie between the plasma membrane and the cell wall. These are called **lomasomes.** Their composition and function are unknown. They have been identified only with the aid of an electron microscope. The second unique structure is the cell wall, which is distinctively composed of microfibrils of chitin or cellulose or mixtures of the two. Like cellulose, chitin forms an insoluble, rigid cell wall that is freely permeable to water and dissolved nutrients.

A final ultrastructural feature of all fungi that should be reemphasized is the complete lack of chloroplasts containing chlorophyll. Therefore, fungi are never autotrophic.

REPRODUCTION

The reproductive potential of most fungi is far greater than that of most other organisms. The number of spores produced by individuals in some genera is in the billions. Who has not seen a piece of bread, left almost anywhere, that has become "moldy" within a short period of time? Consider where the spores for this "mold" have come from and how they have reached the bread.

Cellular reproduction among the fungi is accomplished through regular mitotic divisions. However, the process is different from that observed in most eukaryotes, because the nuclear membrane remains intact during the entire division. The furrowing process of the nuclear membrane during telophase causes the production of two separate nuclei within a common cytoplasm. Observation of mitotic division is hindered by the small size of the nuclei and chromosomes. Heath (1978) has compiled ultrastructural information on mitosis and meiosis in fungi.

Spore Production

Asexual spores are nearly always produced in large numbers by fungi. They usually develop in special structures, the sporangia. Sexual cells (gametes) are generally produced in structures called gametangia. The diversity and architecture of the sporangia and gametangia provide a tool for the identification and classification of many fungi. The reproductive cells (spores and gametes) produced in these structures may be motile or nonmotile, providing another tool for classification and identification.

Asexual spores are formed by three different methods. First, spores may develop within sporangia borne on upright hyphae. These special sporangia-bearing hyphae are called sporangiophores. Upright sporangiophores are common in bread molds (see Fig. 16-1). The nonmotile spores produced are dispersed by air currents. Nonmotile spores of this type are called sporangiospores or aplanospores and are characteristically produced by terrestrial fungi. Aquatic forms produce submerged sporangia from which emerge motile zoospores (planospores) or "swarmers." A good example of this type is found in the common water mold *Saprolegnia* (see Fig. 15-3). Second, spores may be formed by individual vegetative cells contained within a filament. The cells secrete a thick cell wall that separates the spore from adjacent spores or cells within the filament. One spore per cell is produced. Such spores are transformed hyphal cells. They are produced in enormous numbers by some of the basidiomycetes (e.g., the smuts) and are called **chlamydospores.** A third type of spore is produced at the growing tips of hyphae where thin-walled cells are cut off from the hyphal filament. These spores are very lightweight and are easily dispersed by wind currents and drafts. Called conidia (sing. conidium), these spores are produced on modified hyphae called condiophores. They are easily seen in species of such genera as *Aspergillus* or *Penicillium* (see Fig. 17-6).

Another method of asexual reproduction may occur through fragmentation of the individual thallus. Pieces of most fungi are capable of developing into an entirely new reproductive thallus under the proper conditions.

Sexual Reproduction

Sexual reproduction in fungi may be isogamous, anisogamous, or oogamous. (The terminology here is the same as that used to describe algae; see Fig. 1-10.) Isogamy is characteristic of the more primitive genera, whereas oogamy is found among the derived forms. Oogamy often resembles that which occurs in the algal genus *Vaucheria* (Xanthophyta, Fig. 10-1).

The sexual process in the ascomycetes and the basidiomycetes is cytologically complex because of the failure of the haploid nuclei from the two gametes to fuse after the gametes have united. The net result of this peculiarity is the production of two distinct types of hyphae: (1) those in which the fungal nuclei are haploid with one nucleus per cell (monokaryotic condition), and (2) those in which the fungal nuclei

are haploid with two nuclei in each cell (dikaryotic). In the second type the two haploid nuclei fuse, forming a single diploid nucleus just before meiosis, thus forming meiospores called ascospores or basidiospores, depending upon which class of fungi is represented. This is accomplished by **crozier formation** in ascomycetes (see Fig. 17-4), and by **clamp connections** in basidiomycetes (see Fig. 18-2).

Some sexual processes in the fungi are unique in the biological world. Normally the sexual process involves the union of two relatively similar lines of genetic information found within haploid gametes, followed by genetic mixing. The union of gametes doubles the chromosome number, and meiosis returns the chromosome number to its haploid condition. This sort of sexual cycle is a universal process among all eukaryotic organisms, including fungi, plants, animals, and protistans. Crozier formation in ascomycetes and clamp connections in basidiomycetes are both mechanisms that ensure isolation of two haploid nuclei within a single cell. Nuclear fusion and meiosis are always preceded by one of these two processes.

The Deuteromycetes (Fungi Imperfecti), however, do not under normal circumstances undergo a regular sexual cycle. Instead, these organisms have developed a unique series of cytological events called the **parasexual cycle. Parasexuality** is a mechanism whereby genetic recombination is accomplished through mitosis rather than meiosis. It involves the fusion of haploid nuclei, gene segregation, and return to the haploid condition without meiosis. Parasexuality was discovered by Pontecorvo and Roper (1952) while working with *Aspergillus nidulans* at the University of Glasgow in Scotland. It is discussed in detail by Roper (1966) and Pontecorvo (1956), who found that haploid nuclei in the hyphal filaments occasionally fuse, producing diploid nuclei in the same filaments and thus forming a heterokaryon (mycelium) containing five distinct nuclear types, $1N+$, $1N-$, two different $2N$ homozygotes, and a $2N$ heterozygote (Fig. 12-2). Heterokaryosis may

FIG. 12-2 Parasexuality in fungi.
Various methods by which fungi are able to accomplish genetic mixing without going through the usual
sexual cycle. Note that heterokaryosis may be accomplished by three basic mechanisms: (1) diploidization,
in which haploid nuclei fuse to produce diploid nuclei, thus forming a hypha with as many as five different nuclear types contained within the common
cytoplasm; (2) mutation of any of the nuclei contained
within the hyphae; and (3) by hyphal fusion or plasmogamy, in which a homokaryotic hypha becomes a
heterokaryotic one.

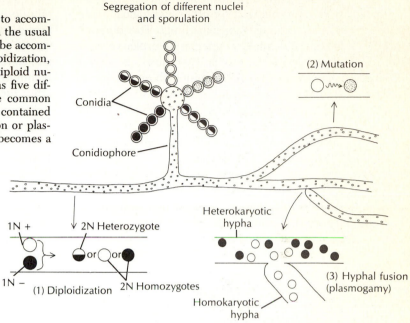

be initiated by three different processes: (1) diploidization, or the fusion of two haploid nuclei
(karyogamy); (2) mutation; or (3) the fusion of
two hyphae, sometimes called plasmogamy.
The process is common in the deuteromycetes,
ascomycetes, and basidiomycetes. It has not
been reported in the lower fungi.

Heterokaryosis is a very important process in
the evolution of fungi because it imparts to the
fungus a genetic potential not encountered with
a single haploid nucleus. If, for example, the
genetically different nuclei in a heterokaryotic
hypha are segregated when spores are formed,
then at least five different hyphae can be isolated by culturing the separate spores. Each
spore gives rise to a hypha that is genotypically
different and that may also be phenotypically
distinct. The evolutionary significance of heterokaryosis becomes clear when one considers
that the heterokaryote is really a functional diploid organism. Experiments with *Neurospora*
strains have shown one haploid strain to be un-

able to synthesize the amino acid arginine,
whereas other haploid strains are unable to synthesize glutamic acid, another amino acid. The
first strain will not grow unless arginine is added
to its environment; the second requires glutamic acid. When the two strains form a heterokaryon, the new hyphae require neither arginine nor glutamic acid; thus, a functional diploid
filament has been generated—the one haploid
strain complements the other. Such hetero-
karyons have obvious adaptive values. The
physiological condition of the heterokaryote is
determined by the interaction of the two genetically different haploid nuclei. Clearly, as
recessive mutations accumulate in some of the
hyphae, they are compensated for by genes that
are present within other nuclei. Since many of
the nuclear lines are unable to compete in certain environmental conditions when they are in
the homokaryotic state, the heterokaryotic condition is favored by natural selection.

Formation of heterokaryotic hyphae is only

the first step in the complete parasexual cycle. It is soon followed by other cytological events that are unique among the fungi. The genetically different nuclear types in heterokaryotic hyphae continue to divide by mitosis, generating coenocytic hyphae containing thousands of nuclei. When the diploid nuclei divide by mitosis, occasional **crossing over** occurs between homologous chromosomes. The frequency of these mitotic crossovers is not especially high, but their occurrence is significant, since they provide the organism with a mechanism for gene recombination without meiosis (Fig. 12-3). It is clear that these crossovers provide an important mechanism in the evolution of fungi.

Heterokaryotic hyphae, in which a series of mitotic crossovers has occurred in the diploid nuclei, contain nuclei with an almost infinite variety of genetic combinations. These hyphae also contain haploid nuclei of at least two types. If mutations have occurred, then there will be other types of haploid or diploid nuclei.

Both diploid and haploid nuclei are randomly dispersed throughout the hyphae in a common cytoplasm. When asexual conidia (spores) are formed, the various nuclear types are randomly distributed within the spores (Fig. 12-2). After the spores are dispersed, they may germinate and develop into hyphae containing the new combinations of genetic information. Tech-

FIG. 12-3 Mitotic crossing over.
Nucleus, **A,** containing a single homologous pair of chromosomes. Nucleus is entering mitotic prophase, and homologous chromosomes, by chance, come to lie near each other. A cross-over between homologous chromatids occurs, with the result that either cells **B** and **C** or **D** and **E** are produced; y = para-aminobenzoic acid (PABA)–requiring; z = biotin-requiring; + = wild-type dominant gene that is able to synthesize both PABA and biotin. Note that parent nucleus **A** requires neither PABA nor biotin, whereas nucleus **C** requires biotin. Nuclear types **B, D,** and **E** require neither PABA nor biotin and their genetic composition is different from nucleus **A.**

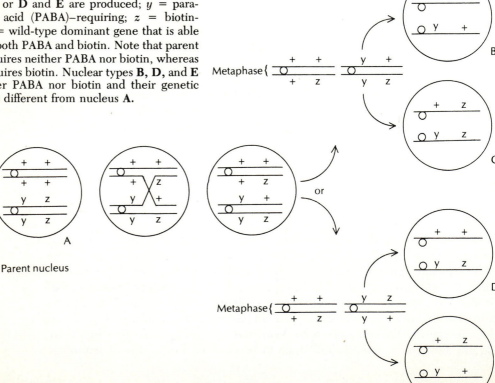

niques for the collection and isolation of individual spores enable mycologists and geneticists to study the individual metabolism of the nuclear types. This is done by culturing individual spores on defined media in a manner similar to that used by bacteriologists for the study of bacteria.

It has been discovered that diploid nuclei are able to revert to the haploid condition without going through meiosis. This process is called **haploidization** to differentiate it from classic meiosis. The mechanism of haploidization is poorly understood although it is probably accomplished through a series of mitoses in which **nondisjunction** of chromosomes takes place at

metaphase (Fig. 12-4). During nondisjunction, mitotic chromosomes fail to separate at the centromere, and the two chromatids of a single chromosome are isolated in one of the two daugher nuclei. This means that one nucleus will contain an extra chromosome $(2N + 1)$, whereas the other will lack a chromosome $(2N - 1)$. Nondisjunction results in a series of **aneuploid** nuclei that lack even numbers of chromosome sets such as N, 2N, 3N, and so on. Aneuploids, therefore, contain chromosome numbers like $2N + 1$, $2N + 2$, and $2N + 3$, or $2N - 1$, $2N - 2$, $2N - 3$, and so on. If the haploid chromosome number is small, then the haploid number might be restored from diploid

FIG. 12-4 Haploidization in fungi.
Diagram illustrates how haploidization may occur in nuclei with low chromosome numbers without going through meiosis. Begin cycle by studying haploid nuclei on upper left and follow arrows counterclockwise. Such a sequence of cytological events depends on the occurrence of a series of nondisjunctions taking place during mitotic anaphase.

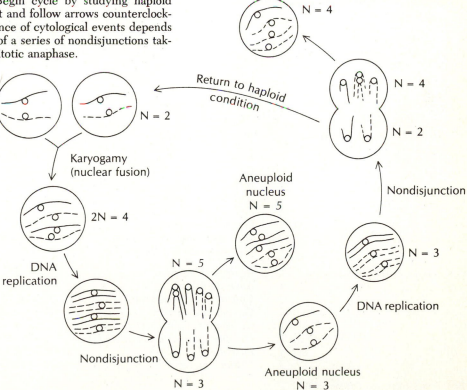

N = 4

Return to haploid condition

N = 2

Karyogamy (nuclear fusion)

2N = 4

DNA replication

Nondisjunction

N = 3

N = 4

N = 2

Nondisjunction

N = 3

DNA replication

Aneuploid nucleus
N = 5

N = 5

Aneuploid nucleus
N = 3

nuclei through repeated nondisjunctions occurring within a single nucleus. A series of such events has been observed by Sinha and Ashworth (1969) in strains of the slime mold *Dictyostelium*. Chromosome numbers were found ranging from the haploid condition (N = 7) in 60% of the nuclei, to the aneuploid condition (N = 8 to 13) in 10% of the nuclei, to the diploid condition (N = 14) in 30% of the nuclei.

It seems clear, therefore, that for haploidization to occur in a significant quantity, there must be a relatively small number of chromosomes in each nucleus. It so happens that this is the case; it is not uncommon for fungi to have low haploid chromosome numbers as is indicated in Table 12-2.

The sequence of the parasexual process in fungi is outlined in Alexopoulos and Mims (1979) as a seven-step series of events (presented here with some modification):

1. Formation of heterokaryotic mycelia
2. Fusion between haploid nuclei; this may occur between genetically identical nuclei (homozygous nuclei)
3. Multiplication (mitoses) of both diploid and haploid nuclei
4. Occasional crossing over during mitosis in diploid nuclei
5. Segregation of diploid and haploid nuclei during spore formation
6. Occasional haploidization of diploid nuclei
7. Segregation of new haploid nuclei during spore formation

TABLE 12-2. Haploid Chromosome Numbers in Fungi

Taxonomic group	Species	Chromosome number
Division Myxomycota (Slime molds)	*Dictyostelium discoideum*	N = 7
	Physarum polycephalum	N = 8
Division Eumycota	*Achlya ambisexualis*	N = 3
Subdivision Mastigomycotina (Flagellate fungi)	*Allomyces arbusculus*	N = 8, 16, 24
	A. macrogynus	N = 14, 28
Subdivision Zygomycotina	*Rhizopus nigricans*	N = 16
Subdivision Ascomycotina	*Aspergillus nidulans*	N = 8
	Ceratocystis fagacearum	N = 4
	Nannizzia spp.	N = 4
	Neurospora crassa	N = 7
	Penicillium expansum	N = 4 or 5
	Peziza vesiculosa	N = 8
	Saccharomyces cerevisiae	N = 8, 16
	Schizosaccharomyces pombe	N = 2
Subdivision Basidiomycotina	*Agaricus campestris*	N = 12
	Amanita fulva	N = 8
	Coprinus atramentarius	N = 6, 16
	C. comatus	N = 14
	C. lagopus, C. micaceus	N = 12
	Lycoperdon pyriforme	N = 6
	Puccinia spp.	N = 3 to 6
	Ustilago violacea	N = 10 to 12

Based on Altman and Dittmer (1972).

SELECTED REFERENCES

Ainsworth, G.C., and A.S. Sussman (eds.). 1965. The fungi: an advanced treatise. Vol. I. The fungal cell. Academic Press, Inc. New York. 748 pp.

Ainsworth, G.C., and A.S. Sussman (eds.). 1966. The fungi: an advanced treatise. Vol. II. The fungal organism. Academic Press, Inc. New York. 805 pp.

Ainsworth, G.C., and A.S. Sussman (eds.). 1968. The fungi: an advanced treatise. Vol. III. The fungal population. Academic Press, Inc. New York. 738 pp.

Ainsworth, G.C., F.K. Sparrow, and A.S. Sussman (eds.). 1973a. The fungi: an advanced treatise. Vol. IV, A. A taxonomic review with keys: Ascomycetes and Fungi Imperfecti. Academic Press, Inc., New York. 621 pp.

Ainsworth, G.C., F.K. Sparrow, and A.S. Sussman (eds.). 1973b. The fungi: an advanced treatise. Vol. IV, B. A taxonomic review with keys: Basidiomycetes and lower fungi. Academic Press, Inc., New York. 504 pp.

Alexopoulos, C.J., and C.W. Mims. 1979. Introductory Mycology, ed. 3, John Wiley & Sons, Inc., New York. 632 pp.

Burnett, J.H., and A.P.J. Trinci (eds.). 1979. Fungal walls and hyphal growth. Cambridge University Press, London, England. 418 pp.

Cole, G.T. 1981. Architecture and chemistry of the cell walls of higher fungi. In Schlessinger, D. (ed.). Microbiology—1981, pp. 227-231. American Society for Microbiology, Washington, D.C.

Griffin, D.H. 1981. Fungal physiology. John Wiley & Sons, Inc., New York. 383 pp.

Heath, I.B. (ed.). 1978. Nuclear division in the fungi. Academic Press, Inc., New York. 235 pp.

Heath, I.B. 1981. Nucleus-associated organelles in fungi. In Internat. Rev. Cytol. **69**:191-221. Bourne, G.H., and J.F. Danielli (eds.). Academic Press, Inc., New York.

Smith, J.E., and D.R. Berry (eds.). 1976. The filamentous fungi. Vol. 2. Biosynthesis and metabolism. Edward Arnold Publishers, Ltd., London, England. 520 pp.

Smith, J.E., and D.R. Berry (eds.). 1978. The filamentous fungi. Vol. 3. Developmental mycology. John Wiley & Sons, Inc., New York. 464 pp.

13

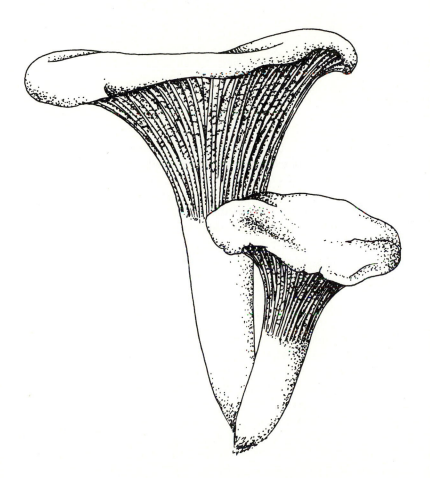

Fungi are essential organisms in the world's ecosystems: their major ecological role is that of decomposer (see Chapter 2). They contain an armory of exoenzymes that degrade many types of complex organic molecules into their inorganic components, which are then utilized and recycled by the producers, green plants. Fungi and bacteria make up the greatest number of decomposers, and although their biomass is small, their impact is enormous. In 1 g of fertile soil there may be 1×10^6 fungi, including the single-celled yeasts (Ascomycotina) and portions of mycelia from other fungal groups. In the top 25 cm of soil there may be 10 to 100 g (dry weight) of fungi and bacteria per square meter. The fungi and bacteria work together to decompose organic material; often the fungi can degrade compounds that bacteria cannot attack. The fungi are especially important in the degradation of wood, which contains the complex carbohydrate cellulose and the complex polymer lignin. Litter containing oak leaves and pine needles will support many fungi because these materials are high in lignin. Not only do the fungi release inorganic nutrients from the breakdown of organic molecules but they also serve directly or indirectly as food for various consumers (see Chapter 2).

In the terrestrial environment much of the primary productivity is channeled into the detritus (dead organic matter) food chain—that is, most of the energy fixed by producers is not consumed by herbivores, so the green plants contribute dead organic matter to the ecosystem in the form of leaves, twigs, or their entire biomass when they die. This dead organic matter is decomposed by fungi, bacteria, and some invertebrates that in turn become food for larger animal consumers. The importance of decomposers may be emphasized by the fact that in some terrestrial environments over 90% of the net primary productivity may be channeled to detritus food chains, and secondary production of the decomposers may be up to 24 times that of animal production (Whittaker, 1975). In the aquatic ecosystem of Silver Springs, Florida (see Fig. 2-2), the decomposers make up only 0.58% of the biomass, but they are responsible for approximately 24% of the energy flow. In the marine environment over 60% of the net primary productivity may be directed to the decomposers.

DISTRIBUTION AND DISPERSAL

Fungi are distributed throughout the world but are much more common in the warm, humid tropics. Their spores are found from the Antarctic and Arctic ice caps to the tropical rain forests and have also been isolated from the atmosphere 9 km above the earth. The fruiting bodies of some ascomycetes and deuteromycetes have been found growing in the sea at a depth of 5,300 m at temperatures of 2.5° to 4° C (Cooke, 1979). Many fungal species are adapted for wind dispersal and may be transported hundreds of kilometers by air currents because of their extreme lightness and their shape. Air currents may carry fungal spores to high altitudes from which transportation over hundreds of kilometers becomes highly probable. Well-documented incidences of long-distance spore transport were observed during the spread of black stem rust of wheat, *Puccinia graminis* f. sp. *tritici,* from the southern United States to the northern states and Canada.

Water may also play an important role in fungal spore dispersal. The spores of some Mastigomycotina and Deuteromycotina are dispersed by water. Flowing rivers and streams carry Mastigomycotina spores many kilometers, as does runoff from fields, forests, and urban areas. Currents in "still" bodies of water such as lakes or ponds also move spores about. Water moving through the soil can transport fungal spores both horizontally and vertically, and raindrop splash can disperse spores many centimeters.

Because of these many mechanisms of spore dispersal, fungal spores are found in the air and water all over the earth. Anyone who suffers from allergies to fungal spores is keenly aware of the ubiquity of these spores. Fungal spores are second only to pollen in causing respiratory allergies, and most of the more than 100,000 fungal species produce airborne spores at some stage of their life cycle (Christensen, 1975). During a wheat rust epidemic in 1925, spore densities of 2×10^{13} spores per hectare were recorded over an infected wheat field (Stakman, 1955). Rust spores, however, do not appear to be as potent respiratory allergens as the spores of the commonly occurring saprophytic fungi that may grow on decaying organic matter of all forms. The spores of the deuteromycetes *Hormodendrum* and *Alternaria* are most commonly found in air samples; these genera appear to be the more common allergens.

When fungal spores occur inside buildings, the spores of **xerophytic** (dryness-loving) fungi may cause many respiratory problems. House dust from a Minnesota home in winter averaged 180,000 spores/g. Most of the spores were from the ascomycetes *Aspergillus* and *Pencillium* (Christensen, 1975). Spores from xerophytic species of *Aspergillus* (e.g., *A. restrictus, A. glaucus*) are common in homes, often producing respiratory symptoms in allergic individuals. These species grow readily on damp overstuffed furniture, on pillows, and on mattresses; they do not need very much moisture. In older buildings spores of the **dry rot** fungus (*Serpula lacrymans*—Chapter 18) may cause structural damage to the supporting wood as well as respiratory difficulties for humans. Knowledge of the occurrence and growth habits of fungi can facilitate the treatment of respiratory allergies. For example, susceptible individuals may suffer severe asthmatic or other allergic reactions following a move to an older building or after camping in a fungus-infested trailer or tent. To add to these problems, the spores can remain viable for up to 16 years, resisting both low temperatures and desiccation.

SAPROPHYTIC FUNGI

Fungi can assume both saprophytic and parasitic roles, and some species can switch from one role to the other depending on conditions. In the recycling of organic matter in ecosystems, fungi are especially important in the decomposition of the recalcitrant molecules of **cellulose** and **lignin.** In a field or forest, fungal mycelia make up the majority of the decomposer biomass. The amount of fungal biomass is extremely low when compared to the biomass of producers and other consumers; however, the decomposers are responsible for a large portion of energy flow. In areas where numbers of fungi have been significantly reduced, decomposition is retarded and dead organic material builds up rapidly (Buchauer, 1971).

Fungi can decompose a wide range of carbon compounds found in nature such as the protein keratin and the polysaccharides cutin, chitin, pectin, starch, cellulose, and the aromatic alcohol polymer lignin. Fats and oils and the complex hydrocarbons of petroleum products can also be attacked. For most compounds produced by the metabolism of living organisms there are also some organisms with the biochemical ability to decompose these compounds. Fungal decomposition is accomplished by the production of various exoenzymes (e.g., proteases, lipases, amylases, and cellulases) that break down large polymers to smaller molecules (e.g., amino acids, monosaccharides, and fatty acids) that can then be absorbed into the cell and further degraded by **endoenzymes.**

The amount of detritus requiring decomposition in an ecosystem can be extremely large. In a temperate deciduous forest, for example, a large elm tree will drop over 180 kg of leaves and branches in one autumn (Gray, 1959). This energy-containing material will become an energy source for the billions of decomposers that in turn will become food for secondary or tertiary consumers. The organic material will be mineralized (degraded to its inorganic components) to be used by subsequent generations of producers and consumers. A large amount of the potential energy in an ecosystem is usually in the form of detritus. The biomass of detritus may be 10 to 100 times that of the living organisms; at any one time it represents an equilibrium between detritus production and decomposition. The rate of decay and percentage of detritus vary greatly from one ecosystem to another.

Many synthetic organic compounds are very resistant to decomposition by either bacteria or fungi. Such materials as the chlorinated hydrocarbon pesticides (e.g., DDT and dieldrin), fungicides, detergents, and plastics (e.g., polyvinyl chloride, polychlorinated biphenyls, polyurethane, and polystyrene) are highly resistant to decomposition. In fact, the major advantages of these products are their longevity and resistance to decay. Over time, however, certain decomposers have developed the biochemical abilities (e.g., enzymes) to degrade these synthetic products. For example, fungi may start to decompose a polyurethane pillow when it is only a few years old and the pillow may crumble soon after that. Deacon (1980) discusses the chemical mechanisms by which synthetic molecules are degraded by fungi.

Mechanisms by which fungi decompose natural compounds are well documented. The biodegradation of cellulose, the major component of plant cell walls, requires the production of at least three enzymes known as the cellulase enzyme complex. First, the straight-chain glucan (composed of thousands of glucose units) is broken into units of **cellobiose** by two depolymerase enzymes. Then the disaccharide cellobiose is cleaved by a glucosidase to produce glucose. However, the decomposition of the three-dimensional aromatic polymer lignin is not completely understood. Many diverse exoenzymes that attack many different structural parts of the molecule are required for its breakdown. (See Kirk et al, 1980, and Crawford, 1981, for further information on the biodegradation of lignin.) A study showed that in a forest 50% of the available glucose was mineralized to carbon dioxide and water in 3 days and 70% in 10 days. Cellulose remained, decomposing slowly, and lignin remained even longer. Up to 25% of the carbon from the lignin may ultimately be found in decay-resistant humus (Mayaudon and Simionart, 1959a, 1959b).

When detritus is fresh, the first group of saprophytic fungi to appear on it are the **sugar fungi,** which can utilize simple, soluble nutrients such as disaccharides or monosaccharides. Often aging leaves are weakly parasitized by fungi even before they fall from the tree. After the leaves fall, the sugar fungi colonize them. These fungi also appear on freshly deposited dung or on animals that have recently died. Growth of the fungi is rapid and the simple nutrients are soon exhausted.

The next group of fungi to appear are the **cellulolytic fungi,** which can produce cellulases and break down both cellulose and hemicellulose (polymers of various simple sugars). Together these two polymers comprise 70% of all plant material. The degradation of cellulose takes many weeks.

The third and last group of fungi to gain energy from the decomposition of plant materials is the **ligninolytic fungi.** These appear much later than the sugar and cellulolytic fungi, are slow growing, and degrade lignin very slowly. Lignin may comprise up to 30% of wood tissues. Other fungi will appear on keratin-containing animal remains (e.g., hair and skin) and will also slowly degrade these compounds. Many fungi can degrade a wide variety of organic substrates and therefore may fit into more than one of the categories just discussed. For example, some fungi, primarily basidiomycetes, can partially degrade both cellulose and lignin.

Fungi also play an important but poorly defined role in the production of humus in the soil. Humus is a dark colloidal substance, composed of dead organic material, that is highly resistant to decomposition. It may contain a complex combination of lignins, waxes, fats, and proteinaceous materials. Humus improves soil quality because it aids both moisture and nutrient retention and increases soil permeability.

PARASITIC FUNGI

Parasitic fungi are responsible for many extremely destructive plant diseases (e.g., late blight of potatoes, Dutch elm disease, black stem rust of wheat, and late corn blight), but only a small percentage of all fungi are parasitic. Agrios (1969) discusses fungal parasites of higher plants. The great majority of fungi are beneficial, even essential, to the functioning of the biosphere. It is important to remember that a discussion of parasitic fungi centers around only a small percentage of the over 100,000 known species of fungi.

Many saprophytic fungi can, under certain conditions, become parasitic on plants or animals. These fungi are opportunistic and are called **facultative parasites.** Other fungi are **obligate parasites**—that is, they require living organic material in order to live and reproduce. Parasitic fungi may be divided into two additional categories: (1) **necrotrophic,** living on dead cells that the fungus has killed, and (2) **biotrophic,** living on live cells, deriving nutrients from them by haustoria, but not killing them. Parasitic fungi may be either endobiotic, living entirely within the cells of their host, or **epibiotic,** living primarily on the surface of the host, but with haustoria penetrating the host cells.

Plant diseases caused by fungi include the wilts, smuts, rusts, rots, and blights of a wide variety of pterophytes, gymnosperms, and angiosperms. Almost every higher plant species is subject to one or more fungal diseases. Even the algae are parasitized by a group of mastigomycetes, the chytrids, as well as by ascomycetes.

A biotrophic fungus may live on its host for a long time without causing much damage or killing cells. The effects of such parasitism on the host may not be readily discernible and the host does not appear to be diseased in any way. However, when a fungal parasite causes damage to the host, cell death, or a disturbance of the host's physiology, the parasite may be said to be "pathogenic," or disease causing. The most highly evolved form of fungal parasitism is considered to be biotrophism, because the fungus not only does not kill or significantly injure its host but is also restricted to a very narrow range of hosts, often just one.

The parasitic fungus may penetrate the host by a combination of mechanical and chemical (exoenzyme) means. The fungus may penetrate the host's surface by invading a wound or growing through natural openings such as a leaf stomata. Once the fungus is located below the epidermis, inducible **lytic** enzymes are produced and the fungus thus gains entrance into the cell. The cellulose cell wall is digested by **cellulases,** and the **pectin** middle lamella may be dissolved by **pectinases.** Haustoria then develop within the host cell, and cellular nutrients are absorbed by the fungus for its nutrition. Certain fungi (e.g., *Ceratocystis ulmi*, which causes Dutch elm disease; see Chapter 17) produce toxins that can alter the permeability of the host's membranes and therefore interfere with the cell's ability to regulate what goes in and out of the cell. Other fungal parasites may produce plant hormonelike substances that cause the host to lose control over the growth of some of its parts, resulting in **hypertrophy** or **hyperplasia** of some tissues. Other fungi produce slimelike carbohydrates that may block the flow of water in the host's vascular tissues.

Fungi may also parasitize animals, including humans. The diseases produced may range from mild skin infections (e.g., athlete's foot) to fatal infections of the internal organs (e.g., **histoplasmosis** in humans, **aspergillosis** in birds and humans). Fungal parasites of animals may either be components of the normal fungal flora of the animal (e.g., *Candida albicans*) or opportunistic fungi that exist primarily as saprophytes (e.g., *Coccidioides immitis*). The conditions under which normally saprophytic fungi invade animal tissues are not completely understood. It is known, however, that humans whose immune systems have been suppressed by drugs or disease are often the victims of fungal attack. (Ainsworth, 1952, Ainsworth and Austwick, 1973,

and Christensen, 1975, discuss fungal diseases of animals, including man.)

Fish may be parasitized by an opportunistic member of the mastigomycetes (*Saprolegnia parasitica*); this has been responsible for many fish deaths, not only in hatcheries but also in lakes and streams.

PREDATORY FUNGI

Two groups of soil fungi can capture and kill minute invertebrates in the soil. These fungi (see Zygomycetes—Chapter 16, and deuteromycetes—Chapter 19) can trap amoebae, other protozoa, rotifers, and nematodes with sticky secretions to which the prey adheres or with loops formed from hyphal filaments. Once immobilized, the fungi penetrate the invertebrate and feed on the cellular contents, eventually killing the host.

NUTRITION

What environmental factors, chemical and physical, promote or discourage the growth of fungi? Fungi have a remarkable ability to use a wide variety of substrates as energy sources. There are few substances, either natural or synthetic, that some fungus cannot decompose. Fungi are heterotrophic and therefore must have preformed organic molecules in order to grow and reproduce. (The broad range of carbon sources that fungi can use are discussed in the sections on saprophytic and parasitic fungi.) Some fungi are very limited in the carbon sources that they will break down (e.g., biotrophs), whereas other fungi will attack a wide range of organic substrates, living or nonliving (e.g., facultative parasites). Almost all fungi must have an organic source of carbon, but a few mastigomycetes can assimilate carbon dioxide on a limited basis (see Chapter 15).

Inorganic nitrogen sources such as ammonia, nitrite, and nitrate can be utilized by most fungi, although some can use only ammonia while others can use only nitrite or nitrate. A few fungi have an obligate requirement for organic nitrogen, and most fungi can obtain their nitrogen requirements from organic sources when necessary. The ability to break down proteins to their component amino acids is widespread among the fungi. Fungi can also break down fats to fatty acids and glycerol, which are then absorbed into the cell.

Many fungi can synthesize the vitamins they require for growth and reproduction; others require exogenous sources of vitamins from the environment or media. Generally the widely distributed saprobes have few if any vitamin requirements, whereas the parasitic forms have complex nutritional requirements, including vitamins, that they may obtain from their host. The vitamin required by most filamentous fungi is thiamine; biotin is required most frequently by ascomycetous yeast. Fungi known to require a certain vitamin may be used as assay organisms to determine the presence or absence of such a vitamin on complex substrates.

Fungi require sulfur and most can reduce the sulfate ion (SO_4^{-2}) for their metabolism. A few fungi cannot reduce sulfate and therefore require sulfur in a reduced form. Some fungi can utilize organic sulfur sources (e.g., sulfur-containing amino acids), but no fungi are known to have an obligate requirement for organic sulfur.

The micronutrients required by most fungi are zinc, iron, manganese, molybdenum, copper, and perhaps calcium. Required concentrations of these essential micronutrients may range from barely detectable levels to 0.5 ppm (parts per million). A large excess of micronutrient may prove toxic. For example, 0.5 ppm of zinc is required for fungal growth but 32.5 ppm is toxic (Weinburg, 1971).

Most fungi require oxygen for vegetative growth or sexual reproduction. A few fungi can grow in environments that are low or lacking in oxygen (e.g., fermentation of sugars), but most fungi evidently require oxygen at some stage of their life cycle, usually during sexual reproduction. The yeasts (Ascomycotina) are the best known fungi that grow anaerobically, metabolizing sugars to ethyl alcohol and carbon dioxide via the fermentative pathway. This ability is used commercially in the production of beer and wine.

Most fungi also require light for both vegetative growth and sexual reproduction. A few fungi can grow under conditions of reduced light; however, light is often necessary for the development of sexual structures. This requirement for light to develop sex organs generally limits soil fungi to the top half-meter of soil. As a general rule the biochemical and environmental requirements for sexual reproduction and spore production are narrower than the requirements for vegetative growth. In addition, the fungal thallus must reach a certain level of maturity (i.e., competence; see p. 277) before sexual structures are formed, no matter what the environmental stimuli.

Most saprophytic fungi occurring in nature have an optimum growth temperature near the mean summer temperature of their natural habitat. Yeasts that grow in the Arctic have an optimum growth temperature of 15° C, whereas tropical fungi exhibit optimum growth near 35° C. Parasites of warm-blooded animals have optimum growth near the body temperature of their host. Facultative parasites that can live as saprobes or parasites have a wide range of optimum growth temperatures. Some fungi can grow on and spoil refrigerated foods (4° C), while others can even grow on frozen foods (−5° C). Thermophilic (heat-loving) fungi that grow in heated compost during its decomposition can withstand temperatures of 60° C.

Moisture in varying amounts is required by all fungi. The aquatic fungi (Mastigomycotina) require moisture at all times; they are extremely susceptible to desiccation, and their motile zoospores require water for dispersal. The terrestrial fungi (zygomycetes, ascomycetes, basidiomycetes, and deuteromycetes) also require adequate moisture at most stages of their life cycles. Fungi need some moisture for their exoenzymes to diffuse into the substrate and for nutrients to diffuse into their cells. Growing fungi that are rapidly metabolizing are especially vulnerable to deficiencies of water. Mycelia are much more susceptible to desiccation than are the spores. Moist environments such as forest leaf litter and rotting logs support a large fungal population, whereas arid environments have a much smaller population. Fungi that grow in low-moisture environments are called xerophytes (dryness loving).

The moist and warm tropics support a very large fungal population, many of which can cause problems by parasitizing humans and by rapidly decomposing synthetic products. Humid, warm weather in temperate regions also encourages the growth and spread of such destructive plant parasites as the late potato blight, wheat rust, and the downy mildew of grapes.

The majority of fungi cannot tolerate high salinities (salt concentrations); therefore relatively few (1% of described species) are found in the marine environment (Kohlmeyer and Kohlmeyer, 1979). The Ascomycotina are the most prevalent fungal group in the oceans and estuaries. These marine organisms are osmophilic (thrust loving) because they can withstand the increased osmotic pressure of salt water. Other osmophilic fungi, mainly yeasts, can live in media that are high in sugar concentration and may be found growing in jams and jellies.

Most fungi grow well at pH levels below 7.0. In acid forest soils, fungi are often more important decomposers than the bacteria. Some fungi, however, can do well at alkaline pHs up to 11.0 and may be found growing in limestone soils or other alkaline environments.

A large percentage of the carbon assimilated by fungi is directed toward the synthesis of the cell wall. Chitin comprises a large portion of the cell wall, although ascomycetous yeast cell walls contain principally glucans (glucose polymers) and **mannans** (mannose polymers). Cellulose occurs occasionally in the cell walls of some fungal groups (e.g., Oomycetes).

The majority of fungi store energy as the polysaccharide glycogen, which is also an energy storage form frequently found in animals. Glycogen granules may comprise approximately 5% of the dry weight of both spores and mycelia. Lipid globules may also be formed as energy storage compounds, as may the sugar alcohol mannitol.

Carbohydrates are usally translocated by fungi in the form of **mannitol** or trehalose, a disaccharide. The fungal ability to convert carbohydrates to mannitol and trehalose gives the parasitic fungi an advantage over their vascular plant hosts. Most vascular plants translocate carbohydrates in the form of sucrose, fructose, or glucose. When fungi convert host carbohydrates to mannitol and/or trehalose, a one-way transfer of nutrients occurs because the vascular plant cannot metabolize these fungal carbohydrates. The fungal haustoria absorb the host sugars, convert them to fungal carbohydrates, and translocate the fungal carbohydrates to other parts of the mycelium where they are needed. Trehalose and mannitol are also found in many nonparasitic fungi.

Fungal growth may be limited by any number of factors that are required for their growth. Saprophytic fungi are widely distributed and often may be limited by the moisture content of their environment. For example, basements may be dry during winters when furnaces are functioning, but when summer arrives with its warm, humid weather, basements become damp and fungi (**mildew**) grow well. The cool basement walls cause moisture from the humid air to condense, and drops of water form on the walls and other objects: a perfect environment for fungal growth is created. Articles stored in the basement may become covered with mildew and be subject to damage. Once the weather becomes cooler and dryer in autumn, the furnace is started and the basement dries out. The mildew-causing fungi cease to grow, but they have already produced spores that may cause problems the following summer after their germination.

Temperature, light, nutrients, or carbon sources may also be limiting factors for fungal growth. If the mycelium does not have the proper conditions for vegetative growth, spores may be produced. Spores are much more resistant to adverse environmental conditions than the growing mycelium, so any effort to eradicate fungi from an area must also include methods for destroying the spores.

PREVENTION OF FUNGAL GROWTH

Fungi are essential to recycling of the inorganic nutrients that are chemically bound in organic material. Fungi will indiscriminately attack a rotting tree, a telephone pole, a picnic table, or stored lumber without regard for the economic value of the attacked structure. It is necessary, therefore, to protect certain objects from fungal decomposition. Telephone poles, railroad ties, and other wooden objects used outside or in contact with the soil must be treated with creosote or pentachlorophenol— phenol-based preservatives that prevent fungal growth for many years. Lumber used for porches, steps, and park benches may be impregnated under pressure with **fungicides** such as chromated copper arsenate. Lumber stored in a lumber yard is carefully seasoned and dried, thereby reducing its moisture content so fungi will not grow.

Cellulose-decomposing fungi attack fabrics in tents, tarpaulins, and ropes, causing these materials to decay. Fabrics and fibers destined for outdoor use may be treated with appropriate copper-based fungicides. Fiber and fabric decay is especially troublesome in the tropics, where fungi grow rapidly and decompose fabrics quickly. The substitution of plastics in tropical areas has partially solved the problem.

Perishable foods must also be protected from fungal growth. Fungi grow readily on bread, fruits, cheese, jellies, and many cereal products. Many such foods are treated with a fungal inhibitor (e.g., sodium propionate) to lengthen their shelf life. Other foods (e.g., milk and fruit juices) may be pasteurized (heated to 62° C for 30 minutes) to kill fungal spores. Manipulation

of the environment may also retard fungal growth; for example, apples may be stored at low temperatures in an atmosphere high in carbon dioxide (8% to 10%) to retard fungal growth and to keep them fresh for several months.

Fungicides are often sprayed on fruit, grain, and vegetable crops to prevent growth or to kill existing fungi. Likewise many trees in both managed forests and residential areas are sprayed with fungicides when fungal diseases threaten. In 1978 total fungicide production in the United States was 7.2×10^7 kg (U.S. Department of Agriculture, 1978). Siegel and Siser (1977) discuss the chemical compounds used to retard fungal growth.

Fungicides, like many chemicals designed to kill insects, are not specific as to the organisms they kill. A fungicide sprayed over many acres will not only kill the target fungus but also many beneficial saprophytic fungi. Fungicide use may also decrease the population of fungi that would parasitize insects and keep the insect populations under control (Tedders, 1981). Decomposition of litter in the sprayed area may be significantly retarded and the rate of nutrient recycling thereby reduced.

Inorganic copper, sulfur, or mercury-based chemicals are the most frequently used fungicides against plant disease. Mercury-based fungicides have been used primarily on grain seeds, and problems have subsequently developed when birds ate the seeds. In Sweden mercury levels high enough to interfere with reproduction resulted and caused other detrimental metabolic changes. Sweden banned the use of methyl mercury in 1966 because of the toxicity to grain-eating birds (Miller and Berg, 1969). Humans have also become seriously ill from eating bread that was mistakenly made from grains contaminated with mercury or from eating livestock that were fed contaminated grain. In 1970 an American family became gravely ill with mercury poisoning from contaminated pork (Curley et al, 1971). In 1972, 450 people died in Iraq after eating mercury-contaminated bread. Since these incidents, the frequency of seed treatment has been reduced and the formulation of the mercury compound used for seed treatment has been changed. These two measures have significantly reduced the contamination of seeds with mercury.

Organic fungicides, such as the phthalimide compounds captan and folpet and the carbamates fenthion and zineb, are used on agricultural crops. Captan is also used as a dip to prevent fungal growth on fruit such as peaches. Concern has arisen over the widespread use of captan and folpet, because they have been shown to produce defects in developing chick embryos (Verrett et al, 1969) and in rabbits (Shea, 1972).

Systemic organic fungicides such as benomyl and other similar compounds (benzimidazoles) can be introduced into the vascular system of the plant and transported to all parts of the plant to kill fungi. This method has been used to protect elm trees from the Dutch elm disease, but its effectiveness is still unknown. Few such systemic fungicides are available for fighting plant diseases.

Other methods for controlling undesirable fungal growth include alteration of the environment to discourage fungi; for example, for mildews in the basement, a dehumidifier would reduce the moisture content of the basement, thereby discouraging such fungi. Properly dried lumber seldom supports growth, but the lumber must be protected with creosote or fungi-retarding paint if used outdoors. The use of plastics and other synthetic materials retards fungal growth on ropes, tents, and other outdoor equipment. Unfortunately, fungi are now developing exoenzymes that can even decompose synthetic materials (e.g., dacron, polyurethane, and nylon).

Some crops do well without protection against fungi in dry sunny seasons. When the weather becomes damp and cool, fungi can cause severe diseases. By a knowledge of weather conditions in a given area, it is possible to predict outbreaks of certain plant diseases such as potato blight. Such predictions can warn farmers to take preventative

steps with fungicides. These predictions enable farmers to use the minimum amounts of fungicides. Another successful method of fungal control is the elimination of alternate hosts for certain parasitic fungi such as wheat rust (alternate host: barberry) and white pine blister rust (alternate hosts: gooseberry and currant). The breeding of crop strains that are resistant to pathogenic fungi has also been successful against many fungal diseases (e.g., downy mildew of grapes, potato blight, and wheat rust). Rotation of crops to prevent reinfection from overwintering spores and sanitation to destroy fungi-infested plant parts are also effective methods of reducing the incidence of fungal disease.

Biological control of some fungi may be accomplished by encouraging or introducing natural predators of the fungi. Some fungi will inhibit or prevent the growth of other fungi (hyphal interference), thereby restricting growth of the unwanted fungus. Interactions between fungi are discussed in the next three sections.

Secondary Metabolites

A large number of fungi produce chemical compounds that have no known function in the metabolism of the organism. These compounds, called **secondary metabolites,** are produced when normal growth is restricted in some manner. Secondary metabolites differ, therefore, from **primary metabolites;** which are compounds produced during normal growth and which have a known function in the structure or metabolism of the fungus (Griffin, 1981). During normal fungal growth, intermediate products of the basic metabolic pathways are directed into primary metabolites. When fungal growth is restricted, these intermediate products are shunted into the production of secondary metabolites. Many fungal secondary metabolites have been chemically characterized, but the survival value of these has not yet been definitely determined. Deacon (1980) proposes that secondary metabolites may be an "escape valve" to direct intermediate products of primary pathways into compounds that have no effect on the fungus producing them. Some fungi have been able to produce secondary metabolites that are beneficial to the fungi themselves. For example, **antibiotics** discourage competition from bacteria.

Many secondary metabolites are species or even strain specific, and many have been exploited commercially. For example, the antibiotic penicillin is a secondary metabolite produced under certain conditions by a specific strain of *Penicillium chrysogenum* (Ascomycotina).

Certain fungi produce secondary metabolites called **mycotoxins** that are toxic to some animals; and their effect may range from mildly to extremely toxic. Several mycotoxins produced by species of *Aspergillus* (Ascomycotina) are **carcinogenic** (cancer-causing) in laboratory animals. Fungi that produce mycotoxins tend to grow on stored grains; such contaminated grains can cause illness or even death when consumed by animals. The ascomycete *Claviceps purpurea*, which parasitizes grasses, especially rye, produces powerful alkaloids that in humans can cause convulsions, hallucinations, and loss of circulation in the extremities.

Additional Interactions Between Fungi and Other Plants and Animals

Fungi interact with many other organisms in a multitude of ways. For example, hyphae from different fungi may meet in the soil, and one fungus will cease growing while the other continues. Such **hyphal interference** is observed most frequently in the basidiomycetes and may be used in the biological control of undesirable fungi. Antibiotics produced by certain fungi prevent the growth of bacteria in regions where the fungi are growing. The parasitic and pathogenic fungi that attack plants and animals may cause mild or severe diseases.

Mutually beneficial interactions between fungi and algae occur in lichens (see Chapter 20). These symbiotic combinations of an alga and a fungus behave as one organism. The alga provides carbohydrates and removes carbon dioxide for the fungus, while the fungus provides protection and moisture for the alga.

Fungi and insects have many interesting interactions. Fungi (Entomophthorales) may parasitize insects and in some instances effectively control insect populations. Weiser (1982) discusses fungi as agents for the control of insect pests. Some insects (e.g., termites and ants) actively cultivate gardens of fungi in their nests. Other insects and fungi have developed complex symbiotic relationships such as the one between scale insects and a basidiomycete (see Chapter 18). Many insects serve as efficient dispersal agents for fungi, carrying spores from one plant to another.

Fungi are also being used experimentally to control undesirable angiosperms. These **mycoherbicides** are used in the control of weeds in soybean fields and waterways. Kenney et al (1979) and Phatak et al (1983) discuss the use of fungi for weed control.

Mycorrhiza

Complex interactions have developed between soil fungi and a majority of forest trees, herbaceous plants, and agricultural crops. These symbiotic associations between vascular plants and fungi are called mycorrhizae (fungus-root), and they are prevalent in soils low in available nutrients. The term mycorrhiza refers to the plant root and its associated fungus. The fungal hyphae invade the root tissue, and a transfer of water, inorganic nutrients, and carbohydrates may occur in both directions. The relationship has been described as one of "physiologically well-balanced reciprocal parasitism" (Hacskaylo, 1972). Both partners in the relationship appear to benefit at one time or another and neither partner is harmed by the parasitism. The host plant evidently has biochemical mechanisms for restricting the growth of the fungus, and the fungus has mechanisms by which it can usually prevent its being destroyed by the host.

Infection of vascular plants with mycorrhizal

fungi is very widespread, occurring in both woody and herbaceous plants. It is estimated that a majority of vascular plants benefit from mycorrhizal associations. Only a few plant families fail to form mycorrhizae, and the incidence of mycorrhizae in the other families may depend on soil fertility or other factors. Benefits to the vascular plant from the fungal association include increased uptake of nutrients (especially phosphorus), protection from pathogens, and production of plant hormones. The fungus benefits from having a ready and plentiful supply of carbohydrates. Identification of the fungus is often very difficult, because many such

TABLE 13-1. Mycorrhizae

Vascular plant family and representative genera	Type of mycorrhiza	Fungus
Pine—Pineaceae Douglas fir, hemlock, spruce, pine, larch	Ectotrophic	Basidiomycetes such as *Amanita* *Russula* *Suillus* *Pisolithus*
Willow—Salicaceae Willow, poplar	Ectotrophic	
Hickory—Juglandaceae Hickory, walnut	Ectotrophic	
Myrtle—Myrtaceae Eucalyptus	Ectotrophic	
Linden—Tiliaceae Basswood	Ectotrophic	
Beech—Fagaceae Oak, beech	Ectotrophic	Ascomycete—*Tuber* Basidiomycetes
Cypress—Cypressaceae Juniper, cypress, arborvitae	Endotrophic	Zygomycetes *Endogone* *Glomus*
Sequoia—Taxodiaceae Sequoia, bald cypress	Endotrophic	
Legume—Leguminosae Peas, beans, soybeans, peanuts	Endotrophic	
Tomato—Solanaceae Tomatoes, tobacco, egg plant	Endotrophic	
Rose—Rosaceae Peach, apple, strawberry	Endotrophic	
Coffee—Rubiaceae Coffee	Endotrophic	
Citrus—Rutaceae Orange, grapefruit	Endotrophic	
Grass—Gramineae Rye, corn, wheat, rice, oat	Endotrophic	
Orchid—Orchidaceae Orchids	Endotrophic	Basidiomycetes *Marasmius, Armillariella, Fomes*
Heath—Ericaceae Cranberry, blueberry, mountain laurel	Endotrophic	Unidentified Fungi

After Harley (1969) and Hacskaylo (1972).

fungi fail to produce fruiting bodies when cultured in the laboratory. Clues to fungal identification must be taken from the regular association of certain fungal species with certain tree species (e.g., *Tuber aestivum*, a truffle, with oak and beech trees; *Suillus americanus*, a bolete basidiomycete, with eastern white pine).

There are three forms of mycorrhizae: ectotrophic, endotrophic, and ectendotrophic. **Ectotrophic mycorrhizae** are widespread in temperate areas on forest trees such as pine, Douglas fir, hemlock, larch, birch, and beech. Most ectotrophic fungi belong to the basidiomycetes, but a few are ascomycetes (Table 13-1).

Ectotrophic fungi form a sheath of mycelium around individual primary roots, and hyphae invade the root cortex by the secretion of **pectolytic** enzymes. The hyphae in the cortex grow intercellularly in a distinctive pattern called the Hartig net (Fig. 13-1). The fungus may secrete such plant hormones as auxin or cytokinin, which cause characteristic morphological changes in the root. The fungus absorbs carbohydrates from the vascular plant and converts the photosynthate (usually sucrose) into fungal carbohydrates,

mainly trehalose, mannitol, and glycogen. The fungus, therefore, becomes a sink for the host carbohydrates, which are then unavailable to the root cells, since the latter cannot utilize the fungal carbohydrates (Lewis and Harley, 1965). Most ectotrophic fungi are "sugar fungi" and cannot decompose the insoluble cellulose and lignin in the soil to an appreciable extent; therefore these fungi must rely on the soluble carbohydrates from the host photosynthate as a carbon source. This type of mycorrhizae develops only when the host is actively photosynthesizing, providing additional evidence that the host photosynthate provides carbon for the fungus. Because the fungus in this partnership obtains most of its carbon from the host, the presence of the fungus must cause a drain on the host's photosynthetic products. The presence of the fungus increases the growth of the host only when soil nutrients are limiting the growth of the host. The fungus can absorb nutrients in excess of its needs, releasing them to the host as needed. The price the host must pay, however, is the loss of some photosynthate to the fungus. If physical or chemical factors (e.g., light or carbon dioxide) other than nutrients are limiting to the host, then the presence of the fungus does not result in increased growth.

Endotrophic mycorrhizae are found in ferns, gymnosperms (except the pine family), and angiosperms, including both herbaceous plants and trees, especially in the tropics (Table 13-1). This type of mycorrhizal association is much more frequently encountered than the ectotrophic form and appears to be very widespread throughout the world. Many economically important agricultural crops (e.g., grains, fruit trees, vegetables, and coffee) are dependent on these fungal associations, which increase the productivity of many commercial food and fiber crops. In the endotrophic types, the hyphae may not only penetrate the root cortex intercellularly but they also penetrate the cells by producing haustoria. Many endotrophic fungi (usually zygomycetes such as *Endogone* or *Glomus*) develop prominent vesicles and extensively

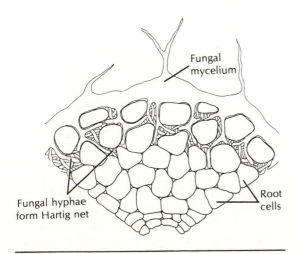

Fungal mycelium

Fungal hyphae form Hartig net

Root cells

FIG. 13-1 Ectotrophic mycorrhiza.
Note netlike appearance of intercellular hyphae, Hartig net, and sheath of mycelia surrounding root.

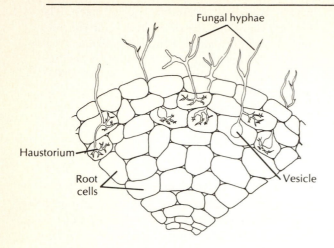

Fungal hyphae

Haustorium

Root cells

Vesicle

FIG. 13-2 Endotrophic mycorrhiza.
Note intracellular, branching haustoria and vesicles produced by the fungus. This form of intracellular haustoria is called a vesicular-arbuscular mycorrhiza.

branched haustoria within the cells of the cortex (Fig. 13-2). These associations are called **vesicular-arbuscular** (treelike) **mycorrhizae.** Intracellular portions of such hyphae are often digested by the host within a few days after their formation, thereby releasing fungal nutrients to the host. The fungi also transfer soil nutrients to their host in a manner similar to that of the ectomycorrhizal fungi.

Both the extent of endotrophic fungal formation and the depth of penetration of the fungus into the root are governed by nutrient levels in the soil. The poorer the soil in mineral nutrients, especially phosphorus, the greater the biomass of fungus and the deeper the penetration into the root. As in the ectotrophic association, the mycorrhizae are formed only when the host is actively photosynthesizing, probably demonstrating a need by the fungus for the host's photosynthate. The positive identification of such fungi has proved difficult because they are biotrophic and cannot be grown in the laboratory unless grown with the host species.

All orchids (Orchidaceae) have mycorrhizae that present a slightly different type of endomycorrhizal relationship. The fungi involved are usually basidiomycetes (Table 13-1) that can decompose insoluble carbohydrates (cellulose and lignin) and translocate soluble carbohydrates into the orchid (Harley, 1969). The orchid may

occasionally digest the fungus. The mycorrhizal relationship is essential to the young orchid, which is saprophytic and dependent on the fungus for carbohydrates. Some orchids remain saprobes when mature, whereas others become autotrophic.

The third type of mycorrhizal association is the **ectendotrophic,** which has been observed in tree species such as pine, fir, and spruce. This type of association combines the ectromycorrhizal and endomycorrhizal characteristics and is more frequently found in nursery-grown gymnosperms than in trees growing in the forest. The fungi involved are septate, but they have not been positively identified (Wilcox, 1971).

Mycorrhizae have evidently evolved independently among several groups of fungi and among several different vascular plant divisions. Mycorrhiza-like aseptate fungi have been seen in the fossils of primitive vascular plants that are over 300 million years old (Hayman, 1980).

Knowledge of the importance of mycorrhizal fungi to crop and forest productivity has been used to aid the establishment and increase the growth of commercially important vascular plants. For example, the successful introduction of pines into Puerto Rico, Australia, Rhodesia, and areas of the Caribbean was accomplished by ensuring that soil containing mycorrhiza-forming fungal spores was introduced at the same time. Without the mycorrhizal fungus, the trees do not grow well and may even die (Fig. 13-3). Mycorrhizae appear to be extremely important, if not essential, to host species in both nutrient-poor and highly competitive environments. This knowledge has been useful in revegetating such areas as anthracite waste piles (Schramm, 1966) and coal mine spoils in arid areas (Aldon, 1975).

FIG. 13-3 Virginia pine *(Pinus virginiana)* **with and without mycorrhizae.**

A, Two 4-month-old Virginia pine seedlings. The one on the left is growing with the ectomycorrhizal fungus *Pisolithus tinctorius* (Basidiomycotina). *P. tinctorius* is widely used in reforestation and is known to be mycorrhizal with other species of *Pinus*, in addition to other important tree species such as *Betula* (birch), *Carya* (hickory), and *Quercus* (oak). Marx (1977) discusses the ectomycorrhizal role of this fungus. The presence of the mycorrhizae causes the tree to grow faster and healthier. Roots of the seedling on the left, **B,** show extensive ectomycorrhizal roots. The seedlings were inoculated with *P. tinctorius*. **C,** Roots of seedling on the right were *not* inoculated with *P. tinctorius* and lack extensive branching. (Courtesy Donald H. Marx, Institute for Mycorrhizal Research and Development, United States Forest Service, Athens, Georgia.)

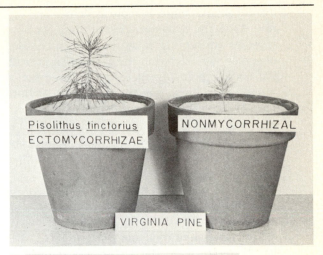

A

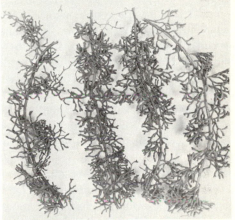

B

Continued research into these extremely beneficial and essential relationships will yield more information into this complex form of symbiosis. Harley (1969), Hacskaylo (1972), Marks and Koslowski (1973), Sanders et al (1975), and Schenck (1982) provide in-depth discussions of these phenomena.

SUCCESSION

The succession of fungi can be easily observed on various substrates. Succession may be viewed as a type of interaction between species where one species "conditions" the substrate for another species. A similar type of "conditioning" was discussed previously in the succession of algal species (Chapter 2). The succession of fungi on fresh dung can be observed in the laboratory. It involves first the sugar fungi, primarily the zygomycetes, which utilize the simple, soluble molecules. Second, the ascomycetes and deuteromycetes appear, utilizing the more resistant cellulose. Last, the basidiomycetes emerge and decompose lignin and any remaining cellulose. A similar species succession may be seen on straw compost. Degradation of straw compost takes much longer, how-

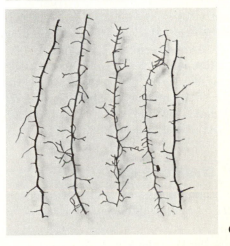

C

ever, because there is more cellulose and lignin. As in the decomposition of dung, basidiomycetes are the last to appear. The succession observed is reflected in the production of fruiting bodies; probably spores of all three fungal groups were present at the start of decomposition. It takes longer to produce a complex fruiting body such as a mushroom than it does to produce a zygomycete sporangiophore. The zygomycete fruiting bodies, therefore, appear first.

Succession of fungi has also been studied on pine needles. Since needles have a high percentage of lignin, decomposition may take as long as 7 to 9 years (Kendrick and Burges, 1962). The basidiomycetes are the last group of fungi to appear, and evidently they outcompete other fungi. Specific basidiomycetes may be found growing under specific gymnosperms, and these fungi may have both a mycorrhiza-forming role and a decomposition role. Similar studies of succession have involved the leaves of deciduous trees. Oak leaves appear very resistant to decomposition and, because they also have a high lignin content, remain on the forest floor for many years.

Fungal succession has also been observed in aquatic ecosystems. The **Hyphomycetes*** (Deuteromycotina) and the Oomycetes (Mastigomycotina) are important in the decomposition of leaves and twigs in aquatic habitats. Willoughby (1977) discusses the succession of such fungi on leaves. The invertebrate amphipod *Gammarus* preferentially feeds upon leaves partially decayed by fungi over newly fallen leaves with sparse fungal populations, indicating that the fungi-enriched leaves are probably more readily metabolized (Kaushik and Hynes, 1971).

FUNGAL COMMUNITIES

Fungal communities are readily observed on fallen trees, on wood piles, on fallen leaves, or

*The aquatic Hyphomycetes have recently been renamed the **Ingoldian fungi** (Webster and Descals, 1981).

on animal dung. Specific associations of species exist that are found generally on straw, on hair, or on dry or alkaline soils. In each instance several species will occur together and are capable of decomposing the dead organic matter. The fungal community on fallen leaves may be observed by bringing some leaves into the laboratory, monitoring their decomposition, and observing the fungi that are present. Assays of protein content and cellulase activity may also be performed.

The remains of warm-blooded animals may also develop a specific fungal community that occurs on this particular substrate containing keratin, a protein found in feathers, hair, and hooves.

Studies of fungal communities may also be done with packs of leaves in streams and lakes. Several different species may be identified and the decomposition rate followed. The aquatic Hyphomycetes (Deuteromycotina) are important in the degradation of coarse organic matter in such environments (Sinsabaugh et al, 1981).

Kohlmeyer and Kohlmeyer (1979) discuss the fungal communities of salt marshes. Over 100 species of fungi, both parasitic and saprophytic, have been isolated from the salt marsh grasses (*Spartina* spp.) alone (Gessner and Kohlmeyer, 1976). Most of the fungi identified are ascomycetes and deuteromycetes. Fungi that live in the harsh environment of the salt marsh must be able to withstand the wide ranges of environmental parameters, especially salinity, that regularly occur.

The fungal communities of sewage treatment facilities have also been studied. Hawkes (1965) studied the relationships between the fungal and bacterial populations on **trickling filters.** Cooke and Pipes (1970) investigated the seasonal distribution of the fungal groups in activated sludge systems. Most fungi isolated from the activated sludge system were filamentous rather than yeastlike. Cooke (1976) further discusses the fungi found in sewage.

Complex interactions take place among fungi and other microorganisms in the soil. For example, an ascomycete may partially decompose cellulose with its exoenzymes, and the resulting disaccharides or monosaccharides are available in the soil not only to the ascomycete, but also to other fungi and bacteria in the immediate area. Various soil microorganisms may compete for the nutrients released by the fungus, but antibiotic production by certain fungi may discourage competing bacteria. Cooke (1979) provides a thorough discussion of fungal communities.

COMMERCIAL USES OF FUNGI

Fungi are cultured commercially for a large variety of products, which may be divided into four classes. These include: (1) the fungal cells themselves, especially the yeasts and mushrooms; (2) compounds with high molecular weights such as exoenzymes, which are synthesized by the fungi; (3) such primary metabolites as organic acids and vitamins; and (4) such secondary metabolites as antibiotics and plant hormones (Emerson, 1973; Demain and Solomon, 1981).

Ethyl alcohol, organic acids (e.g., citric, gluconic), various enzymes, and antibiotics are all produced by fungi. Fungal strains are cultivated and bred for their ability to produce the desired by-products or secondary metabolites. The environment in which the fungus is grown is manipulated to obtain maximum production of the desired end product. Ethyl alcohol thus produced is used as a solvent in industry and in scientific laboratories. Citric acid is used in many foods, especially soft drinks. The enzymes may be used for washing powders, cheesemaking, starch hydrolysis, or medicines.* Onions et al (1982) outline the industrial uses of fungi.

Other medical drugs, including penicillin, are produced by fungi. Several other antibiotics, antifungal compounds, and an antitumor substance are also synthesized by fungi. In addition, steroid hormones may be transformed from inactive to physiologically active compounds. Alkaloids extracted from the sclerotia of *Claviceps purpurea* are used for the production of the drug ergot (or ergotamine), which is used to prevent hemorrhaging during childbirth.

Other yeasts (Ascomycotina) have the ability to ferment sugars to ethyl alcohol under anaerobic conditions. Such yeasts are used widely in the production of beer and wine. Beer is made from fermented barley malt, whereas wine is made from fermented grape or other fruit juices. Fermentation by bakers' yeast is also an anaerobic process resulting in the production of ethyl alcohol and carbon dioxide, which causes bread to rise. During baking, the alcohol is volatilized and the carbon dioxide escapes, leaving air pockets in the bread. Yeasts may also be used in the manufacture of vitamins or as a food source by themselves.

Fungi are also used to age meat and to ripen and give distinctive tastes to cheeses. In the Orient other fungi are used to ferment foods and to increase their nutritive value. Wild fungi (truffles, morels, and assorted mushrooms) are gathered for gourmet additions to meals. Several mushroom species (*Agaricus brunnescens, Lentinus edodes*) are cultivated commercially and sold fresh or canned. The yeasts (primarily Ascomycotina and Deuteromycotina) are used to produce "single-cell protein," which is used as a food source for humans or as fodder. The fungi may be grown on organic waste products from petroleum and papermaking industries. Moo-Young (1976) discusses the different microorganisms used in the production of single-cell protein.

PROBLEMS CAUSED BY FUNGI

The major impact of fungal activity on mankind is the destruction of crops, both agricultural and timber. Plant diseases caused by parasitic fungi have been discussed previously. Fungi are the most important cause of diseases of

---

*An entire issue of *Scientific American* (245[3], 1981) discusses many industrial applications of microbiology, including the use of fungi in these processes.

vascular plants, with over 5,000 fungal species causing disease. Parasitic fungi cause losses of over 20,000 m³ of timber per year in California alone. Fungal pathogens of plants outnumber the bacterial pathogens and the viruses. When listing losses caused by plant diseases, usually no division is made between fungal diseases and bacterial or viral diseases, making it difficult to ascertain how much disease is attributable to each group.

Crop losses between 1942 and 1951 amounted to 13.26% of the total production in dollars (Gray, 1959). Between 1951 and 1960 losses were approximately 10% of the annual crop production (1% of the Gross National Product), with a loss of approximately $4.25 billion (Tippo and Stern, 1977). In addition, between 1951 and 1960, $250 million was spent on the control of fungal diseases. Worldwide losses of crops to fungal diseases probably amounts to billions of dollars per year. Annual crop losses in the United States from fungal diseases are estimated at $2.5 billion.

Saprophytic fungi degrade many products valued by man, in addition to many substances whose nutrients need to be recycled. Problems arise because fungi are indiscriminate decomposers. They will attack the wood in a cooling tower as well as a fallen tree and stored peaches as well as rotting fruit left in the orchard.

Fungi routinely attack foods such as fruit, bread, jellies, dairy products, and meats. Refrigeration retards fungal growth, as do preservatives. It is estimated that millions of kilograms of food are spoiled by fungi every year.

The pulp and paper industry also experiences problems with fungal growth, and fungicides must be used during processing to prevent spotting and slimes on the final product.

In the tropics, fungi are an ever-present problem because they grow rapidly in the favorable warm, moist climate. During World War II much research was devoted to preventing fungal deterioration of fabrics, tents, clothes, ropes, books, and leather. Not only did fungi attack such cellulose-based articles, but paints,

rubber, electrical insulation, optical equipment, and photographic supplies were also degraded. Fungal growth on binoculars, microscopes, and telescope lenses rendered this equipment useless because the mycelium coated the lenses and exoenzymes etched the glass. Plastic panels in radios and the insulation covering wires were degraded also. Damage to valuable equipment was extensive. Research conducted by the U.S. Army Quartermaster Corps led to the discovery of many ways of retarding fungal growth. However, the rapid deterioration of equipment in the tropics remains a significant problem.

In the South Pacific during World War II, the cellulytic deuteromycete *Trichoderma reesei* (= *Trichoderma viride*) caused widespread damage. Currently the fungus' cellulose degrading abilities are being used to convert tons of waste cellulose from newspapers, peanut shells, straw, and other sources into glucose that is usable in foods. Fungi may attack paint, causing "sooty mold" in moist areas such as kitchens and bathrooms. These fungi are usually deuteromycetes utilizing the carboxymethyl cellulose that is added to paints as an emulsifier (Deacon, 1980). Paints for outdoor use often have fungicides added to retard fungal growth. The deuteromycete *Aureobasidium pullulans* not only causes deterioration of painted surfaces but also attacks plastics in subtropical Florida. The fungi that attack plastics are just beginning to be characterized, but it appears that some fungi can partially decompose certain plastics, exposing portions of the plastic molecules to oxidation by light (i.e., **photooxidation**). Lubricating oils, linseed oil, kerosene, and aircraft fuels may also be attacked by fungi. Along with bacteria, fungi can grow in fuel storage tanks, utilizing and decomposing the hydrocarbons. Water in the fuel tanks encourages fungal growth, yet removal of the water is not practical. Many millions of dollars are spent annually to prevent or retard fungal growth; furthermore, the cost of manufacturing and distributing the chemical fungal inhibitors amounts to billions of dollars annually.

BENEFITS OF FUNGI

The principal benefit conveyed by fungi is the decomposition of dead organic matter into its organic components, which are reusable by green plants. Without both fungi and bacteria, dead organic materials would rapidly build up in the environment. Fungi may also provide the biochemical tools to decompose such synthetic, persistent, organic compounds as the insecticide DDT (dichlorodiphenyl trichloroethane), fungicides, plastics, alkyl benzyl sulfonate (ABS) detergents, and polychlorinated biphenyls (PCBs). The degradation and detoxification of these molecules is extremely important to prevent their persistence in the ecosystem and their bioconcentration in living organisms.

In addition to the benefits from the commercial use of fungi, certain fungi serve to biologically control potential plant pests. Some fungi (*Entomophthora, Beauveria*) parasitize insects and keep the population numbers in control.

Fungi are frequently used in the study of genetics. The ascomycete *Neurospora* (see Chapter 17) is used in the laboratory for studies of the inheritance of various biochemical and physical characteristics. Burnett (1975) discusses fungi as tools in genetic research.

An old field or a virgin forest has many fungal mycelia living in its soil. Left undisturbed over time, a large number of fungi grow and pursue their role as decomposers of dead organic matter. Many symbiotic associations develop as these ecosystems mature. For example, lichens will flourish on the rocks and trees and mycorrhizal associations between basidiomycetes, and the resident gymnosperms and angiosperms will increase in frequency (Fig. 13-4). Every spring or fall many different kinds of mushrooms (basidiocarps) will make their appearance in **fairy rings** or under trees. The interactions and associations that develop in an undisturbed ecosystem are very beneficial to the ecosystem, increasing the efficiency of nutrient recycling. As the ecosystem approaches a climax (mature) condition, both the number of species and the number of symbiotic associations between species increase. Harley (1975) states, "The importance of symbiosis in ecosystems has been much underestimated and the extent and effects of diversion and short-circuiting of carbon and nutrients throughout symbiotic systems requires further evaluation." Should man or some coup of nature severely stress the ecosystem, these beneficial, complex interactions that have taken years to evolve will be significantly reduced.

FIG. 13-4 Interaction of woodland mushrooms (Basideomycotina) and forest trees.
Mycorrhizal associations may develop between fungi and gymnosperms and angiosperms. Such associations provide carbohydrates for the fungus and nutrients and expanded absorptive surfaces for the vascular plant.

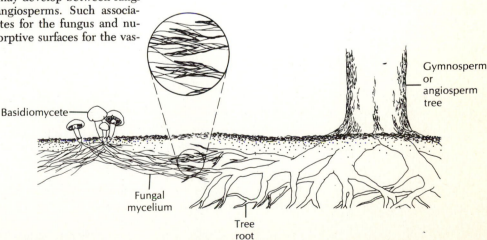

Basidiomycete

Fungal mycelium

Tree root

Gymnosperm or angiosperm tree

EVOLUTION

Fossils of aquatic fungi are among the oldest eukaryotic fossils, dating back to the Precambrian era over 1 billion years ago (Banks, 1970). By the Carboniferous period all major fungal groups were in existence. The geological record of fungi is very sparse, so it is extremely difficult to draw any definite lines of fungal evolution.

The presumably advanced fungi have septate mycelia, no motile zoospores, and complex fruiting bodies (e.g., mushrooms). Biochemical evidence indicates that one fungal class, the Oomycetes (see Chapter 15), evidently diverged very early from the main line of fungal evolution. Relationships among the "lower fungi" are not clear; in fact, the groups may be evolutionarily unrelated. Various relationships between the algae and the different fungal groups have been proposed, but these relationships must remain speculative until additional fossil evidence is found.

It is possible that fungi arose at the same time or soon after organic matter was starting to accumulate in the biosphere. As dead organic matter increased, it tied up essential nutrients. Organisms evolved that used this energy source and filled the decomposer niche by biodegrading the organic material. Then the nutrients contained in the organic matter were made available to the producers (green plants).

SELECTED REFERENCES

Aharanowitz, Y., and G. Cohen. 1981. The microbiological production of pharmaceuticals. Sci. Am. **245**(3):140-152.

Ainsworth, G.C., and P.K.C. Austwick. 1973. Fungal diseases of animals. Ed. 2. Commonwealth Agricultural Bureaux, Slough, U.K.

Christensen, C.M. 1975. Molds, mushrooms and mycotoxins. University of Minnesota Press, Minneapolis. 254 pp.

Cooke, W.B. 1979. The ecology of fungi. CRC Press, Inc. Boca Raton, Florida. 274 pp.

Deacon, J.W. 1980. Introduction to modern mycology. Vol. 7. Basic microbiology. Halsted Press. John Wiley & Sons, New York. 197 pp.

Demain, A.L., and N.A. Solomon. 1981. Industrial microbiology. Sci. Am. **245**(3):66-74.

Gray, W.D. 1959. The relation of fungi to human affairs. Henry Holt & Co., Inc. New York. 492 pp.

Hacskaylo, E. 1972. Mycorrhiza: the ultimate in reciprocal parasitism? BioScience **22**(10):577-583.

Harley, J.L. 1969. Biology of mycorrhiza. Ed. 2. Leonard Hill, London. 334 pp.

Malloch, D. 1981. Moulds: their isolation, cultivation and identification. University of Toronto Press, Canada. 97 pp.

Onions, A.H.S., H.O.W. Eggins, and D. Allsopp. 1982. Smith's introduction to industrial mycology. Ed. 7. John Wiley & Sons, Inc., New York. 416 pp.

Parkinson, D. 1981. Ecology of soil fungi. In Cole, G.T., and B. Kendrick (eds.): Biology of conidial fungi, pp. 277-294, Vol. I. Academic Press, Inc., New York.

Rose, A.H. 1981. The microbiological production of food and drink. Sci. Am. **245**(3):126-138.

14

CLASSIFICATION

Division: Myxomycota: "slime molds"; phagotrophic nutrition; no cell walls, except in spores

Class 1: Acrasiomycetes: cellular slime molds; myxamoebae; pseudoplasmodia; haplontic life cycles; distinctive sorocarps; cellulose in spore walls of some

ORDER: Protosteliales: minute, often multinucleate, cellular "protostelids"; no sex; common on dead plant material; not well known or studied; *Protostelium*

ORDER: Dictyosteliales: the "dictyostelids"; no flagellate cells; uninucleate myxamoebae with one or more nucleoli per nucleus; aggregating myxamoebae forming pseudoplasmodia; *Dictyostelium*

Class 2: Myxomycetes: the acellular slime molds; true multinucleate brightly colored plasmodia; no aggregation phase to life cycle; diplohaplontic life cycles; distinctive sporophores; cytoplasmic streaming; flagellate "swarmers"

ORDER 1. Ceratiomyxales: widely distributed soil organisms; spores borne externally on stalks; no peridium; one genus, *Ceratiomyxa*

ORDER 2. Liceales: distinctive sporophores lacking lime deposits; no capillitium; brown to tan spores; *Dictydium, Lycogala*

ORDER 3. Trichiales: most common slime molds; distinctive sporophores, often brightly colored; *Arcyria, Hemitrichia*

ORDER 4. Physarales: lime-encrusted sporophores, purple-brown spores; *Didymium, Fuligo, Physarum, Stemonitis*

Class 3: Hydromyxomycetes: mostly marine slime molds of uncertain affinity; produce a "net plasmodium"

ORDER: Labyrinthulales: acellular tubes within which spindle shaped cells migrate; heterokont flagellate zoospores; *Labyrinthula* (parasitic or saprophytic on marine and estuarine grasses such as *Zostera* and *Spartina*), *Labyrinthoriza* (a freshwater form)

Division Myxomycota is composed of an assemblage of organisms whose taxonomic relationship with fungi, algae, or protozoans is not very clear. They are commonly known as the "slime molds" because of their viscous nature. They resemble the fungi because they lack photosynthetic pigments and contain similar food reserves; they resemble the protozoa in being phagotrophic and having naked protoplasts during their vegetative existence; and they resemble algae in having cellulosic cell walls in some of their reproductive phases.

Some of the Myxomycota are cellular and are usually referred to as the cellular slime molds (Acrasiomycetes). Others are acellular organisms (Myxomycetes) forming multinucleate **plasmodia.** The division poses an interesting problem in taxonomy and classification because of the diversity of organisms that are usually considered within its realm—mycologists claim them as a branch of fungi, protozoologists usually claim them as a separate branch of unicellular or multinucleate protozoa, and still other scientists place the group as a separate branch of achlorophyllous algae. Most introductory textbooks include the slime molds in discussions of the fungi; this text will follow that tradition.

Perhaps the most characteristic feature of the slime molds is physiological rather than morphological, since they all exhibit phagotrophic nutrition accompanied by amoeboid movement; that is, they all actively pursue and engulf their food. In this respect they are like many protozoans and other protistan organisms. They also possess additional protistan characteristics such as the lack of cell walls during the vegetative aspects of their life cycles. A major exception to this is the cellulosic walls produced during spore formation.

Division Myxomycota is obviously a heterogeneous grouping of organisms that have questionable phylogenetic affinities with the rest of the biological world. In this text we have separated the division into three separate classes based on the existence of individual cells, multinucleate plasmodia, and the presence of a "net plasmodium." The phylogenetic relationships among the three classes are difficult to establish—indeed, they may not be related at all.

CELL STRUCTURE

The organisms of division Myxomycota exhibit various types of cellular organization. The cellular slime molds (Acrasiomycetes) are composed of individual, naked, uninucleate, nonflagellate cells called **myxamoebae** (Fig. 14-1). These cells resemble individual, amoeboid, protozoan cells; they are called myxamoebae to differentiate them from the protozoa, because there are apparently no significant structural differences between the cells of the two groups. Each myxamoeba contains all of the usual ultrastructural characteristics that are encountered within most cells, including a plasma membrane, mitochondria, rough and smooth endoplasmic reticulum, ribosomes, Golgi bodies, various types of vacuoles, and a membrane-bound nucleus containing chromosomes and one or more nucleoli.

FIG. 14-1 Life history of a cellular slime mold.

Life cycle for *Dictyostelium discoideum*, one of the common cellular slime molds. Individual myxamoebae feed on bacteria and other organic matter. When food is exhausted, the cells aggregate, forming a pseudoplasmodium and eventually a multicellular slug. Slugs migrate as a coordinated unit for short distances, then produce sorocarps. Sexuality is expressed by formation of macrocysts, forming the only diploid phase in the life cycle. Myxamoebae may also form haploid microcysts.

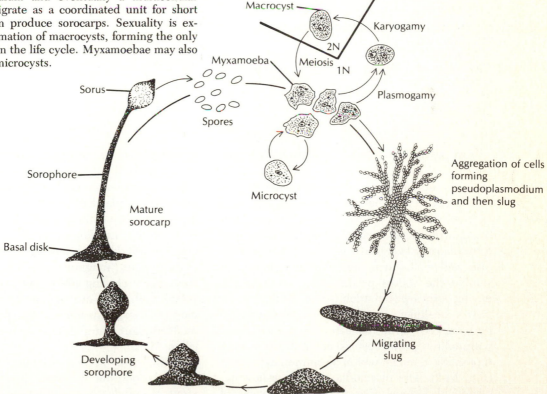

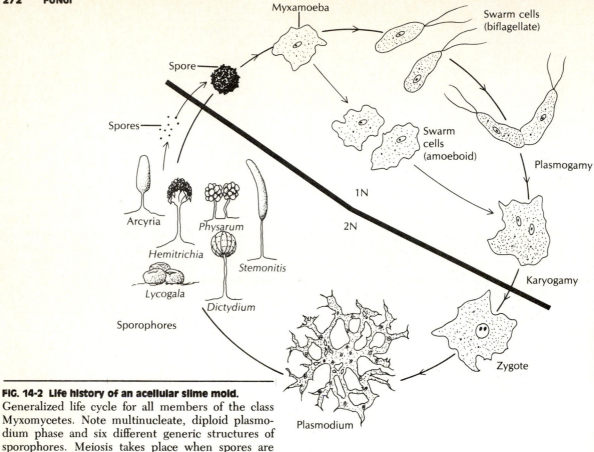

FIG. 14-2 Life history of an acellular slime mold.
Generalized life cycle for all members of the class
Myxomycetes. Note multinucleate, diploid plasmodium phase and six different generic structures of
sporophores. Meiosis takes place when spores are
formed in each sporangium.

Vegetative myxamoebae resemble flattened,
stellate, tissue-culture cells. They move by
cyclosis, producing amoeboid **pseudopodia,** and
feed on bacteria, engulfing and digesting them
in food vacuoles. Under unfavorable environmental conditions, individual myxamoebae may
form **microcysts** (Fig. 14-1) by secreting a cellulosic cell wall. Such individual microcysts resemble the spores produced by myxamoebae
during sorocarp formation (see p. 273).

The acellular slime molds (Myxomycetes)
form naked, multinucleate plasmodia (sing.
plasmodium), which consist of hundreds or even
thousands of nuclei within a common cytoplasm
(Fig. 14-2). The plasmodia move in a flowing,
amoeboid fashion over the moist surfaces of

dead leaves, logs, or organic soils. As the plasmodia glide along, they engulf bacteria, protozoa, and smaller food particles, all of which they
digest. Most plasmodia possess prominent veins
throughout the common cytoplasm, as is easily
seen in the genus *Physarum* (Fig. 14-5, *B*). The
color of individual plasmodia is often indicative
of the species, ranging through all colors of the
spectrum including black and white, although
yellow and white are the most common. Plasmodia of various sizes are easily spotted growing
on dead plant material in moist areas—often on
fallen trees.

Within the membrane-bound cytoplasm of
each plasmodium is found the usual complement of cellular organelles such as mitochon-

dria, Golgi bodies, rough and smooth endoplasmic reticulum, vacuoles, ribosomes, and
microtubules. The membrane-bound nuclei are
found floating free within the cytoplasm. Each
nucleus contains one or more nucleoli and a set
of chromosomes.

REPRODUCTION AND GROWTH

The life history of the cellular slime molds
follows the pattern presented in Fig. 14-1 for
the well-known genus *Dictyostelium.* When
spores germinate on a suitable medium, they
give rise to individual, uninucleate myxamoebae by the process of mitosis. Myxamoebae
move about and feed on bacteria or small organic particles until the organic food sources are
exhausted. When this happens, one or more of
the myxamoebae will begin to produce and secrete a chemical signal that diffuses away from
the myxamoeba, thus setting up a chemical gradient. Other myxamoebae in the vicinity are
stimulated by the chemical signal, and they in
turn produce more of the substance while they
begin to move toward the center of the gradient. As the myxamoebae start to aggregate, they
secrete another chemical that causes their surface properties to change. They become sticky
and adhere to each other. Research suggests
that the chemical that causes adhesion belongs
to a group of proteins called lectins, which bind
carbohydrates (Rosen et al, 1975).

The migration of myxamoebae toward the
center of the chemical gradient is called the **aggregation phase** of the life cycle. The chemical
signal for aggregation has been identified as
cyclic adenosine monophosphate (cAMP) in
Dictyostelium discoideum, and is called **acrasin**
(Bonner, 1969). There are different types of
acrasin for each species of cellular slime mold.

After the myxamoebae have aggregated in
significant numbers, they form a mass of individual cells called a **slug.** Slugs are groups of
individual myxamoebae that migrate for short
distances at a rate up to 2 mm/hr in a well-
coordinated manner toward light (especially

green). Among the slime molds, the slug is a
unique structure, and the cellular slime molds
may be distinguished from other members of
Myxomycota on the basis of its presence. The
slug is sometimes called a **pseudoplasmodium**
(false plasmodium), since it is not multinucleate
as are true plasmodia. Slugs are normally
formed from the aggregation of 100,000 to
200,000 individual myxamoebae. The numbers
of cells involved may be manipulated in culture.

After the migration phase, which may last for
several days, the multicellular slug differentiates into a fruiting body called a **sorocarp.**
Sorocarps are formed in a very specific manner.
Cells forming the anterior one third of the slug
rise from the substrate surface and produce a
stalk containing rigid walls of cellulose fibers,
glucans, and protein. Cells from the posterior
two thirds of the slug ascend through the center
of the stalk and are transformed into spores with
cellulosic cell walls. Thus a mature sorocarp
with spores is produced. Each uninucleate
spore is produced by a single myxamoeba;
therefore, the individuality of each myxamoeba
is retained.

The distinctive structure of sorocarps is used
as a basis for species identification. Sorocarps
may grow to heights of 1 cm, producing wind-
dispersed spores. They do not require a period
of dormancy and therefore will germinate immediately in suitable environments, producing
new myxamoebae.

Sexual reproduction among the cellular slime
molds was undocumented before the early
1970s, when a giant cell called a **macrocyst** (Fig.
14-1) was discovered among aggregating myxamoebae. The giant cell was shown to be the
site of nuclear fusion and meiosis and thus represents the site of sexual activity. Sexuality has
been reported in a number of species (Macinnes
and Francis, 1974; Erdos et al, 1975). When
macrocysts arise, they grow by engulfing nearby
myxamoebae. Nuclear fusion (karyogamy) takes
place within the macrocyst. The enlarged diploid nucleus undergoes meiosis, thus producing
multiple haploid nuclei. When the macrocyst

germinates, the same number of individual uninucleate haploid myxamoebae that entered the macrocyst are liberated (Fig. 14-1). This type of life cycle is very close to the haplontic life cycle outlined in Fig. 1-11 in which the only diploid phase is the zygote. Before the discovery of the macrocyst, it had been established that parasexual genetic recombination occurred during the aggregation stage. The cellular slime molds do not always form macrocysts, and in such instances genetic recombination occurs by parasexual methods. Environmental conditions appear to determine whether a dictyostelid produces a macrocyst or whether it proceeds with the sequence of sorocarp formation.

The acellular or true slime molds follow a somewhat different life cycle than the cellular slime molds. Separate diploid and haploid phases are recognized so that a diplohaplontic type of cycle is produced, and the diploid phase forms a multinucleate plasmodium (Figs. 14-2 and 14-5). It is from the amorphous plasmodium that the slime molds take their name. There is no cellulose associated with plasmodia. The plasmodium is a constantly changing, flowing, creeping mass of protoplasm that is bound by a plasma membrane covered with a gelatinous slime sheath. As the plasmodium creeps over a substrate, portions of the sheath are shed, leaving a visible trail of slime. Individual plasmodia may attain considerable size (up to several square meters), often covering dead logs in their entirety. As the mass creeps over dead organic matter, it engulfs particles of food consisting of bacteria, plant spores, protozoans, yeasts, and bits of organic matter. These particles enter into food vacuoles within the cytoplasm of the plasmodium, where digestion is accomplished by enzymes. The nuclei contained within the common cytoplasm of each plasmodium are diploid and divide mitotically. Mitotic divisions are synchronous (i.e., all nuclei divide at the same time).

Plasmodia are unique, often richly pigmented structures. Most plasmodia are yellow or white, but colors from yellow and red to violet and black have been reported for the various genera. The chemical nature and specific function of the various pigments is not definitely known. Plasmodia are vegetative structures that spend their entire existence feeding and gathering energy. They are capable of initiating two types of reproductive structures, the sclerotium and sporophores (Fig. 14-2).

Sclerotia (sing. sclerotium) are hard, irregular, thick-walled masses that are formed in response to unfavorable environmental conditions in some slime molds. Sclerotia may remain dormant for extended periods—up to 2 years—then germinate when favorable conditions return. Each sclerotium is composed of thick-membraned "spherules" that contain multiple nuclei (from 1 to 14). The spherules are embedded in a hardened mass of secreted polysaccharide material (Zaar and Kleinig, 1975). Not all slime molds produce sclerotia.

The plasmodia may migrate for a considerable distance from the main feeding area before they form **sporophores.** They may climb trees or cars or may cross paved roads. Sporophores are distinctive structures that bear sporangia and spores. The spores are produced by meiosis, and the resulting cells are haploid. The spore walls contain cellulose, chitin, and galactosamines. Spores are wind dispersed and are very resistant to adverse conditions, especially desiccation—some have germinated after 70 years in dried collections.

When individual spores germinate, they release one or a few myxamoebae or biflagellate swarm cells. The nature of the emergent cells depends on the availability of water. With adequate water the cells are anteriorly biflagellate; without plenty of water the cells remain amoeboid.

Sporophore structure varies to some extent from species to species, but in general each sporophore consists of a basal, membranous section (the hypothallus), a stalk, and a sporangium (Fig. 14-3). The sporangium is the site of meiotic nuclear divisions and haploid spore formation. At the same time meiosis is occurring, the

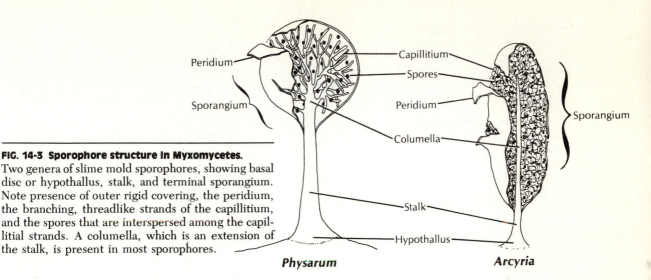

FIG. 14-3 **Sporophore structure in Myxomycetes.**
Two genera of slime mold sporophores, showing basal
disc or hypothallus, stalk, and terminal sporangium.
Note presence of outer rigid covering, the peridium,
the branching, threadlike strands of the capillitium,
and the spores that are interspersed among the capillitial strands. A columella, which is an extension of
the stalk, is present in most sporophores.

capillitium is formed. The capillitium is a network of secreted threadlike material with a distinctive structure. The capillitial threads may
aid in spore dispersal, or they may function in
spore nutrition, since some threads have been
shown to be hollow conduits and thus may carry
food materials. The capillitium is sometimes
covered with a membrane of variable thickness
called the **peridium,** an ephemeral structure
that may be lost during development of individual sporophores. Some species lack a peridium,
whereas some have lime encrustations associated with it.

Spores may also be produced in other structures. Some genera do not form upright sporangia but produce their spores in stalkless sporangiophores called **plasmodiocarps,** which are
formed by the veins of plasmodia. These contain
spores and capillitia and are covered with a
peridium. Plasmodiocarps are found in some
species of *Hemitrichia*. A third spore-producing
structure is called an **aethalium** (pl. aethalia).
Aethalia (Fig. 14-4, *B*) are macroscopic, round,
or flattened oval structures ranging from one to
several centimeters in diameter. They resemble
small yellow, brown, or reddish puffballs (basidiomycetes). They are stalkless and covered

with a rigid peridium. The genus *Fuligo* (Fig.
14-4) may form aethalia that cover many square
centimeters of decaying wood, sawdust, or rich
organic soil. Such aethalia have been commonly
called "flowers of tan."

When the haploid spores germinate, individual myxamoebae or swarm cells emerge and
may divide mitotically and forage on their own
on bacteria or other organic material. Under
certain adverse conditions they may secrete a
thin cellulose wall, forming microspores that
remain viable up to 2 years. They may also fuse
in a sexual process (**plasmogamy**), forming binucleate cells. These cells then undergo karyogamy (nuclear fusion), thus producing a diploid
cell or zygote. The zygotic nucleus then divides
mitotically, producing a multinucleate plasmodium, and the life cycle continues.

PHYSIOLOGY

Both the cellular slime mold *Dictyostelium
discoideum* and the plasmodial slime mold *Physarum polycephalum* have been used extensively as experimental organisms. Consequently, a
large body of knowledge exists concerning their
molecular biology, biochemistry, physiology,

FIG. 14-4 Stages in life history of *Fuligo*.
A, Three pieces of wood on which bright yellow (white in this photograph) plasmodia of *Fuligo* sp. are growing. **B,** Round to oval spore-forming structures called aethalia produced by *Fuligo* sp. growing on dead wood. The aethalia (white in this photograph) are yellow or tan in the field. (Courtesy George Knaphus, Department of Botany, Iowa State University, Ames, Iowa.)

A

growth, and development. The cellular slime molds produce chemical signals that stimulate individual myxamoebae to aggregate in the slug phase, to migrate, and to form a sorocarp (Loomis, 1975; Sussman and Brackenburg, 1976). *P. polycephalum* produces a substance that induces the formation of a plasmodium (Youngman et al, 1977).

Individual Acrasiomycete myxamoebae feed until either the exogenous supply of organic material is exhausted or until the population of cells reaches a certain minimum number. If the food supply is adequate, the myxamoebae may continue feeding. Feeding ceases and aggregation is initiated when a stressed myxamoeba has exhausted the food supply in its immediate environment. This myxamoeba, called the "pacemaker," begins to secrete acrasin in 5-minute pulses. Acrasin acts as a **chemotactic agent** attracting nearby myxamoebae that are also nutri-

tionally stressed. These myxamoebae orient toward the aggregation center that was established by the pacemaker cell. The aggregating myxamoebae become sticky, adhere to each other, and also begin to secrete acrasin. In *D. discoideum*, the pacemaker myxamoeba emits pulses of cAMP at a rate somewhat faster than responding aggregating myxamoebae. The slower respondents amplify the cAMP signal and spread it into the immediate environment, thus establishing a concentration gradient of cAMP. Myxamoebae come together in established streams, flowing toward the aggregation center established by the pacemaker (Fig. 14-1). The concentration gradient of cAMP is maintained by additional secretion by the cells. The concentration of cAMP does not rise above saturation because an enzyme, phosphodiesterase, is secreted periodically by the myxamoebae, and it inactivates cAMP.

Aethalia

B

The aggregation process and the chemical interaction of the myxamoebae have been studied by Darmon and Brachet (1978) and Mato et al (1978). The myxamoebae of *D. discoideum* evidently are constantly responding to varying levels of cAMP. The bacteria upon which the myxamoebae feed secrete low levels of cAMP, which attracts the myxamoebae (Konijn et al, 1969). As the bacterial food supply becomes exhausted, the myxamoebae secrete phosphodiesterase, which breaks down the bacterial cAMP in their immediate environment (Gerish et al, 1974). When the myxamoebae cease all feeding, they secrete a chemical that inhibits extracellular phosphodiesterase; therefore, cAMP from the myxamoeba is no longer broken down and its concentration in the environment starts to rise (Riedel et al, 1973). At this stage the rate of cAMP synthesis by the myxamoebae appears to remain constant, but the amount in the immediate environs varies with the level of phosphodiesterase. The increase in cAMP thus results from a reduction in the rate of cAMP inactivation (Loomis, 1975). It is during the sticky preaggregation stage that phosphodiesterase is inhibited, cAMP starts to build up, and aggregation follows. Olive (1975) chronicles the events in the aggregation of *D. discoideum*.

To respond to cAMP in the environment, myxamoebae must be mature enough to be "competent." Competence involves the organism's reaching a certain size and accumulating enough stored energy so that a fully developed reproductive structure can be differentiated (Ross, 1979). Myxamoebae are only competent at certain stages of their life cycle; the sensitivity to cAMP is not always expressed. The expression of cAMP sensitivity is evidently genetically controlled. When *Dictyostelium discoideum* pacemaker cells begin emitting cAMP pulses, the other competent myxamoebae respond in a like manner, aggregation proceeds, and a slug is formed. Depending upon the species and the environment, slugs may immediately form a sorocarp or may migrate for several hours or days before doing so. As migration continues, the slug deposits a slimy sheath in its wake. Migration rates have been measured at 0.3 to 2.0 mm/hr at 20° C for *D. discoideum* (Bonner et al, 1950). The slug is a closed system of thousands of individual myxamoebae; it does not feed and contains all of the energy and nutrients necessary for sorocarp formation. Several researchers (Wright, 1973; Wright et al, 1977) have attempted to model the process of the conversion of precursor molecules in the slug into the needed products of the sorocarp. Such a model provides information on both the genetic control and the timing of the synthesis of necessary enzymes.

The anterior portion of the slug is both **phototrophic** and **thermotrophic** and will move toward light and heat. A photoreceptive pigment has been identified that may play a role in the phototropism of the slug of *Dictyostelium* (Poff et al, 1974). The slug still consists of individual myxamoebae, but the myxamoebae now coordinate their movements during slug migration and sorocarp formation. If the slug is broken apart with a needle, for example, the individual myxamoebae will commence feeding on bacteria again. The cells of the slug differentiate at an early stage in its development: the cells in the front third of the slug are destined to become stalk cells in the sorocarp, whereas the remaining cells will become sporogenous cells. The anterior cells, which are destined to become stalk cells, have higher cAMP concentrations than the cells toward the posterior. This gradient in cAMP develops as the slug migrates. Bonner (1971, 1983) and Raper (1973) discuss morphogenesis in *D. discoideum* in further detail.

The sequence outlined for *Dictyostelium discoideum* does not hold for all dictyostelids. For many cellular slime molds the detailed biochemistry and morphogenesis have not been worked out. The exact sequence of events depends not only on the species but also on environmental conditions. Acrasin is cAMP in *D. discoideum*, but in some other species acrasin has been shown to be folic acid, whereas in still other species the aggregating compound has not yet been identified.

Plasmodia of acellular slime molds have provided scientists with an excellent system for studying both the synchronization of mitosis and protoplasmic streaming. Plasmodia of *Physarum polycephalum* are easily grown in axenic culture in the laboratory (Fig. 14-5). The plasmodia may be induced to produce sporophores, but no species has completed its life cycle within axenic culture. Myxamoebae have also been grown in axenic culture, and their nutritional requirements have been found to vary considerably from those of the plasmodium. Dove and Rusch (1980) discuss the genetics, growth, and differentiation in *P. polycephalum*.

Plasmodia of most myxomycetes are fan-shaped and contain veins or conduits of protoplasmic gel through which the more-fluid protoplasm streams (Fig. 14-5). Protoplasmic streaming in these conduits is easily observed; it can be both rhythmic and reversible in certain genera (*Physarum*) and irreversible in others (*Ceratiomyxa*). In those genera having reversible flow, the protoplasm streams in one direction for a period of time (1 to 30 minutes), then reverses and flows in the opposite direction. The flow may be observed with a hand lens. Streaming rates up to 1 mm/sec have been recorded; these rates exceed those observed in other fungi, in algae, or in vascular plant cells.

The plasmodium starts to form a fruiting body (sporophore) when it is mature (i.e., competent) and when environmental conditions are conducive to its formation. Several environmental factors appear to control and stimulate sporulation; these include light (310 to 500 nm), temperature, pH, reduction in moisture, and depletion of food sources. Several vitamins and amino acids also appear necessary for sporulation. Apparently all the plasmodia have an obligate re-

FIG. 14-5 Plasmodium of *Physarum polycephalum.*
This plasmodium grew over the entire inside of a
9 cm Petri dish. It was growing on agar and was fed
oat flakes. **A,** Growing plasmodium (2.5×). **B,** Veins
and first stages of sporophore formation are readily
seen (10×). (Courtesy D. Spess, Department of Biology, Lehigh University, Bethlehem, Pa.)

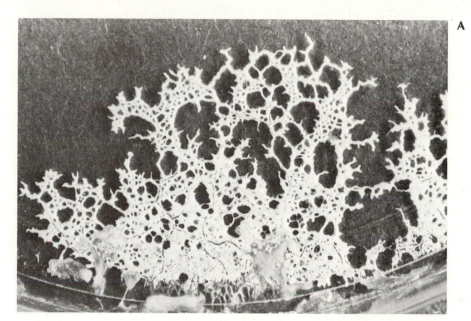

A

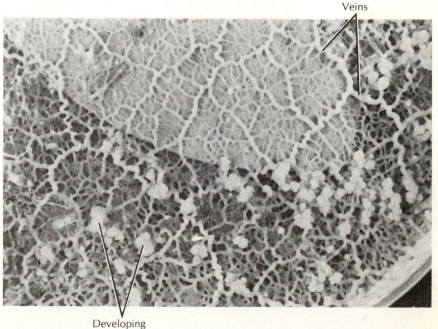

Veins

B

Developing
sporophores

quirement for light in order to sporulate, but the highly colored plasmodia evidently need longer periods of light than the colorless or slightly pigmented ones (Sauer, 1973). Once the plasmodium ceases feeding, it may migrate a considerable distance toward light and to an exposed place before it sporulates. Once sporulation has started, the process is irreversible, and no matter how adverse the environmental conditions, the organism will attempt to produce spores.

As the sporophore develops, the threads of the nonliving capillitium develop into a network of branching tubes (Fig. 14-3). Both the capillitium and peridium may contain calcium carbonate, and the capillitium may serve as a system for transporting the calcium carbonate to the outer peridium (Webster, 1980). The capillitium is often **hygroscopic** (i.e., water absorbing), expanding and rupturing the peridium and releasing the wind-dispersed spores.

In the class Hydromyxomycetes, a **net plasmodium** is formed; it is not a true plasmodium, since it is composed of a network of branching tubes through which spindle-shaped uninucleate cells glide. The network is called an **ectoplasmic net** because it is secreted outside the individual cells. The cells within the ectoplasmic net do not usually exhibit amoeboid motion. The net extends ahead of the spindle cells and over potential food sources such as bacteria, yeasts, or other small cells. It is composed of polysaccharide materials and has the ability to break down (lyse) both plant and animal cells, to absorb the nutrients from these cells, and to transport them to the enclosed spindle cells. The ectoplasmic nets of some members of the Labyrinthulales can even degrade such highly resistant molecules as sporopollenin, a component of pollen grain walls.

The ectoplasmic net is formed from unusual organelles, the **sagenogens,** which are depressions located at the cell's periphery and which contain endoplasmic reticulum. Research by Porter (1972) and Perkins (1976) has increased the knowledge of both the ultrastructure of the ectoplasmic net and its relationship to the sagenogens. Olive (1975) outlines a possible life cycle for *Labyrinthula*. Much additional research is needed on these little-known organisms.

EVOLUTION

Myxomycota provide few clues to their evolutionary history—they have left no fossil record. Members of the group have animal, algal, and fungal characteristics, and most of what is known about their evolution is gathered through inference. On the one hand, the plasmodium moves about and ingests nutrients much like a protozoan (e.g., *Amoeba*), while on the other hand, the same organism produces an aerial, cellulose-containing reproductive structure resembling those of some fungi. The slime molds have historically been studied in conjunction with the fungi, largely because of their heterotrophic nutrition and their ability to grow on dead wood.

One investigator (Olive, 1975) has classified them within kingdom Protozoa as a separate class of "fungus animals," the Mycetozoa. He has proposed an evolutionary sequence for Myxomycota in which the Acrasiomycetes and the Myxomycetes are derived from a common flagellate protistan ancestor. Olive and Stoianovitch (1960), who discovered the protostelids, view them in evolutionary sequence as being the most primitive of the Myxomycota and ancestral to both the cellular and acellular slime molds.

The third class, the Hydromyxomycetes, have an uncertain ancestry. Some investigators do not even consider them to be slime molds. Olive (1975) proposes that they are derived from a flagellate protist like the other classes, although not from the same flagellate protist as the other classes.

Until further research is done, the phylogenetic position of the Myxomycota will remain a matter of both speculation and disagreement. At this time there is not enough information from which to draw definite evolutionary patterns.

ECOLOGY

Although Myxomycota are widely distributed throughout the world, they are rarely noticed because they are quite inconspicuous. Their role in the ecosystem is not well defined, but their common occurrence in decaying vegetation and dung implies a significant role in nutrient recycling. Little data are available, however, to support this thesis.

The cellular slime molds (Acrasiomycetes) are very common in woodland and cultivated soils throughout the world. They comprise one of the main components of the soil flora, but their role in soil ecology is not yet defined. The feeding stage, the single-celled myxamoebae, consume bacteria and decomposing organic matter in the soil. The sorocarps are ephemeral and seldom noticed, so the presence of these organisms is rarely obvious. They have been found in soils wherever investigators have looked, but only further research will provide clues to their role in the soil environment. A new species of *Dictyostelium*, *caveatum*, has been isolated from bat droppings in a cave and has been shown to phagocytize amoebae (Waddell, 1982). Bonner (1959, 1967) discusses the Dictyosteliales in detail.

The protostelids (Protosteliales) are also widely distributed on decaying plant material and dung. However, both their small size and their recent discovery (1960) indicate that they have often been overlooked in studies of ecosystems. Additional research should help to define their role in the ecosystem.

The ecology of the acellular slime molds (Myxomycetes) is better known than that of the previous class. The distribution and ecology of this class is reviewed by Martin and Alexopoulos (1969). The plasmodia of this class may become fairly large and therefore are more conspicuous (Figs. 14-4 and 14-5). *Ceratiomyxa* occurs all over the world on decaying wood in forests, but its presence is not evident until it produces spores. *Lycogala*, with its rounded aethalia that resemble the puffballs of basidiomycetes, is often found on the decaying wood of conifers. *Arcyria* and *Hemitrichia* occur frequently in temperate regions in forests on moist, decaying logs from spring until fall.

The Physarales, which are also widely distributed, produce plasmodia that can grow to large proportions (1 m²), creeping over logs or plants and feeding upon the decomposing bodies of fleshy fungi such as the basidiomycetes. In the colder climates many members of the Physarales produce sclerotia that can overwinter and give rise to a viable plasmodium the following spring. The famous "yellow blobs" that occurred in Texas were the large yellow plasmodia of *Fuligo septica*. *Stemonitis* is also widely distributed throughout the woodlands of the world. A survey of any damp temperate or tropical woodland would reveal many species of the Physarales. Their distribution is limited not only by moisture and temperature but also by light which is necessary for sporulation.

Much remains to be learned about the ecology of the Myxomycota. Very little information is available about their ecological role or their sensitivity to pollutants and other ecosystem stresses. These subjects must be investigated before an adequate understanding of the role of these interesting organisms can be reached.

ECONOMIC IMPORTANCE

The major economic contribution of the Myxomycota is their usefulness as experimental organisms in the laboratory. The two most widely used species are *Dictyostelium discoideum* and *Physarum polycephalum*. Both have been grown frequently in culture. *P. polycephalum* has been grown in axenic culture and in chemically defined media, but *D. discoideum* has not been grown in either. However, *D. discoideum* is easily grown in the laboratory with bacteria as a food source, and complex culture media such as peptone-dextrose or hay-infusion agar are used

(Olive, 1975). The extensive use of these two organisms and some of their close relatives qualifies them as domesticated organisms along with the bacterium *Escherichia coli* and the laboratory rat *Rattus norvegicus*.

D. discoideum has provided important information about its physiology, molecular biology, biochemistry and morphogenesis. Developmental biolgists have also extensively studied the development of the sorocarp in *D. discoideum*. *P. polycephalum* is an excellent experimental system; research with this organism has yielded information concerning cytoplasmic streaming, control of mitosis, and morphogenesis, to name but a few areas of increasing knowledge. The McArdle Laboratory for Cancer Research at the University of Wisconsin has pioneered in the use of *P. polycephalum* for studies of cellular biology, including RNA and DNA synthesis.

The Myxomycete *Fuligo septica* caused considerable concern in 1973 when, for some unknown reason, it invaded many yards in suburban areas of Dallas, Texas. Many people were frightened by the large (70 × 54 × 3 cm) foamy yellow plasmodia—labeled the "yellow blob"—that crept over lawns and produced extremely large aethalia (up to 20 cm). Concern over the "blob" diminished when it was positively identified and pronounced harmless by biologists. Sunhede (1974) provides additional information on *Fuligo*.

Another Myxomycete, *Physarum cinereum*, may become conspicuous on lawns in cities or suburbs. The bluish plasmodium may cover many square centimeters. Bonifacio (1960) discusses this organism in more detail. Other Myxomycetes may occasionally attack cultivated mushrooms.

The Hydromyxomycete *Labyrinthula macrocystis* has been implicated as a possible cause of the **wasting disease** of eel grass (*Zostera marina*—Angiospermae). During the 1930s, in coastal areas on both sides of the Atlantic, eel grass was killed by the wasting disease. The loss of eel grass in estuaries and coastal waters reduced the amount of vegetation that protected not only the eggs and young of marine invertebrates and vertebrates but also the shoreline from eroding waves. In addition, the eel grass provided food for such herbivorous wildlife as ducks and other marine birds. *L. macrocystis* was found on the infected eel grass, but its role as the causative agent was not definitely established (Porkorny, 1967). The organism apparently can be parasitic or saprophytic, depending on the circumstances. The net plasmodium of the Labyrinthulales can break down plant cells, so *Labyrinthula* has the potential for being a destructive plant parasite. *Labyrinthula* has also been isolated from other marine angiosperms— for example, the estuarine cord grass, *Spartina alterniflora*.

SUMMARY

Myxomycota (slime molds) constitute a division of heterotrophic organisms whose phylogenetic relationships have not been established. They possess characteristics that appear to relate them to algae, fungi, and protozoans. The division is separated into three classes based upon the nature of their cellular structure: the Acrasiomycetes or cellular slime molds, the Myxomycetes or acellular slime molds, and the Hydromyxomycetes. All three classes are nonphotosynthetic, heterotrophic organisms that obtain nutrition through phagotrophic activity accompanied by amoeboid movement.

The cellular slime molds are composed of individual, uninucleate myxamoebae that live an independent existence. When their nutrient supplies are exhausted, myxamoebae migrate and aggregate, forming a multicellular slug (pseudoplasmodium). Slugs in turn develop into distinctive fruiting bodies (sorocarps) that produce spores by mitosis. The life cycles are haplontic.

The acellular slime molds are composed of multinucleate plasmodia that flow over, engulf, and digest food particles. When the food supply is depleted, plasmodia produce distinctive fruiting bodies in the form of upright sporangia, stalkless plasmodiocarps, or macroscopic ovoid aethalia. Meiospores are produced in these fruiting bodies. Capillitial threads form a skeletal network in many sporangia. The life cycles are diplohaplontic. Included as members of the acellular slime molds are the relatively unknown, yet ubiquitous, "protostelids," the brightly colored, true plasmodial slime molds, and a small group of marine slime molds that produce a "net plasmodium."

The nature of the life cycle in cellular slime molds has made them an ideal organism for studying cellular aggregation, cellular differentiation, and cellular growth and development. The biochemistry and physiology of individual cells can be examined either before or after they aggregate into a slug or differentiate into a sorocarp. The system provides an excellent opportunity to study the synthesis and destruction of chemotactic agents like acrasin (cAMP).

The acellular slime molds provide an equally good system for the study of protoplasmic streaming and synchronous mitotic cell divisions, since all nuclei in any one plasmodium divide at the same time. Study of the life cycles of the acellular slime molds has indicated that spore formation in the form of fruiting bodies (sporangia, plasmodiocarps, and aethalia) are related to plasmodial maturity, available light, temperature, pH, moisture, and food depletion.

Myxomycota are widely distributed organisms in soils and areas of decaying vegetation. However, they are seldom observed because of their minute size and inconspicuous nature. They probably play an important role in nutrient turnover in natural environments.

SELECTED REFERENCES

Aldrich, H., and J.W. Daniel (eds.). 1982. Cell biology of *Physarum* and *Didymium*. Vol. 1, Organisms, nucleus and cell cycle, 488 pp. Vol. 2, Differentiation, metabolism and methodology, 400 pp. The Cell Biology Series, Academic Press, Inc., New York.

Bonner, J.T. 1967. The cellular slime molds. Ed. 2. Princeton University Press, Princeton, New Jersey. 205 pp.

Bonner, J.T. 1983. Chemical signs of social amoebae. Sci. Am. 248(4):114-120.

Dove, W.F., and H.P. Rusch. 1980. Growth and differentiation in *Physarum polycephalum*. Princeton University Press, Princeton, New Jersey. 250 pp.

Loomis, W.F. 1975. *Dictyostelium discoideum:* a developmental system. Academic Press, Inc., New York. 214 pp.

Martin, G.W., and C.J. Alexopoulos. 1969. The myxomycetes. University of Iowa Press, Iowa City. 561 pp.

Olive, L.S. 1975. The mycetozoans. Academic Press, Inc., New York. 293 pp.

Raper, K.B. 1973. Acrasiomycetes. In Ainsworth, G.C., F.K. Sparrow, and A.S. Sussman (eds.). The fungi: an advanced treatise. IVB. Academic Press, Inc., New York. pp. 9-36.

Sauer, H.W. 1973. Differentiation in *Physarum polycephalum*. In Ashworth, J.M., and J.E. Smith (eds.). Microbial differentiation, pp. 375-405. Cambridge University Press, London.

Waddell, D.R. 1982. A predatory slime mold. Nature 298(5873):464-466.

CLASSIFICATION

Division: Eumycota: "true fungi"; assimilative nutrition

  Subdivision 1: Mastigomycotina: "flagellate fungi" or "lower fungi"; centrioles; unicellular or multinucleate (coenocytic) filaments

    Class 1: Chytridiomycetes: single posterior whiplash flagellum; cell walls chitinous; mostly haplontic life cycles; sexual reproduction in some

      ORDER 1. Chytridiales: mostly freshwater algal parasites; isogamous; the "chytrids"; *Chytridium, Synchytrium, Polyphagus, Olpidium, Rhizophlyctis, Chytriomyces, Rhizophydium*

      ORDER 2. Blastocladiales: freshwater and soil molds; membrane-bound nuclear caps; thick-walled resting spores; coenocytic branching filaments; anisogamous; *Allomyces, Blastocladia, Blastocladiella, Coelomomyces*

    Class 2: Plasmodiophoromycetes: mainly higher-plant parasites; heterokont biflagellate zoospores with one anterior and one posterior flagellum; naked plasmodia similar to slime molds; haplontic life cycles

      ORDER 1. Plasmodiophorales: the "endoparasitic slime molds"; *Plasmodiophora, Spongospora*

    Class 3: Oomycetes: laterally inserted heterokont biflagellate zoospores; cell walls containing glucans and cellulose; oogamous; coenocytic branching filaments; diplontic life cycles

      ORDER 1. Saprolegniales: "water molds" or "fish molds"; common freshwater plant and animal parasites and saprophytes; *Achlya, Saprolegnia*

      ORDER 2. Leptomitales: small order of "water molds"; saprophytic, chitin present in cell walls of some species; *Aqualinderella, Leptomitus*

      ORDER 3. Peronosporales: highly specialized obligate and facultative higher-plant parasites; damping-off fungi, white rusts, and downy mildews; economically important; *Albugo, Peronospora, Phytophthora, Plasmopara, Pythium*

      (A few other minor orders)

    Class 4: Hyphochytridiomycetes: small group of aquatic fungi having anterior uniflagellate zoospores; cell walls composed of cellulose and chitin; *Rhizidiomyces*

MASTIGOMYCOTINA

The evolutionarily unrelated fungi of division Eumycota, sometimes referred to as the "true fungi," constitute a large group that possesses, as a common characteristic, an assimilative mode of nutrition. They do not, like the Myxomycota, possess cellular pseudoplasmodia or acellular plasmodia. Most of the familiar fungi belong to this division. They are ubiquitous in the environment and worldwide in distribution, having invaded almost all ecological niches in all climates. Their role in recycling dead organic matter is very important to the ecosystem; in nature they may be found growing saprophytically on plant or animal remains in aquatic or damp terrestrial habitats. They are also important as parasites on and in both plants and animals.

Classification of the Eumycota is complicated by the great diversity of organisms found within the division. The arrangement used in this text follows that established by Ainsworth (1973). Accordingly, there are five subdivisions, as follows:

Subdivision 1: Mastigomycotina*: motile cells
Subdivision 2: Zygomycotina*: no motile cells; zygospores formed
Subdivision 3: Ascomycotina: no motile cells; ascus formed
Subdivision 4: Basidiomycotina: no motile cells; basidia formed
Subdivision 5: Deuteromycotina: no motile cells; no sexual stages; the "imperfect fungi"

Morphological organization among the Eumycota reflects the vast variety of organisms within the division. Most of the "true fungi" are characterized by the presence of a branching, filamentous hypha; however, there are exceptions, since a few species are unicellular, especially among the flagellate Mastigomycotina and the common

*Mastigomycotina and Zygomycotina together were known as the Phycomycetes in older classifications.

yeasts (Ascomycotina). A few fungi even have the ability to exist either as filamentous or unicellular forms. These fungi are said to be dimorphic. Dimorphism is found among some of the pathogenic fungi, where the noninfectious, attenuated form exists as a filament, whereas the virulent, infectious organism exists as a unicellular, yeastlike fungus.

Subdivision Mastigomycotina

The fungi of subdivision Mastigomycotina are characterized by the presence of flagellate cells (zoospores) at some point in their life histories. The subdivision is ubiquitous in damp and aquatic habitats throughout the world. In nature they may be isolated from any habitat that has sufficient water to enable the fungi to survive. Scientists can easily collect, observe, and isolate them by using "baiting" techniques in which pieces of fruit, wood, cellophane, dead invertebrates, wool, dead fish, snake skins, hair, seeds, or other biodegradable matter is placed in shallow dishes of chlorine-free water or damp soil for a few days (Sparrow, 1960).

The subdivision is often referred to as the lower fungi because of the relatively simple structure and the apparent primitive qualities of its members. They are also known as the "flagel-late fungi," since all members of the subdivision produce flagellate cells.

The taxonomy of the subdivision is based on the types of flagella produced by the zoospores (Fig. 15-1). There are at least four distinct flagellar types that form the basis for the four separate classes: (1) those with a single anterior tinsel-type flagellum (Hyphochytridiomycetes); (2) those with a single posterior whiplash-type flagellum (Chytridiomycetes); (3) those with biflagellate, heterokont, apically or laterally inserted flagella, one of which is tinsel-type and the other whiplash-type (Oomycetes); and (4) those having two heterokont whiplash-type flagella, one at the anterior end of the cell and the other at the posterior end (Plasmodiophoromycetes). Electron microscopy of the zoospores is required to accurately make these observations.

The different flagellar types and insertions imply that Mastigomycotina is an artificial grouping of organisms that is not based on any genetic relationship. The only unifying characteristic of the division is the development of motile zoospores. Each of the four classes of this division probably has its own separate origin; it is therefore a **polyphyletic** group. Indeed, many mycologists place the Plasmodiophoromycetes

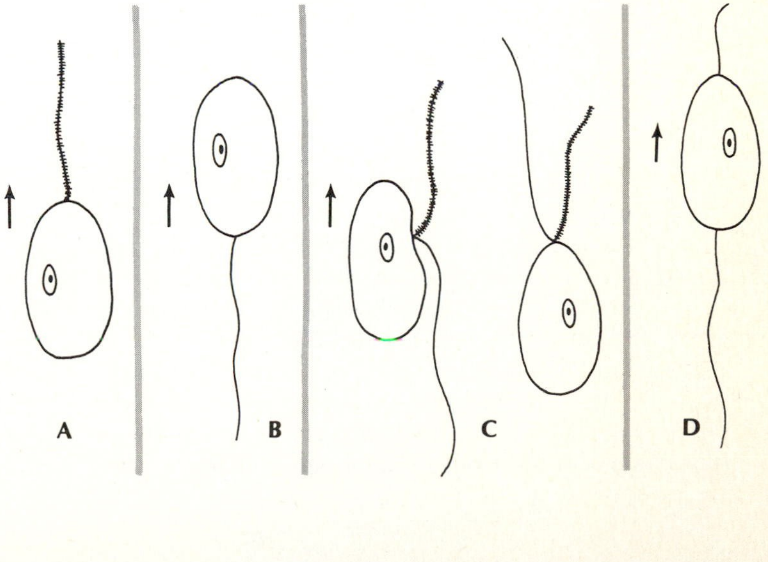

FIG. 15-1 Flagellar types in the Mastigomycotina.
Classes of organisms within this subdivision are based on flagellar types illustrated: **A,** single anterior tinsel-type flagellum of the Hyphochytridiomycetes; **B,** single posterior whiplash flagellum of the Chytridiomycetes; **C,** biflagellate, heterokont flagella of two types of Oomycetes; and **D,** biflagellate heterokont whiplash flagella of the Plasmodiophoromycetes.

in a separate class of the slime molds (Myxomycota), or even as a separate division. Karling (1968) has proposed a close relationship between this class and the protozoa.

In the recent past the subdivision Mastigomycotina has been classified as part of the now-abandoned group, the Phycomycetes. This classification, which no longer has taxonomic credibility, was based on both the resemblance of many members of the group to photosynthetic algae and the presence of sporangia that are borne on a multinucleate (coenocytic) filamentous mycelium. Until more is known about the biochemistry and physiology of the Mastigomycotina, it is clear that the taxonomy of this division will remain uncertain.

CELL STRUCTURE

The Mastigomycotina contain some genera that are unicellular, but most of its members are composed of siphonaceous, multinucleate filaments or hyphae. All species of this division produce flagellate cells at some point in their life histories. The flagella that are produced contain the typical 9 + 2 microfibrillar construction and are similar to the flagella found in higher plants, animals, and algae (see Fig. 1-8).

The individual cells and multinucleate hyphae possess ultrastructural characteristics that are similar in most respects to those found in other eukaryotic cells. Endoplasmic reticulum, ribosomes, and vacuoles are present in abundance. Mitochondria are variable in shape, but have a tendency to be elongate, with only a few tubular or flattened cristae projecting into the matrix. The nature of the cristae has been used to distinguish the various classes (Taylor, 1978). The zoospores of *Blastocladiella* contain a single large mitochondrion located near the base of their flagella. Microtubules may also be identified in most fungal hyphae; they appear to function in connection with cytoplasmic streaming.

The Blastocladiales contain special cellular organelles called **nuclear caps.** These are extensions of the nuclear membrane containing ribo-

FIG. 15-2 Life cycle of *Chytridium*.

Sexual and asexual phases of the life history of a generalized parasitic species of *Chytridium*. Note production of large numbers of zoospores (swarmers) during asexual reproduction. Cysts represented in both sexual and asexual phases are composed of zoospores that settle on and attach to a filamentous alga and then resorb their flagella. Subsequently cysts invade the host algal cell through growth of a tubelike series of rhizoids that form an intracellular haustorium within the algal cells. Resting sporangia are composed of thick-walled, multinucleate structures capable of withstanding adverse environmental conditions. Germination of resting sporangia yields haploid zoospores.

somes that are rich in ribonucleic acid and protein and are therefore important sites of protein synthesis. The nuclear caps are easily seen in electron photomicrographs of zoospores and may also be observed as a refractile covering over the nucleus in phase-contrast and light microscopes.

The cytoplasm of the cells and hyphae is bound by a plasma membrane that is chemically and architecturally similar to other eukaryotic cells and conforms to the fluid-mosaic model of Singer and Nicolson (1972). Outside the plasma membrane, the cell or hypha is covered with a cell wall composed of chitin or cellulose or a mixture of the two (see Table 12-1). Protein substances often form a matrix for chitin or cellulose deposition, and enzymes (protein) also often form an integral part of the cell wall.

The vegetative hyphae of organisms in this subdivision are nonseptate, so that the cytoplasm is one contiguous unit that contains many nuclei. Septa are produced, however, when reproductive structures are formed, as seen in the life cycles of *Saprolegnia* (Fig. 15-3) and *Allomyces* (Fig. 15-4). The amount of differentiation within vegetative thalli is almost totally restricted to the growth and development of the reproductive structures. However, some of the parasitic members produce specialized invasive

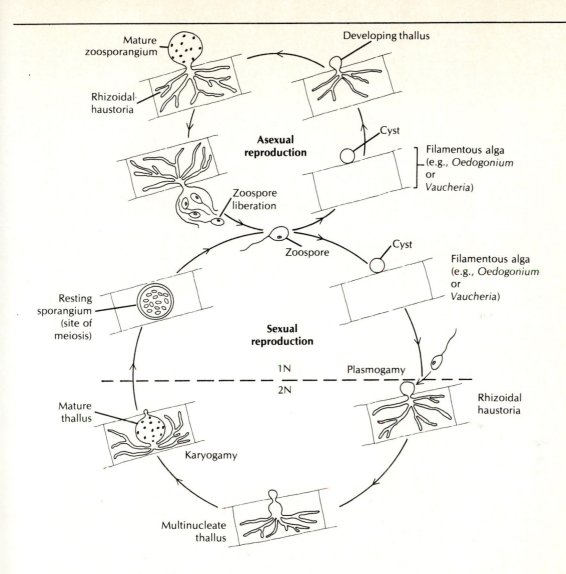

hyphae called haustoria (Figs. 13-2 and 15-2). These specialized branches grow into the cytoplasm of living host cells and tissues, where they function in the absorption of nutrients. Haustoria are particularly well developed in obligate parasitic forms such as the "downy mildews" or "white rusts" of the Peronosporales. It should be understood, however, that not all parasitic forms produce haustoria, since they are absent in some **chytrids** (Chytridiales) and endoparasitic slime molds (Plasmodiophorales).

REPRODUCTION AND GROWTH

Both sexual and asexual reproductive mechanisms for the Mastigomycotina have been described in the literature. Some of the species appear to reproduce by asexual means only, al-

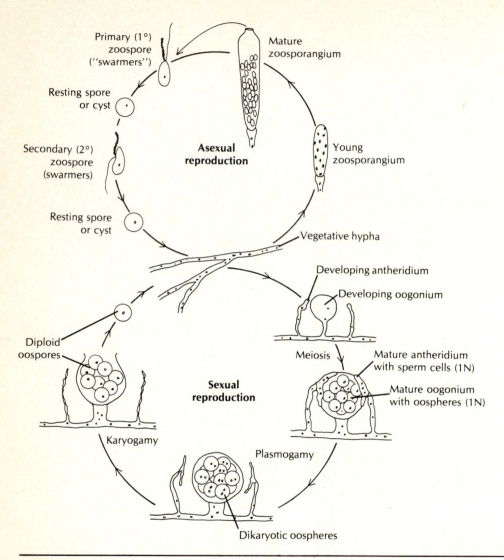

FIG. 15-3 Life cycle of Saprolegnia.

Saprolegnia, the common water mold, lives a diplontic life cycle. The vegetative hyphae live a saprobic life on dead organisms or a parasitic life on some live organisms. Asexual reproduction is accomplished on some live organisms. Asexual reproduction is accomplished through production of elongate zoosporangia, which liberate primary (1°) anteriorly biflagellate zoospores. These 1° zoospores encyst after a few hours, only to germinate some time later, producing secondary (2°) laterally biflagellate zoospores. These swarmers also encyst after a few hours.

Vegetative hyphae may also reproduce sexually through formation of specialized gametangial branches called antheridia (male) and oogonia (female). Meiosis occurs within the gametangia, producing many haploid sperm nuclei and many haploid oospheres. Plasmogamy takes place through special fertilization tubes, forming dikaryotic oospheres. Karyogamy (nuclear fusion) then takes place, forming diploid oospores that are liberated from the oogonium and that germinate to form vegetative multinucleate hyphae.

though it may be that sexual aspects of these species have not yet been observed. The asexual and sexual reproductive strategies are both closely related to the aquatic habit of the individual species. These fungi produce motile uniflagellate or biflagellate zoospores in great numbers when they reproduce asexually. The number, type, and position of the flagella depend on the individual class of the organism and are used as characteristics to help identify them (see p. 287).

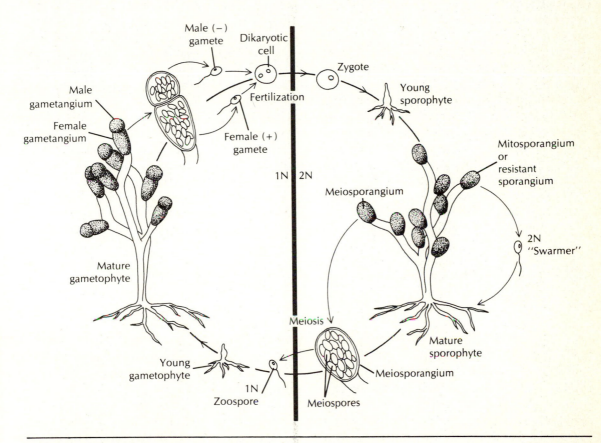

FIG. 15-4 Life cycle of *Allomyces*.

Note that the thalli of both the sporophytic and gametophytic generations are composed of hyphae that lack septa when in the vegetative state but form complete septa when sporangia and gametangia are formed. The various species of *Allomyces* may be isogamous or anisogamous, with motile female gametes being slightly larger than the male gametes. Arrangement of male and female gametangia may be reversed in some species, with the male gametangium being formed beneath the female gametangium. The sporophyte produces two types of sporangia, mitosporangia and meiosporangia. Mitosporangia produce 2N mitospores called "swarmers" or zoospores and are capable of forming another diploid sporophyte. Mitosporangia may also develop into a resistant, resting stage in response to adverse environmental conditions. Meiosporangia produce 1N meiospores by meiosis. These haploid zoospores are capable of forming a new gametophytic generation. This species is isogamous.

The simplest and most primitive members of the division are all isogamous and are represented by the chytrids. Reproduction among the chytrids is difficult to follow because they are parasitic, with part of their life history spent within host algal cells (Fig. 15-2). *Chytridium* is a good representative genus of chytrid that is parasitic on filamentous algal cells. Fig. 15-2, showing a life cycle for *Chytridium* is based on the studies of Koch (1951). Note the manner by which part of the life cycle is spent as a motile swarmer or zoospore, whereas another part is spent as an invasive, parasitic haustorium growing within the host algal cells.

Asexual reproduction via zoospores is particularly well developed in this division. Millions of swarmers or zoospores can easily be demonstrated through the use of the baiting techniques described earlier. Members of the division form zoospores by two different methods: (1) a species is **holocarpic** if the entire thallus, usually a unicellular organism, divides into a number of individual cells that are released as zoospores; a few unicellular chytrids are holocarpic; and (2) a species is **eucarpic** if they produce their zoospores in specialized branches of their hyphae; most fungi are eucarpic. By either method, the number of zoospores produced is indefinite and usually represents a rather large number of functional swarmers, as in *Saprolegnia* (Fig. 15-3) or *Allomyces* (Fig. 15-4).

Sexual reproduction always involves the fusion of gametic nuclei (**karyogamy**), which generally takes place shortly after the union of male and female gametes (plasmogamy). Plasmogamy always results in the formation of a dikaryotic phase, even though it often exists for only a short period. In *Saprolegnia,* when the individual diploid nuclei of the oospores (zygotes) divide, as the oosphere germinates, no septa are formed between them; thus a coenocytic vegetative hypha is produced.

Individual organisms of Mastigomycotina may be dioecious or monoecious. They are monoecious if both male and female gametes are produced on the same thallus. A monoecious species of *Saprolegnia* is represented in Fig. 15-3. Dioecious forms produce male gametes on one thallus and female gametes on another; the thalli are, therefore, separate sexed. The gametes produced are isogamous in the more primitive organisms, but anisogamy and oogamy are found among the derived forms. In this respect the fungi are similar to the algae. An isogamous species of *Allomyces* is illustrated in Fig. 15-4. The isogamous species produce gametes that are morphologically similar but physiologically distinct. Such forms are often indicated as mating types + (plus) and − (minus), instead of the usual female and male designations.

Monoecious fungal thalli are often designated as being **heterothallic** or **homothallic.** Heterothallic forms produce male and female gametes that are self-incompatible, and therefore two compatible thalli (mating types) are required for sexual reproduction. Homothallic forms produce male and female gametes that are self-compatible, and therefore sexual reproduction can occur within a single thallus.

PHYSIOLOGY

Most of the information on the physiology of Mastigomycotina has been gained from studying those species that can be grown easily in the laboratory. Many saprophytes are readily grown and observed under laboratory conditions, but obligate parasites are extremely difficult—sometimes impossible—to culture under controlled conditions. The members of the Blastocladiales have been the subject of extensive experimentation and therefore much is known about their cytology, morphology, and reproduction. Most mastigomycetes are aquatic because they depend on water for their nutrition and spore dispersal. Those species that grow in soils still depend on water in the form of soil moisture for dispersal of zoospores.

The Mastigomycotina have a very wide range of nutritional requirements. They can utilize many diverse sources of carbon, indicating a wide variety of exoenzymes and also reinforcing the probable unrelatedness of the four classes. Carbon may be utilized from such varied organic sources as cellulose, chitin, alcohols, sugars, starch, glycogen, or organic acids, or from such proteinaceous sources as keratin or gelatin. Several species can even use carbon dioxide. Some mastigomycetes can assimilate inorganic nitrogen or sulfur, whereas others require organic sources of these nutrients. Many members of this subclass can synthesize all needed vitamins or only require thiamine.

Mastigomycetes vary in their oxygen requirements from obligately aerobic through **facultatively** aerobic to obligately anaerobic (Oprin, 1975; Gleason, 1976). Certain Chytridiomycetes such as *Blastocladia ramosa* and three species isolated from the rumens of cattle and sheep (Oprin, 1977) are obligately anaerobic, as is the Oomycete *Aqualinderella fermentans* (Emerson and Held, 1969). Several other species (e.g., *Blastocladia pringsheimii*) **grow** well under low oxygen tensions and may be called **microaerobes** (Held et al, 1969). Under these low oxygen conditions the slowly growing fungi appear to be able to synthesize ATP for energy requirements via lactic acid fermentation, but they cannot synthesize such essential cell components as steroids without adequate oxygen (Gleason, 1976). Differentiation of gametangia may also require high concentrations of oxygen.

Several mastigomycetes require high levels of carbon dioxide. For example, the growth of some species of *Aqualinderella, Blastocladiella,* and *Blastocladia* is stimulated by high (5% to 20%) concentrations of carbon dioxide (Held et al, 1969). They probably require carbon dioxide for growth; in fact, carbon dioxide in concentrations below 5% may be required by all fungi for growth (Griffin, 1981). All three genera appear to be able to fix carbon dioxide (Cantino and Horenstein, 1956). In *Blastocladiella emersonii* carbon dioxide levels influence morphogenesis by determining whether the thallus bears a resistant sporangium in a high carbon dioxide con-

centration or a zoosporangium in a low carbon dioxide environment. *B. pringsheimii* remains viable in carbon dioxide levels up to 95% (Cooke, 1976) and in low levels of oxygen (Cantino, 1955). Under such high concentrations of carbon dioxide and low oxygen tension, resting spores are produced and vegetative growth of *B. pringsheimii* is suspended (Willoughby, 1977).

Visible light also influences the growth of several mastigomycetes. *B. emersonii*, for example, can fix more carbon dioxide in light (400 to 500 nm) than in darkness via a process known as **lumisynthesis.** The fungal thallus grows larger, contains more nucleic acid, and synthesizes more proteins and polysaccharides than thalli grown in the dark (Cantino and Horenstein, 1956, 1957); evidently this fungus has some form of a photoreceptor. Furthermore, *B. emersonii* is an unusual fungus in that its metabolism can be influenced by the presence or absence of bicarbonate. Cantino (1966) and Lovett (1975) have worked extensively with this fungus as well as other members of the Blastocladiales and have provided considerable information about the development and differentiation of members of this order.

The Chytridiomycetes are easily grown in the laboratory, and therefore a considerable amount of information is available concerning their physiology. Most chytrids can reduce sulfate and use both nitrate (NO_3^-) and ammonium (NH_4^+) as nitrogen sources. Chytrids play an important role in the decomposition of such stable molecules as cellulose, chitin, and the protein keratin. The chytrids (Fig. 15-2) also parasitize both freshwater vascular and nonvascular plants and microscopic freshwater animals such as rotifers and protozoa. Parasitism on marine algae is less common (Johnson, 1976), but they may parasitize marine diatoms such as *Rhizosolenia, Melosira,* and *Navicula* (Johnson, 1966). Species of *Olpidium* not only parasitize aquatic algae, moss, or vascular plants but also may serve as vectors for transmitting viruses from plant to plant. Other chytrids parasitize algae such as *Oedogonium (Chytridium olla), Vaucheria (Chy-*

tridium sexuale), euglenoids, blue-green bacteria, and diatoms.

Members of the genus *Synchytrium* parasitize flowering plants (Angiospermae), usually causing galls on leaves, stems, or fruits, but not causing major damage. *Synchytrium endobioticum,* however, causes **black wart** of potato, a devastating disease that significantly reduces the yield in areas of potato production, especially in Europe. The cells of the potato tuber or stem are stimulated to divide by the fungus, and galls (warts) are formed. Resting sporangia may remain alive and viable in the soil for years. As in many fungi, the production of asexual or sexual reproductive structures may be controlled by the environment. In *S. endobioticum,* adequate moisture induces zoospore formation, whereas lack of moisture induces gamete formation. Gametes are also formed at the end of the growing season, a phenomenon similar to that which occurs in algae.

Saprophytic chytrids include *Rhizophlyctis rosea,* which may be isolated from many cellulose-rich soils or sediments, where it grows using cellulose as the sole source of carbon. Chitinous insect exuviae (exoskeletons that have been shed) may be decomposed by the saprophytic chytrid *Chytriomyces hyalinus.* Because several saprophytic chytrids feed on declining algal populations, their role as either parasites or saprophytes is not clear. A chytrid that attacks a rapidly growing algal population, however, is undoubedly a parasite (Masters, 1976).

Noteworthy among the Blastocladiales is the obligate parasite of mosquito larvae, *Coelomomyces psorophorae.* This parasite has an alternate host, the crustacean *Cyclops vernalis;* it is the only known parasitic mastigomycete that requires an alternate host for the completion of its life cycle (Whisler et al, 1975). *C. psorophorae* has been considered as a biological control agent for mosquitoes; the recent discovery (1975) of the alternate host may increase its effectiveness in mosquito control. Other members of this genus are obligate parasites of other insects and may be useful as biological control agents for such insect pests.

Research with *Allomyces macrogynus* demonstrated that the female gametangia and gametes secrete a hormone, sirenin, that chemotactically attracts the male gamete (Machlis, 1958). Sirenin is effective at concentrations as low as 10^{-10} molar. See Machlis (1972) for further discussion of fungal hormones.

In the class Oomycetes, oxygen requirements range from being obligately aerobic to obligately anaerobic. Most Oomycetes require the calcium ion for stabilization of their internal membranes. The cell walls contain small amounts of cellulose, but the principal components are long-chain glucans. Chitin has been found only in the order Leptomitales (Ross, 1979). The class is sensitive to the antibiotic streptomycin, which is also effective against bacteria; other fungi are not affected by streptomycin. The method of action of streptomycin on the Oomycetes is proposed to be the blockage of calcium ion uptake at the cell wall, since 95% of the antibiotic remains at the cell wall (Deacon, 1980). Streptomycin has been used to effectively control the downy mildew of hops, the vascular plant vine that is so important in the brewing of beer. The Oomycetes synthesize the amino acid lysine by a metabolic pathway known as the **diaminopimelic acid (DAP)** pathway (LeJohn, 1971). The DAP pathway also is found in bacteria and higher plants. All the other Eumycota, however, use a different pathway for lysine synthesis, the **aminoadipic acid (AAA)** pathway. Another significant difference between the Oomycetes and other fungi is that the Oomycetes do not translocate sugar alcohols (mannitol or arabitol) or trehalose (a disaccharide) as other fungi do. (See Chapter 13 for a discussion of the significance to parasitism of the translocation carbohydrate.) Many Oomycetes, especially in the order Peronosporales, are obligate parasites of commercially important food crops.

Nutritional requirements in the Oomycetes range from general requirements for organic nitrogen (amino acids) and sugars to highly specific requirements for preformed organic materials. The Peronosporales cannot synthesize sterols (Bu'Lock, 1976), which are apparently required for sexual reproduction, so they must obtain sterols from their host.

The Saprolegniales are widespread in soil and fresh and salt water. They occur both as saprobes and parasites. As in the chytrids, it is sometimes difficult to draw the line between a saprophyte and a parasite. Species of *Saprolegnia* (Fig. 15-3) decompose dead animal matter that is high in protein and may often parasitize fish, fish eggs, and occasionally zooplankton. Members of this order require no exogenous vitamins, and they preferentially utilize organic sulfur and organic nitrogen over inorganic forms.

Saprolegnia parasitica may cause death in fish within 24 hours in such environments as crowded hatcheries and may also infect anadromous fish (saltwater fish that ascend to fresh water to·spawn). *Saprolegnia* is an important and ubiquitous fish parasite; the zoospores may be found year round and present a constant threat of infection. Parasitic Saprolegniales can be controlled by increasing the salinity of the water because the fungus cannot reproduce at salinities above 7.5 ‰* (Harrison, 1972). Saprolegniales often follow bacterial infections, thereby increasing the difficulty of determining the role of the fungus as saprophyte or parasite (Wilson, 1976). Species of *Achlya* may also be involved as fish parasites.

Many fungi require specific inducers, often from the environment (pH or exogenous vitamins, for example) to induce differentiation of gametangia on the mature (i.e., competent) thallus. Considerable research has been done on hormones as inducers in *Achlya* (Machlis, 1966; Raper, 1966; Barksdale, 1969; McMorris et al, 1975; Horgen, 1981). Raper (1966), working with heterothallic strains of *Achlya ambisexualis* and *A. bisexualis,* determined that the male and female thalli only produced sexual structures in the presence of thalli of the opposite sex. The female thallus produces a diffusible

steroid, **antheridiol,** which stimulates the male thallus to differentiate antheridia; these in turn produce another diffusible steroid, **oogoniol.** Oogoniol stimulates the differentiation of oogonia on the female thallus, and the oogonia secretes more antheridiol, which attracts the antheridium, and fertilization occurs (Fig. 15-5). This hormonal induction of gametangial growth has been used as a model system for the steroid induction of differentiation (Horgen, 1977). The **sterols** isolated from *Achlya* chemically resemble the sterols of vascular plants and algae more than those found in other Eumycota.

Sporulation in *Achlya* may be induced by removal of all exogenous nutrients. Differentiation of the sporangia then takes place using only the preexisting compounds in the thallus. Work by Timberlake et al (1973), who used a protein synthesis inhibitor, indicated that the proteins in the hyphae were broken down and resynthesized into the proteins needed for sporulation. The fungus was able to utilize the proteins existing before sporulation for spore production, a process of reutilization.

In the Leptomitales, several genera can ferment sugars. *Aqualinderella* is an obligate fermenter, metabolizing glucose via glycolysis and having a very weak cytochrome system. This fungus is considered a microaerobic fungus. *Aqualinderella fermentans* grows well at carbon dioxide concentrations up to 20% (Emerson and Weston, 1967). However, *A. fermentans,* like other fungi, may require oxygen for differentiation of gametangia. *Leptomitus,* however, requires oxygen for vegetative growth and can reduce inorganic sulfate, but cannot utilize inorganic nitrogen. *Leptomitus lacteus,* the "sewage fungus," is usually found in waters containing sewage (Cooke, 1979) and can utilize primarily fatty acids, organic acids, organic nitrogen, and glycerol. *Leptomitus* has a normal cytochrome system, cannot utilize any of the fermentation pathways, and requires oxygen (Ross, 1979).

The order Peronosporales contains many important agricultural crop parasites including the **downy mildews, white rusts, damping off dis-**

---

*Parts per thousand—sea water is usually 35 ‰ in the open ocean.

eases, and the blight of potatoes. The order contains both facultative and obligate parasites, and this group has caused severe and detrimental impacts on agricultural economy.

Pythium, a common inhabitant of soils, can survive saprophytically in either soil or water for years. It can rapidly become parasitic, causing damping off disease of seedlings. The fungus can produce pectinases, which dissolve the middle lamella of the host cells, causing it to collapse and die. *Pythium* is common in cultivated soils and has a rapid growth rate. One species, *Pythium mammilatum*, can remain dormant until the roots of a susceptible species grow into the vicinity, and diffusible substances from the roots can induce germination of the dormant spore (Barton, 1957). The mycelia of *P. mammilatum* are then attracted chemotactically to the roots of

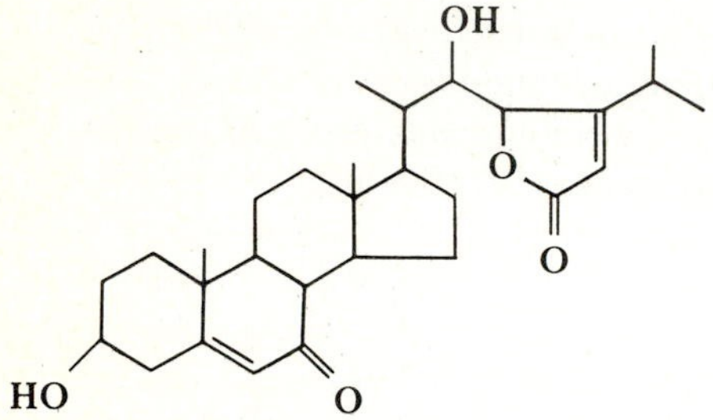

Antheridiol

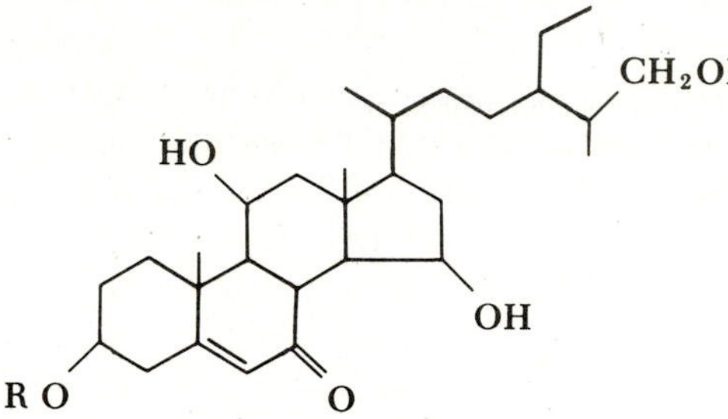

Oogoniols — R may vary

1. ♀ Vegetative thallus

                → Antheridiol →

2. ♂ Vegetative thallus differentiates antheridial initials which produce oogoniol

3. Differentiation of oogonial initials, which produce other hormones or antheridiol in some *Achlya* species

          ← Oogoniol ←

                → Hormones →

4. Antheridial initials chemotactically attracted to oogonial initials and grow toward them

5. Contact between antheridial and oogonial initial
6. Differentiation of antheridium and oogonium
7. Syngamy
8. Formation of zygote
9. Zygote forms thick-walled resting spore

FIG. 15-5 Structure and functioning of the hormones in *Achlya*.
Sterols secreted by the female vegetative thallus induce formation of male sex organ (antheridium) initials, which then secrete another sterol, oogoniol. There are several chemical forms of oogoniol that stimulate development of female sex organ (oogonium) initials. Oogonium initials then produce hormones that chemotactically attract the antheridial initials. Upon contact between male and female initials, sex organs and gametes are formed and syngamy ensues. See Griffin (1981) for detailed information on these hormones.

the host. *Pythium* requires thiamine for its growth and sterols for oospore formation. Control is best accomplished by avoiding both overcrowding and poor drainage.

All species of *Peronospora* are obligate parasites of vascular plants. For example, *Peronospora tabacina* causes blue mold on tobacco and other species of Peronosporales parasitize onions, beets, and spinach. *Plasmopara viticola* is an important parasite of grapes; the sporangiophores and sporangia covering the leaves gives the disease the name "downy mildew." Species of *Albugo* parasitize higher plants, causing white rusts of cabbage and horseradish. The mycelium is intercellular (Fig. 15-6), but the haustoria are intracellular (Fig. 15-2). Sporangiophores bearing sporangia grow out of the leaf between the mesophyll cells, and the sporangia are blown to other suitable host plants. The zoospores escape from the sporangia under appropriate conditions and encyst on the host, then produce a hypha that grows through the stomata and between the cells. Eventually haustoria are produced that invade the cells and extract the host nutrients. During the spring and summer, reproduction is vegetative, with sexual reproduction occurring in the fall.

As with *Pythium*, members of the genus *Phytophthora* also require exogenous sterols for oospore production. Evidently sterols are necessary in this class for sexual reproduction, but they are not needed for vegetative growth. Sterols for sexual reproduction in the Peronosporales may be provided by the host plant because many higher plant sterols can fulfill the role of exogenous sterols. If the host plant does provide the sterols for fungal reproduction, a complex relationship between host and parasite must occur (Ross, 1979).

Zoospores of *Phytophthora infestans* (Fig. 15-6) demonstrate chemotaxis toward the stomata of potato leaves, through which they can gain entrance to the leaf interior and initiate infection. Chemotaxis, then, is a definite physiological advantage, because the zoospores apparently do not have adequate stored energy to penetrate the intact leaf cuticle (Deacon, 1980). Chemotaxis of zoospores toward host plant roots has been demonstrated for several other species of *Phytophthora*.

Species of *Phytophthora* parasitize avocados, eucalyptus and citrus trees, and members of the Solanaceae (potato-tomato) family. The pathogen that causes late blight of potatoes, *Phytophthora infestans*, inflicted economic and cultural havoc on Ireland in the mid-1800s. The fungus killed or spoiled potato plants and tubers resulting in a great famine. In *P. infestans*, genetically compatible hyphae must be present before sexual reproduction can take place. Two mating types, A$_1$ and A$_2$, have been identified in *P. infestans*, and oospores are formed when gametic fusion occurs between the two types. A$_1$

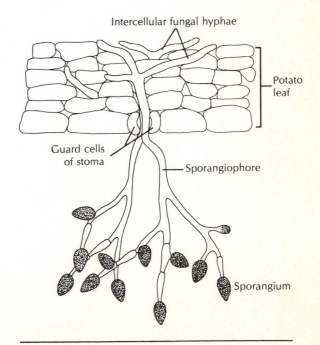

Intercellular fungal hyphae

Potato leaf

Guard cells of stoma

Sporangiophore

Sporangium

FIG. 15-6 *Phytophthora infestans* **habit.**
Mature sporangiophore of *Phytophthora infestans* projecting through a stoma on the ventral surface of a potato leaf. Note mature and developing sporangia and the intercellular fungal hyphae growing among the parenchyma cells of the host potato leaf.

and A_2 are types found only in Central and South America, where both the potato and *P. infestans* evidently originated. Outside of this area, however, only mating type A_2 has been found, but oospores have also been discovered. Various other methods (homothallism, **parthenogenesis**) of genetic mixing and oospore induction have been proposed (Galindo and Gallegly, 1960). Oospores may survive in the soil for many years; the fungus may also overwinter in infected potato tubers. Sporangia are wind dispersed and survive the longest in high humidity.

The Plasmodiophorales are obligate parasites of such vascular plants as cabbage, potato, watercress, and eel grass and such aquatic algae as *Vaucheria*. *Plasmodiophora brassicae*, for example, causes **clubroot** of cabbage and other related plants, producing distortion and swelling of the roots and eventually a reduction in growth rate. Aist and Williams (1971) chronicle the penetration of the root hair by the fungus. When this parasitic fungus modifies the hormonal balance in the host, the infected root becomes a site of renewed cell growth. The cells of the host increase both in number (hyperplasia) and in size via swelling (hypertrophy). Usually this type of growth in vascular plants is stimulated by plant hormones. Butcher et al (1974) have identified a hormone precursor in healthy cabbage plants that is converted to the plant hormone **indoleacetic acid (IAA)** when the host becomes infected with the fungus. The mechanism by which the fungus increases IAA levels in the root cells is not yet definitively established. Another member of this order, *Spongospora subterranea*, has a life cycle similar to *P. brassicae*, but it parasitizes the potato, causing "powdery scabs" on the tubers, and also transmits a pathogenic virus to the host (Kole and Gielink, 1963).

The Hyphochytridiomyces, a small class of Mastigomycotina, contain both saprophytes and parasites on aquatic algae, other fungi, and insects. Morphologically, the members of this class resemble chytrids. *Rhizidiomyces*, a well-known genus of the group, has been grown in the laboratory and its life cycle has been determined (Fuller, 1962).

EVOLUTION

Mastigomycotina is a group of fungi that is apparently unrelated evolutionarily. The division is considered primitive among the fungi because all members produce flagellate cells. The grouping in this division—according to the possession of motile zoospores—is artificial, and no evolutionary relationship between the four classes is implied or evident. The group is probably polyphyletic, with each class arising from different ancestors. Evolutionary trends, therefore, are difficult to ascertain. The Mastigomycotina were formerly included in the now-defunct division, the Phycomycetes.

The Chytridiomycetes appear to be the class with fewest advanced characteristics and may be related to the lower protists. Many exhibit primitive nutritional characteristics such as being able to utilize inorganic sources of sulfate and nitrogen and to synthesize all essential vitamins. Some chytrids, however, require several exogenous vitamins. Sources of carbon utilized by them include sugars, starch, and cellulose.

The Plasmodiophoromycetes, an evolutionary enigma, have been placed by other authors in the Myxomycota (slime molds) because they have a plasmodium (Webster, 1980). Sparrow (1958) places them in their own separate division. In addition, a relationship to phagotrophic protozoans has been proposed for the class, and Ross (1979) places them in the Protista. Waterhouse (1973a) suggests that they bear no clear relationship to other fungal divisions.

The Hyphochytridiomycetes, with their cellulose and chitinous cell walls (Fuller and Barchad, 1960; Aronson, 1965) and anterior tinsel-type uniflagellate zoospores, are also an evolutionary enigma. Until 1952 (Gäumann), they were included with the Chytridiomycetes because of their close morphological resemblance to this group. At this time they are difficult to place in any type of evolutionary sequence. Olive (1975) places them in their own division, the Pantonemomycota, along with the Oomycetes.

The Oomycetes are obviously not closely related to the other Mastigomycotina because of several unusual biochemical, reproductive, and morphological characteristics. Although many Oomycetes have some cellulose in their cell walls, the primary components are long-chain glucans that are found in most fungi. Chitin, however, has been isolated from the cell walls of several species of the Leptomitales. The Oomycetes synthesize the amino acid lysine via a different biochemical pathway (the DAP pathway) than all other members of the true fungi. Other biochemical characteristics that separate the Oomycetes from other Eumycota are: (1) they apparently cannot translocate mannitol and trehalose, and (2) their sterols are more similar to those of vascular plants than to those of other fungi. Diplontic life cycles are observed in the Oomycetes, another characteristic separating them from the other fungi. Morphologically, the insertion of the flagella in the zoospores is different from that observed in other Eumycota.

Gäumann (1952, 1964) states that the Oomycetes may have originated from the siphonaceous green algae. There is indeed a distinct morphological resemblance between sexual reproduction in the Saprolegniales and in the xanthophyte *Vaucheria* (see Fig. 10-1). Olive (1975) proposed the phylum (division) Pantonemomycota to include the Oomycetes and the Hyphochytridiomycetes while all other fungi were placed in the Eumycota. Olive did not believe, however, that there was a close biochemical or structural relationship between the Oomycetes and the hyphochytrids. Shaffer (1975) excludes these two groups from the fungal kingdom. Copeland (1956) proposed a close relationship between the Oomycetes and the heterokont algae. Obviously much more information must be gathered before the evolutionary relationships of the Oomycetes are determined definitively.

It is extremely difficult to draw any conclusions about evolutionary relationships in the Mastigomycotina. Evidently the different classes have had different origins, and current attempts to unravel relationships are mainly speculative. Further research is needed to determine relationships, and as a result of such work, the classification in the division may change along with present-day concepts of fungal evolution.

ECOLOGY

The Mastigomycotina are found primarily in aquatic or moist terrestrial habitats, because excessive moisture favors their growth and reproduction. All members of the division are dependent on water, be it only a thin film, for zoospore dispersal. Members of the group range from saprophytes, which decompose and recycle organic matter, to plant parasites, which are extremely destructive to certain agricultural crops.

The chytrids may easily be found in fresh water and soils and less frequently in salt water. Chytrids are numerous in the "soil flora" and live saprophytically, recycling dead organic material from animal and plant remains. They produce exoenzymes that can decompose chitin, cellulose, and keratin. *Rhizophydium*, the most numerous chytrid, has been labeled the "plankton eater" because it parasitizes phytoplankton, including the diatoms. *Rhizophydium planktonicum* is a common diatom parasite. Other chytrids such as *Olipidium* and *Chytridium* (Fig. 15-2) may parasitize green algae, euglenoids, xanthophytes, blue-green bacteria, and zooplankton. *Polyphagus euglenae* parasitizes euglenoid algae and may have a significant effect on such algal populations in puddles and ponds.

Currently researchers differ in their assessment of the impact of the chytrids on phytoplankton populations. Some scientists (Soeder and Maiweg, 1969; Webster, 1980) believe that parasitic chytrids can severely deplete apparently healthy and rapidly growing algal populations and influence the interspecific competition between algal species. Others such as Canter and Lund (1969) indicate that healthy algal cells may be parasitized, especially when very numerous, but that seasonal peaks of host algae are not affected. Moss (1980) thinks that chytrids only reduce populations when algal densities are very high. Wetzel (1975) feels that chytrid parasitism is especially important in eutrophic waters (see Chapter 2). Canter and Lund (1969), Paterson (1970), and Masters (1976) discuss the effect of chytrids on phytoplankton populations. Further research will aid in clarifying the role of chytrids in regulating algal populations.

An important chytrid that is parasitic on higher plants is *Synchytrium endobioticum*, the cause of black wart disease of potatoes. Control of this parasite is best accomplished by the development of resistant potato strains.

Allomyces (Fig. 15-4) is widely used as an experimental fungus in the laboratory, as are other Blastocladiales. Most species of this filamentous fungus are found in warm climates in either water or damp soils. It is common in the tropics, where it plays an important saprophytic role in decomposition. Most members of the order Blastocladiales are found in tropical areas also, and some genera parasitize vascular plants (e.g., cashew trees). Species have been found in temperate areas, however, including one that parasitizes the Cyanobacterium *Anabaena* (Canter and Willoughby, 1964).

Saprolegnia (Fig. 15-3) is an important decomposer and parasite in aquatic habitats and in damp soils. *Saprolegnia* zoospores are always present in fresh water, including fish hatcheries where densities average 400 zoospores per liter, lakes and reservoirs (1 to 10 zoospores per liter), streams (hundreds of zoospores per liter) and sewage plant effluents (4,000 zoospores per liter) (Willoughby, 1969). The trickling filters used in sewage treatment plants contain many members of the Saprolegniales (Willoughby, 1977), and evidently the zoospores are washed off into the effluent. Whether *Saprolegnia* assumes a sap-

rophytic or parasitic role depends on an as-yet-undetermined complex of environmental conditions, fish susceptibility, and the strain of fungus.

Saprolegnia is evidently a facultative parasite that can establish itself on the external mucus on fish. Infection has been observed in sport fish such as Atlantic salmon, various trout, and perch. The fungus grows rapidly, eventually enzymes are produced that break down body tissues, and the fungal infection becomes systemic. Death results, often within 24 hours, if the fungus invades both the spinal cord and the blood vessels (Nolard-Tintigner, 1973). Sodium chloride added to the water in fish hatcheries increases the salinity, making reproduction of *Saprolegnia* impossible and thereby controlling the disease. Thick-walled oospores, resistant to adverse conditions, remain in fresh water as potential sources of future infection. Rotifers, copepods, and fish eggs may also be parasitized, and occasional parasitism of marine organisms has been recorded (Johnson, 1976).

A dead housefly placed in pond water will exhibit a cottonlike growth of *Saprolegnia*, or occasionally *Achlya*, within two days, and students may readily observe the live fungus using this procedure. Dead fish left in relatively clean water will also be covered with *Saprolegnia* and associated bacteria and other microorganisms within a few days, indicating just how ubiquitous the zoospores are.

In culture, the Saprolegniales can utilize a wide variety of organic carbon sources, inorganic or organic nitrogen, and organic sulfur. The chytrids in culture can utilize inorganic sulfur, but require thiamine (B_1) for growth. The ability of the Saprolegniales to manufacture their own vitamins could give them a competitive advantage over the chytrids in vitamin-poor environments. However, extrapolation of laboratory situations to natural systems is both difficult and risky. So many variables operate in natural systems that it is almost impossible to predict the result of the interactions of such variables upon natural populations.

Leptomitus lacteus (Leptomitales) has been labeled the "sewage fungus" because it often grows in polluted water. *L. lacteus* requires high oxygen concentrations and may also be found in relatively clear, clean waters (Hynes, 1963). Actually what has been called the "sewage fungus" is more accurately designated the "sewage fungus community" because the "community" is composed of fungi, protozoa, and bacteria (Cooke, 1954). The bacterium *Sphaerotilus natans* is an important component of this community and may be misidentified as a fungus because of its filamentous growth form. *L. lacteus* may occasionally, but not always, be an important member of the community, especially in cold weather when oxygen levels in the waters are high (Hynes, 1963). Curtis (1969) considers *L. lacteus* an unimportant member of the sewage fungus community. Apparently, *L. lacteus* does not deserve the label "sewage fungus."

Aqualinderella fermentans is probably a better indicator of pollution than *Leptomitus lacteus* because it can respire anaerobically. *A. fermentans* can live in waters low in oxygen (less than 5.0 mg/l, high in carbon dioxide (26 mg/l), and high in dissolved organic material (Willoughby, 1977). The fungus may commonly be found in warm stagnant waters, an unusual ecological niche for a fungus (Cooke, 1979). Other fungi found in polluted waters are discussed in Cooke (1979). The ability of *A. fermentans* to successfully compete with bacteria under anaerobic conditions demonstrates the biochemical versatility and diversity of the fungi. Several other genera of the Leptomitales can also gain their energy requirements via fermentative pathways.

Some members of the Peronosporales can exist saprophytically in organic soils or can be found as plant parasites. A good example is *Pythium*, the cause of damping off disease of seedlings. When not parasitizing plants, these fungi often exist for many years as dormant oospores that resist adverse conditions. Such a capability makes these diseases difficult to control.

One parasite that caused innumerable problems in France is *Plasmopora viticola* (Fig. 15-7), the downy mildew of grapes. In the mid-1800s the grape vines of southern France were being heavily damaged by root aphids. Because American vines were resistant to the aphid, the French imported grape roots from the United States and grafted French vines on the American roots. The aphid problem was solved. Unfortunately, the American vines also carried the downy mildew fungus (to which they were also resistant) and soon the fungus was destroying the foliage of the French vines. By the late 1870s the French grapes were severely threatened and no solution was apparent. The density of *P. viticola* spores reached 32,000 s/dm^2 in the grape orchards (Christensen, 1975).

One day in 1882, Pierre Millardet, a professor at the University of Bordeaux in southern France, noticed that certain vines along the road were not touched by the mildew and that they had a strange bluish tinge. Upon inquiry Millardet discovered that the farmer had purposely sprayed his grape vines blue to discourage people walking by from eating the grapes. Millardet noted that not only did the spray discourage grape-picking from the road but it also protected the grapes from attack by the mildew. The spray, haphazardly compounded from leftover chemicals, was a mixture of copper sulfate and calcium oxide. From this mixture of chemicals Millardet developed the first effective fungicide **Bordeaux Mixture,** which subsequently was widely used on the grape vines, providing excellent protection against the downy mildew. Bordeaux Mixture was one of the first inorganic pesticides and continues in use today. Many organic fungicides are also used today, but Bordeaux Mixture is still effective when properly applied. Since copper is quite phytotoxic, great care must be used in the application of Bordeaux Mixture.

Millardet's discovery of the fungicidal properties resulting from the combination of copper sulfate and lime came soon after another member of the Peronosporales, *Phytophthora infes-*

FIG. 15-7 *Plasmopora viticola,* **downy mildew on concord grapes.**

This fungus attacks both the leaves and fruit of the European grape. Fungal hyphae grow intracellularly and produce haustoria that penetrate the cells. Sporangiophores grow above the surface and produce whitish sporangia, giving the leaves and grapes a "downy" appearance. (Courtesy William Merrill, Department of Plant Pathology, The Pennsylvania State University, University Park, Pa.)

tans, (Fig. 15-6) had caused extreme economic hardship in another part of the world. *P. infestans,* the fungus causing late blight of potatoes, had destroyed the potato crop in Ireland in the mid-1840s, causing widespread famine and death. At least 2 million people died from starvation; over a million people emigrated from Ireland, many to the United States. The population of Ireland dropped from 8 million in 1846 to about 4.7 million in 1900 (Fig. 15-8). Unfortunately, because the efficacy of copper sulfate and lime against fungi was discovered more than 40 years after the disastrous potato famine, Ireland endured many years of severe hardship before an effective control for the fungus was found. Following the discovery of the effectiveness of Bordeaux Mixture against fungi, it was used successfully against the potato blight.

The Irish famine occurred 20 years before Koch and Pasteur established the relationship between microorganisms and disease. Nobody really knew what caused the disease, and prizes were offered for discovering the cause of the blight. In 1855 a German mycologist, Anton DeBary, proposed the idea that the blight was caused by a fungus, but it was several years later that the causative agent was definitely established.

Why did the fungus *Phytophthora infestans* cause such destruction? The potato grew originally in the dry highlands of South and Central America, where it was too cool and too dry for the fungus to cause much damage. When the potato became established in Europe in the 1500s, it had no resistance to the fungus, because it had not been exposed. When *P. infestans* came in contact with the massive and susceptible potato crop in Ireland, disaster resulted. The cool, damp weather was ideal for the propagation of the fungus. In addition, the

FIG. 15-8 Human population of Ireland, 1750 through 1970.
Following the late blight of potatoes caused by *Phytophthora infestans* in the mid-1800s, millions of people either died during the famine or emigrated to other countries; many Irish emigrated to the United States at this time. The population of Ireland dropped dramatically; the decline lasted more than a century. (From the data of Connell [1941] and Reinhard et al [1968], Macmillan Publishing Co, Inc. In Whittaker, R.H., Communities and ecosystems, Ed. 2. Copyright © 1975 by Robert H. Whittaker. Reprinted with permission.)

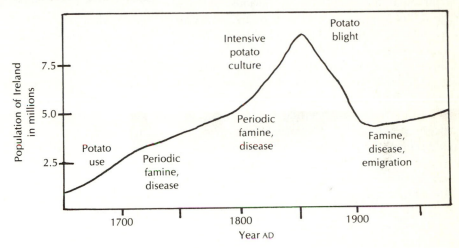

potato had been propagated vegetatively and the Irish crop was genetically uniform, so all the potato plants were equally susceptible to the fungus. In 1845 the weather was cool and wet enough for the fungus to flourish rapidly, and 75% of the total crop was lost (Klein, 1979). Again in 1847 the weather was suitable for the fungus, and within 1 week in summer, 80% of the plants had died.

Great effort has been devoted to breeding resistant potato plants. Niederhauser and Cobb (1959) discuss some of the work sponsored by the Rockefeller Foundation and the Mexican government. Shimony and Friend (1975) have done ultrastructural studies of resistant and susceptible strains (cultivars) of potato.

Obviously *P. infestans* is a highly successful fungus that flourishes in high humidity. The fungus is spread when the sporangia become detached from infected leaves: the sporangiophore bearing sporangia grow out of the leaf through the stomata (Fig. 15-6), and sporangia are blown by the wind to uninfected leaves. The sporangia produce either new mycelia or zoospores, depending on environmental conditions. The zoospores are chemotactically attracted to the leaf stomata, hyphae grow from the zoospore, and the leaf becomes infected. Following infection the leaves eventually die. Such a reduction in the photosynthetic area of the plant reduces yield and tuber weight. Both zoospores and sporangia may encyst and remain potential sources of infection for many years. Erwin et al (1982) discuss the biology of *Phytophthora*.

Infected tubers may be destroyed in storage by the fungus or become vulnerable to bacterial spoilage. Tubers may also serve as a reservoir of potential infection when they are stored over the winter. Recently, pregnant women have been warned that they should not eat or handle infected tubers, since work by a British scientist in 1974 suggested that exposure to the fungus may cause birth defects.

As noted previously, humidity plays an important role in outbreaks of *P. infestans*. Many sporangia are produced in periods of high humidity; in addition, the sporangia survive longer in high humidity (Cochrane, 1958). Great progress has now been made on forecasting outbreaks of the blight based on weather conditions. When suitable weather conditions occur, preventative fungicidal programs may be started. Conditions favoring the fungus (cool, damp weather) also favor high potato yields. Therefore, the infection of a crop is not necessarily disastrous because, if the infestation is not extensive, losses to the blight may be balanced by higher yields.

Willoughby (1977) has outlined the succession of saprophytic fungi on decomposing leaves in the aquatic environment. Species of both *Phytophthora* and *Pythium* are often the initial colonizers, and members of the Saprolegniales take over the decomposition following these organisms. For leaves to decompose, the fungi must be able to break down both cellulose and lignin. The breakdown of cellulose requires the action of at least three enzymes before it is hydrolyzed to glucose and can subsequently enter the fungal cell. Lignin is a complex, three-dimensional polymer that makes up to 5% to 30% of the dry weight of all plant tissue. It is extremely resistant to breakdown and requires many enzymes for its decomposition. It is usually the last molecule to be hydrolyzed. Many logs, stumps, and twigs may remain for a long time in a natural stream or river because such woody parts have a high lignin content and decay very slowly.

Mastigomycotina are extremely important ecologically. On the one hand, they are essential in decomposing plant and animal matter and thereby recycling the nutrients back into the ecosystem. Furthermore, in the process of decomposition they may develop a rather significant biomass and become food for various invertebrates or vertebrates. On the other hand, they may be highly destructive parasites, causing widespread economic and social havoc.

ECONOMIC IMPORTANCE

The Mastigomycotina in general and the Peronosporales in particular have had an enormous impact on various agricultural crops throughout the world and therefore on the economy of certain countries.

The late blight of potatoes (*Phytophthora infestans*) probably had the most significant, documented economic impact of any fungal disease. No accurate dollar figures can be placed on the destruction of the Irish potato crop and the reduction in yield that precipitated the potato famine, nor is it possible to apply monetary values to the number of deaths or emigrations that resulted from the famine. Salaman (1949) states: "History has few parallels to such a disaster—a disaster due to the criminal folly of allowing a single, cheaply produced foodstuff to dominate the dietary of the people."

Why was the potato famine such a disaster? The potato (*Solanum tuberosum*) was introduced into Ireland in the 1500s. It soon became extremely popular in Ireland because it requires little care, is a versatile and nutritious food, and provides a high yield from a small area (6 to 22 metric tons/hectare). The Irish became extremely dependent on the potato as a food staple as the birth rate quickly increased; thus the rapid destruction of the potato crop in fields and storage by the late blight disease devastated the food supply of the country. Famine and death ensued, and many inhabitants (over a million) had to leave the country to find food and jobs. The immigration of the Irish had a profound impact on the economics, the politics, and the culture of the United States. The famine also influenced the British to repeal their corn laws, which restricted the importation of grains into Great Britain. Inexpensive grains (many from the United States) were imported into Great Britain and used to feed the starving Irish. The repeal of the corn laws also encouraged free trade among Britain, Europe, and the United States. For further discussion of the far-reaching effects of the potato blight, see Large (1940, 1958) and Salaman (1949).

The potato continues to be an extremely important food crop for the world. It is second only to rice in tons produced and requires considerably less acreage than rice while yielding more calories per acre (Niederhauser and Cobb, 1959). Fortunately the ability to predict weather conditions favorable to the fungus enables farmers to initiate fungicidal sprays before the disease becomes well established.

The downy mildew of grapes also had a tremendous economic impact on European agriculture. The fungus infected all the vineyards in Europe, seriously jeopardizing the entire European wine industry, especially in France, Spain, and Portugal; the timely discovery of the efficacy of Bordeaux Mixture saved the wine industry. The decrease in wine production, however, forced alcohol lovers to turn to the distilled whiskys of Ireland and Scotland. Distilled whiskys became increasingly popular when wine became difficult to obtain.

Other Mastigomycotina cause considerable agricultural damage. The damping off fungi, white rusts, downy mildews, clubroots, and galls all can be significant pests and can cause important crop losses.

Saprolegnia and its relatives can seriously reduce egg viability and fish survival in hatcheries, often wiping out all the young fish of the year.

The chytrids appear to control algal populations under some conditions. Masters (1976) proposes that parasitic chytrids may present problems in mass outdoor cultures of algae. Such mass cultures have been suggested as a method of increasing the food supply for livestock and humans, especially for the production of protein (see Chapters 2 and 4).

The Oomycete *Ostracoblabe* bores into oyster shells and deforms the shell (Alderman and Jones, 1971). Severe infection of the oyster beds in the Netherlands in the 1930s caused many oyster deaths (Korringa, 1951). The fungus is evidently endemic in European waters, and the 1930 epidemic was caused by using old infected shells for settlement of the larvae.

The development of effective fungicides evolved from the need to control members of the Mastigomycotina. The production of fungicides is an important industry worldwide (see Chapter 13).

Not only are some of the Mastigomycotina destructive plant parasites but some are also extremely beneficial in degrading organic materials to their inorganic components. Their role in recycling recalcitrant molecules of chitin, cellulose, and lignin, for example, in the woods, fields, and waters of the world is incalculable. When the numbers of fungi are reduced in an ecosystem, decomposition is significantly retarded and organic matter collects in the ecosystem. Such a situation exists near a zinc smelter in Pennsylvania, where high levels of zinc in the soil have reduced numbers of fungi. Decomposition is greatly retarded and organic matter accumulates (Buchauer, 1971).

SUMMARY

The Mastigomycotina comprise a large group of flagellate fungi of four separate classes, each of which is evolutionarily unrelated to the other. The division, therefore, is polyphyletic in origin. Among its members are the chytrids, endoparasitic slime molds, water molds, downy mildews, and white rusts. The more primitive forms tend to be unicellular; the others have a filamentous habit composed of siphonaceous, multinucleate hyphae. Most of the Mastigomycotina are aquatic, although the chytrids form an important part of the soil flora, and other species are prominent vascular plant parasites. The cell walls contain chitin/polysaccharides in some (Chytridiomycetes, Hyphochytridiomycetes, and Plasmodiophoromycetes) and cellulose/polysaccharides in others (Oomycetes).

The parasitic forms of this division consist of hyphae that develop into invasive haustoria. Haustoria may grow intracellularly in the adult host tissues of higher plants and in algae; fungi, and protozans or other aquatic invertebrates.

Both sexual and asexual reproductive mechanisms have been observed in the division. For many members it appears that asexual reproduction is their only means of propagation. Asexual spores are produced by both holocarpic and eucarpic mechanisms. In sexual species, both dioecious and monoecious forms have been reported, and both homothallic and heterothallic species exist.

Nutrition in the Mastigomycotina is saprophytic or parasitic. Many individuals require exogenous vitamin supplements in culture. Aerobic, facultative anaerobic, and obligate anaerobic species exist. Some species even require relatively high concentrations of carbon dioxide to grow and reproduce. Visible light is needed by some species to complete their life cycles. Other nutritional requirements range from very general organic nitrogen in the form of amino acids to highly specific low-molecular-weight growth substances such as certain vitamins.

The chytrids are a relatively little-known group of microscopic freshwater algal parasites. The water molds, including *Saprolegnia* and *Achlya,* are a ubiquitous group of freshwater plant and animal parasites and saprophytes that play an important role in recycling nutrients in aquatic habitats. They are also economically significant, since they often infect fish in hatcheries or aquaria, especially when the fish are injured or under crowded conditions.

Members of the order Peronosporales are highly specialized higher-plant parasites, causing such common diseases as downy mildews associated with grapes, white rusts associated with common garden-variety vegetables, damping off diseases associated with developing seedlings, and some blight diseases associated with

potatoes and other economically important crops. Knowledge and understanding of this group is especially important in agriculture, where many of these plant pathogens have been implicated in seriously reducing crop productivity.

SELECTED REFERENCES

Bu'Lock, J.D. 1976. Hormones in fungi. In Smith, J.E., and D.R. Berry (eds.): The filamentous fungi. Vol. 2. Biosynthesis and metabolism, pp. 345-368. Edward Arnold Ltd., London, England.

Erwin, D.C., S. Bartnicki-Garcia, and P.H. Taso (eds.). 1982. *Phytophthora:* its biology, taxonomy, ecology, and pathology. Am. Phytopath. Soc., St. Paul, Minn. 392 pp.

Gleason, F. 1976. The physiology of the lower freshwater fungi. In Gareth Jones, E.B. (ed.): Recent advances in aquatic mycology, pp. 543-572. Elek Science, London, England.

Held, A.A., R. Emerson, M.S. Fuller, and F.H. Gleason. 1969. *Blastocladia* and *Aqualinderella:* fermentative water molds with high carbon dioxide optima. Science **165:**706-709.

Horgen, P.A. 1981. The role of the steroid sex pheromone antheridiol in controlling the development of male sex organs in the water mold *Achlya.* In O'Day, D.H., and P.A. Horgen: Sexual interactions in eukaryotic microbes, pp. 155-178. Academic Press, Inc., New York.

Johnson, T.W., Jr. 1976. The Phycomycetes: morphology and taxonomy. In Gareth Jones, E.B. (ed.): Recent advances in aquatic mycology, pp. 193-211. Elek Science, London, England.

Large, E.C. 1958. The advance of the fungi. Jonathan Cape, Ltd., London, England. 488 pp.

Salaman, R.N. 1949. The history and social influence of the potato. Cambridge University Press, London, England. 685 pp.

Sparrow, F.K. 1973. Mastigomycotina (zoosporic fungi). In Ainsworth, G.C., F.K. Sparrow, and A.S. Sussman (eds.): Vol. IVB. The fungi: an advanced treatise, pp. 61-73. Academic Press, Inc. New York.

Sparrow, F.K. 1976. The present status of classification of biflagellate fungi. In Gareth Jones, E.B. (ed.): Recent advances in aquatic mycology, pp. 213-222. Elek Science, London, England.

Waterhouse, G.M. 1973a. Plasmodiophoromycetes. In Ainsworth, G.C., F.K. Sparrow, and A.S. Sussman (eds.): Vol. IVB. The fungi: an advanced treatise, pp. 75-82. Academic Press, Inc., New York.

16

CLASSIFICATION

Division: Eumycota: "true fungi"; assimilative nutrition

    Subdivision 2: Zygomycotina: multinucleate or partly septate branching filaments; no flagella; spindle pole bodies in cells; sexual reproduction by gametangial fusion, producing a thick-walled zygospore

        Class 1: Zygomycetes: asexual reproduction via sporangiospores; 100 genera, 600 species

            ORDER 1: Mucorales: "black molds"; soil and dung fungi; *Pilobolus* from dung and *Rhizopus*, the common bread mold; *Absidia, Mucor, Phycomyces, Pilaira, Piptocephalis, Syzygites, Zygorhynchus*

            ORDER 2: Endogonales: many forming ectotrophic and endotrophic mycorrhizae; most bear spores singly in soil: *Endogone, Glomus*

            ORDER 3: Entomophthorales: mostly insect parasites; *Basidiobolus, Entomophthora*

        Class 2: Trichomycetes: parasites and/or commensals in arthropod digestive systems—especially centipedes and millipedes

ZYGOMYCOTINA

The Zygomycotina constitute a subdivision of Eumycota characterized by a complete lack of flagellate cells at any time during their life histories and by the formation of a thick-walled zygote (zygospore) during sexual reproduction. The body of members of this class is composed of multinucleate (coenocytic) hyphae. The more primitive members of the class contain hyphae that completely lack cross walls (septa), whereas the derived members possess hyphae with incomplete septa. Therefore, in all the genera, the filaments are coenocytic.

Most Zygomycotina are terrestrial fungi. Under earlier classifications schemes, they, along with the Mastigomycotina, were considered to be members of the "Phycomycetes" (algaelike fungi), so called because they possess coenocytic hyphae and superficially resemble some siphonaceous algae. Nutrition in the zygomycetes is entirely **assimilative,** with absorption of organic molecules taking place after partial digestion of other organisms by production and secretion of exoenzymes. Most forms are saprophytic, gaining their energy through the degradation of dead organic matter, while others employ a parasitic life style. Saprophytic forms are especially common growing in and on animal dung or in soils containing high concentrations of organic matter. Many of the Zygomycetes are familiar to the casual observer, since they include bread molds. A few species are used in the commercial production of organic chemicals, including organic acids and enzymes. The ubiquitous distribution of spores produced by Zygomycetes makes them a nuisance to scientists, housekeepers, and food handlers; it is difficult to keep unprotected foods without their being contaminated by a member of Zygomycetes.

Perhaps the most distinguishing feature of the class is the sexual production of a sculptured, thick-walled resting zygospore (Fig. 16-

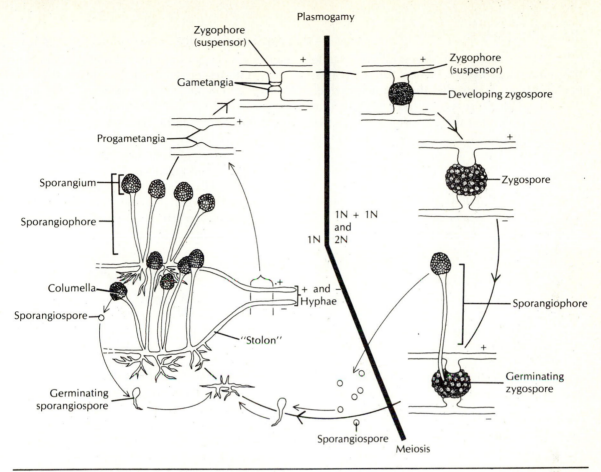

FIG. 16-1 Life cycle of *Rhizopus*, the common bread mold.
Sexual and asexual reproduction in the common bread mold *Rhizopus stolonifer*. Asexual reproduction (shown at lower left) occurs when sporangia produce multinucleate sporangiospores, which germinate to form more haploid hypha. Sexual reproduction is initiated by prohormone synthesis (see Fig. 16-3) and formation of progametangia. Progametangia fuse, forming gametangia that are isolated from the parent hypha by septae. Gametangial contents fuse (plasmogamy), forming a zygospore that develops into a multinucleate, thick-walled cell. Zygospores may act as resting spores, thus delaying germination for a period of time (up to 3 months). During this time nuclear fusion (karyogamy) occurs, followed by meiosis just before zygospore germination. Zygospore germination results in formation of a sporangiophore with its sporangium, and the life cycle is completed.

1). Zygospores result from fusion (plasmogamy) between two sexually compatible hyphae (sometimes called **stolons***). The life histories are haplontic, since meiosis is zygotic (Fig. 16-1). Asexual reproduction is by fragmentation of individual hyphae or by production of haploid, nonmotile **sporangiospores** (also known as aplanospores), which are produced in great numbers. They are borne on top of special reproductive filaments called sporangiophores. The spores are wind dispersed.

*The term "stolon" is used by mycologists to describe fungal structures that resemble runners or rootstocks found in certain grasses and other flowering plants such as strawberries.

CELL STRUCTURE

The zygomycetes are structurally similar to the other fungi discussed thus far (i.e., Mastigomycotina), except that they do not produce flagellate cells. Most zygomycetes produce a coenocytic hypha that develops into a distinctive mycelium, although the mycelium may not be as compact as in the more advanced "higher fungi." The coenocytic hyphae possess the ultrastructural organelles characteristic of eukaryotic cells. The typical fungal endoplasmic reticulum, ribosomes, mitochondria, and membrane-bound vesicles and nuclei can be isolated from the hyphae. Golgi complexes are not common; centrioles are absent. There is little about the internal structure of the hyphae that is distinctive or easily demonstrable without electron microscopy.

The growing apices of hyphae, when viewed with the electron microscope, show some accumulation of membrane-bound vesicles (Grove and Bracker, 1970). The content of the vesicles has not been determined, but it is assumed that they contain building materials and enzymes necessary for extension of the cell wall. Some evidence for this is presented by Bracker et al (1976), who demonstrated a special structure, the **chitosome,** associated with growing apices. Chitosomes are packets of an enzyme, chitin synthetase, which is necessary for the synthesis of chitin, the major material of the cell wall.

The cell walls of the Zygomycotina are chemically complex and composed primarily of chitin, **chitosan,** and other polysaccharides (see Table 12-1). Chitosan (glucosamine) is often the most abundant component. Also present in varying amounts are different polysaccharides, proteins, lipids, and mineral substances. No cellulose is present. The composition of the cell wall may vary at different stages of the life cycle. Such variation has been observed in *Mucor rouxii* (Bartnicki-Garcia, 1968).

An interesting feature of this group is the dimorphic nature of some genera—that is, some of the zygomycetes have the ability to exist in two forms, a unicellular yeastlike form and a coenocytic mycelial or hyphal form. Such dimorphism is common among animal pathogens. The genus *Mucor,* for instance, can cause a human infection known as **zygomycosis** or **mucormycosis** (Fig. 16-7) and can shift from unicellular to hyphal form, depending upon its environment. Under anaerobic conditions *M. rouxii* can be maintained indefinitely as a yeast-like, single-celled organism. Under aerobic conditions the cells form a hyphal complex (Lara and Bartnicki-Garcia, 1974). Changes in the morphological structure of these organisms make disease diagnosis difficult.

REPRODUCTION AND GROWTH

Both sexual and asexual reproductive structures are distinctive and easily identified in most members of this subdivision. The name Zygomycotina means "zygote-fungi," and the entire group is so called because of the distinctive nature of the resting zygote, called a zygospore. Most of the taxonomy within the subdivision, however, is based upon the structure of asexual reproductive hyphae, the sporangiophores, which bear sporangia and sporangiospores (Figs. 16-1, 16-2 and 16-4).

Sporangiophores (fr. Greek *phoreus* = bearer) are specialized upright hyphae. Each sporangiophore bears a terminal sporangium containing thousands of small sporangiospores, which are formed by cell wall formation around the many nuclei within the sporangium. Within each sporangium is a clear central area called the **columella,** which is an extension of the hyphal stalk forming the sporangiophore. Spores formed within the sporangium are distinctive and vary from species to species according to their surface sculpturing, shape, size, and color. Dissemination of spores is generally accomplished by their being wafted away by air movements. In some genera (e.g., *Pilobolus*) the entire sporangium is actively discharged by hydrostatic forces from within the sporangiophore (Fig. 16-2). These forces are strong enough to eject the entire sporangium 2 m verti-

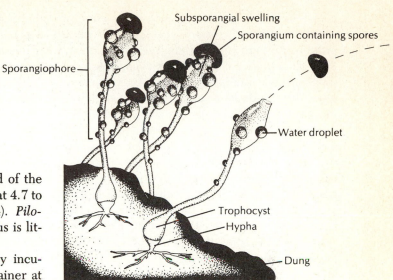

cally and 2.5 m horizontally. The speed of the ejected sporangium has been measured at 4.7 to 27.5 m/sec (Page and Kennedy, 1964). *Pilobolus* means "hat thrower," as the fungus is literally capable of "blowing its top."

Pilobolus hyphae are easily grown by incubating fresh horse dung in a glass container at room temperature for a few days. The sporangiophores possess a basal swelling, the **trophocyst,** which contains high concentrations of carotenes that color it yellow, orange, or red. The trophocyst appears to be active in phototropic mechanisms, even though it develops below the surface of the dung substrate in which the fungus grows. Sporangiophores also possess a fluid-filled **subsporangial swelling.** Pressures built up within the subsporangial swelling are responsible for the ejection of the sporangium cap. Sporangiophore development and sporangium discharge have been described in detail by Buller (1934) and updated by Page and Curry (1966).

Sporangiophores of most other members of Zygomycotina are not as distinctive as those of *Pilobolus. Mucor,* a common soil and dung fungus, produces stout rounded sporangiophores occurring singly, while *Rhizopus,* a common bread or fruit mold, produced long slender sporangiophores occurring in groups. *Rhizopus* also produces vegetative hyphae, called stolons, which connect two asexual reproductive areas where sporangiophores and rhizoids emerge from a hypha (Fig. 16-1). In other genera, asexual conidia may be produced at the growing tips of the hyphae (see also p. 241, Spore Production).

FIG. 16-2 *Pilobolus* **sporangiophores.**
Five sporangiophores are shown here, one of which is seen discharging its black-colored sporangium; direction of discharge is toward light. Discharge is accomplished by turgor pressures in the vacuole of the subsporangial swelling. When pressure reaches the breaking point, membranes of the sporangium rupture, sending the entire sporangium for a distance of 2 m; sporangiophore then collapses. The sporangium adheres to some substrate such as a blade of grass. When the grass is eaten by some animal, spores are passed, unharmed, through the digestive tract of the animal and then deposited in the animal's excrement, where the cycle begins again.

The trophocyst is a basal swelling of the sporangiophore. It is rich in carotenes, which give it a characteristically bright yellow-orange or red color.

Since most members of the Zygomycotina are dioecious, sexual reproduction involves separate individual hyphae. Such hyphae are usually designated as + or − strains, since they are morphologically identical but biochemically distinct. Sexual union of two sexually compatible hyphae is accomplished when the two hyphae grow toward each other and produce specialized hyphal branches called **suspensors** (or more correctly **zygophores** [Fig. 16-1], which means "zygote-bearer"). Zygophore growth is initiated by an inducing agent, **trisporic acid (TSA),** which is produced by growing hyphae (see next section, Physiology). TSA both induces zygophore production and inhibits sporangiophore production. After induction and growth, zygophores produce septa, thus isolating part of the multinucleate cytoplasm in separate structures called **gametangia.** The wall between adjacent gametangia is then broken down and the + and − mating types mix in a common cytoplasm (plasmogamy), thus forming a multinucleate cell or zygospore. Septa isolate the zygospore from the parent hyphae. A thickened cell wall, which bears characteristic spines, develops around the cell. The zygospore thus formed becomes a true "resting spore," since most of these spores require extended periods of time before they will germinate (up to 3 months).

Nuclear fusion (karyogamy) occurs at any time during the resting period. Timing of karyogamy appears to depend on the individual species. Meiosis of the resulting diploid nuclei may also occur at any time during the resting period and it too is species dependent. In *Mucor,* karyogamy and meiosis occur shortly after the zygospore is formed. In *Rhizopus,* karyogamy occurs at various times, but meiosis does not take place until the zygospore germinates. Variation in the nuclear cytology of the various species is reviewed by Webster (1980).

PHYSIOLOGY

The Zygomycotina are principally terrestrial, saprophytic organisms that can utilize simple soluble organic molecules such as glucose and other sugars; a few species can decompose starch and cellulose. Thiamine is a requirement for growth for the majority of Zygomycetes, and often additional exogenous vitamins are needed. Certain zygomycetes are parasites of humans, lower animals, mushrooms (Basidiomycotina), vascular plants, and even other zygomycetes.

Much information is available on the physiology of *Rhizopus* spp. and *Mucor* spp. because they are easily grown in the laboratory and have been well studied. In addition, species of the two genera have been used to produce various industrial products such as organic acids (lactic, gluconic, fumaric, and citric), steroid hormones, and ethyl alcohol. In industrial production both the nutrient levels and environmental parameters such as oxygen and carbon dioxide concentration may be manipulated to obtain maximum production of the desired chemical. For example, ethyl alcohol and lactic acid production are greatest at low oxygen concentrations because the glycolytic and fermentative pathways are used. When oxygen is abundant, lactic acid and ethyl alcohol will be oxidized via the oxidative pathway to carbon dioxide and water. Most species that can ferment sugars anaerobically need adequate oxygen for spore development (Cochrane, 1958). Research with various species of *Mucor* and *Rhizopus* indicates that certain species can produce exoenzymes to decompose such widely diverse carbon sources as proteins, fats, fatty acids, pectins, and starch.

The phenomenon of dimorphism can be induced in some Mucorales by altering their environment. The normally filamentous mycelium may be reversibly converted to a single-celled yeastlike phase in response to changing or unfavorable environmental conditions. In 1860 Pasteur first observed that *Mucor rouxii* would change from the mycelial to the yeast form under conditions of low oxygen concentrations. Increased carbon dioxide, nitrogen, and sugar levels will also induce the yeast form in *M. rouxii.* The single-celled yeast form not only has different enzymes and different vitamin requirements but also different proportions of cell wall components (Griffin, 1981). Cell walls in yeasts are thicker, containing up to six times as much mannan (polysaccharides containing mannose) as the cell walls of the mycelial form (Bartnicki-Garcia and Nickerson, 1962). When environmental conditions become favorable again, the fungus will revert to the mycelial form.

Some species of *Mucor* and *Rhizopus* are not only tolerant of high temperatures but also show optimum growth at temperatures up to 46° C (Cochrane, 1958). Other zygomycetes can withstand very cold temperatures (4° C) and can grow on refrigerated meat.

The hormones of the Mucorales have been intensively studied. Both heterothallic and homothallic forms produce diffusible (in air or water) hormone precursors or **prohormones.** In homothallic forms the production of prohormones is restricted to that area on the thallus where the gametangia will be located. In heterothallic forms the + strain produces low levels of the prohormone, which diffuses through the substrate to the − strain. The − strain in turn adds a chemical group to the prohormone, thereby synthesizing the active hormone, trisporic acid (TSA; Fig. 16-3). The − strain also produces low levels of the prohormone, which diffuses to the + strain and is converted, by the addition of a chemical group, to the active hormone, the same trisporic acid. Both strains now contain TSA and this stimulates further prohormone production. TSA levels in each strain and in the surrounding medium increase as the prohormone is converted to active hormone (Bu'Lock, 1976). The hormone concentrations

reach levels that induce both the formation of zygophores and the directional growth of the zygophores toward each other (i.e., **chemotropism**). Gametangia form on the zygophores, which are now close enough for plasmogamy to occur, and zygospores are formed (Fig. 16-3).

This prohormone and trisporic acid system, which effects morphogenesis and directional growth of zygophores, is easily observed under laboratory conditions. At this time, however, no evidence shows that such a system is common under natural conditions (Burnett, 1976). Not only are zygospores rarely found in nature but + and − minus strains of the same species are also not often found. By producing very low levels of prohormones, the fungus may be conserving energy until a compatible partner is nearby (Ross, 1979). When two compatible partners are stimulated to produce TSA and zygophores, energy is directed to sexual reproduction and gamete attraction, thereby enhancing the probability of fertilization. Should the fungus produce high levels of TSA at all times, a considerable amount of energy would be wasted on hormone production if no compatible strains were nearby.

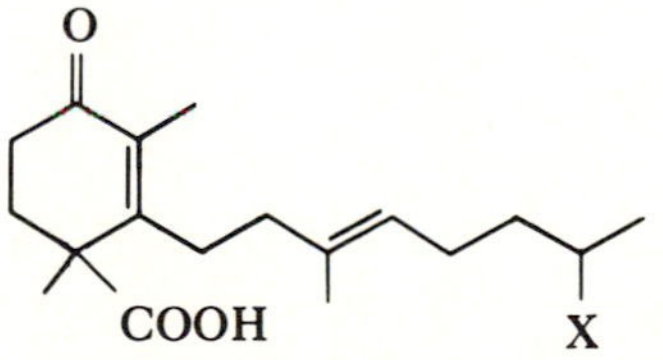

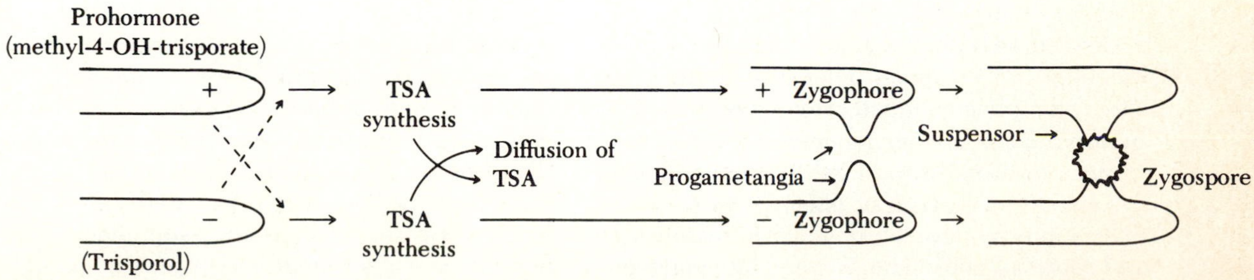

FIG. 16-3 Hormonal Induction of zygospore formation. Structure of trisporic acid (TSA) and method for induction of synthesis of TSA from the prohormone precursors, methyl-4-OH-trisporate in + hyphae, and trisporol in − hyphae. TSA has two effects: (1) induction of zygophore formation, and (2) repression of sporangiophore formation.·

TSA (trisporic acid)

X may be $=O$ or $-OH$

Prohormone (methyl-4-OH-trisporate)

(Trisporol)

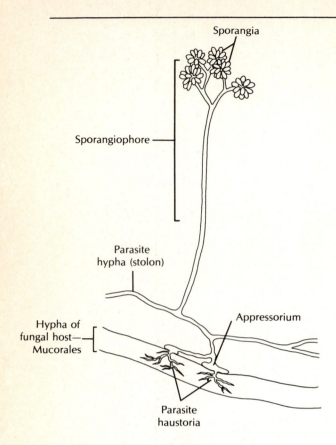

Sporangia

Sporangiophore

Parasite
hypha (stolon)

Appressorium

Hypha of
fungal host—
Mucorales

Parasite
haustoria

FIG. 16-4 Habit of *Piptocephalis* **sporangiophore.**
Piptocephalis is a mycoparasite, usually parasitizing other members of the Mucorales. Spores of *Pipto-cephalis* germinate near hyphae of other fungi and form a broad, flattened hyphal segment called an **appressorium.** The appressorium hugs closely to the host hypha and develops fine haustoria internally, inside the host hypha, and externally as "stolons." The stolons develop other appressoria, sporangiophores for asexual reproduction, or they fuse with receptive hyphae, producing a zygospore. *Piptocephalis* spp. are homothallic, so any stolon may fuse with any other stolon. The sporangiophores produce cylindrical sacs of nonmotile spores. Such an arrangement is called merosporangium and is common among some species of the Mucorales.

Some Mucorales parasitize members of their own kingdom and are called **mycoparasites.** For example, *Piptocephalis* sp. (Fig. 16-4) may parasitize other members of the Mucorales (especially *Mucor*), some ascomycetes, and some deuteromycetes. Members of this genus may be observed growing on *Mucor* following its fruiting phase (i.e., sporangiophore production) on dung or other suitable substrates.

Several members of the Zygomycetes (*Basidiobolus, Phycomyces, Pilobolus, Entomophthora*) have been studied extensively in the laboratory. Their sporangiophores exhibit phototropism (growing toward light); therefore, they must have some type of photoreceptive molecule, either a carotene or a flavoprotein (Page and Curry, 1966). The 5 mm-long sporangiophore of *Phycomyces* responds most actively to light in the blue spectrum, with a maximum response at 430 nm. Bergman et al (1969) have intensively studied the growth of the sporangiophore.

Most members of the Entomophthorales are obligate parasites of insects and other animals, including humans. The members of this order, like most obligate parasites, have complex nutritional requirements. In the laboratory these parasites grow well on meat and fish but poorly on synthetic media, indicating their need for complex organic molecules. Several species also parasitize algae and vascular plants. One species, *Entomophthora coronata* (= *E. empusae*) is a facultative parasite not only growing saprophytically in the soil but also parasitizing aphids and termites. Some Entomophthorales are highly specific as to what host they will parasitize, whereas other species have a broad range of hosts. The hyphae of this order are unusual because they are generally short and somewhat restricted in growth form. A mycotoxin has been isolated from *E. coronata* that causes blood cell damage and eventually death in the insect host (Prasertphon and Tanada, 1961).

Entomophthora muscae (Fig. 16-5) parasitizes houseflies (*Musca domestica*) and, like many fungi, is more prevalent in damp weather. The

FIG. 16-5 *Entomophthora muscae* habit.
A, Dead housefly (*Musca domestica*) adhering to a pane of window glass. Its tissues have been invaded by the fungus *Entomophthora muscae*. Note the mass of conidiophores *(con)* projecting from beneath the exoskeleton, and the circle of discharged conidia around the fly. **B,** Enlargement of the conidiophore area. Note the multinucleate conidiophores *(con)*, and the exoskeleton of the host fly *(ex)*.

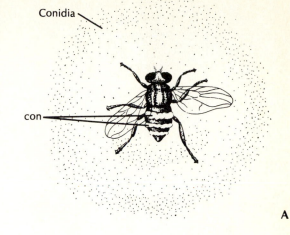

A

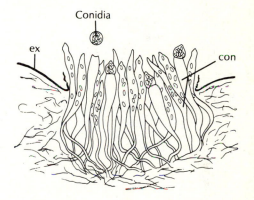

B

infection starts when the fly either ingests the spores or the spores contact and penetrate the exoskeleton. The spores germinate inside the insect, and the mycelium grows throughout the soft body tissues, digesting them. Eventually sporangiophores project out through the exoskeleton. The fly may still be flying about, somewhat erratically, with its internal organs partially digested and riddled with mycelia. When it can no longer fly, it usually lands on a flat surface, often a windowpane, sticks to the surface, and dies. The sporangiophores, now growing through the exoskeleton, forcibly eject their spores, coated with sticky mucus, creating a 2 cm halo of spores about the dead fly (Fig. 16-5). The fly usually dies within a week of infection, and the cycle begins again when a sticky spore contacts another housefly. Spores of the parasitic Entomophthorales may remain viable for up to 2 years. Other Entomophthorales parasitize other insects; for example, *Entomophthora americana* attacks blowflies and *Entomophthora greesa* parasitizes the spruce bud worm. Some Entomophthorales can even be predacious and capture protozoa.

Several species of *Basidiobolus* (Entomophthorales) spend part of their life cycle within the alimentary tracts of amphibians or reptiles. *Basidiobolus ranarum* is a common isolate from frog excrement. In the frog intestine, *B. ranarum* conidia form large, round uninucleate cells that pass out undigested with the feces. The fungal cells germinate and produce a septate hyphal mass (mycelium) that then forms conidiophores (Fig. 16-6). Development of the conidiophores is stimulated by blue light (420 to 470 nm), and the structures are phototrophic. The conidia are ejected forcibly by hydrostatic pressures that build up in the conidiophore vesicle. The discharged conidia are ingested by foraging beetles that in turn are eaten by frogs, thus completing the cycle. Sexual reproduction occurs with zygospore formation by older hyphae that grow on the frog excrement.

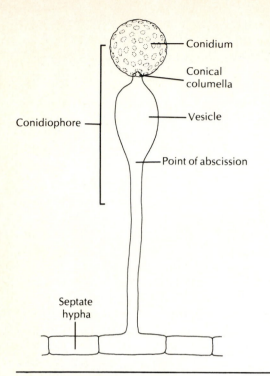

FIG. 16-6 Conidiophore of *Basidiobolus*.
Septate hypha that forms part of a mycelium for *Basidiobolus ranarum*, which grows on dung excreted by frogs. The hypha has produced an upright conidiophore that bears a terminal conidium. The conidia are forcibly ejected when hydrostatic pressure within the swollen vesicle causes the vesicle wall to rupture at the point of abscission. Conidia may be ejected for distances up to 2 cm. Projectiles of this sort resemble miniature rockets.

Members of the Endogonales grow symbiotically with the roots of herbaceous and woody plants in both endotrophic and ectotrophic mycorrhizal associations (Saunders et al, 1976) (see Chapter 13). The sporangia grow underground; the spores germinate close to host roots and hyphae, then penetrate root epidermal cells. In this order the most common type of mycorrhizal association is the endotrophic, vesicular-arbuscular type. The hyphae invade the root, grow between the cells, and then produce branched haustoria that invade the cells (see Fig. 13-2). Species of *Endogone* and *Glomus* have been isolated from *Eucalyptus* trees and poinsettias, but evidently this type of beneficial association is widely distributed, not only on many agricultural plants but also on trees and shrubs in the natural environment. Spores of the Endogonales have been isolated from the guts of rodents, giving a clue to a potential method of spore dispersal (Silver-Dowding, 1955).

Many zygomycetes can alter their internal osmotic pressure in response to varying osmotic pressures in their environment. Such fungi are called osmophilic and can grow in solutions of high osmotic pressure. For example, *Mucor heimalis* can alter its osmotic pressure by transforming the osmotically inactive storage compound glycogen to the osmotically active glucose as the external osmotic pressure rises (Adebayo et al, 1971).

Members of the class Trichomycetes live in the digestive tract of both arthropods and nematodes. However, this relationship does not appear to be parasitic, because the fungi do not penetrate the host tissues. The Trichomycetes obtain their nutrition from the partially digested food in the gut, an apparent instance of **commensalism**. Lichtwardt (1976) discusses the relationship between the fungus and its host.

EVOLUTION

For many years the Zygomycotina as well as the Mastigomycotina were included in the now-defunct group, the Phycomycetes. The Zygomycetes were removed from this group because they lacked the motile characteristic of the other members. The Zygomycetes appear to be a natural grouping of morphologically and functionally similar fungi. Vitamin requirements provide few evolutionary clues because each genus has such a large range of these requirements (Cochrane, 1958). However, biochemical characteristics such as cell wall components and energy storage compounds are conservative features and should give indications of both their ancestry and present-day relationships.

The fossil record reveals few clues to the evolutionary lines of fungi. Fungal fossils, at 600 million years old, are some of the oldest eukaryotic fossils known (Cloud, 1965; Martin, 1968). Fungi are extremely variable morphologically, depending on their environments, and the few extant fossils are difficult to classify. Not only is there no point at which to begin a phylogenetic scheme but there is also a lack of definitive geological record for determining which fungal characteristics have been stable over time.

Fossils of primitive vascular plants, the Psilophytales, from the lower Devonian Rhynie chert of Scotland, contain mycorrhizal-like structures in their rhizomes. These apparent fossil mycorrhizae resemble present-day members of the Endogonales (Webster, 1980). Vascular plant fossils from the Upper Devonian period also contain fossils of what appear to be nonseptate fungi. In addition, mycelia resembling the Mucorales have been found in fossils of vascular plants from the Carboniferous period (Arnold, 1947). Fungi have evidently been growing inside vascular plant tissues for millions of years.

Many investigators think that the Zygomycetes are ancestral to the Ascomycotina. Until additional evidence is discovered, the position of Zygomycetes in fungal evolution will remain controversial.

The relationship of the Trichomycetes to the other Zygomycotina is questionable. Little is known about the evolution and affinities of the Trichomycetes because they have not been extensively researched. Some workers think they are an artificial group and that they contain fungi from several groups.

ECOLOGY

The Zygomycotina play an important role in the ecosystem by reducing organic molecules to inorganic molecules that are utilized by green plants. Most zygomycetes are terrestrial and live saprophytically in soil, on dung, or on decaying organic matter. Generally the zygomycetes cannot utilize the larger, complex organic molecules but instead use the smaller ones produced following the breaking apart of large molecules such as starch and cellulose usually by mastigomycetes, ascomycetes, or basidiomycetes. A few species of zygomycetes parasitize plants, arthropods, other fungi, and humans. The airborne spores of the Mucorales, especially, are ubiquitous and *Rhizopus* and *Mucor* are common contaminants of both bacterial and fungal cultures in the laboratory.

Certain Mucorales can damage stored fruits and vegetables. *Rhizopus stolonifer* (= *nigricans*), for example, can cause rotting of sweet potatoes, strawberries, apples, tomatoes, and bananas. Usually *R. stolonifer* is growing as a saprophyte, because the growing mycelium secretes exoenzymes that first kill the fruit or vegetable cells and then dissolve the cell walls. As the mycelium grows through the killed cells it absorbs the soluble organic molecules for its nutrition. The fungus has killed the cells with its exoenzymes, but no haustoria are formed and nutrients are obtained from dead tissue; therefore, the fungus is technically a saprophyte. This fungus also grows well on bread, where it is known as "black bread mold" because of the production of black spores massed in abundant sporangia.

The Zygomycetes are common saprophytes on animal dung (i.e., **coprophilous** fungi) and interesting studies have been conducted on the succession of species on this substrate. The Zygomycetes are usually the first decomposers to produce sporangia, with the fruiting bodies of *Pilaira* being produced within 2 days and *Pilobolus* sporangia appearing in 4 days. These organisms require simple sugars and use such molecules in the dung. The zygomycetes are usually followed by various Ascomycotina and then by Basidiomycotina (Harper and Webster, 1964). All such fungi probably were present as spores when the dung was deposited, but the succession observed is one reflected by the production of the sporangia (Webster, 1970).

Sporangium discharge in *Pilobolus* (Fig. 16-2) has already been discussed, but the dispersion of the fungus has also been widely studied. The sporangia are forcibly ejected toward the light. If they fall on grass, they may eventually be eaten by grazing animals. In the vertebrate gut, gastric juices induce the release of the spores from the sporangium, and the spores are passed out with the feces. The spores germinate to form mycelia, which subsequently decompose the dung and produce sporangia in about 4 days. Ammonia, a common breakdown product during dung decomposition, stimulates sporangial production in *Pilobolus* (Webster, 1980). Similar studies on succession have been done involving wheat straw compost. On this substrate, thermophilic Mucorales may grow, because high temperatures (up to 70° C) may result from microbial exothermic decomposition. Fungi other than zygomycetes also grow on decomposing compost.

Some spores of the Mucorales can remain dormant for years. The spores of *Mucor mucedo,* for example, are extremely resistant to degradation because they contain **sporopollenin,** a polymer of carotene. Spores of this species, therefore, remain dormant but viable in the soil for a very long time (Gooday et al, 1973).

Other members of the Mucorales are also important saprophytes. *Zygorhynchus* has been isolated from soil at fairly great depths (Hesseltine et al, 1959). *Phycomyces* can decompose fatty materials and may be found growing on oily substances, bread, and dung. *Syzgites* grows on the decomposing fruiting bodies of basidiomycetes.

Another important role played by species of the Mucorales in the ecosystem is that of soil improvement. A common farming practice is that of "green manuring," in which a crop is plowed into the soil to increase the organic content. The organic material, "green manure," provides raw material for the soil fungi, primarily *Rhizopus* spp. and *Mucor* spp. From the raw material, these fungi synthesize complex polysaccharides that cement together inorganic particles in the soil. The polysaccharides, then, aid in the formation of "soil aggregates" that greatly improve the structure of the soil. To synthesize such polysaccharides, the fungi must have the green manure as raw material. Other microorganisms also are involved in soil aggregate formations, but the Mucorales are extremely important in this process (Gray, 1959).

ECONOMIC IMPORTANCE

Certain Zygomycotina have been cultured commercially to produce industrial products such as organic acids, ethyl alcohol, enzymes, and steroids. Other members of this subdivision have also been used to nutritionally enrich foodstuffs (e.g., soybeans). Economic losses may be caused by some species that spoil food or textiles, while still other zygomycetes cause diseases of humans, domestic animals, and plants.

Mucor and *Rhizopus* have been used widely in the laboratory as experimental organisms. Consequently many strains have been developed that engage in various metabolic activities; these strains are used for commercial production of chemicals and foods. In the industrial production of ethyl alcohol, *M. rouxii* may be used to decompose starch to glucose before anaerobic fermentation by ascomycetous yeasts proceeds. Unlike most zygomycetes, *M. rouxii* can synthesize amylases that break down starch to glucose. Yeasts subsequently ferment glucose to ethyl alcohol and carbon dioxide (see Chapter 17). Still other Mucorales can produce ethyl alcohol directly from various substrates.

Other species of *Mucor* and *Rhizopus* are used to produce organic acids—for example, gluconic, lactic, citric, fumaric, oxalic, and succinic acids. Manipulation of both the substrate composition and the environment of these fungi enables industrial growers to achieve maximum production of the desired acid. The soluble calcium salt of gluconic acid is used widely as a nontoxic medium for administering calcium. Calcium gluconate can be injected directly into tissues without adverse effects; it is also used as a method for administering calcium to pregnant women or others in need of extra calcium (Gray, 1959). Another use for gluconic acid is in dishwasher detergents, where it prevents the precipitation of spot-causing calcium and magnesium salts on glassware (Eveleigh, 1981).

Species of *Rhizopus* and *Mucor* are used to hydroxylate available steroid compounds to make them clinically effective. *Rhizopus arrhizus* and *Rhizopus stolonifer*, among others, have been used industrially to transform steroids into physiologically active compounds (Peterson and Murray, 1952; Phaff, 1981). Zygomycetes have also been used for antibiotic production. *Mucor ramanmanus* is used in the production of the antibiotic **fusidic acid,** which is active against gram positive bacteria (Berdy, 1974).

Species of *Rhizopus* and *Mucor* have been used for centuries in the Orient to enrich and upgrade soybean products (Hesseltine and Wang, 1980). The majority of soybean protein is unavailable because it is associated with both fats and a trypsin inhibitor that prevents digestion of the protein. When soybeans are cooked and fermented with *Rhizopus oligosporus*, the resulting product is called "tempeh," and it is more palatable, more digestible, and has more available protein than the original soybeans. Other forms of tempeh may be made from fermented coconuts. Species of *Mucor* are used in China to ferment soybean curd; the resulting food is called "sufu" (Rose, 1981).

Certain zygomycetes such as *M. mucedo* grow on stored meat and improve its quality in a process called "aging" (Gray, 1959).

Incalculable benefits are derived from symbiotic zygomycetes in mycorrhizal association with many commercially important green plants. Such mycorrhizae benefit the plant in various ways, including increased absorptive root surface and increased phosphate uptake (see Chapter 13).

The insect parasites Entomophthorales have been evaluated as potential biological control agents for insect pests. Investigators have not been able as yet to infect enough insects to effectively control pest populations. Recent work, however, indicates that some species of *Entomophthora* may naturally control some insect pest populations such as the clover leaf weevil and the alfalfa weevil under certain environmental conditions (Watson et al, 1981).

On the detrimental economic side, the Zygomycotina cause considerable damage both to stored fruits and vegetables and to bakery products. Destruction of baked goods by fungi (Zygomycotina and Ascomycotina) has necessitated the incorporation of preservatives into bakery products to preserve shelf life. Textiles and paper products may also be damaged by the zygomycetes by their decomposition of fabrics and their spotting of finished papers.

Several zygomycetes may cause both mild and severe diseases in humans. Members of the Mucorales cause diseases known as mucormycoses (Fig. 16-7), which are especially virulent in patients whose immune systems have been suppressed by drugs, as in leukemia, in organ transplants, or in steroid drug therapy. These conditions predispose the patient to fungal diseases. Fungi that have been isolated from lesions in humans, in order of the frequency of their isolation, are *Rhizopus, Mucor,* and *Absidia* (Baker, 1977). Three hundred cases of human mucormycosis have been reported throughout the world. Many of the pathogenic zygomycetes are dimorphic, with the yeastlike stage being parasitic and the mycelial stage saprophytic. Certain species grow well at body temperature (37° C) and are therefore tolerant of relatively high temperatures. These disease-causing species are **opportunistic** fungi that normally live saprophytically but may cause either localized lesions or generalized infections. When systemic infections invade internal organs, the disease may be fatal. *Basidiobolus* (Entomophthorales) has also been implicated in some human mycoses (Bartell et al, 1974). Additionally, Zygomycetes have been implicated in a few diseases of domestic animals and agricul-

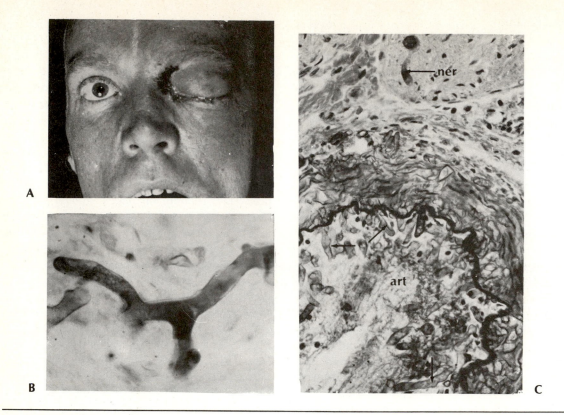

FIG. 16-7 Mucormycosis of the eye.

The patient's eye, **A,** is infected with *Rhizopus* spp. This patient is diabetic and therefore more susceptible to fungal infection than most individuals. **B,** The infectious form of the fungal hypha is coenocytic. **C,** A histological section taken from the eye orbit of this fatal case of mucormycosis. Arrows indicate fungal hyphae growing within an artery *(art)* and a nerve *(ner)*. (Adapted from photographs from Baker, R.D. 1977. Fungal, actinomycetic and algal infections. In Anderson, W.A.D., and J.M. Kissane (eds.): Pathology, Ed. 7. The C.V. Mosby Co., St. Louis. p. 516.)

tural crops. Ajello (1976) discusses other diseases caused by the Zygomycotina.

The economic value of the decomposition of huge amounts of dead organic matter by zygomycetes is inestimable. In this process the zygomycetes do not work alone but are joined by an army of other fungi and bacteria that contain an armory of enzymes. These enzymes can break down those complex organic molecules synthesized in nature to the inorganic molecules that are usable by green plants. The zygomycetes have their place in the sequence of decomposition and recycling. Primarily they break down the small organic molecules such as dissaccharides and monosaccharides to smaller, soluble organic molecules such as citric acids, lactic acid, and ethyl alcohol. They are important not only in dung decomposition but also in the biological breakdown of sewage sludge and in the improvement of soil structure.

SUMMARY

The organisms of subdivision Zygomycotina, division Eumycota, are characterized by a lack of flagellate cells at any time during their life histories. They represent a relatively simple group of fungi that lack active motility. The group has some of the characteristics of Mastigomycotina such as coenocytic hyphae, but they produce mycelia with a much higher degree of cellular differentiation and thus resemble some of the higher fungi such as Ascomycotina. The subdivision contains only two classes, with the Zygomycetes being by far the more important. Included among the Zygomycetes are the universal black molds associated with soils, dung, and spoiled foods. The common bread mold, *Rhizopus*, is a member of this class.

The hyphae of this group develop into distinctive mycelia, although the amount of differentiation and structural integrity of the mycelia does not reach the level of complexity encountered in the higher subdivisions of the Eumycota. The hyphae are coenocytic, and cross walls are usually formed only during development of sporangia or gametangia.

The cell walls of the multinucleate hyphae do not contain cellulose, although they do possess chitin and other complex organic substances. The composition of the cell wall may vary at different times in the life history of the organism. Some zygomycetes exhibit dimorphism and may exist as filamentous forms or yeastlike unicellular forms. The particular form exhibited at any one time appears to be environmentally induced.

The various species of Zygomycotina produce distinctive sexual and asexual structures. Sporangiophores are specialized, asexual, upright hyphae that bear sporangia, the site of asexual sporangiospore formation. Some sporangiophores attain a rather complex organization and are actively involved in spore dispersal (e.g., *Pilobolus*).

Distinctive, thick-walled zygospores are formed following plasmogamy between two sexually compatible mycelia. In some species zygophore growth and development is initiated by a hormone, trisporic acid (TSA). Prohormonal inducers are also known for some other species. Plasmogamy occurs when the cell wall between two adjacent gametangia dissolves. Karyogamy and meiosis take place before the germination of the resting zygospore.

In some species, hyphal growth has been demonstrated to exhibit a positive phototropism.

While many zygomycetes are well represented as saprophytic organisms in soil, on dung, or on spoiling food, they also have significant representatives that grow symbiotically or as parasites on other fungi, higher plants, invertebrates, and vertebrates. *Entomophthora muscae* parasitizes houseflies, while others (e.g., *Mucor*) are responsible for diseases (mucormycoses) of humans and invertebrates.

SELECTED REFERENCES

Benjamin, R.K. 1979. Zygomycetes and their spores. In Kendrick, W.B. (ed.): The whole fungus, pp. 573-616. National Museums of Canada, Ottawa.

Bu'Lock, J.D. 1976. Hormones in fungi. In Smith, J.E., and D.R. Berry, (eds.): The filamentous fungi. Vol. 2. Biosynthesis and metabolism, pp. 345-368. John Wiley & Sons, Inc., New York.

Cole, G.T., and R.A. Samson. 1979. Patterns of development in conidial fungi. Pitman Publishing, Ltd., London, England. 190 pp.

Griffin, D.H. 1981. Fungal physiology. John Wiley & Sons, Inc. New York. 383 pp.

Hesseltine, C.W., and J.J. Ellis, 1973. Mucorales. In Ainsworth, G.C., F.K. Sparrow, and A.F. Sussman, (eds.): The fungi. Vol. IVB: A taxonomic review with keys: Basidiomycetes and lower fungi. pp. 187-217. Academic Press, Inc., New York.

Lockwood, L.B. 1975. Organic acid production. In Smith, J.E., and D.R. Berry (eds.): The filamentous fungi. Vol. I. Industrial mycology, pp. 140-157. Arnold, Great Britain; John Wiley & Sons, Inc., New York.

Waterhouse, G.M. 1973b. Entomophthorales. In Ainsworth, G.C., F.K. Sparrow, and A.S. Sussman (eds.): The fungi. Vol. IVB: A taxonomic review with keys: Basidiomycetes and lower fungi. pp. 219-229. Academic Press, Inc., New York.

17

CLASSIFICATION

Division: Eumycota: "true fungi"; assimilative nutrition

    Subdivision 3: Ascomycotina: "sac fungi"; presence of a saclike ascus containing ascospores; partially septate hyphae; zygotic meiosis

        Class 1: Hemiascomycetes: no ascocarps; often yeastlike unicells

            ORDER 1. Endomycetales: zygote transformed directly into an ascus; the yeasts; *Ashbya, Dipodascus, Eremascus, Saccharomyces, Schizosaccharomyces*

            ORDER 2. Taphrinales: plant parasites causing galls, blisters, leaf curl, and growth deformation such as "witches broom"; *Taphrina*

        Class 2: Plectomycetes: closed ascocarps (cleistothecia) containing ephemeral asci

            ORDER 1. Onygenales: stalked ascocarps; important animal parasites causing skin disorders (ringworm, athlete's foot, jock itch, etc.) and deep mycoses; *Ajellomyces* (blastomycosis), *Arthroderma* (= *Trichophyton*) causing dermatitis, *Emmonsiella* (histoplasmosis), *Nannizzia* (= *Microsporum*) causing dermatitis, *Onygena* (diseases of hoofs, claws, and keratin-containing structures)

            ORDER 2. Eurotiales: microscopic cleistothecia; some with imperfect life cycles: *Aspergillus* ("black molds"), *Penicillium* ("blue" or "green molds")

            ORDER 3. Microascales: a few important plant parasites with specialized asexual reproductive mechanisms; *Ceratocystis* (Dutch elm disease)

        Class 3: Pyrenomycetes (Hymenoascomycetes): closed ascocarps (cleistothecia) or open ascocarps (perithecia); ascospores without septa

            ORDER 1. Erysiphales: cleistothecial ascocarps; perfect life cycles; plant parasites causing "powdery mildews"; *Erysiphe* (grass powdery mildew), *Microsphaera* (on lilacs), *Sphaerotheca* (rose powdery mildew), *Uncinula* (on grapes)

            ORDER 2. Xylariales: perithecial ascocarps; mostly saprophytic; *Chaetomium, Neurospora* (red bread mold or bakery mold), *Oceanitis, Sordaria* (a common dung fungus)

            ORDER 3. Clavicipitales: perithecial ascocarps produced in a mat of mycelia (stroma); threadlike ascospores; important grain parasites and drug (ergot) producers; *Claviceps purpurea*

            ORDER 4. Diaporthales: perithecia with spores forcibly expelled; serious garden plant and hardwood tree parasites. *Diaporthe citri* (citrus melanose); *Endothia parasitica* (American chestnut blight); *Gibberella* (diseases of corn, coffee, and rice)

        Class 4: Discomycetes (Hymenoascomycetes): wide open, disklike ascocarps (apothecia)

            ORDER 1. Helotiales: minute to moderately sized cup-shaped apothecia; spores forcibly discharged; plant parasites or saprobes; *Monilinia* (brown rot of stone fruits such as peaches or plums), *Naemacyclus* (diseases of pine trees), *Sclerotina* (garden vegetable and flower diseases)

ORDER 2. Phacidiales: dark colored apothecia developing within a stroma; spores forcibly discharged; mostly plant parasites; *Rhytisma* (tar-spot of maple trees)

ORDER 3. Pezizales: large, open, colorful apothecia; mostly saprophytes on dead wood and leaves; *Helvella* (saddle-shaped apothecia), *Morchella* (morels), *Peziza* (with white to brown large apothecia), *Scutellinia* (with large red apothecia)

ORDER 4. Tuberales: subterranean fungi with large round or oval closed ascocarps; *Tuber* (truffles)

Class 5: Laboulbeniomycetes: obligate parasites of arthropods, mostly insects; some parasites of algae

ORDER 1. Laboulbeniales: arthropod parasites

ORDER 2. Spathulosporales: algal parasites

Class 6: Loculoascomycetes: ascus divided into two halves at maturity; asci borne in locules located in a mycelial mat or stroma; ascocarps with prominent paraphyses (sterile hairs); mostly saprobes on wood, but some important plant parasites; *Venturia* (apple and pear scab)

The fungi of subdivision Ascomycotina are commonly referred to as the ascomycetes, or "sac-fungi," since they all produce a saclike structure called an ascus (pl. asci) at some point in their life histories. Asci contain a small, finite number of spores (ascospores) produced by meiosis and subsequent mitoses. The number of ascospores per ascus is usually a multiple of four—most frequently eight. Ascospores are often arranged in a linear fashion within the ascus, as in *Neurospora* (Fig. 17-1). Other ascomycetes, however, may produce other spore multiples arranged in various patterns.

The Ascomycotina, together with the Basidiomycotina and Deuteromycotina, are often called the **higher fungi,** since they are the most structurally and biochemically complex of the kingdom. Most ascomycetes are unicells or filaments that require the use of a microscope for positive identification. However, some are macroscopic and are identifiable from observation of their gross structure and color as exhibited by their distinctive **ascocarps** or mycelia.

Ascomycetes have haplontic life cycles that vary in some respects from the typical haplontic cycle presented in Chapter 1. The major difference is the nature of the individual cells: ascomycetes possess monokaryotic and dikaryotic cells in their hyphae during their "haploid" phase.

Most of the blue-, green-, red-, or rust-colored molds associated with food spoilage are ascomycetes. This includes the salmon-colored bread mold *Neurospora,* which has become an important tool in the study of genetics. Also included are the powdery mildews, which are most often found growing on harvested fruits and vegetables and which are responsible for large economic losses in the food industry. Ascomycetes also include species that cause significant plant diseases such as the American chestnut blight *(Endothia parasitica),* Dutch elm disease *(Ceratocystis ulmi),* citrus melanose *(Diaporthe citri),* and many more. Human and animal diseases are also caused by ascomycetes.

FIG. 17-1 *Neurospora* **ascus.**
Ascus from genus *Neurospora.* Contained within ascus are eight ascospores arranged in a linear fashion so that spores emerge from ascus one spore at a time through opening at its upper end.

Many cause fungal infections called **mycoses** (sing. mycosis). Both deep internal mycoses (e.g., lung infections) and superficial mycoses (e.g., skin diseases), which are common among humans throughout the world, are a major concern of the World Health Organization. *Penicillium* is a genus of ascomycete that synthesizes the antibiotic penicillin.

Other important ascomycetes include the **yeasts** *(Saccharomyces),* which are the major fermenters in the brewing industry and carbon dioxide producers in the baking industry. Gourmet cooks go to great lengths to procure ascocarps of truffles *(Tuber* spp.) or morels *(Morchella)* to serve with very special meals.

CELL STRUCTURE

Ascomycetes are primarily composed of branching hyphae that, unlike the fungi in the previous chapters, contain cross walls or septa. The septa generally possess pores through which the cytoplasm may stream, carrying nuclei and cytoplasmic organelles from one cell to another. In the more complex ascomycetes, hyphae develop into a three-dimensional, complicated, intricate mycelium, as in *Peziza* and *Morchella* (Figs. 17-8 and 17-21). The simpler ascomycetes never progress beyond the single-celled stage, as in most yeasts. Septa of individual hyphae divide them into discrete cells that contain one or two nuclei. If there is only one nucleus per cell, it is called a monokaryotic hypha; if there are two nuclei per cell, it is dikaryotic. Multinucleate cells have been observed in young hyphae.

As previously stated, hyphal septa of ascomycetes contain pores that provide a passage for cytoplasmic contents from one cell to another.

These pores are lined with a continuous plasma membrane, and they appear to become clogged with membranous materials as the cells age. Special organelles called woronin bodies, have been observed in electron micrographs near the septal pores. These are round, crystalline, electron-dense bodies that may play a role in formation of material used in plugging the pores (Camp, 1977).

The cells also contain the usual complement of organelles associated with eukaryotic organisms, including mitochondria, smooth and rough ER, microtubules, and ribosomes. In some cells, vacuoles and food storage vesicles are prominent. Golgi bodies have not been reported.

Just outside the nuclear membrane is an electron-dense structure called the spindle pole body (SPB). The SPB divides in two when mitosis occurs and migrates to opposite sides of the outer nuclear membrane. Its behavior is reminiscent of centrioles, and it is believed that the SPBs are derivatives of centrioles. It has not been definitely established whether SPBs have a role in the production of the intranuclear spindle apparatus that appears during mitosis or meiosis.

The hyphae of ascomycetes are enclosed within a cell wall composed primarily of chitin and insoluble glucans (polysaccharides containing glucose). Peptides, proteins, and lipids also constitute a significant part of the cell walls (Cole, 1981). The protein complement may be structural or enzymatic. The external surface of the cell walls possesses enzymes that play an important role in the biochemical activity of individual organisms, including transport of biochemical substances across the fungal membranes.

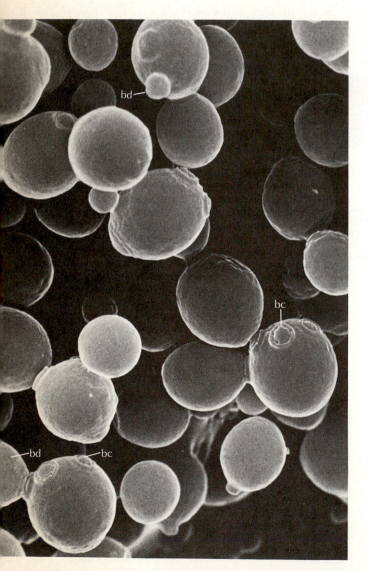

FIG. 17-2 Yeast cells.
Scanning electron microscope view of *Saccharomyces cerevisiae*, the organism commonly known as bakers' or brewer's yeast. Note budding scars *(bc)* on some of the cells, as well as smaller cells or buds associated with some cells *(bd)* (10,000×). (Courtesy Dr. Alastair T. Pringle, Department of Microbiology, University of California at Los Angeles.)

REPRODUCTION AND GROWTH

Ascomycetous fungi have well-developed methods of both sexual and asexual reproduction. Asexually, most of these organisms have the capability of reproducing by fragmentation, mitotic cell divisions, budding, and production of different types of spores, including chlamydospores and conidia. Fragmentation involves the mechanical breaking of individual hyphae with release of pieces of hypha capable of regenerating new fungal organisms. **Budding** is a special type of cell division common to yeast cells (Fig. 17-2). It involves the asymmetrical partition of the cytoplasm of a single cell, forming "buds" or cells of unequal size. Nuclear division (mitosis) occurs within the larger parent cell, and a single nucleus migrates into the bud along with other cytoplasmic organelles. The entire mitotic process is intranuclear—the nuclear membrane constricts during telophase, thus forming two separate nuclei, each of which contains the same chromosome number. (A complete description of the process is given by Fuller, 1976.) One may observe the budding operation with the light microscope by using cultures of brewer's or bakers' yeast *(Saccharomyces cerevisae)*. Not all yeast cells reproduce by budding, however; it should be noted that the "fission yeasts" (e.g., *Schizosaccharomyces*) do not "bud" but divide by normal fungal mitosis and cytokinesis, producing two daughter cells of equal size.

Spore production in ascomycetes usually involves the formation of conidia, which are asexually produced spores borne on a wide variety of conidiophore types. The structure of the conidiophore and the shape and surface sculpturing of individual conidia provide significant characteristics for the identification of individual species.

Hyphae of individual ascomycetes often develop special structures called **sporocarps** in which conidiophore formation takes place (Fig. 17-3). The two most common sporocarps are the **pycnidium** (pl. pycnidia) and the **acervulus** (pl. acervuli). Pycnidia (fr. Greek *pyknon* = concentrated) are pear-shaped or flask-shaped hol-

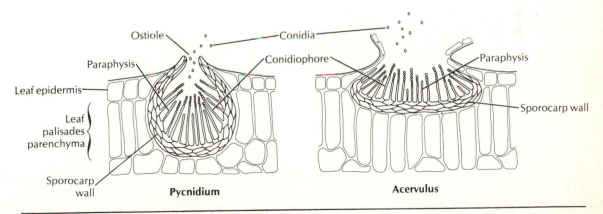

FIG. 17-3 Sporocarps (asexual reproductive structures). Two types of sporocarp that are common among the ascomycetes and deuteromycetes. Pycnidia are flask-shaped structures that emerge through the epidermis of the host leaf and liberate conidia to the environment. Acervuli are flattened structures with a similar function. Note sterile hairlike structures (paraphyses) that develop among the conidia.

low forms that are lined with conidiophores bearing conidia. Scattered sterile hairs called paraphyses also occur within pycnidia.

Acervuli (fr. Latin *acervus* = heap) are flattened groups of conidiophores bearing conidia that develop just beneath the epidermal layer of the host's leaf or stem (Fig. 17-3). At maturity the conidia erupt through the epidermal host tissues and are subsequently dispersed by various means. Brown, red, or yellow spots on the leaves, stems, and other surfaces of higher plants are often outward manifestations of a developing acervulus. Pycnidia, acervuli, and other asexual **fructifications** are common among the ascomycetes and the deuteromycetes.

Sexual reproduction among the ascomycetes involves a complicated, difficult-to-observe series of cytological events. The pattern of the events, however, is always similar; a generalized life cycle is represented in Fig. 17-4. A

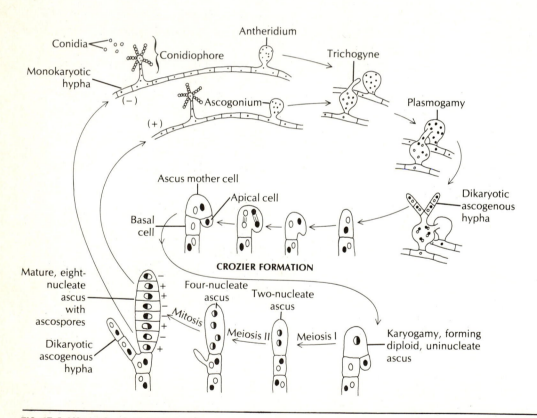

FIG. 17-4 Life cycle of an ascomycete.

Life history pattern followed by most members of Ascomycotina. Note that the cycle is haplontic, with the only diploid phase occurring during karogamy when the uninucleate ascus is formed. Dikaryotic cells also occur after fusion of compatible hyphae. During ascus formation the single diploid nucleus in the ascus cell divides by meiosis, producing a two-nucleate and then four-nucleate ascus. In some ascomycetes the four nuclei divide again, this time by mitosis, producing the classic eight-nucleate ascus. These nuclei then become separated by cell walls, producing eight ascospores.

Asexual reproduction is accomplished through production of conidia borne on conidiophores produced by vegetative hyphae.

major exception to this type of life cycle is found in the unicellular yeasts (Fig. 17-5). In both instances, however, the end result of the sexual process is always the formation of the saclike ascus that contains a small but definite number (usually eight) of haploid ascospores. Asci, with their contained ascospores, are easily observed with the light microscope. The presence of the ascus is the most prominent unifying feature of subdivision Ascomycotina and provides the reason for its name, the sac fungi.

In some ascomycetes, the ascus is only rarely produced, which means that reproduction is mainly accomplished through asexual methods via the formation of conidia. Those fungi that reproduce sexually are referred to as "perfect," and the stages and structures associated with sexual reproduction are called the **perfect stages.** The asexual reproductive phases are "imperfect" and are referred to as **imperfect stages.** The fungi of one subdivision, Deu-

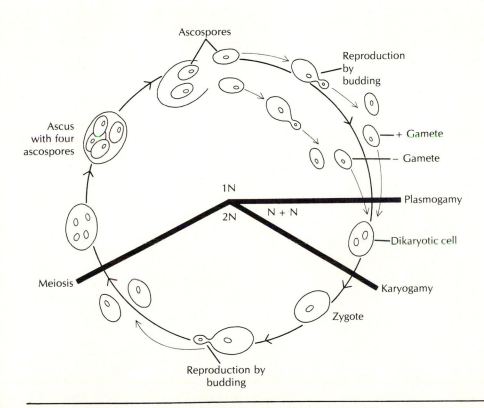

FIG. 17-5 Diplohaplontic life cycle of the yeast *Saccharomyces cerevisiae.*

The life cycle of unicellular yeasts differs from other ascomycetes in that it lacks the usual crozier formation, production of asexual conidia, and production of separate antheridium and ascogonium. Furthermore, other ascomycetes do not reproduce by budding. Yeasts themselves produce variations on the above cycle for *S. cervesiae.* In *Saccharomyces ludwigii,* ascospores fuse within the ascus, forming diploid ascospores. In *Schizosaccharomyces octosporus,* the haploid ascospores divide mitotically while they are still within the ascus, producing an eight-celled ascus. Still other yeasts carry out karyogamy and ascus formation immediately following plasmogamy, thus producing a haplontic life cycle (e.g., *Saccharomyces elegans*).

teromycotina, do not produce sexual stages and are therefore known only by their imperfect stages (and are called the "imperfect fungi").

The cytological events leading up to the formation of the ascus are intricate, and some variation exists among individual species. Usually, however, the sequence of events follows the pattern diagrammed in Fig. 17-4. Haploid ascospores germinate in an appropriate environment and develop into hyphae containing monokaryotic (single-nucleus) cells. Because of

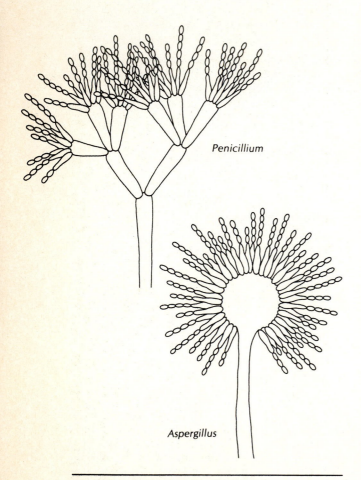

Penicillium

Aspergillus

FIG. 17-6 Conidiophores bearing conidia.
Two types of conidiophores: one from a *Penicillium* sp. and one from an *Aspergillus* sp. Each conidiophore bears numerous conidia arranged in a linear fashion.

hyphal fusion and nuclear migration, some of these cells may become dikaryotic (containing two nuclei) or even multinucleate. Mutation and **diploidization** may complicate the situation (see discussion on parasexuality, pp. 242-243), producing heterokaryotic hyphae. Mitotic crossing over and haploidization of diploid nuclei may also occur. It is best to temporarily disregard the effects of heterokaryosis and parasexuality and concentrate on a "normal" series of events.

The monokaryotic hyphae reproduce asexually through the reproduction of conidiophores bearing conidia (Fig. 17-6). At certain times chemical or physical signals cause these same hyphae to form sexual structures called **ascogonia** if they are female in nature and antheridia if they are male in nature. Both structures are multinucleate and contain many haploid nuclei. The ascogonium, in response to a nearby antheridium, produces a long, tubelike projection, the trichogyne, which grows toward the antheridium. The structure of the ascogonium and antheridium may be otherwise identical. Instead of producing antheridia, some ascomycetes produce pycnidium-like structures that are called **spermogonia** and that in turn produce small uninucleate nonmotile cells called **spermatia.** Spermatia are transported to **receptive trichogynes** (special hyphae) by insect, wind, or water vectors. Spermogonium production, common among the parasitic fungi in both the ascomycetes and the basidiomycetes, is illustrated in *Puccinia* (see Fig. 18-8), a basidiomycete.

The tubular trichogyne fuses with the antheridium or with spermatia that settle upon the trichogyne. The nuclei from the antheridium or spermatium and those from the ascogonium mix in the common cytoplasm, forming pairs of nuclei that eventually migrate into the ascogonium. Shortly thereafter, in most ascomycetes, ascogenous hyphae develop from the wall of the ascogonium, and the paired nuclei migrate into the ascogenous hyphae, where dikaryotic cells are produced as the hyphae become septate. Simultaneous mitotic divisions of the apical cell nuclei ensure that each additional cell will contain a pair of nuclei, one from the female (−) hypha and one from the male (+) hypha.

After only a few cell divisions the apical cell of the dikaryotic hypha divides in a characteristic manner, which leads to ascus formation. The lead portion of the apical cell folds over, forming a hook that resembles a shepherd's crook and is therefore called a crozier (fr. Germanic *crosse* = crutch or crook) (Fig. 17-4). During crozier formation, the two leading pairs of haploid nuclei divide in such a way that their intranuclear spindles are parallel with each other. Two septa subsequently divide the crozier into three cells—a binucleate cell known as the **ascus mother cell,** an apical uninucleate cell, and a basal uninucleate cell. Karyogamy (nuclear fusion) takes place in the ascus mother cell, forming the only sexually produced diploid cell in the life history of ascomycetes. Meiosis soon follows karyogamy, with four haploid nuclei produced within the young ascus. Usually one mitotic division follows, resulting in an eight-nucleate ascus. Each nucleus then becomes surrounded by cytoplasm during cleavage and develops into an ascospore.

The two uninucleate haploid cells developed during ascus mother cell formation often fuse while meiosis is taking place in the ascus mother cell. The dikaryotic cell thus fashioned may continue to grow, forming another dikaryotic hypha, which repeats crozier formation. This process may continue, producing many asci in any given location.

In *Neurospora crassa* (Fig. 17-1) and other species within the genus, the eight ascospores are arranged in a linear fashion. One spore at a time emerges from the apical pore of the ascus. Such spores can be collected and cultured separately so that individual phenotypic and genotypic traits of each spore can be determined, thus providing a major tool for the analysis of meiosis through genetic studies. This technique was first developed by Shear and Dodge (1927). Since that time, ascomycetes have been used by geneticists to unravel the perplexities of genetic recombination and gene segregation during the first and second meiotic cell divisions.

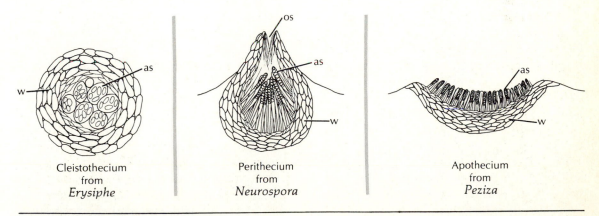

Cleistothecium from *Erysiphe*

Perithecium from *Neurospora*

Apothecium from *Peziza*

FIG. 17-7 Ascocarps of Ascomycetes.
The three types of ascocarps found within the Ascomycetes. Notice that each ascocarp contains a thick wall of fungal hyphae (*w*), and numerous asci (*as*). Each ascus contains a small number of ascospores, usually eight. Between the asci, numerous sterile, hairlike paraphyses develop. Cleistothecia possess asci that are entirely enclosed with thick hyphal walls. Perithecia are open to the atmosphere through an opening called the ostiole (*os*). Apothecia are saucer-shaped disks that produce asci and paraphyses on their open surface.

Multiple asci often develop within definite hyphal layers called **hymenia** (sing. **hymenium**). Hymenial layers containing many asci are located within definite fruiting bodies called ascocarps (fr. Greek *askos* = sac + *karpos* = fruit). The architecture of individual ascocarps may provide structures that are sufficiently distinct to be used in fungal identification. Fig. 17-7 illustrates three common types of ascocarps: (1) a **cleistothecium,** which is completely enclosed within a special wall of fungal hypha (e.g., *Erysiphe*); (2) a **perithecium,** which is pear-shaped, hollow, and possesses an opening at one end (e.g., *Neurospora, Claviceps*); and (3) an **apothecium,** which is a broad, open structure lacking a cover (e.g., *Peziza, Morchella*). Apothecia structurally resemble saucers or disks and form a distinctive feature of the class Discomycetes.

The more primitive ascomycetes do not produce asci within ascocarps; their asci are formed free of any type of fruiting body. The yeasts are good examples of this naked type of ascus formation.

PHYSIOLOGY

The Ascomycotina can utilize a wide range of carbohydrates, from hexose sugars to starch and cellulose. Although some members of this subdivision are important organisms in the industrial production of antibiotics and biochemicals, many others are highly destructive plant and animal parasites. Most ascomycetes are terrestrial, although some are found in both salt and fresh water.

The physiology of the Endomycetes or yeasts will be discussed first, because many experimental studies have been conducted with them; consequently a large body of knowledge exists concerning yeast biochemistry, genetics, cell growth, and reproduction.

Most yeasts are saprophytic, although a few are parasitic. Their cell walls contain mannans and proteins. Yeasts have an extremely wide variety of metabolic patterns. Many yeasts metabolize hexose sugars via glycolysis to pyruvate and then anaerobically, via fermentation, to lactic acid or ethyl alcohol and carbon dioxide. This fermentative pathway can proceed until lactic acid or ethyl alcohol accumulates to toxic levels. If oxygen is then added to a culture, the yeast will employ the **oxidative pathway** and oxidize the alcohol or lactic acid to carbon dioxide and water. Evidently alcohol is produced by yeast even when oxygen is present, but the alcohol is oxidized immediately. Lowered oxygen levels favor alcohol production.

Saccharomyces cerevisiae (Figs. 17-2 and 17-5), the "brewer's" and "bakers'" yeast, and *Saccharamyces ellipsoideus*, the wine yeast, can both live anaerobically, breaking down sugars incompletely. Various strains of *S. cerevisiae* are used in beermaking, where alcohol is the desired product. Other strains are used in baking, where carbon dioxide is the desired product, because the gas causes the bread to "rise." The alcohol produced during the "rising" (i.e., fermentation) of bread dough is vaporized during the high temperatures of baking.

The production of yeast for sale as bakers' yeast is accomplished by heavily aerating the culture. This process encourages the oxidative pathway and the growth of yeast biomass and discourages alcohol production via the fermentative pathway.

Yeasts cannot ferment sugars anaerobically for an indefinite period, because they need adequate oxygen at some stage of their life cycle, usually during sexual reproduction. More energy in the form of adenosine triphosphate (ATP) is available to the organism via the oxidative pathways. However, many yeasts can switch readily from oxidative metabolism to anaerobic pathways when oxygen becomes limiting. Domesticated strains of the Endomycetales require many vitamins for growth, but they are also able to synthesize many vitamins. Vitamin requirements vary from strain to strain, but most yeasts require biotin. Very few, if any, require riboflavin. Free-living species appear to be able to synthesize the essential vitamins. The requirements for vitamins may be "absolute"—that is, the organism will not grow without the vitamin—or "relative"—the organism grows very slowly without it. Yeasts with absolute growth requirements for a certain vitamin are used as assay organisms to determine if complex substrates contain the required vitamin.

In addition to vitamins, the yeasts also need a carbon source (usually a hexose sugar), organic or inorganic nitrogen, and trace elements (Phaff et al, 1966). All these nutritional requirements can be supplied by malt extract, which is produced from sprouted barley. Malt extract is used as the substrate for the initiation of fermentation in beermaking.

Most yeasts will grow at a temperature of 20° to 25° C, but free-living yeasts in the arctic grow best at 15° C. Parasitic yeasts in warm-blooded animals have an optimum growth temperature of 30° to 35° C. The yeasts obviously have adapted well to growing in each of these specialized habitats.

Certain yeasts are osmophilic; that is, they thrive on high osmotic pressures. These yeasts, such as *Saccharomyces mellis* and *Saccharomyces rouxii*, grow well in media high in sugar such as honey, jellies, and jams. High sugar concentrations, however, are not required, since they can adapt to lower concentrations (facultative osmophiles). *Eremascus albus* is an obligate osmophile that requires at least 50% sugar in the substrate for growth. This organism may spoil foods with high sugar contents. The marine yeasts, also osmophiles, can withstand the sodium chloride levels found in sea water (35° $^o/_{oo}$), but they do not appear to have an absolute requirement for high salt concentrations. Their distribution in the marine environment does not seem to be limited by salinity (Kohlmeyer and Kohlmeyer, 1979).

In the filamentous ascomycetes, considerable research has been done with the form genera *Aspergillus* and *Penicillium*. These are referred to as **form genera** because sexual (perfect) stages in many species have not been observed; classification, therefore, is based on their asex-

ual stages seen in conidial development (Fig. 17-6).

One of the reasons that "perfect" or sexual stages of some ascomycetes have not been found is that many fungal thalli must reach a certain chronological or physiological age (i.e., competence) before they will develop sexual reproductive structures. Once that certain age is reached, some form of chemical or environmental induction must take place before growth of the sexual structures is initiated.

A great deal of information is available regarding the physiology of the filamentous Ascomycotina. Many species can be considered "domesticated organisms," because selected strains of *Aspergillus* and *Penicillium* are used in the commercial production of organic acids, antibiotics, and other economically important materials. They are able to produce a wide variety of enzymes and can use many different substrates. Both genera also occur very frequently in the environment.

Domesticated strains of *Aspergillus niger* metabolize sugars aerobically to citric, oxalic, gluconic, or other organic acids. This ability is used commercially to produce large amounts of citric acid from sucrose, for example. Citric acid is frequently used in soft drinks. Some strains of *Aspergillis* spp. are used in the production of enzymes (e.g., proteases and amylases) for commercial or scientific use, while other strains produce substances with antibiotic activity.

Some species of *Aspergillus* can utilize substrates of leather and natural fabrics, causing degradation problems with these materials. *Aspergillus* spp. also attack fruits, cheese, and bakery products. *Aspergillus flavus*, a common inhabitant of soil, may grow on stored foods such as peanuts and rice and produce a powerful carcinogenic toxin. Other ascomycetes such as *Chaetomium* can decompose cellulose and may utilize this substrate in moist environments (Phaff, 1981).

Penicillium is as common in nature as *Aspergillus* (see Fig. 17-6). Wild strains (blue or green molds) not only degrade citrus and other fruits but also decompose leather and cellulose-based fabrics. *Penicillium* may also be used in the commercial production of organic acids and cheeses. The genus is best known, however, for the production of the antibiotic penicillin, a secondary metabolite.

A large amount of experimental work with *Neurospora* has provided much information about the reproduction and metabolism of this genus. *Neurospora* is easily grown and manipulated in the laboratory and is an important tool in genetic research (Dodge, 1927, 1935; Moreau-Froment, 1956). The laboratory strain is haploid and the spores are arranged linearly in the ascus (Fig. 17-1). It is possible, therefore, to dissect out individual spores, culture them, and determine instances of recombination, crossing over, and other processes and to ascertain the assortment of alleles during the meiotic divisions. The work of Beadle (1959) and Tatum (1959) with *Neurospora* led to their "one-gene, one-enzyme" hypothesis of gene regulation and activity, for which they won the Nobel Prize in 1958. The hypothesis has since been modified as more information has become available. The related genus *Sordaria* has also been used frequently in genetic research. In nature *Neurospora* and *Sordaria* occur as saprobes on dung or decaying plant material. *Neurospora*, called the pink, salmon, or red bread mold, frequently attacks baked products.

EVOLUTION

The evolution of the ascomycetes is neither clear nor without controversy. As with other fungi, there are few extant fossils providing clues to their evolutionary ascent. Current hypotheses must therefore be based on studies of existing species and attempts to ascertain biochemical, physiological, and morphological similarities among these species. Trying to understand evolutionary processes by looking at extant species is analagous to trying to accurately draw a large tree while viewing only tips of its branches. Most of the discussion of fungal evolution is based on the view of tips of the branches represented by extant organisms and therefore is subject to much conjecture and disagreement.

Two main theories exist to explain ascomycete evolution. The first postulates that these fungi arose from the Zygomycotina and assumes that the first ascomycetes were filamentous at some stage in their life cycle (Moore, 1971). The evidence cited for a relationship between the Zygomycotina and the Hemiascomycetes is the dimorphism observed in some zygomycetes (e.g., *Mucor*) and the similarity between zygospore formation and ascus formation in some Hemiascomycetes (Ross, 1979). The Taphrinales have been proposed as the most ancient order, because they not only have a yeastlike thallus and naked asci but also a short dikaryotic mycelial stage (Mix, 1949; Raper, 1968). The unicellular yeasts and the complex ascocarp-producing mycelial forms are considered derived forms that may have diverged from a common ancestor (Bartnicki-Garcia, 1970). Some researchers think that the Taphrinales may be a transitional form between the lower and more advanced ascomycetes. Other investigators, however, propose the filamentous Endomycetales *Eremascus* and *Dipodascus* as transitional genera between the primitive ascomycetes (filamentous with naked asci) and the advanced groups with well-developed dikaryotic hyphae and complex ascocarps.

A second theory holds that the ascomycetes evolved from parasitic red algae (Rhodophyta). This theory was developed by Kohlmeyer (1973) as a result of studies of marine fungi and by Demoulin (1974). Parasitic members of the red algal class Floridiophyceae bear considerable resemblence to the marine, parasitic ascomycetes in both structure and reproductive mechanisms. The discovery of cellulose in the cell walls of some ascomycetes (Jewell, 1974) and the presence of esters of choline sulfate in some higher marine fungi (Catalfomo et al, 1972-73) and red algae lends biochemical evidence in favor of this theory. Morphological evidence is found in the complete lack of flagellate cells in both the red algae and the ascomycetes.

According to the second theory, the ascomycete orders of obligate parasites, the Laboulbeniales and Spathulosporales, are the most primitive forms, with all other orders derived from them. Kohlmeyer and Kohlmeyer (1979) have summarized the theory: the two obligate parasitic orders are closely related to a proposed marine ancestor that closely resembled parasitic red algae. Identification of that proposed marine ancestor of the ascomycetes has not been established.

Placement of the various orders of the ascomycetes on a theoretical evolutionary tree depends entirely upon which theory is followed. The listing of orders given under the classification of the ascomycetes (pp. 323-324) follows the first theory more closely than the second.

Fortunately little controversy exists over the direction in which advanced ascomycetes have evolved. The similarity between the dikaryotic hyphae and clamp connections in the Basidiomycotina and crozier formation in the Ascomycotina seems to indicate a close relationship between the two. Most mycologists agree that the basidiomycetes are derived from the ascomycetes.

ECOLOGY

Ascomycotina occur in a great variety of habitats; they may be found growing saprophytically in soil, in fresh or salt water, on plants, and on all forms of dead or decaying organic material. They also grow on and decompose many substances used and valued by humans (e.g., leather, cellulose-containing fabrics, petroleum products, and even synthetic materials such as plastics and polyesters). Members of this subdivision also may grow on molluscs in the marine environment or may parasitize insects. As with most fungi, the most important role of the ascomycetes in the ecosystem is that of decomposing complex organic molecules and releasing the inorganic components, which can then be reused by the green plants.

The yeasts are abundant in fresh and salt water, in soils, on the surface of fruits, and in exudates from injured plants. Most yeasts are saprophytic but a few are parasitic. The yeasts serve as important food sources for protozoans, insects, and other invertebrates.

Plants may have a variety of yeasts growing on them. Some leaves exude sugary liquids that serve as food sources for the yeasts; other yeasts may grow in the syrupy nectar of flowers. The sap flowing from tree wounds usually become infested with various bacteria and yeasts (e.g., *Dipodascus*), giving the sap a slimy consistency—thus the designation slime flux. Insects are often the vectors that carry yeasts to uninfected plants.

Yeasts may parasitize agricultural crops such as cotton, lima beans, coffee beans, nuts, and citrus trees, especially in the warm climates of the subtropics and tropics. Yeasts also constitute part of the normal intestinal flora of warm-blooded animals. These thermophilic fungi may be obligate or facultative parasites, but they seldom cause disease, since they live on the partially digested food in the intestine. Several yeast species may, however, become pathogenic in humans or domestic animals. In addition, pathogenic yeasts have been isolated from two arthropods: the crustacean *Daphnia* and the brine shrimp *Artemia*.

The filamentous ascomycetes are as widely distributed as the yeasts. They are not only important saprobes but also plant and animal parasites. Many have been isolated from the marine environment. One ascomycete, *Oceanitis scuticella*, has been found growing at an ocean depth of 4,000 m and at temperatures of 2° to 4° C. Marine ascomycetes are often found growing on living or dead algae or vascular plants. Decomposing wood is another habitat for marine ascomycetes; 76 species of ascomycetes have been found growing on these cellulose and lignin-containing substrates (Kohlmeyer and Kohlmeyer, 1979). Marine ascomycetes, in conjunction with other marine fungi, may "condition" wood by predigestion with cellulolytic enzymes, with the result that wood-boring organisms such as shipworms may readily invade the wood (Kampf et al, 1959). As with other fungal decomposers, this form of decomposition recycles nutrients into the environment. However, this is not always beneficial: when wood pilings or docks are involved, preservatives must be used to prevent fungal degradation.

Freshwater ascomycetes are usually found on the submerged, dead remains of swamp plants such as reeds and cattails (Ingold, 1976). The species found in fresh water are usually related to well-described terrestrial genera.

The Helotiales, the Pezizales, and the Tuberales are frequently observed in woodlands growing saprophytically on buried twigs, rotting logs, and animal dung (Fig. 17-8). Many members of these orders (e.g., truffles and some cup fungi) form mycorrhizal associations with forest trees.

The Eurotiales, especially *Aspergillus* and *Penicillium* (Fig. 17-6), are ubiquitous and may be isolated from soil, decmposing materials, or even house dust. Both of these form genera produce secondary metabolites that have antibiotic activity and therefore kill bacteria. At least 248 substances with antibiotic activity have been isolated from these two form genera (Berdy, 1974). The question has often been raised of the selective value of antibiotic production, because little evidence exists that antibiotics are functional in natural environments. The antibiotic producers do not appear to be more successful in the environment than non-

producers, nor are antibiotic concentrations in nature high enough to be effective against competing microorganisms. Most researchers think that antibiotic production is beneficial to the organism in restricted microhabitats very close to the site of their production by the fungus. The antibiotics may accumulate in this microhabitat, thus discouraging competitors.

Spores of *Aspergillus restrictus* are the most prevalent in household dust. This fungus and *Aspergillus glaucus* are adapted to fairly low levels of moisture; they can grow in relatively low humidities and also in a wide variety of materials (Christensen, 1975). Moisture will condense on cool surfaces such as basement walls or windows in cold weather. The condensing moisture raises the humidity in the immediate area, thereby providing adequate moisture for fungal growth. Certain fungi, commonly called "mildew," can then grow on such surfaces.

Many free-living ascomycetes (e.g., *Onygena*) contain enzymes that decompose the protein keratin. Such fungi may be found on keratin-containing feathers, hair, and horns, where they play a significant role in the breakdown of this protein. Fungi that parasitize the skin of warm-blooded animals also contain keratinases.

Ascomycetes are very instrumental in the recycling of dung, appearing after the zygomycetes and before the basidiomycetes in the succession of fungal fruiting bodies on this substrate (Harper and Webster, 1964). Some ascomycetes such as *Aspergillus fumigatus* are also able to decompose straw compost and can withstand the heat (60° to 80° C) generated during the decomposition process.

—Apothecium

FIG. 17-8 *Peziza* habit.
Cup fungus, *Peziza* spp., growing on a dead log. Asci are produced within the inner membrane of the cup-shaped apothecium. Cup fungi grow in a variety of habitats including manure, humus, soil, and sand. One species, *Peziza proteana*, grows well in areas burned over by forest fires. Most cup fungi produce asci during the winter months.

ECONOMIC IMPORTANCE
Parasitic Fungi of Warm-blooded Animals, Including Man

Fungal diseases often occur in people whose immune systems have been suppressed. Medically these patients are called "compromised," because they are in a highly susceptible condition with little resistance. Many disease-causing fungi are opportunistic and usually plague only those patients with predisposing conditions such as diabetes and leukemia or those undergoing steroid therapy or organ transplantation. As these types of medication regimens become more common, more research will have to be done to prevent the increasing numbers of fungal diseases that are likely to occur.

Many human fungal diseases are contracted by inhaling the spores, which are very small (2 to 4 μm); these can bypass the defense mechanisms of the respiratory system and infect the lungs. Most fungi that infect humans also live saprophytically on either decomposing plant matter or on the dung of birds or mammals. Fungal diseases are usually not widespread; they occur infrequently in isolated incidences. Airborne spores from the saprophytic fungi appear to be the source of infection for both man and domestic animals. Human-to-human disease transmission rarely occurs among the systemic (internal) diseases.

Species of *Aspergillus*, often *Aspergillus fumigatus*, may cause a pulmonary disease called aspergillosis in both humans and domestic animals. The spores are inhaled into the lungs, where they germinate and cause serious lesions. The disease may attack young chicks, causing **brooders pneumonia.** *A. fumigatus* is unusual in that both the parasitic and the saprophytic forms are mycelial. Many pathogenic fungi are dimorphic, assuming the yeast form when they are in their host.

Histoplasmosis is an occasionally fatal disease of the reticuloendothelial system caused by *Emmonsiella (Histoplasma) capsulatum.* The fungus grows saprophytically on droppings from bats and birds, especially chickens. At human body temperature (37° C) the fungus is yeast-like in form; however, at the lower temperatures generally encountered in nature, the dimorphic fungus converts to the mycelial form.

Another group of fungi that parasitizes birds and mammals (including humans) is the **dermatophytes.** These attack dead keratinized tissues (e.g., skin, hair, and nails). It was recently discovered that their sexual stages occur in the soil, with only the asexual stages occurring on the host. The group was placed in the Deuteromycotina until the sexual stages were discovered. This group includes the troublesome fungal infections "ringworm" (Fig. 19-18) and "athlete's foot" caused by *Arthroderma* and *Nannizzia* (form genera = *Microsporum* and *Trichophyton*). These dermatophytes have been isolated from a wide range of warm-blooded hosts, including humans and domestic and wild animals. Susceptibility of the host is important in determining the course of infection. Ringworm of the scalp may persist for several years in the affected individual's childhood, but it often clears up during puberty, because fatty acids secreted from the scalp have fungistatic properties (Deacon, 1980). After the individual reaches puberty, the site of ringworm infection may shift to the underarm or groin area.

Keratin, a protein found in skin, contains many disulfide bonds and therefore is not readily degraded by proteases of bacterial origin. The dermatophytes produce exoenzymes that can break the disulfide bonds, thus initiating skin diseases. These fungi do not appear to be dimorphic; the mycelial form is found in skin lesions. Transmission of the dermatophytic fungi can occur from host to host or via shared shower or bathing facilities. Fungal diseases of humans and lower animals are discussed in detail by Conant et al (1971), Ainsworth and Austwick (1973), and Emmons et al (1977).

Ergotism

One of the most horrendous diseases of the Middle Ages, ergotism, was caused by the ingestion of sclerotia (see Chapter 12) of the ascomycete, *Claviceps purpurea* (Fig. 17-9). Some evidence indicates that epidemics of **ergotism** date back as far as 430 BC; the first recorded epidemic occurred in 857 AD. This fungus primarily infects rye and forms dark, hard sclerotia that replace the ovaries of the rye flowers. During harvesting of the rye, the sclerotia may not be noticed and subsequently are ground into flour along with uninfected grain. When the flour is made into bread and the bread is eaten, a serious illness or even death may result. Flour containing 2% ergot can cause the disease. In 994 AD, 40,000 people are reported to have died from ergot poisoning. In 1720 the cavalry of Czar Peter the Great of Russia was struck down by ergotism after consuming contaminated bread; his planned invasion of Turkey was cancelled when 20,000

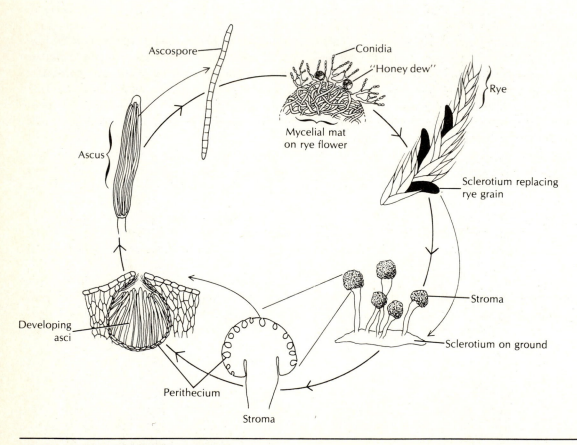

Fig. 17-9 Life history of *Claviceps purpurea*.

This fungus overwinters as a hard sclerotium. In the spring, sclerotia germinate to produce stromata in which develop perithecia with asci and ascospores. Karyogamy and meiosis take place within the stroma during the spring. Needlelike ascospores are forcibly discharged and dispersed by the wind. These germinate on the developing rye grain and grow into a mycelial mat that replaces the developing rye ovary.

The mat produces a sweet "honey dew" that attracts insects. Conidia from the mat adhere to the insects and spread the fungus to other rye plants. As the mycelial mat matures near the end of the growing season, it becomes firm and dark, forming a sclerotium. Sclerotia may be larger than individual rye grains. They fall to the ground during harvest, and spend the winter in a resting phase.

men and horses became ill. Ergotism among the soldiers also contributed to Napoleon's defeat by Russia and his retreat from Moscow. Between 1580 and 1900, 65 epidemics occurred; both humans and lower animals were involved.

The sclerotia, also called ergots, contain up to 20 identified alkaloids. Several of these are highly toxic to humans and cause fever, convulsions, hallucinations, and loss of circulation in the extremities. The hallucinations are apparently caused by one or several of the **alkaloids,** related to **lysergic acid,** a precursor of the potent hallucinogenic drug **lysergic acid diethylamide** (Fig. 17-10), popularly known as LSD. Caporael (1976) has proposed that the 20 young women executed as witches in 1692 in Salem, Massachusetts, may have been the victims of hallucinations caused by ergot-contaminated bread. The probability of ergotism's being responsible for the actions of the Salem "witches" is further debated by Spanos and Gottleib (1976) and Matossian (1982).

Ergotism has been frequent in southern France, where rye bread is a staple of the diet. In the eleventh century a hospital for victims of ergotism was built in the Rhone Valley, where the patients were cared for by monks of the order of St. Anthony. The disease became known subsequently as "St. Anthony's Fire." Patients often improved because they were fed uncontaminated wheat bread rather than rye bread, even though the cause of the disease was still unknown. In 1670 a French physician proposed that the ergots were causing the disease. It took years, however, before farmers were convinced that ergot was the causative agent, and the disease was rampant for a long time after its cause was first proposed. The history and influence of ergotism is discussed by Carefoot and Sprott (1967).

Ergotism has been well under control since its etiology was established and home milling decreased. Modern baking practices have virtually eliminated ergotism. Occasionally, when wild birds or other animals eat grasses that are parasitized, they have been reported to act strangely and may be victims of ergot poisoning. A suspected epidemic of ergotism in France in 1951 is described by Fuller (1968). The incident is described in fictional form, but many scientists think the account describes a real epidemic. During this suspected epidemic 7 people died and 300 experienced weeks of harrowing hallucinations. If ergotism was the cause, the hallucinations were probably the result of the hallucinogenic alkaloid lysergic acid, the precursor of LSD. LSD is hallucinogenic in minute concentrations, and reactions are totally unpredictable.

Concentrations of the various alkaloids in *Claviceps purpurea* may vary with the strain, the location, or the season. Some of the alka-

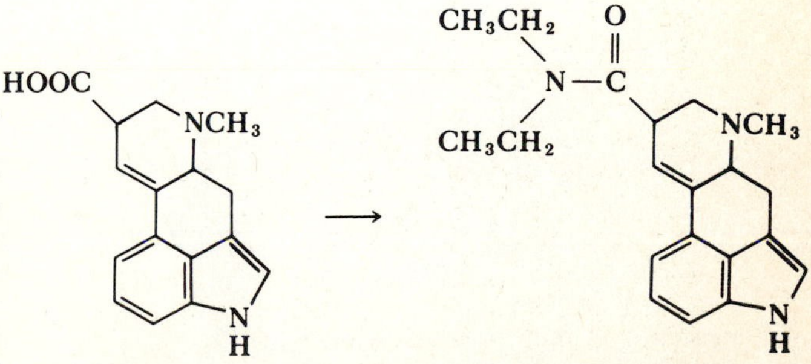

FIG. 17-10 Lysergic acid and lysergic acid diethylamide. Lysergic acid is derived from the amino acid tryptophan. Ergot toxins, including lysergic acid, cause hallucinations in addition to other effects. Hallucinations may be caused by the chemical similarity between these toxins and the neurotransmitter serotonin (Griffin, 1981). Lysergic acid is used to manufacture the hallucinogenic drug lysergic acid diethylamide (LSD).

Lysergic acid   **D-lysergic acid diethylamide**

loids from the sclerotia of *C. purpurea* have been extracted and used as drugs (ergots, ergotamine) to constrict blood vessels (Mantle, 1975). In the midwestern United States in 1945, 100 tons of ergot were produced from purposely infected rye fields (Klein, 1979). In the 1940s the United States also imported 40 metric tons of ergot, mostly from Europe, where rye fields are similarly infected. The drug is still frequently used. Lysergic acid has also been isolated and used in the manufacture—legal or illegal—of LSD.

Tree and Crop Diseases

The brown rot of peaches and other stone fruits is a disease caused by the ascomycete *Monilinia fructicola* (Fig. 17-11). The fungus may attack the flowers, leaves, twigs, or fruit, depending on the season, and may inflict light to severe damage in peach-growing areas, especially in the humid southeastern United States. A fungal pectinase dissolves the middle lamella of the fruit cells, the tissues soften, and the mycelium invades the fruit. Infected fruits become dried and hard (i.e., "mummified") and composed of a compact fungal stroma (Fig. 17-11). Mummified fruit left in the soil may produce apothecia and ascospores for several years. Another genus of the Helotiales, *Naemacyclus*, includes fungi that attack trees of the pine (Pinaceae) family (Fig. 17-12). A related species, *Rhytisma acerinum*, causes "tar spot" on maple trees.

A widespread disease of peach trees, peach leaf curl, is caused by a yeastlike organism, *Taphrina deformans*. The fungus stimulates host cells to multiply, causing blistering on the stem surface and curling and arching of the leaves (Fig. 17-13). The leaves are eventually killed, the tree is partially defoliated, the photosynthetic area is reduced, and the yield of peaches is consequently low. Still other species of *Taphrina* attack fruit trees such as cherry and plum.

Apples and crab apples often become infected with the "apple scab" fungus, *Venturia inequalis*. As with most fungal infections, the disease is most severe in humid weather. The fungus may attack the leaves, flowers, or fruit. When fruit is attacked, the characteristic "scab" is seen on the skin of the apple. Control is accomplished by the use of fungicides or resistant varieties of apple trees. *Venturia inequalis* is one of the significant diseases of apple trees with which orchard owners must contend in most apple-growing areas of the world. *Venturia pyrina* causes a similar scab of pear trees.

Powdery mildews (Erysiphales—Fig. 17-14) cause fluffy white coverings on infected leaves and stems of many different plants. The white

Apothecium

"Mummified" peach

FIG. 17-11 *Monilinia fructicola* **or brown rot of peaches.** This fungus invades the leaves or twigs of trees that bear stone fruits. Its mycelia then produce conidia that germinate on mature fruit, where it grows into a dense mycelium throughout the tissues of the fruit, causing it to shrivel and become hard and dry (i.e., mummified). Mummified fruits may produce apothecia, as seen in this illustration, within a year or two. Apothecia in turn produce asci with ascospores.

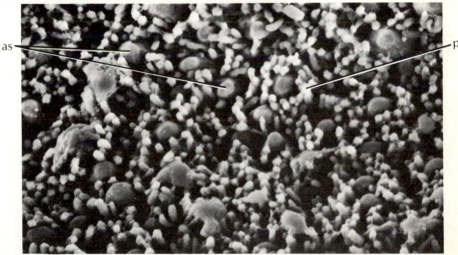

FIG. 17-12 *Naemacyclus minor, a parasite of Pinus.*

The upper photograph (190×) shows apothecium of *N. minor* rupturing through epidermis of a needle of a Scotch pine (*Pinus sylvestris*). The lower photograph (1700×) shows inner surface of apothecium with the swollen tips of the asci *(as)* protruding above the paraphyses *(p)*. Following discharge of the ascospores the asci shrink below the paraphyses. Members of the Helotiales have numerous paraphyses distributed in their apothecia among the asci. (Courtesy William Merrill, Department of Plant Pathology, The Pennsylvania State University, University Park, Pa.)

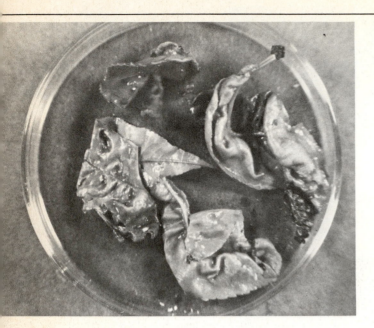

FIG. 17-13 *Taphrina deformans* **causing leaf curl.**
Peach tree leaves that have been infected by *T. deformans,* causing the leaves to curl. Ascospores land on the leaves and germinate, producing mycelia, which invade the leaf tissues and cause the deformation. (Courtesy D. Spess, Department of Biology, Lehigh University, Bethlehem, Pa.)

FIG. 17-14 Cleistothecium of a powdery mildew.
This cleistothecium (200×) is from one of the powdery mildews (order Erysiphales), which have closed ascocarps. During summer these fungi produce numerous conidia, giving the host plant a powderlike covering. Cleistothecia are usually produced in the late summer or early fall. In this species the cleistothecal appendages (A) are dichotomously branched and may serve to anchor the cleistothecium to the leaf surface of the host. The asci (B) mature and burst forth from the cleistothecium, bearing the ascospores (C). Such cleistothecia may be resistant to cold temperatures and aid the fungus in surviving winter. (Courtesy William Merrill, Department of Plant Pathology, The Pennsylvania State University, University Park, Pa.)

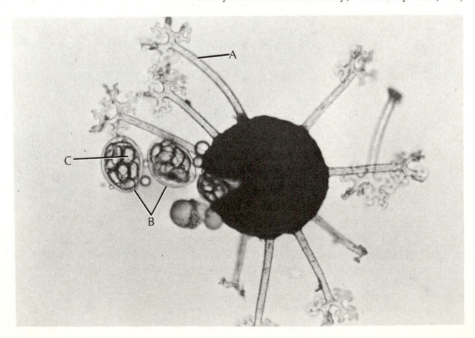

FIG. 17-15 Perithecium of _Ceratocystis ulmi_, the Dutch elm disease fungus.
Scanning electron photomicrograph (approx. 200×) of a perithecium of _C. ulmi_, which may be found in bark-beetle galleries (see Fig. 17-16). When the perithecium ruptures, ascospores are released and easily dispersed by bark beetles. (Courtesy of R.J. Scheffer, Phytopathological Laboratory, "Willie" Commelin Scholten, Baarn, The Netherlands.)

powder is caused by the abundant conidia that are produced by the invading parasite. Powdery mildews are obligate parasites of such plants as grape (parasite: _Uncinula necator_), cucumber family (parasite: _Erysiphe cichoracearum_), lilacs (parasite: _Microsphaera alni_), roses (parasite: _Sphaerotheca pannosa_), and wheat (parasite: _Erysiphe graminis_). Some powdery mildews may be highly specific as to the hosts they attack, whereas others attack a wide range of hosts. Damage done by the fungus may range from light to very heavy.

Diaporthe citri, commonly known as **citrus melanose,** has been a problem wherever citrus trees are grown. Other species of _Diaporthe_ attack oak, dogwood, and pear trees and vegetables such as egg plant and various beans.

The **Dutch elm disease** caused by _Ceratocystis ulmi_ is discussed below. However, other species of _Ceratocystis_ may also cause considerable crop and tree damage. _Ceratocystis fagacearum_ causes wilt of oak trees, _Ceratocystis fimbriata_ attacks sweet potatoes, and _Ceratocystis minor_ or _Ceratocystis pilifera_ causes "blue stain" on stored lumber, reducing its value.

Dutch Elm Disease. Dutch elm disease is caused by a fungus, _Ceratocystis ulmi_ (Fig. 17-15), which was first introduced to the United States in the early 1930s with some wood from England. The term "Dutch" elm disease is really a misleading name—it was Dutch scientists who first isolated and identified the fungus; the disease did _not_ come from the Netherlands but originated in the Orient. Ironically, the contaminated wood was sent to the United States in partial payment of debts from World War I. The disease was first noticed in Ohio in the mid-1930s and thereafter spread through eastern Canada and the eastern and southern United States, reaching California in the 1970s. The disease rapidly kills the American elm (*Ulmus americana*) and several species of European elms. Billions of elms have been killed in North America and Europe. American elm trees are highly prized as majestic shade trees; they were frequently planted on city streets, college campuses, and village greens, but the fungus has killed almost all such trees, changing the landscape drastically in many areas of the eastern and midwestern United States.

The fungal mycelium grows into the xylem vessels of the tree, blocking them and prevent-ing the movement of water and nutrients into the leaves. A toxin may also be involved. Fungal spores may also travel through the xylem, spreading the disease to other parts of the tree. Bark beetles, which inhabit cracks and dig out "galleries" (Fig. 17-16) beneath the bark to lay their eggs, may carry spores from one tree to the next. The fungal spores are introduced into the vascular system of a healthy tree when the bark beetle starts digging its galleries. Spread of the fungus may also occur when the roots of adjacent trees grow together and spores are transferred from the infected tree to a healthy one. Death of the tree is rapid, usually within 1 to 3 years after infection (Strobel and Lanier, 1981).

Control of the disease has been attempted by several methods. Insecticides, especially di-

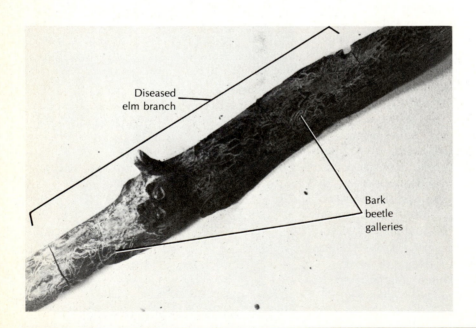

FIG. 17-16 Bark-beetle galleries in elm wood.
Galleries produced by elm tree bark-beetles (*Scolytus* spp). Such galleries are produced when female beetles lay eggs, which hatch into larvae that subsequently tunnel beneath the bark, forming feeding galleries. They feed on the sap carried in phloem tissues. Adult bark beetles may fly to other elm trees and infect them with the fungal spores picked up in the galleries. (Courtesy D. Spess, Department of Biology, Lehigh University, Bethlehem, Pa.)

chloro-diphenyl-trichloroethane (DDT) and methoxychlor, to kill the bark beetle were frequently used. Unfortunately the extensive use of DDT in the Midwest contaminated Lake Michigan to the extent that fish reproduction (in coho salmon, especially) was greatly reduced. The beetles also developed resistance to the insecticides. Further control efforts included attempts to breed a resistant tree using hybrids of the American elm with resistant species from Asia. In 1972 a fungicide, benomyl (a **benzimidazole**), proved effective when injected from a pressurized canister into the vascular system of infected elms. This systemic fungicide has also been tried as a method of protecting valuable, healthy trees from the fungus; however, the treatment is extremely expensive ($100 per tree)—it requires extreme precision in determining the dosage and should be repeated every 2 years. Edington and Peterson (1977) discuss the use of systemic fungicides. Sanitation is another method of control and includes the rapid destruction of infected trees and logs to prevent spore dispersal. Stroebel and Lanier (1981) have experimented with pheromone-mediated behavioral modification in the bark beetle. They propose an integrated pest management program that employs a combination of the methods mentioned above for controlling the spread of the fungus.

A toxin that has been isolated from the pathogen interferes with the movement of water in the xylem (Stroebel and Lanier, 1981). The toxin affects the permeability of the membranes covering the xylem pit connections and thereby restricts the horizontal flow of water to branches and leaves. Movement of the injected fungicide to infected portions of the tree may also be impeded by the toxin, thereby decreasing the fungicide's effectiveness.

Webber (1981) reports that the spread of *C. ulmi* in the United Kingdom has been retarded by another fungus, *Phomopsis oblonga* (Deuteromycotina), which interferes physiologically with the successful reproduction of bark beetles. To date *Phomopsis* has been found primarily in the wych elm, *Ulmus glabra*. There is hope that *P. oblonga* or other similar fungi may reduce populations of the bark beetle and therefore the spread of Dutch elm disease.

Chestnut Blight. Earlier in this century, the **chestnut blight** fungus (*Endothia parasitica*) devastated American chestnut trees (*Castanea dentata*) in the eastern United States. Spores of the fungus were carried on Japanese chestnut trees imported from the Orient in 1904. The Japanese chestnut is resistant to the disease. Before the 1900s there were approximately 1 billion chestnut trees covering 9 million acres in the northeastern United States and eastern Canada (Klein, 1979). In less than 25 years almost all mature chestnut trees east of the Mississippi River were dead. By 1930 the American chestnut was considered lost in the United States. The fungus does not kill the roots, however, and new shoots sprout from the roots. Young chestnut trees may often be seen in northeastern forests. Some have even borne nuts, but when they reach 10 m or taller in height and up to 15 cm in diameter, the omnipresent blight hits and the tree dies back to its roots.

Spread of the disease may occur as spores are blown by the wind or carried by insects, birds, rodents, or rain. Asexual spores are shed in damp weather; sexual ascospores are formed in dry weather. Both types of spores germinate in wounds or cracks in the bark, and the mycelium grows into the tree, producing toxins that kill both phloem and cambium cells.

The effects of chestnut blight meant the loss of trees that could have provided 30 billion board feet of exceptionally durable lumber. In past years chestnut wood was widely used for homes, fences, flooring, furniture, and utility poles. It is greatly prized because it is very resistant to decay-causing organisms. Furthermore, the elimination of mature chestnut trees in the forests of the eastern United States has had far-reaching effects on the flora and fauna of the local ecosystems. The nuts alone were not only food for many wild animals but were also prized as a delicacy by humans.

Currently efforts to save the chestnut include a search for naturally blight-resistant trees growing wild in the United States and the hybridization of *Castanea dentata* with European or Oriental chestnut trees that are resistant to the fungus. The European chestnut (*Castanea sativa*) became infected with the blight in the 1930s, but the trees recovered. Researchers isolated a **hypovirulent** strain of the fungus that aided the tree in recovery from the virulent strain of the pathogen. Evidently the hypovirulent strain carries a viruslike agent that kills virulent *E. parasitica* strains. Such a strain provides hope for a solution to the chestnut blight problem, but unfortunately the hypovirulent strains spread more slowly and are less vigorous than the virulent ones (Klein, 1979). Efforts are being made in the United States to effectively disseminate the hypovirulent fungal strains containing the viruslike agent. No successful method of spreading these strains has yet been discovered (Anaghostakis, 1982).

Mycotoxins and Toxic Ascomycetes

Mycotoxins. The common *Aspergillus flavus* and *Aspergillus parasiticus* are known to produce potentially toxic secondary metabolites known as mycotoxins. Although they are not very toxic in their native state, they may be converted to toxic molecules in vertebrate livers. Mycotoxins produced by these species, called **aflatoxins** (Fig. 17-17), were first discovered in 1960. Several different chemical forms of such toxins exist; some are more toxic than others. In 1960 approximately 100,000 turkey pullets died in Great Britain after eating peanut meal contaminated with aflatoxin. Deaths of calves and ducklings, following their consumption of peanut meal, have also been reported. *A. flavus* and *A. parasiticus*, commonly known as **storage fungi,** grow readily in moist climates on stored foods such as peanuts, rice, and cottonseed. High moisture content in stored products encourages fungal growth.

Because aflatoxins are powerful carcinogens, considerable research is currently directed to-

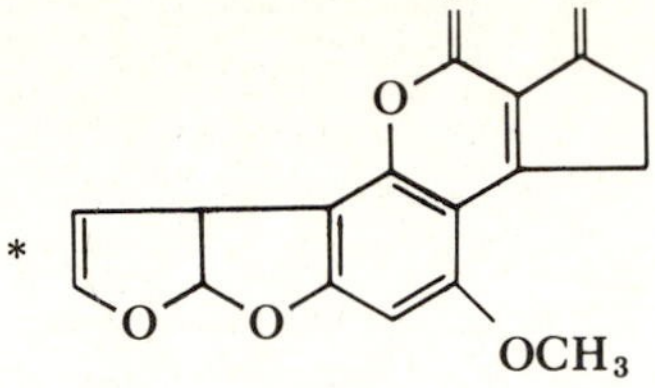

FIG. 17-17 Aflatoxin.
There are several different aflatoxins that are produced by *Aspergillus flavus* or *A. parasiticus*. Some aflatoxins are more toxic than others. The aflatoxins are some of the most potent carcinogenic (cancer-causing) chemicals known. Toxicity may be caused by the double bond at the asterisk (*) where an epoxide may be formed.

FIG. 17-18 *Helvella* habit.
Helvella (Gyromitra) infula, one of the "saddle fungi" or "false morels." Genus *Helvella* is difficult to separate morphologically from genus *Gyromitra*, and the species may be listed under both genera in some manuals. The common names stem from the bifurcate nature of the ascocarp or cap that bears the asci, and from their resemblance to the true morels, *Morchella* (Fig. 17-21). Note that the cap does not possess the many convoluted apothecia associated with the true morels, and that the stalk is slender and irregularly ridged.

ward preventing their occurrence in foods. The target organ of the toxin is usually the liver. Edible products that are susceptible to *Aspergilus* growth (e.g., peanut butter) are regularly checked for aflatoxin content. Improved storage and handling procedures for susceptible foods should reduce fungal toxin contamination. Goldblatt (1969), Stoloff (1977), and Heathcote and Hibbert (1978) provide thorough discussions of aflatoxins. Some fungi such as *A. flavus* will fluoresce in ultraviolet light; therefore, in some food processing plants, foods are screened under an ultraviolet light to check for fungi that might produce mycotoxins. Recent research suggests that a mycovirus may be involved in the production or inhibition of aflatoxin. Virus particles that have been found in a strain of *A. flavus* that did not produce a toxin, and no virus particles were detected in a strain that did produce a toxin. A toxin-producing strain of *A. parasiticus* likewise shows no evidence of virus particles (Mackenzie and Adler, 1972; Wood et al, 1974).

Additional species of *Aspergillus* and *Penicillium* can also produce mycotoxins. These fungi may grow on improperly stored feed and cause illness when the feed is consumed by livestock. Kadis et al (1971, 1972), Christensen (1975), Ciegler and Bennett (1980) and Shank (1981) discuss mycotoxins in greater detail.

Toxic Ascomycetes. The saprophytic "saddle" fungus *Helvella (Gyromitra)* (Fig. 17-18) may be poisonous if eaten. Some people are highly susceptible to the toxin and become ill after ingesting the fungus, whereas others can eat it with no apparent undesirable effects. *Helvella esculenta* and *Helvella brannea* have proved poisonous to some people, even though *esculenta* means "edible" in Latin. Seventeen people in Normandie, France, became very ill after eating *H. esculenta* (Denis, 1961). The variation in human reactions to this potentially poisonous fungus has led to many theories about its toxicity. Christensen (1975) discusses hypotheses on the causes of its toxicity. Included among the hypotheses are: the fungus is toxic only when decaying; the fungus is misindentified; the fungal toxin is destroyed by cooking; and only certain individuals are sensitive. Christensen concludes that he personally will not eat this fungus.

Decomposition of Synthetic Products

The subdivision Ascomycotina contains many fungi that produce a wide array of exoenzymes that can decompose both complex carbohydrates, such as cellulose and starch, and simple sugars. They also produce many primary and secondary metabolites during decomposition. These capabilities are used to human advantage in many industrial production processes, but large problems arise when fungi start to degrade synthetic products.

Certain species of ascomycetes cause deterioration of fabrics, leather, radio parts, photographic equipment, paper, and rope. Many foods are also attacked by ascomycetes, especially *Penicillium* spp., *Aspergillus* spp., and *Saccharomyces* spp., which produce blue, green, or black molds on citrus fruits, jellies, vegetables, and breads.

Chemicals such as copper or boron-based chemicals, sodium propionate, alcohols, or hydroxybenzoic acid are added to retard fungal growth in fabrics, paper products, and breads. The production of petroleum-based plastics has solved many such problems with fabrics, especially in the tropics, where fungal degradation is rampant because of the high humidity. Species of fungi are presumably evolving now, however, that can even decompose such synthetic molecules as those in polyurethane plastic.

Unwanted yeasts may grow in fruit juices or cheeses, imparting unpleasant tastes and odors. Some yeasts even ferment sugar-containing foods at refrigerated temperatures (0° to 4° C). Spoilage of fruit juices (e.g., apple cider) may occur fairly rapidly in the refrigerator. To preserve the quality of fruit juices for a long time, pasteurization is necessary to kill the wild yeasts that naturally occur in the juice. Several yeasts can ferment the lactose in milk products, and others grow in the high-salt liquids in which olives, pickles, or sauerkraut are preserved (Phaff et al, 1966).

Some ascomycetes can grow in and decompose petroleum-based aircraft fuels such as kerosene. Boron or ethanol derivatives are usually mixed with such fuels to prevent fungal growth and subsequent deterioration. An additional problem may be presented by supersonic aircraft fuel, because the fuel tanks are much hotter than in conventional aircraft, and thermophilic fungi such as *Aspergillus fumigatus* may be able to grow.

Important Drugs Produced by Ascomycetes

In ancient Egypt and China, cheese, bread, and meats covered with green or blue molds were used as a folk medicine for infected wounds. The reason moldy foods were effective in preventing or curing infection was not understood until the antibiotic properties of the secondary metabolites of *Penicillium* spp. were discovered. In 1929 Sir Alexander Fleming (1881-1955) observed that a fungus *(Penicillium notatum)* that contaminated his bacterial cultures of *Staphylococcus* had killed many of the bacteria. Fleming received the Nobel Prize in 1945 for the discovery of the antibiotic activities of *Penicillium.*

It was not until the second World War, however, when a substance was desperately needed to kill bacteria in wounds, that Howard Florey of Oxford, England, purified the antibiotic penicillin (Fig. 17-19), originally isolated by Fleming. Florey and several colleagues came to the United States in 1941 and initiated a search for a strain of *Penicillium* that would produce higher amounts of the antibiotic. The search produced a "moldy" cantaloupe from a Peoria, Illinois, market, and the strain of *Penicillium chrysogenum* that grew on this melon yielded 500 times the antibiotic of Florey's original strain (Emerson, 1952). The high-yielding strain has now been improved upon and distributed worldwide. Industrial strains of *Penicillium* today produce 10,000 times more penicillin per unit volume than Fleming's original culture (Demain

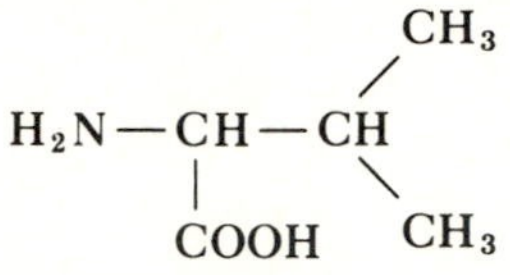

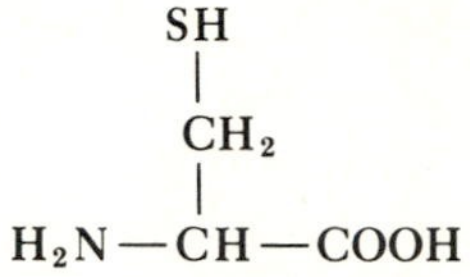

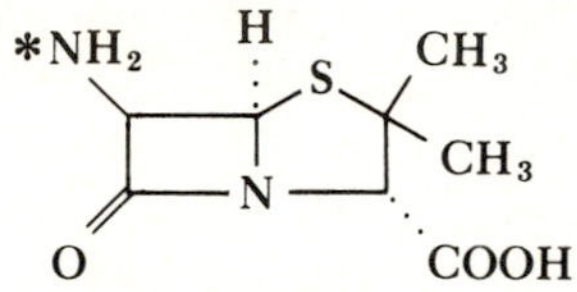

6-Aminopenicillanic acid (6-APA)

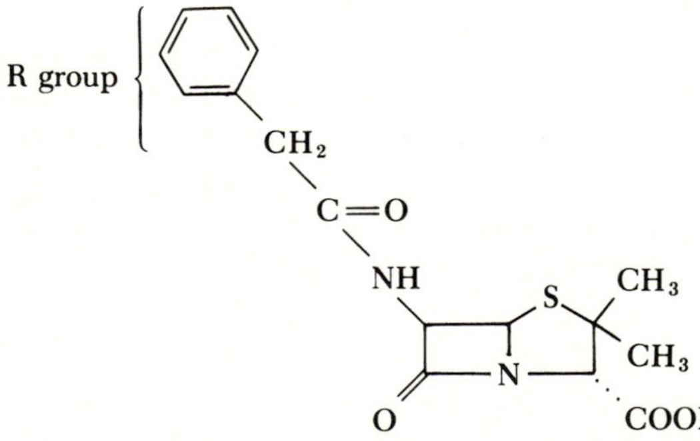

Penicillin G (also known as benzylpenicillin)

FIG. 17-19 6-Aminopenicillanic acid (APA) and Penicillin G.
The aminopenicillanic acid molecule is derived from the amino acids valine and cysteine. Various strains of *Penicillium* spp. produce over a dozen different types of penicillin. Many other semisynthetic forms are created in the laboratory from the fungal-produced 6-APA. The term penicillin refers to the sodium, potassium, or calcium salt of penicillin G. Other semisynthetic penicillins such as ampicillin and carbenicillin have a different chemical group at (*).

and Solomon, 1981). In 1979 sales of penicillins of various types in the United States exceeded $200 million and in 1978 world sales exceeded $1 billion (Aharowitz and Cohen, 1981).

Penicillin is not a single compound but a group of related compounds with 6-amino penicillanic acid as their base (Fig. 17-19). Natural penicillins are active against the gram positive bacteria that cause *Streptococcus* infections, anthrax, pneumonia, meningitis, and endocarditis. The drug's mechanism of action is the inhibition of bacterial cell wall synthesis. Specific side chains may be added to penicillin, increasing its activity to include gram negative bacteria.

Penicillium griseofulvin produces a fungistatic drug, griseofulvin, which is effective against some skin diseases caused by fungi, especially athlete's foot. Only chitin-containing fungi are affected by this antibiotic; it is not effective against the pathogenic yeasts (Deacon, 1980).

The drug ergot, which is obtained from wheat that has been infected with *Claviceps purpurea*, is used to induce contractions in childbirth and to control hemorrhaging following birth. In addition, it is used to treat migraine headaches, high blood pressure, and varicose veins. The method of action of the drug is the contraction of blood vessels (i.e., vasoconstriction). Various physiologically active alkaloids derived from the *C. purpurea* sclerotia are used as drugs (e.g., ergotamine, ergonovine). Lysergic acid diethylamide (LSD) was first synthesized by the Swiss chemist Albert Hofmann in 1938 from lysergic acid (Fig. 17-10), an alkaloid isolated from *C. purpurea*. Although the drug has been widely misused by untrained people, many psychiatrists and other medical personnel think it may have great potential in the treatment of the mentally ill.

Industrial Uses of Ascomycetes

Under acid conditions, *Aspergillus* spp. synthesize large amounts of citric acid, an organic acid that is widely used in soft drinks, medicine,

candy, and in printing processes. In 1952, 9,000 metric tons of citric acid were produced from fungi, replacing the citric acid previously obtained from fruits from Italy (Emerson, 1952).

Strains of the form genera *Aspergillus* and *Penicillium* are often used in the industrial production of organic acids such as gluconic acid, citric acid, gallic acid, and oxalic acid. The controlled reactions that produce these acids are all aerobic processes. Smith (1968) and Lockwood (1975) discuss the production of organic acids by fungi. Other industrially produced metabolic products of *Aspergillus* spp. and *Penicillium* spp. are alcohols, esters, and compounds containing chlorine or sulfur.

Ethyl alcohol is an important industrial solvent, extractant, and antifreeze, to mention only a few uses. Ethanol is also used as a substrate for the synthesis of other commercially important organic compounds. In 1980 in the United States, 550,000 metric tons of ethanol were produced, with sales totaling $297 million. This does *not* include the sales of alcohol-containing beverages (Eveleigh, 1981). Food and drug laws in the United States require that all alcohol used in beverages be produced by fermentation. In the late 1970s, about 70% of the industrial ethanol in the United States was produced by chemical processes using petroleum products rather than by biological processes. With the large increases in petroleum prices, more ethanol will probably be produced by biological processes in the future. Brazil, for example, has substantially reduced its consumption of gasoline by substituting an ethanol-gasoline mixture. The ethanol is fermented from locally abundant sugar cane and cassava (Baker, 1978).

Ethanol has also been added to gasoline in the United States. "Gasohol," compounded of nine parts gasoline to one part ethanol, is meeting with good public acceptance. Much of the ethanol used comes from fermentation of renewable, abundant substrates (e.g. corn and wood). The corn, wood, or other substrate must first be broken down to sugars by some process or series of processes before fermentation by yeast can occur. Wood is less expensive and more abundant than corn, but decomposing the cellulose and lignin requires fungi and/or bacteria with the proper enzymes. To have a significant impact on current gasoline consumption in the United States, the production of gasohol would have to increase substantially. Maximum ethanol production from excess grain is estimated to be 7.4 billion liters per year. To replace gasoline with gasohol in the United States would require at least 44 billion liters per year, so other substrates in addition to grain would have to used (Eveleigh, 1981).

Yeasts are used extensively in the production of vitamins. The yeast *Ashbya gossypi* is used commercially for the production of the vitamin riboflavin, B_2, and very large yields can be obtained (Demain and Solomon, 1981). Yeasts are also high in ergosterol, the vitamin D precursor, which may be converted to vitamin D by irradiation with ultraviolet light. Irradiated yeasts are commonly used as a source of vitamin D. Bakers' yeast is grown commercially, often in molasses, and sold for use in home baking.

Other fungi are used in the commercial production of enzymes, which may then be used for both the breaking or making of specific chemical bonds in industrial and commercial processes or in scientific research.

Gibberella fujikuroi (Diaporthales), a fungal parasite of rice, is grown commercially for its production of the plant growth hormone **gibberellic acid**. This hormone is used by commercial and home gardeners to promote growth, flowering, fruit formation, and seed germination.

Beer, Wine, and Distilled Alcoholic Drinks

Beer. Beer has been used to assuage thirst for thousands of years. Drawings on an Egyptian tomb indicate that beer was consumed during the building of the tomb in 2400 BC (Demain and Solomon, 1981). Beer evidently was used as early as 6000 BC by the Sumarians and Babylonians. Beer not only quenches thirst, but before

water was purified for drinking, it was also safer to drink than water, because it is relatively free from bacterial contamination.

Yeast cells were first seen in a drop of fermenting beer by Anton van Leeuwenhoek through his microscope in 1680. He did not perceive, however, the relationship among the yeast, fermentation, and the taste of beer and wine. When Louis Pasteur observed yeast in 1859, he discerned that yeast is living and growing and that when sugar is fermented by its metabolic activities, alcohol is produced.

Beer is produced from malt (germinated barley), which is boiled and extracted. The malt is rich in amylases, enzymes that break down starches to sugar. The starch must first be broken down to sugars before the yeast can commence their fermentation. Dried flowers of the hop vine, *Humulus lupulus*, are added for flavor, and fermentation is allowed to proceed in a cool, dark place. The end result is a deliciously flavored but slightly bitter beverage containing up to 15% ethyl alcohol and carbon dioxide. In the process of brewing beer considerable yeast is produced, since during the fermentation process the yeasts are growing and increasing in numbers. This extra yeast is removed and sold as "brewer's yeast," a highly nutritious by-product rich in proteins, amino acids, and vitamins.

Wine. Many years ago, before grape growers knew that ethyl alcohol production from grapes was caused by yeasts, grape juice would ferment naturally because of the presence of naturally occurring yeasts, often *Saccharomyces ellipsoideus*, on the skin of the grapes. The skin of a single grape may contain up to 10×10^6 yeast cells. Most of these yeast cells are "wild" wine yeasts and are capable of fermenting the sugars in grape juice. The juice of grapes may contain up to 24% fermentable sugars, especially fructose and glucose. Wine, "the oldest of medicines," has been used medically for hundreds of years as an analgesic, sedative, vasodilator, and diuretic (Amerine, 1964). The fermented juice has been consumed for at least 4,000 years.

Modern-day winemakers have improved on

the "wild" yeasts by culturing those with desirable characteristics and adding them to the grape juice. Undesirable yeasts are usually suppressed by chemicals. Subsequently the juice, along with the pulp, seeds, and skins, is placed in a vat and fermentation is allowed to proceed. Temperatures are carefully controlled, because these regulate both the rate of fermentation and the type of wine produced. The vats must be cooled constantly, because the temperatures of fermentations may exceed 38° C, the lethal temperature for wine-producing yeast. The pulp, seeds, and skins float to the top, forming a thick layer that reduces the amount of oxygen entering the vat, creating semianaerobic conditions. When the alcohol concentration reaches 15% to 18%, the yeast is inactivated, and, after further aging and processing, the wine is ready for sale.

The process of fermentation of sugars in the grape juice to ethyl alcohol requires 22 enzymes and 6 or more coenzymes (Amerine, 1964). Several other compounds (acetaldehyde, glycerol, aromatic compounds) are produced during fermentation. Winemakers must chemically or physically suppress the formation of these unwanted compounds because they may interfere with the desired taste of the wine. The amount of alcohol produced and the flavor and aroma of the wine are all determined by the complex interactions of temperature, the strain of yeast, and the initial sugar content.

In China and Japan species of *Aspergillus* are used to ferment rice starch in the production of the popular oriental wine **sake**. *Aspergillus oryzae* is often used because it produces amylases that break down the rice starch to sugar. After starch digestion a special strain of *Saccharomyces cerevisiae* is used for fermentation. Approximately 10% of the rice produced in Japan is fermented for sake.

To increase the alcoholic content above 15% to 18% in beverages, the alcohol produced from fermentation must be distilled to separate the alcohol from water. Distilled whiskys such as scotch and rye are produced from fermented corn, rye, wheat, barley, or a combination of these, and the alcohol from the fermentation is then distilled to concentrate it.

Ascomycetes as a Food Source

One of the great gourmet delicacies that come from the Ascomycotina is the truffle (*Tuber aestivum;* Fig. 17-20). Truffles are subterranean ascocarps of fungi that live in a mycorrhizal association with oak or beech trees. The most delectable truffles, with an odor of spice, cheese, and garlic, are found in France and Italy. Extensively trained dogs or pigs root out the underground ascocarp by smell. Pregnant sows are reputed to be the best truffle-snuffers because they have a heightened sense of smell. In the evening, when the air is still, the pig or dog is taken to wooded areas during the truffle season to begin the hunt. The pig may often be carried or pushed in a wheelbarrow to avoid overexerting the animal. When the pig or dog starts to dig out the truffle, the animal is forcibly restrained by a leash and the hunter proceeds to dig out the fungus with a shovel. The pig or dog is rewarded with a small bit of food—but *not* the truffle—which may sell for up to $200 a pound

FIG. 17-20. *Tuber* **habit.**
Tuber aestivum, the delicious truffle, is a prized delicacy throughout the world. This species is indigenous to Europe, and its subterranean ascocarp may measure up to 7 cm across. It is dark brown with white inner flesh. They are hunted in the summer with the aid of truffle-hunting pigs and dogs. Other European truffles that are in great demand are *T. magnatum,* the white truffle, *T. melanosporum,* the black truffle, and *T. brumale,* the winter truffle.

(1975 dollars). The truffle *Tuber melanosporum* contains a steroid pheromone that is also secreted by boars during premating rituals. The presence of the steroid may explain the enthusiasm with which sows dig up truffles (Claus et al, 1981). Unfortunately no North American species of *Tuber* have been discovered that are suitable for human food, although many species do occur, especially on the West Coast of the United States.

The ascocarps of *Tuber* remain closed underground and the spores are not discharged into the environment. How, then, are the spores dispersed? One theory is that the odor attracts rodents, which unearth and eat the ascocarp,

FIG. 17-21 *Morchella esculenta* **(morel) habit.**
Morels, which grow in the spring from April through June, are especially common in apple orchards and pine or oak forests. The apothecium may reach sizes up to 14 cm. It is considered a gastronomic delicacy. The cap resembles a sponge, but is actually an apothecium with an hymenial layer containing asci lining the indentations. The entire morel is an ascocarp. Many morels are thought to be mycorrhizal.

thereby dispersing the spores in their feces. Another theory is that the tissue surrounding the asci and ascospores decomposes, leaving the spores behind unharmed.

Another gourmet delicacy belonging to the ascomycetes is the common morel (*Morchella esculenta*). This organism (Fig. 17-21) resembles a mushroom; in fact, it is called the "sponge mushroom." Close inspection, however, reveals it to be the fruiting body (i.e., ascocarp) of an ascomycete. Morels may be found for several weeks in the spring, often under pine and oak trees. Several species of *Morchella* are consumed as delicacies. See Canby (1982) for a description of a hunt for morels in the midwestern United States.

Another product, the popular soy sauce used in oriental dishes, is produced by fermenting a mixture of soybeans, wheat flour, and salt with *Aspergillus oryzae* in a multistep process. This fungus is also used in Japan to ferment whole soybeans, resulting in the product called "natto."

The flavor of cheeses such as Camembert and Roquefort is the result of the growth and enzymatic digestion of milk curds by two species of *Penicillium*, *Penicillium roquefortii* and *Penicillium camemberti*. *P. roquefortii* was first isolated in southern France, and to this day the cheese that bears the name "Roquefort" is produced from ewe's milk in controlled environments in southern France (Wernick, 1983). Many cheese connoisseurs believe that the local environment is very influential in cheesemaking and taste. Similar cheeses, Gorgonzola (Italy), Stilton (England), and blue cheese (Denmark, United States), are produced with similar fungi in different geographical locations. The cheese (from goat's, cow's, or sheep's milk, usually) is inoculated with the fungus, and the blue fungal mycelium and spores give the cheese its characteristic blue-veined appearance. The metabolites of the fungi add the taste, zest, and aroma. *P. camemberti*, in a similar way, is used to give the characteristic flavor and texture to Camembert cheese.

Yeasts apparently have great potential for providing inexpensive single-celled protein (SCP) to increase the world's supply of protein food. Not only can many convert carbohydrates to protein very efficiently and with little energy loss, but yeast cells are also high in vitamin content, especially the B vitamins. The conversion of carbohydrate to protein by yeast is a much more efficient process than the conversion of grain carbohydrate to animal protein by the feeding of livestock.

Brewer's yeast as a food tends to be bitter and unpalatable. When attempts are made to "de-bitter" it, many vitamins are lost. Many yeasts grown specifically as food sources are also bitter and unpalatable. The consumption of too much yeast can also cause digestive upsets. Yeast is probably best used as a food supplement, added to bread or other foods. It takes only 42 g of yeast to provide the food value of ¼ lb (110 g) of beef. Australian citizens consume large amounts of concentrated yeast extract, which they use on bread as a highly nutritious spread. Yeast is also frequently used as an animal food additive to increase nutritive value.

In both World Wars I and II, the Germans grew yeasts (mostly brewer's) on a large scale, adding it to foods. The use of yeast as a nutritious food supplement helped considerably to ease the food shortages during these wars (Rose, 1981).

British and Dutch scientists imprisoned in Japanese prison camps during World War II observed that the prisoners developed vitamin deficiency diseases as a result of the highly restricted diet. The scientists, familiar with the nutritive value of yeasts, started their own yeast cultures from leftover rice and spores from the soil. After the prisoners ate the yeast brew for several weeks, the deficiency disease improved in many prisoners (Emerson, 1952).

Experimentation with yeast as a food source is currently being carried out where sugary waste liquids are abundant. Such wastes from sugar beets, cassava, or sweet potatoes can be converted to protein by certain yeast species. Large-scale conversions of these tropical and subtropical crops to protein could significantly increase the protein intake of people in tropical countries.

SUMMARY

Subdivision Ascomycotina is composed of four classes. They constitute a rather large, diverse group of some well-known "higher fungi," ranging from unicellular yeasts to the mushroomlike morels and the tasty truffles. Also included are many animal parasites and disease-causing forms (e.g., ringworm), plant pathogens (e.g., ergot) and the common blue, green, red, and black molds (e.g., *Penicillium, Aspergillus,* and *Neurospora*). The major unifying characteristic of the subdivision is the presence of a sac-like ascus containing ascospores, which occur in small numbers, usually multiples of four, since they are a product of meiosis. Ascospores are often arranged in a linear fashion in groups of eight.

Life cycles of ascomycetes are haplontic, with both monokaryotic and dikaryotic hyphae represented in the haploid phase. The only diploid phase occurs at the uninucleate stage of ascus formation. Hyphae are partially septate, with uninucleate or binucleate cells. They may form extensive mycelia, which produce both sexual (asci) and asexual reproductive structures. The cell walls of the hyphae are composed of chitin and insoluble polysaccharides (glucans) in addition to proteins, peptides, and lipids.

Asexual reproduction is accomplished by hyphal fragmentation, budding, and production of various spore types, including chlamydospores and conidia. Conidia are borne on conidiophores, whose structure is often distinctive. Conidiophores may be produced in specialized mycelial structures known as sporocarps, such as pycnidia and acervuli. The nuclei of ascomycetes are small and difficult to see with light microscopy. Mitosis is intranuclear and is directed by a centriolar type of structure called the spindle pole body (SPB).

Sexual reproduction is always accompanied by nuclear fusion (karyogamy) and zygotic meiosis during ascus formation. It involves a series of complex cytological events resulting in crozier formation and eventual ascus production. Asci are sometimes formed in special fruiting bodies called ascocarps, the most common of which are the cleistothecium, perithecium, and apothecium. Parasexuality is also known among members of the subdivision.

The yeasts are an especially important group among the ascomycetes, since they play a key role in ecology and have significant human economic impact. Yeasts are used to produce ethyl alcohol and are therefore employed in producing alcoholic beverages (beer, wine, etc.). Another by-product of yeast metabolism is carbon dioxide, and yeasts are used for this purpose in the baking industry (bakers' yeast). Yeasts are also used in other manufacturing processes, including vitamin synthesis. Other ascomycetes are important because they produce antibiotic substances (e.g., *Penicillium*) and various organic molecules. Still others are used in food fermentations and diverse industrial processes (e.g., *Aspergillus*).

Many vascular plant diseases are caused by ascomycetes, and some have had a notable impact on history and agriculture throughout the world. In North America the chestnut blight and Dutch elm disease have changed the character of the forests and landscapes over the past 100 years. In medicine the ascomycetes are important as pathogens (ergotism, skin diseases) and in disease prevention and cure (penicillin).

Most ascomycetes are saprophytic. Many grow in places where they are not wanted, thus spoiling food, contaminating grains, and decomposing fabrics. They are important to energy flow in our ecosystems, where they decompose organic matter in forests and fields, thus transforming organic molecules into nutrients that are more readily available to other organisms. They may grow wherever there is a suitable substrate and will indiscriminately attack dead organic matter, whether or not it is useful to humans.

SELECTED REFERENCES

Anagnostakis, S.H. 1982. Biological control of chestnut blight. Science **215**:466-471.

Carefoot, G.L., and E.R. Sprott. 1967. Famine on the wind: man's battle against plant disease. Rand McNally & Co., Chicago. 299 pp.

Ciegler, A., and J.W. Bennett. 1980. Mycotoxins and mycotoxicoses. BioScience **30**(8):512-515.

Emmons, C.W., C.H. Binford, J.P. Utz, and K.J. Kwon-Chung. 1977. Medical mycology. Lea & Febiger, Philadelphia. 592 pp.

Heathcote, J.G., and J.R. Hibbert. 1978. Aflatoxins: chemical and biological aspects. Elsevier North-Holland, Amsterdam. 212 pp.

Hesseltine, C.W., and H.L. Wang. 1980. The importance of traditional fermented foods. BioScience **30**(6):402-404.

Kohlmeyer, J. 1973. Fungi from marine algae. Bot. Mar. **16**:201-215.

Matossian, M.K. 1982. Ergot and the Salem witchcraft affair. Am. Sci. **70**:355-357.

Strathern, J.N., E.W. Jones, and J.R. Broach (eds.) 1981-1982. The molecular biology of the yeast Saccharomyces. Monograph IIA, 751 pp., and IIB, 672 pp. Cold Spring Harbor Laboratory, Cold Spring Harbor, New York.

Strobel, G.A., and G.N. Lanier. 1981. Dutch elm disease. Sci. Am. **245**(2):56-66.

18

CLASSIFICATION

Division: Eumycota

   Subdivision 4: Basidiomycotina: "club fungi"; spores borne externally on small bases (the basidia); septate hyphae; zygotic meiosis

     Class 1: Hymenomycetes: spores forcibly discharged (ballistospores); septate or nonseptate basidia; basidia developed on a hymenial layer

       Subclass 1: Holobasidiomycetidae: single-celled, club-shaped basidia (holobasidia)

         ORDER 1. Aphyllophorales (Polyporales): leathery or woody basidiocarps; parasitic or saprophytic; *Amylostereum, Cantharellus, Clavaria* (coral fungus), *Digitatispora, Fomes, Ganoderma, Halocyphina, Heterobasidium, Hydnum* (tooth fungus), *Inonotus, Peniophora, Piptosporus, Polyporellus, Polyporus, Schizophyllum, Serpula, Stereum*

         ORDER 2. Agaricales: saprophytic, parasitic or mycorrhizal mushrooms; the gill fungi and the fleshy tube fungi; *Agaricus, Agrocybe, Amanita, Armillariella, Boletus, Clitocybe, Coprinus, Flammulina, Galerina, Lentinus, Marasmius, Mycena, Omphalotus, Pleurotus, Psilocybe, Termitomyces, Volvariella*

         ORDER 3. Dacrymycetales: small, gel-like, usually orange or yellow basidiocarps on dead wood; *Calocera; Dacryopinax*

         ORDER 4. Tulasnellales: saprophytic, parasitic and mycorrhizal with orchids; waxy basidiocarps on roots, foliage and fruits; some notable plant pathogens; *Rhizoctonia*

       Subclass 2: Phragmobasidiomycetidae: linear, four-celled septate basidia (phragmobasidia)

         ORDER 1. Tremellales: saprophytic "jelly fungi" with jellylike basidiocarps; *Tremella* (trembling fungus or "witches' butter")

         ORDER 2. Auriculariales: saprophytic and parasitic "ear-shaped" fungi with gelatinous basidiocarps; *Auricularia*

         ORDER 3. Septobasidiales: parasitic on scale insects; crusty, dry, or spongy fruiting bodies; *Septobasidium*

     Class 2: Gasteromycetes: basidiospores passively dispersed; single-celled, club-shaped basidia (holobasidia); hymenium not clearly present at maturity

         ORDER 1. Sclerodermatales: small, hard, warty puffballs; *Pisolithus, Scleroderma*

         ORDER 2. Lycoperdales: saprophytic, soft leathery puffballs and earthstars; *Calvatia, Geastrum, Lycoperdon*

         ORDER 3. Phallales: foul, odorous "stinkhorns"; basidiocarps resembling a phallus; *Mutinus, Phallus*

         ORDER 4. Nidulariales: "bird's nest" fungi; saprophytic, especially on dung; *Crucibulum, Cyathus, Sphaerobolus*

Class 3: Teliomycetes: thick-walled, binucleate, resting teliospore formation; vascular plant parasites; linear septate basidia; no basidiocarp formation

ORDER 1. Uredinales: rusts; all plant parasites, many on commercially important plants; complicated life cycles with special sex bodies, the spermatia and receptive hyphae; *Cronartium ribicola* (white pine blister rust), *Gymnosporangium, Hemileia, Puccinia graminis* (wheat rust or black rust), *Uredinopsis*

ORDER 2. Ustilaginales: smuts; mostly plant parasites that produce black spore masses; complex life cycles but lacking sex organs; *Filobasidiella, Melanotaenium, Tilletia, Urocystis cepulae* (onion smut), *Ustilago maydis* (corn smut)

The Basidiomycotina and the Ascomycotina from which they were apparently derived constitute the "higher fungi." Their degree of structural complexity far exceeds that of organisms we have studied earlier. They have haplontic life cycles, with the only diploid phase being the product of nuclear fusion during the formation of the basidium. This diploid cell (zygote) immediately undergoes meiosis, producing four haploid nuclei that are destined to become the four basidiospores.

The basidiomycetes comprise some of the most common and recognizable fungi, including mushrooms, puffballs, shelf fungi, earthstars, stinkhorns, coral fungi, and many vascular-plant pathogens. The major unifying feature of the entire group is the existence of a small base or **basidium** upon which the sexual spores (**basidiospores**) are formed following meiosis (Figs. 18-1 and 18-2). All basidiomycetes produce basidia. Another microscopic feature is "clamp connections" produced in the dikaryotic mycelia during cell division (Fig. 18-2); they occur in all the basidiomycetes except some important plant parasites belonging to the subclass Teliomycetes. A third microscopic structure, usually detectable only with electron microscopy, is the **dolipore septum.** Mature dolipore septa of the sort seen in Fig. 18-3 are characteristic of dikaryotic hyphae of all basidiomycetes except for some members of the Teliomycetes (the rusts and smuts).

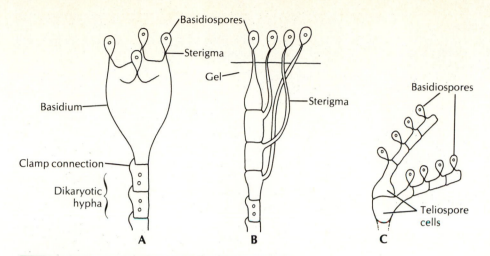

FIG. 18-1 Types of basidia.
Representative basidial types showing basidiospore formation in three groups of basidiomycetes: **A,** Single-celled, club-shaped basidium (holobasidium) from common mushroom (Agaricales); **B,** linear, four-celled septate basidium (phragmobasidium) from jelly fungus (Tremellales); **C,** linear septate basidium developing directly from resting teliospores of parasitic rust (Uredinales).

FIG. 18-2 Clamp connection and basidium formation.
Development of clamp connections, **A** through **F,** as the terminal cell in the dikaryotic hypha undergoes mitosis. Note how process ensures that each new cell will contain one light and one dark nucleus from the parent cell. **G,** Nuclear fusion or karyogamy with the resultant diploid nucleus, which divides by two successive meiotic divisions, **H** and **I.** The result of meiosis II is four separate haploid nuclei that form the four basidiospores found in mature basidia, **J.**

Identifying Characteristics

Most basidiomycetes produce macroscopic spore-bearing fruiting bodies called **basidiocarps,** which become quite complex in structure. The common mushroom is one example of an elaborate basidiocarp (Fig. 18-4). The use of macroscopic characteristics such as color, size, and shape in the scientific classification of basidiomycetes is considered by many mycologists to be too artificial, although their use presents a generally workable and convenient system for identification (Table 18-1).

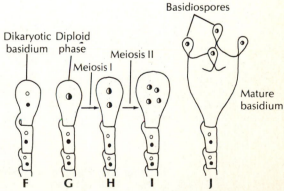

FIG. 18-3 Dolipore septum.

Septa between adjacent cells in basidiomycete hyphae resemble the neck of a large jar and are called dolipore septa. Each septum is traversed by a single pore that is covered by a septal pore cap composed of endoplasmic reticulum.

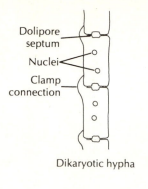

Dikaryotic hypha

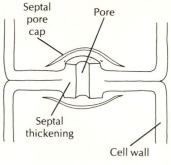

Dolipore septum

Large jar
(Latin: dolium)

TABLE 18-1. Features of Common Basidiomycetes and Confusing Ascomycetes

Fungus	Description	Common name	Order
Fungi with cap, with or without stem	Presence of gills beneath cap	Agarics, gilled mushrooms	Agaricales
	Presence of tubes easily separable beneath cap	Boletes	Agaricales
	Usually stalkless woody cap with tubes beneath	Polypores, shelf fungi	Aphyllophorales
	Funnel- or vase-shaped cap with gills beneath	Chanterelles	Aphyllophorales
	Presence of spikes or teeth beneath cap	Hydnums, teeth fungi	Aphyllophorales
Fungi lacking cap	Fruiting body coral-like or shrublike	Clavarias, coral fungi	Aphyllophorales
	Fruiting body globular or pear shaped	Puffballs	Lycoperdales and Sclerodermatales
	Fruiting body star shaped	Earthstars	Lycoperdales
	Fruiting body resembling small bird's nest with eggs	Bird's nest fungi	Nidulariales
	Fruiting body resembling phallus with foul smell	Stinkhorns	Phallales
	Fruiting body jellylike, yellow, orange, or red, growing on dead wood	Trembling or jelly fungi, "witches' butter"	Tremellales and Dacrymycetales
	Fruiting body ear shaped, gelatinous to leathery	Ear fungi	Auriculariales
	Fruiting body in form of chambered sponge, often resembling pine cone	Morels	Ascomycetes (Pezizales)
	Fruiting body stalkless, resembling small, flattened cup	Cup fungi or pezizas	Ascomycetes (Pezizales)

FIG. 18-4 Representative forms of basidiocarps and ascocarps.
Major forms of basidiomycetes and ascomycetes (listed in Table 18-1). Note that morels and cup fungi, seen at lower right, are ascomycetes, whereas all others are basidiomycetes.

Gilled mushroom

Bolete

Shelf fungus

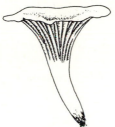

Chanterelle

Hydnum
or
tooth fungus

Coral fungus

Puffball

Earth star

Bird's nest fungus

Stinkhorn

Jelly or trembling fungus

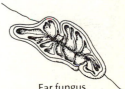

Ear fungus

Morel

Cup fungus

CELL STRUCTURE

Unlike flagellate fungi and zygomycetes, which have coenocytic hyphae, and ascomycetes, which have partially septate hyphae, the basidiomycetes have septate hyphae that contain uninucleate or binucleate cells. Therefore they are composed of monokaryotic hyphae with uninucleate cells during the primary mycelial stage and dikaryotic hyphae during the secondary and tertiary mycelial stages (Fig. 18-5). The structural features of individual cells include nuclei, various organelles, and other cellular inclusions that are similar in most respects to those that were discussed in the chapters on ascomycetes and other fungi. This includes the nucleus-associated organelle (NAO), which appears in the position occupied by centrioles in flagellate cells.

The septa associated with the hyphae have a distinctive structure. At maturity they possess a small central pore similar to that associated with ascomycetes, but unlike the ascomycetes, the pore is surrounded by a doughnut-shaped thickening or lip that resembles the mouth of a large water jar (Fig. 18-3)—thus the name dolipore septum (fr. Latin *dolium* = large jar). Dolipore septa are especially characteristic of mature dikaryotic hyphae. They appear to be evolutionary elaborations on the septal structure of ascomycetes, from which they were presumably derived.

The pores in the dolipore septa are small, but movement of mitochondria and other organelles has been observed through them. Even nuclei have been known to pass through the pores at

FIG. 18-5 Life cycle of an agaric mushroom.
Most of the life history is spent underground as a mycelium. The primary mycelial stage *(A)* is composed of individual hyphae designated (+) and (−). Fusion of compatible hyphae produces a dikaryotic secondary mycelium *(B)*, which may exist for extended time periods. The familiar mushroom or tertiary mycelium *(C)* is differentiated from organized masses of secondary mycelia. A uninucleate diploid cell is formed when individual nuclei from a dikaryotic cell fuse within individual basidia *(D)*, and this stage is quickly followed by meiosis producing four haploid nuclei, which migrate into the four basidiospores *(E)*.

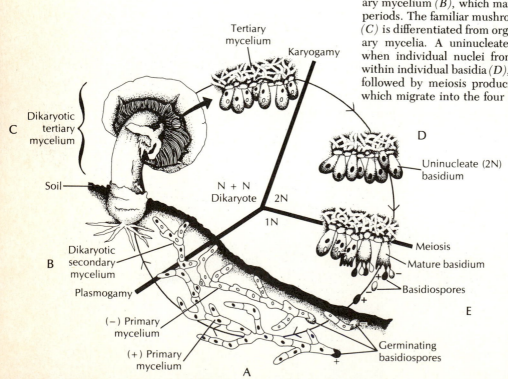

certain early stages in the development of the hyphae (Janszen and Wessels, 1970). Each central pore is covered with a membranous cap of endoplasmic reticulum known as the **septal pore cap.** During hyphal development, changes occur in the dolipore septal apparatus. Early in their development the septa permit a free exchange of nuclei and cellular organelles from one cell to another. Later in their ontogeny, the openings through the septa become occluded, and movement of substances from one cell to another is impeded or completely stopped (Flegler et al, 1976). The precise role of dolipore septal structures still has not been explained to everyone's satisfaction, but it would appear that they certainly are involved in the movement of nuclei and other organelles from cell to cell during the early stages of mycelial development.

Small channels of cytoplasm called clamp connections join adjacent cells of many dikaryotic basidiomycete hyphae (Figs. 18-1, 18-2, and 18-3). Their main function is to ensure the segregation of the two genetically similar nuclei found in the dividing cells at the tips of dikaryotic hyphae.

REPRODUCTION AND GROWTH

Although both asexual and sexual reproductive strategies play an important role in basidiomycetous fungi, the asexual methods generally play a less common role than in other groups. Fragmentation of hyphae, conidia, chlamydospores, and a special spore type, the **oidium,** are all common methods of asexual reproduction. Conidia are most prevalent in the class Teliomycetes, where specialized conidia are seen in the form of **aeciospores** and **urediniospores** (Fig. 18-8). Hyphal fragmentation is carried out by most genera to some extent, and many basidiomycetes form thick-walled chlamydospores within their hyphal filaments. A fourth form of asexual reproduction is found in the production of specialized hyphal filaments, **oidiophores,** which bear oidia (fr. Greek *oidion* = egg, diminutive). Oidia may be produced by both mono-

karyotic or dikaryotic hyphae, thus forming uninucleate or binucleate oidia. These cells act as spores and may germinate to form a new hypha, or, if they are uninucleate, may act as a spermatium by fusing with a compatible uninucleate hyphal filament, thus producing a new dikaryotic hypha. Oidia are dispersed by water, air current, or insect vectors.

Sexual reproduction in basidiomycetes is found in every species and always results in the formation of a basidium. The shape and form of the basidium may vary (Fig. 18-1). However, only the rusts (order Uredinales) produce specialized sex organs for production of sex-associated cells. These are the spermatia and **receptive hyphae,** both of which play a role in bringing together sexually compatible nuclei (Fig. 18-8). The rest of the basidiomycetes carry out the sexual process through fusion of compatible monokaryotic hyphae, thus forming a dikaryotic hypha, the secondary mycelium. The fusion of these hyphae, therefore, brings about the process known as plasmogamy, or the fusion of protoplasm without the fusion of nuclei. Although each nucleus in dikaryotic cells remains haploid, the cells act as functional diploid cells, since they contain the nuclear materials of both + and − strains of the same species. Karyogamy, or nuclear fusion, is accomplished later during formation of the basidium by the tertiary mycelium (Fig. 18-5).

We shall describe, by way of example, the life cycle of a common agaric field mushroom from the genus *Agaricus*. The hyphae of *Agaricus* and other basidiomycetes form mycelia that develop in three separate phases during the life cycle of the fungus; these are called the primary, secondary, and tertiary mycelia (Fig. 18-5). When haploid, uninucleate basidiospores germinate, they form the primary mycelium composed of uninucleate or monokaryotic hyphae, which contain genetically identical nuclei (and are therefore homokaryotic). In many basidiomycetes, these hyphae form separate + and − strains (sometimes referred to as male and female strains). The monokaryotic phase

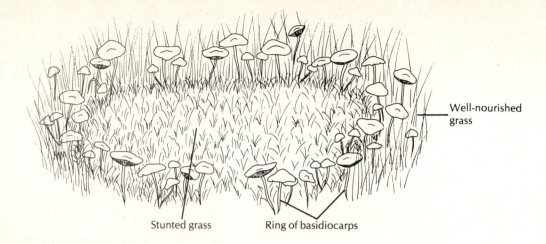

Well-nourished grass

Stunted grass

Ring of basidiocarps

FIG. 18-6 "Fairy ring" formed by *Marasmius* sp.
Fungal spore germinates into primary and secondary mycelia, which grow outward in a circle. Basidiocarps are formed at the edges of the secondary mycelial growth, forming a ring of basidiocarps. Similar environmental conditions induce simultaneous formation of basidiocarps. Such fairy rings are often seen in lawns, fields, and parks in summer and autumn.

FIG. 18-7 Mushroom growth and development.
Manner in which a common agaric mushroom develops from a secondary mycelium. Each drawing represents an older stage of the developing, mature basidiocarp (mushroom), seen on extreme right.

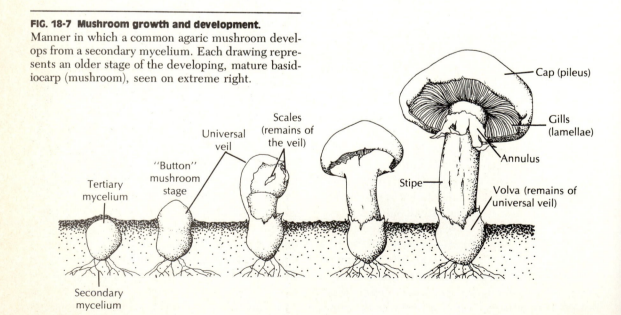

Cap (pileus)

Gills (lamellae)

Annulus

Stipe

Volva (remains of universal veil)

Scales (remains of the veil)

Universal veil

"Button" mushroom stage

Tertiary mycelium

Secondary mycelium

generally exists for only a short time. Fusion of compatible hyphae produces a secondary mycelium that is made up of binucleate cells and forms dikaryotic hyphae (mycelia). The dikaryotic secondary mycelium often forms the most extensive growth for any given species, although it is seldom observed, since in saprophytic forms it develops beneath the surface in organically rich soils. Secondary mycelia may, however, be observed without a microscope as slimy, stringy, white, yellow, or orange growths around the roots of grass or trees, beneath dead wood, or in humus-containing soils. The secondary mycelium may persist for years in any given location, producing tertiary mycelia called **mushrooms** (basidiocarps) year after year. The result of this growth pattern may be observed in the large circles of mushrooms that grow on grass lawns, golf courses, and forest floors. Such developments are known as "fairy rings," since folklore has it that these mushrooms were the seats for fairies and gnomes when they danced and played during the night (Fig. 18-6).

Under certain conditions of light, temperature, humidity, and nutrient concentrations, secondary mycelia differentiate into tertiary mycelia (the basidiocarp), thus forming the familiar mushroom or the fruiting body of some other basidiomycetes. The tertiary mycelium is actually a specialized dikaryotic group of hyphae whose major function is the sexual recombination of genetic factors. Most basidiomycetes produce their basidia within specialized layers called hymenia (sing. hymenium), which occur within the basidiocarp. In mushrooms, hymenia line individual gills or lamellae beneath the cap or **pileus** (Fig. 18-7). The number of basidiospores produced by any individual is often quite high (Table 18-2).

The tertiary mycelium (basidiocarp) of many basidiomycetes is a complex structure with organ-level differentiation. The manner by which mushrooms develop into a functional basidiocarp is presented in Fig. 18-7. Other forms of basidiocarps found among the basidiomycetes are illustrated in Fig. 18-4. Note especially the common polypores or shelf fungi, coral fungi, puffballs, earthstars, and stinkhorns. Basidiocarps are not produced by members of the Teliomycetes, a class of Basidiomy-

TABLE 18-2. Basidiomycete Spore Numbers

Fungus	Total number of spores	Spore drop period	Number of spores dropped per day
Calvatia gigantea (Giant puffball)	7×10^{12}	—	—
Ganoderma applanatum (Artist's shelf fungus)	5.46×10^{12}	6 months	3×10^{10}
Polyporus squamosus (Shelf fungus)	5×10^{10}	14 days	3.5×10^{9}
Agaricus brunnescens (bisporus) (Commercial mushroom)	1.6×10^{10}	6 days	2.6×10^{9}
Coprinus comatus (Shaggy-mane mushroom)	5.2×10^{9}	2 days	2.6×10^{9}
Puccinia graminis (Wheat rust)	—	14 days	10^{6}/uredium

From Buller, A.H.R. 1922. Researches on fungi, Vol. 2. Longmans, Green & Co., Ltd., London. 492 pp.
Blanks indicate information not available.

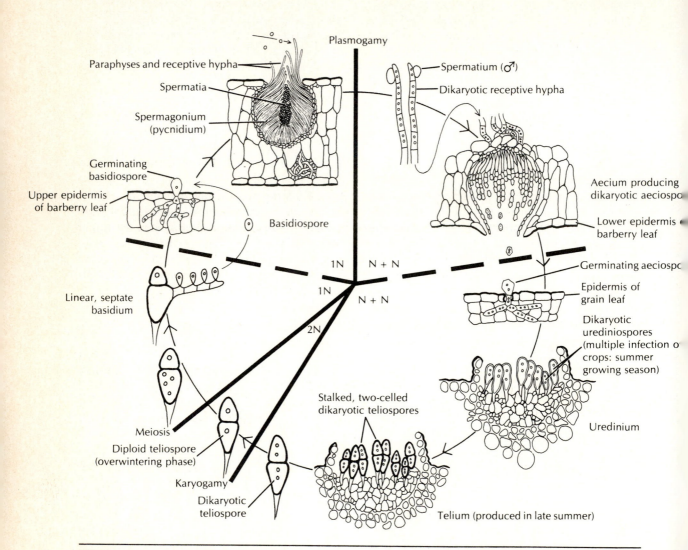

FIG. 18-8 Life cycle of *Puccinia graminis*, a rust.

Uredinales (rusts) all have complicated life cycles with one or two intermediate hosts. *P. graminis* spends part of its life as a parasitic fungus in barberry leaves and part as a parasite in the leaves of various grains. Linear septate basidia develop in the spring, producing four basidiospores, two of which are (+) and two (−). These infect barberry leaves and produce (+) or (−) spermagonia on leaf's upper surface. Spermagonia produce (+) or (−) receptive hyphae and (+) or (−) spermatia. Cross-fertilization is ac-complished by insects carrying spermatia; (+) spermatia fuse with (−) receptive hyphae, producing dikaryotic hyphae that in turn form aecia on the leaf's lower surface. Aeciospores are then wind dispersed and infect the leaves of cereal grains, producing uredia. Uredinia create multiple infections through-out the summer. In late summer, telia are formed that produce thick-walled teliospores. Teliospores overwinter and germinate the following spring, form-ing linear septate basidia.

cotina that is composed entirely of vascular plant parasites including the smuts (Ustilaginales) and rusts (Uredinales). These orders produce basidia directly upon germination of dikaryotic resting spores called **teliospores** (fr. Greek *telos* = end). Teliospores represent the end of the life cycle, and their germination is the beginning of a new generation (Fig. 18-8).

Reproduction and growth within the class Teliomycetes differ considerably from that of other basidiomycetes. Moore (1972) has even placed the smuts within a separate division, the Ustomycota, based on such differences. However, we shall retain the group as a separate class of the Basidiomycotina until further developments indicate otherwise. The Teliomycetes retain their taxonomic position among the basidiomycetes because they form basidia and basidiospores. The thick-walled teliospores are considered to be resting basidia; when they germinate, they form a septate basidium, which normally produces four basidiospores (Fig. 18-1) in the Uredinales (**rusts**) and multiple basidiospores in the Ustilaginales (**smuts**).

Besides basidiospore formation, the Teliomycetes differ from other basidiomycetes in a number of respects. First, as previously mentioned, they never produce large, macroscopic basidiocarps. Second, although their hyphae are septate with a small central pore in each septum, they do not normally produce a dolipore type of septum with a septal cap as described previously for the Basidiomycotina. An exception to this rule is found in the genus *Tilletia*, a common parasite of wheat known as **bunt,** one of the smuts found on various cereals. Third, most of the Teliomycetes are haustorial parasites of vascular plants. In general the rusts are confined to angiosperm (flowering plant) hosts, whereas the smuts are parasites of angiosperms, gymnosperms, and ferns. And finally, unlike most basidiomycetes, the Teliomycetes have complex life cycles involving one or two unrelated alternate vascular plant hosts.

A good example of this is the economically important species *Puccinia graminis,* which parasitizes both barberry (dicotyledonous angiosperm) and cereal grains (monocotyledonous angiosperms) during its life history (Fig. 18-8). The life cycles of *P. graminis* and other rusts are complicated not only by their reliance on one or two alternate hosts but also by production of different types of asexual spores as well as specialized sexual hyphae. The cycle begins with production of haploid basidiospores in the spring. These are dispersed by the wind to leaves of common barberry plants (*Berberis vulgaris*), where they germinate and penetrate the leaf tissues, producing haploid, monokaryotic hyphae with haustoria. At maturity these hyphae appear as yellow pustules on the upper surface of the leaf; these pustules are called spermagonia. (They are often referred to as pycnidia because of their shape; compare with the asexual reproductive pycnidium of Fig. 17-3.) Mature spermagonia are designated as + or −, since they produce + or − receptive hyphae and + or − spermatia. The receptive hyphae are colored orange-yellow. Receptive hyphae and sterile, hairlike paraphyses emerge from the upper surface of the barberry leaf and secrete a sweet-smelling ooze that attracts insects. This sets up a symbiotic relationship analagous to the insect-plant symbiosis that is responsible for pollination in flowering plants. Insects are attracted to and eat the nectar; spermatia stick to them and are passed to a receptive hypha. When positive (+) spermatia are transferred to negative (−) receptive hyphae, or when negative (−) spermatia are transferred to positive (+) receptive hyphae, nuclear transfer by plasmogamy is accomplished. Nuclear migration and mitoses follow, resulting in dikaryotization of the infecting hyphae. Shortly thereafter (about 3 days), dikaryotic aeciospores appear in the early stages of blisterlike structures called **aecia,** which occur on the ventral surface of the barberry leaf. Several cup-shaped, orange-colored aecia are often bunched together on the ventral leaf surface beneath a single spermagonium.

Aecia in this condition are referred to as the "cluster cup" of infection. They are clearly visible to the naked eye.

Dikaryotic aeciospores are wind dispersed. They can infect only certain monocotyledonous plants such as wheat and other cereal grasses, including oats, barley, and rye. Germinating aeciospores penetrate the leaf tissues of the cereal host, producing haustorial hyphae that in turn may form two types of spores—urediniospores and teliospores. Urediniospores are single-celled, dikaryotic, specialized conidia. They are produced throughout the cereal-growing season, are wind dispersed, and may continually reinfect other plants during the course of a summer. Urediniospores are produced in large, rust-colored blisters called **uredinia** (sing. **uredinium**); thus the common name "rust" has been given to these fungi.

Teliospores are dikaryotic spores produced in special pustules called **telia** (sing. **telium**). Telia form only in late summer during senescence of individual cereal plants. Teliospores have thicker cell walls than urediniospores and appear black. Teliospores are a type of resting spore, since the fungus survives the winter in this condition. When spring arrives, karyogamy occurs in the teliospores and is immediately followed by meiosis. Teliospores then produce germination tubes that become septate basidia bearing four haploid basidiospores. The cycle then begins again (Fig. 18-8).

The life cycles of members of the Ustilaginales (smuts) are somewhat simpler than that just described. No sex organs are produced and therefore no spermagonia, spermatia, or receptive hyphae are created. Aecia and uredinia are also lacking. Plasmogamy is accomplished through fusion of two compatible hyphae, basidiospores, or conidia. All hyphal elements may be converted to spores, or they may form buds much like those observed in yeasts. In fact, asexual reproduction of the Ustilaginales is so much like that of the yeasts that they are sometimes referred to as the "basidiomycetous yeasts," along with certain members of the imperfect fungi (Deuteromycotina).

The Ustilaginales are capable of producing large numbers of spores in a dark-colored powdery mass that is usually associated with the flower or fruit of the infected angiosperm. The spores are teliospores (although they are often called chlamydospores in older texts). The teliospores have characteristic wall ornamentation that is often used to identify the various smuts.

Most rusts and smuts are heterothallic, which means that the thalli are not self-fertile. The incompatibility of hyphae from the same thallus is controlled by multiple genes within each species. The haploid phase or monokaryotic phase of smuts is often a weak parasite. However, once plasmogamy between two compatible hyphae has taken place, the infectiousness of the dikaryotic phase is increased. Dikaryotic hyphae may cause the formation of large tumors or galls because of increased growth (hypertrophy and hyperplasia) within the host-plant tissues and organs such as flowers or seeds. The galls in turn may produce more dikaryotic teliospores.

In some instances the nuclei of teliospores may fuse, forming diploid teliospores that may be dispersed when the fungal mass or gall causes the tissues of the host to rupture. The "bunt" or "stinking smut" of *Tilletia caries*, a major parasite of wheat, produces hyphae that invade the wheat seeds in autumn. The hyphal segments lie dormant until germination of the seed after planting the following spring. When the young wheat plant begins growing, it is already infected with fungal hyphae. It is interesting that the name "stinking smut" originates from the foul odor produced by the by-product triethylamine, which is formed by the fungus.

PHYSIOLOGY

The Basidiomycotina are primarily saprophytes, but the division includes some economically important plant pathogens. Species from various genera such as *Polyporus*, *Fomes*,

FIG. 18-9 Wood-rotting bracket fungi.

A, *Polyporus* spp. (Aphyllophorales), **B,** *Pleurotus* spp. (Agaricales), and **C,** *Fomes* spp. (Aphyllophorales) may grow on living trees or dead wood. These genera produce exoenzymes that can degrade both cellulose and lignin. Identification of the different bracket fungi is difficult; spore prints may aid in the identification of both the Aphyllophorales and the Agaricales. Spore prints are made by cutting off a mature pileus and placing it, gills or pores down, on white paper. Covering the pileus with a bowl will prevent disturbance of the spores by air currents. Spores will be shed and a spore print can be obtained in 6 to 8 hours. Both the color of the spores and the pattern they create on the paper help to positively identify the basidiomycete. See Arora (1979) for further information about spore prints.

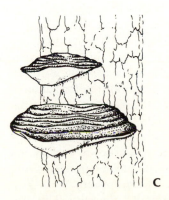

Stereum, and *Pleurotus* (Fig. 18-9) can degrade the complex and recalcitrant carbohydrates cellulose and lignin and are therefore often found in woodlands on living or dead trees. Many common woodland basidiomycetes can decompose cellulose, and some such as *Marasmius, Armillariella,* and *Ganoderma* can attack both cellulose and lignin, which is extremely resistant to decomposition (Crawford, 1981). Many basidiomycetes form mycorrhizal associations with commercially valuable trees and crops, while other members of the subdivision are very destructive obligate parasites of crops.

The requirements for vitamins vary considerably among the different species of the basidiomycetes. Many members of this subdivision appear to be able to use ammonium nitrogen as a nutrient source. *Coprinus* spp. can utilize both nitrate and urea.

Several basidiomycetes have been grown successfully in the laboratory: *Schizophyllum commune, Polyporellus brumalis* (= *Polyporus brumalis*), *Agaricus brunnescens* (= *A. bisporus*), *Flammulina velutipes, Coprinus* spp., and *Rhizoctonia solani* have all been grown and studied. Considerable information is available on the biochemistry, genetics, physiology, ultrastructure, and ecology of these species. *Tremella* (trembling or jelly fungus, Fig. 18-4) has also been grown in the laboratory, and the presence of diffusible hormones has been demonstrated during sexual reproduction (Reid, 1974). Sakagami et al (1978) have isolated one of these hormones called **tremerogen.**

Some of the teliomycetes—although normally obligate parasites—have also been cultured in the laboratory. The devastating *Puccina graminis* f. sp. *tritici*, black stem rust of wheat (Fig. 18-8), has been grown in axenic culture and much has been learned about its life cycle. Coffey (1975) discusses the culture of rusts. The spores of many Ustilaginales have also been cultured in the laboratory and the knowledge of their biology has thereby been increased.

Basidiocarp Development

Much of the information on the induction and differentiation of basidiocarps has come from the large-scale culture of mushrooms sold commercially for food, especially *Agaricus brunnescens* (= *A. bisporus*). Bonner et al (1956) and Deacon (1980) discuss the development of *A. brunnescens*. Fruiting may be stimulated by various bacteria in the soil (Park and Agnihotri, 1969; Hayes, 1972).

Schizophyllum commune, a species widely distributed on decomposing wood, has been used extensively in studies of basidiomycete genetics and basidiocarp development. Wessels (1965) described four stages in the basidiocarp development: (1) undifferentiated growth, (2) **primordia** initiation, (3) primordial growth, and (4) pileus formation. Various nutrients are required at the first three stages, but no exogenous nutrients are needed for the formation of the pileus (Griffin, 1981). The initiation of the primordia is induced by a reduction in exogenous nitrogen. If carbon becomes a limiting factor, the fungus can break down its own cell wall and recycle the component nutrients into pileus development (Niederpruem and Wessels, 1969. Rusmin and Leonard (1975) isolated a low-molecular-weight substance, named **fruiting-inducing substance** (**FIS**), that caused localized fruiting on certain sections of the mycelium of *S. commune*.

Reduction in available nutrients including carbohydrates also induces fruiting in *Coprinus macrorhiza*, whereas a high level of glucose inhibits fruiting (Uno and Ishikawa, 1974). Considerable research has also been done on the development of basidia and basidiospores in other *Coprinus* species—for example, *Coprinus cinereus* (= *C. lagopus*) (Mathews and Niederpruem 1972; 1973) and *Coprinus congregatus* (Ross, 1979).

Differentiation of the basidiocarp varies within this subdivision. Generally environmental factors such as moisture, light, temperature, and aeration combine with such physiological variables as nutrition and maturity to regulate the development of the basidiocarp. Once the primordium is formed it may grow very rapidly—up to 10 times its original size. The basidiocarps of some *Coprinus* and *Agaricus* species grow so quickly that they can crack an asphalt pavement, occasionally with a loud noise. Certain *Coprinus* species produce basidiocarps that can appear above ground, grow, mature, and start to digest themselves—all in the course of a few hours.

In some mushrooms—for example, *Coprinus cinereus* (= *C. lagopus*)—the rapid appearance of the fruiting body—overnight in some instances—appears to be caused by the uptake of water and the concomitant increase in size of the hyphal segments. Such fruiting body development is therefore designated determinate, because the increase in size results primarily from the expansion of cells already formed in the primordia. In *Schizophyllum commune*, however, the basidiocarp apparently differentiates as it enlarges (Ross, 1979), and the development is termed indeterminate.

Regulatory compounds, analagous to hormones, evidently move from the gills to the stipe in *Agaricus brunnescens* and *Schizophyllum commune* (Gruen, 1963, 1969), but in *Co-*

prinus cinereus elongation of the stipe appears to be regulated autonomously without chemical signals from other portions of the basidiocarp (Gooday, 1974).

The dikaryotic secondary mycelium grows extensively and assimilates many nutrients before the fungus forms a fruiting body. As in other subdivisions, the fungus must reach a certain physiological or chronological competence (see Chapter 13) before the basidiocarps develop. In culture the mycelium of *Coprinus congregatus* must be at least 4 days old before basidiocarps will start to form. This species appears to require light to induce formation of the fruiting body (Ross et al, 1976). A small molecule such as glycine or a trauma such as cutting the mycelium will also induce fruiting in *C. congregatus* if the fungus is mature. Ross (1979) suggests the higher fungi may respond to several different fruiting-body inducers, provided the mycelium is sufficiently mature.

Light is also required for pileus development in some Agaricales such as *Agaricus*, *Flammulina*, and *Pleurotus* and in some Aphyllophorales such as *Polyporus*. In some Hymenomycetes whose mycelia grow in dark areas, light may stimulate not only the directional growth of the stipe but also the development of the pileus.

Decomposition of Wood

The growth of Basidiomycotina such as *Schizophyllum*, *Stereum*, and *Fomes* causes degradation of the heartwood (i.e., dead interior wood) and the sapwood (i.e., live outer wood) in forest and shade trees. When heartwood is attacked the result is called "heart rot," and the tree is seriously weakened structurally and subject to blowing over. The timber value of the tree is also reduced. Two kinds of heart rot are recognized: **white rot** and **brown rot.** The former is caused by fungi that decompose both lignin and cellulose, leaving the wood a whitish color; the latter is caused by fungi that decompose the cellulose primarily and leave the brownish lignin. In both types of rot the cellulose content is reduced, but the lignin content is reduced only in white rot (Findlay, 1940).

The decomposition of cellulose and lignin is a complex process involving many different exoenzymes. The cellulose microfibrils may either be sheathed or complexed with lignin, making them inaccessible to the cellulolytic enzymes. Wood-decomposing fungi may either modify the ensheathing lignin, gaining access to the cellulose, or they may degrade the lignin and cellulose simultaneously. The ligninolytic fungi produce extracellular **phenoloxidases,** which probably aid in the decomposition of lignin (Kirk, 1971). Kirk et al (1977, 1978, 1980) and Crawford (1981) discuss the degradation of lignin in detail. Most of the hundreds of species of white-rot fungi are members of the Aphyllophorales or the Agaricales. The various species differ as to which wood components are at-

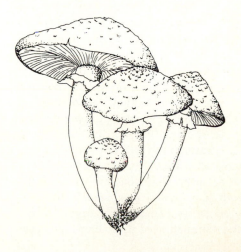

FIG. 18-10 *Armillariella mellea* (= *Armillaria mellea*), **the honey mushroom.**
Armillariella mellea may be both a saprophyte and a parasite. It may be mycorrhizal with an orchid and simultaneously parasitize a nearby tree, causing root rot. Both the mycelium and the rhizomorphs may exhibit luminescence. Basidiocarps are usually found in fall or winter on the ground or on dead or living trees.

tacked and the rate at which decomposition proceeds.

Fungi that degrade heartwood are technically saprophytes because they break down only dead wood. Species of *Stereum* and *Fomes* (Fig. 18-9) invade the living sapwood and cause considerable damage to or even the death of certain trees. The mycelia of some Aphyllophorales may grow in heartwood for years before their presence is detected by the appearance of a basidiocarp on the bark or on a root some distance from the stem of the tree.

The honey mushroom, *Armillariella mellea* (=*Armillaria mellea*) (Fig. 18-10), may be either saprophytic or parasitic, causing root rot of dead or living gymnosperms and angiosperms (Fig. 18-25). This fungus has also been called the "shoestring" or "bootlace" fungus because it produces **rhizomorphs,** which are strong, fast-growing, thick filaments composed of up to 1,000 hyphae. These may be found under the bark of fallen trees or in wounds of living trees.

Spore Discharge and Dispersal

The basidiospores of many Holobasidiomycetidae, Phragmobasidiomycetidae, and all Teliomycetes are violently discharged in sequence from the basidium. Such spores have been labeled **ballistospores.** The mechanisms by which the spores are forcibly discharged are controversial. Immediately before discharge a drop of an unidentified liquid substance appears and grows in size. When the drop reaches a certain volume, the spore is ejected from the basidium a distance of 0.10 to 0.20 mm. Ingold (1971) discusses the mechanisms of spore discharge.

Spore dispersal mechanisms in the Basidiomycotina are as varied as they are interesting. As previously noted, in the Aphyllophorales and the Agaricales the basidiospores are forcibly discharged, fall through the gills or pores, and may be dispersed by wind or occasionally by water or animals. The very lightweight spores of the rusts (Uredinales), often produced by the billions, are easily and widely dispersed by the wind. The combination of the billions of spores produced and easy wind dispersal is perhaps partly responsible for the widespread occurrence of these devastating plant diseases, especially wheat rust. Some wheat rust urediniospores have been blown as far as 3,300 km from Mexico to Canada on the prevailing southern winds. A concentration of 100 urediniospores per m³ of *Puccinia graminis* has been recorded 4,500 m above central Canada, and hundreds of urediniospores per m³ have been found 3,000 m above the North Atlantic (Ross, 1979).

The Lycoperdales and Sclerodermatales produce spores that may be discharged through the ostiole in *Lycoperdon* (Fig. 18-11) or through the cracks in the outer covering (i.e., peridium) in puffballs such as *Calvatia* and *Scleroderma* (Fig. 18-12). Raindrops striking the *Lycoperdon* basidiocarp cause the strong and elastic peridium to compress, and the spores are ejected in a

FIG. 18-11 *Lycoperdon gemmatum* (= *Lycoperdon perlatum*), **the gem-studded puffball.**
This puffball is common on the edges of woods and usually grows in clumps of three or more. At maturity an ostiole forms at the tip. Spores are released in a "puff" through the ostiole when the basidiocarp is disturbed by an animal or pounded by rain. This species is the most frequently encountered puffball in North America.

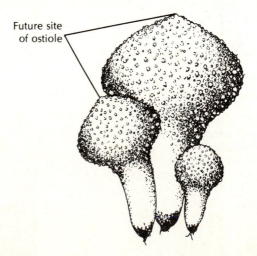

Future site of ostiole

FIG. 18-12 *Scleroderma* sp., the earthball.
This fungus grows in woods and fields. Basidiocarps are usually produced in late summer or early fall. Spores are released when tough outer covering, the peridium, is ruptured. Dark gleba inside contains the spores. When mature basidiocarp is disturbed, spores are released in a pufflike cloud.

FIG. 18-13 *Calvatia gigantea*, the giant puffball.
This puffball measured 50 cm by 20 cm and weighed 3.5 kg. It was found growing in a field in eastern Pennsylvania in October. At maturity the peridium will crack open and release millions of spores. Some *Calvatia* species are not firmly attached to the substrate and may be knocked or blown free; they then roll about, distributing the basidiospores through the cracked peridium. (Courtesy *The Express*, Easton, Pa., Midic Castelletti, photographer.)

FIG. 18-14 Spore dispersal in bird's nest fungus, *Cyathus* sp.

In *Cyathus* the basidiocarp is composed of a "nest-like" peridium containing several oval peridioles or "eggs." The peridioles are a portion of the basidiospore-containing gleba surrounded by a covering. Peridioles are attached to the peridium by a cord, the funiculus, which is composed of hyphae. When a raindrop falls into the peridium ("splash cup"), it displaces the mature peridioles with their attached funiculi. Peridioles may travel up to 1 m before the funiculus, with its adhesive end, catches on nearby vegetation and winds around it. The vegetation may then be eaten by an animal, and *Cyathus* spores will be deposited with the animal dung. Bird's nest fungi are common in fields, lawns, and forests, especially where manure has been deposited.

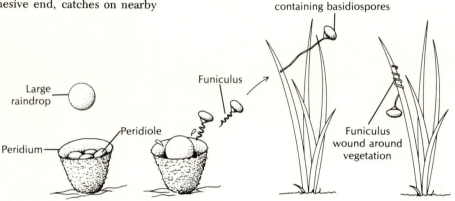

FIG. 18-15 Spore dispersal in *Sphaerobolus* sp.

Basidiocarp of *Sphaerobolus* is initially shaped like a ball. Peridium consists of several layers surrounding the basidiospore-containing peridiole. At maturity, peridium splits open and the large peridiole is ejected for distances up to 6 m (Buller, 1933). Ejection is caused by the layers of peridium separating and the inner layers increasing in osmotic pressure and expanding outward, thereby ejecting the peridiole. *Sphaerobolus* grows on both animal dung and dead wood.

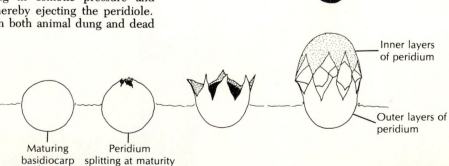

puff of "smoke." The spore-release mechanism in *Lycoperdon* has been likened to a bellows. The giant puffball, *Calvatia gigantea* (Fig. 18-13), can produce about 7×10^{12} spores over a 6-month period.

In the bird's nest fungi (Nidulariales), the **gleba** or spore chamber is differentiated into several spherical structures *(Cyathus)* or a single spherical structure *(Sphaerobolus)*. These glebal spheres are called **peridioles.** The spore-containing peridioles of *Cyathus* (Fig. 18-14) are forcibly ejected by water droplets falling into the peridium, which serves as a splash cup (Brodie, 1951). *Sphaerobolus* (Fig. 18-15) means "sphere thrower"; it is aptly named, since the single, round, large peridiole may be ejected for a distance of 560 cm by a raindrop (Brodie, 1975, 1978). The immature peridium contains osmotically inactive glycogen, but at maturity the glycogen is enzymatically converted to osmotically active glucose. The rising concentration of glucose causes the cells of the peridium to take up more water, thereby increasing the turgor pressure (Engel and Schneider, 1963). The peridium inner layers swell and

the peridiole is discharged toward the light. Peridioles containing basidiospores may remain viable for several years (Ingold, 1972).

Another interesting spore dispersal mechanism observed in the basidiomycetes is seen in *Coprinis comatus* (Fig. 18-16), the "shaggy-mane" mushroom. The basidia of *Coprinus* mature progressively from the lower edge of the gills upward. As the basidiospores are discharged, the surrounding tissues undergo **autolysis,** changing to a black fluid and giving the mushroom another of its popular names, the "inky cap." Following autolysis the next layer of mature basidiospores is discharged, and spores are dispersed into the air without encountering the gills.

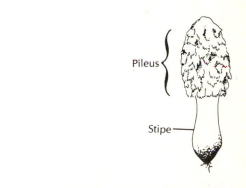

Pileus

Stipe

FIG. 18-16 *Coprinus comatus,* **the "shaggy-mane" mushroom.**

"Shaggy-mane" mushroom is also called the "inky cap," because the cap becomes black after spores are released. When the basidiocarp first emerges from the ground, the pileus resembles the shaggy mane of a horse. Basidiospores mature and are released from outer edge of the pileus inward. Following spore release the outer portions of the cap decompose (i.e., autolyse) and turn black or inky. By the time all basidiospores are released, the pileus is greatly reduced in size. *Coprinus* often grows in lawns and fields rich in organic matter in early autumn. *C. comatus* can burst through an asphalt pavement, often overnight.

Autolysing black edge of pileus

In the stinkhorns (*Phallus, Mutinus*) the basidia are enclosed until the basidiocarp matures. At maturity the sterile tissue surrounding the gleba (i.e., fertile part) dissolves, forming a sticky, odorous fluid that attracts insects (Fig. 18-17). The basidiospores are contained in the fluid and dispersed by insects that come to feed on the foul-smelling material. The spores may attach to the insect legs or carapaces or, if ingested, may pass intact through the insect gut and thereby be dispersed.

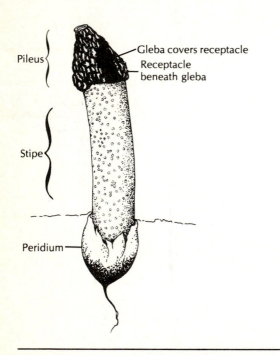

FIG. 18-17 *Phallus impudicus*, the stinkhorn.
As *Phallus impudicus* starts to mature underground in its oval or round peridium, the stipe elongates and ruptures the peridium. The stipe breaks through the ground and may reach a height of 15 cm. The pileus contains a sticky, foul-smelling (hence the name stinkhorn) gleba in which the basidiospores develop. The receptacle lies under the gleba. The foul smell attracts insects, which then disperse spores. Members of Phallales are common in the tropics but also grow in temperate woodlands in early fall.

Optimum pH and Temperatures

Many woodland basidiomycetes (e.g., *Marasmius*, Fig. 18-6) grow well at pHs below 7.0 and may not be able to grow under alkaline conditions. However, some species of *Coprinus* will grow only in alkaline conditions and may be found growing on limestone soils.

Woodland mushrooms (e.g., *Polyporus vaporarius*) generally have optimum growth at temperatures near 21° C, the mean temperature of late summer or early fall in temperate areas. *Coprinus* spp., however, have an optimum growth temperature of 30° to 35° C and may be labeled **thermotolerant,** although sporulation in *Coprinus* does not occur above 30° C. The optimum temperature for growth of the timber-rotting fungus *Serpula* (= *Merulius*) *lacrymans* is 23° C, with a maximum of 26° C. These temperature ranges limit this destructive fungus to cool, damp areas in the temperate zones (Cartwright and Findlay, 1958).

Balanced Physiology

Many parasitic smuts achieve a temporary, balanced physiological relationship with their hosts. Some, such as *Tilletia caries* and *Ustilago nuda,* infect the host when the plant is young but cause little damage. Sometimes the host may even grow larger and faster than uninfected plants would. When the plant matures and flowers, the host physiology apparently changes and the fungus then takes over, invading or replacing the flower and infecting the seeds. In corn smut, *Ustilago maydis,* the fungus invades the developing ovary, causing hypertrophy and producing many black, sooty fungal spores (Fig. 18-18).

Toxic and Intoxicating Basidiomycotina

Approximately 30 to 40 species of basidiomycetes are known to produce powerful—often deadly—toxins. The most notorious genus is *Amanita,* and the species *Amanita phalloides, Amanita verna,* and *Amanita virosa* cause most mushroom poisonings. *A. phalloides* (Fig. 18-19) is responsible for 90% to 95% of the deaths

FIG. 18-18 *Ustilago maydis*, corn smut.
Ustilago maydis fungus grows in the developing corn ear and causes hypertrophy of the kernels. Infected kernels rupture and appear black from production of black basidiospores. This fungus may cause extensive damage to corn crops, especially during periods of high humidity.

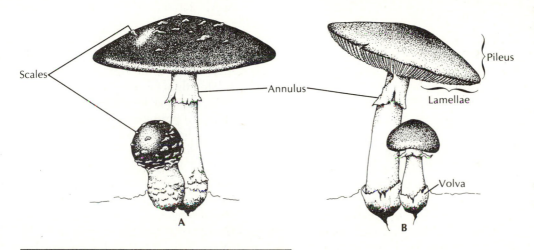

A

B

FIG. 18-19 Three species of *Amanita*—*muscaria*, *phalloides*, and *caesarea*.
A, *Amanita muscaria*, the "fly agaric," contains hallucinogenic toxins. **B,** *A. phalloides*, the "death angel," is extremely poisonous and is often pure white. **C,** *A. caesarea* is edible and delicious and has a red to orange pileus and yellow gills. Note the scales, remnants of the universal veil, on *A. muscaria*, which usually has a red or yellow pileus. Identification of the amanitas is difficult. Most species are mycorrhizal and are therefore found near trees in pastures or in woods. *A. muscaria* is common near conifers but may also be mycorrhizal with birch or oak trees. It fruits during the cool weather of late fall through early spring. *A. phalloides* usually fruits in late fall to early winter in leaf litter in close proximity to oak trees. *A. caesarea* grows in forests in southern Europe and southern North America.

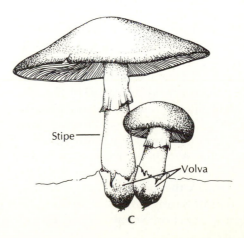

C

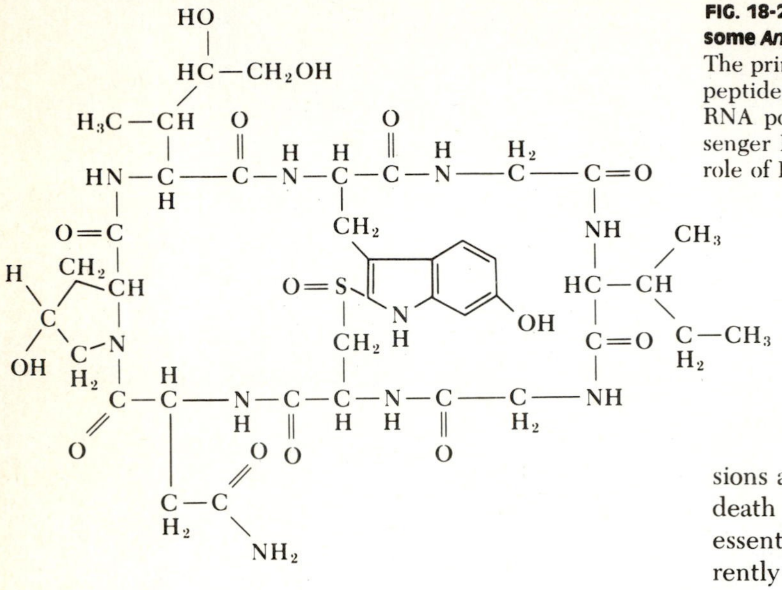

FIG. 18-20 α-**Amanitin, an important amatoxin found in some *Amanita* spp.**

The principle amatoxin, α-Amanitin, is a cyclic oligopeptide. Its method of action is the inhibition of the RNA polymerase that directs the synthesis of messenger RNA; α-Amanitin is used in the study of the role of RNA polymerase in living systems.

resulting from mushroom ingestion recorded in Europe. These three species have also been implicated in 90% of such deaths by mushroom poisoning in the United States. Some *Amanitas* are pure white or olive green and are called "destroying angel," "death cup," or "death angel." There has been confusion in the United States over the classification of some *Amanitas* (Litten, 1975). These species contain three types of toxins, the cyclic oligopeptides **amatoxins** and **phallotoxins** and the tryptophan-derived **bufotenine.** The amatoxins (Fig. 18-20) are the most toxic and may cause death if enough is ingested and medical treatment is unavailable (Lindcoff and Mitchell, 1977). The amatoxins inhibit the synthesis of messenger RNA (Griffin, 1981). One toxic genus, *Galerina* (Agaricales), contains amatoxins but no phallotoxins.

Human reaction to these toxins begins 6 to 16 hours after ingestion of even minute amounts. The toxins attack the kidneys, causing lesions, and may cause necrosis of the liver as well. Severe intestinal cramps, vomiting, and convulsions also occur. In 50% of the cases, coma and death follow in 2 to 3 days. Prompt diagnosis is essential to save the patient. No antidote is currently known, although the U.S. Food and Drug Administration is now trying an experimental antidote, **thioctic acid,** which was first used in Czechoslovakia in 1958. Some recoveries from *Amanita* poisonings (in which thioctic acid was used) have been recorded. Litten (1975), Lincoff and Mitchell (1977), and Rumack and Salzman (1978) discuss phallotoxins and amatoxins and their effects in greater detail.

Wasson and Wasson (1957) proposed that the Roman emperor Claudius Caesar was poisoned by his fourth wife by the substitution of *Amanita phalloides* (Fig. 18-19) for the delicious and nontoxic *Amanita caesarea* (Fig. 18-19). The two species appear quite different when growing in nature, since *A. phalloides* is pure white, yellow, or olive green, whereas *A. caesarea* is yellow to orange, but a substitution in cooking would hardly be noticed. Following the death of Claudius, the infamous Nero assumed the throne and the Roman Empire began its decline.

Another species of *Amanita* that is toxic but not usually lethal is the yellow- to red-capped *Amanita muscaria* (Fig. 18-19). This species is called the "fly agaric," because flies that feed on it are killed. It was used many years ago as a vegetable insecticide. This species is a mycorrhizal form that grows in deciduous and coniferous forests and old pastures of temperate zones of Asia, Europe, and North America. The toxin of this species is hallucinogenic, causing inebriation, euphoria, illusions of strength, and visual hallucinations. The extent of intoxication depends on the dose. When the mushrooms are eaten in moderation, they raise one's spirits and increase one's strength. One toxin from *A. muscaria* is an alkaloid that affects the nervous system and is called **muscarine.** The other toxins (**muscimol** and others) that have been isolated are derivatives of **ibotenic acid,** which is related to amino acids, and appear to cause the intoxication. Many folktales about the use of this mushroom have evolved, including one that maintains that it was ingested by the ancient Scandinavian "berserkers," who went on periodic rampages of violence and killing. Wasson recounts that Siberian tribesmen have eaten *A. muscaria* in rituals for centuries. Evidently the intoxicating chemicals from the mushroom pass through the kidneys intact, and the urine may be collected and used again as an intoxicant during ritualistic ceremonies. *A. muscaria* has also been used in India in religious rituals. (See Wasson, 1967, 1978, for the folklore and history of the use of this basidiomycete.)

Toxic proteins have been isolated from two edible mushrooms, *Volvariella volvacea* (**volvatoxin**) and *Flammulina velutipes* (**flammutoxin**). These toxins are called **cardiotoxic** because they affect heart function. Flammutoxin causes blood pressure to fall in conjunction with other toxic reactions. This toxin is heat labile, and toxicity can be eliminated by heating at 100° C for 20 minutes (Lin et al, 1974). *F. velutipes* is canned in the Orient for both local use and export.

The Indians of Central America and Mexico have used several species of *Psilocybe* (Fig. 18-21) (e.g., *Psilocybe mexicana*) in religious ceremonies for thousands of years. Ingestion of these mushrooms causes hallucinations, and people eat them, thinking they can commune with the gods. Evidently mushroom worship in these areas has been practiced for at least

FIG. 18-21 *Psilocybe* sp.
Several species of the agaric *Psilocybe* contain hallucinogenic agents; the basidiocarps may be eaten during religious ceremonies. Psilocybes are inconspicuous and grow on substrates rich in organic matter in lawns, fields, or woods.

3,000 years, because frescoes and statues from 1000 BC depict the practice of mushroom-eating and worship. Wasson and Wasson (1957) and Christensen (1975) discuss the history of religious use of mushrooms. The psychoactive compounds have been isolated and are identified as the tryptamine derivatives **psilocybin** and **psilocin.** These compounds affect the central nervous system, causing distortions of touch and vision similar to the effects of LSD, mescaline, and marijuana. Lincoff and Mitchell (1977) discuss the hallucinogenic effects in detail. Psilocybin was synthesized in 1958 and like LSD (see Chapter 17) may have some value in the treatment of psychiatric disorders.

Hyphal and Spore Interference

Several interesting physiological interactions between basidiomycetes and other fungi have been reported. Some members of the group tend to exclude other fungi, as well as other basidiomycetes, from the immediate habitat in which they are growing. The abilities of some species to exclude others are well known and predictable, in spite of a lack of isolated substances from the antagonistic species. These interactions, called hyphal interference, may be observed on sterilized dung when it is inoculated with various fungal spores. The zygomycetes *Pilaira* and *Pilobolus* continue fruiting for some time unless *Coprinus heptemerus* is introduced, at which time the fruiting stops. On contact the hyphae of *C. heptemerus* cause the hyphae of the ascomycetes to stop growing—a case, apparently, of contact inhibition. Only one fungus has been found, the deuteromycete *Stilbella* spp., that will limit the growth of *C. heptemerus*, probably by the production of an antibiotic. Other documented hyphal interferences are used in biological control of fungi such as *Heterobasidium annosum*. Both hyphal interference and antibiotic production convey a competitive advantage to the species that use them. The very localized contact inhibition observed in some basidiomycetes, however, would appear to require less energy than the more widely dispersed antibiotic production.

Interaction between basidiospores has been observed in the inhibition of spore germination in the Uredinales. Two types of compounds have been isolated from various rust spores: compounds that inhibit and compounds that stimulate spore germination. The spores will germinate when the stimulators of germination counteract the inhibitors. A large concentration of spores, for example, has an inhibitory effect on spore germination. Only a few spores germinate, thereby reducing competition. Many of the spores remain dormant, a condition in which they are more resistant to unfavorable environmental conditions.

Antibiotics

Several basidiomycetes produce compounds with antibiotic activity. The bird's nest fungus, *Cyathus helenge*, produces secondary metabolites that are antifungal, antibacterial, and **antiactinomycetal** (Allbutt et al, 1971). The agarics *Clitocybe diatreta* and *Agrocybe dura* produce antibiotics with antifungal properties (Cochrane, 1958).

FIG. 18-22 *Omphalotus olearius* (= *Clitocybe illudens*), the jack-o'-lantern mushroom.
Under proper physiological conditions, the mycelium and basidiocarps of this mushroom may glow in the dark. It is often orange and may grow in clumps of 15 to 20 basidiocarps in late summer or early fall. This agaric fungus is toxic.

Bioluminescence

At least 17 species of Basidiomycotina contain enzymes and chemicals that enable them to glow in the dark under certain conditions, a phenomenon called bioluminescence. The jack-o'-lantern mushroom, *Omphalotus olearius* (= *Clitocybe illudens*), has orange caps and luminescent orange gills and often grows at the base of oak trees (Fig. 18-22). These basidiocarps contain phosphorous and mycotoxins and are toxic if ingested. The basidiocarp of *Mycena luxcoeli* also is bioluminescent and may emit enough light so that one may photograph it at night without an additional light source.

The mycelia of some basidiomycetes such as *Armillariella mellea* (Fig. 18-10) also bioluminesce. As the mycelium proliferates through decaying wood, the wood glows, a phenomenon known as **fox fire.** Bioluminescence in basidiomycetes is an enzymatic process requiring oxygen. The light emitted is in the 520 to 530 nm range in those genera studied—for example, *Polyporus, Pleurotus, Agaricus melleus, Armillariella mellea,* and *Omphalia flavida* (Harvey, 1952).

Genetic Diversity

One problem in attempting to control the rusts is their genetic plasticity and their ability to rapidly produce new **physiological races.** One of the reasons such races arise is the abundance of spores produced with a myriad of different genomes. First, several wheat plants may become resistant to wheat rust, and agricultural researchers save the seeds for planting the following season. The resistant wheat strain produces several years of good crops, but then a race of the fungus produces a genome that can successfully parasitize the resistant wheat strain. Eventually another strain of wheat may evolve that is resistant to the new fungal race, and so the cycle goes. More than 200 races of *Puccinia graminis* (Fig. 18-8) exist. Plant breeders work very hard to keep several years ahead of the rapidly mutating physiological races of *P. graminis.*

Hopes were high that a 1918 campaign to eradicate the secondary host, barberry (*Berberis* spp.), of the black stem wheat rust (*P. graminis* f. sp. *tritici*) would reduce the damage and the incidence of infection. However, in the Western Hemisphere, not only does barberry have a wide geographical range but it also occurs in remote areas where it may grow unnoticed. The campaign did bring about significant reduction in damage. The eradication of barberry eliminated the host where plasmogamy and genetic recombination takes place, resulting in a slowing in the evolution of new genotypes (physiological races). Unfortunately the eradication of barberry did not eliminate the wheat rust, because both the urediniospores and the dikaryotic mycelium can survive winters in warm areas of the southern United States. The spores produced can then reinfect the wheat in the spring, thus bypassing the stage on barberry. Urediniospores can also be blown north for thousands of miles, infecting crops in the northern states. Evidently, genetic exchange (e.g., parasexuality) can also occur between nuclei of the dikaryotic hyphae growing on wheat plants; thus barberry is unnecessary in the cycle. (Walker, 1969, documents the history of the barberry eradication program.)

Mycoviruses

The basidiomycetes, like most fungi, are host organisms for many types of **mycoviruses.** Viral infection has been an ever-present problem in the commercial cultivation of mushrooms and continues to harass those who attempt to cultivate fungi.

Infection with a virus may cause biochemical changes in the fungal host. For example, the presence of a mycovirus in some stains of the corn smut, *Ustilago maydis*, induces the secretion of a toxin by the fungus. Such a toxin-secreting strain is called a "killer strain" because the toxin is lethal to susceptible strains of *U. maydis.* However, the "killer strain" is immune to damage from the toxin. The mycovirus in the killer strains is a double-stranded RNA virus that controls the production of the toxin (Berry and Bevan, 1972; Bozarth et al, 1981).

Mycoviruses have a significant effect on the physiology of their hosts (e.g., toxin production). Lemke and Nash (1974) have indicated that mycoviruses may also play an important role as a selective force in fungal evolution. Fungi have a number of mechanisms for increasing genetic diversity, including parasexuality and heterokaryosis, but in spite of this they still tend to interbreed and are often self-fertile. The selective forces governing the tendency to interbreed are unknown; however, inbreeding reduces viral infection and spread among the individual fungal species. It is possible that the course of fungal evolution has been influenced by the presence of mycoviruses, since experiments have shown that random breeding among the fungi fosters the spread of viruses, whereas inbreeding limits viral dissemination.

EVOLUTION

Evolution among the various groups of fungi is a much-debated subject, and the position of the Basidiomycotina is no exception. The definite shortage of fossil evidence means that most phylogenetic relationships among the fungi are based on structural and biochemical characteristics. Most mycologists believe that the Basidiomycotina arose from the Ascomycotina, basing their opinions on the structural and reproductive similarities between the ascus and the basidium and between the crozier formation and clamp connections. Another structural similarity is found in the septal walls. The ascomycete septum (see Chapter 17) could easily have given rise to the basidiomycete dolipore septum. Some investigators feel that the more primitive ascomycetes—the Endomycetales and the Taphrinales—were the progenitors of the basidiomycetes. Cain (1972), however, has speculated that the two groups Ascomycotina and Basidiomycotina evolved from two different lines of autotrophic ancestors when they moved from an aquatic to a terrestrial environment.

Among the Basidiomycotina the "jelly fungi" (subclass Phragmobasidiomycetidae, Fig. 18-4) are considered by some researchers to be the most primitive group because of the septate nature of their four-celled basidia (Fig. 18-1, *B*). The Holobasidiomycetidae with their single-celled club-shaped basidia (Fig. 18-1, *A*) and the Teliomycetes with their linear septate basidia (Fig. 18-1, *C*) are both derivatives of the jelly fungi. However, there appears to be no universal agreement on this matter. From a structural viewpoint such an evolutionary tree would seem logical, even if it were not genetically correct. Further information about basidiomycete evolution may be found in Petersen (1971).

Some theorists consider the most highly developed basidiomycetes, the Teliomycetes, to be derivatives of separate ancestors, thus making them a polyphyletic group. If this is true, then the rusts (Uredinales) and smuts (Ustilaginales) are probably not closely related phylogenetically. Laundon (1973) thinks the rusts with the complex **heteroecious** cycles are primitive, whereas the simpler **autoecious** rusts are more advanced. Savile (1976, 1978) discusses the evolution of the rusts. The study of the complicated rust life cycles causes scientists to wonder what selective advantage these complex life cycles convey. The rusts have been extremely successful with a complicated and specialized (in terms of hosts) life cycle. If specialization often leads to extinction, why then do the rusts continue to be so successful? The answer seems to lie in their genetic plasticity and their ability to produce innumerable different genotypes that can adapt to parasitizing the many different strains of the host.

The rusts have some of the oldest fossil records of any of the basidiomycetes, reportedly dating from the Carboniferous period, when they were parasites of ancient tree ferns. Inference suggests that tree ferns were parasitized shortly after they evolved. Some investigators believe the geological age of a host plant is a good indicator of the phylogenetic age of its parasite. If this is true, then the rust genus

Uredinopsis represents the most ancient extant basidiomycete, since it is the cause of a rust disease associated with *Osmunda*, one of the oldest fern genera. Tiffney and Barghoorn (1974) reported that Holobasidiomycetidae fossils have also been found in Carboniferous rocks, indicating that this subclass was also extant at that time. As more fossils are found, the course of basidiomycete evolution will become clearer.

ECOLOGY

The Basidiomycotina, especially the Aphyllophorales and the Agaricales, are essential in the reduction and recycling of the nutrients entrapped in cellulose and lignin (Frankland et al, 1982). These orders are probably the most important decomposers of lignin especially, and the large populations observed in woodlands testify to their dominance of this particular ecological niche. There are at least 1,600 species of woodrotting basidiomycetes (Gilbertson, 1980). Fungal mycelia may be found in almost any moist place, such as under rotting logs, in leaf litter, and under the bark of fallen trees; some basidiomycetes even occur in the desert and under the snow (Gilbertson, 1980).

Most woodlands or fields where there is organic material and adequate moisture will have a large biomass of basidiomycete mycelia. Their presence, however, is generally not evident until the basidiocarps appear. The dikaryotic mycelium usually grows under leaf litter most of the year, producing basidiocarps for only a short period when genetic exchange and spore dispersal take place.

The Tremellales and the Dacrymycetales are both saprophytic orders that occur regularly in woodlands. The Tremellales have jellylike basidiocarps (Fig. 18-4) and are often found on decaying logs. Other saprophytic orders such as Lycoperdales, Nidulariales, and Auriculariales are also commonly found in forests or overgrown fields. In temperate areas most basidiocarps are formed in the fall, though quite a few

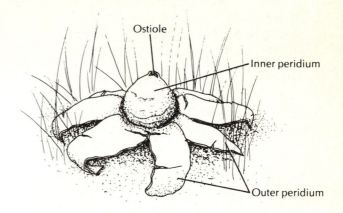

FIG. 18-23 *Geastrum* **sp., the earthstar.**
Geastrum grows in woodland areas rich in organic matter. The peridium is several layers thick, and at maturity the outer layers peel back, forming a stellate design. Inner peridium layers contain an ostiole through which the spores are dispersed. Following discharge of basidiospores, inner peridium collapses.

are formed in the spring and summer. Mushrooms such as the "shaggy mane," *Coprinus comatus* (Fig. 18-16), may occur in lawns or open fields. The saprophytic earthballs (*Scleroderma*, Fig. 18-12), puffballs (*Calvatia*, Fig. 18-13) and earthstars (*Geastrum*, Fig. 18-23) also can be found in forests, woodland edges, or fields. The Phallales and Nidulariales are also saprophytic and have habitats similar to the other saprophytic basidiomycetes.

Many Basidiomycotina form mycorrhizal associations with gymnosperms and angiosperms. *Pisolithus tinctorius* (Sclerodermatales), for example, is ectomycorrhizal with *Pinus* spp. (Fig. 13-3) and some angiosperms (see Chapter 13). This fungus has been introduced with *Pinus* into such unfavorable environments as strip-mined areas or infertile, overused soils. The presence of the fungus not only enhances the pine's chance of becoming established in such harsh environments but also helps the trees grow faster. Berry and Marx (1976) discuss the use of this fungus with *Pinus* for revegetation.

Some basidiomycetes, mostly Agaricales, form mycorrhizal associations with orchids, some with larch, fir, apple, willow, oak, birch, beech, and hickory trees, and still others with food crops. Most mycorrhizal associations in this subdivision are ectomycorrhizal, and some associations are more beneficial to the partners than others. Some fungi form mycorrhizae with specific vascular plants, others form them only with certain families, and still others will form with a broad range of hosts. Northern white pine (*Pinus strobus*) and Norway spruce (*Picea abies*), for example, may form associations with 12 fungal species from such genera as *Amanita*, *Boletus*, and *Lycoperdon* (Haynes, 1975). The association is usually mutually beneficial, but the balance is delicate, and should the vascular plant become weakened or damaged, the bal-

ance shifts and the fungus may then assume a parasitic role. If the host dies, the fungus becomes saprophytic. If the fungus becomes weakened, however, the vascular plant may digest the fungus and use the component nutrients.

The woodland angiosperm Indian Pipe (*Monotropa uniflora*) contains no chlorophyll and is therefore white. How, then, does this plant obtain energy? Its roots form mycorrhizae with a species of *Boletus* (Fig. 18-24); the angiosperm appears to be a parasite upon the fungus, absorbing food from the fungus. The fungus in turn is also mycorrhizal with neighboring trees, and sugars from the tree (as demonstrated with radioactively labeled samples) may ultimately end up in the Indian Pipe (Bjorkman, 1960). A similar situation apparently occurs in some orchids, which are frequently dependent on mycorrhizae. One genus of white orchid, *Gastrodia* sp., for example, is dependent on the honey mushroom, *Armillariella mellea* (Fig. 18-10), for nutrition and will not flower without the presence of this fungus. The relationship between the orchid and the fungus is mycorrhizal, but the fungus is parasitic upon a nearby tree, which ultimately provides nutrients for the orchids via the fungus (Hamada, 1940).

Another interesting relationship has developed between *Septobasidium* (Septobasidiales) and colonies of scale insects in the family Coccidae. The fungal spores germinate on several adult insects and the hyphae invade their bodies. Only a few insects in the colony are parasitized. The infected insects continue to suck sap from the host tree, and some nutrients from the tree are transferred to the fungus via the insect. The parasitized insect becomes a virtual conduit for the movement of nutrients from tree to fungus. The fungus grows, forming a thick mat of mycelial threads over the insects, protecting them from predators. These mats persist for years, with tunnels and chambers developing in the mat as the insect colony enlarges (Couch, 1938). The parasitized insects continue to feed and are not killed by the fungus, but they are immobilized and rendered sterile. Ross (1979) thinks

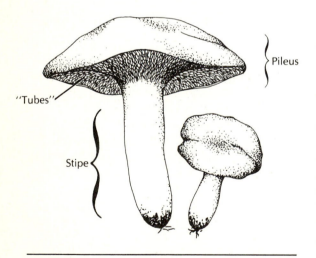

FIG. 18-24 *Boletus* sp., a bolete.
"Boletes" or "tube fungi" are a major group of Agaricales. They are called tube fungi because the undersurface of the pileus is composed of tubes rather than gills. Basidiospores are borne on the hymenium, which lines the tubes. The stipe is thick and fleshy. Boletes grow in late spring and summer but are most common in early fall, especially under conifers. Some boletes are mycorrhizal; some are delicious but others are quite toxic. See Snell and Dick (1970) for further information about the boletes.

the relationship is beneficial to both partners. The fungus gains nutrients and the insects gain protection; both benefits are gained at the expense of the tree.

Another insect-fungus relationship occurs between the wood wasp *(Sirex noctillo)* and *Amylostereum* spp. (Aphyllophorales) in the forests of Australia, Canada, New Zealand, and Japan (Gilbertson, 1980). The wasp, a serious pest, bores holes in trees and injects not only its eggs but also spores of the fungus. The fungus grows and starts to decompose the wood (brown rot), and the larvae feed upon the rotting wood. The relationship is so well developed evolutionarily that the wasp even has special pouches in which it carries fungal conidia. Other wood wasps have developed relationships with still other bracket fungi (e.g., *Stereum*) in which the wasp acts as a dispersal agent. After the fungus predigests the wood, the wasp larvae are able to use the nutrients thus made available.

Termites also develop beneficial associations with fungi. In tropical Asia and Africa, termite colonies have "fungus gardens" in their enormous nests. Large fungal masses of various genera develop in these nests, but their role in termite nutrition is not definitely known. The fungus may aid in the breakdown of cellulose. Parts of the mycelium are moved about the "garden" by the termites, and portions of the fungus are also eaten by the termites. The most abundant fungus in termite nests is the agaric *Termitomyces*. Batra and Batra (1967) discuss this as well as other fungal-insect relationships.

The basidiocarps of many saprophytic Agaricales tend to occur in large circles called "fairy rings" (Fig. 18-6). Rings develop as the mycelium starts growing from a central point and spreads out in all directions. Each year, usually in the autumn, basidiocarps are formed around the circular growing edge of the mycelium. The grass immediately inside the fairy ring is usually stunted, since the fungus is using the nutrients that would otherwise be available to the grass. The grass outside the ring, however, is greener and longer than the surrounding grass, for, as the mycelium grows outward, it releases exoenzymes that break down dead plant material, releasing nutrients that the grass absorbs. *Marasmius oreades* often forms fairy rings and may be parasitic on the grasses. Some fairy rings are reputed to have occurred in the same location for centuries, enlarging in circumference a little bit each year (Parker-Rhodes, 1955).

In the succession of fungi on dung or straw compost, the Basidiomycotina basidiocarps are usually the last to appear. Their fruiting bodies are larger and require more energy for production than those of the other fungi (i.e., oomycetes, zygomycetes, ascomycetes, deuteromycetes) that usually precede the basidiomycetes in succession. Also, when the basidiomycetes appear, they tend to exclude other fungi (i.e., hyphal interference). As the basidiomycetes start growing, the chief materials remaining are cellulose and lignin; it is the basidiomycetes primarily that possess the exoenzymes to use these carbon sources. Some ascomycetes and deuteromycetes can also degrade these recalcitrant molecules (see Chapters 17 and 19). *Coprinus* will appear on dung in 9 to 13 days (Harper and Webster, 1964) and on straw compost in about 30 days. When the compost is being used commercially to culture *Agaricus brunnescens*, steps must be taken to exclude *Coprinus* because of antagonistic interactions between the two species. Sinden (1971) discusses the ecological control of both unwanted fungi and pathogens in the cultivation of *A. brunnescens*.

The naturalist who combs the fields and forests in search of mushrooms soon notices that cultivated and fertilized farm fields seldom contain mushrooms. Areas rich in dead organic materials that are relatively untouched by chemical fertilizers or pesticides should bear many mushrooms if the hunter goes at the right time of year. Even the home garden fertilized with manure may yield a few mushrooms. Coastal areas usually produce few mushrooms, because the salt from sea spray seems to discourage their growth; however, some species may be found on sand dunes near the sea.

Several Basidiomycotina are found in aquatic environments. Two genera, *Digitatispora marina* (Aphyllophorales) and *Nia vibrissa* (Gasteromycetes), have been found growing on wood in the marine environment (Ingold, 1976). These organisms have passive spore release and **tetraradiate** (i.e., four-armed) conidia (Fig. 19-7), a form of spore that occurs frequently in the aquatic environment. Kohlmeyer and Kohlmeyer (1979) list two additional marine genera—*Halocyphina* and *Melanotaenium*. Several freshwater basidiomycetes have been identified from streams in Nigeria and in Queensland (Australia). One organism, *Ingoldiella hamata*, has tetraradiate conidia and clamp connections (Ingold, 1976). Nawawi (1973) discusses freshwater basidiomycetes from Malaysia.

Many of the basidiocarps of the shelf or bracket fungi such as *Fomes* (Fig. 18-9) and *Ganoderma* are perennial, growing every year as new sporogenous tissue is added. One basidiocarp of *Fomes* is reportedly 80 years old. At one time people in rural areas sometimes collected these large fruiting bodies and burned them for fuel. *Ganoderma applanatum* (= *Fomes applanatus*) produces spores in inner layers of tubes. The pure white undersurface of the large basidiocarp is ideal for painting or drawing pictures. The fruiting body is perennial, so it gets large enough for a good-sized painting. These basidiocarps were used by American Indians for both writing and drawing surfaces.

ECONOMIC IMPORTANCE
Degradation of Timber

The Aphyllophorales have been responsible for the destruction of many valuable timber trees. Heart rot will seriously weaken trees and reduce their value as lumber. Sapwood rot, of course, will eventually kill the tree. These fungi are extremely important decomposers of cellulose and lignin, but they destroy timber stands and shade trees along with fallen and rotting trees. In North America approximately 10% of the annual timber harvest is destroyed by wood-

FIG. 18-25 Consequences of root rot in *Catalpa*, caused by *Armillariella mellea*. *Armillariella mellea* invaded the roots of this catalpa tree and caused them to rot. The tree then was weakened and fell, crushing the car. (Courtesy William Merrill, Department of Plant Pathology, The Pennsylvania State University, University Park, Pa.)

rotting fungi (Gilbertson, 1980). Lignin, the most persistent compound, makes up 15% to 25% of the woody tissue and decomposes very slowly, requiring at least three complex steps to completely break down the three-dimensional polymer. Some Aphyllophorales and Agaricales (Figs. 18-9 and Fig. 18-10) are indiscriminate in their decomposition and do not distinguish among rotting logs, valued timber, or stored lumber. The damage to standing trees, felled trees, lumber, and building supports by these fungi has been estimated at $9 billion a year (Tippo and Stern, 1977). Fergus (1960) and Gilbertson (1980) discuss the fungi that decay wood.

Heterobasidium annosum (= *Fomes annosus*) establishes itself rapidly on the newly cut stumps of southern pines, causing rot of such plants in the southeastern United States. From the stumps, the fungal spores are blown by the wind to live trees where they begin degradation. Chemical control by putting borax or creosote on the stumps was not proved effective. Rishbeth (1963) found that another bracket fungus, *Peniophora gigantea*, could colonize pine stumps and suppress the growth of *H. annosum* in an apparent instance of hyphal interference (see Chapter 13). The use of *P. gigantea* as a means of biological control was proposed, and currently spores of *P. gigantea* are added to the oil on the chain saws used to fell the trees. The stump is thus inoculated with spores of the antagonistic fungus, and damage from *H. annosum* has been reduced substantially (Artman, 1972).

Important decomposers of wood are *Polyporus*, *Heterobasidium*, *Ganoderma*, *Stereum*, *Fomes*, and *Armillariella* (Figs. 18-9, 18-10, and 18-25). Some trees are more resistant to fungal decay than others. Redwood, juniper, and Douglas fir are the most resistant gymnosperms, while black locust, black walnut, chestnut, and red mulberry are highly resistant angiosperms. The degree of resistance is related to the amount of metabolites such as resins and tannins found in the heartwood. For this reason

these woods are useful for outdoor furniture, fence posts, and porches.

Mushroom Cultivation

Most mushrooms cultivated for human consumption are members of the Agaricales. These are mushrooms of woodland and field and as such are responsible for a significant amount of nutrient recycling. Basidiocarps are the temporary fruiting bodies of an extensive underground mycelium that may be up to a hundred years old, producing basidiocarps every year.

Agaricus brunnescens (= *A. bisporus*) is grown commercially for sale to markets and restaurants. The United States and France lead the world in *A. brunnescens* production, having produced about 40% of the world's total production of 600,000 metric tons in 1978 (Chang, 1980). More than 1,000 metric tons of mush-

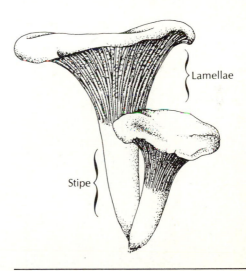

FIG. 18-26 *Cantharellus* **sp., a chanterelle.**
Chanterelles are some of the most delicious mushrooms known. They resemble funnels in their morphology. The lamellae branch and are shallower than in other agarics. Protein-splitting enzymes from *Cantharellus cibarius* are used commercially in food processing. *C. cibarius* is yellow to orange, fruits in cool weather, and may grow in close proximity to conifers and oak trees.

FIG. 18-27 Cultivation of *Agaricus brunnescens*.
Two stages in the development of mushrooms. The basidiocarps first appear above the substrate as initials, *A*, which quadruple in size, forming "pins," *B*. The pins continue to enlarge to the button stage, *C*, which then develops to the harvestable, fully grown mushroom, *D*. (Courtesy Paul J. Wuest, Department of Plant Pathology, The Pennsylvania State University, University Park, Pa.)

rooms are consumed annually in the United States (Atlas and Bartha, 1981). *A. brunnescens* is not the most delicious mushroom available, but it is the only species that has been grown commercially on such a large scale in the temperate zone. In terms of taste, the chanterelles (e.g., *Cantharellus cibarius*, Fig. 18-26), some boletes—even *Amanita caesarea* (Fig. 18-19)—and the ascomycetous truffles and morels (Figs. 17-20 and 17-21) are reportedly much more delicious than *A. brunnescens*. These species, however, are mycorrhizal and, therefore, cannot be cultivated. In Europe edible wild mushrooms are gathered and sold fresh in markets for a high price or they may also be dried or canned.

A. brunnescens is grown on compost, a mixture of straw and horse manure (Fig. 18-27). The compost is placed in piles and decomposition proceeds. Internal temperatures of the pile may reach 70° C in 4 to 6 days because of exothermic microbial decomposition. In the next month, temperature will gradually fall to ambient temperatures. *A. brunnescens* is usually introduced into the compost as temperatures cool, because it cannot tolerate high temperatures. It must be introduced, however, before *Coprinus* spp. gain a foothold, because *Coprinus* will suppress the growth of *A. brunnescens*. Often the compost is sterilized to kill *Coprinus* spores or other competitors. The **spawn** (i.e., young mycelium) of *A. brunnescens* is introduced at the proper time and growth proceeds. The spawn is grown on special media in laboratories under controlled conditions. The commercial cultivation of *A. brunnescens* is the most technologically refined method of mushroom production. The large buildings in which the mushrooms are grown must have a carefully controlled environment. Ventilation is especially important, because an atmospheric concentration of carbon dioxide exceeding 1% will inhibit basidiocarp development. With the maintenance of favorable growing conditions, two or three crops per year may be produced in the northeastern United States. Further discussion

of mushroom cultivation may be found in Chang and Hays (1978) and Atlas and Bartha (1981).

A great deal of information about the life cycles of mushrooms, basidiocarp development, and other processes has been learned through commercial attempts to increase mushroom growth. *A. brunnescens* growers have been plagued by a virus that causes malformations and reduced yields. Insects can also cause many problems during mushroom cultivation. Cantelo and San Antonio (1982) discuss methods of controlling invertebrate pests in mushroom culture.

In Japan the "shiitake" mushroom, *Lentinus edodes*, a wood-rotting species, and the "matsutaki" mushroom, *Armillariella ponderosa*, are cultivated commercially and sold in dried or powdered form. In 1978, 110 metric tons of *L. edodes* were produced in Japan (Chang, 1980). The straw mushroom, *Volvariella volvacea*, is also grown commercially in parts of the Orient.

Dry Rot

A very economically important basidiomycete, *Serpula* (= *Merulius*) *lacrymans*, causes "dry rot" of timbers. The term "dry" is misleading because the fungus needs at least 20% humidity to become established. Once the fungus starts growing, however, it can transport water along its hyphae for many centimeters and thus it can invade dry timbers. This fungus breaks down cellulose to glucose and ultimately to carbon dioxide and water. This metabolic water that is produced may be seen as droplets on infected wood, giving the fungus its name, "lacrymans," which means "crying" in Latin. Infected wood is brownish, crumbly, and covered with hairline cracks. Ships and older homes are often the targets of this extremely destructive fungus; many fungal spores may be isolated from the basements of homes. Once a timber is infected it must be replaced. *S. lacrymans* and other wood-rotting fungi thrive under conditions of low oxygen levels. Dry rot, therefore, often starts in poorly ventilated basements of old buildings. The fungus merely needs moisture to start its invasion, and once established, it can spread to drier areas, even the upright supports of the building.

Dry rot, or wood-rotting of some sort, was the scourge of the British navy for nearly 400 years. Many parts of the ships were built from oak, which decayed rapidly. The ever-increasing demand for oak timbers denuded the English countryside of oaks. British ships virtually rotted at anchor, thereby earning the name of "Rotten Row" for the warships at their docks. During the American Revolution so many British ships were damaged by fungi that, according to Ramsbottom (1937), "It was necessary to shovel away the toadstools and filth from rotting planks and timbers." Some ships evidently were decayed before they were completely built and launched. These problems continued until the hulls were clad in iron and the holds were better ventilated. Emerson (1952) and Gray (1959) discuss these problems further.

Prevention of fungal decomposition and the preservation of lumber are accomplished by several procedures or combination of procedures. These include steam sterilizing, proper drying (even dry rot requires moisture to become established), and treating the wood with a phenolic preservative such as creosote or pentachlorophenol. Railroad ties and telephone poles are usually impregnated with creosote under pressure and a new process impregnates lumber with pentachlorophenol. Use of lumber from trees resistant to fungi is effective but expensive. Further discussion of the use of fungicides against wood-decaying fungi is found in Levi (1977).

Plant Parasites

Wheat rust has been destroying crops for thousands of years, probably since ancient Babylonia (2000 to 1800 BC). Fungal remains of this rust have been found that are 3,300 years old (Kislev, 1982). The migrations of the Jews, as told in the Bible, were probably caused by crop failures resulting from wheat rust. The ancient Romans told of a "red material" that destroyed their wheat fields. They thought their crop failure was caused by the god Robigus as punishment. In order to placate the god they held annual festivals in his honor.

The rust fungi have caused billions of dollars in damage to grain crops over the world. There are approximately 4,000 species of rusts (Uredinales) that are obligate parasites on ferns, gymnosperms, and angiosperms. Several require two different vascular plants to complete their life cycle.

On the Great Plains of the United States, a spore density of 2×10^{13}/ha was recorded in the early 1900s during a wheat rust epidemic. These vast numbers of spores, each bearing different genetic information, make control difficult. The plant geneticists who breed for resistant strains of wheat can only stay about 3 years ahead of the rapidly changing rusts. One race, 15B, spread over 8 million km^2 in 1 year. This race, first discovered in Texas in 1939, caused widespread damage in the United States in 1953 and 1954, when it was well established and the weather conditions were favorable (Ross, 1979).

Wheat rust epidemics have caused severe losses in Russia, Europe, Africa, Canada, and the United States. The losses of wheat per year throughout the world are estimated at $500 million (Haynes, 1975). In the United States in the late 1930s, 100 million bushels valued at $50 million were destroyed in a devastating epidemic that greatly exacerbated the problems of a nation struggling with the Depression.

Ustilago maydis, corn smut (Fig. 18-18), is an important disease of corn. In bad years yields may be reduced by 60%. The teliospores overwinter in the soil or corn rubble and may remain dormant but viable for 10 to 15 years. The disease causes the formation of galls on the corn ear. Some people slice and fry these galls for eating before the spores are formed. A recipe from Mexico calls for cooking the fungus with garlic and chili. Also in Mexico, citizens may collect the mature, black teliospores from the galls, mix them with olive oil, and use them as a substitute for "caviar."

The bunt or stinking smut of wheat, *Tilletia caries*, is almost as damaging as *Puccinia graminis*. During threshing of the wheat, clouds of smut spores may accumulate in the machine and explode, causing flash fires. The spores may also produce allergic reactions in susceptible individuals, as do many fungal spores. Other smuts attack onion, oats, wheat, and various cereals. Damage by smuts to agricultural crops and ornamental shrubs and flowers may total many millions of dollars.

The **white pine blister rust,** *Cronartium ribicola*, has caused considerable damage to five-needle pine trees, especially in the United States. Wild currant and gooseberry *(Ribes)* are the primary hosts of this rust, and it is on these plants that urediniospores and teliospores are formed. The spermogonia and aecia are formed on the white pine, where they injure the tree by girdling the trunk, thereby reducing its timber value and sometimes killing the tree. To control the rust, currant and gooseberry were eradicated in the vicinity of pines; this method has been quite successful. In the northwestern United States another fungus has been fairly effective in biologically controlling the blister rust (Kimmey, 1969). The blister rust was introduced into the United States from Russia in the late 1800s, another instance of a pest introduced to an area where the native flora had little or no resistance to the parasite.

Another important disease is the cedar apple rust *(Gymnosporangium juniperi—virginianae)*, which parasitizes apple and juniper trees. Most trees commonly called "cedar" are actually juniper *(Juniperus)*. The rust produces reddish-brown galls called "cedar apples" on junipers. The galls are composed of masses of mycelia. The fungus overwinters as dikaryotic hyphae in the gall, and teliospores are usually produced in the spring. Spermagonia and aecia are produced on the secondary hosts, apple or crab apple.

Control of this rust is partially accomplished by isolating junipers far away (over 3 km) from apple orchards. Fungicides may also be applied to either or both hosts.

The agaric *Mycena citricolor* (= *Omphalia flavida*) parasitizes coffee tree leaves, causing premature leaf drop and reduction in yield. This fungus produces an enzyme that inactivates the plant hormone auxin. Adequate levels of auxin prevent premature leaf drop. When auxin is inactivated, levels are reduced and the coffee tree leaves fall as a result of premature abscission. The fungus grows only vegetatively on the host. The sexual stage, the basidiocarp, occurs when the fungus is growing saprophytically on dead leaves and is induced by the presence of other fungi such as the ascomycete *Penicillium* (Salas and Handcock, 1972). This pattern of vegetative reproduction on the host and sexual reproduction on decaying matter has been observed in many other parasitic fungi.

In Asia, coffee trees are parasitized by a rust, *Hemileia vastatrix*, which plagues coffee-growing areas of the Middle East and India. Coffee trees originated in Ethiopia and were transplanted to these two regions. The Dutch introduced coffee trees into Ceylon, where they flourished. The British took over Ceylon in the late 1700s, and coffee became an important export from that colony. A severe coffee rust epidemic hit Ceylon in the mid 1800s, and by 1892 most trees were dead. Coffee beans were still available, however, because in the early 1700s trees were transplanted to Brazil and Central America. Following the death of the trees in Ceylon, the Western Hemisphere became the center of coffee production in the world. When the coffee trees were killed by *H. vastatrix* in Ceylon, the resourceful British planted tea, which rapidly replaced coffee as the favored drink of Great Britain. The coffee rust has not yet crossed the Atlantic in significant concentrations to plague the trees of the Western Hemisphere. At last reports it had been found on the west coast of Africa (Haynes, 1975). Christensen

(1975), however, reports that fungal spores now have been found in Brazil. The spores probably did not blow across the southern Atlantic, because the spores appear adapted for dispersal by splashing raindrops and not by wind. They have probably been introduced on some type of imported plant.

Poisonous Mushrooms

Human poisonings from mushrooms do occur but, fortunately, infrequently. Most botanists are extremely cautious and will not eat a mushroom unless the identification has been verified by a competent mycologist. Nevertheless, tragic poisonings still occur. *Amanita* species (Fig. 18-19) are responsible for most documented poisonings. In 1975, for example, the newspapers reported that ten people died and hundreds were hospitalized in Italy from the consumption of poisonous mushrooms. No particular species was implicated. It was a warm and rainy September; Italy had a bumper crop and many people were picking mushrooms from the wild.

The consumption of certain mushrooms may temporarily upset human digestive systems. If susceptible individuals eat basidiocarps of *Coprinus atramentarius* and consume alcohol in the same meal, extreme digestive upset may result. The reaction is similar to the effect of the drug antabuse (**disulfiram**) which is used to discourage alcoholics from drinking alcohol. The illness-causing compound from *Coprinus* has been isolated and labeled **coprin.**

Many folktales about mushrooms claim ways of telling whether they are poisonous. There is *no* foolproof method of detecting a poisonous mushroom from an edible one—except by a trained mycologist! Poisonous mushrooms will *not* necessarily turn a silver spoon or coin black when they are cooked! Laymen tend to call edible basidiocarps "mushrooms" and nonedible or poisonous basidiocarps "toadstools." To the mycologist no technical distinction exists.

Followers of the "drug culture" seek out mushrooms with psychedelic properties—for example, *Psilocybe* (Fig. 18-21) and its relatives. These genera grow in Mexico and the northwestern United States. In November 1976, the news media reported that hundreds of young people were seeking psychedelic mushrooms in the Seattle, Washington, area, and at least one teenager was hospitalized with a violent reaction to the mushrooms.

Mushrooms as a Food Source

Do mushrooms have any food value beyond their delectable taste? Actually they are mostly water (85% to 92%) and are low in calories. Their nutritive value appears to be low—3% to 4% protein, 5% to 7% carbohydrate, and 1% to 2% minerals. Some of the protein may be indigestible by humans. However, they do contain vitamins (e.g., vitamin C, niacin, thiamine, and riboflavin) that are not denatured by cooking. Chang (1980) and Royce and Schisler (1980) point out that mushrooms are higher in protein than many vegetables and that they contain essential amino acids. Mushrooms not only grow readily on such waste products as straw, sawdust, and cotton, but they can also convert the cellulose and lignin of these wastes into edible food.

Human Pathogens

The human pathogen *Cryptococcus neoformans* (perfect state = *Filobasidiella neoformans*) has recently been placed in the Teliomycetes, Ustilaginales, because its dikaryotic mycelium has clamp connections (Kwon-Chung, 1975, 1976). The perfect state produces basidiospores in chains. Formerly classified in the Ascomycotina, this fungus causes **cryptococcosis** in humans. In its pathogenesis the fungus causes pulmonary lesions and may spread to the bones, central nervous system, and skin. Occasionally the disease is fatal. The fungus may also exist as a free-living saprophyte on pigeon droppings. In both the saprophytic and parasitic state the fungus is yeastlike, hence a basidiomycetous yeast.

Medical Uses

An analog of muscimol, one of the intoxicating constituents of the toxic *Amanita muscaria*, produces an analgesic effect in humans. The chemical is low in toxicity, crosses the blood-brain barrier, and appears to be as effective as morphine in reducing pain (Maugh, 1981a).

Inonotus obligius grows on live birch trees in Russia; the fungus is reputed to contain anticancer properties. Reid (1976) discusses the medical qualities of this fungus.

SUMMARY

The Basidiomycotina constitute the most advanced members of the "higher fungi" associated with the division Eumycota. Their life cycles are characterized by the presence of monokaryotic primary mycelia, dikaryotic secondary mycelia, and dikaryotic, structurally complex, tertiary mycelia, which are organized into basidiocarps. Basidiocarps often occur in the form of mushrooms. Primary and secondary mycelia often develop mycorrhizae, which are symbiotic relationships with the roots of various trees and other plants.

Three characteristic morphological features of the subdivision are found on the cellular level: (1) the basidium or small base, which is a structure bearing a definite number of basidiospores (typically four), and is formed as a result of karyogamy and meiosis in terminal branches of a dikaryotic hypha; (2) clamp connections, which are associated with nuclear divisions in dikaryotic hyphae; and (3) dolipore septa, with their characteristic central pores.

Sexual reproduction is characteristic of every group in the subdivision; however, only the rusts (Uredinales) form specialized sexual structures. Genetic recombination is accomplished by karyogamy and meiosis within individual basidia following fusion of monokaryotic hyphae, forming dikaryotic hyphae. The life history of most basidiomycetes is similar to that of the common mushroom *Agaricus*. The parasitic forms of the Teliomycetes, however, have complicated life cycles involving one or two intermediate hosts and the production of various types of spores. These latter organisms are mostly plant parasites, many of which are important economic crop pathogens, including the rusts and smuts.

Basidiocarp (tertiary mycelium) development depends on a number of nutritional and environmental factors, including the presence of certain organic compounds, vitamin availability, light, temperature, pH, and in some instances, the presence of low-molecular-weight–inducing substances. Many basidiomycetes produce enzymes that are capable of decomposing wood. Spore dispersal is generally passive and is accomplished by wind; however, a few species have active dispersal mechanisms.

The basidiomycetes have long been known to produce biochemical toxins and hallucinogens. The genus *Amanita* is especially well known for its toxic principals, the phallotoxins and the amatoxins. *Psilocybe* is noted for its hallucinogenic compounds, psilocybin and psilocin. Antibiotic substances are produced by some genera (e.g., *Clitocybe*). Bioluminescence occurs in some genera (e.g. *Omphalotus*). Mycoviruses have been isolated from some species, and they may have promise as biological control agents in certain basidiomycete diseases of plants.

Ecologically the basidiomycetes are significant in terrestrial habitats because of their role in recycling nutrients that are bound in the cellulose and lignin of woody plants. They also form important symbiotic associations with higher plants through mycorrhizal connections. In this respect they are especially important to terrestrial woodlands and somewhat less important to aquatic habitats.

Economically the basidiomyctes are significant because of their role in the degradation of timber and lumbers, as a food source, as commercial plant parasites, and as human pathogens.

SELECTED REFERENCES

Baker, K.F., and R.J. Cook. 1974. Biological control of plant pathogens. W.H. Freeman and Co., San Francisco. 433 pp.

Brodie, H.J. 1978. Fungi—delight of curiosity. University of Toronto Press, Canada. 331 pp.

Chang, S.T. 1980. Mushrooms as human food. BioScience **30**(6):399-401.

Frankland, J.C., J.N. Hedger, and M.J. Swift (eds.). 1982. Decomposer basidiomycetes: their biology and ecology. Brit. Mycol. Soc. Symp. 4. Cambridge University Press, New York. 355 pp.

Gilbertson, R.L. 1980. Wood-rotting fungi of North America. Mycologia **72**(1):1-49.

Kirk, T.K., T. Higuchi, and H. Chang (eds.). 1980. Lignin biodegradation: microbiology, chemistry and potential applications. Vols. I and II. CRC Press, Inc., Boca Raton, Fla. Vol. I, 256 pp.; Vol. II, 272 pp.

Miller, O.K. 1972. Mushrooms of North America. Dutton & Co., New York. 460 pp.

Phillips, R. 1981. Mushrooms and other fungi of Great Britain and Europe. Pan Books, Ltd., London, England. 288 pp.

Rumack, B.H., and E. Salzman (eds.) 1978. Mushroom poisoning: diagnosis and treatment. CRC Press, West Palm Beach, Fla. 263 pp.

Singer, R. 1975. The Agaricales in modern taxonomy. Ed. 3. J. Cramer, Weinheim, Germany. 912 pp.

Wasson, R.G. 1978. Soma brought up to date. Harvard University Bot. Mus. Leaflet **26**(6):211-223.

19

CLASSIFICATION

Division: Eumycota: "true fungi"; assimilative nutrition

    Subdivision 5: Deuteromycotina: "imperfect fungi" or "Fungi Imperfecti"; no sexual stages; parasexuality common

    Class 1: Blastomycetes: yeastlike organisms; many apparently related to the Basidiomycotina

        ORDER 1. Sporobolomycetales: conidiophores with ejectile mechanisms for conidia (ballistospore) dispersal; *Tilletiopsis*

        ORDER 2. Cryptococcales: reproduction primarily by budding; no ballistospores; *Candida, Cryptococcus, Torulopsis*

    Class 2: Coelomycetes: spores produced in pycnidia or acervuli

        ORDER 3. Sphaeropsidales: spores formed in pycnidia; many plant leaf and stem parasites; *Phoma, Phomopsis, Septoria*

        ORDER 4. Melanconiales: spores formed in acervuli; many plant parasites causing anthracnoses; *Colletrotrichum, Cylindrosporium, Glomerella*

    Class 3: Hyphomycetes: some with spores produced directly on special branches or sporophores but not in pycnidia or acervuli; some with sterile mycelia

        ORDER 5. Moniliales: large group of over 10,000 form species; some important soil fungi and plant and animal pathogens; some predaceous forms; *Acontium, Alternaria, Anguillospora, Arthrobotrys, Aureobasidium, Beauveria, Blastomyces, Botrytis, Cephalosporium, Cercospora, Cladorrhinum, Cladosporium, Clavariopsis, Coccidioides, Dactylaria, Dactylella, Drechslera, Epidermophyton, Flagellospora, Fusarium, Heliocondendron, Heliscus, Helminthosporium, Hormodendrum, Lemonniera, Microsporum, Monacrosporium, Monilia, Nematoctonus, Oidium, Paracoccidioides, Philaophora, Pyricularia, Sporothrix, Sporotrichum, Stachybotrys, Trichoderma, Trichophyton, Tricladium, Verticillium*

        ORDER 6. Agonomycetales: no production of conidia; reproduction mainly by hyphal fragmentation; *Rhizoctonia, Sclerotium*

Subdivision Deuteromycotina, or "Fungi Imperfecti," is perhaps better understood as the "imperfect fungi," since organisms in this group all lack described methods for classic sexual genetic recombination. Reproduction is accomplished entirely through asexual processes, usually by formation of distinctive spores (conidia) borne on special structures (conidiophores). Imperfect fungi are ubiquitous in the environment. Most genera are classified and identified by their conidial structure and color. Perfect stages of members of this subdivision are either nonexistent or unknown; genetic mixing and recombination are accomplished through the parasexual cycle described in Chapter 12. Over 22,000 species of imperfect fungi have been described in the literature. This chapter will attempt to present the most important biological aspects of the deuteromycetes.

Organization and classification of the imperfect fungi differ in some respects from the other groups of fungi because they are not based on evolutionary relationships; instead they are grouped together based on the form of the individual conidia that they produce. Such a system of classification is called "artificial," since it lacks any foundation in a phylogenetic sequence as in all "natural" systems of classification. Attempts to develop "natural" relationships among the deuteromycetes have not been entirely successful (see Booth, 1978, and Kendrick, 1979).

The Deuteromycotina are grouped into three classes—the Blastomycetes, the Coelomycetes, and the Hyphomycetes—based on the nature of their asexual reproductive structures. Most members of the deuteromycetes produce mycelia and conidia that resemble imperfect (asexual) stages of known members of Ascomycotina; thus many have been reclassified as ascomycetes after their sexual cycles have been observed (e.g., *Penicillium, Aspergillus, Neurospora*).

Some deuteromycetes appear to belong to Basidiomycotina based on the presence of a dolipore-like septum and clamp connection associated with dikaryotic hyphae (even though they have not been observed to produce basidia). A very few of the deuteromycetes have been found to be members of Zygomycotina based on the discovery of their methods of sexual reproduction with characteristic zygospores.

Many of the imperfect fungi are ecologically important soil organisms and economically important producers of antibiotics, food supplements, and fats. Others produce skin diseases, allergies, and various respiratory problems in humans. Some deuteromycetes are even predaceous, capturing and devouring various invertebrates such as nematodes, rotifers, and protozoans. This is accomplished by two basic methods: (1) production of sticky secretions on the surface of short hyphal knobs to which the invertebrates stick and are subsequently devoured by exoenzymes; and (2) production of hyphal loops that swell very rapidly, closing like a noose around long, thin invertebrates such as nematodes when they come in contact with the inner surface of the loop.

CLASSIFICATION INTO "FORM GENERA"

The taxonomy and subsequent classification of the deuteromycetes has always posed a major problem, because their sexual stages have either not yet been discovered or no longer exist. The taxonomy of the other groups of fungi is based largely on descriptions of their sexual or perfect stages. The result of this dilemma is the development of taxonomic criteria whose foundation lies in the structure of asexual reproductive conidiophores and conidia rather than on phylogenetic relationships. The result of this artificial system is a series of **form taxa** such as form order, form genus, and form species. For example, the form genus of an organism is a name given to that genus based on the description of a particular part of the organism. Among imperfect fungi, that particular part is the spore-producing structure. This means that on occasion two phylogenetically unrelated organisms may be grouped as a single genus, since their asexual reproductive structures resemble each other.

An artificial classification scheme of this sort is not based on the genetic relationships of the organisms involved; it is a classification of convenience based upon similarity in structure. The use of form taxa is common in paleobotany, where they are used to describe parts of organisms found in the fossil record. In such instances fossilized roots of a vascular plant might be assigned to one form genus, fossil leaves to another, and fossil stems to yet another—for example, *Stigmaria* (root), *Lepidophyllum* (leaf), and *Lepidodendron* (stem). After each of these has been described, it sometimes happens that a fossil is discovered in which the root, stem, and leaf are found joined on the same plant! They must then all be placed in the genus to which the stem has been assigned. However, the form genera remain in the literature.

A similar situation exists in describing the imperfect fungi. Thus certain fungi, before their sexual stages were discovered, were grouped in a form genus for similar conidial stages. This is true for the species *Neurospora crassa* (class Ascomycetes; order Xylariales) and *Sclerotina frutigens* (class Ascomycetes; order Helotiales). Both species have morphologically similar conidial stages. After their sexual stages were discovered, it became clear that both organisms were ascomycetes belonging to two separate orders. Before the sexual stages of these two species were described, they were both placed in the form genus *Monilia* (class Deuteromycetes; order Moniliales).

The result of the use of form taxa is that perfect fungi (those having described sexual stages) and imperfect fungi (those without described sexual stages) often have two different generic names that appear in the literature. When one considers that some ascomycetes develop sexual structures only once a year, it becomes a matter of convenience to use the form genus name that is based on conidial morphology. A good example of this practice is the use of the form species *Oidium tuckeri* to refer to certain powdery mildews found on grapes, although the actual perfect stage of the fungus is really *Uncinula necator,* an ascomycete. When two generic names are known for a single organism, it is current practice to use the name that describes the sexual phase as the taxonomically valid name; the other becomes the form genus and applies to its conidial stage. Thus the fungus causing certain skin diseases in humans belongs to the genus *Nannizzia,* whereas its form genus is *Microsporum* (Fig. 19-16). This situation is generally written as: *Nannizzia* (= *Microsporum*). Another dermatitis-producing fungus is the ascomycete *Arthroderma* (= *Trichophyton*). The example cited previously for powdery mildews in grapes would be written: *Uncinula necator* (= *Oidium tuckeri*).

Furthermore, some genera of ascomycetes such as *Penicillium* and *Aspergillus* contain species that are known to produce sexual stages, whereas other species of these genera do not. In such instances it is not uncommon to find these genera listed as either ascomycetes, deuteromycetes, or both.

CELL STRUCTURE

The features of the cells in Deuteromycotina as seen by the light microscope and the electron microscope are not significantly different from those cells already discussed for the other subdivisions. This is not surprising, since the imperfect fungi comprise an artificial classification of members of the zygomycetes, ascomycetes, and basidiomycetes whose sexual phases either do not exist or have not yet been described. Many of the Deuteromycotina have cells whose ultrastructural characteristics are unmistakably ascomycetous but whose hyphae lack described sexual stages. In some instances these fungi, on further study, have been shown to produce sexual stages bearing asci, at which time they were reclassified. A few deuteromycetes possess ultrastructural features resembling the basidiomycetes, and they too are reclassified after their basidia are discovered. The same follows for those that have proved to be zygomycetes.

REPRODUCTION

Reproduction among the deuteromycetes is largely a matter of hyphal fragmentation and conidial formation, although the parasexual cycle evidently plays an important role in genetic recombination. Parasexuality was first discovered by Pontecorvo and Roper (1952) while working with deuteromycetes; it is discussed in some detail in Chapter 12. Evidently parasexuality provides an important mode of genetic mixing in those fungi that do not readily undergo classic sexual processes accompanied by regular meiosis and zygote formation.

Asexual reproduction through formation of conidia is the most common means of spore production in imperfect fungi. Most of the classification is based on types of conidiophores and conidia that are produced. The manner by which various conidia develop has been the subject of a number of studies of ultrastructure, and different aspects of development have been described and classified (Ellis, 1971). The developmental processes involve the differentiation of conidial cells and the degree of involvement for inner and outer cell walls of those parts of the hyphae that produce the conidia. (The variations in conidial development are described in Chapter 12.) The number of developmental types of conidia can be reduced to a few ultrastructural patterns.

The conidial forms produced by the various imperfect fungi exhibit a wide spectrum of patterns, both for individual spores and for the conidiophores. Conidia may consist of one, two, or many cells and may be round, elliptical, cylindrical, sigmoid, branched, crescent shaped, tetraradiate, or helically coiled (Fig. 19-1). They may range from colorless to brightly colored (usually pink or green). The nature of individual conidia may be observed with the light micro-

Unicellular, round
Cladorrhinum

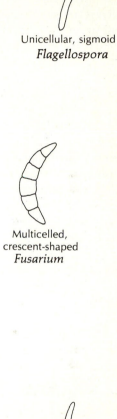

Unicellular, oval
Monacrosporium

Unicellular, sigmoid
Flagellospora

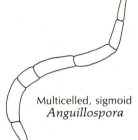

Three-celled, oval
Pyricularia

Multicelled,
tetraradiate
Lemonniera

Multicelled,
crescent-shaped
Fusarium

Multicelled, sigmoid
Anguillospora

Multicelled, spiral
Helicodendron

Multicelled, oval
Alternaria

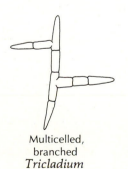

Multicelled,
branched
Tricladium

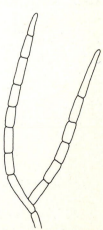

Multicelled, elongate
cylindrical
Cercospora

FIG. 19-1 Conidial shapes among various imperfect fungi.
Representative sample of unicelled and multicelled conidia in various configurations, as encountered among the Deuteromycotina and as viewed with the light microscope. Genus identification may be made on the basis of conidial structure.

scope, and a representative group can be found easily in the foam of any stream flowing through wooded areas, especially after leaves have fallen in the autumn.

Conidia are borne on conidiophores and together they are often called fructifications or "fruiting bodies." The fructifications of deuteromycetes come in a variety of shapes and forms. Some are small and nearly undifferentiated from the hyphal segments that produce them; others are larger and clearly differentiated from the hyphae.

The position and number of conidiophores is also significant. Some are produced singly on hyphal branches or on terminal hyphae; others are produced in small or large groups or even in specialized hyphal structures such as pycnidia or acervuli. The latter help to designate fungi belonging to the class Coelomycetes.

One order, Sporobolomycetales, produces specialized conidiophores that actively eject spores from their branches into the air. Such forcibly discharged spores are called ballistospores.

PHYSIOLOGY

The range of physiological abilities observed in Deuteromycotina is similar to that discussed in other fungal groups. Several species of *Fusarium* can grow anaerobically and ferment carbohydrates via glycolysis to either ethyl alcohol and carbon dioxide or lactic acid. Growth will proceed until the alcohol or lactic acid reaches toxic levels.

Many deuteromycetes are dimorphic, and the growth form—yeast or mycelial—often depends on temperature and nutritional state. Generally the mycelial or hyphal state exists at lower temperatures, whereas the yeast state occurs at higher temperatures. Most parasites of warm-blooded animals occur in the yeast form, except for the dermatophytic fungi, which occur chiefly in the mycelial form. The deuteromycetes are also quite **pleomorphic;** that is, they change shape and morphology in response to varying environmental conditions.

Mycotoxins, compounds secreted during fungal growth on livestock feed or human food, are produced by several genera of this group. A

FIG. 19-2 Conidia of *Fusarium* sp.
This widely distributed form genus produces two types of conidia—macroconidia and microconidia. Crescent-shaped septate macroconidia are represented here. Microconidia are usually one celled.

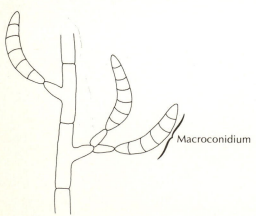

FIG. 19-3 Conidia and conidiophores of *Trichoderma* sp.
This widely distributed form genus may be found in many different kinds of environments. The conidiophores branch and give rise to conidia in bunches.

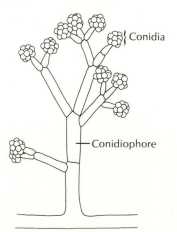

class of **terpenoid** mycotoxins called **trichothecenes** are synthesized as secondary metabolites by species of *Fusarium* (Fig. 19-2), *Stachybotrys*, *Trichoderma* (Fig. 19-3), and *Cephalosporium* (Fig. 19-4). These toxins may be **phytopathogenic** (causing disease in plants) or antibiotic. Other terpenoid toxins are also produced by *Cephalosporium* and by *Helminthosporium* (Fig. 19-5). When ingested by warm-blooded animals, these compounds produce intestinal upsets, convulsions, hemorrhage, or even death. These have caused illnesses in humans, fowl, and livestock such as pigs, horses, and cows. One mode of action of the toxins is the inhibition of protein synthesis (Ciegler and Bennett, 1980). Other deuteromycetes produce phenolic toxins such as **zearalenone** (F-2) (Fig. 19-6), which causes **estrogenic** effects in animals.

Various other mycotoxins found in Deuteromycotina may be derived from amino acids, cyclic peptides, aromatics, terpenoids, polysaccharides, or glycoproteins. Plant hormones, which act as growth regulators, are also produced by Deuteromycotina as well as by other groups of fungi. Ciegler and Bennett (1980) and Griffin (1981) discuss mycotoxin production and physiological effects.

Several deuteromycetes are mycoparasites. *Trichoderma reesei* (= *T. viride*) (Fig. 19-3), for example, may parasitize other fungi such as the honey mushroom, *Armillariella* (see Fig. 18-10), and some "lower" fungi. The *T. reesei* hypha may coil around the host hyphae, resulting in the death of the host. This cellulolytic fungus, found in soils throughout the world, is highly successful because not only does it grow rapidly but it also produces compounds with antibiotic activity. *T. reesei* is able to kill *Armillariella mellea* (Garrett, 1958), but the mechanism or the significance of such a necrotrophic mycoparasite in the natural environment is unknown at this time.

Although it is a common practice to steam-sterilize or fumigate greenhouse soil to kill pathogenic fungi, *T. reesei*'s ability to grow rapidly and outcompete other fungi enables it to

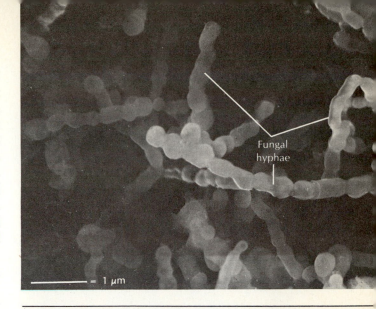

FIG. 19-4 Scanning electron photomicrograph of *Cephalosporium acremonium* mycelium.
This deuteromycete is grown commercially for the production of the antibiotic cephalosporin. This fungus is also widely distributed in the environment. (Courtesy Jonathan King and Erika Hartwieg, Electronmicroscopy Facility, Department of Biology, Massachusetts Institute of Technology, Cambridge, Mass.)

FIG. 19-5 The conidiophores and conidia of *Helminthosporium* sp.
This form genus may have four or more conidia grouped on the conidiopore. Many species of this form genus are saprophytic, but several species are important plant parasites.

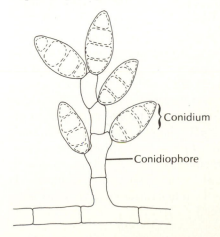

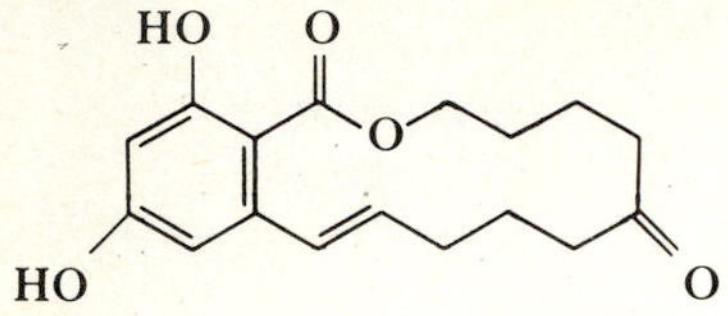

FIG. 19-6 Chemical structure of zearalenone (F-2).
This aromatic mycotoxin is estrogenic in mammals. Several species of *Fusarium* may produce this toxin while growing on agricultural products. The compound also induces the formation of fruiting in the fungus (Inaba and Mirocha, 1979).

quickly colonize such sterilized or partially sterilized soil. The spores of *T. reesei* are quite resistant to steam and fumigants, and the organism will grow in sterile soil and outgrow any pathogenic competitors that might damage young or delicate plants. Therefore, its growth is encouraged.

Several deuteromycetes can degrade cellulose. The cellulase complex of *T. reesei,* which contains three different types of enzymes, has been outlined by Griffin (1981). *T. reesei* produces large quantities of cellulose-splitting enzymes and has been used commercially for the breakdown of cellulose-containing wastes to glucose and other monosaccharides.

Some deuteromycetes are acid tolerant and can live at low pHs, although their maximum growth may occur at higher pHs (e.g., 5.5 to 6.0). For example, *Acontium velatum* is an **acidophile** (acid-loving) that can grow at pHs below 1.0. This fungus has the biochemical ability to reduce the pH of the medium in which it grows, thereby eliminating most competitors. In acidophilic organisms, the cell contents remain at or near neutral pH, thereby preventing inactivation or degradation of vital enzymes and other cell constituents. The successful acidophile must be able to maintain a near-neutral pH within the cell while withstanding very low pHs outside the cell.

Some pectolytic exoenzymes of fungi, especially those of fruit-rotting species, have optimum pHs between 5.0 to 5.5, enabling them to be active in acidic fruits. Other fungi are capable of altering the pH of their immediate surroundings by the production of such metabolites as organic acids or carbon dioxide or by the selective uptake of certain ions. *Sclerotium rolfsi*, a parasite of many different plants, secretes oxalic acid, which lowers the pH of surrounding plant tissue to 4.0, close to the optimum pH for its own pectolytic exoenzymes but not conducive to competing fungi.

Other deuteromycetes are **psychrophilic** (cold-loving) or **psychrotolerant** (cold-tolerant) and can grow at temperatures near 0° C. The snow mold *Fusarium nivale* may cause extensive damage to agricultural crops such as winter wheat or lawn grasses during periods of heavy and long-lasting snow cover.

EVOLUTION

No evolutionary patterns in Deuteromycotina can be discerned because of the lack of sexual structures. The grouping is composed of form genera only, and no evolutionary relationship exists or is implied between or within the artificial form classes. The taxonomy of the deuteromycetes is based on conidial morphology only. As sexual stages of the individual deuteromycetes are discovered, the organisms will be removed from this "catchall" group and placed in the appropriate "perfect" fungal group, based on sexual reproductive structures.

ECOLOGY

The deuteromycetes are important decomposers in both the terrestrial and aquatic environment. On land they grow on leaf litter and other decomposing plant material, including grains and straw. They can even grow inside desert rocks (Staley et al, 1982). They may be saprophytes or facultative or obligate parasites. Many are opportunists and can readily adopt a

TABLE 19-1. Deuteromycotina That Cause Disease in Humans

Disease	Causative fungus	Organs infected
Coccidioidomycosis	*Coccidioides immitis*	Lungs, other internal organs
North American blastomycosis	*Blastomyces (= Ajellomyces) dermatitidis*	Lungs
South American blastomycosis (paracoccidioidomycosis)	*Paracoccidioides (Blastomyces) brasiliensis*	Mouth, lungs
Cryptococcosis	*Cryptococcus (= Filobasidiella) neoformans*	Lungs, central nervous system
Candidiasis	*Candida albicans*	Mouth, vagina, systemic infection
Mycetoma (maduromycosis)	*Monosporium apiospermum (Allescheria boydii), Cephalosporium* spp.	Subcutaneous mycoses, skin lesions
Sporotrichosis	*Sporothrix schenckii*	Skin lesions
Tinea capitis ("ringworm")	*Microsporum audouini*	Itching, scaling of scalp
Tinea pedis ("athlete's foot")	*Trichophyton mentagrophytes*	Itching, scaling of foot
Other "ringworms"	*Trichophyton rubrum*	Various diseases of skin

Compiled from Briody, B.A., and Z.C. Kaminski, 1974; Shovlin, F.E., and R.E. Gillis, 1974; Baker, R.D., 1977; and Deacon, J.W., 1980.

saprophytic or parasitic role, depending on environmental conditions and the physiological condition of the potential host. Several species are extremely destructive plant parasites (Table 19-2), whereas others cause serious diseases of humans (Table 19-1). Deuteromycetes may also assume mycorrhizal roles with various flowering plants. For example, *Phoma* forms mycorrhizal associations with members of the heath family (Ericaceae) and *Rhizoctonia* with members of the orchid family (Orchidaceae) (see Chapter 13). The trickling filters found in sewage treatment plants consist of beds of small stones covered with fungi, especially *Fusarium*, algae, and bacteria. These organisms use—and thereby remove—organic material in the sewage.

As leaves start to senesce, weakly parasitic deuteromycetes called leaf-surface fungi may colonize the leaves. Such fungi as *Alternaria tenuis*, *Botrytis cinerea*, and *Cladosporium herbarum* can use both starch and simple sugars on and in the leaves. These fungi must not only be resistant to the high light intensity and ultraviolet light on unshaded tree leaves high above the ground but they must also be tolerant of the fluctuating moisture levels on the leaf surfaces. When the leaves fall to the ground, the leaf-surface fungi cease their growth and the saprophytic sugar fungi take over. After the sugar fungi have used the available monosaccharides and disaccharides, the cellulolytic fungi begin growth, and many of these, such as *Tricho-*

TABLE 19-2. Deuteromycotina—Plant Parasites

Genus	Host	Disease
Alternaria	Wheat, mustards, potatoes, tomatoes	Leaf blight, leaf spots
Cercospora	Lettuce, celery, beets, potatoes, tobacco, tomatoes	Leaf spot
Cladosporium	Peaches	Scab
Colletotrichum	Cereal, beans, cotton, alfalfa, coffee, trees	Anthracnose
Cylindrosporium	Apples, hops	Leaf spot
Drechslera	Barley, corn	Leaf blight, ear blight, seedling blight
Fusarium	Tomatoes, sweet potatoes, cabbage, alfalfa, cucumbers, bananas, cotton, peas	Vascular wilt
Helminthosporium	Potatoes, rice, oats	Scurf
Phoma	Beets, peas, crucifers	Black leg
Pyricularia	Cereals, rice, grasses	Rice-blast, leaf spots
Rhizoctonia	Many plants	Damping-off, root rot, canker
Sclerotium	Onions, garlic, other plants	White rot
Septoria	Celery, oats, wheat, pears, tomatoes, blackberries	Leaf spot, blight
Verticillium	Cotton, tomatoes, potatoes, eggplants, peaches	Wilt

Based on Gray, W.D., 1959; Haynes, J.D., 1975; Alexopoulos, C.J., and C.W. Mims, 1979; Deacon, J.W., 1980; and Webster, J., 1980.

derma and *Stachybotrys,* are deuteromycetes. As the cellulose is decomposed the ligninolytic fungi take over, and a few of these, such as *Fusarium* spp., may also be deuteromycetes. Ligninolytic deuteromycetes, however, are more common in the aquatic environment than in the terrestrial.

The Hyphomycetes* are extremely important in the degradation of plant material in the aqua-

tic environment and have been reported from streams all over the world. Conidia of the numerous aquatic deuteromycetes may be found in the foam of well-aerated streams. These conidia have two prevalent shapes, tetraradiate (four-armed) and sigmoid, which aid in their flotation. Ingold (1976) provides an extensive description of this group. A common hyphomycete with tetraradiate conidia is *Clavariopsis* (Fig. 19-7), while sigmoid conidia are produced by *Flagellospora* (Fig. 19-1). Other hyphomycetes have helical conidia and are called **aeroaquatic fungi** because their conidia develop above the water's

*The aquatic Hyphomycetes have been renamed the "Ingoldian fungi" in honor of C.T. Ingold, who studied these fungi (Webster and Descals, 1981).

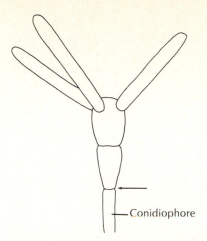

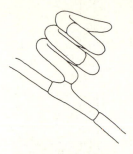

FIG. 19-7 Conidium of a hyphomycete, *Clavariopsis* **sp.**
The conidium of this aquatic deuteromycete develops
three apical arms, and, when mature, the conidium
breaks away from the mycelium (at the arrow). In this
manner a four-armed (tetraradiate) conidium is
formed. The mycelium is benthic, and the conidiophore projects up from the substrate into the water.
The tetraradiate shape aids the flotation of the conidia.

FIG. 19-8 Conidia of *Helicodendron* **sp.**
This genus is an "aeroaquatic" fungus, because the
conidia develop above the water's surface. The helical
conidia can trap air inside, which then keeps the
conidia floating on the surface.

surface. These organisms are often found in
stagnant waters; the helical structure retains air
inside it so that the conidia float. *Helicodendron*
(Fig. 19-8) is a common aeroaquatic fungus that
colonizes fallen leaves of oak, beech, and maple
trees (Gareth Jones, 1976). Taxonomy of these
fungi is discussed by Tubaki (1975a, b).

The aquatic hyphomycetes play an important
role in the cycling of nutrients in running water.
Most headwater streams receive the majority of
their carbon from **allochthonous** materials that
are produced outside the stream, such as leaves
and twigs. For the nutrients from this plant material to become available to the detritus-pro-
cessing invertebrates, the lignin and cellulose
especially must be broken down to easily assimilated smaller molecules. In the degradation process, the fungal biomass provides additional
food, including protein, to the detritivores. The
fungi concentrate the nitrate in the water into
their protoplasm. Aquatic hyphomycetes grow

well in waters that have a near-neutral or alkaline pH and that are high in oxygen and free of
pollutants. They appear to be extremely sensitive to pollution, but they tolerate low water
temperatures and actively decompose plant material during the winter. The peak of spore production in water occurs in late autumn in temperate areas, following the deciduous leaf fall
(Iqbal and Webster, 1973). Bärlocher (1982) discusses the ecology of these fungi, and the succession of hyphomycetes on leaves is reported
by Suberkropp and Klug (1976). These aquatic
fungi are less common in lakes because of the
poor aeration of most lake bottoms.

Aquatic hyphomycetes are also important
decomposers in peat (*Sphagnum*—Bryophyta)
bogs where the prevalent invertebrates, spiders
and springtails, consume them readily (Smirnov, 1958). In estuaries these fungi play a vital
role in the ecosystem by rapidly colonizing the
remains of cord grass (*Spartina* spp.), thereby

increasing the protein content of the *Spartina* detritus (Bärlocher and Kendrick, 1976). The Deuteromycotina appear to be the dominant fungal group in the marine environment and have been isolated from wood, sediments, and dead organic material (Gareth Jones, 1976). Over 50 species of deuteromycetes have been isolated from the marine environment, and over half of these were found growing on submerged wood (Kohlmeyer and Kohlmeyer, 1971).

Certain marine deuteromycetes can degrade some of the hydrocarbons found in petroleum. *Cladosporium* spp. (Fig. 19-9) and *Cephalosporium* spp. (Fig. 19-4) can break down complex hydrocarbon molecules, and *Candida* (Fig. 19-10), *C. lipolytica*, and *C. tropicalis* can oxidize straight-chained alkanes (Ahearn and Meyers, 1976).

The deuteromycetes, along with other fungi,

are important in the degradation of synthetic organic molecules, which may persist for long periods in the environment. For example, the chlorinated hydrocarbon insecticide DDT (dichloro-diphenyl-trichloro-ethane) is partially degraded in vitro by species of *Heliscus*, *Trichoderma* (Fig. 19-3), *Fusarium* (Fig. 19-2), and *Cladosporium* (Fig. 19-9). Many fungi, however, are adversely affected by DDT, and exposure to levels found in natural environments where DDT has been used causes detrimental changes in their growth and reproduction (Hodkinson and Dalton, 1973). Fungi may also concentrate and retain DDT in the ecosystem as they incorporate the chemical into their protoplasm. The exact role of fungi in DDT degradation is not known. Even though several species can break down DDT in vitro, the extent of fungal degradation in the natural environment

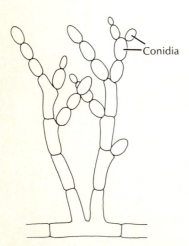

FIG. 19-9 Conidia and conidiophores of *Cladosporium* sp. The conidia of this deuteromycete are produced in chains that may branch. This fungus has been isolated from many habitats, including estuaries and the open ocean.

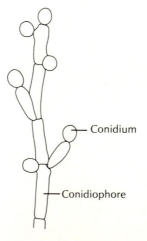

FIG. 19-10 Conidia of *Candida* sp. This widespread form genus has many species with a wide variety of environmental requirements. Some species are part of the internal flora of mammals, whereas other species can degrade petroleum fractions in the marine environment, where they are common inhabitants.

has not yet been determined (Hodkinson, 1976).

Several predatory hyphomycetes attack nematodes (roundworms) in the soil. Nematodes may live saprophytically in decaying vegetation or dung, or they may parasitize vascular plants, especially root and tuber crops. Nematodes also attack citrus and coconut trees, forest tree seedlings, grains, pineapple plants, and even commercially grown mushrooms (Christensen, 1975). Deuteromycetes may ensnare fungi with one of two mechanisms. One group of fungi has rings formed from the mycelium that trap the nematode. These rings are produced by *Arthrobotrys robusta* and *Dactylaria brochopaga*, for example (Fig. 19-11). The loops may be either single or interconnected. The several cells of the loop are stimulated to swell by contact with a nematode. As osmotic pressure in the loop cells increases and water uptake occurs, the

loop constricts almost instantaneously, usually killing the worm. Should the nematode wiggle free from the mycelium, it will take the loop with it, and eventually the fungus will invade the nematode and kill it. The nonconstricting loops act as mechanical traps, ensnaring the worm and eventually invading the body, even if the nematode escapes carrying a ring.

A second trapping mechanism seen in other nematode-trapping fungi is the adhesive knob or knobs on the hyphae to which the nematode adheres. Upon contact the nematode sticks to the knobs and becomes immobilized, and eventually the fungus invades and kills the worm. After the death of the worm, the fungus produces conidia, thereby insuring its dispersal. Fig. 19-12 shows the fungus *Monacrosporium* (= *Dactylella*) with adhesive knobs on the ends of short hyphal filaments.

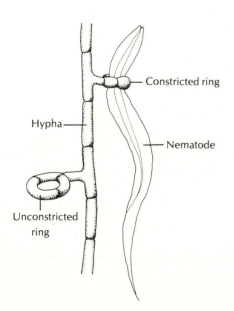

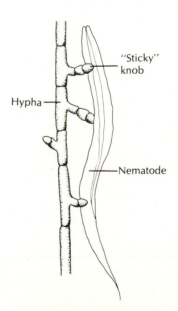

FIG. 19-11 The nematode-capturing *Dactylaria brochopaga*.
This deuteromycete forms rings when a nematode is close-by. When the nematode wiggles into the ring, the ring rapidly constricts, thereby trapping the worm. Not all species of *Dactylaria* produce constricting rings.

FIG. 19-12 The nematode-capturing *Monacrosporium* (= *Dactylella*).
This fungus produces knobs that secrete a sticky material. The nematode adheres to these sticky knobs and is thereby trapped.

Because the predacious fungi are never very prevalent in soils, the question arises: how do the fungi and the nematodes come into contact? Most investigators doubt that contact is made by chance alone. Evidence indicates that the nematodes are chemically attracted and move toward the traps (Field and Webster, 1977). The nematode-trapping fungal species (the same fungus may produce more than one kind of trap) apparently produce the trapping hyphae only if nematodes are nearby. If no nematodes are close, the fungus probably exists as a saprobe. Chemicals, probably peptides, exuded by nematodes evidently induce the fungi to form traps. Certain predatory fungi also produce a toxin that can be extracted either from parasitized worms or from the fungal mycelia, depending on the species.

Other species of Hyphomycetes are endoparasites of nematodes. Infection occurs either by the adhesive conidia attaching to the nematode and penetrating the host or by ingestion of the conidia, which then grow in the intestinal tract. *Nematoctonus*, an example of such an endoparasitic genus, is thought to have basidiomycete affinities, because the hyphae have clamp connections (Fig. 19-13).

Nematode-trapping fungi can easily be grown in the laboratory by incubating several grams of leaf mold in corn meal agar at room temperature. Nematodes should be evident in a few weeks, and, shortly thereafter, the nematode-trapping fungi. The fungi can be grown in culture as well, but they will produce traps only if nematodes are present. The trapping mechanisms can then be observed through the microscope. Duddington (1957) and Barron (1977) provide thorough discussions of nematode-destroying fungi.

These fungi have been proposed as biological control agents for crop-damaging nematodes; however, success has been limited in such use because it is difficult to sustain a large enough fungal population to significantly alter the numbers of nematodes.

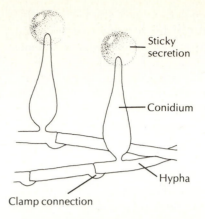

FIG. 19-13 Conidia of *Nemactoctonus* sp.
This fungus is endoparasitic on nematodes. The tip of each tapered conidium has a sticky secretion that helps the conidium to adhere to a worm. The fungus invades the cuticle of the worm, consumes the internal organs, and eventually kills the worm. *Nemactoctonus* has clamp connections, but basidium formation has not been observed.

Several deuteromycetes produce antibiotics that give them a competitive advantage over other microorganisms. Over 50 antibiotic substances have been detected in *Fusarium* and *Trichoderma* alone (Berdy, 1974). In addition, broad-spectrum antibiotics, the cephalosporins, were first isolated from *Cephalosporium acremonium* (Fig. 19-4).

The seeds of gymnosperms and angiosperms provide a specialized ecological niche for many facultative parasitic deuteromycetes. Commonly found in the angiosperm flower parts are *Aureobasidium* (Fig. 19-14), *Alternaria* (Fig. 19-15), and *Cladosporium* (Fig. 19-9). Pathogenic species such as *Phoma* are also found in seeds, providing not only a dispersal mechanism for these fungi but also a ready source of nutrition for the fungus when the seed germinates. Neergaard (1977) discusses seed-borne fungi in detail.

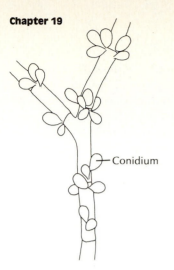

FIG. 19-14 Conidiophore and conidia of *Aureobasidium* sp.
Several conidia may develop from a single point along the conidiophore. *Aureobasidium pullulans* is common in soil but may also decompose paints and plastics. This fungus has also been isolated from marine habitats.

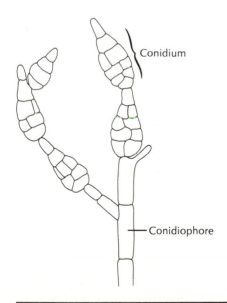

FIG. 19-15 Conidia of *Alternaria* sp.
This widespread and large genus grows on dead and dying plant material. Its airborne spores may cause allergic reactions in susceptible individuals. Some species of *Alternaria* may be parasitic on wheat or other food crops.

ECONOMIC IMPORTANCE
Diseases in Mammals, Including Humans

Many species of Deuteromycotina may cause disease in humans (Table 19-1, p. 401). These diseases may range from mild but annoying skin infections to systemic and life-threatening infections of vital internal organs. Most serious fungal infections occur in individuals whose immune system has been suppressed in some manner, either by disease or drugs. Certain fungal parasites of humans are endemic to particular regions of the world. These diseases occur as mild, seldom-detected infections in a large portion of the population, but such infections may occasionally become severe. An example of such a disease is "San Joaquin Valley (California) Fever," also known as **coccidioidomycosis,** which is prevalent in the hot, dry, desert climates of the southwestern United States, Mexico, and Central America.

The causative fungus of coccidioidomycosis, *Coccidioides immitis*, exists saprophytically in the soil, often 20 to 30 cm below the surface. It grows rapidly and produces many spores during the short rainy seasons (Christensen, 1975). Infection occurs following the inhalation of spores, which are numerous in the dust and soil. About 60% of those infected with the fungus have no symptoms, 40% develop a "flulike" illness accompanied by fever, and in about 0.2% of affected individuals the fungus becomes systemic (Fiese, 1958). Between 1950 and 1955 in California, close to 400 cases of systemic coccidioidomycoses were reported, with over 240 deaths (Fiese, 1958). If these figures represent only 0.2% of the cases, the incidence of the infection with the fungus in the desert areas of California must be very high. Indeed, the high incidence of the disease is documented by the large number of people in the area who exhibit immunological evidence of infection (Briody and Kaminski, 1974).

North American blastomycosis is caused by *Blastomyces dermatitidis* (= *Ajellomyces*) and produces lung lesions resembling those of tu-

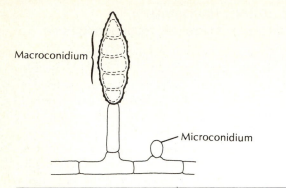

FIG. 19-16 Macroconidia and microconidia of form genus *Microsporum* (= *Nannizzia*).

This fungus causes diseases of mammalian skin, hair, and nails and is called a dermatophytic fungus. This fungus is mycelial on the host and produces microconidia and spindle-shaped macroconidia. This fungus does not reproduce sexually on the host. Sexual reproduction may take place in the soil, and the sexual stage, *Nannizzia*, produces a cleistothecium and is an ascomycete. The eight known species of *Nannizzia* all have conidia that resemble *Microsporum*.

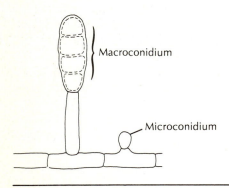

FIG. 19-17 Macroconidium and microconidium of form genus *Trichophyton* (= *Arthroderma*).

This fungus is also a dermatophyte and causes skin diseases such as "ringworm" in mammals. Infection appears to be restricted to dead keratin-containing tissues and very seldom becomes serious or systemic. This fungus may reproduce sexually on soil litter, producing cleistothecia. Eleven of the 15 known *Arthroderma* species have conidia resembling those of the *Trichophyton* form genus. The macroconidia are blunt at the end and thin walled.

berculosis. The infection is usually asymptomatic, but occasionally it will spread to other parts of the body. The incidence of infection is not known, but the fungus is apparently endemic to the southeastern United States, the Mississippi River valley, and Africa, where it lives in the soil (Briody and Kaminski, 1974).

South American blastomycosis, also called paracoccidioidomycosis, may cause lesions either in the oral mucous membranes or in the lungs. The disease may remain localized or may spread to the skin or intestinal tract. The disease is caused by *Paracoccidioides brasiliensis* (= *Blastomyces brasiliensis*). The pathogen lives in the soil or on dead plant material in Central and South America.

A serious mycotic meningitis may be caused by *Cryptococcus* (= *Filobasidiella*) *neoformans*. This organism enters the body through the lungs, producing lesions. If the disease becomes systemic, the central nervous system in general and the meninges in particular may be involved. Treatment for this disease and for other systemic fungal diseases is with the **polyene antibiotic** amphotericin B, which is produced by various species of *Streptomyces*, an actinomycete bacterium. Usually amphotericin B is given intravenously, but it may produce undesirable and serious side effects, especially in the kidneys. A great need exists for additional antifungal drugs, because the number of effective drugs is very limited (Maugh, 1981b).

Many deuteromycetes are dermatophytes that live on keratin-containing tissues such as nails, skin, and hair, causing diseases of these tissues. The three most common genera causing skin problems are *Epidermophyton*, *Microsporum* (Fig. 19-16), and *Trichophyton* (Fig. 19-17). These genera are widespread throughout the world and contain about 30 species that cause infections in humans. Others cause infections in domestic animals and livestock. The infecting fungi are usually mycelial when on the skin, in contrast to those causing systemic infections, which are usually in a yeastlike form inside their host. Spores may enter the skin through minor cuts or abrasions to initiate infection. The general medical term given to skin infections, or **dermatomycoses,** is "tinea"; thus "tinea pedis" is an infection of the foot (e.g., athlete's foot) and "tinea capitis" is a head or scalp infection (e.g., ringworm of the scalp; Fig. 19-18). Many species of the ascomycete genus *Arthroderma* have imperfect stages classified in the form genus *Trichophyton*, while most species of the ascomycete *Nannizzia* have the form genus *Microsporum.* Ajello (1977) reviews the dermatophytes and their perfect and imperfect states.

Baker (1977) lists the dermatomycoses as one of the three great endemic mycoses. Humans are usually the source of infection: person-to-person transmission occurs between children and in shared bathing facilities such as locker rooms. Dogs and cats may also be hosts for the ringworm fungus *Microsporum canis*, which frequently infects pet owners (Fig. 19-18). Many skin diseases are more severe and more widespread in the tropics because of the moist, warm climate, which is conducive to rapid fungal growth. Dermatomycoses are usually treated with the polyene antibiotic nystatin or the *Penicillium*-produced griseofulvin (see Chapter 17). Administration of nystatin may be either topical or internal, whereas griseofulvin is administered internally.

Candida albicans is a deuteromycete that is part of the normal flora of the human mouth, skin, rectum, and vagina. When an individual is stressed by antibiotic or steroid therapy or by diabetes, *C. albicans* may overgrow the other flora, causing soreness and itching. The disease is called "thrush" when it occurs in the mouth, and it may cause epidemics in nurseries for newborns if nursing bottles or nipples become contaminated (Shovlin and Gillis, 1974). *C. albicans* may spread throughout the body, causing a severe disease. When the dimorphic fungus does become systemic, it may occur either in the mycelial or the yeast form, in contrast to most other serious fungal pathogens, which occur in the yeast form within their host.

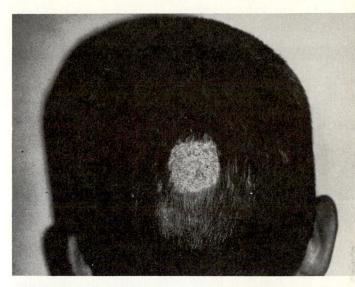

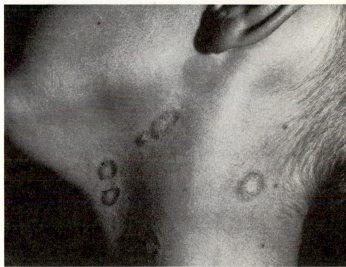

FIG. 19-18 Ringworm of the scalp and body.
Two areas of skin infection caused by *Microsporum canis*. This ringworm fungus is often called the "kitten" or "puppy" fungus, since cats and dogs are sometimes the vectors for it. Top photograph shows ringworm of the scalp or "tinea capitis," while the bottom photo is ringworm of open skin or "tinea corporis." The common term "ringworm" refers to the circular patches on the skin which are produced by growth of the fungus. (From Stewart, W.D., J.L. Danto, and S. Maddin. 1978. *Dermatology*. Ed. 4. The C.V. Mosby Co., St. Louis. p. 268.)

Both dermatomycoses and the internal mycoses can be extremely persistent, resisting antibiotic therapy, which often has to be continued over months or even years. The systemic mycoses often go into remission, but severe recurrences may happen after weeks or even months of apparent good health. Parasitic fungi can cause a broad spectrum of diseases in humans whose resistance has been lowered by disease or drug therapy. Endocarditis, pneumonia, and meningitis may all be caused by fungi, some of which are mere opportunists, occurring as part of the normal, endogenous, human flora. Human fungal diseases are reviewed by Rippon (1982).

High concentrations of fungal spores in the air may cause severe allergic reactions such as asthma in susceptible individuals. The widespread deuteromycetes *Alternaria* (Fig. 19-15) and *Hormodendrum* grow on decaying or dying plant material in late summer and early autumn in temperate areas. These fungi produce myriads of conidia to which many people are allergic. Farmers are exposed to large concentrations of fungal spores during the harvesting and storing of crops. In susceptible individuals the inhalation of large numbers of spores may cause "farmer's lung," an inflammation of the alveoli of the lungs (Bartell et al, 1974).

Mycotoxins Causing Diseases in Humans and Livestock

When humans or livestock eat food contaminated with mycotoxins they may become ill. The severity of the illness is related to the amount of toxin ingested. For example, when *Fusarium tricinctum* (= *Fusarium sporotrichioides*) grows on a grain such as millet, the trichothecene mycotoxin T-2 may be produced, causing illness in those who consume the contaminated food. **Alimentary toxic aleukia (ATA)** results from the ingestion of the toxin. Symptoms of the disease are a reduction in the number of leukocytes (type of white blood cell), resulting in severe illness and occasionally death.

An epidemic of ATA in the USSR in 1944 killed thousands of people (Christensen, 1975).

Fusarium is widespread throughout the world, existing usually as a saprophyte in decaying plant material. The fungus may also parasitize plants (Table 19-2). Corn grown in the United States has been found occasionally to be contaminated with T-2 toxin from *Fusarium*. **Moldy corn toxicosis,** a disease of cattle, results from livestock ingesting this contaminated corn. The presence of the toxins may also cause livestock to refuse to eat the corn (Ueno, 1977).

Stachybotrys alternans (= *Stachybotrys atra*) may grow on hay or straw and produce a trichothecene toxin. This toxin causes stachybotryotoxicosis, a disease that was first recognized in the 1930s, but the cause of the disease was not definitively established until the 1940s. Forgacs (1972) discusses this disease in the first well-documented case of **mycotoxicosis** in livestock. The disease was extremely important in Russia and eastern Europe in the 1930s, because it caused the illness and death of many horses when contaminated hay was fed to the horses during the winter. The death of the horses caused economic distress in the Ukraine especially, where horses were essential to the economy of the agricultural region (Ciegler and Bennett, 1980). In his memoirs Nikita Khrushchev (1970) recognized the potentially detrimental effect of the toxin and discussed the need for uncontaminated feed for livestock.

Certain species of *Fusarium* (Fig. 19-2) produce a phenolic mycotoxin, zearalenone or F-2 (Fig. 19-6), which not only produces estrogenic effects in mammals but also induces fruiting in the fungus (Inaba and Mirocha, 1979). Zearalenone, when ingested by pigs, produces a hyperestrogenic syndrome that may include vaginal prolapse, ovarian atrophy, abortion, and enlarged uteri, mammary glands, and vulva (Ciegler and Bennett, 1980). This estrogenic syndrome has been observed in swine in Europe, the United States, and Russia. In male swine, feminization and testicular atrophy may

occur following consumption of F-2-contami-
nated feed (Christensen, 1975). Because in low
doses zearalenone causes increased growth in
both cattle and sheep, its use as an additive in
livestock feed has been proposed (Mirocha et al,
1977).

Fusarium species may produce other toxins;
some have been implicated in a phenomenon
known as the **refusal factor corn** in which contaminated corn is refused by pigs. The compound involved in the refusal factor has not
been isolated. In the corn belt of the United
States, pigs will starve rather than eat the contaminated corn (Christensen, 1975). *Fusarium
roseum (= Gibberella zeae)* was found growing
on the corn. The refusal factor and other contaminating mycotoxins are difficult to isolate because the toxin evidently persists after all microscopic or macroscopic evidence of fungal growth
has disappeared.

In 1981 the U.S. State Department made
public a suspicion that mass-produced trichothecene mycotoxins (Fig. 19-19) were being
used and causing illness in the peoples of southeast Asia (Wade, 1981a). The trichothecenes
implicated were ones produced by the ubiquitous form genus *Fusarium*. Symptoms typical of
trichothecene poisoning have been reported by
refugees following the dispersal of a substance
called **yellow rain** by crop-dusting airplanes.
Samples of foliage from southeast Asia were
analyzed for trichothecene, and levels up to 160
ppm were found (Wade, 1981a). Subsequently
several parts per billion of T-2 were found in the
blood of victims of a yellow rain attack in Kam-
puchea (Cambodia), suggesting exposure to the
toxins (Marshall, 1982a; Mirocha, 1982). There
is a paucity of information concerning both the
distribution of *Fusarium* in southeast Asia and
the production of trichothecenes by the fungus
under various environmental conditions (Wade,
1981b).

Jarvis et al (1981) reported the first study of
trichothecene levels in a common shrub in Brazil. Levels ranged from 200 to 300 ppm. The

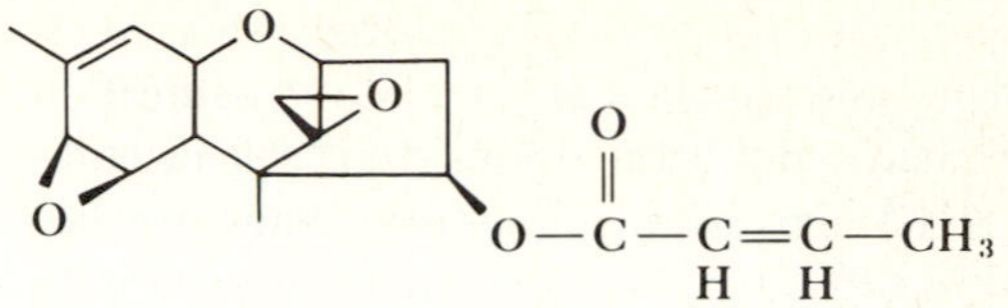

**FIG. 19-19 Chemical structure of a mycotoxin, tricho-
thecene (T-2).**
This terpenoid is produced by various species of
Fusarium. Other deuteromycete genera such as
Cephalosporium, *Stachybotrys*, and *Trichoderma*
may also produce trichothecene mycotoxins. Some
trichothecene toxins inhibit protein synthesis in animals (Wei and McLaughlin, 1974).

shrub evidently absorbed, chemically modified,
and concentrated the mycotoxins that were produced by soil fungi. Cattle that grazed on the
shrub became seriously ill. Other fungi in addition to *Fusarium* produce trichothecenes. Until
further research elucidates the extent of trichothecene production under varying environmental conditions, especially in the jungles of
southeast Asia, the question of whether potent
mycotoxins are being introduced to cause illness
or death to the local populace will remain unan-
swered. Until additional evidence is forthcom-
ing, some scientists remain skeptical about the
very serious charges of mycotoxin use, which is a
violation of biological and chemical warfare
treaties (Marshall, 1982b).

Further information on the effects of mycotoxins on both humans and livestock may be
found in Rodricks et al (1977), Uraguchi and
Yamazaki (1978), and Tamm and Breitenstein
(1980). The alleged use of mycotoxins in southeast Asia is discussed by Seagrave (1981).

Plant Diseases

Although many serious diseases of economically important gymnosperms and angiosperms
are caused by members of Deuteromycotina,
classification of some of these pathogenic fungi

is extremely difficult. Sexual stages have been found in some instances, but the classification still remains doubtful because of lack of sufficient evidence. Many deuteromycetes are believed to belong to Ascomycotina and a few with obvious clamp connections to Basidiomycotina. Until more convincing information is available, they remain among the imperfect fungi.

Several economically important plant diseases caused by the deuteromycetes are listed in Table 19-2 (p. 402). The list is by no means complete, because there are probably thousands of disease-causing imperfect fungi. Many of the disease-causing organisms can live saprophytically in the soil for long periods, as can the fungi that produce diseases in humans. The conditions that induce a normally saprophytic fungus to become parasitic or pathogenic are not known. Such fungi are opportunistic and are ready to parasitize when the proper combination of environmental factors and host susceptibility occurs. A fungus that only causes mild infections may become a serious threat to crops when conditions favor its growth or when it mutates to become a more successful parasite.

Several extremely important plant diseases are worth emphasis here. *Pyricularia oryzae* causes **rice blast,** one of the most destructive parasites of rice in Asia; damage from the fungus was reported as far back as 1637. Antifungal antibiotics have been used in Japan to control this fungus. The pathogen primarily attacks leaves. The fungus flourishes on rice grown in nitrogen-rich soils; thus the addition of nitrogen-containing fertilizers may do more harm than good. In one report, yields of rice were increased, but at the same time losses because of the fungus also increased (Chrispeels and Sadava, 1977).

Another serious parasite of rice, *Helminthosporium oryzae,* caused a widespread famine in Bengal in 1943. Two million people died of starvation following the destruction of a large portion of the rice crop (Deacon, 1980).

A very destructive disease in the corn belt of the United States is the southern corn blight

Drechslera (= *Helminthosporium*) *maydis* (Fig. 19-20). In 1970 this fungal disease caused the destruction of about 15% of the U.S. corn crop, and the damage approached $1 billion. The disease had been present in low densities in the southern United States for many years. In 1970 the combination of mutations in the fungus, weather favorable for fungal growth, and a genetically uniform (and therefore equally susceptible) corn crop made conditions conducive to the development of an epidemic. The blight spread rapidly via windborne spores to the northern portion of the corn belt—Indiana, Illinois, and Iowa. The ecologically unsound practice of planting genetically similar crop species over such large areas was again underscored, as it was during the potato blight (see Chapter 15). Fortunately seed companies reacted rapidly, and by the 1971 growing season seed varieties resistant to corn blight had become available

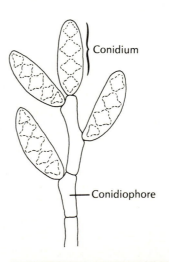

FIG. 19-20 Conidia of *Drechslera* sp.
Many species of form genus *Drechslera* were previously classified in form genus *Helminthoporium.* The club-shaped conidia of *Drechslera* develop apically from the conidiophore tip. Many *Drechslera* species cause disease in plants and may be found in the seeds of monocotyledonous angiosperms. Species of *Drechslera* attack corn, barley, sorghum, oats, and wheat.

and corn blight damage was greatly reduced. The seed companies had maintained reserves of many genetically different types of crops, thereby enabling them to quickly find blight-resistant varieties among their reserves (Walsh, 1981). The potential for a disease to rapidly attack and destroy genetically identical major food crops was emphasized by the 1970 epidemic. The devastation from the blight caused the National Academy of Sciences (1972) to study the problems of genetic uniformity and resulted in their recommending the maintenance of stocks of many genetically different strains of food crops.

Corn blight fungus attacks the lower leaves of the corn plant first, killing them. As the fungus rapidly progresses to the top of the plant, many more leaves are killed, yield is reduced, and death of the plant may ensue. *Drechslera maydis* can overwinter in debris left in the field; thus elimination of the fungus is difficult. The development of resistant varieties holds the best promise for the control of the disease. Tatum (1971) and Roberts and Boothroyd (1972) discuss the corn blight epidemic in detail.

Fusarium oxysporum causes diseases in many important crop plants. The fungus causes the host to wilt, probably because of the production of a toxin that affects membrane permeability and therefore water uptake. The form genus *Fusarium* is difficult to define taxonomically, because it not only exhibits great variability but also is distributed worldwide, and many species from remote areas probably have not been identified. One variety, *F. oxysporum* var. *cubense*, causes a highly destructive wilt of banana trees in Central America. The disease has been called the "Panama disease" because of its prevalence in that country. The fungus grows rapidly on the banana tree, the leaves wilt, and the tree dies soon thereafter. A plantation of banana trees can be ruined by the fungus in several months. Many hundreds of hectares of banana plantations have had to be abandoned in Central America. Manipulations of the environment such as silting and flooding of the fields have been attempted as a means of controlling the

fungus, but the planting of resistant varieties has been the best defense thus far.

The species *Fusarium nivale* is cold-tolerant and may kill grass that has been covered for a long period by snow or other covers such as tarpaulins. Homeowners may be shocked in spring to see large patches of dead grass, especially where snow has lain for several months. British cricket players and fans were appalled to find their cricket field devoid of grass because a tarpaulin had protected the field from heavy rains for several days (Deacon, 1980). *F. nivale*, the "snow mold," had grown rapidly under the ideal conditions of high moisture and low light, and the grass had been quickly killed. The fungus may also cause destruction of winter-sown wheat, especially when snow cover is persistent.

Other widespread and destructive plant parasites occur in the following form genera: *Rhizoctonia*, *Sclerotium*, *Cercospora*, *Alternaria*, *Phoma*, *Colletotrichum*, *Phomopsis*, *Verticillium*, *Septoria*, and *Cylindrosporium*. Further information concerning Deuteromycotina-caused plant diseases may be found in Stevens (1913), Agrios (1969), Roberts and Boothroyd (1972), Dickinson (1977), and Holliday (1980).

Decomposition of Synthetic Products

Wherever there is a source of carbon, no matter how complex the molecule, there seems to be a fungus that can utilize it. This ability can be both beneficial and detrimental to humans. *Cladosporium* sp. (Fig. 19-9) can utilize the carboxymethylcellulose used in paints as a carbon source, causing the growth of dark "sooty mold" on paints used in such moist areas as kitchens and bathrooms. Another species, *Cladosporium herbarum*, can attack and decompose meats at temperatures of −6° C. *Aureobasidium pullulans* (Fig. 19-14) not only attacks painted surfaces but can also partially decompose certain plastics, especially in tropical areas where growth conditions for fungi approach the ideal.

Cellulose-containing fabrics such as wool and cotton may be decomposed by species of *Alter-*

naria, Stachybotrys, or *Fusarium.* Paper mills are plagued by various deuteromycetes such as *Cladosporium* and *Trichoderma,* which can decompose or stain the paper in different stages of manufacturing. Several deuteromycetes can also grow on stored lumber, causing stains and resulting in downgrading of the lumber. Formalin may be added to wood products to discourage such fungal growth.

Deuteromycetes have been isolated from petroleum-based fuels. *Candida* spp. (Fig. 19-10), *Torulopsis* spp., and *Cladosporium* spp. have the biochemical ability to biodegrade and use various hydrocarbons in jet and conventional aircraft fuels (Ahearn and Meyers, 1976). Walters and Elphick (1968) document the various fungi that degrade synthetic materials.

Production of Food and Drink

When weather conditions are favorable, *Botrytis cinerea* (Fig. 19-21), a parasite, grows well on ripening grapes. The growth of this fungus results in "botrytized" grapes, which produce sweeter wines. Not only does the presence of the fungus increase the sugar content but it also reduces the acid content of the resulting wine. Many vineyards, therefore, encourage the growth of the weak parasite *B. cinerea,* whereas other agriculturists discourage other species of *Botrytis* because they are destructive parasites of broad beans, onions, and ornamental flowers (see Table 19-2).

Several deuteromycetes have been grown as food sources. *Candida utilis, Candida arborea,* and *Fusarium* show promise as food sources. Both fungi and bacteria have been used to produce single-celled protein (SCP), which can serve as a substitute for conventional protein sources. Well-aerated wastes from the paper industry have been used to culture *C. utilis.* The fungus grows well and is rich in protein (up to 50%), vitamins, and fats. The yeastlike fungus is pleasantly flavored and, after drying, may be eaten directly or added to less-nutritious foods. Fungi can convert carbohydrate to protein much more efficiently than cows, pigs, or even

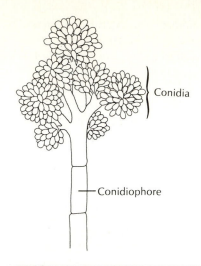

FIG. 19-21 Conidiophore and macroconidia of *Botrytis* sp.
One species of this genus, *Botrytis cinerea,* is a weak parasite that attacks dying or aging plant tissues. Other species are virulent plant parasites. The conidia are easily dispersed by the wind. *Botrytis* may also produce microconidia.

poultry can. Deuteromycetes that were cultured in Germany during World War II contributed to the supply of nutritious food during a time of food shortage (Rose, 1981). The ability of some deuteromycetes to grow on by-products of the petroleum industry such as straight-chain hydrocarbons can result in the conversion of these waste hydrocarbons to nutritious food or fodder. Gray (1970) and Rose (1981) discuss the commercial use of fungi for food and fodder.

Trichoderma reesei (= *T. viride*) is used for the commercial production of cellulolytic enzymes that hydrolyze cellulose. These enzymes (cellulases) are used to convert various cellulose-containing substrates into glucose, which in turn may be fermented by Ascomycotina to ethyl alcohol or synthesized into food, fodder, or organic chemicals by other fungi or bacteria. Substrates used include wastes from carbohydrate crops such as sugar cane and corn, wood chips, waste paper, algae, and aquatic angiosperms. Eveleigh and Montenecourt (1977) review the different cellulase enzymes produced by fungi. Bungay (1981) discusses the methods of conversion of biomass to food and fuels.

Commercially Important Products

Deuteromycetes are cultivated commercially for their production of fats, antibiotics, and single-celled protein. Deuteromycotina were used in the 1930s for fat production. Several deuteromycetes have the biochemical ability to metabolize sugars to fats under the proper conditions. The production of glycerol, a fat precursor, in Germany peaked at about 1,000 metric tons per month during the 1930s.

Cephalosporium acremonium (Fig. 19-4) is used for the commercial production of cephalosporin antibiotic drugs. For over 30 years *C. acremonium* and *Penicillium chrysogenum* (see Chapter 17) were the major sources of the β-lactam antibiotics, which are not only low in toxicity but also effective against a broad spectrum of bacteria. Recently antibiotic manufacturers have been able to add chemically certain side groups to the cephalosporin molecule, thereby increasing its activity against gram-negative bacteria. Sales of the cephalosporins throughout the world in 1978 totaled half a billion dollars (Aharonowitz and Cohen, 1981). Methods are being worked out to increase the activity spectrum of the cephalosporins to include those antibiotic-resistant gram-negative bacteria which are prevalent in hospitals. The current annual cephalosporin sales of $1 billion are expected to increase as the range of susceptible bacteria continues to increase (Maugh, 1981b).

Biological Control Of Insect and Plant Pests

Several deuteromycetes have been successfully used as control agents for insect pests. Fungi that parasitize insects are called **entomogenous.** For fungal parasites to be successful, the climate must be conducive to the rapid dissemination of the fungus. Such conditions exist in the U.S. Gulf states and in other areas (e.g.,

Trinidad) where fungal control of several insects has been successful (Gray, 1959). Species of *Beauveria* have been successfully used in Georgia for the control of the pecan weevil (Tedders, 1981). A species of *Phomopsis* that has been isolated in Great Britain interferes with reproduction in the elm bark beetle. With further research this particular fungus may be able to help in the control of the Dutch elm disease (see Chapter 17), which is spread by the beetle (Webber, 1981).

Cercospora rodmanii (Fig. 19-1) shows some potential for controlling populations of the troublesome water hyacinth (Angiospermae), which is overgrowing and blocking waterways in the southern United States (Conway, 1976).

SUMMARY

The Deuteromycotina are known as the "imperfect fungi" because their sexual (perfect) stages are unknown. These stages either do not exist or have not yet been observed and described. Most of the imperfect fungi are ascomycetes whose sexual reproduction is unknown, although a few have been shown to be basidiomycetes or even zygomycetes, based on their cell structures and the discovery of their sexual stages. Reproduction in imperfect fungi is entirely asexual by fragmentation or by production of distinctive conidia borne on conidiophores. Taxonomy of the group is based solely on the structure of asexual reproductive cells; this results in an artificial classification based on transient names called "form genera" or "form species."

The physiology and biochemistry of Deuteromycotina is much the same as that found in other members of division Eumycota, from which they are all derivatives. Some exhibit dimorphic forms existing either as hyphae or unicellular, yeastlike fungi. Some produce mycotoxins and some are mycoparasites that grow at particular temperatures and pH ranges. Their evolution is closely associated with other aflagellate fungi.

Like the other members of the division, the imperfect fungi play an important role in the decomposition of organic materials in terrestrial and aquatic habitats. They are commonly found growing on decaying plant materials such as leaves, fruit, and stems. They are especially important in the breakdown of cellulose and lignin associated with woody stems. Their enzyme systems can also degrade petroleum products and various synthetic materials and products, including DDT. A few genera are even predaceous on soil nematodes and have evolved ingenious prey-catching devices.

Economically the Fungi Imperfecti have had a significant effect on human development. Many genera cause diseases in humans and thus have had a significant impact on the medical arts (Table 19-1). They have had an even greater effect on agriculture (Table 19-2) and the study of plant pathology. Still other genera have been used to produce food (single-celled protein [SCP]) and to adjust the sweetness of wines. Some are even used in the production of chemicals and commercially important organic molecules.

SELECTED REFERENCES

Bärlocher, F. 1982. On the ecology of Ingoldian fungi. BioScience 32(7):581-586.

Barnett, H.L., and B.H. Hunter. 1972. Illustrated genera of imperfect fungi. Ed. 3. Burgess Publishing Co., Minneapolis. 241 pp.

Barron, G.L. 1977. The nematode-destroying fungi. Topics in mycology. No. 1, Canadian Biological Publications, Ltd., Guelph, Ontario. 144 pp.

Cole, G.T., and B. Kendrick (eds.). 1981. Biology of conidial fungi. Vols. 1 and 2. Academic Press, Inc., New York. Vol. 1, 486 pp; Vol. 2, 660 pp.

Gray, W.D., 1966. The relation of fungi to human affairs. Henry Holt & Co., New York. 510 pp.

Heitefuss, R., and P.H. Williams (eds.). 1976. Physiological plant pathology. Springer-Verlag Inc., New York. 890 pp.

Kendrick, B. (ed.). 1971. Taxonomy of Fungi Imperfecti. University of Toronto Press, Canada. 309 pp.

Rippon, J.W. 1982. Medical mycology. Ed. 2. W.B. Saunders Co., Philadelphia. 842 pp.

Uraguchi, K., and M. Yamazaki. 1978. Toxicology, biochemistry and pathology of mycotoxins. John Wiley & Sons, Inc. New York. 288 pp.

LICHENS AND BRYOPHYTES

20

CLASSIFICATION

Lichens are classified according to the morphology and the fruiting bodies of the mycobiont.

Ascolichens: fruiting bodies contain asci; approximately 500 genera

ORDER 1. Arthoniales:* thallus crustose, foliose, or fruticose; mostly tropical; phycobiont often chlorophyte *Trentepohlia*

ORDER 2. Dothideales:* thallus crustose

ORDER 3. Verrucariales: thallus crustose or foliose; perithecia; on rocks, trees, or soil; some aquatic: *Dermatocarpon, Endocarpon, Verrucaria*

ORDER 4. Pyrenuales: thallus crustose; perithecia; phycobionts usually chlorophyte *Trentepohlia*, subtropical to tropical; *Strigula*

ORDER 5. Caliciales: thallus crustose, foliose or fruticose; apothecia; on bark or wood; green algae phycobionts; *Calicium* (phycobiont = *Chlorella*)

ORDER 6. Ostropales:* thallus crustose; phycobiont usually chlorophyte *Trentepohlia;* in warm climates

ORDER 7. Graphidales:* thallus crustose; green algae phycobionts; in warm climates

ORDER 8. Lecanorales: thallus crustose, foliose, or fruticose; well-developed apothecia with paraphyses; on rocks, trees, soil, man-made substrates;

A. With blue-green phycobiont: *Collema* (a gelatinous form), *Ephebe, Lobaria, Peltigera, Sticta*

B. With green phycobionts: *Bacidia, Lecania, Lecanora, Lecidea, Rhizocarpon*

C. With chlorophyte *Trebouxia* phycobiont: *Alectoria, Bryoria, Cetraria, Cladina, Cladonia, Evernia, Lasallia, Omphalodium, Parmelia, Pertusaria, Physica, Ramalina, Stereocaulon, Umbilicaria, Usnea, Xanthoria*

Basidiolichens: fruiting bodies contain basidia; 6 or more genera

ORDER 1. Aphyllophorales: fruiting body crustose to bracket; phycobiont blue-green; tropical; *Dictyonema, Cora*

ORDER 2. Agaricales: foliose thallus; arctic to temperate; *Omphalina* (grows on humus)

Deuterolichens: no ascocarps; crustose on soil, rocks, trees; arctic to tropical; 6 genera; *Lepraria*

LICHENS

---

*See Poelt (1973) for further information about these orders.

A lichen is an autotrophic, symbiotic organism that is composed of an alga and a fungus. Both partners may derive benefit from this association, which therefore may also be labeled **mutualistic.** The morphology of the lichen is distinct from the morphology of either partner when grown separately. Lichens produce chemical compounds, the **lichen substances,** that neither symbiont produces when growing alone. The lichen is thus an autonomous organism with morphological and biochemical characteristics that are different from either the algal or the fungal partner.

The algal component of the lichen is called the **phycobiont** and the fungal component the **mycobiont.** The form of the mature lichen thallus is usually determined by the fungus, so many algal characteristics may become obscure.

The two components of lichens are easily observed under a compound microscope, with fungal hyphae making up the bulk of the organism's biomass and algae forming a much smaller portion. Most lichenized fungi are classified as Ascomycotina and most lichen-forming algae are Chlorophyta. However, a few mycobionts have been identified as Basidiomycotina or Deuteromycotina, and several nitrogen-fixing Cyanobacteria may be phycobionts.

Much research and speculation has focused on the symbiotic relationship between the phycobiont and the mycobiont. Lichens have been studied for over 200 years, and the dual nature of the organism has been known for over 100 years. The exact nature of the symbiosis, however, still remains unknown. It has been shown experimentally that carbohydrates and—in cyanobacterial phycobionts—nitrogenous compounds are transferred from the alga to the fungus. In some lichens the fungi produce haustoria that invade the algal cell wall but rarely penetrate the cell membrane. No experimental evidence has documented the benefits of the association to the alga,

algal: phycobiont
* fungal: mycobiont

420

but several benefits have been proposed, such as protection from high light intensity, protection from desiccation, and the transfer of inorganic nutrients. The relationship may be one of **controlled parasitism,** in which the host alga has developed some resistance to the fungal parasite (Ahmadjian, 1982a). Whatever the nature of the relationship, it is a delicate one; any environmental factor that favors the growth of one of the symbionts may destroy this precarious balance, causing the demise of one partner and ultimately of the lichen association.

Lichens are able to live in many unfavorable terrestrial habitats throughout the world, including the Arctic, the Antarctic, and deserts. They are often considered pioneer species in some habitats, because they are frequently the first organism to invade newly exposed smooth rock or soil. Following the colonization of a substrate, they may promote soil formation by adding organic material and dissolving minerals from the rock.

Lichens grow readily on rocks, soil, trees, or man-made structures. They occur in forests, on the tundra, in deserts, on rocky ocean shores, and even within rocks or wood. A few species are aquatic, but most are terrestrial. Lichens grow luxuriously in unpolluted habitats, but in areas of polluted air their growth, numbers, and diversity are greatly reduced. Certain lichen species have been used as indicators of clean air; their presence or absence is mapped to indicate areas of air pollution.

Lichens are classified according to the mycobiont. The present system of classification used by Poelt (1973) and Henssen and Jahns (1974), based on the type of fungal fruiting structure, indicates the current theories of evolutionary relationships.

Morphologically, lichens may be divided into three groups: (1) crustose lichens, which adhere very closely to the substrate; (2) **foliose** lichens, which have leaflike margins detached from the substrate; and (3) **fruticose** lichens, which grow perpendicular to the substrate, forming upright or pendulous thalli that are sometimes branched

FIG. 20-1 Three different lichen growth forms.
A, Crustose lichen is intimately associated with the substrate. This lichen is *Lecanora atra*, which grows on rocks and walls in Europe. It may also cause deterioration of stone works of art (Richardson, 1974). Other *Lecanora* species occur in North America. **B,** Foliose lichen is leaflike and its edges are separate from the substrate. This foliose lichen is *Parmelia* sp. The many species of *Parmelia* are widely distributed in both Europe and North America. **C,** Fruticose lichen has erect portions and is attached to the substrate at the base. This lichen, *Cladonia cristatella*, is popularly known as "British soldiers" because the apothecia are bright red; it grows in eastern North America on soil or dead wood. The erect portions, called **podetia,** are 1 to 2 cm high. The small, prostrate, leaflike portions are called **squamules.**

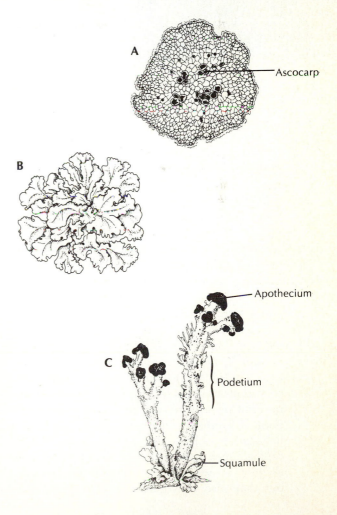

Apothecium

FIG. 20-2 Pendent, fruticose lichen, *Usnea trichodea*.

This fruticose lichen, which grows on trees in southeastern North America, may measure up to 30 cm in length. Apothecia may be seen with a hand lens. (Courtesy Martin B. Berg, Department of Biology, Lehigh University, Bethlehem, Pa.)

(Fig. 20-1). The distinctions among the three types are artificial and often unclear because the gradation is continuous from one group to the next.

The simplest crustose thallus is a powdery crust that grows on a rock and is composed of loosely associated algae and fungi. The most complex fruticose thallus may be either upright or **pendent**, hanging down from trees, and they may branch. Such a fruticose form, *Usnea* (Fig. 20-2), may be found on forest trees in temperate and subtropical areas.

Many lichens are extremely resistant to such adverse conditions as freezing, subzero temperatures, desiccation, and high light intensities. The ability to survive under such harsh conditions has enabled lichens to colonize many nutrient-poor corners of the earth where competition from other green plants is limited. The lichens have physiological mechanisms for "escaping" such extreme environments; not only do they lose water and become dormant rapidly, but they also rehydrate and very quickly be-

come physiologically active. While they are dormant and physiologically inactive, they are very insensitive to extremes of temperature and moisture.

MORPHOLOGY

Both the algal and fungal cells of the lichen contain the usual complement of cellular organelles. The structure of these cells has been discussed previously. The fungal hyphae form a close network around the algal cells. Crustose and **unstratified** lichens with poorly defined thalli may have intracellular haustoria that penetrate the algal cell membrane, whereas the foliose and fruticose forms with well-defined or **stratified** thalli may have haustoria that penetrate only the algal cell wall. Other lichens have no fungal haustoria penetrating the algal cells, but the cell walls of the phycobionts in these lichens are extremely thin and permeable, thus facilitating the movement of organic materials from alga to fungus. The presence or absence of

FIG. 20-3 Cross sections of unstratified and stratified lichen thalli.

The unstratified thallus (at left) has algal cells distributed throughout the thallus. A cortex is present in this lichen, and it is only one cell thick. Not all unstratified lichens have a cortex. In the stratified lichen, an upper cortex composed of compressed fungal tissue is present and is usually several cells thick.

The algal layer immediately below the upper cortex is thin but fairly densely packed with algal cells. Below the algal layer is the medulla, which is composed of interwoven hyphae with many air spaces. A lower cortex may or may not be present.

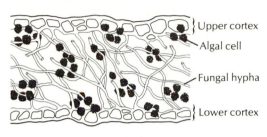

Unstratified thallus

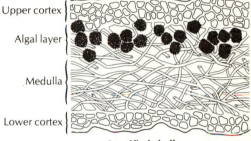

Stratified thallus

FIG. 20-4 Scanning electron photomicrograph of cross section of stratified lichen thallus.

Section through the thallus of *Cladonia cristatella*. Note the upper cortex, algal layer (*Trebouxia erici*), and the medulla with intercellular spaces. This lichen has no lower cortex. (Courtesy Vernon Ahmadjian and Jerome B. Jacobs, Clark University, Worcester, Mass.)

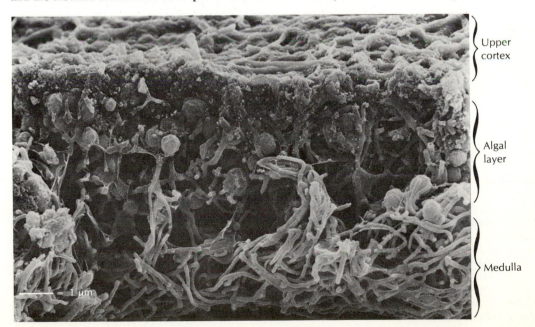

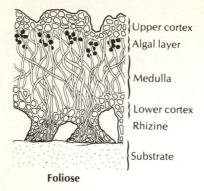

Foliose

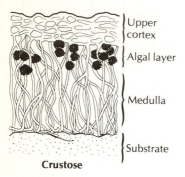

Crustose

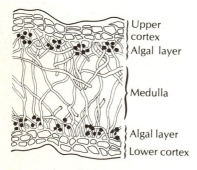

Fruticose

FIG. 20-5 Cross sections of foliose, crustose, and fruticose lichen thalli.
The foliose thallus is stratified and has rhizines that attach the thallus to the substrate. The crustose thallus is stratified, but there is no lower cortex or rhizines in this species. The fruticose thallus is stratified, and the cortex and the algal layer surround a central core of the medulla.

haustoria is variable, as is the extent of haustorial penetration. For example, there may be haustoria in older parts of the thallus, but not in younger portions (Geitler, 1963), or they may be present in a species under xeric conditions, but not under conditions of adequate moisture (Ben Shaul et al, 1969).

In cross section the lichen thallus may be unstratified (i.e., **homoiomerous**) or stratified (i.e., **heteromerous**). The unstratified thallus (Fig. 20-3) consists of the fungal hyphae forming a network throughout which the algal cells are dispersed. The stratified thallus (Fig. 20-3) shows a layering of the algal cells and the fungal cells with compressed hyphae, forming an upper and sometimes a lower cortex. The thin algal layer is located directly below the upper cortex, and the fungus forms the bulk of the thallus in the medulla. Most lichens have stratified thalli.

In the stratified thalli, the cortex (upper and sometimes lower) functions like an epidermis and may be covered by a thin cuticle-like material that retards water loss. The algal layer, located immediately below the upper cortex, is in a position to receive solar energy that has been filtered through the cortex. The fungal hyphae in the medulla are loosely arranged, with large intercellular spaces (Fig. 20-4), thereby increasing the water-retaining capacity of the thallus. The lower cortex may often develop **rhizines,** which attach the thallus to the substrate. Rhizines are composed of compact strands of hyphae.

In cross section the crustose thallus (Fig. 20-5) may have a very thin cortex, a thin algal layer and medulla if stratified, and an extremely thin lower cortex. Occasional hyphae from the lower cortex may act as rhizines for attachment. The crustose thallus is intimately associated with the substrate and is usually very difficult to dislodge intact (Fig. 20-6). The foliose thallus in cross section (Fig. 20-5) is generally stratified with a well-developed upper and lower cortex sometimes having branched or unbranched rhizines. A cross section of a cylindrical fruticose thallus (Fig. 20-5) reveals a cortex surrounding the algal layer, which in turn surrounds the medulla that forms the central core. Research with the scanning electron microscope (Peveling, 1970) revealed small pores between the hyphae of the cortex in a foliose lichen. The function of these **aeration pores** is presumed to be gas exchange. Other lichens have small openings in the cortex that may also function in gas exchange. In *Sticta* these rounded pores have a definite structure and are designated **cyphellae.** Similar structures in other lichens are called **pseudocyphellae** or tubercules (Jahns, 1973).

The phycobiont may be modified so much from its free-living form that it is barely recognizable. Morphological changes are evident even at the ultrastructural level. To identify the alga definitely, one must isolate and culture it individually. The most common phycobiont, found in over 50% of lichens, is the single-celled green alga *Trebouxia* (Tetrasporales). Fig. 20-7 shows a closely related alga, *Pseudotrebouxia*. The filamentous green alga *Trentepohlia* (Ulo-

trichales) is the second most frequently found green alga. A total of 21 genera of green algae have been isolated from different lichens. Among the blue-green phycobionts the nitrogen-fixing *Nostoc* (Fig. 3-1) is the one most commonly isolated, with *Calothrix* the second most frequently isolated. Eleven blue-green bacterial genera have been identified from lichens. Different genera and species of lichens may have the same phycobiont; for example, Santesson (1952) reported that *Trebouxia* had been isolated from 83% of the lichens examined in Scandinavia.

The mycobiont is also changed considerably. The fungus usually determines the morphology of the lichen, and the lichen form may be the same, even if the fungus combines with different phycobionts. Exceptions to this are the gelatinous lichens such as *Collema* that have cyanobacterial phycobionts; in these the phycobiont usually determines the lichen morphology. The fungi that enter into lichen associations have never been found in the free-living form.

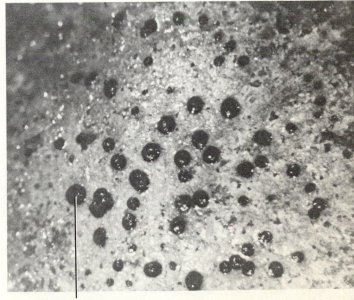

FIG. 20-6 Crustose lichen growing on a rock.
The crustose thallus is intimately associated with the rock substrate, and the separation of the two is extremely difficult. This lichen appears as a greenish-gray patch on the dark gray sandstone. Apothecia often occur in concentric rings.

Apothecium

FIG. 20-7 Transmission electron photomicrograph of *Pseudotrebouxia* **sp.**

Pseudotrebouxia sp. is the phycobiont of the umbilicate lichen *Omphalodium arizonicum*. This lichen grows on rocks at high elevations in southwestern North America. Note the algal nucleus *(AN)*, the fungal protoplast *(FP*-top right) and the pyrenoid *(P)* with lipid-containing, dark-staining pyrenoglobuli *(PG)*.

The fungal cell wall *(FCW)* and algal cell wall *(ACW)* can also be seen. Note also that the fungal hypha does not penetrate the cell wall of this algal cell. Most of the algal cell is filled with chloroplast endoplasmic reticulum *(CER)* (5300×). (Courtesy Vernon Ahmadjian, Clark University, Worcester, Mass.)

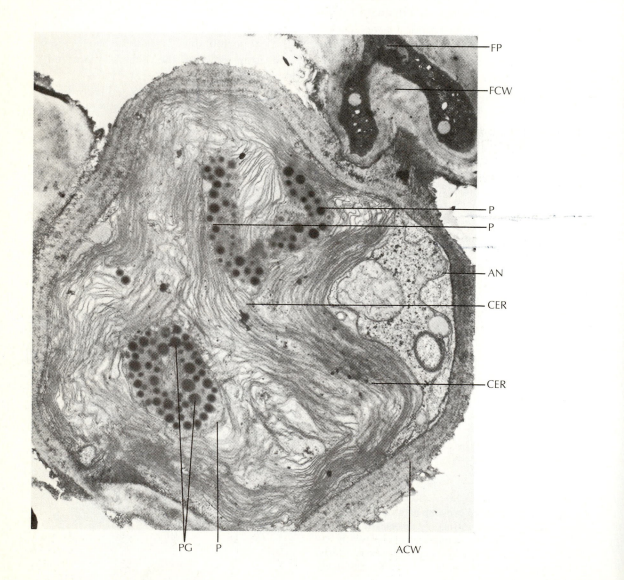

FIG. 20-8 Asexual reproductive structures—soredium and isidium.
Soredia appear on the lichen surface as particles of "dust" that are actually fragments of hyphae and associated algae. Isidia grow perpendicular to the surface of the lichen and are covered by the cortex.

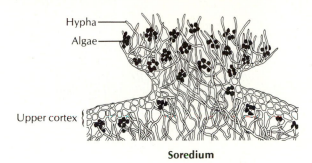
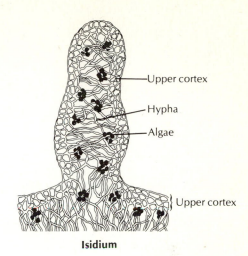

Soredium

Isidium

Some lichens have outgrowths or swellings of their thalli that contain species that are different from the phycobiont in the algal layer. These outgrowths, called **cephalodia**, usually contain the blue-green *Nostoc*. These "foreign" cyanobacteria can fix atmospheric nitrogen and contribute nitrogenous materials to the thallus.

REPRODUCTION

Lichens inhabit many terrestrial habitats all over the earth; this attests to their effective reproduction and dispersal. However, their methods of reproduction, especially sexual, are not completely understood. Many lichens form asexual reproductive structures such as **soredia** (sing. **soredium;** Fig. 20-8), which are filaments of the fungus with associated algae that break through the cortex. Soredia may be seen as dust-like particles on lichen surfaces, and observation under the microscope reveals their algal and fungal composition. Foliose and fruticose lichens may produce protuberances covered by the cortex that are called **isidia** (sing. **isidium;** Fig. 20-8) and that contain algae and fungi. The isidia are easily broken off from the crustose

thallus, but they are not as easily dislodged from foliose or fruticose forms. Isidia are relatively rare in crustose forms but are common in foliose or fruticose species, where they may function to increase the surface area, resulting in a thallus with a greater assimilative capacity (Jahns, 1973).

Both soredia and isidia, collectively called **propagules** or **diaspores,** are effective methods of dispersal because they contain both algae and fungi and may be wind-, water-, or animal-borne to new habitats. The development of isidia and soredia is chronicled in Henssen and Jahns (1974). Fragmentation of the thallus may also serve as a means of asexual reproduction.

Both the alga and the fungus may individually reproduce by mitosis or by the production of asexual spores. The green algae may produce aplanospores or, rarely, planospores. The blue-green bacteria may reproduce by akinetes, heterocysts, or hormogones. The fungus may produce conidia in pycnidia (Fig. 20-9). For the lichen to form in a new habitat, the right fungus and the right alga must come together. The manner in which this union is accomplished is the subject of much conjecture.

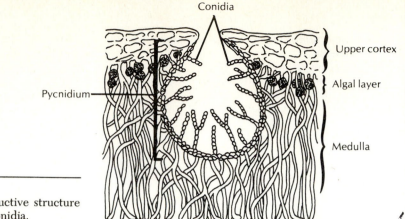

FIG. 20-9 Pycnidium of *Lecanora* sp.
The pycnidium is an asexual reproductive structure
produced by the mycobiont. Note conidia.

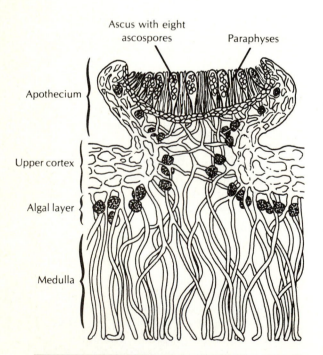

FIG. 20-10 Apothecium of *Lecanora* sp.
The mycobiont may also reproduce sexually. This
drawing shows the apothecium of *Lecanora* sp. with
associated asci and ascospores. Note paraphyses in
the apothecium.

The isolated alga may be induced to produce
gametes in culture, but no documented record
of sexual reproduction by a phycobiont has been
observed. Slocum et al (1979) observed plano-
spores from *Trebouxia* in one lichen, but no
syngamy occurred. The fungus, however, does
reproduce sexually, producing ascospores in Ascomycotina and basidiospores in Basidiomycotina. These sexual spores are produced in ascocarps (Fig. 20-10) or in basidiocarps, which
aid in the identification of those lichens that produce them. The lichen must be several years old
before ascospores are formed, and these may be
produced over several years in the same ascocarp. When the mycobiont is cultured individually in the laboratory, it very seldom produces
ascospores. Letrouit-Galinou (1973) discusses
sexual reproduction in both ascolichens and
basidiolichens.

If sexual reproduction takes place only in the
mycobiont, what does this indicate about genetic diversity in the lichens as a whole? At this
time little is known about the frequency or efficiency of sexual reproduction in lichens. The
crustose lichens do not have as many asexual
reproductive methods as the foliose and fruticose forms; therefore, investigators hypothesize
that sexual methods are more prevalent in these
species. The indirect evidence indicating sexual
reproduction among lichens includes (1) the

growth of fungi that are genetically different (chemically or morphologically) from ascospores from the same lichen and (2) the existence of lichen hybrids, forms intermediate between two species (Hale, 1974). The evidence indicates that sexual or parasexual reproduction does take place, but its importance and contribution to lichen success and dispersal remains to be determined. Genetic diversity is occurring among the lichens, but the mechanisms are still unknown.

The asexually produced soredia and isidia ensure that the lichen alga and fungus will be transported to a new habitat together and therefore will be able to establish a new lichen immediately. When an ascospore, a conidium, or an aplanospore reach a new habitat, how does the alga find the compatible fungus or vice versa? No evidence shows that the hyphae actively grow toward the algae. During its first stages of growth, which may be heterotrophic, the fungus may come in contact with several different algal species. Will it only form a lichen with the "right" alga? Perhaps it incorporates several species of algae initially but only one alga is partially resistant to parasitism by the fungus, resulting in a "balanced parasitism" of the alga by the fungus (Ahmadjian, 1982b).

The green alga *Trentepohlia* and the blue-greens *Nostoc* and *Calothrix* are found free-living in the environment, so they are available for lichenization with the right fungus. *Trebouxia*, however, the phycobiont in the majority of lichens studied, is rarely found free-living in the environment. The source of *Trebouxia* for a new lichen thallus is not known, but the alga evidently has become quite specialized and has developed a dependence on forming a symbiotic association. Examination of tree bark occasionally reveals *Trebouxia* growing in small groups among other algae (Ahmadjian, 1967). Other potential sources for *Trebouxia* are soredia or isidia from different lichen species. A microscopic examination of alga-covered tree bark usually shows many fungal hyphae growing among the algae. From this mixture of potential lichen-forming algae and fungi many combinations may be tried, but only a few result in viable lichens that grow and flourish.

If the recently germinated fungal hyphae do not immediately contact an algal cell and start a new lichen, the fungus may be capable of living saprophytically for some time until a compatible alga arrives (Scott, 1973). Fungal spores may also lie dormant until stimulated to germinate by the proximity of the proper algal partner.

Asexual reproduction is probably the most frequent method of reproduction in lichens. Those species that produce propagules are more numerous in most lichen communities and are more widely distributed than lichens that do not produce propagules (Hale, 1955, 1974). For example, in the widespread genus *Parmelia*, 26 species occur on all continents, and most of these species produce soredia or isidia (Hale, 1965). In lichens whose mycobionts form ascospores, there may be algal cells in close proximity to or within the ascocarp so that as the ascospores are discharged algal cells may be carried along with them, thus ensuring the propagation of the lichen in a new habitat. Additional research should increase our understanding not only of the role of sexual reproduction in lichen biology but also of the mechanisms by which a new lichen thallus becomes established.

PHYSIOLOGY

The physiology of a lichen may be examined in three ways—the physiology of (1) the phycobiont, (2) the mycobiont, or (3) the composite organism. As previously stated, the lichen assumes both morphological and physiological properties that are not observed in either partner. Some information about the physiology of the individual partner has been gained from laboratory cultures, but the data gathered from studying individual cultures cannot usually be extrapolated to lichens as a whole.

The most common phycobiont, *Trebouxia*, has been grown in culture with an organic source of nitrogen in the light and with glucose in the dark. Exogenous vitamins are not required. Growth is generally slow, and vitamins (especially biotin and thiamine) and polysaccharides may be excreted into the growth medium. The alga can be induced to produce aplanospores in culture, but aplanospores have rarely been observed in the lichenized alga. Ahmadjian (1967; 1973) and Hale (1974) discuss the culture of *Trebouxia* and other phycobionts.

When the blue-green phycobiont *Nostoc* is grown in culture, it fixes atmospheric nitrogen, and nitrogenous compounds are excreted into the growth medium. *Nostoc* also excretes glucose into the medium.

Growth of the mycobiont in culture is much slower than other common laboratory fungi such as *Aspergillis*, and the fungus assumes a compact, amorphous shape. The growth of the mycobiont may be stimulated by algal extracts, indicating that the alga produces some form of growth-stimulating compound. There is a question as to whether the mycobiont can grow saprophytically in nature, as it does on artificial media. The fungus culture is started from an asexual (conidium) or sexual spore (ascospore). Vitamins, most often biotin and thiamine, must be supplied in the medium. Most artificially cultured mycobionts do not produce any sexual fruiting structures, but a few have produced conidia. The mycobiont of the rock-inhabiting lichen *Lecidea erratica* did produce conidia when grown in culture. The conidia closely resembled those of the free-living deuteromycete, *Aureobasidium pullulans* (see Fig. 19-14), which

---

FIG. 20-11 A stage in the early synthesis of *Cladonia cristatella*.

This scanning electron photomicrograph shows the early envelopment of the phycobiont *Trebouxia erici* by the *C. cristatella* fungal hyphae. (Courtesy Vernon Ahmadjian and Jerome B. Jacobs, Clark University, Worcester, Mass.)

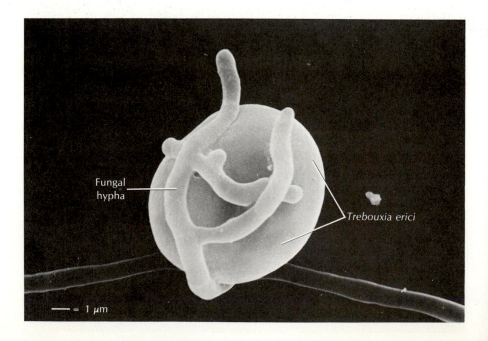

Fungal hypha

Trebouxia erici

— = 1 µm

is widely distributed (Cooke, 1959). Under certain circumstances this fungus may have physiologically adapted to lichenization.

In culture the mycobiont seldom produces ascospores. The mycobiont of *Cladonia cristatella* (Fig. 20-1) produced ascocarps, ascogonia, ascogenous hyphae, and pycnidia, but no ascospores were seen (Ahmadjian, 1966). Lallemant (1977) cultured the mycobiont of *Pertusaria pertusa*, and not only were ascospores formed, but the fungus also produced an organized thallus rather than an amorphous one.

Most mycobionts grown in culture do not produce the unique lichen chemical substances, but one instance of the production of two lichen compounds by a mycobiont has been recorded (Kurokawa, 1971).

Once the lichen components have been grown individually in artificial culture, the next step is to attempt to resynthesize the lichen, and in a few instances resynthesis has been accomplished. The first spore-to-spore synthesis was reported by Ahmadjian and Heikkilä (1970), who used the components of the soil lichen *Endocarpon pusillum*. The researchers used ascospores to first grow the fungus, then combined it with its phycobiont, and subsequently viable ascospores were produced in perithecia. The phycobiont in *E. pusillum* is the filamentous green alga *Stichococcus* (Ulotrichales).

Early stages of lichen synthesis have been achieved in vitro with *Lecidea albocaerulescens* (Ahmadjian et al, 1978). Scanning electron photomicrographs of this synthesis revealed that the

---

FIG. 20-12 A later stage in the synthesis of *Cladonia cristatella*.
This scanning electron photomicrograph shows the phycobiont *Trebouxia erici* almost totally enveloped by the mycobiont hyphae. Note the spaces on the alga through which water and nutrients may diffuse directly into the alga. (Courtesy Vernon Ahmadjian and Jerome B. Jacobs, Clark University, Worcester, Mass.)

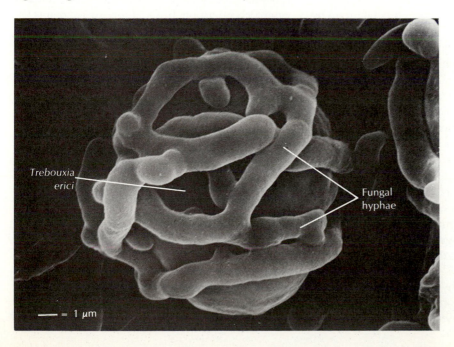

alga *Trebouxia* produced a veillike extracellular substance, presumably a polysaccharide, that bound the algal cells with the fungal hyphae. This algal sheath may be a mechanism by which the symbionts recognize a compatible partner. As the lichen association becomes established, the fungi either produce haustoria that penetrate the algal cell wall or the cell wall becomes thinner and more permeable. The transfer of carbohydrates from alga to fungus may then take place. In *L. albocaerulescens* about 40% of the algal cells were penetrated by haustoria (Ahmadjian et. al, 1978). Other algal cells were partially covered by fungal appressoria (sing. appressorium), which are flattened, thin-walled hyphal strands. The algae cells were not completely enveloped by hyphae, so there was free surface area through which water and minerals could diffuse directly into the alga. The early stages of the synthesis of *Cladonia cristatella* are shown in Figs. 20-11 and 20-12.

Carbohydrates and, in nitrogen-fixing cyanobacteria, nitrogenous compounds move from alga to fungus. When radioactively labeled carbon (^{14}C) in the form of sodium bicarbonate was supplied to *Nostoc*, the phycobiont in *Peltigera polydactyla*, the labeled carbon was taken up by the cyanobacterium and converted to glucose. Labeled mannitol, a sugar alcohol and common fungal storage compound, was found in the fungal medulla within 15 minutes (Smith and Drew, 1965). The glucose had passed through the permeable phycobiont cell wall and had been changed enzymatically to mannitol by the mycobiont. A concentration gradient was established and carbohydrate continued to flow from *Nostoc*. The sugar alcohols mannitol and arabitol are abundant in lichens and may make up to 5% of the dry weight (Pueyo, 1960). Green alga phycobionts evidently may excrete fixed carbon in the form of other sugar alcohols such as ribitol, sorbitol, and erythritol, in contrast to *Nostoc*, which excretes glucose. Whatever the form in which the phycobiont excretes fixed carbon (glucose or alcohol), the mobile carbohydrate is changed enzymatically to a form such as mannitol or arabitol, which cannot be utilized by the alga. It is common for the carbohydrate of a heterotrophic symbiont to be chemically different from the carbohydrate produced by the autotrophic partner. The fungal carbohydrate accumulates, is stored, and cannot be utilized by the alga. This phenomenon is also observed in parasitic fungi, which convert host carbohydrates to mannitol or other storage products (see Chapter 13). Transfer of water from the fungus to the alga has been proposed but not demonstrated experimentally. Nutrients such as nitrate and phosphate are dissolved in water, so there could also be some transfer of these along with the water.

For lichen synthesis to occur, the partners must be in a nutrient-poor medium. If conditions are conducive to the growth of either partner, lichenization will not occur. Synthesis in the laboratory has been successful when the partners are provided with a nutrient-deficient substrate, a slow removal of water, and alternate wetting and drying (Ahmadjian, 1967). Evidently similar conditions must also occur in nature for lichenization to be accomplished. Lichens usually grow in nutrient-poor habitats where competition from other green plants is limited.

Lichens are difficult to grow in culture, because the partnership exists only as long as environmental conditions are unfavorable for the growth of either partner alone. Any environmental change that stimulates or retards the growth of one of the components leads to unbalanced growth and eventual dissolution of the symbiotic relationship. To keep a lichen growing in the laboratory, one must duplicate as closely as possible the environmental conditions under which it grows in the field.

Many changes take place in the lichenized partners. In the phycobiont, cell walls are thinner or completely absent, permitting ready transfer of carbohydrates to the fungus. Changes in membrane permeability of the algae permit both rapid absorption and excretion of various materials. Changes also may occur in the pigmentation of the alga. For example, *Trente-pohlia* produces large amounts of β-carotene when free living and it appears orange or yellow, but when lichenized, the alga has very little pigmentation (Ahmadjian, 1967).

The morphological configuration of the mycobiont is entirely different in culture from its structure in the lichen. In some manner the presence of the alga induces a significant change in the differentiation of the fungus. Generally the fungus determines both the lichen morphological characteristics and fruiting bodies, but the alga probably exerts some influence. In gelatinous lichens such as *Collema*, the blue-green phycobiont *Nostoc* usually determines the morphological characteristics of the thallus. The lichenized fungi require vitamins, especially biotin and thiamine, and these are probably supplied by the alga.

Lichens have an exceptional ability to withstand desiccation. Their structure resembles a **hydrophilic gel** that absorbs water very rapidly and loses water almost as quickly. The absorption and loss of water from the thallus appears to be entirely a physical rather than a biological process. Lichens are very well adapted to alternating periods of drying and wetting; in fact, some lichens may even require wet and dry periods for growth and reproduction. Many do not grow well under conditions of constant high humidity. When a dried thallus is flooded with water, saturation may occur within several minutes. The water is held extracellularly, especially in the intercellular spaces of the spongy medulla (Fig. 20-4). When the lichen thallus is saturated with water, following a rain, for example, the water may increase the weight of the thallus by 100% to 300%. Under arid or drought conditions the water content may drop rapidly to 15% or less of the weight, and the lichens then become dormant and highly resistant to damage.

Lichens, especially crustose ones, are extremely tolerant of arid conditions and are one of the few life forms that can endure the intense insolation and heat of the desert. By becoming

dormant the lichen can survive these harsh conditions, which would severely damage a water-saturated thallus. When the thalli dry out, the rates of most metabolic processes are greatly reduced. When moisture is again available, the respiration of the thallus increases rapidly to several times the normal rate, followed by a slow decline to the normal rate. The photosynthetic rates, however, rise slowly to the normal rate. The algae make up a small portion (5% to 10% of dry weight) of the thallus biomass, and chlorophyll content is low so that photosynthetic rates per unit surface area are considerably lower than those of vascular plant leaves. Respiration rates, however, are similar to those of vascular plants; therefore, net carbon fixation is lower in lichens. Photosynthetic rates, respiration rates, and ultimately growth rates are generally correlated with water content or rainfall (Kärenlampi, 1971). In Europe photosynthetic rates are usually highest in winter when water is plentiful, but respiration rates show little seasonal variability, so net carbon fixation is highest in winter. Translating these factors into biomass added is difficult, but Hale (1974) thinks that lichen growth in temperate zones may be bimodal, with spring and fall maxima. Growth varies greatly with the amount of light and moisture available and with the ambient temperature. Seasonal growth occurs in temperate and arctic areas, but growth is fairly constant in tropical areas.

Growth rates in lichens are often extremely slow, from 0.5 to 3.8 mm/year in crustose forms (Richardson, 1973). The low photosynthetic rates, in comparison with those of vascular plants, are probably responsible for the slow growth. The annual growth rates of foliose lichens (0.5 to 27.0 mm) are faster than crustose; the fruticose (1.5 to 90 mm) have the fastest rates. In conjunction with slow growth, the crustose lichens are also very long lived. One crustose lichen, *Rhizocarpon geographicum*, in Greenland is reported to be 4,500 years old, while other arctic lichens are estimated to be at least 1,000 years old (Beschel, 1961). Many lichens first observed in the early 1800s are still alive today, and their growth rates and patterns have been documented (Richardson, 1974).

Both growth rates and longevity depend on the environment. In the humid tropics where moisture is plentiful, lichens will grow much faster than in the desert, for example, where the only moisture available is from dew or a rare rainstorm. Many temperate lichens grow fastest at temperatures of 15° to 20° C (Henssen and Jahns, 1974) and high humidities. Longevity depends not only on environmental conditions remaining favorable for growth but also on the substrate remaining undisturbed over a long time period. Growth of the crustose and foliose lichens is marginal, taking place at the edges, thus forming a circular thallus that may attain a diameter of 1m (Jahns, 1973). As a crustose lichen grows older, the center tissues may form fruiting bodies or may die, but the margins keep growing. The center may then be colonized by another lichen of the same species, resulting in two thalli of different ages. Fruticose lichens usually grow apically (in length), and each branch grows independently of the others. Hale (1973, 1974) provides additional information on lichen growth and methods of measurement.

Another environmental factor that must be considered is light. Lichens will flourish in a wide range of light intensities but they are generally considered to require high light intensity. Some grow well in lightly shaded forests, whereas others thrive in the intense light and heat of the desert. The phycobiont *Trebouxia* usually occurs in those lichens adapted to lower light intensities, but when it occurs in areas of high light intensity, the cortical layer is thick and heavily pigmented, thereby reducing the intensity of light reaching the algae.

In addition to being adapted to high light intensities, lichens can also grow at a wide range of temperatures. On the one hand, the ability of crustose forms to live and grow in deserts indicates their tolerance (often in a dormant, dehydrated state) of both high temperatures and high light intensities. Desert lichens can grow at temperatures of up to 70° C. On the other hand, the growth of lichens under arctic conditions (−25° C) with high light intensities demonstrates their ability to survive very low temperatures. In a laboratory experiment, respiration continued after an arctic lichen was exposed for several hours to temperatures of −268° C (Llano, 1944).

Lichens usually grow under conditions of inorganic nutrient scarcity. The phycobiont supplies the mycobiont with carbohydrates. The mycobiont, evidently, can absorb dissolved organic nutrients and such inorganic nutrients as nitrates and phosphates from the substrate or rainwater (Smith, 1962) and could pass these on to the phycobiont. However, such a transfer of nutrients from mycobiont to phycobiont has not been demonstrated experimentally. Lichens are extremely efficient in absorbing inorganic nutrients and may accumulate substances such as phosphorous in great excess of their immediate needs (Syers and Iskandar, 1973).

Lichens may accumulate and can tolerate levels of heavy metals such as zinc, cadmium, and lead that are lethal to most vascular plants. Levels of 13,000 ppm zinc and 3,000 ppm lead have been found in lichens growing on rocks or soils containing high quantities of these metals (Ahmadjian, 1967). In areas where both air and soils are polluted with zinc and cadmium by a smelter, *Verrucaria* was found growing with zinc levels of 25,000 ppm and cadmium levels of 334 ppm (Nash, 1971). The lichens secrete acidic compounds, lichen substances that are unique to lichens, and these can dissolve and combine with metals in the substrate, possibly by chelation. The metals are then absorbed by the lichen, possibly by a cation exchange mechanism (Puckett et al, 1973). Because knowledge of the mineral requirements of lichens is sparse, little is known about the types and quantities of minerals they require. Some lichens that grow on limestone apparently need the calcium and magnesium found in the limestone, whereas others that do not grow on limestone do very well without these minerals.

The intimate physiological relationship between the alga and the fungus enables the two organisms together to produce as secondary metabolites lichen substances that neither can produce individually. These extracellular compounds are often used in the classification of lichens. The substances have been characterized by many researchers. They consist mainly of a series of weak phenolic acids, especially the diphenyl **depsides** and **depsidones** (Fig. 20-13) and several pigments. Over 200 lichen substances have been chemically characterized (Henssen and Jahns, 1974). Culberson (1969) and Huneck (1973) discuss additional lichen substances that have been discovered. These substances are deposited either as granules or crystals on the hyphal surface or extracellularly in the medulla or cortex. Pigment colors range from deep red through orange to pale yellow, and these often color the lichen. When lichen substances are used in lichen identification or taxonomy, certain chemical tests are used to determine which substances are present. Several keys to lichen identification include chemical reactions of the species (Alvin, 1977; Hale, 1979).

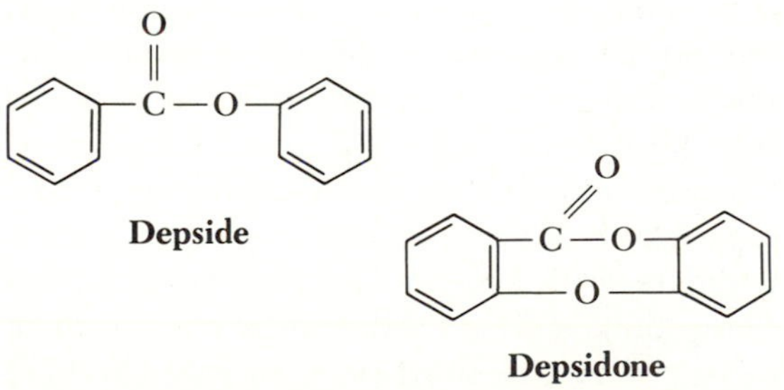

The function of the lichen substances is not known, but several functions have been postulated. Some researchers think these substances may protect the slow-growing thallus from decomposition by other fungi or bacteria or from insect grazing. Many microorganisms and insects may be found in the lichen thalli, so the protection function is questionable. Other researchers propose that the substances protect the lichen from competing bryophytes or algae. Several lichen substances do exhibit antibiotic activity, and they may therefore inhibit the growth of certain microorganisms. Other proposed functions of these compounds are: (1) assistance in the extraction of metallic elements via chelation from the substrate, and (2) increase in algal cell permeability so that carbohydrates are passed to the mycobiont. Mosbach (1973) thinks that the excess carbohydrates produced by the phycobiont may be channeled into the production of lichen substances. Because of the slow growth rate, the fungus may not utilize the carbohydrates as rapidly as the alga produces them. Usually the alga does provide the carbohydrate precursors for the lichen substances.

The lichen substances produced can vary from individual to individual within the same species; species may then be divided into different physiological or chemical strains. Further, the physiological strains may vary geographically or ecologically, and there may be different physiological strains of both the phycobiont and mycobiont, resulting in many different combinations of lichens. In fact, there may be two or three different strains of the phycobiont *Trebouxia* in one lichen thallus (Ahmadjian, 1967).

The fungus produces **lichenin** or **isolichenin,** which is found in the hyphal walls. These linear polymers of glucose are evidently energy storage compounds. Their molecular weights fall between starch and cellulose. Little information exists on the chitin content of lichenized fungi, but chitin is an important component of the cell wall of most fungi.

FIG. 20-13 Chemical structure of the lichen substances depside and depsidone.
These compounds produced by lichens are weakly acidic. The depsides are the largest group of lichen substances and are colorless. The depsidones are the second largest group and are probably derived from the depsides. These lichen substances and their derivatives are usually deposited in the medulla.

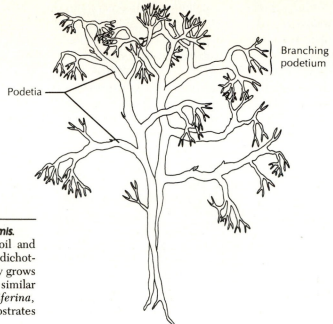

Branching podetium

Podetia

FIG. 20-14 A "reindeer" lichen, *Cladonia rangiformis.*
This fruticose lichen covers large areas of soil and grass in Europe and North America. Note the dichotomously branching podetia. This lichen usually grows on dry, grassy areas with alkaline soils. A similar "reindeer" lichen, *Cladina* (= *Cladonia*) *rangiferina*, grows at higher altitudes on more acidic substrates and has a distinct main stem.

In addition to the ability to concentrate heavy metals, lichens will also concentrate radioactive elements. These elements may be concentrated to high levels that would be toxic to higher plants. As with the heavy metals, lichens have a very high tolerance for radioactive elements, and evidently they have no metabolic mechanisms for excluding or excreting these materials. The bioconcentration of the long-lived artificial radionuclides ^{137}cesium and ^{90}strontium by reindeer lichens in Alaska and northern Scandinavia is well documented (Liden, 1961; Weichhold, 1962). The lichens (Fig. 20-14), an important link in the far-north food chain, are eaten by reindeer or caribou that in turn are eaten by Eskimos or Laplanders. ^{137}Cesium and ^{90}strontium are closely related to the physiologically important elements potassium and calcium, respectively, and are taken up by living organisms in the same manner as potassium and calcium. The radioactive elements, therefore, enter the body and are utilized in the same manner as potassium and calcium. ^{90}Strontium is deposited in bone in a manner similar to the deposition of calcium. Following atmospheric testing of nuclear bombs in the late 1950s and early 1960s and the resulting release of artificial radionuclides, high levels of both ^{137}cesium and ^{90}strontium appeared in some peoples of the far north; these levels caused great concern for their health, which is being carefully monitored to detect any deleterious effects of the excess radioactivity. The lichens of these areas appear to be reservoirs of radioactive elements, because they still showed substantial levels of these radioactive substances up through 1970 (Tuominen and Jaakkola, 1973). The half-life of both ^{137}cesium and ^{90}strontium is approximately 30 years. Lichens growing on rocks or soils that are naturally high in radioactive elements will also concentrate these elements in their thalli.

In addition to concentrating metallic elements and radioactive elements, lichens may also accumulate compounds of nitrogen, phosphorous, and sulfate. By concentrating these necessary plant nutrients, the lichens, upon their death and decomposition, enrich their immediate environment, and these nutrients are then available to the bryophytes and vascular plants that may succeed them on the same substrate (Syers and Iskandar, 1973).

Further discussion of the physiology of lichens may be found in Ahmadjian and Hale (1973), Henssen and Jahns (1974), and Brown et al (1976).

EVOLUTION

The lichens, like many algae and fungi, have delicate thalli and thus have left very few fossils. The first fossils are from the Mesozoic era, and many lichen fossils are found in rocks of the Cenozoic era (Smith, 1921). Lichens appeared relatively late in the evolutionary history of nonvascular plants; evidently both the algae and especially the fungi had to be quite highly evolved before lichenization could take place (Salisbury, 1962). A free-living fungus may have become parasitic upon an alga under conditions of nutrient scarcity.

Ahmadjian (1967) proposes that the lichen association probably started when a fungus parasitized an alga and that the parasitism eventually developed into a mutually beneficial association as the two organisms became progressively dependent on each other. The alga may have developed some resistance to parasitism by the fungus (Ahmadjian, 1982b). Natural selection may have favored for lichenization those algae that could withstand either haustorial penetration or increased permeability of their cell walls and those fungi that grew very slowly. The most advanced lichens have mycobionts that lack haustoria and phycobionts without cell walls. In these species the permeability of the algal cell membrane is increased, permitting the passage of materials, especially carbohydrates, to the fungus; the need for haustoria is thereby eliminated (Gilbert, 1973).

As competition for food and substrate became intense, those algae and fungi survived that formed an association, crustlike at first, in which both partners benefited. This association enabled lichens to survive in nutrient-poor environments where neither partner could survive alone. For this reason lichens have been characterized by Hawksworth and Rose (1970) as "two organisms united in adversity." Because lichens developed physiological mechanisms for "escaping" harsh environmental conditions, they successfully colonized such inhospitable environments as deserts and polar regions. Competition from other plants in these extreme conditions was negligible.

Evidently lichenization evolved independently in several different algal groups (Cyanobacteria, Chlorophyta, and Xanthophyta) and in several different fungal groups (Ascomycotina and Basidiomycotina). Scott (1973) proposes that the evolutionary progression was from free-living algae and fungi through facultative symbiosis to obligate symbiosis.

The Ascomycotina are the most numerous lichenized fungi and the Chlorophyta the most frequently lichenized algae. Over 50% of the known ascomycetes (primarily Discomycetes) are found only as lichens. The number of lichenized algal genera (30) is small compared to the number of lichenized fungal genera (500 or more). The green alga *Trebouxia* is apparently the most highly evolved phycobiont, because it rarely occurs as a free-living form. It appears to be totally dependent on lichenization for its existence (i.e., obligate symbiosis). Other green phycobionts such as *Trentepohlia* and *Chlorella* can still live independently. The blue-green phycobionts *Calothrix* and *Nostoc* can also live independently and are faculative symbionts. Most lichenized fungi do not occur as free-living organisms, so there is a gap in the record of lichen evolution, because no extant fungi are facultative symbionts and can exist either independently or as a mycobiont (Scott, 1973). Lichen-forming fungi may have descended from free-living forms that are now extinct.

The crustose lichens with unstratified thalli are considered to be the most primitive lichens. The stratified fruticose lichens with highly differentiated thalli are considered the most advanced and the foliose forms intermediate. Gilbert (1973) notes the similarity of the structure of a heteromerous lichen to that of an angiosperm leaf, especially in the location of the photosynthetic cells beneath a light-absorbing layer—an instance of convergent evolution. Thick or highly pigmented cortices that protect the phycobiont from high light intensities and water loss are also considered advanced characteristics. Scott (1973) concludes that mechanisms for controlling the light that reaches symbiotic algae are advanced. Such control mechanisms also evolved in three different animal groups—Protozoa, Coelenterata, and Mollusca—in which there are symbiotic algae.

Asexual reproduction by soredia or isidia appears to be the most common method of lichen reproduction. Sexual reproduction by ascospores probably plays a minor role in reproduction. The more advanced lichens such as *Sticta* and *Peltigera* produce mycobiont spores that will germinate only in the vicinity of the proper alga. The most advanced lichens such as *Endocarpon* release cells of the phycobiont *Stichococcus* with the ascospores, thereby ensuring that upon germination the fungi will be in close proximity to a suitable alga (Scott, 1973). Because of the predominance of asexual reproduction in lichens, the rate of genetic exchange is slow compared to other fungi and vascular plants, and the rate of evolution is therefore relatively slow. Lichen symbiosis generally seems to be a stable, slow-to-change association.

ECOLOGY

Very few corners of the earth have not been invaded by lichens. Lichens are found on desert soil or rock, on lava from volcanoes, at the top of tall fir trees, on tropical plant leaves, on arctic ice, on gravestones, on asbestos shingles, on tree bark, on mountain tops, and on rocky seashores.

Lichen communities may be divided into three major types—the **corticolous** communities that grow on tree bark, the **saxicolous** communities that grow on rock, and the **terricolous** communities that grow on soil. A minor community would be the animal-inhabiting community, which encompasses the few lichens that grow on the Galapagos turtles (Hendrickson and Weber, 1964) and on flightless weevils of New Guinea (Gressitt et al, 1965). The lichens within each community show definite preferences for substrate type (e.g., tree bark or calcareous rock), directional orientation (e.g., north or south) and vertical location (e.g., high or low on a tree). Both the acidity and moisture-retaining capacity of the substrate influence the type of lichen community that develops. Those lichens, usually crustose, that are in closest contact with the substrate have the most specific substrate requirements, while some foliose and fruticose species, in less intimate contact with the substrate, may be substrate-indifferent species (Brodo, 1973). Hale (1974) provides additional discussion of lichen communities.

The only environment that has not been successfully invaded by lichens is water. There are, however, a few freshwater and marine forms that are always submerged. A species of *Verrucaria*, for example, lives permanently under water or ice in Antarctica (Kappen, 1973). Other species of *Verrucaria* inhabit marine and freshwater habitats, whereas others are terrestrial. Often lichens and bacteria are the only exposed organisms living in such harsh environments as the Sahara Desert or the Antarctic ice. Lichens that are unclassified at this time have recently been found growing inside sandstone rocks in the Antarctic. These lichens obtain moisture and nitrates from snow that seeps in through the porous rock; adequate light penetrates the first centimeter or so of the rock, and other essential minerals are absorbed from the rock substrate (Young, 1981). Although many lichens are endolithic, especially in limestone, their occur-

FIG. 20-15 An umbilicate lichen, *Lasallia papulosa*, covering a boulder.
This light-brown, saxicolous, foliose lichen may cover boulders in open spaces in northeastern and southwestern North America. Note the blisterlike pustules on the surface and the black apothecia (small dots).

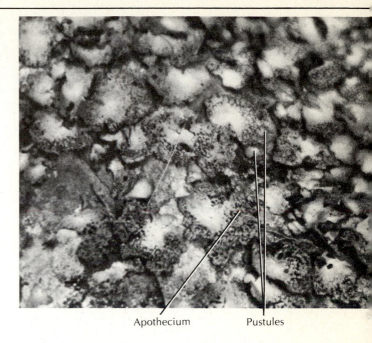

Apothecium   Pustules

rence in the extreme environment of Antartica is unusual.

Lichens may be pioneer species in some habitats, colonizing bare, smooth rocks or newly exposed soils. The **umbilicate** species such as *Lasallia* (Fig. 20-15) or *Umbilicaria* are attached to the substrate at their center by a single strand of rhizines and often cover exposed rock surfaces. The lichen substances that are secreted are often acidic and can therefore slowly dissolve rock minerals, making them available to the lichens and causing a very gradual weathering of the rock surface. As fragments of the rock are broken up and as lichens decompose, soil begins to be formed in the rock crevices, and then bryophytes or vascular plants, which require some soil, may become established. The lichens thus condition the substrate for the next step in plant succession. A thorough discussion of the role of lichens in rock weathering and soil formation may be found in Syers and Iskandar (1973).

In many harsh environments, such as arctic climates and deserts, the rock weathers very slowly and a lichen community becomes established. Lichen growth and formation proceeds so slowly that successional species do not have enough soil in which to become established. Under these circumstances lichen communities will persist for as many years as the substrate and environment remain relatively unchanged. Lichen communities in the Arctic and on high mountain tops may be several centuries old.

Lichens form extensive pastures that are an important component of the ecosystem in the far northern treeless areas of Europe and North America. During the long winter months these **reindeer "mosses,"** which are actually lichens, are the major winter food of the reindeer and caribou. The reindeer mosses are primarily *Cladonia* species (Figs. 20-14 and 20-16), also called *Cladina* by some investigators, with the terricolous *C. rangiferina* being one of the most prominent. These lichens are low in nutritive value, but they evidently contain enough nourishment to sustain the grazing animals and contribute to the primary productivity of the Arctic tundra. Richardson (1974) lists the nutritive value of several of these lichen species compared to dry grass. *Cladonia* spp. have the highest percentage (94%) of carbohydrate of the northern terricolous lichens. However, the lichens are considerably lower in protein (2.7% to 7.5%) than dried grass (15.8%). The caribou and reindeer preferentially graze those lichens having the highest complex carbohydrate content (Richardson, 1974).

Podetia

FIG. 20-16 A "reindeer" lichen, *Cladina* (= *Cladonia*) *subtenuis.*
This fruticose lichen is widely distributed in unpolluted areas of northeastern North America. It often grows in open areas of pine forests. Podetia may range up to 8 cm tall.

FIG. 20-17 Two fruticose lichens, *Cladonia chlorophaea* **and** *Cladonia coniocraea.*
These two lichens grow readily in areas free of air pollution in North America and Europe. *C. chlorophaea*, a cup lichen, has cuplike podetia that may be covered with soredia. Red ascocarps develop along the edges of the cup. This lichen is widely distributed on damp rocks and soil. *C. coniocraea* has long (to 2 cm), pointed podetia that may be covered with green soredia. It grows on humus or dead wood in forests. These species were found growing on moss-covered soil in the Pocono Mountains of eastern Pennsylvania.

C. chlorophaea podetia

C. coniocraea podetia

The number of lichen species in the far north is quite large: 375 species have been recorded in Alaska (Krog, 1968) and 350 species in Antarctica (Rudolph, 1967). These polar species probably contribute substantially to the primary productivity of these regions; for example, there are approximately 3.4 metric tons (dry weight) of *Alectoria* per hectare in Alaskan timberline forests (Edwards et al, 1960). In temperate and tropical areas the contribution of lichens to primary productivity is probably quite small.

Land snails and slugs are major lichen predators, while water bears (Tardigrades), springtails (Insecta—Collembola), mites (Arachnida), and various microorganisms also use lichens as a food source (Richardson, 1974). Bark lice (Insecta—Psocid) may feed upon corticolous lichens. One species of bark louse can be reared in the laboratory on a diet of the crustose lichen *Lecanora conizaeoides* (Hale, 1974). Many invertebrates evidently secrete **lichenase,** an enzyme that breaks down the polysaccharide lichenin into its component glucose molecules, thereby enabling the invertebrates to gain energy from the lichenin.

Those lichens that have nitrogen-fixing blue-green phycobionts may play an important role in providing fixed nitrogen to the ecosystem. In addition, nitrogen-fixing phycobionts enable the lichen to thrive in nitrogen-poor environments. The corticolous lichen *Lobaria*, living on Douglas fir trees in the United States Northwest, provides large amounts of fixed nitrogen to the forest ecosystem (Rhodes, 1978). Other lichens of the arctic and subarctic heaths also contribute substantial nitrogen to these ecosystems. Several crustose, terricolous species found in the desert and saxicolous species in the Antarctic also contribute fixed nitrogen to these harsh environments by the action of their nitrogen-fixing phycobionts.

Lichens have been used to map air pollution in many areas of the world (Houten, 1969). Certain pollutant-sensitive lichens may be used as indicators of unpolluted areas (Fig. 20-17), whereas pollution-tolerant species may be used as indicators of polluted air. A considerable amount of work with lichens as indicators of air pollution has been done in Great Britain. In polluted areas, sensitive species disappear first from trees, then from sandstone rocks, and finally from asbestos roofs and calcareous rocks. The reason that sensitive species persist longest on asbestos or calcareous substrates is the tendency of the alkaline substrates to neutralize the effect of acidic pollutants on the lichen. Tree bark has no such neutralizing capability, so corticolous species disappear first.

By mapping the incidence of certain indicator lichen species around a suspected air pollution source such as a smelter or a city, scientists can note zones of lichen occurrence or absence. The sensitive species will be found on basic substrates such as asbestos and limestone closer to the polluting source than they will on acidic substrates such as many tree barks. Very close to pollution sources lichens may be totally absent, thus resulting in a **lichen desert.**

The lichen flora of the London and Stockholm areas have been the most thoroughly studied. Lichen flora have also been mapped in Zurich, Caracas, Belfast, Montreal, New York, and Christchurch, New Zealand. Biological scales for estimating air pollution have been devised by Gilbert (1970) and Hawksworth and Rose (1970) using the available information about the presence or absence of certain indicator species.

One of the main reasons for the disappearance of lichens in urban regions is the presence of the air pollutant sulfur dioxide. The gas acidifies substrates by producing sulfuric acid when it comes in contact with moisture. The gas also contributes to the acidity of rain, which can also lower the pH of suitable lichen substrates. As the pH of tree bark becomes very low (pH 3.0 to 4.0) because of sulfur dioxide or acid rain in urban areas, the bark becomes unsuitable for lichens. First the lichen biomass decreases, then fruiting is depressed, and finally at the lower limit of the lichen's tolerance it becomes small and sterile. Laboratory experiments with sulfur dioxide show that chlorophyll *a* is degraded to phaeophytin *a* in the phycobiont *Trebouxia* when it is exposed to 5 ppm of sulfur dioxide (Rao and LeBlanc, 1966). Nash (1971) reported decreased chlorophyll content in lichens exposed to 2 ppm sulfur dioxide.

Similar zones of lichen disappearance have been mapped around pollution sources isolated in rural areas. Areas around rural pollution sources such as zinc smelters (Nash, 1972), aluminum smelters (LeBlanc et al, 1971), iron smelters (Rao and LeBlanc, 1967), nickel smelters (Richardson and Puckett, 1971), and an oil-shale factory (Skye, 1958) all had depauperate lichen communities or lichen deserts in the immediate vicinity of the factories.

Several pollution-tolerant crustose forms have now colonized the lichen deserts in urban centers of Great Britain and other areas. On acid substrates such as rock and tree bark, *Lecanora conizaeoides* now is widespread in cities, and the lichen increases in abundance as urban areas are approached. On alkaline substrates such as asbestos, limestone, or concrete, *Lecanora dispersa* and *Lecania erysibe* thrive in cities. These lichens are now considered indicator species of polluted air. In extremely polluted areas not even the pollution-tolerant crustose lichens can survive. The crustose lichens are tolerant of pollution because they are very slow growing and therefore have a very slow metabolic rate and take up pollutants slowly.

Lecanora conizaeoides did not occur in studies of polluted areas of Montreal, New York City, and Pennsylvania, but another lichen, *Bacidia chlorococca*, evidently replaces it as a highly pollution-resistant lichen. Lichens vary in their sensitivity to various pollutants, so the form of pollution (e.g., sulfur dioxide, heavy metals, and particulate matter) will determine which lichen species thrive in the most heavily polluted areas. For example, in a study in Scotland in the vicinity of an aluminum smelter that produces airborne fluorides, the saxicolous *Stereocaulon pileatum* was the most pollution-tolerant species (Gilbert, 1971).

The size and shape of an area influenced by air pollution can be delineated by mapping indicator lichen species. Several such maps have been published, including those by Hawksworth and Rose (1970) for Great Britain and O'Hare (1973) for western Scotland. Another method that can be used to evaluate lichen communities is a comparison of the communities in an allegedly polluted area and in a clean-air area. Nash (1971) and Bradt et al (1978) used such comparisons to delineate air pollution near a zinc smelter and near an urban area, respectively.

Some of the most pollution-tolerant lichens have been mentioned (e.g., *Lecanora*, *Bacidia*, and *Lecania*). The most pollution-sensitive lichens are usually the fruticose ones such as *Usnea* (Fig. 20-2), a pendent lichen that is known as "old man's beard," *Ramalina*, which may be either pendent or erect, and *Evernia*, a corticolous species.

There was considerable controversy over whether the lichen deserts in urban centers were caused by toxic air pollutants, the pollution hypothesis, or by a lack of humidity, the drought hypothesis. The controversy has been discussed at length in scientific journals. Recent evidence appears to favor the pollution hypothesis. Le Blanc and Rao (1973) discuss these hypotheses in detail.

As air pollution has abated in many areas, sensitive lichens are beginning to reappear in the vicinity of pollution sources. Skye and Hallberg (1969) document such recovery near the oil-shale factory where Skye surveyed lichens in 1958. Showman (1982) documents lichen recovery near a coal-fired power plant in the United States and found that some lichens were approaching the normal distribution seen in unpolluted areas after 8 years of decreasing air pollution. Recovery of lichen communities can and will take place, but it will take many years because of the slow growth of lichens. Such recovery studies help to document the reduction in air pollutants.

ECONOMIC IMPORTANCE

Lichens have been used as folk medicine cures for hundreds of years, at least from the Middle Ages. Extracts of lichens were used to treat such diseases as rabies and lung ailments. In modern times some lichens, especially *Usnea* (Fig. 20-2), are still used as folk remedies by East Indians, American Seminole Indians, the Chinese, and the Swedish. The use of lichens in disease treatment was quite logical, because many lichens produce lichen substances that exhibit antibiotic activity. A survey by Burkholder et al (1945) of 100 species of lichens in eastern North America revealed that extracts from 52 different lichen species inhibited the growth of bacteria. The lichen substances that display antibiotic activity usually inhibit gram-positive bacteria such as *Escherichia coli* and *Salmonella*.

The most promising antibiotic lichen substance is the yellow pigment usnic acid (Fig. 20-18), which is found in many lichens. Usnic acid is not only effective against the tuberculosis bacterium, *Mycobacterium tuberculosis*, but also against a broad spectrum of other gram-positive bacteria (Vartia, 1973). Chemically, usnic acid does not fall into any convenient category of lichen substances. Usnic acid is not widely used because it has a low solubility in water, so it is difficult to dissolve into an aqueous solution. It is most often used in skin ointments in Europe for application to superficial wounds and burns. The source of the commercially used usnic acid in Europe is the reindeer lichens, *Cladonia* spp., found in northern Europe and Scandinavia, especially. Other sources of usnic acid are the fruticose lichens *Alectoria*, *Cetraria*, and *Ramalina*. Usnic acid also has antifungal properties and may also be used to treat fungus diseases of the skin.

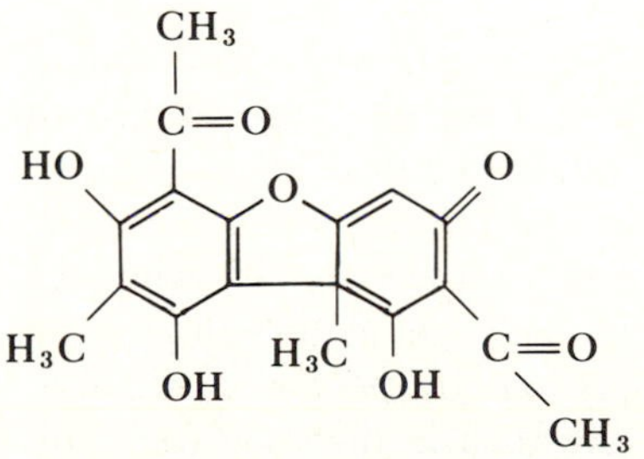

To extract usnic acid from lichens, one must collect large numbers. This is a time-consuming and labor-intensive chore. It is extremely difficult to culture lichens in the laboratory in axenic culture, so they must be collected in the field.

Many other lichen substances exhibit antibiotic properties. Lichens may live 50 to 100 years in nature, and they are rarely attacked by either bacteria or fungi, attesting to their antibiotic activity. The preservation of herbarium specimens is seldom a problem because of the prevalence of antibiotic substances. Lichens, however, are eaten by insects, and grazing insects may cause problems with herbarium specimens. Several lichen substances are also effective against pathogenic bacteria and viruses that cause plant diseases.

Several lichen substances from *Cladonia*, *Parmelia*, *Umbilicaria*, *Usnea*, and others have been investigated for antitumor activity. The effectiveness against tumors appears to be related to the polysaccharide components of these lichens. Vartia (1973) discusses in detail both the antibiotic and antitumor properties of lichens.

Lichens have had limited use as a food source by peoples of the far north and Japan. Lichens often contain acidic substances, making the lichens quite bitter. The lichens may be boiled in

FIG. 20-18 Chemical structure of usnic acid.
Usnic acid is a lichen substance with an unusual chemical structure. It is found in many fruticose lichens, especially the yellow *Cladonias*, and has both antibacterial and antifungal properties. This substance may make up to 4% of the dry weight of the lichen thallus and is deposited only in the cortex.

soda to counteract this bitterness. Their food value is quite low, consisting mainly of the carbohydrates lichenin and isolichenin, with some protein, fat, minerals, and vitamins. After treatment to remove the bitterness, lichens are occasionally used in Scandinavia in soups or other foods. In Japan, rock tripe, *Umbilicaria* sp., is considered a delicacy and is eaten raw in salads or fried in deep fat. Some arctic explorers have eaten umbilicate lichens such as *Umbilicaria* to reduce the incidence of scurvy in their expedition parties. Usually lichens are eaten only when food is extremely scarce, but some populations do consume lichens as part of their daily diet (Richardson, 1974).

Caribou and reindeer of the far north will graze on a dozen or more different species of reindeer lichens, and these may form a major portion of their diet. Many of these reindeer lichens are various species of *Cetraria*, *Cladina*, and *Cladonia* (Figs. 20-14 and 20-16), which form extensive mats on soil in the Canadian subarctic and the Scandinavian birch forests. The reindeer lichens are more conspicuous, however, in Antarctica than in the Arctic, because bryophytes and vascular plants share the far north landscape with the lichens. The resident caribou and reindeer use this abundant food source in fall and winter, even though it is relatively low in food value. If snow cover is exceptionally deep, the animals may even feed on the corticolous lichens or the pendent fruticose lichens *Alectoria*, *Bryoria*, and related forms. In severe winters reindeer lichens comprise over 90% of the diet of the reindeer and caribou, but in a milder winter, lichens make up less than 65% of their diet. The Laplanders practice controlled grazing by their reindeer herds to preserve the lichen pastures. Other mammals of the northern tundra such as lemmings may also feed on lichens.

In the Libyan desert a crustose lichen, *Lecanora esculenta*, which is abundant on rocks and soil, provides food for grazing sheep. This desert species is also eaten by humans in Iran.

The pigments from lichen substances provide dyes for coloring assorted fabrics (Bolton, 1960). The dyes used in the Harris tweed wools are derived from lichens native to the British Isles. Lichens provide purple, red, mustard yellow, and brown dyes and have been used to color fabrics since the times of the ancient Greeks. In present-day Scandinavia, locally abundant lichens such as the fruticose *Evernia* and the foliose *Parmelia* (Fig. 20-19) are the basis of a small dye industry. The amphoteric (e.g., able to react as an acid or base) dye litmus is derived from certain lichen depsides. Litmus is widely used as an acid-base indicator for chemical solutions.

Several lichens such as *Evernia*, *Parmelia* (Fig. 20-19), and *Ramalina* contain pleasantly scented oils that are extracted and used to scent perfumes and soaps in certain sections of Europe. Additional information concerning the use of lichens for perfume oils may be found in Gruenther (1952).

Certain lichens may be used as indicators of various geological formations; for example, some species of the bright yellow *Cetraria* are usually associated with limestone and marble formations and serve to indicate the location of such rocks. Other saxicolous lichens may be used as indicators of certain rare mineral deposits.

A technique called lichenometry is used to estimate the age of rock surfaces (Brown et al, 1976). The annual growth rate of lichen species is estimated from existing structures of known age such as buildings or gravestones. The assumption is made that the lichen colonized the structure soon after its erection. Knowing the annual growth rate, the age of a lichen growing on a glacier or rock may be estimated, together with the age of the rock in question (Beschel, 1961). For example, the age of the lichen-covered stone structures called megaliths on Easter Island in the South Pacific was estimated at 430 years by lichenometry (Richardson, 1974).

Lichens may occasionally cause problems by their growth on young trees, older fruit trees, tropical plant leaves, and stained-glass church windows. An overgrowth of corticolous lichens

FIG. 20-19 A foliose lichen, *Parmelia sulcata*.
This corticolous species is widely distributed in unpolluted areas of North America and Europe. Soredia usually develop along ridges. This specimen was growing on a white oak, *Quercus alba*, in the eastern Pennsylvania Pocono Mountains. In these mountains, *P. sulcata* species may be 25 to 30 cm in diameter.

on very young trees may interfere with their growth, because the lichen rhizines often penetrate the cambium, restrict the supply of nutrients, and block lenticels (Porter, 1917). A heavy lichen cover, however, does not appear to interfere with the growth of older trees. In Europe and the southern United States many orchard owners destroy the corticolous lichens on fruit trees with fungicides (Hale, 1974). These orchard owners fear that the lichens may interfere with the health of the trees and that a heavy lichen cover may provide a habitat for insect pests (Fry, 1926). In the tropics, *Strigula* produces thick growths on angiosperm leaves. Some of the trees upon which these lichens grow are economically important, so there is concern that such a heavy lichen cover may decrease photosynthesis and therefore growth. Crustose lichens may also grow on old stained-glass church windows. Some beautiful and highly prized windows dating back to the thirteenth century have been etched and damaged by lichen growth, probably from rhizine growth, but some chemical action by lichen substances may also be involved (Mellor, 1923).

Exposure to soredia-containing dust from *Evernia* and *Usnea* has caused skin rashes on lumberjacks in Canada. These workers are exposed to lichen propagules as they fell lichen-covered trees in the forests. Spouses of lumberjacks have also developed rashes when they were exposed to the soredia-laden dust on the worker's clothing.

Probably the most significant use of lichens is as indicators of polluted or clean air. Comparisons of lichen-covered trees in mountain or rural areas and urban trees devoid of lichens yield considerable information about both environments. The mountain or rural areas have clean air, free of the phytotoxic gases that plague many urban areas. The lichens cannot excrete toxic substances that they may absorb from the air or substrate, so they concentrate the pollutants. When pollution-sensitive lichens accumulate enough pollutants they die, and the tree bark in polluted areas becomes bare. A glance through *The Observer's Book of Lichens* for Great Britain and Europe (Alvin, 1977) will show how many lichen species are severely restricted in their distribution to a few areas where the air is very clean. The sensitivity of many lichens to air pollution emphasizes how much information on environmental quality can be gained from a close observation of a few key ecosystem components.

SUMMARY

Lichens are autotrophic organisms composed of an alga or a cyanobacterium, the phycobiont, and a fungus, the mycobiont. The form of the relationship has been described in several ways: symbiosis, mutualism, or balanced parasitism. A lichen can live in extreme environments and produce extracellular lichen substances, which neither partner can do alone. Most lichen-forming fungi are members of Ascomycotina, whereas most phycobionts are members of Chlorophyta.

The classification of lichens is based on the morphology of the sexual fruiting bodies of the fungus. Chemical tests for lichen substances are also used in classification. The mycobiont may reproduce sexually, but there is no evidence of sexual processes in the phycobiont. Most reproduction in lichens is asexual, by propagules such as soredia or isidia, which contain both algae and fungi.

The lichen thallus may take a crustose, foliose, or fruticose form. The three forms are not distinct and many intermediary forms exist. In cross section the thallus may be stratified or unstratified, and a cortex is usually present on the upper surface and sometimes on the lower surface. Rhizines may be present to anchor the thallus to the substrate. Large intercellular spaces are often present in the medulla and are filled with water when the thallus is saturated.

The mycobiont may form haustoria that penetrate the phycobiont cell wall and occasionally the cell membrane. Haustorial production may vary with the age of the lichen or with environmental conditions. The phycobiont cell walls are thin and very permeable. Fixed carbon moves from the alga to the fungus and is enzymatically converted to a fungal metabolite, mannitol. When the phycobiont is a nitrogen-fixing cyanobacterial species, fixed nitrogen often moves from phycobiont to mycobiont.

Both lichen partners are morphologically and physiologically changed from their free-living form or their form in culture, even at the ultrastructural level. The common phycobiont *Trebouxia* may be so modified in the lichen thallus that a definite identification can be made only by isolation and culture. Both partners have been individually cultured and some nutritional requirements determined. Attempts to resynthesize lichens from the individually cultured components have been occasionally successful. *Endocarpon* and *Lecidea* species have been successfully resynthesized in the laboratory with the use of a nutrient-poor medium. For lichenization to take place in the laboratory, and, by extrapolation, under natural conditions, environmental conditions must not be conducive to the growth of either partner.

Lichens can dehydrate and rehydrate very rapidly. This ability is physiologically advantageous, because it enables the lichen to go dormant rapidly under xeric conditions. While dormant the lichen is highly resistant to damage from intense insolation and extremely high or low temperatures. When adequate moisture becomes available, the lichen quickly absorbs water and begins respiration and photosynthesis.

Crustose lichens grow very slowly and may live for hundreds or even thousands of years. Foliose and fruticose lichens have faster growth rates than the crustose. The rate of growth is generally correlated with available moisture, so lichens indigenous to moist tropical areas grow much faster than species in the desert or the frozen arctic tundra.

Lichens may concentrate inorganic nutrients such as phosphorous, heavy metal such as zinc, and radionuclides such as ^{137}cesium from rain, the substrate, or the atmosphere. Many lichen substances are acidic and may dissolve or combine with metals in the substrate. The lichen thallus, in turn, absorbs and accumulates the metals.

Fossils of lichens have been found in Mesozoic rocks. Researchers propose that both partners had to be quite advanced in the free-living forms before lichenization could take place. The evolutionary sequence suggested is from free-living forms through facultative symbiosis to obligate symbiosis or balanced parasitism. The lichens considered the most advanced are those in which the relationship is obligate and in which there is an absence of haustoria, and phycobiont cell walls are either very thin or absent altogether. *Trebouxia*, the phycobiont in over 50% of the lichens, is considered advanced because it is rarely found in the free-living form. None of the lichen-forming fungi have been found in free-living form, so they all are considered advanced and descended from extinct free-living forms.

Unstratified crustose lichens are the most primitive forms, whereas fruticose forms are the most advanced. The rate of genetic exchange in lichens appears to be very slow, possibly because of the infrequency of sexual reproduction. Consequently, the rate of evolution is also slow when compared to both fungi and vascular plants.

Based on their substrate, lichen communities may be divided into three types: corticolous, saxicolous and terricolous. Those lichens, usually crustose, that are most intimately associated with the substrate have the most specific substrate requirements.

In the far north lichens are important primary producers, providing food during winters for grazing reindeer and caribou. In nitrogen-limited habitats those lichens with nitrogen-fixing phycobionts, often *Nostoc*, provide fixed nitrogen to the ecosystem.

Many foliose and fruticose lichens are extremely sensitive to air pollutants, and zones of polluted air have been mapped with the use of the presence or absence of sensitive lichens. Such mapping studies can also document the recovery of lichen communities following air pollution abatement.

Lichens are used as sources of antibiotics, fabric dyes, and perfumes. Because of their substrate specificity, certain lichens are used as indicators of various geological formations or mineral deposits. Lichenometry is used to determine the age of rocks or glaciers.

SELECTED REFERENCES

Alvin, K.L. 1977. The observer's book of lichens. Frederick Warne & Co., Ltd., London. 188 pp.

Ahmadjian, V. 1967. The lichen symbiosis. Blaisdell Publishing Co., Waltham, Mass. 152 pp.

Ahmadjian, V. 1982b. The nature of lichens. Natural History 91(3):30-36.

Ahmadjian, V., and M.E. Hale (eds.). 1973. The lichens. Academic Press, Inc., New York, 697 pp.

Ahmadjian, V., B. Jacobs, and L.A. Russell. 1978. Scanning electron microscope study of early lichen synthesis. Science **200**:1062-1064.

Hale, M.E. 1974. The biology of lichens. Ed. 2. Edward Arnold (Publishers) Ltd., London. 181 pp.

Hale, M.E. 1979. How to know the lichens. Ed. 2. Wm. C. Brown Co., Publishers, Dubuque, Iowa. 246 pp.

Hale, M.E., and W.L. Culberson. 1970. A fourth checklist of the lichens of the continental United States and Canada. Bryologist **73**:499-543.

Hawksworth, D.L., and F. Rose. 1976. Lichens as pollution monitors. Edward Arnold (Publishers) Ltd., London. 60 pp.

Houton, J.G. Ten. 1969. Air Pollution: Proceedings, First European Congress on Influence of Air Pollution on Plants and Animals, Wageningen 1968. Centre for Agricultural Publications and Documents, Wageningen, Netherlands. 415 pp.

Richardson, D.H.S. 1974. The vanishing lichens. Hafner Press, New York. 231 pp.

21

CLASSIFICATION

Division: Bryophyta: moss plants; 5,000 genera, 30,000 species; creeping, mostly terrestrial plants with distinct heteromorphic alternations of generations; dominant photosynthetic gametophytes; subordinate, often nonphotosynthetic sporophytes; lack of specialized water-conducting cells

Class 1: Hepaticae: liverworts; parasitic, nonphotosynthetic sporophytes; gametophyte flat and thallose or upright and leafy; sporophytes usually with elaters among the spores

ORDER 1. Jungermanniales: the leafy liverworts; many epiphytic and tropical; gametophytes with upright stems and rhizoids; leaves in three ranks on the stem; resembling the true mosses (Musci); often occurring in dense mats in damp places; *Lejeunea, Porella*

ORDER 2. Metzgeriales: thalloid gametophytes with stalked sporophytes attached; mostly tropical epiphytes, although some *(Pellia)* may be temperate or northern

ORDER 3. Sphaerocarpales: small thalloid gametophytes with antheridia and archegonia enclosed in bottlelike sacs; mostly subtropical; 3 genera; *Sphaerocarpos*

ORDER 4. Marchantiales: flat, thallose, dichotomously branched gametophytes clearly differentiated into dorsal and ventral surfaces; some with upright gametophores bearing parasitic sporophytes; *Marchantia, Riccia*

Class 2: Anthocerotae: hornworts; small, simple, thallose gametophytes that bear elongate, split, hornlike, photosynthetic sporophytes; cells with a single chloroplast containing a single pyrenoid (unlike other bryophytes)

ORDER: Anthocerotales: sporophytes bearing stomata and producing cuticles; gametophytes often with symbiotic colonies of *Nostoc* (Cyanobacteria); *Anthoceros*

Class 3: Musci or Bryophytae: true mosses; two distinct growth phases, the filamentous "protonema" and the upright gametophytic stem bearing spirally arranged leaves and producing an attached sporophyte; multicellular rhizoids; sporophytes with a complex spore capsule and no elaters

ORDER 1. Sphagnales: peat mosses; robust leafy mosses, mostly confined to acid peat bogs; complex gametophyte and sporophyte structures; a single genus, *Sphagnum*, with 135 species

ORDER 2. Andreacales: the granite mosses; mostly small, leafy, black, brown, or reddish gametophytes growing on siliceous rocks in mountainous cool to cold climates; *Andreaea*

There are 14 more orders listed under the subclass Bryidae by Crum and Anderson (1981), and referred to as the "true mosses." Selected orders are listed on p. 450. Classification is based principally on the structure of the sporophyte capsule and the microscopic features exhibited by the **peristome,** with its characteristic teeth. Two basic growth habits exist, the tuft-forming habit (**acrocarpous**) and the carpet-forming habit (**pleurocarpous**).

BRYOPHYTA

ORDER 3. Pottiales: mostly acrocarpous gametophytes; peristome with 16 elongate and twisted teeth; *Tortula*

ORDER 4. Funariales: acrocarpous gametophytes, peristome lacking; single or double with 16 teeth; *Funaria*

ORDER 5. Bryales: acrocarpous gametophytes; peristome double with 16 teeth; *Leptobryum, Mnium*

ORDER 6. Isobryales: acrocarpous or pleurocarpous gametophytes; peristome single or double with 16 teeth; order includes the strictly aquatic genus, *Fontinalis*, of the water-moss family Fontinalaceae

ORDER 7. Polytrichales: acrocarpous gametophytes that are coarse and robust; found growing on soil; peristome single with 16 to 64 small teeth; *Atrichum, Polytrichum*

Division Bryophyta comprises a group of nonvascular plants whose evolutionary position lies somewhere between the algae and vascular plants. They are commonly known as mosses, even though some of them do not fit the classic "moss" appellation of those low-growing, green, leafy, carpetlike plants that cover moist forest floors, rocks, stream banks, and trees. Included among the Bryophyta are three distinct groups, the **liverworts,** the **hornworts,** and the mosses. The three occupy separate classes, as presented in this chapter.

Bryophyta are sometimes difficult to separate taxonomically from the algae and the higher plants on the basis of their gross morphology. In their simplest form they resemble filamentous algae, whereas in their most complex form they may resemble certain higher plants. On the cellular level the degree of differentiation is far greater than that encountered among the algae. However, the degree of cellular complexity falls far short of that found among the many cell types and differentiated tissues and organs of higher plants. Generally speaking, the bryophytes are the most structurally simple of the land plants, even though they are all vegetatively multicellular, three-dimensional organisms with a heterotrichous habit and a truly parenchymatous nature. They also produce multicellular sex organs that are unlike any of the algae and are similar to many vascular plants. Finally, they are all characterized by diplohaplontic life cycles in which there is a dominant gametophytic generation.

The Bryophyta resemble higher vascular plants in having structures that resemble roots, stems, and leaves; however, these organs lack the characteristic internal structure and vascular tissues of "true" roots, stems, and leaves. The Bryophyta are usually creeping or low-growing, photosynthetic organisms found in damp terrestrial habitats, although a few grow in desert regions and some are aquatic. They are common plants of the woodlands, stream banks, acid soils, and any relatively cool and moist habitat. In some areas they may become the dominant flora of the low-growing plants. In some northern latitudes and higher elevations, the mosses are the dominant terrestrial vegetation, creating a unique habitat. In the moist tropics these plants are often a dominant epiphyte, growing on the bark and leaves of vascular plants. They may also be found growing on coastal sand dunes, where they play an important role in establishment of vegetation on shifting sands. In hot desert areas mosses are less common, although they may be found growing in the shade. Bryophytes are worldwide in distribution, ranging from pole to pole and from sea level to the highest mountains and invading most habitats, especially under moist, acid conditions. In many ways they are similar to both the Protista and the Tracheophyta (Table 21-1).

Biochemically the bryophytes are similar to the green algae (Chlorophyta) and the higher vascular plants. They contain both chlorophyll a and b, have cellulosic cell walls, and store energy in the form of starch. Their accessory photosynthetic pigments include various carotenes and xanthophylls that are chemically similar to those in the green algae and vascular plants.

TABLE 21-1. Comparison of the Protista and the Metaphyta

Protista	Bryophyta	Tracheophyta
Mostly aquatic	Mostly terrestrial	Mostly terrestrial
No embryology*	All develop embryos	All develop embryos
No vascular tissue*	Cells resembling conducting elements only	Xylem and phloem tissues
No water-conducting elements	Simple water-conducting cells	Xylem tissue
No special absorbing tissues (rhizoids and haustoria in some fungi)	Rhizoids as anchoring and absorbing structures	Roots
Mostly filamentous	Parenchymatous	Parenchymatous
Definite alternation of generations in some	Alternation of generations in all	Alternation of generations in all
Sporophyte and gametophyte independent	Sporophyte parasitic on gametophyte	Gametophyte parasitic on sporophyte
Sporophyte frequently large and complex	Sporophyte small and simple	Sporophyte always large and complex
Gametangia single cells or groups of single cells not encased in jacket of sterile cells	Gametangia always protected in jacket of sterile vegetative cells	Gametangia always protected in jacket of sterile vegetative cells

*There are some exceptions.

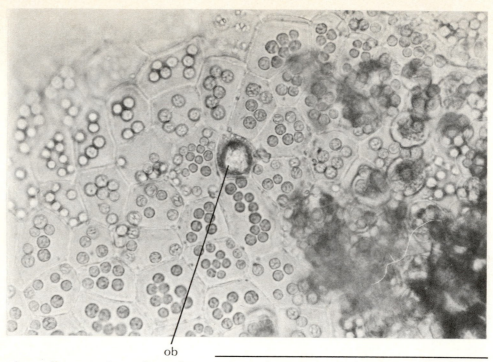

ob

FIG. 21-1 Oil bodies.
This hand-cut, razor blade section of a *Marchantia* spp. thallus shows an oil body *(ob)* among the parenchyma cells, which possess chloroplasts (400×).

Ultrastructural similarities also relate the three groups. The biflagellate sperm cells of the bryophytes have two anterior whiplash flagella that resemble those of some algae (e.g., *Coleochaete, Chara,* and *Nitella*) and some flagellate cells from primitive vascular plants (Pickett-Heaps, 1975). The chloroplasts contain thylakoids that occur in stacks (up to 20 per stack), forming grana. Finally, the cell divisions of bryophytes employ a phragmoplast that is similar in every respect to that found in higher plants, the Charophyta, and the organisms associated with the bryophytan line of green algal evolution (see Fig. 4-8).

CELL STRUCTURE

Bryophyte cell structure is remarkably similar to that found in certain green algae and in the higher plants. The biochemical characteristics are the same for all three groups, with chlorophyll *a* and *b,* cellulose, and starch being present. The ultrastructural characteristics are also similar. They all possess the usual organelles

such as chloroplasts, mitochondria, endoplasmic reticulum, ribosomes, various vacuoles, and other inclusion bodies. The chloroplasts always occur as many small oval green disks floating free in the cytoplasm, where they are moved about by cytoplasmic streaming (cyclosis). The ultrastructure of each individual chloroplast closely resembles that found in higher plants. They are membrane-bound structures containing thylakoids, arranged in stacks of 2 to 20, forming a granum (pl. grana). Many grana occur in each chloroplast. The thylakoids are the site of photosynthesis. The chloroplast structure appears to represent a transition from that of the green algae to that of the higher plants. The nucleus of the bryophytes is membrane bound and contains nuclear sap, chromosomes,

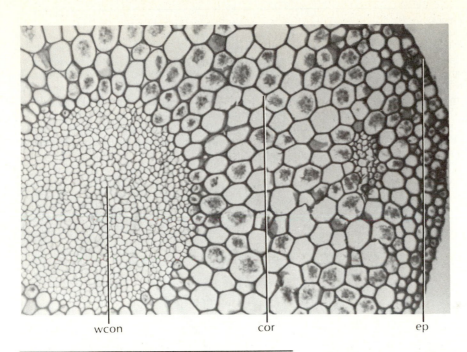

wcon   cor   ep

FIG. 21-2 Moss gametophyte stem.
Low magnification photograph of the cell structure of a moss gametophyte stem. Note tissue-level cellular differentiation into separate structures, including epidermis *(ep)*, cortex *(cor)*, and area of water-conducting cells *(wcon)* (125×).

and nucleoli. Nuclear division (mitosis and meiosis) involves the formation of a phragmoplast (see sections on Growth and Reproduction). Centrioles are found in the biflagellate sperm cells and are located just outside the nuclear membrane. Within the nucleus in some members of the Bryophyta are found special **microchromosomes** (also known as **m-chromosomes** or accessory chromosomes). Microchromosomes are small chromosomes that are less than $1/10$ the length of normal **autosomes.**

The cells of some liverworts contain membrane-bound organelles called **oil bodies** (Fig. 21-1). These distinctive bodies, which are very diverse in form and easily recognizable under the light microscope, are absent in hornworts and mosses. The oil bodies often impart a dis-

tinctive odor to individual liverwort species. The chemistry of the oils and their physiological significance are not known. They are discussed in some detail by Schuster (1966).

Bryophyta are all three-dimensional, multicellular, parenchymatous organisms, and as such they exhibit a considerable degree of cell and organ differentiation in which the plant body of each individual contains an association of a number of cell types. The different cell types are expressed in both the gametophytic and sporophytic generations. However, the degree of cellular complexity of the bryophytes falls far short of that characteristic of the vascular plants. The tissues formed do not include true xylem and phloem, even though some of the conducting cells in a few mosses (e.g., Polytrichales) closely resemble conducting cells of vascular plants in structure and function (Hebant, 1979).

Tissue-level differentiation in mosses includes formation of an outer covering of cells composed of a single layer called the epidermis (Fig. 21-2). Beneath the epidermis lies the other

FIG. 21-3 Life cycle of a moss.

Generalized diplohaplontic life cycle of a moss, showing the alternations of haploid and diploid generations. The gametophyte generation begins when a haploid spore germinates producing a filamentous protonema stage which bears colorless rhizoids. The gametophytic generation in this example is dioecious, with separate male and female upright leafy shoots. The biflagellate sperm released by the antheridium swim through water to the archegonium, where both plasmogamy and karyogamy occur within the venter, producing a zygote. The zygote soon develops into an embryo and finally into a mature sporophyte. Mature sporophytes are parasitic upon the gametophyte. They are composed of a foot, a stalk, and a capsule. The foot anchors the sporophyte to the gametophyte and absorbs nutrients. The capsule is an elaborate structure with an operculum or lid (b), a series of hygroscopic teeth called the peristome (c), a sporangium containing meiospores (d), and an enlarged base (e).

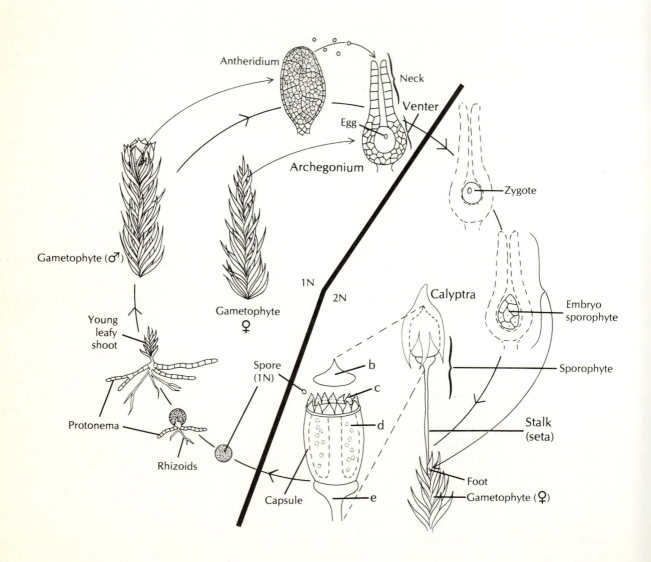

major tissue, the cortex, which is composed of thin-walled parenchyma cells. The cortex may contain, in some moss species, a central, poorly defined cylinder of water-conducting cells called **hydroids.** These cells do not become functional until they die. The cortex may also contain vascular cells called **leptoids** located between the central cylinder and the epidermis. Both hydroids and leptoids may be found in mosses belonging to the Polytrichales. The hydroids resemble tracheids of higher plants, but they lack the pit connections associated with the end walls of tracheids. Their function is the conduction of water (Scheirer and Goldklang, 1977). The leptoids function as conductors of the products of photosynthesis, mainly sucrose. These cells structurally resemble the sieve tube elements found in the phloem tissues of higher vascular plants, especially the ferns. (Behnke, 1975).

GROWTH: GAMETOPHYTE

The gametophytic generation is most conspicuous in the Bryophyta. It is initiated when meiospores, produced in the sporangium of the sporophyte, germinate after falling into a suitable environment. The meiospores grow into photosynthetic filaments known as protonemata (sing., protonema). The protonemal stage varies from a few chloroplast-containing cells in the peat mosses to many chloroplast-containing cells in the true mosses. In the liverworts, only a rudimentary protonema develops, consisting of only a few cells that soon divide into a multicelled thallus. The protonemal stage, then, is characteristic of all Bryophyta and primitive vascular plants such as ferns that produce independent gametophytic generations. The protonema of mosses resembles an unbranched filamentous green alga, except that the end walls of its cells are sharply angled rather than perpendicular to the main axis. In most instances the filamentous protonema soon branches, producing an extensive series of green filaments that closely resemble certain branching green algae such as *Fritschiella* and other members of the Ulotrich-

ales that have heterotrichous habits. However, since it has been established that ulotrichalean algae do not produce phragmoplasts (they form phycoplasts), and their flagellar roots are cruciately arranged, it is improbable that they are the direct phylogenetic ancestors of bryophytes (see Chapter 4, Fig. 4-8). The protonema also initiates branches of colorless or brownish filaments called rhizoids that are negatively phototropic (Fig. 21-3).

The young, branching gametophyte of mosses soon develops a heterotrichous habit with many upright shoots, each of which bears many **microphyllous leaves** (leaves lacking vascular tissues) (Fig. 21-4) arranged in a spiral fashion.

FIG. 21-4 Moss leaves.
Cross section through the microphyllous leaves of a moss (*Polytrichum* spp.) gametophyte. Many moss leaves are only one cell thick; however, *Polytrichum* leaves show considerable differentiation, including filamentous rows of lamellae *(lam)* emerging from the upper surface of the leaf. The cells of the lamellae are photosynthetic, and the spaces between the lamellae function as storage areas for water (125×).

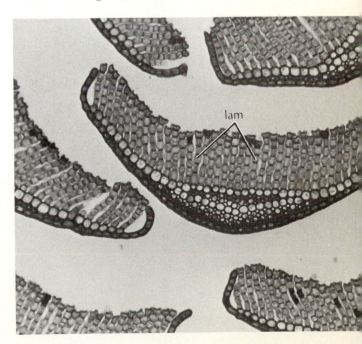

FIG. 21-5 Life cycle of a liverwort.

Diplohaplontic life cycle of *Marchantia* spp., with the alternations of haploid and diploid generations. The gametophyte generation begins when a haploid spore germinates and develops into the flattened prostrate gametophyte. *Marchantia* is dioecious, with male gametophytes producing specialized antheridiophores bearing antheridia and female gametophytes forming archegoniophores bearing archegonia. Both male and female gametophytes repro-duce vegetatively by forming gemmae cups, which form gemmules. Biflagellate sperm fertilize the egg cell within the archegonium, and sporophyte development begins. Mature sporophytes are parasitic on the gametophytic archegoniophore. They are composed of a foot, stalk, and sporangium. Sporangia contain spore mother cells, which undergo meiosis, producing haploid spores. Hygroscopic elaters develop among the spores.

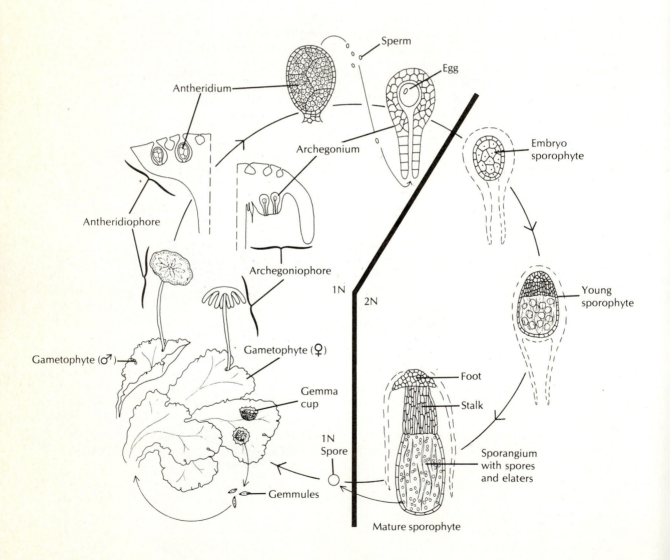

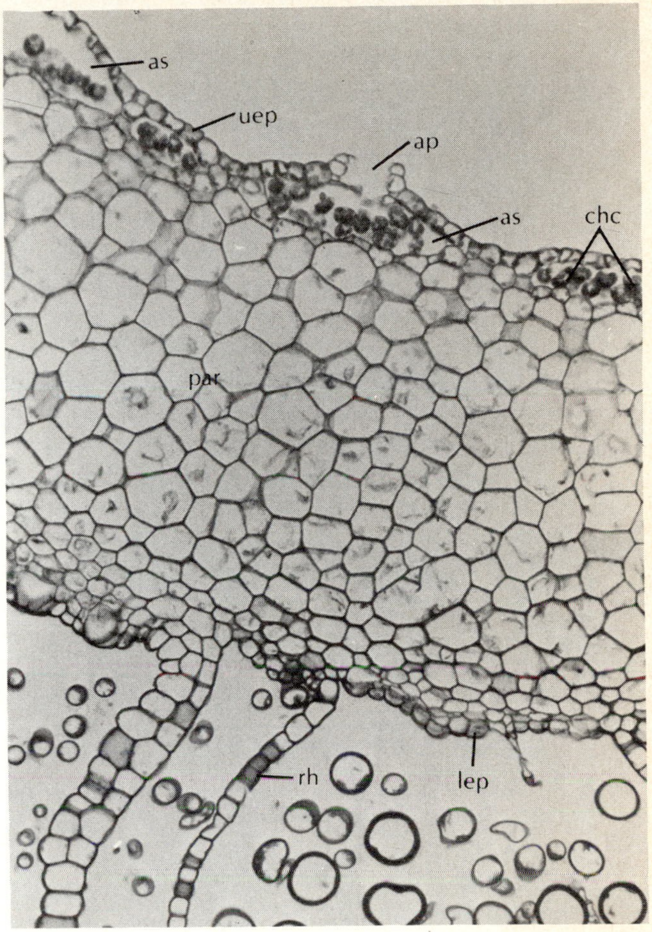

FIG. 21-6 *Marchantia* thallus.
Microscopic section through a thallus of *Marchantia* spp., showing various cell types and structures. Note air spaces *(as)* associated with branched filaments of chlorophyllous cells *(chc)* near the upper epidermis *(uep)* through which an air pore *(ap)* is clearly seen. Below the chlorophyllous cells is a thick layer of compact, thin-walled parenchyma cells *(par)*. Emerging from the lower epidermis *(lep)* are numerous filamentous rhizoids *(rh)* (125×).

Under certain light conditions, the growing apex of the upright gametophytic shoots will differentiate sex organs in the form of archegonia or antheridia (Fig. 21-3). Some moss species produce both archegonia and antheridia on the same apex (**monoecious** species) whereas others are separate sexed (**dioecious** species).*

The developing young gametophyte also continues to produce rhizoids; these may be unicellular filaments in some species or multicellular filaments in others. In both types their function appears to be that of an anchoring device, since there is no good evidence that they act in nutrient absorption from the underlying substrate. Rhizoids are characteristic of some algae, some fungi, and the independent photosynthetic gametophytes of terrestrial plants.

Among the Hepaticae, the leafy liverworts (Jungermanniales) produce protonemata and gametophytes that closely resemble those of the true mosses. The other liverworts develop into a flattened gametophyte with a body (thallus) that is clearly differentiated into dorsal and ventral surfaces (Fig. 21-5). Both sides are covered with a cellular epidermis. The cells between the two epidermal layers attain various degrees of differentiation, depending on the species. In *Marchantia* (Fig. 21-6), a number of cell types

*The terms "dioicous" and "monoicous" are used by some bryologists to differentiate these growth forms in the Bryophyta, while "dioecious" and "monoecious" are reserved for higher vascular plants and algae.

are clearly formed. A variety of **chlorophyllous cells** and storage cells can be identified. Air spaces, pores, and reproductive structures (archegonia and antheridia) are found in association with the upper epidermis, while many rhizoids emerge from the lower one. In the genus *Marchantia* and certain other dioecious liverworts, the reproductive structures are borne on special umbrella-like structures called **gametophores.** Archegonia are produced on the **archegoniophore** and antheridia are produced on the **antheridiophore** (Fig. 21-5).

Asexual reproductive structures may also be associated with bryophyte gametophytes. **Gemmae** (sing. **gemma**) are composed of groups of specially differentiated cells affiliated with the epidermal layers. They may be produced by many different bryophytes but are particularly well developed in the Marchantiales (Hepaticae). In *Marchantia*, gemmae are formed in special structures called **gemmae cups** (Fig. 21-5). They may contain many multicellular reproductive green disks or gemmae that are capable of breaking away from the parent thallus, settling in a different location, and differentiating into an entirely new gametophyte.

The Anthocerotae (hornworts) have relatively simple gametophytes that consist of a series of parenchyma cells located between an upper and lower epidermis. They are similar in form to the thalloid gametophytes of some Hepaticae, but they lack the internal cellular differentiation. The archegonia and antheridia are borne on the upper surface of the flat gametophyte.

GROWTH: SPOROPHYTE

The life cycles of all the Bryophyta are remarkably uniform, following a diplohaplontic pattern with clearly delineated heteromorphic alternations of generations. The sporophyte is invariably dependent upon the more conspicuous and independent gametophyte for its nutrition and support. It is sometimes a nonphotosynthetic structure that grows attached to or buried in the gametophytic tissues. It is anchored to the gametophyte by a special organ

called the **foot** (Fig. 21-3). In liverworts (Hepaticae) the sporophyte is a much-reduced, inconspicuous, parasitic structure consisting of a foot, **stalk** (seta), and **capsule** (sporangium) (Fig. 21-5). Its only function is production of meiospores. The capsule contains spore mother cells, spores, and hygroscopic **elaters.** In the genus *Marchantia* and certain other liverworts, the sporophytes are borne on the underside of umbrella-like structures called gametophores, since they bear the gametangia produced by the gametophyte.

The sporophytes of the Anthocerotae are more complex. They are elongate, macroscopic,

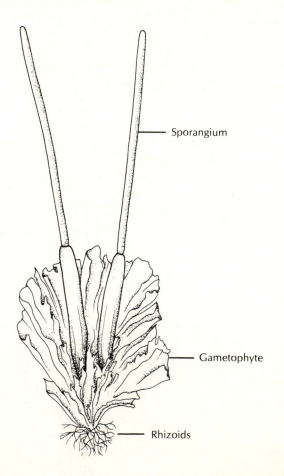

— Sporangium

— Gametophyte

— Rhizoids

FIG. 21-7 *Anthoceros* habit.
Both the gametophyte and sporophyte of the hornwort *Anthoceros*. Note the elongate sporangium, which is the spore-containing part of the sporophyte. The sporangium will split longitudinally as it matures, giving the appearance of horns growing out of the gametophyte.

photosynthetic structures that grow attached to the thalloid gametophyte (Fig. 21-7). The growth of the sporophyte is indeterminate, so that it continues to produce new sporogenous tissue throughout its existence. Furthermore, the tissues of the sporophyte are produced by an intercalary meristem located at the base of its stalk near its origin from the gametophyte. Sporophytes bear capsules (sporangia) that are unique among the Bryophyta. The slender capsule **dehisces** as it matures and dries, thus releasing its spores for wind dispersal. Like the true mosses (Musci), the sporophyte contains photosynthetic cells in its epidermal tissue. It also has stomata scattered among the epidermal layers of the stalk, and it produces a waxy **cuticle** that aids in prevention of water loss. All these factors serve to make the sporophyte of the Anthocerotae unique among the Bryophyta.

The sporophyte of the true mosses (Musci) often is structurally more complex than those previously described (Fig. 21-3). It consists of an upright, usually elongate structure that is anchored deep into gametophytic tissue by means of a special organ, the foot. Emanating from the foot is a long stalk (seta) that is capped by a structurally complex sporangium called the capsule. Within the capsule are located sporogenous tissues that produce spore-mother cells. These cells divide meiotically, forming functional haploid spores. The capsule also has, at the opening of the sporangium, a distinctive circular structure known as the peristome (Fig. 21-3), which bears rows of hygroscopic teeth whose ability to move with changes in humidity helps in spore dispersal. The peristome structure often forms the basis for moss classification. The capsule is topped by another structure, the **operculum,** which in turn is covered by the **calyptra.** The calyptra is formed from gametophytic tissue that originally covered the sporophyte during its embryonic development within the archegonium. The morphological structure of the calyptra is often useful in identifying the various moss species.

REPRODUCTION

Nuclear reproduction within the Bryophyta is accomplished by mitosis and meiosis. Cytoplasmic division is accomplished through cytokinesis. The process of cytokinesis is interesting in the bryophytes, because it provides a major phylogenetic connection between them, certain green algae, and the higher plants (Pickett-Heaps, 1975). Bryophyte cells all produce a phragmoplast, as is true in all the higher vascular plants and certain members of the bryophytan line of green algal evolution, including Charophyta, *Coleochaete*, and a few others (see Chapter 4, Fig. 4-8).

The phragmoplast is composed of a series of microtubules arranged perpendicular to the plane of cytokinesis that form a well-defined spindle apparatus. A cell plate is formed midway between the daughter nuclei, and a new cell wall develops from the center of the cell toward the periphery. The phragmoplast is in sharp contrast to the phycoplast, which is found in green algal members of the chlamydomonad line of evolution.

As in other nonvascular plants, reproduction among bryophytes may be viewed from both asexual and sexual aspects. The methods of asexual reproduction include fragmentation (especially among the gametophytes), meiospore production, and formation of specialized epidermal cells that produce gemmae.

Formation of meiospores always takes place in the sporangium of the sporophytic generation. The sporogenic tissues produce meiospore mother cells that divide by meiosis, forming **tetrads** of functional spores. In liverworts the spores are interspersed with hygroscopic elaters, which are specialized cells that help to disperse the spores when the cell wall of the sporangium breaks down. In mosses the spores are produced in a structurally complex capsule (Fig. 21-3). Meiospore mother cells within the sporogenic tissues of the capsule produce functional spores through meiotic divisions. In many mosses the spores are dispersed by specialized toothlike structures within the capsule, forming the peristome. The teeth of the peristome respond to changes in humidity and aid in actively moving the spores out of the capsule.

Sexual reproduction in all the Bryophyta involves the union of an egg and sperm cell, each of which is produced in a special multicellular organ associated with the gametophyte. Eggs are formed in archegonia, whereas sperm are formed in antheridia. The multicellular nature of these structures separates Bryophyta from the algae and the fungi.

Archegonia (Fig. 21-3) are complex, multicellular organs containing a neck, neck canal, and an enlarged basal area, the **venter,** which contains an egg cell. Archegonia are remarkably similar in structure in all the Bryophyta.

Antheridia (Fig. 21-3) are also multicellular, complex organs containing a jacket of sterile cells within which develop the sperm-mother cells and the sperm.

Fertilization (syngamy) takes place when biflagellate sperm cells are liberated in an aqueous environment and swim to the archegonium. The sperm then enters the neck canal and fuses with the haploid egg, forming a diploid zygote. Fertilization occurs in situ within the archegonium of the female thallus. This is not true for most algae, since fertilization occurs after the release of the gametes into an aquatic habitat; only the charophytes and some red algae have in situ fertilization. The zygote of the Bryophyta develops in a very specific manner, with the first cell division produced by a transverse cell wall so that the apex of the filamentous embryo emerges through the neck of the archegonium (an **exoscopic** development). Cellular polarity within the young embryo is thus established at a very early stage, perhaps within the zygote itself. An embryonic filamentous organism consisting of two to nine cells is soon formed; the precise number of cells depends on the species. Wardlaw (1955) reports two-celled filamentous embryos in some species of *Mnium* and nine-celled filaments in *Atrichum (Catharinea)*. The embryo filament develops a single apical cell

that directs the growth and development of the sporophytic generation. The establishment of **apices** early in development is characteristic of all bryophytes and higher plants. The **apex** of the bryophytes exhibits some of the physiological characteristics exhibited by all vascular plants, including apical dominance.

PHYSIOLOGY

The Bryophyta are all low-growing plants whose life history is inextricably attached to an aquatic environment—even though most species are terrestrial—because of the nature of the sexual process, in which biflagellate sperm cells require a water film through which they swim to reach an archegonium containing an egg cell. Development of sex organs is environmentally determined. Archegonia and antheridia formation is primarily under the influence of temperature. These structures will not differentiate at cooler temperatures but will develop at warmer temperatures in some species (e.g., *Polytrichum* spp.). In others the reverse is true, as in *Leptobryum pyriforme*, a species that is very useful for the study of temperature effects, since it is easily cultured and maintained in a sterile condition at 25° C. When transferred to 10° C, gametophytes form both archegonia and antheridia within an 8-day period (Chopra and Rawat, 1977).

At least one moss species, *Sphagnum plumulosum*, is known to produce sex organs under the influence of light. This species remains sterile during long days (16 hours of light in 24 hours) but produces both archegonia and antheridia when days are short (8 hours of light in 24 hours) (Benson-Evans, 1964).

Sporophytes of mosses have some special mechanisms adapted to spore dispersal. The teeth cells of the peristome in the capsule of many mosses are hygroscopic, absorbing water in times of high humidity and releasing it as the air around them becomes drier. In so doing the elongate teeth twist and turn, causing the contained spores to move. The *Sphagnum* mosses disperse their spores by creating air pressure within the spore capsule, thus causing the operculum (lid) of the capsule to be forcibly discharged (Maier, 1974).

The sporophytes of the liverworts and hornworts produce elaters, whose hygroscopic nature aids in the dispersal of their spores (Fig. 21-5). Elaters are not produced by the Musci.

Pigments associated with bryophyte cells are identical with those found in vascular plants and green algae. They include chlorophyll *a* and *b*, beta-carotene and various xanthophylls, and **phytochrome** (Wareing and Phillips, 1978). Phytochromes are pigments that have been shown to absorb and release light energy in the red range of the visible spectrum. Thus two forms of a single phytochrome may exist—a red form (P_{660}), which absorbs light at 660 nm and is transformed to the far-red form (P_{730}), which decays back to the red form during periods of extended darkness. Such a mechanism has been implicated in the phenomenon known as photoperiodism in flowering plants (Ting, 1982). Phytochromes have also been implicated in growth responses and cellular differentiation in the higher plants and Bryophyta. It is interesting to note that the phytochrome molecule is composed of a light-absorbing moiety that is a phycobilin and a second moiety that is a high-molecular-weight protein. The phycobilins, you will recall, are components of some algae, including the blue-greens (Cyanobacteria), the reds (Rhodophyta), and the fire algae (Pyrrhophyta).

Another group of pigments associated with the Bryophyta are the **anthocyanins.** These are red, blue, or purple pigments composed of simple sugars in association with cyclic flavonoid molecules called **anthocyanidins.** They are present in some bryophytes and all the higher plants but never in the algae or fungi. Anthocyanins are water soluble and are usually found in association with the watery portions of the cell, often within vacuoles. The specific color expressed by the molecule is a function of the concentration of the pigment, the proportion of the

different anthocyanins present, and the pH of the immediate environment. Most anthocyanins are red when under acid conditions and change to shades of purple or blue as the pH becomes more alkaline. The presence of anthocyanins in cells appears to be related to the accumulation of sugars by the cells involved. The anthocyanins, for instance, are responsible for the colors expressed in some flowers, fruits, and leaves of many higher plants. In *Sphagnum* and some other mosses the anthocyanins are often present in high enough concentrations to cause the living plant to take on a reddish-purple hue.

Most of the growth and reproductive physiology of bryophytes is closely tied to the terrestrial environment of the particular organism. The particular adaptations are all associated with the invasion of dry land. These include the development of stomata on sporophytes of mosses and hornworts and the formation of water-impervious cuticles on the same sporophytes (Proctor, 1979). Other mechanisms include the ability of many moss plants, as well as the lichens (Chapter 20), to endure prolonged drought and extreme cold: many mosses can be reduced to near dryness and can survive. Indeed, a few moss species have been revived after 5 years of dark dryness as herbarium specimens (Richardson, 1981). Under natural conditions some mosses can survive under desert conditions. Studies with desert mosses indicate that these mosses have the ability to perform photosynthesis at low water tensions; this is not true in most moss species (Alpert, 1979). Indeed, the dearth of moss species under dry conditions appears not to be caused by the lack of water but by the reduced amount of time for useful photosynthesis and subsequent assimilation of nutrients. In studies with the dune moss *Tortula*, it was discovered that a period of increased respiration occurred following rehydration. This process was in turn followed by a slow return to normal photosynthesis over the next 24 hours. The implication is that a period of time is required to repair the metabolic mechanisms necessary for photosynthesis, which involves energy and the synthesis of enzyme systems (Bewley, 1979). Many mosses are irreversibly damaged by desiccation, even for short periods. Aquatic and tropical mosses require nearly 100% humidity at all times.

Effects of desiccation are often associated with effects of changing temperatures, as might be expected, because many mosses grow exposed on rocks and soils. The dune moss *Tortula* has been known to survive temperatures of up to 71° C, while granite mosses (Andreacales) regularly withstand temperatures of 50° C (Dilts and Proctor, 1976). In laboratory experiments dried mosses may survive immersion in liquid nitrogen at temperatures lower than −100° C (Sakai and Yoshida, 1967). When moist, however, mosses do not survive these temperatures. In experiments with the rate of freezing, it has been shown that slow freezing (3° C/hour) has little effect on moss metabolism, whereas fast freezing (60° C/hour or faster) caused extensive damage (Brewley et al, 1974). It would apear that the rate of freezing has a significant effect on mosses, rather than the rate of thawing, which has long been considered to be the case. These experiments help to explain how mosses can survive high altitude and arctic conditions. They also help our understanding of how mosses are able to adjust and acclimate to various temperature conditions.

EVOLUTION

Evolution of the Bryophyta is shrouded by the lack of good scientific evidence regarding the history of the group. The liverworts are known from a single fossil dating from the late Devonian period (some 390 million years ago) and the mosses from fossils dating to the same time (Remy, 1982). There is no evidence that the bryophytes were ever a dominant component of the terrestrial vegetation, which would certainly be true if they preceded the vascular plants onto land. However, the fossil record indicates that vascular plants were on land during the Silurian period, 75 to 100 million years be-

fore the mosses. It would appear from this evidence that the bryophytes represent a separate dead-end branch of plant evolution, having developed either from highly modified degenerate vascular plants, or from separate species of green algae as secondary invaders of the terrestrial environment.

Classically the Bryophyta are viewed as evolutionary developments that lie somewhere between the relatively simple algae and the complex vascular plants. They consist of a single line of ascent that has diverged into three classes— liverworts, hornworts, and mosses (Bower, 1908). More recently botanists have proposed that they belong to three separate and relatively unrelated lines and have divided them into separate divisions—the Hepatophyta, the Anthocerotophyta, and the Bryophyta. Evidence for this theory is discussed by Crandall-Stotler (1980). Her reasons are based primarily on developmental patterns found in the three groups. The implication of this reclassification is that each of the three lines of Bryophyta has evolved from a separate ancestor (thus forming a polyphyletic group).

This text retains the classic view elaborated by Bower. The implication of his classification is that the three lines of Bryophyta have evolved from a single ancestor (thus forming a monophyletic group). We have retained this classification on the basis of similar biochemical characteristics, heteromorphic alternations of generations, multicellular gametangia with a dominant gametophyte, and a similar, low-growing habit.

The relationship between bryophytes and the green algae appears to be well established when one considers the similarity in their biochemistry, cell walls, and energy storage products. Further evidence is found in the morphological and ultrastructural studies of Pickett-Heaps (1975). The heterotrichous habit of certain algae in the bryophytan line of green algal evolution leads us to the nature of three characteristics that both alga and bryophytes have in common: first, they both exhibit an oogamous condition;

second, they both produce a phragmoplast when dividing mitotically; and third, they both have flagellar roots occurring in two bundles (see Fig. 4-8). An algal organism that has all these characteristics is *Coleochaete* (Chlorophyta); the implication here is that a *Coleochaete*-like organism is very close to the ancestral alga that gave rise to some bryophyte-like organism.

ECOLOGY

Bryophyta are worldwide in distribution and are common plants in moist, acid, terrestrial habitats. Usually each species has a rather sharply defined, narrow ecological range; consequently the bryophytes, like the lichens, are sometimes used as environmental indicator organisms. However, not all bryophytes are restricted to moist terrestrial habitats; some can be found growing in desert regions and others (*Fontinalis*) are strictly aquatic. None of them are marine.

The Bryophyta are considered to be pioneer plants in the succession of organisms in areas that have no soil. Along with lichens, they have the ability to grow and reproduce in regions of bare rock, eroding the rock and forming soil on which other organisms can grow. Frequently the first colonizers of bare rock are blue-green bacteria and various algae, followed by a succession of lichens and bryophytes. Some mosses are found growing in quite arid environments (e.g., the granite mosses), whereas others are indigenous to more moist areas. During plant succession the bryophytes are generally replaced by plants that require a greater degree of soil, including ferns, grasses, shrubs, and other higher plants. However, in certain ecosystems such as peat bogs, tundra, and higher elevations, mosses may be the dominant plant cover.

Some mosses have even been instrumental in rock building as well as in rock erosion. In areas where limestone springs are located, mosses are important in the deposition of **travertine,** a porous, easily cut limestone rock with an attractive tan or rusty color. Such rock is often used in decorative architecture. The Roman Colosseum is built of travertine.

Some investigators have divided the bryophytes into categories based on the chemical nature of the bryophytes' substrates. Thus the **calciophiles** are mosses that thrive on calcium-yielding substrates, whereas the **silicophiles** are those found in silica-yielding substrates. Other researchers divide the bryophytes according to the physical-chemical nature of their substrates. Thus acidophilic (acid-loving), neutral, and **basophilic** (base-loving) bryophytes are identified (Bodenberg, 1954). Such designations are useful when one attempts to identify and classify environments. These categories offer a good example of the specificity and narrow range of ecological conditions under which certain bryophyte species will grow. Our knowledge of these environmental factors is not very extensive, although it is understood that light, temperature, water, and the availability of nitrogen often have significant effects on bryophyte populations.

Most bryophytes are extremely tolerant of low light intensity and are able to grow and reproduce in heavily shaded areas or even in near darkness inside caves. In regions of high light intensity, mosses usually have adapted devices to reduce the light. Anthocyanins are common in the red-colored mosses that are found growing on mountain tops, where they apparently function as shading devices for the light-sensitive chlorophyll. Also, mosses from exposed areas tend to avoid high light intensity by growing in crevices.

Temperature tolerance is also a bryophyte characteristic. Some mosses apparently have two temperature optima, one that induces spore formation in the sporophyte and one for vegetative growth of the gametophyte. Some mosses are able to grow at high temperatures near hot springs, whereas others are indicative of cold temperatures attained at high altitudes (the alpine mosses) and the Arctic regions of the earth. In cold Arctic regions where vascular plants are unable to survive, the mosses flourish. In warmer climates and tropical areas, such as those of moist tropical rain forests, there are relatively few bryophytes, except for areas where the soil is temporarily exposed and for the many epiphytic species that grow on the trunks and leaves of trees. Liverworts, especially common in these areas, are important epiphytes in tropical climates.

Mosses that tolerate the warm temperatures of hot springs may be found growing on surfaces at temperatures near 45° C in Iceland. Some species of *Sphagnum* can tolerate the warm water as well as the high sulfur content of some hot springs (Lange, 1973).

Because bryophytes are very sensitive to small alterations in their microenvironments, they often are good indicators of changes taking place in an environment. As with the lichens, they provide a good system for analysis and evaluation of environmental conditions. A good example of a moss used for this purpose is *Sphagnum*.

Sphagnum moss is acidophilic, and excessive growth of this organism is often indicative of acid conditions. In the last few years concern has increased about the impact of acid precipitation on the biota of poorly buffered streams and lakes. Acid precipitation is defined as rain or snow with a pH below 5.6 (the hypothetical pH of carbon dioxide in equilibrium with pure rainwater). Precipitation in southern Scandinavia and northeastern North America is quite acid, registering pHs between 3.8 and 4.2. The cause of acid precipitation is thought to be the release of the air pollutants sulfur dioxide and oxides of nitrogen, which combine with moisture in the atmosphere to create sulfuric and nitric acid, respectively. Precipitation then becomes a weak solution of these acids.

When acid precipitation falls on poorly buffered waters (low in carbonate), it may lower the pH of the receiving neutral waters to 5.0 to 4.5. These pHs are not compatible with the growth of many aquatic organisms, and sensitive fish, invertebrates, algae, and macrophytes may disappear. *Sphagnum* moss, however, is well adapted to acid conditions and thrives at pHs below 5.0. Under such acid conditions this moss will grow rapidly over lake bottoms and will colonize substrates left vacant by the more sensitive macrophytes. A thick mat of *Sphagnum* on a lake bottom is considered indicative of acid conditions. Such mats of *Sphagnum* have been reported from acid lakes in Norway, Sweden, New York State, and eastern Canada (Hendry and Vertucci, 1980).

When a large portion of a lake benthos is covered by *Sphagnum*, this indicates a habitat rich in carbon dioxide, because the moss uses free carbon dioxide in photosynthesis. *Sphagnum* also has a strong cation-exchange capacity and may bind and concentrate such biologically essential cations as calcium, iron, and potassium (Hendry and Vertucci, 1980). The moss will also bind aluminum ions, which usually are present in acid lakes. Thus essential ions are removed from solution and are unavailable to other biota. As *Sphagnum* binds some biologically important cations, it releases hydrogen ions in order to maintain electroneutrality. The addition of more hydrogen ions to the water accelerates the acidification process and further reduces the lake pH (Grahn, 1976). Thick *Sphagnum* mats also cover large areas of sediment, thereby preventing the exchange of nutrients between the waters and the sediment. *Sphagnum* is not a preferred habitat for most benthic invertebrates, so their numbers decrease as the area of the *Sphagnum* mats increases (Grahn et al, 1974). In several studies of acid lakes, the abundance of *Sphagnum* was negatively correlated with the pH of the water. The increase in *Sphagnum* in acidified lakes is considered to be detrimental to both the water chemistry and the biota of the lake.

Mosses also provide scientists with a very useful tool for monitoring air pollution, since mosses are very sensitive to a few common industrial pollutants such as sulfur dioxide (SO_2), lead, and various hydrocarbons. Richardson (1981) has pointed out three reasons for mosses being useful in gas pollution biology. First, they are metabolically most active during cool, moist periods of the year when SO_2 exposure is most likely to interfere with their growth and reproduction; second, moss gametophytes do not possess waxy cuticles or stomata that help to exclude toxic gases from vascular plants; and third, most mosses are physiologically active in a moist aquatic environments, and since SO_2 is highly soluble in water, its effect is easily seen in moss growth and development (LeBlanc and Rao, 1974).

ECONOMIC IMPORTANCE

The bryophytes do not have a very large economic impact on humans. A few liverworts have been used for medicinal purposes in the treatment of burns and other conditions, but none have proved to be effective. Some mosses, however, have received some attention throughout history as building materials, fuel, surgical dressings, and insulators. In some parts of the world (especially Ireland) mosses have formed extensive peat moss bogs. They are composed primarily of *Sphagnum* moss, and the material is harvested, compressed, dried, and used as fuel. At the lowest depths of the older peat bogs, the carbonized *Sphagnum* forms a soft brown coal called **lignite.** It is used in much the same manner as coal is in the rest of the world.

In the United States *Sphagnum* bogs are harvested, dried, and sold as peat moss, a mulch used as a ground cover around acid-loving shrubbery. *Sphagnum* plants make an excellent mulching agent because the moss's cellular structure enables it to reduce water evaporation, prevent erosion, control weeds, and enrich the soil. It also helps to maintain an acid pH in the soil.

The cell structure of *Sphagnum* leaves enables the moss to have special water-retentive properties (Fig. 21-8). The leaves are composed

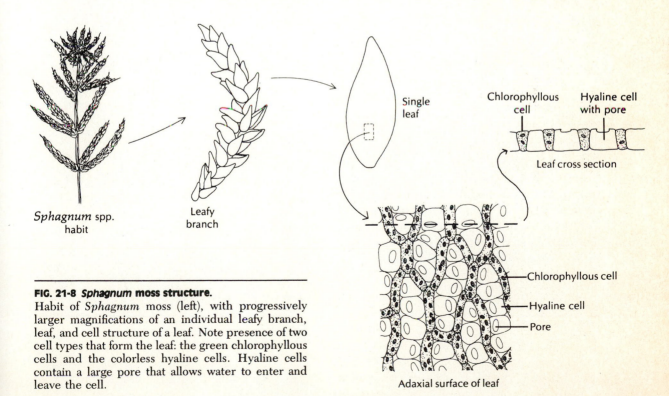

FIG. 21-8 *Sphagnum* moss structure.
Habit of *Sphagnum* moss (left), with progressively larger magnifications of an individual leafy branch, leaf, and cell structure of a leaf. Note presence of two cell types that form the leaf: the green chlorophyllous cells and the colorless hyaline cells. Hyaline cells contain a large pore that allows water to enter and leave the cell.

of two cell types: (1) the green (chlorophyllous) cells, which contain chlorophyll and perform photosynthesis, and (2) the colorless (**hyaline**) cells, which function as water-storage cells. The hyaline cells contain a single pore through which water enters and leaves the cell. The water-retaining properties of the hyaline cells enable *Sphagnum* mosses to absorb and retain up to 90% of their dry weight in water, whether or not the moss is living. Such a property makes *Sphagnum* an excellent plant where insulation and water absorption are required. As every good camper knows, *Sphagnum* bogs are good places for keeping food and drink cold and free from spoilage even during the heat of summer. Fishermen often line their creels with *Sphagnum* mosses to absorb moisture and keep fish cool and fresh, and put *Sphagnum* in worm boxes to keep bait alive and active. *Sphagnum* bogs are natural iceboxes.

The water-retentive properties of *Sphagnum* have also been applied to medicine. During the nineteenth and early twentieth centuries, sterile *Sphagnum* mosses were used in Europe and Asia as dressings for wounds. Even when sterility is a problem, *Sphagnum* is effective, since its low pH inhibits bacterial growth. The healing properties of *Sphagnum* evidently result from a combination of its acid pH and its ability to absorb fluids while allowing free passage of oxygen and other gases into and out of the wound area.

Sphagnum mosses have also been used by modern industry as a packing material for shipping fragile merchandise, foods, and nursery plants. Also, florists and plant growers use it as part of soil mixtures, especially for acid-loving plants. *Sphagnum* will not only keep the pH low but will improve the soil structure and help the soil to retain water.

SUMMARY

The Bryophyta are composed of photosynthetic, mostly terrestrial, plants that have diplohaplontic life cycles with a dominant gametophytic generation and a subordinate, mostly parasitic sporophytic generation. Their structure and biochemistry place them on an evolutionary tree somewhere between the algae and the vascular plants. They possess characteristics of both, including ultrastructural, cellular, and reproductive features.

The bryophytes are ubiquitous in the world, having invaded all climates and habitats. In some northern latitudes and higher elevations, they may constitute the major part of the flora. During the course of their evolution, bryophytes have remained restricted to habitats where water is available, since it is required for fertilization of eggs by swimming sperm. As a result the bryophytes are all low-growing plants, seldom reaching heights of more than a few centimeters. They are all, however, three-dimensional, multicellular, parenchymatous organisms. Internally the cells of bryophytes exhibit much more cellular differentiation than the algae but much less than the tracheophytes. Some have evolved a primitive type of conducting system for water and the products of photosynthesis. Still others have evolved mechanisms to cope with light and temperature changes and variation in desiccation.

Gametophytes of bryophytes develop from a protonema stage into a leafy upright plant or a flat thallus, depending on the genus. The gametophytes produce archegonia and antheridia, both of which are multicellular structures containing gametes. They may be dioecious or monoecious. Some produce specialized asexual reproductive organs called gemmae.

Sporophytes are always attached to gametophytes. Many of them are nonphotosynthetic parasites on the gametophyte (especially the Hepaticae). The structure of moss sporophytes is complex, with a foot, stalk, and capsule. Much of the identification and classification of mosses is based on the structure of the sporophytic capsule and its associated structures, the peristome and calyptra.

Bryophyte pigments are identical to those of green algae and tracheophytes. They also contain a phytochrome complement that is active in photoperiodic responses.

Moss-type plants date back to the Devonian period and were reasonably abundant during the Pennsylvanian period. However, they have never been a dominant component of the terrestrial vegetation, except for a few isolated ecological niches in the far northern and southern latitudes and at higher altitudes.

Ecologically the mosses, together with the lichens, are pioneer plants in the succession of organisms. They are capable of growing on barren soil and bare rock and are excellent soil builders. There are no marine mosses and only a few purely aquatic forms. Some can grow at high temperatures near hot springs, and others are common in arctic tundra. *Sphagnum* is characteristic of acid environments.

Bryophytes have been used as environmental indicator organisms, because they are sensitive to various environmental poisons such as sulfur dioxide. They have also found use in commercial gardening.

SELECTED REFERENCES

Benson-Evans, K. 1964. Physiology of the reproduction of Bryophytes. Bryologist **67**:431-445.

Bewley, J.D. 1979. Physiological aspects of desiccation tolerance. Ann. Rev. Plant Physiol. **30**:195-238.

Bodenberg, E.T. 1954. Mosses, a new approach to the identification of common species. Burgess Publishing Co., Minneapolis. 264 pp.

Clark, G.S.C., and J.G. Duckett (eds.). 1979. Bryophyte systematics. Systematics Association Special Vol. No. 14. Academic Press, Inc., New York. 582 pp.

Conard, H.S., and P.L. Redfearn. 1979. How to know the mosses and liverworts. Ed. 3. Wm. C. Brown, Co., Dubuque, Iowa. 302 pp.

Crum, H.A., and L.E. Anderson. 1981. Mosses of eastern North America, Vols. 1 and 2. Columbia University Press, New York. 1328 pp.

Grout, A.J. 1924. Mosses with a hand-lens. Ed. 3. Published by the author, New York. 340 pp.

Richardson, D.H.S. 1981. The biology of mosses. John Wiley & Sons, Inc., New York. 220 pp.

Schuster, R.M. 1966, 1969, 1974, 1980. The Hepaticae and Anthocerotae of North America east of hundredth meridian. Vols. 1-4. Columbia University Press, New York.

Smith, G.M. 1955. Cryptogamic botany, Vol. II. Bryophytes and Pteridophytes. Ed. 2. McGraw-Hill Book Co., Inc. New York. 399 pp.

abyssal plain Deepest part of the ocean basin, characterized by a smooth, uninterrupted expanse lacking in relief. Fig. 2-14.

accessory photosensitive pigment (also called antenna pigment) Colored organic compound capable of absorbing light and transferring the energy to chlorophyll *a*.

acervulus (pl. acervuli) Type of asexual reproductive structure in ascomycetes and deuteromycetes that contains conidiophores bearing conidia located on an open flat mass of hyphae; a type of sporocarp. Fig. 17-3.

acidophile (adj. acidophilic) Substance, tissue or organism having an affinity for acids.

acrasin Substance that induces the aggregation phase of cell migration in cellular slime molds.

acrocarpous Descriptive term for mosses that produce a tuft-forming habit.

aeciospore Binucleate spore produced by some parasitic basidiomycetes; produced in blisterlike structures called aecia. Fig. 18-8.

aecium (pl. aecia) Blisterlike structure on leaves of host plants infected by some parasitic basidiomycetes; it is composed of dikaryotic hyphae and produces dikaryotic aeciospores. Fig. 18-8.

aeration pore Microscopic pore in the upper or lower surface of stratified lichens; includes both cyphellae and pseudocyphellae.

aeroaquatic Term applied to aquatic fungi that produce conidia above the water's surface.

aerobic Living or taking place in the presence of oxygen.

aethalium (pl. aethalia) Rather large, cushionlike masses of spores covered with a crust; a type of asexual fruiting body in some myxomycetes (slime molds). Fig. 14-4, *B*.

aflatoxin Type of mycotoxin produced by members of the ascomycete genus *Aspergillus*.

agar Complex polysaccharide of red algal origin, liquid at high temperatures and forming a gel at cooler temperatures.

agarophytes Those red algae that are harvested to extract agar; they include *Gelidium*, *Gracilaria*, and others.

aggregation phase State in the life cycle of some cellular slime molds in which individual cells migrate to a center under the influence of acrasins.

akinete Large, thick-walled, nonmotile, reproductive cell of some Cyanobacteria. Fig. 3-6.

algal bloom Significant increase in planktonic or benthic cell members in a body of water.

algicide Any chemical substance that is lethal to algae.

algin Soluble salt (usually sodium) of alginic acid.

alginic acid Complex phycocolloidal substance, a polymer of 5-carbon acids; absorbs large amounts of water; found in cell walls of the Phaeophyta.

alimentary toxic aleukia (ATA) Serious disease of humans and livestock caused by ingestion of grain contaminated with mycotoxin from *Fusarium tricinctum* (Deuteromycotina); characterized by a reduction in the numbers of white blood cells.

alkaloids Large group of naturally occurring, heterogeneous organic plant extracts that usually have an alkaline reaction because of the presence of a nitrogenous heterocyclic ring; many have pronounced biological activity and are used as narcotics, medicinals, tranquilizers, and anesthetics.

allochthonous From outside the immediate ecosystem; of foreign origin; not native.

allophycocyanin Bluish-green phycobilin found in some algae and Cyanobacteria.

alternation of generations Type of life cycle characterized by fluctuations between adult haploid and adult diploid phases. Fig. 1-11.

amatoxin Cyclic oligopeptide toxin produced by some species of *Amanita* (basidiomycete) mushrooms; responsible for some deaths from eating mushrooms. Fig. 18-20.

aminoadipic acid Characteristic intermediary in the α-aminoadipic acid (AAA) pathway of the biosynthesis of the amino acid lysine; the pathway of lysine synthesis found in most fungi. See *diaminopimelic acid pathway (DAP)*, the other lysine pathway.

amylopectin Highly branched component of starch; stains violet-red with iodine.

anaerobic Living or taking place in the absence of oxygen.

androsporangium (pl. androsporangia) Structure in the Oedogoniales that produces androspores. Fig. 4-11.

androspore Zoospore from the Oedogoniales that produces dwarf male filaments. Fig. 4-11.

aneuploid Condition in which the normal chromosome complement of a cell is increased or decreased by one or more chromosomes.

anisogamy (adj. anisogamous) Condition in which one gamete (the male) is somewhat smaller than the other (the female). Fig. 1-10.

antenna pigments Accessory photosynthetic pigments.

antheridiol Organic sterol hormone secreted by female hyphae; induces formation of antheridial branches in certain fungi (Mastigomycotina). Fig. 15-5.

antheridiophore Structure or branch of certain liverwort gametophyte thalli that bears male sex organs; an antheridium bearer. Fig. 21-5.

antheridium Male gametangium or sperm-producing structure; multicellular in charophytes and bryophytes. Fig. 21-5.

anthocyanidin Cyclic flavonoid molecules that, when linked to simple sugars, form the group of pigments called anthocyanins.

anthocyanin Water-soluble, red, blue, or purple pigments found in vacuoles of some bryophytes and higher plant cells.

antiactinomycetal Substance that kills or inhibits the growth of the prokaryotic microorganisms Actinomyocetes.

antibiotics Any of many chemical substances, both natural and synthetic, that inhibit or destroy microorganisms.

apex (pl. apices; adj. apical) tip or angular summit of anything.

apical meristem Growing point that occurs at the apex of a plant. In algae, fungi, and bryophytes, apical meristems are unicellular.

aplanospore Nonmotile reproductive cell or spore in certain algae and fungi.

apothecium (pl. apothecia) Type of ascocarp in ascomycetes that is wide open, resembling a disk and lacking a cover. Fig. 17-7.

appressorium (pl. appressoria) Hyphal portion of a parasitic fungus that is flattened and closely pressed against the host cell. Fig. 16-4.

aragonite Type of crystalline calcium carbonate found in some calcareous algae (Table 1-2).

archegoniophore Structure or branch of certain liverwort thalli that bear female sex organs; an archegonium bearer. Fig. 21-5.

archegonium (pl. archegonia) Multicellular female gametangium forming an organ containing a single egg cell in bryophytes. Fig. 21-5.

articulated Possessing joints and composed of segments.

ascocarp Sexual fruiting structure of ascomycetes bearing asci; various types of ascocarps include cleistothecia, perithecia, and apothecia.

ascogonium (pl. ascogonia) Female reproductive structure of an ascomycete. Fig. 17-4.

ascospore Spore produced in an ascus and found in ascomycetes.

ascus (pl. asci) Saclike structure in which ascospores are produced by ascomycetes. Fig. 17-1.

ascus mother cell Binucleate cell that is a product of crozier formation in ascomycetes. Fig. 17-4.

aspergillosis Respiratory disease of humans and domestic animals caused by *Aspergillus fumigatus* (ascomycete).

assimilate (adj. assimilative) To absorb nutrients and incorporate them as part of the organism or cell.

astaxanthin Red xanthophyll pigment associated with some euglenoids and dinoflagellates; also called hematochrome.

autoecious Parasitic fungus (usually a rust) that can complete its life cycle on one host.

autogamy Sexual reproduction in which two haploid nuclei from one individual fuse to produce a zygote.

autolysis Self-digestion, producing enzymes that digest the structures that produced them.

autosome Chromosome that governs characteristics other than sex determination.

autotroph (adj. autotrophic) Organism that is capable of manufacturing its own food supply, usually by photosynthesis.

auxiliary cell Cell located adjacent to the zygote in some red algae; the zygote nucleus is transferred to the auxiliary cell, which then gives rise to the carposporophyte.

auxins Group of plant hormones that regulate plant growth and development.

auxospore Special type of "growth spore," sexually produced by diatoms (Chrysophyta). Fig. 8-9.

auxotroph (adj. auxotrophic) Organism that requires at least one organic growth factor (usually a vitamin).

axenic culture Culture containing organisms of only one species.

axoneme Axial complex or 9 + 2 protein microfibrillar complex of a flagellum together with its surrounding membrane.

ballistospore Spore that is forcibly ejected by some fungi such as ascomycetes, basidiomycetes, and deuteromycetes.

basidiocarp Fruiting structure of basidiomycetes; it is composed of a tertiary mycelium bearing basidia and basidiospores; a mushroom.

basidiospore Spore produced on basidia by basidiomycetes.

basidium (pl. basidia) Club-shaped structure on which basidiospores are produced by basidiomycetes. Fig. 18-1.

basophile (adj. basophilic) Substance or structure having an affinity for basic (alkaline) materials.

benthos (adj. benthic) Nonplanktonic organisms that grow attached to a substrate, usually the bottom of a body of water.

benzimidazole Organic compound containing carbon, hydrogen, and nitrogen; used as a systemic fungicide in some vascular plants.

bioluminescence Production of light by living organisms.

biomass Total weight of all living organisms or of a species population in any given area or volume.

biotroph (adj. biotrophic) Organism, often a fungus, that is parasitic but does not kill its host.

black tide Large population of marine phytoplankton, often Pyrrhophyta, that colors the water dark brown or black.

black wart Disease of potatoes caused by *Synchytrium endobioticum* (Mastigomycotina).

blade Flattened, broad, leaflike part of some of the larger algae.

Bordeaux mixture Fungicidal spray containing copper sulfate and calcium oxide.

brooder's pneumonia Lung disease of young chickens caused by *Aspergillus fumigatus* (ascomycete).

brown rot Condition in wood in which an invading fungus breaks down primarily cellulose but leaves the brown-colored lignin intact. Compare *white rot*.

bryo- Prefix derived from the Greek word *bryo* = moss.

bryophytan line Line of evolutionary ascent followed by certain green algae having phragmoplasts and flagellar roots arranged in two bundles, culminating in the Bryophyta. Fig. 4-8.

budding Cell division process whereby the parent cell retains its identity, while the offspring is produced with a newly synthesized cell wall attached to the parent cell; common in yeasts (ascomycetes). Fig. 17-2.

bufotenine Cyclic oligopeptide toxin produced by some species of *Amanita* (basidiomycete).

bunt Fungus disease of wheat and other important grains.

calciophile (adj. calciophilic) Plant, often a moss, that grows in areas possessing calcium-yielding substrates; a calcium-loving plant.

calcite Type of crystalline calcium carbonate found in some calcareous algae (Table 1-2).

calyptra Haploid tissue that forms a hood covering the operculum of the capsule in some mosses. Fig. 21-3.

capillitium Network of threadlike structures among which spores are formed in certain slime molds (Myxomycota). Fig. 14-3.

capsule An enclosure; in nonvascular plants, refers to a spore case or sporangium in the bryophytes; in prokaryotic organisms, refers to the mucilaginous slime layer surrounding cells.

carcinogen (adj. carcinogenic) Any substance that produces cancerous growths.

cardiotoxin (adj. cardiotoxic) Any substance that has a significant effect on the function of the heart in vertebrates (e.g., flammutoxin, produced by certain mushrooms).

carnivore Organism whose diet consists of flesh from other animals.

carotenes Group of oxygen-free hydrocarbons with a tetraterpenoid structure and colored yellow to orange; a class of carotenoid pigment, usually associated with light-absorbing structures such as chloroplasts; beta-carotene is the most common.

carotenoids Group of organic compounds including the carotenes and their oxidation products, the xanthophylls; they are all linear tetraterpenoid pigments found especially in chloroplasts and other light-absorbing structures.

carpogonium (pl. carpogonia) Egg cell or oogonium with its attached trichogyne; seen in the red algae. Fig. 11-4.

carpospore Diploid spore produced by the carposporophyte in certain red algae.

carposporophyte First of two diploid generations in some red algae; it is parasitic on the 1N gametophytic generation. Fig. 11-5.

carrageenan Sulfated galactan molecule found in the cell walls of some Gigartinales (Rhodophyta).

cellobiose Disaccharide produced by the hydrolysis of cellulose and composed of glucose units with a β-1,4 linkage.

cellulases Group of enzymes that catalyze the hydrolysis of cellulose.

cellulolytic fungus Fungus that has the capability of producing cellulases and therefore breaks down cellulose.

cellulose (adj. cellulosic) Branched polymer of glucose. Fig. 1-3.

central nodule Opening in the frustule of diatoms (Chryosphyta). Fig. 8-6.

Centrales Order of diatoms (Chrysophyta) based on organisms having radial symmetry.

centric Adjective for diatoms belonging to the Centrales.

centriole Cellular organelle usually located near the nuclear envelope and composed of two segments, each with a 9 + 0 complex of protein microfibrils; common to animal cells and some protistans.

centromere Point of attachment of chromatids on a chromosome from a eukaryotic cell; lacking in chromosomes of mesokaryotic cells.

cephalodium (pl. cephalodia) Minute swelling on a lichen thallus that contains Cyanobacteria that are different from the phycobionts in the algal layer of the lichen.

chemoautotroph (adj. chemoautotrophic) Autotrophic, prokaryotic organism that obtains energy from oxidation of inorganic compounds. Compare with *photoautotroph*.

chemotactic agent Chemical substance that attracts motile cells.

chemotaxis Attraction of motile cells (often male gametes) by chemical substances emitted by other cells (often female gametes).

chemotropism Ability of organisms to move in a certain direction in response to specific chemical stimuli.

chestnut blight Disease of chestnut trees (*Castanea dentata*) caused by the ascomycete *Endothia parasitica*.

chitin Structural polysaccharide common to the cell walls of many fungi; glucosamine; an amino sugar. Fig. 1-3.

chitosan Chitin molecules that are polymerized with other amino sugars.

chitosome Organelle associated with the growing tips of some zygomycetes; it contains packets of the enzyme chitin synthetase.

chlamydomonad line Line of evolutionary ascent followed by certain green algae having a phycoplast and flagellar roots cruciately arranged. Fig. 4-8.

chlamydospores Segments of a hypha that become packed with food reserves, develop thick cell walls, and are passively disseminated; a type of resting spore especially common in rusts and smuts (basidiomycetes).

chlorinated hydrocarbon Class of organic molecules containing one or more chlorine molecules; some have pesticidal or toxic properties and persist in the environment for long periods (e.g., DDT).

chloromonads Class (Chloromonadophyceae) of the Xanthophyta (yellow-green algae).

chlorophyll *a* Green photosynthetic pigment, a magnesium-porphyrin complex, that absorbs light energy in the 450 and 660 nm ranges and that is found in most algae and vascular plants.

chlorophyll *b* Green photosynthetic pigment similar to but differing slightly chemically from chlorophyll *a*; absorbs light in the 480 and 650 nm ranges; found in Chlorophyta, Charophyta, Euglenophyta, Bryophyta, and vascular plants.

chlorophyll *c* Photosynthetic pigment that differs chemically from chlorophyll *a*; absorbs light in the 450 nm range; found in the Chrysophyta, Xanthophyta, Phaeophyta and Pyrrhophyta.

chlorophyll *d* Photosynthetic pigment found in small concentrations in some members of the Florideophycidae (Rhodophyta); chemically similar to but not identical to chlorophyll *a*.

chlorophyllous cells Green cells that carry on photosynthesis, especially in the Bryophyta. Fig. 21-8.

chloroplast endoplasmic reticulum (CER) Single or double membrane system that is an extension of the endoplasmic reticulum; it encloses individual chloroplasts and, in some instances, the nucleus as well. Figs. 1-7 and 9-3.

chromatic adaptation Ability of an alga to alter its pigment ratios in response to changing light wavelengths.

chrysolaminarin Energy storage molecule associated with certain members of the brown line of algal evolution (Table 1-3).

chytrids Common name given to some unicellular flagellate fungi belonging to the Mastigomycotina.

ciguatera Poisoning of humans or fish caused by eating toxin-containing fish or shellfish.

circadian rhythm Repeating cycle of events based on 1 day (about 24 hours).

citrus melanose Fungal disease of citrus trees caused by *Diaporthe citri* (ascomycete).

clamp connection Cell wall structure produced by mitotic cell divisions in dikaryotic hyphae of basidiomycetes. Fig. 18-2.

cleistothecium (pl. cleistothecia) Type of ascocarp in ascomycetes that is completely enclosed. Fig. 17-7.

clubroot Disease of members of the cabbage family causing root enlargement and decreased growth rate; caused by *Plasmodiophora brassicae* (Mastigomycotina).

coccidioidomycosis Disease (San Joaquin Valley fever) of the respiratory tract or other organs caused by the deuteromycete *Coccidioides immitis*.

coccolith Calcareous disk produced as part of the cell covering by coccolithophorids (Prymnesiales; division Chrysophyta). Fig. 8-4.

coccolithophorid Common name given to some members of the Prymnesiales of the division Chrysophyta. Fig. 8-5.

coenobium Colony of cells having a definite number and arrangement of cells in the adult organism; especially common to members of the Volvocales (Chlorophyta). Fig. 4-4.

coenocyte (adj. coenocytic) Multinucleate cell.

columella Sterile part of a sporangium, usually representing an extension of the stalk into the interior of the sporangium in some fungi (Fig. 16-1) and bryophytes.

commensalism Association between two organisms living together in such a manner that one organism benefits and the other neither benefits nor is harmed.

compensation depth Depth of light penetration into water at which oxygen produced by photosynthesis is equal to the oxygen consumed by aerobic respiration.

competence (adj. competent) Level of developmental maturity that an organism reaches before it can differentiate spore-producing or gamete-producing organs; often depends on the organism's reaching a certain physical size or level of stored energy.

conceptacle Cavity containing gametangia located on receptacles of members of the Fucales. Figs. 7-5, 7-6.

conchocelis stage Special diploid, filamentous, microscopic generation of some red algae, formerly thought to be a separate genus, *Conchocelis*. Fig. 11-3.

conchospore Spore produced singly by conchocelis stage of Bangiophycidae. Fig. 11-9.

conditioning Process by which the activities of a preceding species renders the environment more favorable (or less favorable) for a succeeding species; may refer to chemical, physical, or biological processes.

conidiophores Upright structures or stalks on which conidia are produced in some fungi. Fig. 17-6.

conidium (pl. conidia) Asexual, nonmotile spore produced at the tip of growing hyphae; especially common in ascomycetes and deuteromycetes. Fig. 17-6.

conjugation Process whereby two compatible cells or filaments fuse their cytoplasms by producing tubes through which amoeboid gametes move and join together in a sexual process.

conjugation tube Structure through which amoeboid gametes move in order to accomplish conjugation. Fig. 4-13.

consumer Organism found in any ecosystem that gains its energy by ingesting other organisms.

continental shelf Gradual slope of geological land masses from the point of high tide to the abrupt increase in ocean depth at the beginning of the continental slope. Fig. 2-14.

continental slope Drop that occurs between the end of the continental shelf and the abyssal plain at the bottom of the ocean. Fig. 2-14.

controlled parasitism Type of parasitism in which the host has developed some resistance to the parasite so that the host continues to live in spite of the presence of the parasite.

coprin Organic compound isolated from the genus *Coprinus* (basidiomycete) that, when consumed with alcohol, causes stomach upset in some humans.

coprophil (adj. coprophilous) Organism that grows on dung or that has an affinity for dung as a habitat.

cortex In lichens (Fig. 20-3), the outer layer of the thallus, composed of compressed fungal hyphae; in kelps and other plants, any tissue lying between the epidermis and the central core area.

corticolous Growing on trees or dead wood.

costa (pl. costae) Ridge or ridgelike part; used especially to describe diatom frustules.

crista (pl. cristae) Folds produced by the inner membrane of a mitochondrion.

crossing over Interchange of sections of homologous chromosomes during synapsis of meiosis (or mitosis in some fungi; Fig. 12-3).

crozier formation Hooklike formation in dikaryotic hyphae of some ascomycetes in which a + and − nucleus is passed on to the ascus mother cell. Fig. 17-4.

crustose Crustlike.

cryptococcosis Fungal disease of humans; causes lung lesions. Caused by *Cryptococcus neoformans* (Basidiomycotina).

Cryptogamia Older primary division of organisms composed of plants such as ferns, mosses, and thallophytes, all of which lack true flowers and seeds.

cryptomonads Common name given to a class of pyrrhophytes.

cultural eutrophication Condition in aquatic systems in which the waters are high in nutrients because of human activities such as the discharge of sewage or silt.

cuticle Waxy external layer in some bryophytes that serves to prevent water loss.

cyclosis Cytoplasmic streaming.

cyphella (pl. cyphellae) Sunken pit on the lower surface of a lichen; characteristic of the lichen *Stricta*.

cystocarp In the red algae, refers to the diploid carposporophyte (gonimoblast filaments) with its surrounding layer of haploid cells, the pericarp. Fig. 11-4.

cytokinesis Separation of the cytoplasm into two or more parts during cell division.

damping-off disease Fungal disease of plant seedlings caused by the genus *Pythium* (Mastigomycotina).

daughter colonies Asexually produced colonies, especially common in the Volvocales (Chlorophyta). Fig. 4-4.

dawn fish kills Death of fish caused by oxygen depletion; respiration by large numbers of algae and aquatic animals during the dark hours depletes the oxygen in the water. No photosynthesis occurs in the dark, so no oxygen is added to the water.

decomposer Organism that obtains its energy through the enzymatic breakdown of dead organic matter; usually fungi or bacteria.

dehisce To burst open at maturity, discharging contents.

depsides Colorless substances produced by lichens that are esters of phenolcarboxylic acids with a phenyl benzoate skeleton; over 43 have been chemically characterized. Fig. 20-13.

depsidones Colorless lichen substances that are derived from depsides by phenoloxidation; over 23 have been identified. Fig. 20-13.

dermatomycosis Skin infection caused by certain fungi belonging to the Ascomycotina or Deuteromycotina, usually *Arthroderma* (= *Trichophyton*) or *Nannizzia* (= *Microsporum*).

dermatophyte (adj. dermatophytic) Fungal parasite that grows on the skin of mammals.

determinate Type of growth that has defined limits.

detritus Organic and inorganic debris that is in the process of decomposition.

diadinoxanthin Xanthophyll molecule that is characteristic of some members of the Xanthophyta.

diaminopimelic acid Characteristic component of cell walls of prokaryotic organisms; structurally related to the amino acid lysine.

diaminopimelic acid pathway (DAP) Metabolic pathway by which most organisms synthesize the amino acid lysine. The Oomycetes are the only fungi that use this pathway; most fungi use the aminoadipic acid (AAA) pathway.

diaspore Vegetative reproductive structure in lichens that contains both algal and fungal components; includes isidia, soredia, or other fragments of the thallus.

diatom Common name given to those primarily unicellular organisms with siliceous cell walls belonging to the Bacillariophyceae (Chrysophyta).

diatomaceous earth Fine siliceous geological deposit composed chiefly of fossil diatom frustules. Fig. 8-17.

diatomite Name given to diatomaceous earth.

diatoxanthin Xanthophyll molecule that is characteristic of some members of Xanthophyta.

dictyosome Another name for the Golgi body; used especially in plant cell descriptions.

diffuse meristem Type of meristem that is generalized throughout the plant body, with no particular region given to meristematic activity.

dikaryotic hypha Septate fungal filament in which each cell contains two nuclei.

dimorphism (adj. dimorphic) Occurrence of two different forms of the same organism; especially common in some fungi.

dinoflagellate Biflagellate, unicellular member of the Dinophyceae (Pyrrhophyta). Fig. 9-1.

dinoxanthin Type of xanthophyll pigment that is characteristic of certain pyrrhophytes.

dioecious Term used to describe organisms that have separate individuals that produce the male and female reproductive structures.

diplo-diplohaplontic life cycle Type of life cycle associated with certain red algae in which two successive diploid generations are followed by a haploid generation. Fig. 11-2.

diplohaplontic Type of life cycle characterized by the alternation of haploid and diploid generation and sporogenic meiosis. Fig. 1-11.

diploidization Process whereby a haploid nucleus becomes diploid, especially as pertains to certain fungi.

diplontic Type of life cycle characterized by gametogenic meiosis and diploid adults. Fig. 1-11.

disulfiram Sulfur-containing organic compound, used as a drug.

dolipore septum Specific type of cell dividing wall found in most basidiomycetes.

downy mildew Plant disease caused by members of the Mastigomycotina and characterized by a whitish growth.

dry rot Wasting condition of wood caused by a xerophytic fungus, *Serpula (Merulius) lacrymans*, a basidiomycete.

Dutch elm disease Fungal disease of American elm trees caused by the ascomycete *Ceratocystis ulmi*.

ecosystem Natural unit of living and nonliving components that interact to form a system in which the living and nonliving components are under constant change, producing cycles of supply and demand.

ectendotrophic mycorrhizae Type of mycorrhiza that combines characteristics of both the ectotrophic and endotrophic mycorrhizae; usually associated with nursery-grown gymnosperms.

ectocrine (also called exocrine) Chemical substance, often a metabolite excreted by an organism, which may inhibit or stimulate the growth or reproduction of other organisms; an "environmental hormone."

ectoparasite External parasite.

ectoplasmic net Externally secreted network of tubules, called a net plasmodium, produced by slime molds of the class Hydromyxomycetes.

ectotrophic mycorrhizae Type of mycorrhiza that forms a sheath of mycelium around individual roots. Fig. 13-1.

ejectosomes Needlelike projectiles located within the periplast of cryptomonads (Pyrrhophyta).

elater Hygroscopic cell that aids in spore dispersal in some sporangia, especially in the Hepaticae (Bryophyta). Fig. 21-5.

Emerson enhancement effect Effect observed in some algae when one or more accessory photosynthetic pigments occur in consort with chlorophyll *a*, thus increasing photosynthetic efficiency.

endobiotic Term describing parasites that live entirely within the cells of their host.

endoenzymes Enzymes that are produced within a cell and that function within the same cell.

endogenous rhythm Regular periodic events, such as circadian rhythms, that originate from within an organism or cell.

endolithic Living within hard rock.

endoparasite Internal parasite.

endosome Permanent nucleolus of mesokarotic cells, such as those found in euglenoids and pyrrhophytes.

endospore Spore produced within the parent cell, especially in the Cyanobacteria. Fig. 3-4.

endosymbiont (adj. endosymbiotic) Organism living symbiotically within the cells of another organism.

endosymbiosis State in which two organisms live together with one, the endosymbiont, living within the cells of the symbiotic partner.

endotoxin Toxic substance, usually a protein, that is produced within an organism and released only when the cell dies or disintegrates.

endotrophic mycorrhizae Type of mycorrhiza in which fungal hyphae penetrate the root cortex cells, producing intracellular haustoria. Fig. 13-2.

entomogenous Growing in or on insects, often parasitic.

epibiotic Term describing parasites that live primarily on the surface of their host but that possess haustoria that penetrate the host cells.

epicone Epitheca of dinoflagellates. Fig. 9-1.

epidermis Outermost layer of cells covering an organism.

epilimnion Upper layer of a body of water that usually varies significantly, chemically and physically, from the lower layer (hypolimnion).

epilithic Occurring on rocks.

epiphyte (adj. epiphytic) Plant that grows on the surface of another plant.

epitheca Outer valve of diatom frustules (Fig. 8-6) or the thecal plates, which occur anterior to the girdle in dinoflagellates. (Fig. 9-1).

epizoic Plant that uses a member of the animal kingdom as a substrate on which to grow.

equatorial pore Opening in the theca of dinoflagellates through which the two flagella emerge.

ergot Disease of rye and other cereal grasses caused by the ascomycete *Claviceps purpurea*; also, the sclerotium produced by this fungus on cereal grasses or the drug obtained from the fungus.

ergotism Poisoning caused by ingestion of ergot or eating food made from rye or wheat infected with the ergot-causing fungus *Claviceps purpurea*, an ascomycete.

estrogenic Describes biochemical substances that promote or produce estrus and other effects of estrogen in mammals.

eucarpic Describes fungi that produce zoospores in specialized branches of their otherwise vegetative hyphae.

eukaryote Organism that has cells that contain a well-defined, membrane-bound nucleus.

eutrophic Describes a body of water that is excessively fertile and very productive and that contains a large supply of nutrients, especially nitrates and phosphates.

exocrine See **ectocrine.**

exoenzymes Enzymes secreted outside of the cell that synthesized them, thus performing their function externally as with many enzymes produced by fungi.

exoscopic Developmental process in some mosses and ferns whereby the embryo shoot apex emerges through the neck of the archegonium.

exospore Spore produced outside of the parent cell, especially in the Cyanobacteria. Fig. 3-5.

eyespot (also called stigma) Concentration of red or orange lipoid pigments within the cell of some motile unicellular and colonial algae. Fig. 6-1.

facultative Having the capacity to exist or live under more than one specific set of environmental conditions—as in a cell that can exist with or without oxygen or an organism that can live either a parasitic or nonparasitic life. See *obligate.*

facultative parasite Organism that has the capacity to live either as a parasite or a saprophyte.

fairy rings Circles of mushrooms that appear annually and that represent the circumferential limit of growth for the secondary mycelium of some basidiomycetes. Fig. 18-6.

fall overturn Process in temperate waters in which the upper layer of water (epilimnion) cools and sinks to the bottom; the water column then becomes physically and chemically uniform. Fig. 2-9.

fermentative pathway Biochemical series of anaerobic biochemical reactions that result in the production of ethyl alcohol or lactic acid from carbohydrate.

flammutoxin Type of protein cardiotoxin that effects human heart function; it has been isolated from the basidiomycete *Flammulina velutipes*, an edible mushroom.

floridean starch Energy storage molecule of red algae, a branched molecule of α 1-4, 1-6–linked glucose molecules closely resembling glycogen.

floridorubin Carmine-purple pigment that does not contain protein; found in some members of the red algal order Ceramiales.

foliaceous Leaflike.

foliose Flattened, leaflike form.

food pyramid Movement of energy and nutrient materials from one feeding group or trophic level to another, generally progressing from photosynthetic plants to carnivorous animals.

foot One of the three basic parts of a sporophyte in bryophytes that anchors the sporophyte to the gametophyte.

form genus Name given to an organism based on structure rather than phylogenetic relationship; a practice common in describing fossils and imperfect fungi (Deuteromycotina).

form taxa Artificial system of classification based on establishment of form orders, form families, form genera, and so on.

fox fire Bioluminescence in decaying wood caused by some basidiomycete mycelia such as in *Armillariella mellea*.

fragmentation Type of reproduction in many nonvascular plants whereby pieces of the parent plant are capable of growing into an adult organism.

fructification Fruit or fruiting body of plants; in fungi and lichens it refers to structures containing spores.

fruiting inducing substance (FIS) Chemical substance that induces basidiocarp formation in *Schizophyllum commune*, a basidiomycete.

frustule Siliceous cell wall of a diatom.

fruticose Growth form of lichens that is upright or pendent and often branched. Fig. 20-1.

fucoidin Brown alga (Phaeophyta) cell wall component formed from various sulfated polysaccharides.

fucosan granules Refractile vesicles located in the cytoplasm of some brown algal (Phaeophyta) cells and containing organic acids.

fucoxanthin Brownish-colored xanthophyll pigment characteristic of the divisions in the brown line of algal evolution.

fungicide Inorganic or organic compound that prevents, retards, or stops the growth of fungi.

fusidic acid Antibiotic substance produced by some fungi; active against gram-positive bacteria.

galactans Polymers of galactose molecules.

gall Abnormal vegetative growth on plants caused by various agents including chemicals, mechanical injury, invertebrates, and viruses; in red algae, irregular masses of cells are called galls.

gametangium (pl. gametangia) Structure containing gametes.

gamete Haploid male or female sex cell that normally must undergo fusion with another compatible sex cell, thus forming a zygote, before it is capable of further development.

gametogenic meiosis Meiotic process whereby the product is gamete formation.

gametophore Upright structure or branch bearing gametes; a gamete bearer.

gametophyte Name given to the generation of a plant that produces gametes; the haploid generation.

gemma (pl. gemmae) Asexual reproductive structure composed of cells differentiated from the epidermal layer of the parent plant; especially well developed in some liverworts (bryophytes). Fig. 21-5.

gemmae cup Epidermal structure in some liverworts (bryophytes) that produce gemmae. Fig. 21-5.

geotropism Property exhibited by an organism or object that causes it to turn toward the force of gravity.

gibberellic acid Plant growth hormone produced in abundance by the ascomycete *Gibberella fujikuroi*; also occurs naturally in many higher plants.

girdle Transverse, circumferential groove in the thecae (cell walls) of dinoflagellates (Pyrrhophyta). Fig. 9-1.

gleba Spore chamber of certain Gasteromycetes (basidiomycetes). Fig. 18-17.

globule Structure containing male gametes in the Charophyta. See *antheridium*.

glucans Polymers of glucose molecules.

glycogen Energy storage molecule of fungi; a branched molecule of α 1-4, 1-6–linked glucose molecules; sometimes called "animal starch." Fig. 1-3.

glycolytic pathway Biochemical series of reactions involving the anaerobic respiration of carbohydrates to pyruvic acid.

gonidium (pl. gonidia) Specialized asexual reproductive cells in *Volvox* (Chlorophyta).

gonimoblast filaments Series of diploid cell filaments that make up the carposporophyte and produce carpospores; found in certain red algae. Fig. 11-4.

granum (pl. grana) Stack of thylakoid lamellae or sacs that are the site of photosynthesis; especially as referred to in higher plants.

gross primary production Rate at which organic material is produced by photosynthesis.

halophile (adj. halophilic) Organism requiring salt, usually as sodium chloride, for normal growth.

haploidization Process of going from the diploid to the haploid condition without undergoing meiosis, as with some fungi. Fig. 12-4.

haplontic Type of life cycle characterized by zygotic meiosis and haploid adults. Fig. 1-11.

haptonema Appendage that emerges from the cell between the two flagella in the Prymnesiophyceae (Chrysophyta) and that has a unique ultrastructure. Figs. 1-8 and 1-9.

haustorium (pl. haustoria) Specialized portion of a parasite that penetrates host cells; in fungi these organs are specialized hyphal segments. Fig. 13-2.

hematochrome Synonym for the xanthophyll astaxanthin, a red pigment found in some euglenoids.

herbivore Organism that feeds on plants.

heterocyst Colorless, thick-walled cell of some filamentous Cyanobacteria; the site of nitrogen fixation.

heteroecious Refers to a parasite that requires two hosts to complete its life cycle.

heterokaryon Fungal mycelium or hypha in which genetically different nuclei are associated in a common cytoplasm.

heterokaryosis Condition in fungal mycelia or hyphae in which genetically different nuclei occur within a common cytoplasm.

heterokont Having flagella of unequal length. Fig. 1-9.

heteromerous Having parts that differ in quality or number of components; in lichens, a stratified thallus. Fig. 20-3.

heteromorphic Different in form; used to describe alternations of generations when the haploid and diploid phases are morphologically distinct.

heterothallic Term applied to the condition of some organisms that produce both male and female gametes on the same individual, but they are not compatible. In others, male and female gametes are produced in different individuals; therefore, two compatible mating types are required for sexual reproduction.

heterotrichy (adj. heterotrichous) Condition in a branching filament in which both upright, erect portions and horizontal, prostrate portions occur.

heterotroph (adj. heterotrophic) Organism that is not capable of manufacturing its own food and therefore requires preformed organic molecules as a food source.

higher fungi Fungal group that includes Ascomycotina, Basidiomycotina, and Deuteromycotina.

histone Basic protein that is high in lysine and arginine content, often associated with DNA and chromosomes in eukaryotic organisms.

histoplasmosis Systemic respiratory disease of humans caused by the ascomycete *Emmonsiella (Histoplasma) capsulatum*.

holdfast Unicellular or multicellular structure that functions as a point of attachment for many algae.

holocarpic Term describing fungi that convert their entire thallus into one or more reproductive structures.

homoiomerous Having parts that are identical in quality or number of components; in lichens, an unstratified thallus. Fig. 20-3.

homokaryon Fungal mycelium or hypha in which all nuclei are genetically identical.

homothallism (adj. homothallic) Condition of some organisms that produce both male and female gametes on the same individual and that are self-compatible.

hormogonium (pl. hormogonia) Segment of a cyanobacterium filament; sometimes motile; usually formed between two heterocysts or separation disks.

hornwort Common name given to thallose plants that have split, hornlike sporophytes and that belong to the class Anthocerotae (Bryophyta). Fig. 21-7.

humus (adj. humic) Dark-colored amorphous or colloidal material, resistant to microbial decomposition, found in some soils; produced by the partial decomposition of proteins, phenols, and polysaccharides.

hyaline cell Colorless, nonphotosynthetic cell of some bryophytes that functions as a water storage cell. Fig. 21-8.

hydroid Water-conducting cell of some mosses.

hydrophilic gel Colloidal substance having a high affinity for water and easily made wet by water.

hygroscopic Readily absorbing and retaining moisture.

hymenium (pl. hymenia) Layer of hyphae that produces asci or basidia, depending on the division of the organism.

hyperplasia Growth caused by excessive proliferation of normal cells.

hypertonic Describes a situation in which the environment has a higher concentration of osmotically active molecules than the cells in that environment, thus causing the cells to lose water to the environment by osmosis.

hypertrophy Increase in size or bulk, but not necessarily caused by an increase in numbers of cells.

hypha (pl. hyphae) Unit or filamentous structure of which most fungi are composed.

hyphal interference Interaction between growing hyphae of different species of fungi.

Hyphomycetes Class of Deuteromycotina that do not produce pycnidia or acervuli.

hypnospores Thick-walled, resting zygotic (2N) cysts of some pyrrhophytes. Fig. 9-6.

hypocone Hypotheca of dinoflagellates. Fig. 9-1.

hypolimnion Vertical zone in a body of water that lies below the thermocline; may vary chemically and physically (especially temperature) from the shallower epilimnion. Fig. 2-9.

hypotheca Inner valve of diatom frustules (Fig. 8-6) or the thecal plates that lie posterior to the girdle in dinoflagellates (Fig. 9-1).

hypotonic Describes a situation in which the environment has a lower concentration of osmotically active molecules than the cells in that environment, thus causing the cells to gain water form the environment by osmosis.

hypovirulent Less than infective; especially as it applies to strains of viruses, bacteria, and fungi that cause diseases.

hystrichosphere Fossil hypnospore possessing elongate spines, found in some pyrrhophytes. Fig. 9-4.

ibotenic acid Organic material closely related to amino acids that is the base molecule for some toxins produced by *Amanita muscaria*, a poisonous mushroom.

imperfect stage Asexual stage in the life history of a fungus.

indeterminate Type of growth in which cells continue to reproduce and differentiate without having a defined limit.

indoleacetic acid (IAA) One of the plant hormones known as auxins.

ingestion rods Elongate, rigid organelles that occur in pairs and that aid in feeding in *Peranema* and other phagotrophic euglenoids.

Ingoldian fungi Aquatic Hyphomycetes (Deuteromycotina).

intensity adaptation Ability of an alga to alter its pigment concentrations in response to changing light intensities.

intercalary Term describing the position of an object inserted between two other objects, as in an intercalary cell located between the tips of a filament or an intercalary meristem.

internode (adj. internodal) Part of an organism that occurs between two nodes, as in the structure of charophytes.

intertidal zone That area of the marine environment that occurs between high and low tides. Fig. 2-14.

iridescent body Refractile structure that causes some red algae to appear to change colors in changing light patterns.

isidium (pl. isidia) Upright, narrow projection from a prostate lichen thallus. Fig. 20-8.

isogamy (adj. isogamous) Condition in which male and female or + and − gametes are identical in shape and form. Fig. 1-10.

isokont Having flagella of equal length. Fig. 1-9.

isolichenin Starch found in lichens that is composed of glucose residues with α 1-3 and α 1-4 glucosidic linkages; tests positively with iodine.

isomorphic Identical in form; used to describe alternations of generations when the haploid and diploid phases are morphologically indistinguishable.

karyogamy Fusion of compatible haploid nuclei in fungi; preceded by plasmogamy.

kelp Large marine algae, usually members of the Laminariales (Phaeophyta).

keratin Protein substance associated with the external layers of animal skin and that forms nails, feathers, hair, horns, and hoofs.

kieselgühr German word for diatomaceous earth or siliceous earth.

laminarin Energy storage molecule associated with the brown algae (Phaeophyta) (Table 1-3).

leptoid Cells of a vascular nature located within the cortex tissue of some mosses that function in conduction of the products of photosynthesis.

leucosin See *chrysolaminarin*.

lichen Organism composed of a fungus and an alga living together in such a manner that they form a distinct and different organism that has characteristics of its own that are not exhibited by either the algal or fungal partner.

lichen desert Geographic area in which no lichens are found.

lichen substances Compounds produced by and found in lichens; often weak phenolic acids.

lichenase Enzyme produced by organisms that catalyzes the breakdown of the lichen carbohydrate lichenin into glucose.

lichenin Linear polymer of glucose containing α 1-3 and α 1-4 linkages; found in the walls of the mycobiont of lichens; gives a negative test with iodine.

lignin Complex organic substance composed of a polymer of phenylpropanoid units; forms one of the most important parts of higher plant secondary cell walls.

ligninolytic fungus Fungus having the capability of breaking down lignin.

lignite Soft brown coal formed from deposits of plant materials in acid bogs; often high in *Sphagnum* moss content.

limiting factor That chemical factor, essential to the growth of an organism, that is present in the lowest concentration or that physical factor that is critical to the life of the organism.

limnetic zone Shallow-water area of lakes or oceans where light penetrates to the bottom.

littoral zone In fresh water, area where light penetrates to the bottom, permitting plant growth; in marine environments the area between high and low tides; also called the intertidal zone.

liverwort Common name given to mosslike or thallose plants belonging to the class Hepaticae (Bryophyta). Fig. 21-5.

lomasomes Series of small vesicles that lie between the plasma membrane and the cell wall of some fungi; composition and function are unknown.

lorica Hard or gelatinous envelope enclosing algal protoplasts; superficially resembles cell walls; however, unlike most cell walls, loricas are not totally in contact with the plasma membrane, and they do not contain cellulose. Fig. 8-1.

lower fungi Fungal group including Mastigomycotina and Zygomycotina.

luciferase Enzyme that catalyzes the degradation of luciferin, producing light (bioluminescence).

luciferin Protein that, when acted upon by luciferase, produces light (bioluminescence).

lumisynthesis Special process of some species of *Blastocladiella* (Mastigomycotina) that grow faster in light than in darkness; light-assisted growth.

lutein Xanthophyll pigment found in the Chlorophyta, Charophyta, and Rhodophyta.

lysergic acid Toxic metabolite of the ascomycete *Claviceps purpurea*; derived from the amino acid tryptophan. Fig. 17-10.

lysergic acid diethylamide (LSD) Hallucinogenic drug that can be readily synthesized from lysergic acid found in the sclerotia of *Claviceps purpurea* (Ascomycotina). Fig. 17-10.

lysis (v. lyse; adj. lytic) Decomposition or disintegration of organic material in living systems.

macandrous In members of the Oedogoniales (Chlorophyta), this term describes the situation in dioecious species that have male filaments that are the same size as the female filaments. Fig. 4-11.

macroalga (pl. macroalgae) Alga that is multicellular and clearly visible to the unaided eye; often called "seaweed;" also, extremely large algae such as the giant kelps (Phaeophyta).

macroconsumer Organism, visible to the naked eye, that utilizes preformed organic molecules.

macrocyst Giant cell that is the site of karyogamy and meiosis in some slime molds. Fig. 14-1.

mahogany tide Waters containing large populations of microorganisms that turn the water dark red.

mannan Polymer of mannose molecules.

mannitol Alcohol derived from the 6-carbon sugar mannose; one of the organic forms by which carbohydrates are usually translocated in fungi.

mariculture Cultivation of marine organisms; for example, seaweeds such as kelp or invertebrates such as clams and oysters.

marl Hard, unconsolidated deposit formed in freshwater lakes by deposition of calcium carbonate mixed with inorganic impurities.

mastigonemes Hairlike appendages found along the shaft of flagella.

m-chromosome See *microchromosome*.

medulla (adj. medullary) Inner portion of a lichen thallus that is composed of loosely interwoven hyphal filaments, Fig. 20-3; also region found in some red and brown algae as well as higher plants.

meiospore Spore that is produced by meiotic cell divisions.

meristem Point of growth in a plant where mitotic cell divisions are initiated.

meristoderm Meristematic layer of epidermal cells found in members of the Laminariales (Phaeophyta).

mesokaryotic nucleus Nuclear-type structure having characteristics of both prokaryotic and eukaryotic cells; found in euglenoids and dinoflagellates (Euglenophyta and Pyrrhophyta).

mesotrophic Describes a body of water that has moderate primary production and moderate levels of plant nutrients (such as nitrate and phosphate).

metaboly Expanding, contracting, and changing of form that characterizes euglenoids during cell movement.

metachromatic granules Polyphosphate bodies or volutin granules found in cells of the Cyanobacteria and containing stored phosphate.

microaerobe Organism type that is able to grow under low oxygen tensions.

microalga (pl. microalgae) Alga that is microscopic and often single celled.

microchromosome Small chromosomes that are less than $0.1 \times$ the length of normal chromosomes; found in some bryophytes.

microconsumer Microorganism that utilizes preformed organic molecules for energy; usually bacteria and fungi.

microcyst Small cysts containing a cellulose cell wall; formed under adverse environmental conditions by some slime molds (Myxomycota). Fig. 14-1.

micronutrient Trace elements or minor elements required in very minute quantities; often important limiting factor.

microphyllous Term applied to a small leaf in primitive plants whose leaf trace does not produce a leaf gap.

microplankton Organisms that are planktonic and whose size is less than 1 mm but greater than 60 μm. Table 2-2.

mildew Fungal growths that cover the surface of substrates, giving them a whitish appearance; a general term for fungal growth, parasitic or saprophytic.

mineralization (v. mineralize) Process by which organic compounds are converted to inorganic substances.

mitospore Spore that is produced by mitotic cell division.

mixotrophic Pertaining to organisms that are facultative heterotrophs and that can live as autotrophs or heterotrophs, depending on their individual habitats.

moldy corn toxicosis Disease of livestock resulting from ingestion of corn plants infected with the deuteromycete *Fusarium*.

monoecious Producing both male and female reproductive structures on the same individual.

monokaryotic hypha Septate fungal filament in which each cell contains a single nucleus.

monosporangium (pl. monosporangia) Sporangium in which monospores are produced in red algae. Fig. 11-3.

monospore Asexual spore, haploid or diploid, that germinates to reproduce its parent plant; in red algae (Rhodophyta).

morphogenesis Structural development of an organism.

muciferous body Organelle near the surface of euglenoid cells that produces mucilaginous materials secreted by the cell.

mucopolysaccharides Polymers of glucosamine or galactosamine that form a complex with proteins, creating slimy mucoid substances.

mucormycosis Fungal disease of mammals caused by members of the genus *Mucor* and related genera (Zygomycotina).

mud algae Community of algal organisms that live in or on mud.

multiaxial See *multiseriate*.

multiseriate Composed of many filaments of cells.

muramic acid Characteristic cell-wall component of prokaryotic organisms, chemically known as *N*-acetylmuramic acid, a sugar derivative.

muscarine Alkaloid extract that is the toxic substance in the mushroom *Amanita muscaria.*

muscimol Toxic organic substance that is a derivative of ibotenic acid; produced by the mushroom *Amanita muscaria.*

mushroom Fleshy basidiocarp or tertiary mycelium that produces basidiospores; a fruiting body of the subdivision Basidiomycotina. Fig. 18-5.

mutualism (adj. mutualistic) Relationship between two organisms in which both partners benefit from the association.

mycelium (pl. mycelia) Mass of compact hyphal filaments that compose the body of a fungus. Fig. 18-4.

myco- Prefix derived from the Greek word *mykes* = fungus.

mycobiont Fungal component of a lichen.

mycoherbicides Fungi that have deleterious effects on the growth of higher plants, and that may be used to control the growth of "pest" plants.

mycoparasite Fungi that parasitize other fungi.

mycorrhiza (pl. mycorrhizae) Symbiotic relationship between a fungus and the roots of a higher plant.

mycosis (pl. mycoses) Any disease induced by a fungus.

mycotoxicosis Disease caused by toxins produced by fungi.

mycotoxins Organic substances, produced by some fungi, that are toxic to other organisms.

mycovirus Virus that attacks a fungus.

myxamoeba (pl. myamoebae) Amoeboid cell of the division Myxomycota. Fig. 14-1.

myxophycean starch Energy storage molecule of the Cyanobacteria; a branched molecule of α 1-4, 1-6–linked glucose molecules closely resembling glycogen.

nannandrous In members of the Oedogoniales (Chlorophyta), this term describes the situation in dioecious species having small male filaments that are much smaller than the female filaments and that are called dwarf males. Fig. 4-11.

nannoplankton Small planktonic organisms ranging in size from 5 to 60 μm in diameter.

natural eutrophication Process by which a body of water becomes rich in plant nutrients by natural processes. Compare with *cultural eutrophication.*

necrotrophic parasite Parasite that kills the cells of its host and continues to feed upon the dead cells.

nematocysts Structurally complex projectile organelle found in a few dinoflagellates (Pyrrhophyta).

neritic zone Shallow portions of the seas that cover the continental shelves. Fig. 2-14.

net plankton Planktonic organisms that are retained in a net with openings of 60 μm; organisms with a size greater than 60 μm.

net plasmodium Special type of acellular, intersecting, tubular plasmodium produced by certain slime molds (Hydromyxomycetes).

net primary production Rate at which organic matter is stored by photosynthetic organisms *after* the subtraction of the amount of organic matter used in respiration by these organisms (i.e., producers). The amount of energy (in the form of organic material) available to the next trophic level (i.e., the herbivores).

neurotoxin Toxin that has an effect on nerve cells.

nitrogenase Enzyme complex that is required, along with ATP and a source of electrons, for nitrogen fixation; needs anaerobic conditions.

node (adj. nodal) Joint in a stem, or the part of a stem that bears an appendage, as in the Charophyta.

nonarticulated Not jointed; as with the encrusting red algae that lack joints.

nondisjunction Failure of homologous chromosomes to migrate to opposite poles during meiosis or mitosis. Fig. 12-4.

nonhistone chromosomal protein Protein content of eukaryotic cell chromosomes that is not histone.

nuclear cap Special intracellular organelle in zoospores of members of the Blastocladiales (Mastigomycotina).

nucleus-associated organelle (NAO) Electron-dense structure located just outside the nuclear membrane in some fungi in a position where most animal cells contain a centriole.

nucule Structure containing female gametes in the Charophyta. See *oogonium.*

nullipore Any of the coralline red algae having a crustlike plant body and therefore seeming to lack pores.

obligate Restricted to a particular set of conditions, as with a cell that must have oxygen or a parasite that must live in close association with its usual host to survive.

obligate parasite Parasite that must live in close association with its usual host in order to survive.

oceanic zone Area of the marine environment that is beyond the continental shelf. Fig. 2-14.

oidiophore Specialized hypha bearing oidia in certain basidiomycetes.

oidium (pl. oidia) thin-walled hyphal segment or spore that is formed on oidiophores; it may be either dikaryotic or monokaryotic; a special spore type of the basidiomycetes.

oil body Special membrane-bound organelle that contains a high lipid content; found in some liverworts. Fig. 21-1.

oligopeptide Molecule composed of a small number of amino acids.

oligotrophic Adjective describing a body of water that is low in nutrients and primary productivity.

oogamy (adj. oogamous) Condition in which the female gamete is relatively large and nonmotile, whereas the male gamete is small and motile (except in the red algae). Fig. 1-10.

oogoniol Organic sterol hormone secreted by antheridial branches of certain fungi (Mastigomycotina); induces oogonia to synthesize more antherdiol. Fig. 15-5.

oogonium (pl. oogonia) Female gametangium that produces eggs, especially in Charophyta (Fig. 5-3) and Xanthophyta (Fig. 10-1).

oospore Fertilized egg cell of *Vaucheria* and other xanthophytes.

operculum Structure that serves as a lid covering the capsule of mosses. Fig. 21-3.

opportunism (adj. opportunistic) Practice of adapting action to expediency or circumstances, especially as dictated by environmental conditions.

osmiophilic Term used to describe objects or substances that readily absorb osmium. Used extensively in electron microscopy.

osmophile (adj. osmophilic) Organism that lives or thrives in media having high osmotic pressures, as with certain fungi that grow on sugar or syrup.

osmoregulator Organelle, organ, or compound that acts to regulate internal osmotic pressure within a cell or organism (e.g., a contractile vacuole).

oxidative pathway Series of biochemical reactions that involve the removal of hydrogen or the addition of oxygen; the conversion of pyruvic acid to carbon dioxide and water, utilizing molecular oxygen.

palmella (adj. palmelloid) Stage in which many nonmotile cells are embedded in a common gelatinous matrix.

paraflagellar swelling Crystalline enlargement at the base of a flagellum in euglenoids.

parallel evolution Independent development of similar structures or functions by organisms that are distantly related or genetically unrelated; often exhibited by organisms living in similar environmental conditions; also called convergent evolution.

paralytic shellfish poisoning (PSP) Condition in birds and mammals caused by ingestion of shellfish that contain toxic dinoflagellates and/or their toxins.

paramylon Energy storage molecule associated with the Euglenophyta and Xanthophyta.

paraphysis (pl. paraphyses) Sterile cells or filaments found in association with reproductive structures.

parasexual cycle Life cycle involving plasmogamy, karyogamy, and haploidization without meiosis and zygote formation.

parasexuality Process whereby genetic mixing and gene recombination are accomplished without the formation of specific gametes by meiotic cell division and subsequent fusion of the gametes to form a zygote.

parasite (adj. parasitic) Organism that derives its nourishment from another organism.

parasporangium (pl. parasporangia) Sporangium in the red algae (Rhodophyta) that produces a small number of haploid or diploid monospores.

parenchyma (adj. parenchymatous) Plant tissue composed of thin-walled, photosynthetic or storage cells. A fundamental tissue of all multicellular plants.

parthenogenesis Development of an organism from an unfertilized egg cell.

pectic acids Common polysaccharide complexes included in cell walls and slime sheaths of many algae; usually includes galacturonic acid, plus mixtures of various sugar polymers.

pectin Polymer of galacturonic acid methyl ester, associated with plant cell walls and slime sheaths.

pectinase Enzyme that has the ability to catalyze the breakdown of pectins.

pectolytic Adjective referring to the ability to hydrolyze pectin (polymer of methyl *D*-galacturonate) found between and in plant cell walls.

pelagic zone Areas of open ocean (Fig. 2-14); also applies to open-water areas of fresh water.

pellicle Proteinaceous cell covering located beneath the plasma membrane in euglenoids. Fig. 6-2.

pendant Hanging down from trees or other objects; suspended.

Pennales Order of diatoms (Chrysophyta) based on organisms having bilateral symmetry.

pennate Adjective for diatoms belonging to the Pennales.

perfect stage Sexual stage in the life history of a fungus.

pericarp Envelope of haploid cells that surrounds the gonimoblast filaments in some red algae. Figs. 11-4 and 11-5.

peridinin Carotenoid pigment that is characteristic of some pyrrhophytes.

peridioles Entire spore chamber of some basidiomycetes belonging to the Nidulariales. Figs. 18-14 and 18-15.

peridium Outside covering of a spore-producing or fruiting structure in some fungi. Figs. 14-3 and 18-15.

periplast Cell covering in cryptomonads (Pyrrhophyta) that lies just beneath the plasma membrane and contains ejectosomes.

peristome Single or double circular rows of pointed, toothlike, hygroscopic appendages around the opening of the capsule in mosses. Fig. 21-3.

perithecium Type of ascocarp in ascomycetes that is pear shaped with an opening at one end. Fig. 17-7.

phaeophytin Primary degradation product of chlorophyll that results from the loss of the magnesium atom from the tetrapyrole portion of the chlorophyll molecule.

phagocytosis Process of engulfment of food such as bacteria or other cells by another cell, a common practice in protozoa.

phagotroph (adj. phagotrophic) Organism capable of utilizing particulate food substances.

phallotoxins Cyclic oligopeptide toxins produced by some species of *Amanita* (basidiomycete) mushrooms.

phenoloxidase Enzyme capable of catalyzing the oxidation of phenolic compounds.

photic zone Depth in a body of water to which light is able to penetrate so that active photosynthesis can occur. Figs. 2-14 and 2-15.

photoautotroph (adj. photoautotrophic) Autotrophic organism that uses light as its primary energy source.

photooxidation Oxidation that occurs under the influence of radiant energy, often sunlight.

photoperiod (also photoperiodism) Physiological response to various periods of light and dark, especially as exhibited by flowering plants.

photoreceptive molecule Compound that is sensitive to stimulation from light.

photosynthate Product of photosynthesis.

phototaxis (adj. phototactic) Orientation movement by an organism in response to light; may be a positive phototaxis, moving toward light, or negative phototaxis, moving away from light.

phototropism (adj. phototropic) Growth or orientation response of an organism to changes in light.

phragmoplast Name given to the arrangement of microtubules when they are parallel to the spindle fibers (or perpendicular to the plane of cytokinesis) during mitotic telophase in some green algae and higher plants. Fig. 4-8.

phyco- Prefix derived from Greek word *phycos* = seaweed.

phycobilins Group of water-soluble accessory photosynthetic pigments found in some algae and Cyanobacteria.

phycobiliprotein Water-soluble red or blue pigments associated with the phycobilins in some algae and Cyanobacteria.

phycobilisome Cellular organelle containing phycobiliproteins that are attached to the membrane surface of thylakoids in the Rhodophyta and Cyanobacteria.

phycobiont Algal component of a lichen.

phycocolloids Substances, extracted from algae, that are colloidal in nature and important economically as food additives; they include carrageenan, alginic acid and agar.

phycocyanin Blue phycobilin pigment found in some algae and Cyanobacteria.

phycoerythrin Red phycobilin pigment found in some algae and Cyanobacteria.

phycoplast Name given to the arrangement of microtubules when they are perpendicular to the spindle fibers (or parallel to the plane of cytokinesis) during mitotic telophase in some green algae. Fig. 4-8.

physiological races Genetic populations of given species that have various biochemical and physiological capabilities.

phytochrome Pigment in some plants that is the causative agent for photoperiodism and some growth responses in bryophytes and higher plants.

phytopathogen (adj. phytopathogenic) Organism that causes diseases in plants.

phytoplankton Planktonic organisms that photosynthesize.

pileus Cap of some basidiocarps (mushrooms).

pit connection Structurally complex cytoplasmic connection between adjacent cells in some red algae. Fig. 11-1.

plankton (adj. planktonic) Community of organisms that live suspended in water.

planospore Motile spore or zoospore.

planozygote Motile zygote.

plasmodesma (pl. plasmodesmata) Fine cytoplasmic connection that passes through the cell wall and joins adjacent cells, thus forming a continuum from one cell to the next.

plasmodiocarp Stalkless sporangiophore or sporangium that contains a spore, capillitial threads and is covered with a peridium; found in some slime molds (Myxomycota).

plasmodium (pl. plasmodia) Naked, multinucleate mass of cytoplasm that moves in an amoeboid fashion; common to some slime molds (Myxomycota). Fig. 14-2.

plasmogamy Fusion of the cytoplasm of two cells or the fusion of two hyphae in the more advanced fungi, producing heterokaryotes.

plasmolysis Contraction of the cytoplasm of a cell from loss of water because of osmosis.

plastid Specialized membrane-bound organelle in the cytoplasm of eukaryotic cells; the site of such activities as food manufacture and food storage.

pleomorphic Having many different shapes.

pleurocarpous Descriptive term for mosses that produce a carpetlike habit.

plurilocular sporangium Many-chambered sporangium producing mitospores in the Phaeophyta. Fig. 7-3.

pneumatocyst Gas-filled air bladder or float associated with the plant body of some brown algae (Phaeophyta). Figs. 7-5, 7-10, and 7-11.

podetium (pl. podetia) Upright, hollow structure seen in species of the lichen *Cladonia;* often branched. Fig. 20-1.

polar nodules Opening in the frustule of diatoms (Chryophyta). Fig. 8-6.

polyene antibiotic Organic compound such as amphotericin B that restricts the growth of or kills certain organisms, especially fungi.

polymorphism Capacity to exist in many different forms.

polyphyletic Derived from more than one ancestral line.

powdery mildew Disease of vascular plants such as roses and grasses that is caused by members of the order Erysiphales (Ascomycotina); causes a powdery appearance on surfaces of the host because of the numerous conidia produced.

primary metabolites Compounds produced during normal growth and development of fungi that have a known function in the organism's metabolism.

primary pit connections Pit connections formed by adjacent cells in a filament in some red algae. Fig. 11-1.

primary production (productivity) Rate at which light energy is stored as chemical energy (organic compounds) by photosynthesis in green plants or other autotrophs.

primordium Early development or origin of a particular organ or organ part.

producers In ecology, those organisms, usually green plants, that produce chemical energy in the form of organic materials from the light energy of the sun.

profundal zone Dark region of a body of fresh or salt water that is below the depth of the effective penetration of light. Fig. 2-15.

prohormone Organic molecule that is a hormone precursor.

prokaryote Organism that lacks a well-defined, membrane-bound nucleus within its cells.

propagules Multicellular, asexually produced structures that function as dispersal units, especially in some red algae and lichens.

proplastid Early stage in the development of cytoplasmic organelles called plastids.

protonema (pl. protonemata, adj. protonemal) Primary, filamentous stage of a germinating zygote (spore) in bryophytes, charophytes, and related plants. Figs. 5-3 and 21-3.

protothecosis Skin disease of humans caused by the colorless, heterotrophic green alga *Prototheca.*

pseudocyphella (pl. pseudocyphellae) Aeration pore found on the upper or lower surface of lichens, resembling, but not identical to, a cyphella.

pseudoparenchyma (adj. pseudoparenchymatous) False parenchyma composed of densely packed filaments of cells that superficially resemble true parenchyma.

pseudoplasmodium (pl. pseudoplasmodia) False plasmodium composed of many uninucleate cells that aggregate, forming a "slug" in some cellular slime molds (Myxomycota). Fig. 14-1.

pseudopodium (pl. pseudopodia) Armlike projection or "false foot" produced by some cellular slime molds (Myxomycota) and amoeboid protozoa.

pseudoraphe Groove in the frustule of some diatoms that does not go entirely through the frustule. Fig. 8-6.

psilocin Hallucinogenic compound derived from the amino acid tryptophan; produced by certain species of *Psilocybe* (Basidiomycotina).

psilocybin Hallucinogenic compound, the phosphoryl ester of psilocin, produced by certain species of *Psilocybe* (Basidiomycotina).

psychrophile (adj. psychrophilic) Organism that thrives at a relatively cold temperature.

psychrotolerant Describing organisms that can tolerate relatively cold temperatures.

punctum (pl. puncta) Pore or depression in a diatom frustule.

pycnidium (pl. pycnidia) Type of asexual reproductive structure in ascomycetes that contains conidiophores bearing conidia in a flask-shaped, hollow sporocarp. Fig. 17-3.

pyrenoid Proteinaceous portion of a chloroplast associated with synthesis and storage of various polysaccharides in many algal cells. Fig. 4-1.

raphe Longitudinal groove in the frustule of diatoms that passes completely through the frustule. Fig. 8-6.

receptive hypha Specialized hyphae in rusts (basidiomycetes) that receive compatible spermatia, bringing about a sexual process. Fig. 18-8.

receptive trichogyne Special hairlike hypha in some ascomycetes that will accept and fuse with spermatia.

red tide Large populations of planktonic dinoflagellates that color the water red.

refusal factor corn Feed corn that is refused by animals, presumably because of metabolic by-products produced by infecting fungi in the corn.

reindeer "moss" Actually fruticose lichens; grazed by reindeer and caribou; covers large areas in the far north; also occurs in temperate regions.

resting cyst Type of thick-walled cell produced by some dinoflagellates (pyrrhophytes) in response to environmental stress or as a result of sexual union of gametes. Fig. 9-6.

rhizine Rootlike structure in lichens that is composed of fungal hyphae; serves to anchor the lichen to the substrate.

rhizoid Colorless, hairlike filaments of cells that act as anchoring devices in many nonvascular plants; in bryophytes rhizoids emerge from the lower epidermis of gametophytes.

rhizomorph Thick strand composed of many hyphal filaments; the strand behaves as a unit; often in woodland basidiomycetes.

rice blast Disease of rice plants caused by the deuteromycete *Pyricularia oryzae.*

rockweed Common name given to many of the brown algae (Phaeophyta) found growing in the rocky intertidal zone.

rust Common name given to the parasitic, reddish-brown species of the Uredinales (basidiomycetes).

sagenogens Organelles, unique in the Hydromyxomycetes, that aid in formation of ectoplasmic nets.

sake Oriental wine formed through fermentation of rice by species of *Aspergillus* (ascomycete).

saprophyte (adj. saprophytic) Organism that obtains its energy through the breakdown of dead organic matter.

saxicolous Capable of growing on rocky substrates.

saxitoxin (**STX**) Neurotoxin produced by *Gonyaulax catinella*, a dinoflagellate (Pyrrhophyta).

sclerotium (pl. sclerotia) Hard, compact, hyphal masses of some fungi that are formed as a resistant stage to adverse environmental conditions.

secondary metabolites Compounds produced by some fungi when their normal growth and development is hindered in some way; these compounds have no known function in the organism's metabolism.

secondary pit connections Pit connections formed by cells in adjacent filaments in some red algae. Fig. 11-1.

secretory cell Specialized cell from the epidermis of some red algae that plays a role in halogen metabolism; also cell lining the mucilaginous ducts in the Phaeophyta. Fig. 7-2.

separation disk Dead cell in cyanobacterial filaments.

septal pore cap Membranous cap, composed of endoplasmic reticulum, that covers the central pore of the dolipore septum. Fig. 18-3.

septum (pl. septa) Dividing membrane or wall.

shield cells External sterile protective layer of cells in the male reproductive structure of charophytes. Fig. 5-3.

sieve cell Name sometimes applied to the trumpet cells of some large kelps (Phaeophyta) because of their resemblance to the sieve cells (phloem tissue) of higher vascular plants. Fig. 7-1.

sieve plate Common wall between two adjacent trumpet cells (sieve cells) in some large kelps (Phaeophyta). Fig. 7-1.

silicoflagellate Unicellular biflagellate member of the Dichtyochales (Chrysophyta) that produces delicate siliceous skeletal structures. Fig. 8-2.

silicophile Moss plant that grows in areas possessing silica-yielding substrates; a silica-loving plant.

single-cell protein (SCP) Protein produced by microorganisms such as bacteria, fungi (especially yeasts); or algae; often cultured under artificial conditions.

siphonaceous (also siphonous) Term describing the multinucleate condition of many algae and fungi.

slime flux Tree sap, usually flowing from wounds, that contains sugars and is infested with various bacteria and yeast giving it a slimy consistency.

slug Pseudoplasmodium stage in the life cycle of cellular slime molds (Myxomycota). Fig. 14-1.

smut Common name for the black spore masses produced by species of the Ustilaginales (basidiomycetes); fungal disease of vascular plants.

snow algae Algae that grow on snow and glaciers.

soredium (pl. soredia) Microscopic, dustlike particles on the surface of a lichen, composed of hyphal fragments and algal cells; a mechanism of vegetative reproduction. Fig. 20-8.

sorocarp Spore-producing structure of cellular slime molds (Myxomycota). Fig. 14-1.

sorus Group or cluster of sporangia. Fig. 7-4.

spawn Young mycelium used to inoculate media for agricultural production of mushrooms.

spermatium (pl. spermatia) Nonmotile male sex cell in some ascomycetes, basidiomycetes (Fig. 18-8), and red algae (Fig. 11-4).

spermocarp Multicellular structure in *Coleochaete* (Chlorophyta) that encloses the zygote.

spermogonium Structure in some ascomycetes and basidiomycetes (Fig. 18-8) that produces spermatia.

spindle pole body (SPB) See *nucleus-associated organelle* (NAO).

sporangiophore Upright hypha or stalk that bears a sporangium.

sporangiospore Spore that is produced in a sporangium.

sporangium (pl. sporangia) Structure in which spores are formed.

spore Haploid or diploid asexual reproductive cell that is capable of directly undergoing further development without fusing with another cell.

sporocarp Asexual reproductive structure in ascomycetes, composed of masses of hyphae and containing conidiophores bearing conidia.

sporogeneic meiosis Meiotic process whereby the product is spore formation.

sporophore Any structure bearing spores.

sporophyte Generation of a plant that produces meiospores; the diploid generation.

sporopollenin Polymer of carotene that is very resistant to decay; found in fungal spores and vascular plant pollen.

spring diatom increase (SDI) Annual increase in numbers of diatoms in the temperate areas; population increase caused by increased light, higher temperatures, and plentiful nutrients. Fig. 2-8.

spring overturn Process in fresh and salt water in which the upper layer (epilimnion) of water warms to 4° C and sinks to the bottom. In northern temperate areas this occurs in the spring.

squamule Prostrate thallus in lichens that is composed of small lobes; lacks a cortex and rhizines; seen often in *Cladonia*. Fig. 20-1.

stachybotryotoxicosis Disease of horses caused by the ingestion of grain contaminated with a mycotoxin produced by the deuteromycete *Stachybotrys alternans*.

stalk One of three parts of a moss sporophyte.

standing crop Amount of organisms in one particular place at a particular time; may be expressed in numbers of organisms, weight, or energy content; can be used for all organisms or an individual group of organisms.

statoliths Membrane-bound inorganic substance located in the apex of growing rhizoids in the Charophyta. Fig. 5-4.

statospore Special type of cyst or resting spore common to the Chrysophyceae (Chrysophyta).

stephanokont Having multiple flagella arranged in a ring or crown around one end of the cell. Fig. 1-9.

sterigma (pl. sterigmata) Hornlike process projecting from a basidium and bearing a spore (basidiospore) in the basidiomycetes. Fig. 18-1.

sterol Organic compound of the steroid group, having a hydroxyl group at position 3 and a hydrocarbon side chain at position 17.

stigma (pl. stigmata; also called eyespot) Concentration of red or orange lipoid pigments forming an organelle in some motile unicellular or colonial algae.

stipe A stalk, especially as seen in the larger algae (kelps), Figs. 7-4 and 7-5; and in mushrooms, Fig. 18-7.

stolon Horizontal stem or runner characteristic of some flowering plants (especially strawberries); term also used to designate the horizontal vegetative hyphae in zygomycetes. Fig. 16-1.

storage fungi Fungi that grow on such stored foodstuffs as grains and hay.

stratified In lichens, a thallus in which the alga and fungus are in distinct layers; also heteromerous. Fig. 20-3.

stria (pl. striae) Row of puncta or small pores in a diatom frustule.

stroma (pl. stromata) Foundation supporting part of an organelle, cell, tissue or organ; also, a compact mass of fungal hyphae.

stromatolite Concentric or eccentric deposits of calcium carbonate forming geological formations attributed to Cyanobacteria.

subapical Located just beneath the apex.

sublittoral zone Area of the marine environment between the low tide mark and the depth to which light penetrates; also called the subtidal zone. Fig. 2-14.

subsporangial swelling Enlargement of the hypha located just beneath the sporangium on some sporangiophores. Fig. 16-2.

subtidal zone Area of the marine environment between the low tide mark and the depth to which light penetrates; in this text called the sublittoral zone.

succession Sequence of species changing with time as an area progresses from relatively unstable pioneer stage to more stable climax stage.

sugar fungus First group of saprophytic fungi to appear on dead organic matter; so called because they utilize simple sugars as an energy source.

sulcus Groove in the cell wall (theca) of dinoflagellates that extends posteriorly from the girdle and houses the posterior flagellum.

sulfated polysaccharides Long-chain sugars to which sulfate groups have been added; in the red algae these are water-soluble mucilaginous compounds of some commercial importance.

suspensor Specialized hyphal branches that suspend zygotes between two adjacent compatible hyphae in zygomycetes; a zygophore.

swarmer A motile spore; also called a zoospore or planospore.

symbiosis (adj. symbiotic) Association between two organisms living together; relationship may be parasitic, commensalistic, or mutualistic.

syngamy Union of gametes during sexual reproduction.

tannin granules See *fucosan granules*.

teliospore Thick-walled resting spore produced by rusts and smuts (basidiomycetes). Fig. 18-8.

telium Binucleate cell that gives rise to teliospores (Teliomycetes, Basidiomycotina).

temporary cyst Type of thick-walled cell produced by some dinoflagellates (pyrrhophytes) in response to environmental stress. Fig. 9-6.

terpenes (adj. terpenoid) Group of simple (nonsaponifiable) lipids that are multiples of a 5-carbon hydrocarbon, isoprene; many terpenes are found in plants.

terricolous Living on soil, earth.

tetrad Group of four; in bryophytes, refers to the groups of four cells produced by meiosis during sporogenesis.

tetraradiate Having four branches or arms originating from the center, as in spores from aquatic Hyphomycetes (Ingoldian fungi, Fig. 19-7).

tetrasporangium Sporangium in which four spores are produced, generally by meiosis. Fig. 11-5.

tetraspore Any of four cells produced by meiosis in a tetrasporangium of tetrasporophytes (Rhodophyta).

tetrasporine line Line, abandoned in this text, of green algal evolution characterized by uninucleate cells.

tetrasporophyte Second diploid generation following the carposporophyte in some rhodophytes. Fig. 11-5.

Thallophyta Phylum of plants in which the plant body (thallus) lacks true roots, stems, and leaves; term now discarded.

thallus (adj. thalloid) Body of a nonvascular plant, especially when it is horizontally flattened and lacks true roots, stems, or leaves.

theca (pl. thecae) Cellulose plates that make up the cell covering of dinoflagellates (Pyrrhophyta).

thermal stratification In bodies of water, the stratification of layers of water that are at different temperatures.

thermocline Depth(s) in a thermally stratified body of water at which the temperature changes dramatically over a very short distance.

thermophile (adj. thermophilic) Organism that is capable of living at high temperatures.

thermotolerant Capable of withstanding relatively high temperatures.

thermotropism (adj. thermotropic) Orientation movement or response to changes in temperature.

thigmotaxis Movements of an organism toward or away from an object that provides a mechanical (touch) stimulus.

thioctic acid Experimental antidote to poisoning caused by mushroom ingestion.

thylakoid Membranous sac or lamella within chloroplasts that is associated with chlorophyll *a*.

transcription Process whereby messenger RNA is synthesized from a DNA template.

transformers Organisms, often microorganisms, that decompose the organic matter in ecosystems into inorganic components that can then be utilized by the producers (e.g., green plants).

travertine Porous, brownish, limestone rock often used in decorative architecture.

trehalose Disaccharide composed of two glucose molecules with an α 1-1 linkage.

tremerogen Diffusable hormone isolated from *Tremella* (basidiomycete).

trichocyst Threadlike projectile organelle found in some dinoflagellates (pyrrhophytes) and chloromonads (xanthophytes).

trichogyne Hairlike or tubelike projection on the female reproductive structure of red algae and ascomycetes.

trichome Hairlike filament of cells usually surrounded by a gelatinous sheath, as in the Cyanobacteria.

trichothallic growth Growth form produced by an intercalary meristem at the base of a hairlike uniseriate filament, producing a filament of cells in one direction and a multicellular thallus in the other. Fig. 7-7.

trichothecene Terpenoid toxin produced by several deuteromycetes such as *Fusarium* and *Trichoderma*.

trickling filter In sewage treatment plants, a bed of stones covered with algae, fungi, and bacteria; water to be treated is trickled over this bed and organic matter metabolized by the organisms.

triphasic life cycle Type of life history in some red algae that consists of three separate generations.

trisporic acid (TSA) Morphogenic agent that induces formation of suspensors (zygophores) in zygomycetes. Fig. 16-3.

trophic level Feeding level; place in the food web where organisms feed; for example, green plants (producers) are on the first trophic level, herbivorous animals are on the second trophic level, carnivores on the third, and so on; trophic level is a designation of function, and one species could occupy more than one trophic level. Fig. 2-1.

trophocyst Basal swelling in sporangiophores of *Pilobolus* (zygomycete).

trumpet cell Conducting cell in some Laminariales (Phaeophyta).

tubercule Small projection or growth.

umbilicate In lichens, attached to the substrate by a central filament composed of hyphae.

uniaxial See *uniseriate*.

unilocular sporangium Sporangium having a single chamber, produced by meiosis and therefore bearing meiospores; in some brown algae (Phaeophyta). Figs. 7-3, 7-4.

uniseriate Composed of a single linear filament of cells.

unstratified In lichens, a thallus in which algae and fungi are distributed vertically throughout the thallus; not in definite layers; also homoiomerous. Fig. 20-3.

upwelling Physical process in oceans in which the deep waters are brought to the surface. Fig. 2-10.

urediniospore Binucleate spore produced by the rusts (Uredinales: Basidiomycotina).

uredinium Macroscopic rust-colored blisters or pustules containing urediniospores and produced by members of the Uredinales (Basidiomycetes) that are parasitic on cereal grasses.

valve One of two overlapping halves that compose a diatom frustule.

vaucheriaxanthin ester Xanthophyll accessory pigment found in the Eustigamatophyceae (Xanthophyta).

venter Enlarged basal section of an archegonium containing an egg cell.

vesicular-arbuscular mycorrhiza Mycorrhizal fungi that produce swollen, treelike extensions of the hyphae within the cells of the host. Fig. 13-2.

violaxanthin Accessory photosynthetic pigment; a xanthophyll found in Phaeophyta and Eustigmatophyceae (Xanthophyta).

volutin granules See *metachromatic granules.*

volvatoxin Type of protein cardiotoxin that effects human heart function; it has been isolated from the basidiomycete *Volvariella volvacae.*

volvocine line Line of colonial green algal evolution represented by members of the Volvocales. Fig. 4-4.

wasting disease Disease of eel grass (*Zostera marina*).

water bloom Large population of algae. See also *algal bloom.*

white pine blister rust Disease of white pine (*Pinus strobus*) and related five-needle pines caused by *Cronartium ribicola* (Basidiomycotina); alternate hosts of the fungus are currants and gooseberries.

white rot Condition in wood in which an invading fungus breaks down both the cellulose and the lignin associated with the cell walls. Compare *brown rot.*

white rust Disease of angiosperms caused by several mastigomycetes; also called "white blisters;" white, powdery spot formed by sporangiophores rupturing the epidermis.

woronin body Crystalline organelle found in cells of septate fungal filaments in ascomycetes and deuteromycetes.

xanthophyll Group of oxygenated derivatives of carotenes, colored orange to red, and having a tetraterpenoid structure; a class of carotenoid pigment.

xerophyte (adj. xerophytic) Plant adapted for growth under arid conditions.

yeast Unicellular ascomycetes (Endomycetales) that reproduce primarily by budding and that produce little or no mycelium. Figs. 17-2 and 17-5. Also used to describe unicellular forms of the Basidiomycotina and Deuteromycotina.

yellow rain Yellowish substance of unknown chemical composition that has been aerially sprayed on areas of southeast Asia, reportedly causing illness in humans.

zearalenone (F-2) Aromatic toxin produced by several species of *Fusarium* (Deuteromycotina) that produces estrogenic effects in mammals. Fig. 19-6.

zoochorellae Green algae (Chlorophyta) that live in symbiotic association inside a host animal such as the coelenterate hydra.

zoospore Motile spores.

zooxanthellae Yellow to brown algae which live symbiotically inside an animal. Often Pyrrhophyta or Chrysophyta.

zygomycosis Diseases of humans caused by zygomycetes such as *Mucor* and *Rhizopus.*

zygophore See **suspensor.**

zygospore Thick-walled resting spore resulting from conjugation of amoeboid gametes in some green algae (Zygnematales) and fungi (zygomycetes).

zygotic meiosis Meiotic process that occurs when a zygote divides by meiotic cell divisions, usually producing four haploid meiospores.

Abbott, I.A., and G.H. Hollenberg. 1976. Marine algae of California. Stanford University Press, Stanford, Calif. 827 pp.

Adebayo, A.A., R.F. Harris, and W.R. Gardner. 1971. Turgor pressure of fungal mycelia. Trans. Brit. Mycol. Soc. **57**:145-151.

Agrios, G.N. 1969. Plant pathology. Academic Press, Inc., New York. 629 pp.

Aharonowitz, Y., and G. Cohen. 1981. The microbiological production of pharmaceuticals. Sci. Am. **245**:(3):140-152.

Ahearn, D.G., and S.P. Meyers. 1976. Fungal degradation of oil in the marine environment. In Gareth Jones, E.B. (ed.): Recent advances in aquatic mycology, pp. 125-133. Elek Scientific Books, Ltd., London.

Ahmadjian, V. 1966. Artificial reestablishment of the lichen *Cladonia cristatella*. Science **151**:199-201.

Ahmadjian, V. 1967. The lichen symbiosis. Blaisdell Publishing Co., Waltham, Mass. 152 pp.

Ahmadjian, V. 1973. Methods of isolating and culturing lichen symbionts and thalli. In Ahmadjian, V., and M.E. Hale (eds): The lichens, pp. 653-659. Academic Press, Inc., New York.

Ahmadjian, V. 1982a. Algal/fungal symbioses. In Round, F.E. and R.L. Chapman (eds.), Progress in phycological research. Vol. I, pp. 179-233. Elsevier Biomedical Press BV, New York.

Ahmadjian, V. 1982b. The nature of lichens. Natural History **91**(3):30-36.

Ahmadjian, V., and M.E. Hale (eds.). 1973. The lichens. Academic Press, Inc., New York. 697 pp.

Ahmadjian, V., and H. Heikkilä. 1970. The culture and synthesis of *Endocarpon pusillum* and *Staurothele clopima*. Lichenol. **4**:259-267.

Ahmadjian, V., B. Jacobs, and L.A. Russell. 1978. Scanning electron microscope study of early lichen synthesis. Science **200**:1062-1064.

Ainsworth, G.C. 1952. Medical mycology: an introduction to its problems. Pitman Publishing, Ltd., London. 105 pp.

Ainsworth, G.C. 1973. Introduction and keys to higher taxa. In Ainsworth, G.C., F.K. Sparrow, and A.S. Sussman (eds.): The fungi: an advanced treatise. Vol. IVB, pp. 1-7. Academic Press, Inc., New York.

Ainsworth, G.C., and P.K.C. Austwick. 1973. Fungal diseases of animals. Ed. 2. Commonwealth Agricultural Bureaux, Slough, England. 216 pp.

Ainsworth, G.C., F.K. Sparrow, and A.S. Sussman (eds.). 1973a. The fungi: an advanced treatise, Vol. IVA: A taxonomic review with keys: Ascomycetes and Fungi Imperfecti. Academic Press, Inc., New York. 621 pp.

Ainsworth, G.C., F.K. Sparrow, and A.S. Sussman (eds.). 1973b. The fungi: an advanced treatise. Vol. IVB: A taxonomic review with keys: Basidiomycetes and lower fungi. Academic Press, Inc., New York. 504 pp.

Ainsworth, G.C., and A.S. Sussman (eds.). 1965. The fungi: an advanced treatise, Vol. I: The fungal cell. Academic Press, Inc., New York. 748 pp.

Ainsworth, G.C., and A.S. Sussman (eds.). 1966. The fungi: an advanced treatise. Vol. II: The fungal organism. Academic Press. Inc., New York. 805 pp.

Ainsworth, G.C., and A.S. Sussman (eds.). 1968. The fungi: an advanced treatise. Vol. III: The fungal population. Academic Press, Inc., New York. 738 pp.

Aist, J.R., and P.H. Williams. 1971. The cytology and kinetics of cabbage root hair penetration by *Plasmodiophora brassicae*. Can. J. Bot. **49**:2023-2034.

Ajello, L. 1976. The zygomycete *Saksanaea vasiformis* as a pathogen of humans with a critical review of the etiology of zygomycosis. Mycologia **68**:52-62.

Ajello, L. 1977. Taxonomy of dermatophytes: a review of their imperfect and perfect states. In Iwata, K. (ed.): Recent advances in medical and veterinary mycology, pp. 289-297. University Park Press, Baltimore.

Alderman, D.J., and E.B. Gareth Jones. 1971. Shell disease of oysters. Fish. Investig. (London) Ser. II, **26**(8):1-18.

Aldon, E.F. 1975. Endomycorrhizae enhance survival and growth of four wing saltbush on coal mine spills. U.S. Forestry Service Res. Note RM 294.

Aldrich, H., and J.W. Daniel (eds.). 1982. Cell biology of *Physarum* and *Didymium*. Vol. 1: Organisms, nucleus and cell cycle, 488 pp. Vol. 2: Differentiation, metabolism and methodology, 400 pp. The Cell Biology Series, Academic Press, Inc., New York.

Alexopoulos, C.J., and C.W. Mims. 1979. Introductory mycology. Ed. 3. John Wiley & Sons, Inc., New York. 632 pp.

Allanson, B.R. 1973. The fine structure of the periphyton of *Chara* sp. and *Potamogeton natans* from Wytham Pond, Oxford, and its significance to the macrophyte-periphyton metabolic model of R.G. Wetzel and H.L. Allen. Freshwat. Biol. **3**:535-541.

Allbutt, A.D., W.A. Ayer, H.J. Brodie, B.N. Johri, and H. Taube. 1971. Cyathin, a new antibiotic complex produced by *Cyathus helenae*. Can. J. Microbiol. **17**:1401-1407.

Allen, M.B. 1969. Structure, physiology and biochemistry of the Chryosphyceae. Annu. Rev. Microbiol. **23**:29-46.

Almer, B., W. Dickson, C. Ekström, and E. Hörnström. 1978. Sulfur pollution and the aquatic system. In Nriagu, J.O. (ed.): Sulfur in the environment, Part II. Ecological Impacts, pp. 271-311. John Wiley & Sons, Inc., New York.

Alpert, P. 1979. Desiccation of desert mosses following a summer rain storm. Bryol. **82**:65-71.

Altman, P.L., and D.S. Dittmer. 1972. Biology data book. Ed. 2, Vol. 1. Fed. Amer. Socs. for Exptl. Biol., Bethesda, Md. 633 pp.

Alvin, K.L. 1977. The observer's book of lichens. Frederick Warne, Ltd., London. 188 pp.

Amerine, M.A. 1964. Wine. Sci. Am. Aug. 1974. Reprint No. 190.

Anagnostakis, S.L. 1982. Biological control of chestnut blight. Science **215**:466-471.

Anderson, O.R. 1977. Fine structure of a marine ameba associated with a blue-green alga in the Sargasso Sea. J. Protozool. **24**:370-376.

Anderson, W.A.D., and J.M. Kissane (eds.). 1977. Pathology, Ed. 7. The C.V. Mosby Co., St. Louis. Vol. I, 1051 pp.; Vol. II, 1174 pp.

Antia, N.J., T. Bisalputra, J.Y. Cheng, and J.P. Kalley. 1975. Pigment and cytological evidence for reclassification of *Nannochloris oculata* and *Monallantus salina* in the Eustigmatophyceae. J. Phycol. **11**:339-343.

Arnold, C.A. 1947. Paleobotany. McGraw-Hill Book Co., New York, 433 pp.

Arnold, D.E. 1971. Ingestion, assimilation, survival and reproduction by *Daphnia pulex* fed seven species of blue-green algae. Limnol. Oceanogr. **16**:906-920.

Aronson, J.M. 1965. The cell wall. In Ainsworth, G.C., and A.S. Sussman (eds.): The fungi, Vol. 1: The fungal cell, pp. 49-76. Academic Press, Inc., New York.

Arora, D. 1979. Mushrooms demystified. Ten Speed Press, Berkeley, Calif. 669 pp.

Artman, J.D. 1972. Further tests in Virginia using chain saw–applied *Peniophora gigantea* in lob-lolly pine stump inoculation. Pl. Dis. Rep. **56**:958-960.

Atlas, R.M., and R. Bartha. 1981. Microbial ecology: fundamentals and applications. Addison-Wesley Publishing Co., Inc., Reading, Mass. 560 pp.

Baker, H.G. 1978. Plants and civilization. Ed. 3. Wadsworth, Inc., Belmont, Calif. 198 pp.

Baker, K.F., and R.J. Cook. 1974. Biological control of plant pathogens. W.H. Freeman & Co., San Francisco. 433 pp.

Baker, R.D. 1977. Fungal, actinomycetic and algal infections. In Anderson, W.A.D., and J.M. Kissane (eds.): Pathology. Ed. 7, Vol. I, pp. 497-521. The C.V. Mosby Co., St. Louis.

Banks, H.P. 1970. Evolution and plants of the past. Wadsworth, Inc., Belmont, Calif. 170 pp.

Barghoorn, E.S., and S.A. Tyler. 1965. Microorganisms from the Gunflint Chert. Science **147**:563-577.

Barksdale, A.W. 1969. Sexual hormones of *Achlya* and other fungi. Science **166**:831-837.

Bärlocher, F. 1982. On the ecology of Ingoldian fungi. BioScience **32**(7):581-586.

Bärlocher, F., and B. Kendrick. 1976. Hyphomycetes as intermediaries of energy flow in streams. In Gareth Jones, E.B. (ed.): Recent advances in aquatic mycology, pp. 435-446. Elek Scientific Books, Ltd., London.

Barnes, R.S.K., and K.H. Mann (eds.). 1980. Fundamentals of aquatic ecosystems. Blackwell Scientific Publications, Ltd., Oxford, England. 229 pp.

Barnett, H.L., and B.H. Hunter. 1972. Illustrated genera of imperfect fungi. Ed. 3. Burgess Publishing Co., Minneapolis. 241 pp.

Barron, G.L. 1977. The nematode-destroying fungi. Topics in Mycology No. 1. Canadian Biological Publishers, Ltd., Guelph, Ontario. 144 pp.

Bartell, P.F., M.N. Schwalb, and Z.C. Kaminski. 1974. Characteristics of microorganisms and other infectious agents. In Briody, B.A. (ed.), Microbiology and infectious disease, pp. 23-42. McGraw-Hill Book Co., New York.

Bartnicki-Garcia, S. 1968. Cell wall chemistry, morphogenesis, and taxonomy of fungi. Annu. Rev. Microbiol. **22**:87-108.

Bartnicki-Garcia, S. 1970. Cell wall composition and other biochemical markers in fungal phylogeny. In Harbone, J.B. (ed.), Phytochemical Phylogeny, pp. 81-103. Proceedings Phytochemical Society Symposium. 1969. Academic Press, Inc., New York.

Bartnicki-Garcia, S., and W.J. Nickerson. 1962. Isolation, composition and structure of cell walls of filamentous and yeast-like forms of *Mucor rouxii*. Acta Biochim. Biophys. **58**:102-119.

Barton, R. 1957. Germination of oospores of *Pythium mamillatum* in response to exudates from living seedlings. Nature **180**:613-614.

Batra, S.W., and L.R. Batra. 1967. The fungus garden of insects. Sci. Am. **217**(5):112-120.

Beadle, G.W. 1959. Genes and chemical reactions in *Neurospora*. Science **129**:1715-1719.

Bean, B. 1979. Chemotaxis in unicellular eukaryotes. In Haupt, W., and M.E. Feinleib (eds.): Physiology of movements, pp. 335-354. Encyclop. Plant Physiol. N.S. Vol. 7. Springer-Verlag, Berlin.

Behnke, H.D. 1975. Phloem tissue and sieve elements in algae, mosses and ferns. In Aronoff, S., J. Dainty, P.R. Gorham, L.M. Srivastava, and C.A. Swanson (eds.): Phloem transport, pp. 187-210. Plenum Press, New York.

Ben-Amotz, A. 1975. Adaptation of the unicellular alga *Dunaliella parva* to a saline environment. J. Phycol. **11**:50-54.

Benjamin, R.K. 1979. Zygomycetes and their spores. In Kendrick, W.B., (ed.): The whole fungus, pp. 573-616. National Museums of Canada, Ottawa.

Ben-Shaul, Y., H.T. Epstein, and J.A. Schiff. 1965. Studies of chloroplast development in *Euglena*. 10: The return of the chloroplast to the proplastid condition during dark adaptation. Can. J. Bot. **43**:129-136.

Ben-Shaul, Y., N. Paran, and M. Galun. 1969. The ultrastructure of the association between phycobiont and mycobiont in three ecotypes of the lichen *Caloplaca aurantia* var. *aurantia*. J. Microsc. **8**:415-422.

Benson-Evans, K. 1964. Physiology of the reproduction of Bryophytes. Bryol. **67**:431-445.

Berdy, J. 1974. Recent developments of antibiotic research and classification of antibiotics according to chemical structure. Adv. Appl. Microbiol. **18**:309-406.

Bergman, K., P.V. Burke, E. Cerdá-Olmedo, C.N. David, M. Delbrück, K.W. Foster, E.W. Goodell, M. Heisenberg, G. Meissner, M. Zolokar, D.S. Dennison, and W. Shropshire, Jr. 1969. *Phycomyces*. Bacteriol. Rev. **33**:99-157.

Berry, C.R., and D.H. Marx. 1976. Sewage sludge and *Pisolithus tinctorius* ectomycorrhizae: their effect on growth of pine seedlings. For. Sci. **22**:351-357.

Berry, E.A., and E.A. Bevan. 1972. A new species of double-stranded RNA from yeast. Nature New Biol. **239**:279-280.

Beschel, R.E. 1961. Dating rock surfaces by lichen growth and its application to glaciology and physiography (lichenometry). In Raasch, G.O. (ed.): Geology of the Arctic, pp. 1044-1062. University of Toronto Press, Toronto.

Bewley, J.D. 1979. Physiological aspects of desiccation tolerance. Annu. Rev. Plant Physiol. **30**:195-238.

Bewley, J.D., J. Derek, and T.A. Thorpe. 1974. On the metabolism of *Tortula ruralis* following desiccation and freezing: respiration and carbohydrate oxidation. Physiol. Plant Path. **32**:147-153.

Bicknell, W.J., and D.C. Walsh. 1975. The first "red tide" in recorded Massachusetts history: managing an acute and unexpected public health emergency. In LoCicero, V.R. (ed.): Proceedings of the First International Conference on Toxic Dinoflagellate Blooms, pp. 447-458. Massachusetts Science and Technology Foundation, Wakefield, Mass.

Biebl, R. 1962. Seaweeds. In Lewin, R.A. (ed.): Physiology and biochemistry of algae, pp. 799-815. Academic Press, Inc., New York.

Bjorkman, E. 1960. *Monotropa hypopitys* L.—an epiparasite on tree roots. Physiol. Plant **13**:308-327.

Boalch, G.T., and D.S. Harbour. 1977. Unusual diatom off the coast of southwest England and its effect on fishing. Nature **269**:687-688.

Bodenberg, E.T. 1954. Mosses, a new approach to the identification of common species. Burgess Publishing Co., Minneapolis. 264 pp.

Bold, H.C., and M.J. Wynne. 1978. Introduction to the algae: structure and reproduction. Prentice-Hall, Inc., Englewood Cliffs, N.J. 706 pp.

Bolton, E.M. 1960. Lichens for vegetable dyeing. Studie Books, London. 63 pp.

Boney, A.D. 1970. Scale-bearing phytoflagellates: an interim review. Oceanogr. Mar. Biol. Annu. Rev. **8**:251-305.

Bonifacio, A. 1960. Su di una alterazione causata da *Physarum cinereum* et tapeto erboso di un giardino. Riv. ortoflorofrutticolt. Ital. **44**:326-331.

Bonner, J.T. 1959. The cellular slime molds. Princeton University Press, Princeton, N.J. 150 pp.

Bonner, J.T. 1967. The cellular slime molds. Ed. 2. Princeton University Press, Princeton, N.J. 205 pp.

Bonner, J.T. 1969. Hormones in social amoebae and mammals. Sci. Am. **220**:78-91.

Bonner, J.T. 1971. Aggregation and differentiation in the cellular slime molds. Annu. Rev. Microbiol. **25**:75-92.

Bonner, J.T. 1983. Chemical signals of social amoeba. Sci. Am. **248**(1):114-120.

Bonner, J.T., W.W. Clark, Jr., C.L. Neely, and M. Slifkin. 1950. The orientation to light and the extremely sensitive orientation to temperature gradients in the slime mold *Dictyostelium discoideum*. J. Cell. Comp. Physiol. **36**:149-158.

Bonner, J.T., K.K. Kane, and R.H. Levey. 1956. Studies on the mechanics of growth in the common mushroom, *Agaricus campestris*. Mycologia **48**:13-19.

Booth, C. 1978. Do you believe in genera? Trans. Brit. Mycol. Soc. **71**:1-9.

Bouch, G.B. 1972. Architecture and assembly of mastigonemes. Adv. Cell. Molec. Biol. **2**:237-271.

Bower, F.O. 1908. The origin of a land flora. Macmillan & Co., London. 727 pp.

Bozarth, R.F., Y. Koltin, M.B. Weissman, R.L. Parker, R.E. Dalton, and R. Steinlauf. 1981. The molecular weight and packaging of dsRNAs in the mycovirus from *Ustilago maydis* killer strains. Virology **113**:492-501.

Bracker, C.E., J. Ruiz-Herrera, and S. Bartnicki-Garcia. 1976. Structure and transformation of chitin synthetase particles (chitosomes) during microfibril synthesis *in vitro*. PNAS (U.S.) **73**:4570-4574.

Bradt, P.T. 1974. The ecology of the benthic macroinvertebrate fauna of the Bushkill Creek, Northampton County, Pa. Doctoral dissertation. Lehigh University, Bethlehem, Pa. 182 pp.

Bradt, P.T., S. Bieling, and S. Tancin. 1978. A comparison of lichen species richness and abundance: South Mountain and Hawk Mountain, Pennsylvania. Proc. Pa. Acad. Sci. **52**:117-120.

Bradt, P.T., and G.E. Wieland III. 1978. The impact of stream reconstruction and a gabion installation on the biology and chemistry of a trout stream. PB 280-543. NTIS, Springfield, Va. 61 pp.

Branning, T.G. 1976. Giant kelp: its comeback against urchins, sewage. Smithsonian **7**(6):102-109.

Briody, B.A., and Z.C. Kaminski. 1974. Systemic infections involving the respiratory tract. In Briody, B.A., and R.E. Gillis (eds.): Microbiology and infectious disease, pp. 535-546. McGraw-Hill Book Co., New York.

Bristol-Roach, B.M. 1919. On the retention of vitality by algae from old stored soils. New Phytol. **18**:92-107.

Brodie, H.J. 1951. The splash-cup mechanism in plants. Can. J. Bot. **29**:224-234.

Brodie, H.J. 1975. The bird's nest fungi. University of Toronto Press, Toronto. 199 pp.

Brodie, H.J. 1978. Fungi—delight of curiosity. University of Toronto Press, Toronto. 331 pp.

Brodo, I.M. 1973. Substrate ecology. In Ahmadjian, V., and M.E. Hale (eds.): The lichens, pp. 401-441. Academic Press, Inc., New York.

Brongersma-Saunders, M. 1948. The importance of upwelling to vertebrate paleontology and oil geology. Kon. Ned. Ak. Wet., Verh. Afd. Nat. (Tweede Sectie) **45**(4):1-112.

Brook, A.J. 1965. Planktonic algae as indicators of lake types, with special reference to the Desmidaceae. Limnol. Oceanogr. **10**:403-411.

Brown, D.H., D.L. Hawksworth, and R.H. Bailey (eds.). 1976. Lichenology: progress and problems. Academic Press, Inc., New York. 561 pp.

Buchanan, R.E., and N.E. Gibbons (eds.). 1974. Bergey's Manual of Determinative Bacteriology, Ed. 8. The Williams & Wilkins Co., Baltimore. 1246 pp.

Buchauer, M.J. 1971. Effects of zinc and cadmium pollution on vegetation and soils. Doctoral dissertation. Rutgers University, The State University of New Jersey, New Brunswick, N.J. University Microfilms 72-9610. Ann Arbor, Mich. 328 pp.

Buetow, D.E. (ed.). 1968. The biology of *Euglena*, Vols. I and II. Academic Press, Inc., New York. Vol. I, 361 pp.; Vol. II, 417 pp.

Buller, A.H.R. 1922. Researches on fungi, Vol. 2. Longmans, Green & Co., Ltd., London. 492 pp.

Buller, A.H.R. 1934. Researches on fungi, Vol. 7. Toronto University Press, Toronto. 458 pp.

Bu'Lock, J.D. 1976. Hormones in fungi. In Smith, J.E., and D.R. Berry (eds.): The filamentous fungi, Vol. 2: Biosynthesis and metabolism, pp. 345-368. Edward Arnold, Ltd., London.

Bungay, H.R. 1981. Energy, the biomass options. John Wiley & Sons, Inc., New York. 347 pp.

Burkholder, P.R., A. Evans, I. McVeigh, and H. Thornton. 1945. Antibiotic activity of lichens. PNAS (U.S.) **30**:250-255.

Burnett, J.H. 1975. Mycogenetics: an introduction to the general genetics of fungi. John Wiley & Sons, Inc., New York. 375 pp.

Burnett, J.H. 1976. Fundamentals of mycology, Ed. 2. Edward Arnold, Ltd., London. 673 pp.

Burnett, J.H., and A.P.J. Trinci (eds.). 1979. Fungal walls and hyphal growth. Cambridge University Press, Cambridge, England. 418 pp.

Burns, R.L., and A.C. Mathieson. 1972. Ecological studies of economic red algae. 3: Growth and reproduction of natural and harvested populations of *Gigartina stellata* (Stackhouse) Batters in New Hampshire. J. Exp. Mar. Biol. Ecol. **9**:77-95.

Butcher, D.N., E.T. Sayadat, and D.S. Ingram. 1974. The role of indole glucosinolates in the club root disease of Cruciferae. Physiol. Plant Path. **4**:127-141.

Butcher, R.W. 1967. An introductory account of the smaller algae of British coastal waters. Part IV: Cryptophyceae. Ministry Agric., Fish and Food, Fish. Invest., Ser. IV. 54 pp.

Cain, R. F. 1972. Evolution of the fungi. Mycologia **64**(1):1-14.

Cairns, J., Jr., S.P. Almeida, and H. Fujii, 1982. Automated identification of diatoms. BioScience **32**(2):98-102.

Camp, R.R. 1977. Association of microbodies, woronin bodies and septa in intercellular hyphae of *Cymodothea trifolii*. Can. J. Bot. **55**:1856-1859.

Canby, P. 1982. Of men and morels. Audubon **84**:(3):88-93.

Cantelo, W.W., and J.P. San Antonio. 1982. Effect of mushroom mycelium growth on population development of *Lycoriella mali*, nematodes and mites in compost. Environ. Entomol. **11**(1):227-230.

Canter, H.M. and J.W.G. Lund. 1969. The parasitism of planktonic desmids by fungi. Öst. bot. Z. **116**:351-377.

Canter, H.M., and L.G. Willoughby. 1964. A parasitic *Blastocladiella* from Windermere plankton. J. Royal Microscop. Soc. **83**:365-372.

Cantino, E.C. 1955. Physiology and phylogeny in the water molds—a reevaluation. Q. Rev. Biol. **30**:138-149.

Cantino, E.C. 1966. Morphogenesis in fungi. In Ainsworth, G.C., and A.S. Sussman (eds.): The fungi, Vol. 2, pp. 283-337. Academic Press, Inc., New York.

Cantino, E.C., and E.A. Horenstein. 1956. The stimulatory effect of light upon growth and carbon dioxide fixation in *Blastocladiella*. I: The S.K.I. cycle. Mycologia **48**:777-799.

Cantino, E.C., and E.A. Horenstein, 1957. The stimulatory effect upon growth and carbon dioxide fixation in *Blastocladiella*. II: Mechanism at an organismal level of integration. Mycologia **49**:892-894.

Caporael, L.R. 1976. Ergotism: the satan loosed in Salem? Science **192**:21-26.

Carefoot, G.L., and E.R. Sprott. 1967. Famine on the wind: man's battle against plant disease. Rand McNally & Co., Chicago. 229 pp.

Carpenter, E.J., and C.C. Price. 1976. Marine *Oscillatoria* (*Trichodesmium*): explanation for aerobic nitrogen fixation without heterocysts. Science **191**:1278-80.

Carpenter, E.J., and C.C. Price. 1977. Nitrogen fixation, distribution and production of *Oscillatoria* (*Trichodesmium*) spp. in the western Sargasso and Caribbean Seas. Limnol. Oceanogr. **22**(1):60-72.

Carr, N., and B. Whitton. 1973. The biology of the blue-green algae. University of California Press, Berkeley. 676 pp.

Cartwright, K.St.G., and W.P.K. Findlay. 1958. Decay of timber and its prevention. Ed. 2. H.M.S.O., London. 332 pp.

Cassin, P.E. 1974. Isolation, growth and physiology of acidophilic chlamydomonads. J. Phycol. **10**:439-447.

Catalfomo, P., J.H. Block, G.H. Constantine, and P.W. Kirk, Jr. 1972-73. Choline sulfate (ester) in marine higher fungi. Mar. Chem. **1**:157-162.

Chang, S.T. 1980. Mushrooms as human food. BioScience **30**(6):399-401.

Chang, S.T., and W.A. Hayes (eds.). 1978. The biology and cultivation of edible mushrooms. Academic Press, Inc., New York. 819 pp.

Chapman, A.R.O. 1979. Biology of seaweeds: levels of organization. University Park Press, Baltimore. 134 pp.

Chapman, A.R.O., and E.M. Burrows. 1970. Experimental investigations into controlling effects of light conditions on the development and growth of *Desmarestia aculeata* (L.) Lamour. Phycologia **9**:103-108.

Chapman, A.R.O., and J.S. Craigie. 1977. Seasonal growth in *Laminaria longicruris*: relations with dissolved inorganic nutrients and internal reserves of nitrogen. Mar. Biol. **40**:197-205.

Chapman, D.V., J.D. Dodge, and S.I. Heaney. 1982. Cyst formation in the freshwater dinoflagellate *Ceratium hirudinella* (Dinophyceae). J. Phycol. **18**:121-129.

Chapman, V.J. 1970. Seaweeds and their uses. Ed. 2. Methuen, Ltd., London. 304 pp.

Chapman, V.J., and D.J. Chapman. 1980. Seaweeds and their uses. Ed. 3. Methuen, Ltd., London. 334 pp.

Chave, K.E. 1954. Aspects of biogeochemistry of magnesium. I: Calcareous marine organisms. J. Geol. **62**:266-283.

Cheng, T.H. 1969. Production of kelp—a major aspect of China's exploitation of the sea. Econ. Botany **23**:215-236.

Chernush, K. 1980. Seaweed: part of a blue revolution. Agenda **3**:(2):2-4.

Chiang, Y.-M. 1981. Cultivation of *Gracilaria* (Gigartinales, Rhodophycophyta). Taiwan Proc. Int. Seaweed Symp. **10**:569-574.

Chopra, R.M., and M.S. Rawat. 1977. Studies on the initiation of sexual phase in the moss *Leptobryum pyriforme*. Beitr. Biol. Pflanz. **53**:353-357.

Chrispeels, M.J., and D. Sadava. 1977. Plants, food and people. W.H. Freeman & Co., San Francisco. 278 pp.

Christensen, C.M. 1975. Molds, mushrooms and mycotoxins. University of Minnesota Press, Minneapolis. 254 pp.

Ciegler, A., and J.W. Bennett. 1980. Mycotoxins and mycotoxicoses. BioScience **30**(8):512-515.

Clark, G.S.C., and J.G. Duckett (eds.). 1979. Byrophyte systematics. Systematics Association Special Vol. No. 14. Academic Press, Inc., New York. 582 pp.

Clark, W. 1980. China's green manure revolution. Science '80 **1**(6):68-73.

Clarke, G.L. 1939. The utilization of solar energy by aquatic organisms. In Moulton, E.R. (ed.): Problems in lake biology, pp. 27-38. Pub. 10, Amer. Assoc. Adv. Sci., Washington, D.C.

Claus, R., H.O. Hoppen, and H. Karg. 1981. The secret of truffles: a steroidal pheromone? Experientia **37**:1178-1179.

Cleave, M., and E. Percival. 1973. Carbohydrates of the freshwater alga *Tribonema aequale*. II: Preliminary photosynthetic studies with ^{14}C. Br. Phycol. J. **8**:181-184.

Clem, D.J. 1975. Management of the paralytic shellfish poison problem in the United States. In LoCicero, V.R. (ed.): First International Conference on Toxic Dinoflagellate Blooms, pp. 459-471. Massachusetts Science and Technology Foundation, Wakefield, Mass.

Clemons, G.P., D.V. Pham, and J.P. Pinion. 1980. Insecticidal activity of *Gonyaulax* (Dinophyceae) cell powders and saxitoxin to the German cockroach. J. Phycol. 16(2):303-305.

Clendenning, K.A. 1964. Photosynthesis and growth in *Macrocystis pyrifera*. Proc. Intern. Seaweed Symp. 41:55-65.

Cloud, P.E. 1965. Significance of the Gunflint (Precambrian) microflora. Science 148:27-35.

Cloud, P., M. Moorman, and D. Pierce. 1975. Sporulation and ultrastructure in a late Proterozoic cyanophyte: some implications for taxonomy and plant phylogeny. Q. Rev. Biol. 50(2):131-150.

Cochrane, V.W. 1958. Physiology of fungi. John Wiley & Sons, Inc., New York. 524 pp.

Coffey, M.D. 1975. Obligate parasitism and the rust fungi. In Symposia of the Society for Experimental Biology XXIX: Symbiosis, pp. 297-323. Cambridge University Press, London.

Cohen, Y., B.B. Jørgensen, E. Padan, and M. Shilo. 1975. Sulphide-dependent anoxygenic photosynthesis in the cyanobacterium *Oscillatoria limnetica*. Nature 257:489-491.

Cole, G.A. 1983. Textbook of limnology. Ed. 3. The C.V. Mosby Co., St. Louis. 401 pp.

Cole, G.T. 1981. Architecture and chemistry of the cell walls of higher fungi. In Schlessinger, D. (ed.): Microbiology—1981, pp. 227-231. American Society for Microbiology, Washington, D.C.

Cole, G.T., and B. Kendrick (eds.). 1981. Biology of conidial fungi. Vols. 1 and 2. Academic Press, Inc., New York. Vol. 1., 486 pp.; Vol. 2., 660 pp.

Cole, G.T., and R.A. Samson. 1979. Patterns of development in conidial fungi. Pitman Publishing, Ltd., London. 190 pp.

Colman, J.S., and A. Stephenson. 1966. Aspects of the ecology of a "tideless" shore. In Barnes, H. (ed.): Some contemporary studies in marine science, pp. 163-170. George Allen and Unwin (Publishers) Ltd., London.

Conant, N.F., D.T. Smith, R.D. Baker, and J.L. Callaway. 1971. Manual of clinical mycology. Ed. 3. W.B. Saunders Co., Philadelphia. 755 pp.

Conard, H.S., and P.L. Redfearn. 1979. How to know the mosses and liverworts. Ed. 2. Wm. C. Brown Co., Dubuque, Iowa. 302 pp.

Connell, K.H. 1950. The population of Ireland, 1750-1845. Clarendon Press, Oxford, England. 293 pp.

Conway, K.E. 1976. *Cercospora rodmanii*, a new pathogen of water hyacinths with biological control potential. Can. J. Bot. 54:1079-1083.

Cook, A.H., and J.A. Elvidge. 1951. Fertilization in Fucaceae: investigations on the nature of the chemotactic substance produced by eggs of *Fucus serratus* and *F. vesiculosus*. Proc. R. Soc. B. 138:97-114.

Cook, L.L., and S.A. Whipple. 1982. The distribution of edaphic diatoms along environmental gradients of a Louisiana salt marsh. J. Phycol. 18:64-71.

Cooke, W.B. 1954. Fungi in polluted water and sewage. Sewage Ind. Wastes 26:539-549.

Cooke, W.B. 1959. An ecological life history of *Aureobasidium pullulans* (DeBary) Arnaud. Mycopathol. Mycol. Appl. 12:1-45.

Cooke, W.B. 1976. Fungi in sewage. In Gareth Jones, E.B. (ed.): Recent advances in aquatic mycology, pp. 389-434. Elek Scientific Books, Ltd., London.

Cooke, W.B. 1979. The ecology of fungi. CRC Press Inc., Boca Raton, Fla. 274 pp.

Cooke, W.B., and W.O. Pipes. 1970. The occurrence of fungi in activated sludge. Mycopathol. Mycol. Appl. 40:249-270.

Copeland, H.F. 1956. The classification of lower organisms. Pacific Books, Publishers, Palo Alto, Calif. 302 pp.

Corner, E.D.S., and A.G. Davies. 1971. Plankton as a factor in the nitrogen and phosphorous cycles in the sea. Adv. Mar. Biol. 9:101-204.

Couch, J.N. 1938. The genus *Septobasidium*. University of North Carolina Press, Chapel Hill. 480 pp.

Craigie, J.S. 1974. Storage products. In Stewart, W.D.P. (ed.): Algal physiology and biochemistry, pp. 206-235. University of California Press, Berkeley.

Craigie, J.S., J. McLachlan, R.G. Ackman, and C.S. Tocher. 1967. Photosynthesis in algae. III: Distribution of soluble carbohydrates and dimethyl-β-propiothetin in marine unicellular chlorophyceae and Prasinophyceae. Can. J. Bot. 45:1327-1334.

Crandall-Stotler, B. 1980. Morphogenetic designs and a theory of bryophyte origins and divergence. BioScience 30(9):580-585.

Crawford, R.L. 1981. Lignin biodegradation and transformation. John Wiley & Sons, Inc., New York. 154 pp.

Creutz, C., and B. Diehn. 1976. Motor response to polarized light and gravity sensing in *Euglena*. J. Protozool. 23:552-556.

Crum, H.A., and L.E. Anderson. 1981. Mosses of eastern North America. Vols. 1 and 2. Columbia University Press, New York. 1328 pp.

Culberson, C.F. 1969. Chemical and botanical guide to lichen products. University of North Carolina Press, Chapel Hill. 628 pp.

Curley, A., V.A. Sedlak, E.F. Girling, R.E. Hawk, W.F. Bathel, P.E. Pierce, and W.H. Likosky. 1971. Organic mercury identified as the cause of poisoning in humans and hogs. Science 172:65-67.

Curtis, E.J.C. 1969. Sewage fungus: its nature and effects. Water Res. 3:289-311.

Czygan, F. 1970. Blutregen und Blutschnee: Stickstoff-mangel-Zellen von *Haematococcus pluvialis* und *Chlamydomonas nivalis*. Arch. Mikrobiol. **74**:69-76.

Darden, W.H, Jr. 1966. Sexual reproduction in *Volvox aureus*. J. Protozool. **13**:239-255.

Darley, W.M. 1974. Silicification and calcification. In Stewart, W.D.P. (ed.): Algal physiology and biochemistry, pp. 655-675. University of California Press, Berkeley.

Darmon, M., and P. Brachet. 1978. Chemotaxis and differentiation during the aggregation of *Dictyostelium discoideum* amoebae. In Hazelbauer, G.L.: Taxis and behavior, pp. 101-139. John Wiley & Sons, Inc., New York.

Davies, R.R., and J.L. Wilkenson. 1967. Human protothecosis: supplementary studies. Ann. Trop. Med. Parasitol. **61**:112-115.

Davis, R.A. 1973. Principles of oceanography. Addison-Wesley Publishing Co., Inc., Reading, Mass. 434 pp.

Dawes, C.J. 1981. Marine botany. John Wiley & Sons, Inc., New York. 632 pp.

Dawes, C.J., J.M. Lawrence, D.P. Cheney, and A.C. Mathieson. 1974a. Ecological studies of Floridian *Eucheuma* (Rhodophyta, Gigartinales). III: Seasonal variation of carrageenan, total carbohydrates, protein and lipid. Bull. Mar. Sci. **24**:186-199.

Dawes, C.J., A.C. Mathieson, and D.P. Cheney. 1974b. Ecological studies of Floridian *Eucheuma* (Rhodophyta, Gigartinales). I: Seasonal growth and reproduction. Bull. Mar. Sci. **24**:235-273.

Dawson, E.Y. 1966. Marine botany. Holt, Rinehart & Winston, New York. 371 pp.

Dayton, P.K. 1975. Experimental studies of algal canopy interactions in a sea-otter dominated kelp community at Amchitka Island, Alaska. Fish. Bull. **73**:230-237.

Deacon, J.W. 1980. Introduction to modern mycology. Vol. 7: Basic microbiology. Halsted Press Div., John Wiley & Sons, Inc., New York. 197 pp.

Demain, A.L., and N.A. Solomon. 1981. Industrial Microbiology. Sci. Am. **245**(3):66-74.

Demoulin, V. 1974. The origin of ascomycetes and basidiomycetes: the case for a red algal ancestry. Bot. Rev. **40**:315-345.

Denis, A. 1961. Une intoxication collective par *Gyromitra esculenta* en Normandie. Bull. Soc. Mycol. France **77**:64-67.

Denison, W.C., and G.C. Carrol. 1966. The primitive Ascomycete: a new look at an old problem. Mycologia **58**:249-269.

Desikachary, T. 1959. Cyanophyta. Indian Council of Agricultural Research, New Delhi. 686 pp.

Dickinson, C.H. 1977. Plant pathology and plant pathogens. Blackwell Scientific Publications, Ltd., Oxford, England. 161 pp.

Diehn, B. 1969. Two perpendicularly oriented pigment systems involved in phototaxis of *Euglena*. Nature **221**:336.

Diehn, B., and B. Kint. 1970. The flavin nature of the photoreceptor molecule for phototaxis in *Euglena*. Physiol. Chem. Phys. **2**:483-488.

Digby, P.S.B. 1977a. Growth and calcification in the coralline algae, *Clathromorphum circumscriptum* and *Corallina officialis*, and the significance of pH in relation to precipitation. J. Mar. Biol. Assoc. (U.K.) **57**:1095-1109.

Digby, P.S.B. 1977b. Photosynthesis and respiration in the coralline algae *Clathromorphum circumscriptum* and *Corallina officinalis* and the metabolic basis of calcification. J. Mar. Biol. Assoc. (U.K.) **57**:1111-1124.

Dillard, G.E. 1966. Seasonal periodicity of *Batrachospermum macrosporum* Mont. and *Audouinella violacea* (Kuetz) Ham. in Turkey Creek, Moore County, North Carolina. J. Elisha Mitchell Sci. Soc. **82**:204-207.

Dillon, L.S. 1978. Evolution: concepts and consequences. Ed. 2. The C.V. Mosby Co., St. Louis. 288 pp.

Dilts, T.J.K., and M.C.F. Proctor. 1976. Seasonal variation in desiccation tolerance in some British bryophytes. J. Bryol. **9**:239-247.

Dixon, P.S. 1973. The biology of the Rhodophyta. Hafner Press, New York. 285 pp.

Dixon, P.S., and L.M. Irving. 1977. Seaweeds of the British Isles. Vol. 1:Rhodophyta. Part 1: Introduction, Nemaliales, Gigartinales. British Museum (Natural History), London. 252 pp.

Dodge, B.O. 1927. Nuclear phenomena associated with heterothallism and homothallism in the ascomycete *Neurospora*. J. Agric. Res. **35**:289-305.

Dodge, B.O. 1935. The mechanics of sexual reproduction of *Neurospora*. Mycologia **27**:418-438.

Dodge, J.D. 1971. A dinoflagellate with both a mesokaryotic and a eukaryotic nucleus. I: Fine structure of the nuclei. Protoplasma **73**:145-157.

Dodge, J.D. 1973. The fine structure of algal cells. Academic Press, Inc., London. 261 pp.

Dodge, J.D. 1979-80. The Phytoflagellates: fine structure and phylogeny. In Levandowsky, M., S. Hunter, and L. Provasoli (eds.): Biochemistry and Physiology of the Protozoa. Ed. 2, pp. 7-57. Academic Press, Inc., London.

Dove, W.F., and H.P. Rusch. 1980. Growth and differentiation in *Physarum polycephalum*. Princeton University Press, Princeton, N.J. 250 pp.

Drew, K.M. 1949. *Conchocelis*—phase in the life history of *Porphyra umbilicas* (L.) Kütz. Nature **164**:748.

Drew, K.M. 1956. Reproduction in the Bangiophycidae. Bot. Rev. **22**:553-611.

Drews, G. 1973. The fine structure and chemical composition of the cell envelopes. In Carr, N.G., and B.A. Whitton (eds.): The biology of the blue-green algae, pp. 99-116. Blackwell Scientific Publications, Ltd., Oxford, England.

Dring, M.J. 1974. Reproduction. In Stewart, W.D.P. (ed.): Algal physiology and biochemistry, pp. 814-837. University of California Press, Berkeley.

Dring, M.J. 1981. Chromatic adaptation of photosynthesis in benthic marine algae: an examination of its ecological significance using a theoretical model. Limnol. Oceanogr. **26**(2): 271-284.

Dring, M.J. 1983. The biology of marine plants. Edward Arnold, Baltimore. 208 pp.

Droop, M.R. 1974. Heterotrophy of carbon. In Stewart, W.D.P. (ed.): Algal physiology and biochemistry, pp. 530-559. University of California Press, Berkeley.

Drouet, F. 1968. Revision of the classification of the Oscillatoriaceae. Monogr. Acad. Nat. Sci. (Philadelphia) **16**:1-341.

Drouet, F. 1973. Revision of the Nostocaceae with cylindrical trichomes (formerly Scytonemataceae and Rivulariaceae). Hafner Press, New York. 292 pp.

Drouet, F. 1978. Revision of the Nostocaceae with constricted trichomes. J. Cramer, Vadu 2, Liechtenstein. 258 pp.

Drouet, F., and W. Daily. 1956. Revision of the coccoid Myxophyceae. Butler University Bot. Studies **10**:1-218.

Drum, R.W., and J.T. Hopkins. 1966. Diatom locomotion: an explanation. Protoplasma **62**:1-33.

Dubinsky, Z., T. Berner, and S. Aaronson. 1978. Potential of large-scale algal culture for biomass and lipid production in arid lands. In Scott, C.D. (ed.): Biotechnology in energy production and conservation, pp. 51-68. John Wiley & Sons, Inc., New York.

Duddington, C.L. 1957. The friendly fungi: a new approach to the eelworm problem. Faber & Faber, Ltd., London. 188 pp.

Duggins, D.O. 1981. Sea urchins and kelp: the effects of short-term changes in urchin diet. Limnol. Oceanogr. **26**(2):391-394.

Earle, S.A. 1980. Undersea world of a kelp forest. Nat. Geog. **158**(3):411-426.

Edington, L.V., and C.A. Peterson. 1977. Systemic fungicides: theory, uptake and translocation. In Siegel, M.R., and H.D. Sisler (eds.): Antifungal compounds. Vol. 2, pp. 51-89. Marcel Dekker, Inc., New York.

Edmundson, W.T. 1970. Phosphorous, nitrogen and algae in Lake Washington after diversion of sewage. Science **169**:690-691.

Edwards, P. 1969. Field and cultural studies on the seasonal periodicity of growth and reproduction of selected Texas benthic algae. Contrib. Mar. Sci. Univ. Texas **14**:59-114.

Edwards, R.Y., J. Soos, and R.W. Ritcey. 1960. Quantitative observations on epidendric lichens used as food by caribou. Ecology **41**:425-431.

Ellis, B.K., and J.A. Stanford. 1982. Comparative photoheterotrophy, chemoheterotrophy, and photolithotrophy in a eutrophic reservoir and an oligotrophic lake. Limnol. Oceanogr. **27**(3):440-454.

Ellis, M.B. 1971. Dematiaceous Hyphomycetes. Commonwealth Mycological Institute, Kew, Surrey, England. 608 pp.

Ely, T.H., and W.H. Darden. 1972. Concentration and purification of the male-inducing substance from *Volvox aureus*. M5. Microbios. **5**:51-56.

Emerson, R. 1952. Molds and men. Sci. Am. Jan. 1952. pp 1-6.

Emerson, R. 1973. Mycological relevance in the nineteen-seventies. Trans. Brit. Mycol. Soc. **60**:363-387.

Emerson, R., and W. Held. 1969. *Aqualinderella fermentans* gen. et sp. n., a phycomycete adapted to stagnant waters. II: Isolation, cultural characteristics and gas relations. Am. J. Bot. **56**:1103-1120.

Emerson, R., and W.H. Weston. 1967. *Aqualinderella fermentans* gen. et sp. nov., a phycomycete adapted to stagnant waters. I: Morphology and occurrence in nature. Am. J. Bot. **54**:702-719.

Emmons, C.W., C.H. Binford, J.P. Utz, and K.J. Kwon-Chung. 1977. Medical mycology. Lea & Febiger, Philadelphia. 592 pp.

Ende, H., van den. 1976. Sexual interactions in plants: the role of specific substances in sexual reproduction. Academic Press, Inc., New York. 186 pp.

Engel, H., and J.C. Schneider. 1963. Die umwandlung von glykogen in zucker in den fruchtkörpern von *Sphaerobolus stellatus* (Thude) Pers., von ihrem abschluss. Ber. dt. bot. Ges. **75**:397-400.

Englemann, T.W. 1883. Farbe und assimilation. Bot. Ztg. **41**:1-29.

Epp, R.W., and W.M. Lewis, Jr. 1981. Photosynthesis in copepods. Science **214**:1349-1350.

Erdos, G.W., K.B. Raper, and L.K. Vogen. 1975. Sexuality in the cellular slime mold *Dictyostelium giganteum*. PNAS (U.S.) **72**:970-973.

Erwin, D.C., S. Bartnicki-Garcia, and P.H. Taso (eds.). 1983. *Phytophthora*: its biology, taxonomy, ecology, and pathology. Am. Phytopath. Soc., St. Paul, Minn. 392 pp.

Ettl, H. 1978. Xanthophyceae. In Gerloff, H.J., and H. Heynig (eds.): Susswasserflora von Mitteleuropa, Bd. 3.1. Tel. Gustav. Fischer, Stuttgart, Germany.

Eveleigh, D.E. 1981. The microbiological production of industrial chemicals. Sci. Am. **245**(3):155-178.

Eveleigh, D.E., and B.S. Montenecourt. 1977. Review of fungal cellulases. ERDA, Fuels from Biomass Newsletter. Rensselaer Polytechnic Institute, Troy, New York. 20 pp.

Falkowski, P.G. (ed.). 1980. Primary productivity in the sea. Vol. 19. Environmental Science Research Service. Plenum Publishing Corp., New York. 335 pp.

Fan, K.C. 1961. Studies of *Hypneocolax*, with a discussion of the origin of parasitic red algae. Nova Hedwigia **3**:119-128.

Fay, P., W.D.P. Stewart, A.E. Walsby, and G.E. Fogg. 1968. Is the heterocyst the site of nitrogen fixation in blue-green algae? Nature **220**:810-812.

Feldman, G. 1970. Sur l'ultrasture des corps irisants des *Chondria* (Rhodophycées). C.R. Acad. Sci. (Paris) **270**:1244-1246.

Fenical, W. 1975. Halogenation in the Rhodophyta—a review. J. Phycol. **11**:245-259.

Fergus, C.L. 1960. Illustrated genera of wood decay fungi. Burgess Publishing Co., Minneapolis. 132 pp.

Field, J.G., N.G. Jarman, G.S. Dieckmann, C.H. Griffiths, B. Velimirov, and P. Zoutendyk. 1977. Sun, waves, seaweed and lobsters: the dynamics of a west coast kelp-bed. S. Afr. J. Sci. **73**:7-10.

Field, J.I., and J. Webster. 1977. Traps of predacious fungi attract nematodes. Trans. Brit. Mycol. Soc. **68**:467-470.

Fiese, M.J. 1958. Coccidioidomycosis. Charles C Thomas, Publisher, Springfield, Ill. 253 pp.

Findlay, W.P.K. 1940. Studies in the physiology of wood-destroying fungi. III: Progress of decay under natural and controlled conditions. Ann. Bot. **4**:701-712.

Flegler, S.L., G.R. Hooper, and W.G. Fields. 1976. Ultrastructural and cytochemical changes in the basidiomycetes dolipore septum associated with fruiting. Can. J. Bot. **54**:2243-2253.

Fletcher, R.L. and S.M. Fletcher. 1975. Studies on the recently introduced brown alga *Sargassum muticum* (Yendo) Fensholt. I: Ecology and reproduction. Bot. Mar. **18**:149-156.

Fogg, G.E. 1974. Nitrogen fixation. In Stewart, W.D.P. (ed.), Algal physiology and biochemistry, pp. 560-582. University of California Press, Berkeley.

Fogg, G.E. 1980. Phytoplanktonic primary production. In Barnes, R.S.K., and K.H. Mann (eds.): Fundamentals of aquatic ecosystems, pp. 24-45. Blackwell Scientific Publications, Ltd., Oxford, England.

Fogg, G.E., W.D.P. Stewart, P. Fay, and A.E. Walsby. 1973. The blue-green algae. Academic Press, Inc., New York. 459 pp.

Forgacs, J. 1972. Stachobotryotoxicosis. In Kadis, S., A. Ciegler, and S.J. Ajl (eds.): Microbial toxins. Vol. VIII: Fungal toxins, pp. 95-128. Academic Press, Inc., New York.

Forsberg, C. 1965. Nutritional studies of *Chara* in axenic cultures. Physiol. Plant. **18**:275-290.

Fott, B. 1952. Mikroflora oravských rašelin. Preslia **24**:189-209.

Fott, B. 1964. Hologamic and agamic cyst formation in loricate chrysomonads. Phykos **3**:15-18.

Frankland, J.C., J.N. Hedger, and M.J. Swift (eds.). 1982. Decomposer Basidiomycetes: their biology and ecology. Brit. Mycol. Soc. Symp. 4. Cambridge University Press, New York. 355 pp.

Frantz, T.C., and A.J. Cordone. 1967. Observations on deepwater plants in Lake Tahoe, California and Nevada. Ecology **48**:709-714.

Frederick, J.F. 1979. Storage glucan biosynthesis in *Cyanidum, Chlorella* and *Prototheca:* evidence for symbiosis. Phytochem. **18**:1823-1825.

Fredrick, J.F. 1980. The b., e., and Q types of 1,4-α-D-glucan: 1, 4-α-D glucan-6-glucosyl transferase isozymes in algae. Phytochem. **19**:539-542.

French, R.A. 1970. The Irish moss industry. Canadian Atlantic. Current appraisal. Ottawa Department of Fisheries and Forestry, Canada. 230 pp.

Fritsch, F.E. 1945. The structure and reproduction of the algae. Vol. II. Cambridge University Press, Cambridge, England. 939 pp.

Fry, E.J. 1926. The mechanical action of corticolous lichens. Ann. Bot. **40**:397-417.

Fryxell, G.A. 1983. New evolutionary patterns in diatoms. BioScience **3**(2):92-98.

Fuller, J.G. 1968. The day of St. Anthony's Fire. Macmillan, Inc., New York, 310 pp.

Fuller, M.S. 1962. Growth and development of the water mold *Rhizidiomyces* in pure culture. Am. J. Bot. **49**:64-71.

Fuller, M.S. 1976. Mitosis in fungi. Int. Rev. Cytol. **45**:113-153.

Fuller, M.S., and I. Barshad. 1960. Chitin and cellulose in the cell walls of *Rhizidiomyces* sp. Am. J. Bot. **47**:105-109.

Gale, W.F., A.J. Gurzynski, and R.L. Lowe. 1979. Colonization and standing crops of epilithic algae in the Susquehanna River, Pennsylvania. J. Phycol. **15**:117-123.

Galindo, J., and M.E. Gallegly. 1960. The nature of sexuality in *Phytophthora infestans.* Phytopathol. **50**:123-128.

Gantt, E. 1980a. Structure and function of phycobilisomes: light harvesting pigment complexes in red and blue-green algae. In Bourne, G.H., and J.F. Danielli (eds.): Int. Rev. Cytol. 66, pp. 45-80. Academic Press, Inc., New York.

Gantt, E. 1980b. Photosynthetic cryptophytes. In Cox, E.R. (ed.): Phytoflagellates, pp. 381-405. Elsevier North-Holland, New York.

Gantt, E., and S.F. Conti. 1965. Ultrastructure of *Porphridium cruentum.* J. Cell Biol. **26**:365-381.

Gareth Jones, E.B. 1976. Lignicolous and algicolous fungi. In Gareth Jones, E.B. (ed.): Recent advances in aquatic mycology, pp. 1-49. Elek Scientific Books, Ltd., London.

Garrels, R.M., and F.T. Mackenzie. 1971. Evolution of sedimentary rock. W.W. Norton & Co., Inc., New York. 397 pp.

Garrett, S.D. 1958. Inoculum potential as a factor limiting lethal action by *Trichoderma viride* Fr. on *Armillaria mellea* (Fr.) Quél. Trans. Br. Mycol. Soc. **41**:157-164.

Gates, D.M. 1971. The flow of energy in the biosphere. Sci. Am. **225**(3):88-100.

Gäumann, E.A. 1952. The fungi (translated by F.L. Wynd). Hafner Publishing Co., New York. 420 pp.

Gäumann, E.A. 1964. Die Pilze. Birkhäuser Verlag, Basel, Switzerland. 541 pp.

Geesink, R. 1973. Experimental investigations on marine and freshwater *Bangia* (Rhodophyta) from the Netherlands. J. Exp. Mar. Biol. Ecol. **11**:239-247.

Geitler, L. 1963. Über haustorien bei flechten and über *Myrmecia biatorellae* in *Psora globifera*. Österr. Bot. 2. **110**:270-280.

Gerisch, G., H. Beug, D. Malchow, H. Schwarz, and A.V. Stein. 1974. Receptors for intercellular signals in aggregating cells of the slime mold *Dictyostelium discoideum*. In Lee, E.Y.C., and E.E. Smith (eds.): Biology and chemistry of eukaryotic cell surfaces, pp. 49-64. Academic Press, Inc., New York.

Gerwick, W.H., and N.J. Lang. 1977. Structural, chemical and ecological studies on iridescence in *Iridaea* (Rhodophyta). J. Phycol. **13**:121-127.

Gessner, R.V., and J. Kohlmeyer. 1976. Geographical distribution and taxonomy of fungi from salt marsh *Spartina*. Can. J. Bot. **54**:2023-2037.

Gibbs, S.P. 1978. The chloroplasts of *Euglena* may have evolved from symbiotic green algae. Can. J. Bot. **56**:2883-2889.

Gibor, A. 1966. *Acetabularia*: a useful giant cell. Sci. Am. **215**(5):118-124.

Gilbert, O.L. 1970. Further studies on the effect of sulfur dioxide on lichens and bryophytes. New Phytol. **69**:605-627.

Gilbert, O.L. 1971. The effect of airborne fluorides on lichens. Lichenol. **5**:26-32.

Gilbert, O.L. 1973. Lichens and air pollution. In Ahmadjian, V., and M.E. Hale (eds.): The lichens, pp. 443-472. Academic Press, Inc., New York.

Gilbertson, R.L. 1980. Wood-rotting fungi of North America. Mycologia **72**(1):1-49.

Gleason, F. 1976. The physiology of the lower freshwater fungi. In Gareth Jones, E.B. (ed.): Recent advances in aquatic mycology, pp. 543-572. Elek Scientific Books, Ltd., London.

Gleason, F.K., and J.M. Wood. 1977. Ribonucleotide reductase in blue-green algae: dependence on adenosylcobalamine. Science **192**:1343-1344.

Goering, J.J., R. Dugdale, and D.W. Menzel. 1966. Estimates of *in situ* rates of nitrogen uptake by *Trichodesmium* sp. in the tropical Atlantic Ocean. Limnol. Oceanogr. **11**:614-620.

Gojdics, M. 1953. The genus *Euglena*. University of Wisconsin Press, Madison. 268 pp.

Goldblatt, L. (ed.). 1969. Aflatoxin: scientific background, control and implication. Academic Press, Inc., New York. 472 pp.

Gooday, G.W. 1974. Control of development of excised fruit bodies and stipes of *Coprinus cinereus*. Trans. Brit. Mycol. Soc. **62**:391-399.

Gooday, G.W., P. Fawcett, D. Green, and G. Shaw. 1973. The formation of fungal sporopollenin in the zygospore wall of *Mucor mucedo*: a role for the sexual carotenogenesis in the Mucorales. J. Gen. Microbiol. **74**:233-239.

Goodenough, U.W. 1980. Sexual microbiology: mating reactions of *Chlamydomonas reinhardii*, *Tetrahymena thermophila*, and *Saccharomyces cerevisiae*. In Gooday, G.W., D. Lloyd, and A.P.J. Trinci (eds.): The eukaryotic microbial cell, pp. 301-328. Cambridge University Press, Cambridge, England.

Goodwin, T.W. 1974. Carotenoids and biliproteins. In Stewart, W.D.P. (ed.): Algal physiology and biochemistry, pp. 176-205. University of California Press, Berkeley.

Gordon, H.R., D.K. Clark, Jr., J.L. Mueller, and W.A. Hovis. 1980. Phytoplankton pigments from the Nimbus-7 Coastal Zone Color Scanner: comparison with surface measurements. Science **210**:63-66.

Goreau, T.F., N.I. Goreau, and T.J. Goreau. 1979. Corals and coral reefs. Sci. Am. **241**(2):124-136.

Govindjee, R., and B.Z. Braun. 1974. Light absorption, emission and photosynthesis. In Stewart, W.D.P. (ed.) Algal physiology and biochemistry, pp. 346-390. Univeı sity of California Press, Berkeley.

Grahn, O. 1976. Macrophyte succession in Swedish lake caused by deposition of airborne acid substances. In Proceed. Intern. Symp. on Acid Precipitation and the Forest Ecosystem, pp. 519-530. Columbus, Ohio, 1975. NTIS, PB 258-645. Springfield, Va.

Grahn, O., H. Hultberg, and L. Landner. 1974. Oligotrophication—a self-accelerating process in lakes subjected to excessive supply of acid substances. Ambio **3**:93-94.

Gramblast, L.J. 1974. Phylogeny of the Charophyta. Taxon **23**:463-481.

Gray, W.D. 1959. The relation of fungi to human affairs. Henry Holt & Co., Inc, New York. 492 pp.

Gray, W.D. 1966. The relation of fungi to human affairs. Ed. 2. Henry Holt & Co., Inc., New York. 510 pp.

Gray, W.D. 1970. The use of fungi as food and in food processing. CRM Press, Cleveland. 113 pp.

Green, J.C., and M. Parke. 1975. New observations upon members of the genus *Chrysotila* Anand. with remarks upon their relationships with the Haptophyceae. J. Mar. Biol. Assn. (U.K.) **55**:109-121.

Green, P.B. 1964. Cinematic observations on growth and division of chloroplasts in *Nitella*. Am. J. Bot. **51**:334-342.

Green, P.B., R.O. Erickson, and P.A. Richmond, 1970. On the physical basis of cell morphogenesis. Ann. NY Acad. Sci. **175**:712-731.

Gressitt, J.L., J. Sedlacek, and J.J.H. Szent-Ivany. 1965. Flora and fauna on backs of large Papuan moss-forest weevils. Science **150**:1833-1835.

Griffin, D.H. 1981. Fungal physiology. John Wiley & Sons, Inc., New York. 383 pp.

Grout, A.J. 1924. Mosses with a hand-lens. Ed. 3. Published by the author. New York. 340 pp.

Grove, S.N., and C.E. Bracker. 1970. Protoplasmic organization of hyphal tips among fungi: vescicles and Spitzenkörper. J. Bacteriol. **104**:989-1009.

Gruen, H.E. 1963. Endogenous growth regulations in car-pophores of *Agaricus bisporus*. Plant Physiol. **38**:652-666.

Gruen, H.E. 1969. Growth and rotation of *Flammulina velutipes* and the dependence of stipe elongation on the cap. Mycologia **61**:149-166.

Guenther, E. 1952. Essential oils of the plant family Usneaceae: concrete and absolute of oakmoss. In Guenther, E. (ed.): The essential oils. Vol. VI, pp. 179-191. Van Nostrand Co., New York.

Guillard, R.R.L., and P. Kilham. 1977. The ecology of marine planktonic diatoms. In Werner, D. (ed.): The biology of diatoms, pp. 372-469. University of California Press, Berkeley.

Guillard, R., and C. Lorenzen. 1972. Yellow-green algae with chlorophyllide *c*. J. Phycol. **8**:10-14.

Gull, K. 1978. Form and function in septa in filamentous fungi. In Smith, J.E., and D.R. Berry (eds.): The filamentous fungi. Vol. 3: Developmental mycology, pp. 78-93. John Wiley & Sons, Inc., New York.

Habas, E.J., and C. Gilbert. 1975. A preliminary investigation of the economic effects of the red tide of 1973-74 on the west coast of Florida. In LoCicero, V.R. (ed.): Proceedings, First International Conference on Toxic Dino-flagellate Blooms, pp. 499-506. Massachusetts Science and Technology Foundation, Wakefield, Mass.

Hacskaylo, E. 1972. Mycorrhiza: the ultimate in reciprocal parasitism? BioScience **22**(10):577-583.

Hale, M.E. 1955. Phytosociology of corticolous cryptogams in the upland forests of southern Wisconsin. Ecology **36**:45-63.

Hale, M.E. 1965. A monograph of *Parmelia* subgenus *Amphigymnia*. Contr. U.S. Nat. Herb. **36**:193-358.

Hale, M.E. 1973. Growth. In Ahmadjian, V., and M.E. Hale (eds.): The lichens, pp. 473-492. Academic Press, Inc., New York.

Hale, M.E. 1974. The biology of lichens. Ed. 2. Edward Arnold, Ltd., London. 181 pp.

Hale, M.E. 1979. How to know the lichens. Ed. 2. Wm. C. Brown Co., Publishers, Dubuque, Iowa. 246 pp.

Hale, M.E., and W.L. Culberson. 1970. A fourth checklist of the lichens of the continental United States and Canada. Bryol. **73**:499-543.

Halldal, P. 1970. The photosynthetic apparatus of micro-algae and its adaptation to environmental factors. In Halldal, P. (ed.): Photobiology of microorganisms, pp. 17-55. John Wiley-Interscience, London.

Hamada, M. 1940. Physiologisch—morphologische Studien uber *Armillaria mellea* (Vahl.) Quel., mit besonderer Rucksicht auf die Oxalsaurebildung Ein Nachtrag zür Mykorrhiza von *Galeola septentrionalis* Reichb. F. Jap. Bot. **10**:387-463.

Hardy, A.C. 1965. The open sea: its natural history. Part I: The world of plankton. Houghton Mifflin Co., Boston. 335 pp.

Harley, J.L. 1969. Biology of mycorrhiza. Ed. 2. Leonard Hill, London. 334 pp.

Harley, J.L. 1975. Problems of mycotrophy. In Sanders, F.E., B. Mosse, and P.B. Tinker (eds.): Endomycor-rhizas, pp. 1-24. Academic Press, Inc., New York.

Harper, J.L., and J. Webster. 1964. An experimental analysis of the coprophilous fungus succession. Trans. Brit. Mycol. Soc. **47**:511-530.

Harper, M.A. 1969. Movement and migration of diatoms on sand grains. Br. Phycol. J. **4**:97-103.

Harper, M.A. 1977. Movements In Werner, D. (ed.): The biology of diatoms, pp. 224-249. University of California Press, Berkeley.

Harris, T.M. 1939. British Purbeck Charophyta. British Museum (Natural History) IX, London. 83 pp.

Harrison, J.L. 1972. The salinity tolerance of freshwater and marine zoosporic fungi, including some aspects of the ecology and ultrastructure of the Thraustochytriaceae. Doctoral thesis. University of London.

Harrison, W.G. 1976. Nitrate metabolism of the red tide dinoflagellate *Gonyaulax polyedra* Stein. J. Exp. Mar. Biol. Ecol. **21**:199-209.

Harvey, E.N. 1952. Bioluminescence. Academic Press, Inc., New York. 649 pp.

Haselkorn, R. 1978. Heterocysts. Annu. Rev. Plant Physiol. **29**:319-344.

Haupt, W., and E. Schönbohm. 1970. Light-oriented chloroplast movements. In Halldahl, P. (ed.): Photobiology of microorganisms, pp. 283-307. John Wiley-Interscience, London.

Hawkes, H.A. 1965. Factors influencing the seasonal incidence of fungal growths in sewage bacteria beds. Int. J. Air Water Pollut. **9**:693-714.

Hawkes, M.W. 1978. Sexual reproduction in *Porphyra gardneri* (Smith et Hollenberg) Hawkes (Bangiales, Rhodophyta). Phycologia **17**:329-353.

Hawksworth, D.L., and F. Rose. 1970. Qualitative scale for estimating sulfur dioxide air pollution in England and Wales using epiphytic lichens. Nature **227**:145-148.

Hawksworth, D.L., and F. Rose. 1976. Lichens as pollution monitors. Edward Arnold, Ltd., London. 60 pp.

Hayes, W.A. 1972. Nutritional factors in relation to mush-room production. Mushroom Science **8**:663-674.

Hayman, D.S. 1980. Mycorrhiza and crop production. Nature **287**:487-488.

Haynes, J.D. 1975. Botany: an introductory survey of the plant kingdom. John Wiley & Sons, Inc., New York. 562 pp.

Heath, I.B. (ed.). 1978. Nuclear division in the fungi. Academic Press, Inc., New York. 235 pp.

Heath, I.B. 1981. Nucleus-associated organelles in fungi. In Bourne, G.H., and J.F. Danielli (eds.): Int. Rev. Cytol. **69**:191-221. Academic Press, Inc., New York.

Heathcote, J.G., and J.R. Hibbert. 1978. Aflatoxins: chemical and biological aspects. Elsevier North-Holland, Amsterdam. 212 pp.

Hébant, C. 1979. Conducting tissues in bryophyte systematics. In Clark, G.S.C., and J.G. Duckett (eds.): Bryophyte systematics, pp. 365-383. Systematics Assoc. Spec. Vol. No. 14. Academic Press, Inc. New York.

Heelis, D.W., W. Kernick, G.O. Phillips, and K. Davies. 1979. Separation and identification of carotenoid pigments of stigmata isolated from light-brown cells of *Euglena gracilis* strain Z. Arch. Microbiol. **121**:207-211.

Heitefuss, R., and P.H. Williams (eds.). 1976. Physiological plant pathology. Springer-Verlag, Inc., Berlin. 890 pp.

Held, A.A., R. Emerson, M.S. Fuller, and F.H. Gleason. 1969. *Blastocladia* and *Aqualinderella:* fermentative water molds with high carbon dioxide optima. Science **165**:706-709.

Hellebust, J. 1965. Excretion of some organic compounds by marine phytoplankton. Limnol. Oceanogr. **10**:192-206.

Hellebust, J.A. 1974. Extracellular products. In Stewart, W.D.P. (ed.): Algal physiology and biochemistry, pp. 838-863. University of California Press, Berkeley.

Hendry, G.R., and F.A. Vertucci. 1980. Benthic plant communities in acidic Lake Colden, New York: *Sphagnum* and the algal mat. In Drabløs, D., and A. Tollan (eds.): Ecological impact of acid precipitation, pp. 314-315. SNSF Project, Ås-WHL, Norway.

Hendrey, G.R., and R.F. Wright. 1976. Acid precipitation in Norway: effects on aquatic fauna. J. Great Lakes Pres. **2**(Suppl. 1):192-207.

Hendrickson, J.R., and W.A. Weber. 1964. Lichens on Galapagos giant tortoises. Science **144**:1463.

Henssen, A., and A.H. Jahns. 1974. Lichens Eine Einfuhrung in die Flechtenkunde. George Thieme Verlag, Stuttgart, Germany. 467 pp.

Hesseltine, C.W., C.R. Benjamin and B.S. Mehrotra. 1959. The genus *Zygorhynchus*. Mycologia **51**:173-174.

Hesseltine, C.W., and J.J. Ellis. 1973. Mucorales. In Ainsworth, G.C., F.K. Sparrow, and A.F. Sussman (eds.): The fungi, Vol. IVB, pp. 187-217. Academic Press, Inc., New York.

Hesseltine, C.W., and H.L. Wang. 1980. The importance of traditional fermented foods. BioScience **36**(6):402-404.

Heywood, P. 1980. Chloromonads. In Cox, E.R. (ed.): Phytoflagellates, pp. 351-379. Elsevier North-Holland, New York.

Hibberd, D.J. 1976. The ultrastructure and taxonomy of the Chrysophyceae and Prymnesiophyceae (Haptophyceae): a survey with some new observations on the ultrastructure of the Chrysophyceae. J. Linn. Soc. (Bot.) **72**:55-80.

Hibberd, D.J. 1980a. Prymnesiophytes (= Haptophytes). In Cox, E.R. (ed.): Phytoflagellates, pp. 273-317. Elsevier North-Holland, New York.

Hibberd, D.J. 1980b. Xanthophytes. In Cox, E.R. (ed.): Phytoflagellates, pp. 243-271. Elsevier North-Holland, New York.

Hibberd, D.J. 1980c. Eustigmatophytes. In Cox, E.R. (ed.): Phytoflagellates, pp. 319-334. Elsevier North-Holland, New York.

Hillson, C.J. 1977. Seaweeds. The Pennsylvania State University Press, University Park. 194. pp.

Hodkinson, M. 1976. Interactions between aquatic fungi and DDT. In Gareth Jones, E.B. (ed.): Recent advances in aquatic mycology, pp. 447-467. Elek Scientific Books, Ltd., London.

Hodkinson, M., and S.A. Dalton. 1973. Interactions between DDT and river fungi. II: Influence of culture conditions on the compatibility of fungi, and p.p′-DDT. Bull. Environ. Contam. Toxicol. **10**:356-359.

Hoffman, L.R. 1973. Fertilization in *Oedogonium*. I: Plasmogamy. J. Phycol. **9**:62-84.

Hoham, R.W. 1980. Unicellular chlorophytes—snow algae. In Cox, E.R. (ed.): Phytoflagellates, pp. 61-84. Elsevier North-Holland, New York.

Holliday, P. 1980. Fungus diseases of tropical crops. Cambridge University Press, New York. 607 pp.

Holmes, R.W. 1966. Short-term temperature and light conditions associated with auxospore formation in the marine centric diatom *Coscinodiscus concinnus* W. Smith. Nature **209**:217-218.

Holt, C. von, and M. von Holt. 1968. Transfer of photosynthetic products from zooxanthellae to coelenterate hosts. Comp. Biochem. Physiol. **24**:73-81.

Horgen, P.A. 1977. Steroid induction of differentiation: *Achlya* as a model system. In O'Day, D.H., and P.A. Horgen (eds.): Eucaryotic microbes as model developmental systems, pp. 272-293. Marcel Dekker, Inc., New York.

Horgen, P.A. 1981. The role of the steroid sex pheromone antheridiol in controlling the development of male sex organs in the water mold, *Achlya*. In O'Day, D.H., and P.H. Horgen: Sexual interactions in eukaryotic microbes, pp. 155-178. Academic Press, Inc., New York.

Horne, A.J., J.E. Dillard, D.K. Fujita, and C.R. Goldman. 1972. Nitrogen fixation in Clear Lake, California. II: Synoptic studies on the autumn *Anabaena* bloom. Limnol. Oceanogr. **17**(5):693-703.

Houten, J.G. Ten. 1969. Air pollution: Proceed. First Europ. Cong. on Influence of Air Pollution on Plants and Animals, Wageningen 1968. Centre for Agric. Publ. Docmt., Wageningen, Netherlands. 415 pp.

Hovasse, R., J.P. Mignot, and L. Joyon. 1967. Nouvelles observations sur les trichocysts des Cryptomonadanes et les "R bodies" des particules kappa de *Paramecium aurelia* Killer. Protistologica **3**:241-255.

Humm, J.H., and S.R. Wicks. 1980. Introduction and guide to the marine bluegreen algae. John Wiley & Sons, Inc., New York. 194 pp.

Huneck, S. 1973. Nature of lichen substances. In Ahmadjian, V., and M.E. Hale (eds.): The lichens, pp. 495-522. Academic Press, Inc., New York.

Hurst, J.W., Jr. 1975. History of paralytic shellfish poisoning on the Main coast 1958-1974. In LoCicero, V.R. (ed.): First International Conference on Toxic Dinoflagellate Blooms, pp. 525-528. Massachusetts Science and Technology Foundation, Wakefield, Mass.

Hutchinson, G.E 1967. A treatise on limnology. Vol. II: Introduction to lake biology and the limnoplankton. John Wiley & Sons, Inc., New York. 1115 pp.

Hutchinson, G.E. 1975. A treatise on limnology. Vol. III: Limnological botany. John Wiley & Sons, Inc., New York. 660 pp.

Hutchinson, G.E. 1981. Thoughts on aquatic insects. Bio-Science 31(7):495-500.

Hynes, H.B.N. 1963. The biology of polluted waters. Liverpool University Press, Liverpool, England. 202 pp.

Hynes, H.B.N. 1970. The ecology of running waters. University of Toronto Press, Toronto. 555 pp.

Idyll, C.P. 1973. The anchovy crisis. Sci. Am. 228(6):22-29.

Imahori, K. 1954. Ecology, phytogeography and taxonomy of the Japanese Charophyta. Kanazawa University, Japan. 234 pp.

Imahori, K. 1964. Iconograph of the Characeae. In Wood, R.D., and K. Imahori: A revision of the Characeae, Vol. II. Cramer, Weinheim, Germany. 401 pp.

Inaba, T., and C.J. Mirocha. 1979. Preferential binding of radiolabeled zearalenone to a protein fraction of Fusarium roseum Graminearum. Appl. Environ. Microbiol. 37:80-84.

Ingold, C.T. 1971. Fungal spores: their liberation and dispersal. Clarendon Press, Oxford, England. 302 pp.

Ingold, C.T. 1972. Sphaerobolus: the story of a fungus. Trans. Brit. Mycol. Soc. 58:179-195.

Ingold, C.T. 1976. The morphology and biology of freshwater fungi excluding Phycomycetes. In Gareth Jones, E.B. (ed.): Recent advances in aquatic micology, pp. 335-357. Elek Scientific Books, Ltd., London.

Iqbal, S.H., and J. Webster. 1973. Aquatic Hyphomycete spora of the river Exe and its tributaries. Trans. Brit. Mycol. Soc. 61:331-346.

Jahns, H.M. 1973. Anatomy, morphology and development. In Ahmadjian, V., and M.E. Hale (eds.): The lichens, pp. 3-58. Academic Press, Inc., New York.

Janszens, F.H.A., and J.G.H. Wessels. 1970. Enzymic dissolution of hyphal septa in a basidiomycete. Antonie van Leeuwenhoek 36:255-257.

Jarvis, B.B., J.O. Midiwo, D. Tuthill, and G.A. Bean. 1981. Interaction between the antibiotic trichothecene and the higher plant Baccharis megapotamica. Science 214:460-462.

Jarzens, D.M. 1979. Zygospores of Zygnemataceae in the Paleocene of southern Saskatchewan (Canada). Rev. Paleobot. Palynol. 28:21-25.

Jeffrey, S.W. 1976. The occurrence of chlorophyll c_1 and c_2 in algae. J. Phycol. 12:349-354.

Jerlov, N.G. 1951. Rep. Swed. Deep Sea Exped. 3. Physics and Chemistry. No. 1. pp. 1-59.

Jerlov, N.G. 1968. Optical oceanography. Elsevier, Amsterdam, Netherlands. 194 pp.

Jerlov, N. 1976. Marine Optics. Ed. 2. Elsevier, Amsterdam, Netherlands. 231 pp.

Jeschke, L. 1963. Die Wasser- und Sumpfvegetation im Naturschutzgebiet "Ostufer der Muntz." Limnologia 1:475-545.

Jewell, T.R. 1974. A qualitative study of cellulose distribution in Ceratocystis and Europhium. Mycologia 66:139-146.

Johansen, H.W. 1981. Coralline algae, a first synthesis. CRC Press, Inc. Boca Raton, Fla. 256 pp.

Johnson, H.W. 1976. The biological and economic importance of algae. IV: The industrial culturing of algae. Tuatara 22:1-114.

Johnson, T.W. 1966. Chytridomycetes and Oömycetes in marine phytoplankton. Nova Hedwigia 10:579-588.

Johnson, T.W. 1976. The Phycomycetes: morphology and taxonomy. In Gareth Jones, E.B., (ed.): Recent advances in aquatic mycology, pp. 193-211. Elek Scientific Books, Ltd., London.

Jones, O.A., and R. Endean. 1973. Biology and geology of coral reefs. Vol. II: Biology I. Academic Press, Inc., New York. 502 pp.

Jones, O.A., and R. Endean. 1976. Biology and geology of coral reefs. Vol. III: Biology II. Academic Press, Inc., New York. 457 pp.

Jørgensen, E.G. 1957. Diatom periodicity and silicon assimilation. Dansk Bot. Arkiv. 18(1):54 pp.

Kadis, S., A. Ciegler, and S.J. Ajl. 1971. Microbial toxins. Vol. 7. Algal and fungal toxins. Academic Press, Inc., New York.

Kadis, S., A. Ciegler, and S.J. Ajl. 1972. Microbial toxins. Vol. 8. Fungal toxins. Academic Press, Inc., New York.

Kamitsubo, E. 1972. Motile protoplasmic fibrils in cells of the Characeae. Protoplasma 74:53-70.

Kampf, W.D., G. Becker, and J. Kohlmeyer. 1959. Versuche uber das Auffinden und den Befall von Holz durch Larven der Bohrmuschel Teredo pedicellata. Qutrf. Z. Angew. Zool. 46:257-283.

Kappen, L. 1973. Response to extreme environments. In Ahmadjian, V., and M.E. Hale (eds.): The lichens, pp. 310-380. Academic Press, Inc., New York.

Kapraun, D.F. 1977. Asexual propagules in the life history of Polysiphonia ferulacea (Rhodophyta, Ceramiales). Phycologia 16(4):417-426.

Kärenlampi, L. 1971. Studies on the relative growth rate of some fruticose lichens. Rep. Kevo Subarctic Res. Sta. 7:33-39.

Karling, J.S. 1968. The Plasmodiophorales. Hafner Publishing Co., New York. 256 pp.

Kaushik, N.K., and H.B.N. Hynes. 1971. The fate of the dead leaves that fall into streams. Arch. Hydrobiol. **68**: 465-515.

Kelso, J.R.M., R.J. Love, J.H. Lipsit, and R. Dermott. 1982. Chemical and biological status of head water lakes in the Sault Ste. Marie district, Ontario. In Itri, F.M.D. (ed.): Acid precipitation: effects on ecological systems, pp. 165-207. Ann Arbor Science, Ann Arbor, Mich.

Kendrick, B. (ed.). 1971. Taxonomy of Fungi Imperfecti. University of Toronto Press, Toronto. 309 pp.

Kendrick, W.B. (ed.). 1979. The whole fungus: the sexual-asexual synthesis. 2 vols. National Museums of Canada, Ottawa. 793 pp.

Kendrick, W.B., and A. Burges. 1962. Biological aspects of the decay of *Pinus sylvestris* leaf litter. Nova Hedwigia **4**:313-344.

Kenney, D.S., K.E. Conway, and W.H. Ridings. 1979. Mycoherbicides—potential for commercialization. In Underkofler, L.A. (ed.): Developments in industrial microbiology. Vol. 20, pp. 123-130. Society of Industrial Microbiology, Arlington, Va.

Kessel, M. 1977. Identification of a phosphate-containing storage granule in the cyanobacterium *Plectonema boryanum* by electron microscope x-ray microanalysis. J. Bacteriol. **129**:1502-1505.

Kesseler, H. 1966. Beitrag zur kenntnis der chemischen und physikalischen Eigenschaften des Zellsaftes von *Noctiluca miliaris*. Veroff. Inst. Meeresf. Bremerhaven Sb II:357-368.

Keys, V.E. 1975. Management of Florida red tides regarding shellfish harvesting. In LoCicero, V.R. (ed.): First International Conference on Toxic Dinoflagellate Blooms, pp. 483-488. Massachusetts Science and Technology Foundation, Wakefield, Mass.

Khrushchev, N. 1970. Khrushchev remembers. Little, Brown & Co., Boston. 639 pp.

Kilham, P. 1971. A hypothesis concerning silica and the freshwater planktonic diatoms. Limnol. Oceanogr. **16**:10-18.

Kimmey, J.W. 1969. Inactivation of lethal-type blister rust cankers on western white pine. J. Forest. **67**:296-299.

Kirk, T.K. 1971. Effects of microorganisms on lignin. Annu. Rev. Phytopathol. **9**:185-210.

Kirk, T.K., W.J. Connors, and J.G. Zeikus. 1977. Advances in understanding the microbiological degradation of lignin. Recent Adv. Phytochem. **11**:369-394.

Kirk, T.K., T. Higuchi, and H.M. Chang (eds.). 1980. Lignin biodegradation: microbiology, chemistry and potential applications. Vols. I and II. CRC Press, Inc., Boca Raton, Fla. Vol. I, 256 pp.; Vol. II, 272 pp.

Kirk, T.K., H.H. Yang, and P. Keyser. 1978. The chemistry and physiology of the fungal degradation of lignin. Dev. Ind. Microbiol. **19**:51-61.

Kislev, M.E. 1982. Stem rust of wheat 3300 years old found in Israel. Science **216**:993-994.

Klein, R.M. 1970. Relationships between blue-green and red algae. Ann. NY Acad. Science **175**:623-633.

Klein, R.M. 1979. The green world: an introduction to plants and people. Harper & Row Publishers, Inc., New York. 437 pp.

Klein, R.M., and A. Cronquist. 1967. A consideration of the evolutionary and taxonomic significance of some biochemical, micromorphological, and physiological characters in the thallophytes. Q. Rev. Biol. **42**(2):105-296.

Klein, S., J.A. Schiff, and A.W. Holowinsky. 1972. Events surrounding the early development of *Euglena* chloroplasts. II: Normal development of fine structure and the consequences of preillumination. Dev. Biol. **28**:253-273.

Knights, B.A., A.C. Brown, E. Conway, and B.S. Middleditch. 1970. Hydrocarbons from the green form of the freshwater alga *Botryococcus* braunii. Phytochem. **9**: 1317-1324.

Knoll, A.H., and S. Golubic. 1979. Anatomy and taphonomy of a Precambrian Algal stromatolite. Precamb. Res. **10**:115-151.

Koch, W.J. 1951. Studies in the genus *Chytridium*, with observations on a sexually reproducing species. J. Elisha Mitchell Sci. Soc. **67**:267-278.

Kohlmeyer, J. 1973. Fungi from marine algae. Bot. Mar. **16**:201-215.

Kohlmeyer, J. 1975. New clues to the possible origin of the ascomycetes. BioScience **25**(2):86-92.

Kohlmeyer, J., and E. Kohlmeyer. 1971. Synoptic plates of higher marine fungi. Ed. 3. J. Cramer, Lehre, Germany. 68 pp.

Kohlmeyer, J., and E. Kohlmeyer. 1979. Marine mycology: the higher fungi. Academic Press, Inc., New York. 690 pp.

Kole, A.P., and A.J. Gielink. 1963. The significance of the zoosporangial stage in the life cycle of the Plasmodiophorales. Ned. J. Plant Path. **69**:258-262.

Konijn, T.M., J.G.C. van de Meene, Y.Y. Chang, D.S. Barkley, and J.T. Bonner. 1969. Identification of adenosine-3', 5'—monophosphate as the bacterial attractant for myxamoebae of *Dictyostelium discoideum*. J. Bacteriol. **99**(2):510-512.

Koop, H.-U. 1975a. Germination of cysts of *Acetabularia mediterraneae*. Protoplasma **85**:137-146.

Koop, H.-U. 1975b. The site of meiosis in *Acetabularia mediterraneae*. Protoplasma **85**:109-114.

Korringa, P. 1951. Investigations on shell disease in the oyster, *Ostrea edulis* L. Rapports et Proces-verbaux, Conseil Permanent Internat. l'Exploration de la Mer. **128**:50-54.

Krichenbauer, H. 1937. Beitrag zur Kenntnis der Morphologie und Entwicklungeschichte der Gattungen *Euglena* und *Phacus*. Arch. Protistenk. **90**:88-123.

Krog, H. 1968. The macrolichens of Alaska. Nor. Polar Inst., Skr. **144**:1-180.

Kubai, D.F., and H. Ris. 1969. Division in the dinoflagellate *Gyrodinium cohnii* (Schiller). A new type of nuclear reproduction. J. Cell Biol. **40**:508-528.

Kulasooriya, S.A., N.J. Lang, and P. Fay. 1972. The heterocysts of blue-green algae. III: Differentiation and nitrogenase activity. Proc. R. Soc. **81**:199-209.

Kurokawa, S. 1971. Results of isolation and culture of lichen fungi and algae. J. Jap. Bot. **46**(10):297-302.

Kuwabara, J.S. 1982. Micronutrients and kelp cultures: evidence for cobalt and manganese deficiencies in southern California deep seawater. Science **216**:1219-1221.

Kwon-Chung, K.J. 1975. A new genus *Filobasidiella*, the perfect state of *Cryptococcus neoformans*. Mycologia **67**:1197-1200.

Kwon-Chung, K.J. 1976. Morphogenesis of *Filobasidiella neoformans*, the sexual state of *Cryptococcus neoformans*. Mycologia **68**:821-833.

Lackey, J.B. 1938. A study of some ecological factors affecting the distribution of protozoa. Ecol. Monogr. **8**:501-527.

Lackey, J.B. 1967. The microbiota of estuaries and their roles. In Lauff, G.H. (ed.): Estuaries, pp. 291-305. A.A.A.S., Washington, D.C.

Lallemant, R. 1977. Recherches sur le dévelopment en cultures pures *in vitro* due mycobionte du discolichen *Pertusaria pertusa* (L.) Tuck. Rev. Bryol. Lichenol. **43**:255-282.

Lang, N.J. 1963. Electron microscopic demonstration of plastids in *Polytoma*. J. Protozool. **10**:333-339.

Lange, B. 1973. The *Sphagnum* flora of hot springs in Iceland. Lindbergia **2**:81-93.

Lara, S.L., and S. Bartnicki-Garcia. 1974. Cytology of budding in *Mucor rouxii:* wall ontogeny. Archiv. für Microbiol. **97**:1-16.

Large, E.C. 1940. The advance of the fungi. Henry Holt & Co., New York. 488 pp.

Large, E.C. 1958. The advance of the fungi. Jonathan Cape, Ltd., London. 488 pp.

Laundon, G.F. 1973. Uredinales. In Ainsworth, G.C., F.K. Sparrow, and A.S. Sussman (eds.): The fungi, Vol. IVB, pp. 247-279. Academic Press, Inc., New York.

Leadbeater, B.S.C. 1970. Preliminary observations on differences of scale morphology at various stages in the life cycle of "*Apistonema-Syracosphaera*" sensu von Stosch. Br. Phycol. J. **5**:57-59.

LeBlanc, F., G. Comeau, and D.N. Rao. 1971. Fluoride injury symptoms in epiphytic lichen and mosses. Can. J. Bot. **49**:1691-1698.

LeBlanc, F., and D.N. Rao. 1973. Evaluation of the pollution and drought hypotheses in relation to lichens and bryophytes in urban environments. Bryol. **76**:1-19.

LeBlanc, F., and D.N. Rao. 1974. A review of the literature on bryophytes with respect to air pollution. Bull. Soc. Bot. Fr. Colloque Bryol. **121**:237-255.

Lee, R.E. 1977. Evolution of algal flagellates with chloroplast endoplasmic reticulum from the ciliates. S. Afr. J. Sci. **73**:179-182.

Lee, R. 1980. Phycology. Cambridge University Press, Cambridge, England. 478 pp.

Leedale, G.F. 1962. The evidence for meiotic process in the Euglenineae. Arch. Mikrobiol. **42**:237-245.

Leedale, G.F. 1967. Euglenoid flagellates. Prentice-Hall, Inc., Englewood Cliffs, N.J. 242 pp.

Leedale, G.F. 1969. Observations on endonuclear bacteria in euglenoid flagellates. Öst. bot. Z. **116**:279-294.

Leedale, G., and P. Walne. 1971. Bibliography on Euglenophyta. In Rosowski, J., and B. Parker (eds.): Selected papers in phycology, pp. 797-802. Department of Botany, University of Nebraska, Lincoln.

Leeper, E.G. 1976. Seaweed: resource of the 21st century? BioScience **26**:(5):357-358.

Lehringer, A.L. 1975. Biochemistry. Ed. 2. Worth Publishers, Inc., New York. 1104 pp.

LéJohn, H.B. 1971. Enzyme regulation, lysine pathways and cell wall structures as indicators of major lines of evolution in fungi. Nature **231**:164-168.

Lembi, C.A. 1980. Unicellular chlorophytes. In Cox, E.R. (ed.): Phytoflagellates, pp. 5-59. Elsevier/North-Holland, New York.

Lemke, P.A., and C.H. Nash. 1974. Fungal viruses. Bacteriol. Rev. **38**:29-56.

Letrouit-Galinou, M. 1973. Sexual reproduction. In Ahmadjian, V., and M.E. Hale (eds.): The lichens, pp. 59-90. Academic Press, Inc., New York.

Levi, M.P. 1977. Fungicides in wood preservation. In Siegel, M.R., and H.D. Sisler (eds.): Antifungal compounds, Vol. 1. pp. 397-436. Marcel Dekker, Inc., New York.

Levring, T. 1947. Submarine daylight and the photosynthesis of marine algae. Göteborgs Vetensk Samh. Handl., IV Ser., B, **5/6**:1-89.

Levring, T. 1966. Submarine light and algal shore zonation. In Bainbridge, R., G.C. Evans, and O. Rackham (eds.): Light as an ecological factor, pp. 305-318. Brit. Ecol. Soc. Symp., Vol. 6. Blackwell Scientific Publications, Ltd., Oxford, England.

Levring, T., H.A. Hoppe, and O.J. Schmid. 1969. Marine algae: a survey of research and utilization. Cram de Gruyter and Co., Hamburg, Germany. 421 pp.

Lewin, J.C. 1962. Calcification. In Lewin, R.A. (ed.): The physiology and biochemistry of algae, pp. 457-465. Academic Press, Inc., New York.

Lewin, J.C., and R.A. Lewin. 1960. Auxotrophy and heterotrophy in marine littoral diatoms. Can. J. Microbiol. **6**:127-34.

Lewin, R.A. 1976. Prochlorophyta as a proposed new division of algae. Nature **261**:697-698.

Lewin, R.A., and J.A. Robertson. 1971. Influence of salinity on the form of *Asterocystis* in pure culture. J. Phycol. **7**:236-238.

Lewin, R.A., and N.W. Withers. 1975. Extraordinary pigment composition of a prokaryotic alga. Nature **256**:735-737.

Lewis, D.H., and J.L. Harley. 1965. Carbohydrate physiology of mycorrhizal roots of beech. I: Identity of endogenous sugars and utilization of exogenous sugars. New Phytol. **64**:224-237.

Lewis, J.R. 1977. The role of physical and biological factors in the distribution and stability of rocky shore communities. In Keegan, B.F. (ed.): European Marine Biology Symposium, 11th, Galway, 1976. Biology of benthic organisms, pp. 417-424. Pergamon Press, Inc., New York.

Ley, A.C., W.L. Butler, D.A. Bryant, and A.N. Glazer. 1977. Isolation and function of allophycocyanin B of *Porphyridium cruentum*. Plant Physiol. **59**:974-980.

Lichtwardt, R.W. 1976. Trichomycetes. In Gareth Jones, E.B. (ed.): Recent advances in aquatic mycology, pp. 651-671. Elek Scientific Books, Ltd., London.

Liden, K. 1961. Cesium-137 burdens in Swedish Laplanders and reindeer. Acta Radiol. **56**:237.

Liebig, J. 1842. Chemistry in its application to agriculture and physiology. Ed. 2. Taylor & Walton, London. 409 pp.

Lieth, H., and R.H. Whittaker. (eds.). 1975. The primary productivity of the biosphere. Springer-Verlag, New York. 339 pp.

Likens, G.E. (ed.). 1972. Nutrients and eutrophication: the limiting-nutrient controversy. American Society of Limnology and Oceanography, Inc., Lawrence, Kansas. 328 pp.

Likens, G.E. 1975. Productivity of inland aquatic ecosystems. In Leith, H., and R.H. Whittaker (eds.): The primary productivity of the biosphere, pp. 185-202. Springer-Verlag, New York.

Lin, J-Y., Y-J. Lin, C-C. Chen, H-L. Wu, G-Y. Shi, and T-W. Jeng. 1974. Cardiotoxic protein from edible mushrooms. Nature **252**:235-237.

Lincoff, G., and D.H. Mitchel. 1977. Toxic and hallucinogenic mushroom poisoning. Van Nostrand Reinhold Co., New York. 267 pp.

Lind, O.T. 1979. Handbook of common methods in limnology. Ed. 2. The C.V. Mosby Co., St. Louis. 199 pp.

Linskens, H.F. 1963. Beitrag zur Frage der Beziehung Zwischen Epiphyt und Basiphyt bei marinen Algen. Publ. Staz. zool. Napoli. **3**:274-293.

Lipkin, Y. 1977. *Centroceras*, the "missile"-launching marine red alga. Nature **270**:48-49.

Litten, W. 1975. The most poisonous mushrooms. Sci. Am. **232**(3):90-101.

Llano, G.A. 1944. Lichens: their biological and economic significance. Bot. Rev. **10**:1-65.

Lobban, C.S., and M.J. Wynne (eds.). 1982. The biology of seaweeds. Bot. Monog. 17. University of California Press, Berkeley. 798 pp.

LoCicero, V.R. (ed.). 1975. Proceedings, First International Conference on Toxic Dinoflagellate Blooms. Massachusetts Science and Technology Foundation, Wakefield, Mass. 541 pp.

Lockwood, L.B. 1975. Organic acid production. In Smith, J.E., and D.R. Berry (eds.): The filamentous fungi. Vol. I: Industrial Mycology, pp. 140-157. Edward Arnold, Ltd., London.

Loeblich, A.R. III. 1976. Dinoflagellate evolution: speculation and evidence. J. Protozool. **23**:13-28.

Loomis, W.F. 1975. *Dictyostelium discoideum:* A developmental system. Academic Press, Inc., New York. 214 pp.

Lovett, J.S. 1975. Growth and differentiation in the water mold *Blastocladiella emersonii:* cytodifferentiation and the role of ribonucleic acid and protein synthesis. Bacteriol. Rev. **39**:345-404.

Lucas, I.A.N. 1970. Observations on the fine structure of the Cryptophyceae. I: The genus *Cryptomonas*. J. Phycol. **6**:30-38.

Lund, J.W.G. 1964. Primary production and periodicity of phytoplankton. Verh. Internat. Verein. Limnol. **15**:37-56.

Lüning, K., and M.J. Dring. 1973. The influence of light quality on the development of the brown algae *Petalonia* and *Scytosiphon*. Br. Phycol. J. **8**:333-338.

Lyman, H., H.T. Epstein, and J.A. Schiff. 1959. Ultraviolet inactivation and photo reactivation of chloroplast development in *Euglena* without cell death. J. Protozool. **6**:264-265.

Lyman, H., and K. Traverse. 1980. *Euglena:* mutations, chloroplast "bleaching" and differentiation. In Gantt, E. (ed.): Handbook of phycological methods—developmental and cytological, pp. 107-141. Cambridge University Press, Cambridge, England.

Machlis, L. 1958. Evidence for a sexual hormone in *Allomyces*. Physiol. Plant. **11**:181-191.

Machlis, L. 1966. Sex hormones in fungi. In Ainsworth, G.C., and A.S. Sussman (eds.): The fungi, Vol. 2, pp. 415-433. Academic Press, Inc., New York.

Machlis, L. 1972. The coming of age in sex hormones in plants. Mycologia **64**:235-247.

Macinnes, M.A., and D. Francis. 1974. Meiosis in *Dictyostelium mucoroides*. Nature **251**:321-323.

Mackenzie, D.E., and J.P. Adler. 1972. Virus-like particles in toxigenic *Aspergilli*. pp. 68. Abstract, Annual Meeting of the American Society of Microbiologists.

Magne, F. 1969. Meise sans tetrasporocystes chez les Rhodophycees. Proc. Intl. Seaweed Symp. **6**:251-254.

Maier, K. 1974. Ruptur der Kapsewand bei *Sphagnum*. Plant Syst. Evol. **132**:13-24.

Malloch, D. 1981. Moulds: their isolation, cultivation and identification. University of Toronto Press, Toronto. 97 pp.

Malone, T.C. 1971. The relative importance of nannoplankton and net plankton as primary producers in tropical oceanic and neritic phytoplankton communities. Limnol. Oceanogr. **16**:633-639.

Mann, K.J. 1973. Seaweeds: their productivity and strategy for growth. Science **182**:975-981.

Mantle, P.G. 1975. Industrial exploitation of ergot fungi. In Smith, J.E., and D.R. Berry (ed.): The filamentous fungi, Vol. 1, pp. 281-300. John Wiley & Sons, Inc., New York.

Manton, I. 1964. Further observations on the fine structure of the haptonema in *Prymnesium parvum*. Arch. Mikrobiol. **49**:315-330.

Margalef, R. 1968. Présence de *Chattonella subsalsa* Biechler dans le port de Barcelone. Rapp. Comm. Int. Mer. Médit. **19**:581-582.

Margulis, L. 1970. Origin of eukaryotic cells. Yale University Press, New Haven, Conn. 349 pp.

Margulis, L. 1971. Symbiosis and evolution. Sci. Am. **225**(2):48-57.

Margulis, L. 1981. Symbiosis in cell evolution: life and its environment in the early earth. W.H. Freeman & Co., San Francisco. 419 pp.

Marks, G.C., and T.T. Koslowski (eds.). 1973. Ectomyorrhizae: their ecology and physiology. Academic Press, Inc., New York. 444 pp.

Mars, P.W., A.R. Rabson, J.J. Rappey, and L. Ajello. 1971. Cutaneous protothecosis. Br. J. Dermatol. **85**(Suppl. 7): 76-84.

Marshall, E. 1982a. More on yellow rain. Science **216**:967.

Marshall, E. 1982b. Yellow rain: filling in the gaps. Science **217**:31-34.

Martin, D.F., and B.B. Martin. 1973. Implications of metal-organic compounds in red tide outbreaks. In Singer, P.C. (ed.): Trace metals and metal-organic interactions in natural waters, pp. 339-362. Ann Arbor Science Publishers, Inc., Ann Arbor, Mich.

Martin, G.W. 1968. The origin and status of fungi (with a note on the fossil records). In Ainsworth, G.C., and A.S. Sussman (eds.): The fungi, Vol. 2, pp. 635-648. Academic Press, Inc., New York.

Martin, G.W., and C.J. Alexopoulos. 1969. The Myxomycetes. University of Iowa Press, Iowa City. 561 pp.

Marx, D.H. 1977. Tree host range and world distribution of the ectomycorrhizal fungus *Pisolithus tinctorius*. Can. J. Bot. **23**:217-223.

Masters, M.J. 1976. Freshwater Phycomycetes on algae. In Gareth Jones, E.B. (ed.): Recent advances in aquatic mycology, pp. 489-512. Elek Scientific Books, Ltd., Oxford, England.

Mathieson, A.C., and R.L. Burns. 1975. Ecological studies of economic red algae. 5: Growth and reproduction of natural and harvested populations of *Chondrus crispus* Stackhouse in New Hampshire. J. Exp. Mar. Biol. Ecol. **70**(2):137-156.

Mathieson, A.C., and E. Tveter. 1975. Carageenan ecology of *Chondrus crispus* Stackhouse. Aquat. Bot. **1**:25-43.

Mato, J.M., P.J.M. van Haastert, F.A. Krens, and T.M. Konjin. 1978. Chemotaxis in *Dictyostelium discoideum*: effect of concanavalin A on chemoattractant mediated cyclic GMP accumulation and light-scattering decrease. Cell Biol. Internat. Rep. **2**:163-170.

Matossian, M.K. 1982. Ergot and the Salem witchcraft affair. Am. Sci. **70**:355-357.

Matthews, T., and D. Niederpruem. 1972. Differentiation in *Coprinus lagopus*. I: Control of fruiting and cytology of initial events. Arch. Mikrobiol. **87**:257-268.

Matthews, T., and D. Niederpruem. 1973. Differentiation in *Coprinus lagopus*. II: Histology and ultrastructural aspects of developing primordia. Arch. Mikrobiol. **88**:169-180.

Maugh, T.H. II. 1981a. Analgesic from mushrooms begins clinical trials. Science **212**:431.

Maugh, T.H. II. 1981b. A new wave of antibiotics builds. Science **214**:1225-1228.

Mayaudon, J., and P. Simonart. 1959a. Étude de la décomposition de la matiere organique dans le sol au moyen de carbone radioactif. III: Décomposition des substances solubles dialysables des proteines et des hemicelluloses. Plant Soil **11**:170-180.

Mayaudon, J., and P. Simonart. 1959b. Etude de la décomposition de cellulose et de lignine. Plant Soil **11**:181-192.

McConnaughey, B.H. 1983. Introduction to marine biology. Ed. 4. The C.V. Mosby Co., St. Louis. 638 pp.

McMorris, T.C., R. Seshadi, G.R. Weihe, G.P. Arsenault, and A.W. Barksdale. 1975. Structures of oogoniol-1, -2, and -3 steroidal sex hormones of the watermold, *Achlya*. J. Am. Chem. Soc. **97**:2544-2545.

McMurray, L., and J.W. Hastings. 1972. No desynchronization among four circadian rhythms in the unicellular alga *Gonyaulax polyedra*. Science **175**:1137-1139.

Meeks, J.C. 1974. Chlorophylls. In Stewart, W.D.P. (ed.): Algal physiology and biochemistry, pp. 161-175. University of California Press, Berkeley.

Meeuse, B.J.D. 1956. Free sulfuric acid in the brown alga, *Desmarestia*. Acta Biochim. Biophys. **19**:372-374.

Meeuse, B.J.D. 1962. Storage products. In Lewin, R.A. (ed.): Physiology and biochemistry of algae, pp. 289-311. Academic Press, Inc., New York and London.

Meisinger, A.C. 1981. Diatomite. In Mineral Yearbook, 1980. Vol. I, pp. 293-295. U.S. Dept. of the Interior. U.S. Government Printing Office, Washington, D.C.

Mellor, E. 1923. Lichens and their action on the glass and leadings of windows. Nature **112**:299-300.

Michanek, G. 1975. Seaweed resources of the ocean. Food and Agriculture Organization of the United Nations. Fish. Tech. Paper 138. 127 pp.

Miller, M.W., and G.C. Berg (eds.). 1969. Chemical fallout. Charles C Thomas, Publisher, Springfield, Ill. 531 pp.

Miller, O.K. 1972. Mushrooms of North America. Dutton & Co., New York. 460 pp.

Minckley, W.L. 1963. The ecology of a spring stream, Doe Run, Meade County, Kentucky. Wildl. Monogr. Chestertown II. 124 pp.

Mirocha, C.J. 1982. Mycotoxin weapons. Science **217**:776-778.

Mirocha, C.J., S.V. Pathres, and C.M. Christensen. 1977. Chemistry of *Fusarium* and *Stachybotrys* mycotoxins. In Wyllie, T.D., and L.G. Morehouse (eds.): Mycotoxic fungi, mycotoxins, mycotoxicoses: an encyclopedic handbook, Vol. I: Mycotoxic Fungi and Chemistry of Mycotoxins, pp. 365-420. Marcel Dekker, Inc., New York.

Mix, A.J. 1949. A monograph of the genus *Taphrina*. Kansas Univ. Scient. Bull. **33**:3-167.

Moikeha, S.N., and G.W. Chu. 1971. Dermatitis-producing alga *Lyngbya majuscula* in Hawaii. I: Isolation and chemical characterization of the toxic factor. J. Phycol. **7**:4-8.

Moore, G.T., and N. Carter. 1926. Further studies on the subterranean algal flora of the Missouri Botanical Garden. Ann. Missouri Bot. Garden. **13**:101-140.

Moore, R.E. 1977. Toxins from blue-green algae. BioScience **27**:797-802.

Moore, R.T. 1971. An alternative concept of the fungi based on their ultrastructure, pp. 49-64. Tenth International Congress on Microbiology, Mexico.

Moore, R.T. 1972. Ustomycota, a new division of higher fungi. Antonie van Leeuwenhoek **38**:567-584.

Moo-Young, M. 1976. A survey of SCP production facilities. Process. Biochem. **11**(10):32-34.

Moreau-Froment, M. 1956. Les Neurospora. Bull. Soc. Bot. France **103**:678-731.

Mori, M. 1975. Studies on the genus *Batrachospermum* in Japan. Jap. J. Bot. **20**:461-485.

Morris, I. 1981. The physiological ecology of phytoplankton: studies in Ecology, Vol. 7. University of California Press, Berkeley. 625 pp.

Mosbach, K. 1973. Biosynthesis of lichen substances. In Ahmadjian, V., and M.E. Hale (eds.): The lichens, pp. 523-546. Academic Press, Inc., New York.

Moss, B. 1976. The effects of fertilization and fish on community structure and biomass of aquatic macrophytes and epiphytic algal populations: an ecosystem experiment. J. Ecol. **64**:313-342.

Moss, B. 1980. Ecology of fresh waters. John Wiley & Sons, Inc. New York. 315 pp.

Mosser, J.L., A.G. Mosser, and T.D. Brock. 1977. Photosynthesis in the snow: the algal *Chlamydomonas nivalis* (Chlorophyceae). J. Phycol. **13**:22-27.

Müller, D. 1964. Life-cycle of *Ectocarpus siliculosus* from Naples, Italy. Nature **203**:1402.

Müller, D. 1968. Versuche zur Charakterisierung eines Sexual—Lockstoffes bei der Braunalge *Ectocarpus siliculosus*. I: Methoden, Isolierung und gaschromatographischer Nachweis. Planta **81**:160-168.

Müller, D. 1972. Studies on reproduction in *Ectocarpus siliculosus*. Soc. Bot. France, Mém. **1972**:87-98.

Müller, D. 1975. Experimental evidence against sexual fusions of spores from unilocular sporangia of *Ectocarpus siliculosus* (Phaeophyta). Br. Phycol. J. **10**:315-321.

Müller, D.G., and L. Jaenicke. 1973. Fucoserraten, the female sex attractant of *Fucus serratus*. L. (Phaeophyta). FEBS Letters **30**:137-139.

Müller, D.G., L. Jaenicke, M. Donike, and T. Akintobi. 1971. Sex attractant in a brown alga: chemical structure. Science **171**:815-817.

Mulligan, H.F. 1975. Oceanographic factors associated with New England red tide blooms. In LoCicero, V.R. (ed.): Proceedings, First International Conference on Toxic Dinoflagellate Blooms, pp. 23-40. Massachusetts Science and Technology Foundation, Wakefield, Mass.

Munawar, M., and I.F. Munawar. 1975. Some observations on the growth of diatoms in Lake Ontario with emphasis on *Melosira binderana* during thermal bar conditions. Arch. Hydrobiol. **75**:490-499.

Muscatine, L. 1967. Glycerol excretion by symbiotic algae from corals and *Tridacna* and its control by the host. Science **156**:516-518.

Muscatine, L., L.R. McCloskey, and R.E. Marian. 1981. Estimating the daily contribution of carbon from zooxanthellae to coral animal respiration. Limnol. Oceanogr. **26**(4):601-611.

Mynderse, J.R., Moore, M. Kashiwagi, and T. Norton. 1977. Antileukemia activity in the Oscillatoriaceae: isolation of debromoaplysiatoxin from *Lyngbya*. Science **196**: 538-540.

Nash, T.H. III. 1971. Effect of effluents from a zinc factory on lichens. Doctoral dissertation, Rutgers University, New Brunswick, N.J. 114 pp.

Nash, T.H. III. 1972. Simplification of the Blue Mountain lichen communities near a zinc factory. Bryol. **75**:315-324.

National Academy of Sciences. 1972. Genetic vulnerability of major crops. The Academy, Washington, D.C. 307 pp.

National Research Council of Canada (NRCC). 1981. Acidification in the Canadian aquatic environment. Associate Committee on Scientific Criteria for Environmental Quality. Pub. No. 18475. Environmental Secretariat, Ottawa. 369 pp.

Nawawi, A. 1973. Two clamp-bearing aquatic fungi from Malaysia. Trans. Br. Mycol. Soc. **61**:521-528.

Nealson, K.H. 1981. Bioluminescence: current perspectives. IHRDC Publications, Boston, Mass. 165 pp.

Neergaard, P. 1977. Seed pathology, 2 vols. John Wiley & Sons, Inc., New York. 1187 pp.

Neil, J.H., and G.E. Owen. 1964. Distribution, environmental requirements and significance of *Cladophora* in the Great Lakes, pp. 113-121. Great Lakes Research Division, Public. No. 11. University of Michigan, Ann Arbor.

/

Neushul, M. 1972. Functional interpretation of benthic marine algal morphology. In Abbott, I.A., and M. Kurogi (eds.): Contributions to the systematics of benthic marine algae of the North Pacific, pp. 47-73. Japanese Soc. Phycol., Kobe, Japan.

Neuville, D., and P. Daste. 1975. Experiments regarding auxospore production in the diatom *Navicula ostrearia* (Gaillon) Bory cultured *in vitro*. C.R. Acad. Sci. (Paris) **281D**:1753-1756.

Nichols, H.W., and E.K. Lissant. 1967. Developmental studies of *Erythrocladia* Rosenvinge in culture. J. Phycol. 3:6-18.

Niederhauser, J.S., and W.C. Cobb. 1959. The late blight of potatoes. Sci. Am., May 1959. 11 pp.

Niederpruem, D.J., and J.G.H. Wessels. 1969. Cytodifferentiation and morphogenesis in *Schizophyllum commune*. Bacteriol. Rev. 33:505-535.

Nienhuis, P.H., and J. Simons. 1971. *Vaucheria* species and some other algae on a Dutch salt marsh, with ecological notes on their periodicity. Acta Bot. Nederl. **20**:107-118.

Nolard-Tintigner, N. 1973. Étude experimentale sur l'epidemiologie et la pathologenie de la saprolegniose chez *Lebistes reticulatus* Peters et *Xiphophorus helleri* Heckel. Acta Zool. Patho. Antverp. **57**:1-127.

North, W.J. 1971. The biology of giant kelp beds (*Macrocystis*) in California. Nova Hedwigia **32**:1-600.

North, W.J. 1979. The giant kelp *Macrocystis*: a potential producer of marine biomass for energy. Bioresour. Dig. 1:96-102.

Nozawa, K. 1970. The effect of *Peridinium* toxin on other algae. Bull. Misaki Mar. Biol. Inst. **12**:21-24.

Nultsch, W. 1974. Movements. In Stewart, W.D.P. (ed.): Algal physiology and biochemistry, pp. 864-893. University of California Press, Berkeley.

O'Brien, W. 1972. Limiting factors in phytoplankton algae: their meaning and measurement. Science **178**:616-617.

Odum, E.P. 1971. Fundamentals of ecology. Ed. 3. W.B. Saunders Co., Philadelphia. 555 pp.

O'Hare, G.P. 1973. Lichen techniques of pollution assessment. Area **5**:223-9.

Olive, L.S. 1975. The Mycetozoans. Academic Press, Inc., New York. 293 pp.

Olive, L.S., and C. Stoianovitch. 1960. Two new members of the Acrasiales. Bull. Torrey Bot. Club **87**:1-20.

Oltmanns, F. 1892. Über die Cultur-und Lebensbedingungen der Meeresalgen. Jb. wiss. Bot. **23**:349-440.

Onions, A.H.S., H.O.W. Eggins, and D. Allsopp. 1982. Smith's introduction to industrial mycology. Ed. 7. John Wiley & Sons, Inc., New York, 416 pp.

Oohusa, T. 1971. Cultivation of Asakusanori (laver) in Japan. Occas. Pap. Yamamoto Nori Res. Lab., 1971. 2 pp.

Oprin, C.G. 1975. Studies on the rumen flagellate *Neocallimastix frontalis*. J. Gen. Microbiol. **91**:249-262.

Oprin, C.G. 1977. The occurrence of chitin in the cell walls of the rumen organisms *Neocallimastix frontalis*, *Piromonas communis* and *Sphaeromonas communis*. J. Gen. Microbiol. **99**:215-218.

Orme-Johnson, W.H., and W.E. Newton (eds.). 1980. Symbiotic associations and blue-green algae. Vol. II. University Park Press, Baltimore. 352 pp.

Ott, D.W., and R.M. Brown. 1974. Developmental cytology of the genus *Vaucheria*. I: Organization of the vegetative filament. Br. Phycol. J., 9:111-126.

Paasche, E., and I. Østergren. 1980. The annual cycle of plankton diatom growth and silica production in the inner Oslofjord. Limnol. Oceanogr. **25**(3):481-494.

Page, R.M., and G.M. Curry. 1966. Studies on phototropism of young sporangiophores of *Pilobolus kleinii*. Photochem. Photobiol. 5:31-40.

Page, R.M., and D. Kennedy. 1964. Studies on the velocity of discharged sporangia of *Pilobolus kleinii*. Mycologia **56**: 363-368.

Palmer, C.M. 1977. Algae and water pollution. Office of Research and Development, U.S. Environmental Protection Agency, Cincinnati, Ohio. 124 pp.

Palmer, J.D. 1973. Tidal rhythms: the clock control of the rhythmic physiology of marine organisms. Biol. Rev. **43**: 377-418.

Palmer, J.D. 1975. Biological clocks of the tidal zone. Sci. Am. **232**(2):70-79.

Pankratz, H.S., and C.C. Bowen. 1963. Cytology of blue-green algae. I: The cells of *Symploca muscorum*. Am. J. Bot. **50**:387-399.

Park, J.Y., and V.P. Agnihotri. 1969. Bacterial metabolites trigger sporophore formation in *Agaricus bisporus*. Nature **222**:984.

Parke, M., and I. Adams. 1960. The motile *Crystallolithus hyalinus* (Gaarder and Markali) and non-motile phases in the life history of *Coccolithus pelagicus* (Wallich) Schiller. J. Mar. Biol. Assoc. (U.K.) **39**:263-274.

Parker, B.C. 1965. Translocation in the giant kelp *Macrocystis*. I: Rates, direction, quantity of C¹⁴-labelled products and fluorescein. J. Phycol. 1:41-46.

Parker, B.C. 1966. Translocation in *Macrocystis*. III: Composition of sieve tube exudate and identification of the major C¹⁴-labeled products. J. Phycol. 1:38-41.

Parker, B.C., R.E. Preston, and G.E. Fogg. 1963. Studies on the structure and chemical composition of the cell walls of Vaucheriaceae and Saprolegniaceae. Proc. R. Soc. B. **158**:435-445.

Parker, B.C., G.M. Simmons, Jr., F.G. Love, R.A. Wharton, Jr., and K.G. Seaburg. 1981. Modern stromatolites in antarctic dry valley lakes. BioScience **31**(9):656-661.

Parker, B.C., G.M. Simmons, Jr., R.A. Wharton, Jr., and K.G. Seaburg. 1982. Removal of organic and inorganic matter from antarctic lakes by aerial escape of bluegreen algal mats. J. Phycol. **18**:72-78.

Parker-Rhodes, A.F. 1955. Fairy ring kinetics. Trans. Brit. Mycol. Soc. **38**:59-72.

Parkinson, D. 1981. Ecology of soil fungi. In Cole, G.T., and B. Kendrick (eds.): Biology of condial fungi, Vol. I, pp. 277-294. Academic Press, Inc., New York.

Paterson, R.A. 1970. Lacustrine fungal communities. In Cairns, J., Jr. (ed.): The structure and function of freshwater microbial communities. Research Div. Monog. 3. Virginia Polytecnic. Institute and State University, Blacksburg. pp. 209-218.

Patrick, R. 1977. Ecology of freshwater diatoms—diatom communities. In Werner, D. (ed.): The biology of diatoms, pp. 284-332. University of California Press, Berkeley.

Patrick, R., and C.W. Reimer. 1966. The diatoms of the United States exclusive of Alaska and Hawaii. Vol. 1. Monogr. Acad. Nat. Sci. Philadelphia. No. 13. 688 pp.

Patrick, R., and C.W. Reimer. 1975. The diatoms of the United States exclusive of Alaska and Hawaii. Vol. 2, Part 1. Monogr. Acad. Nat. Sci. Philadelphia. No. 13. 213 pp.

Peck, R.E. 1953. The fossil charophytes. Bot. Rev. **19**:209-227.

Pennak, R.W. 1973. Some evidence for aquatic macrophytes as repellents for a limnetic species of *Daphnia*. Int. Rev. ges. Hydrobiol. **58**:569-576.

Pentecost, A. 1980. Calcification in plants. In Bourne, G.H., and J.F. Danielli (eds.): Intern. Rev. Cytol. 62, pp. 1-28. Academic Press, Inc., New York.

Percival, E. 1979. The polysaccharides of green, red and brown seaweeds; their basic structure, biosynthesis and function. Br. Phycol. J. **14**(2):103-117.

Percival, E., and R.H. McDowell. 1967. Chemistry and enzymology of marine algal polysaccharides. Academic Press, Inc., New York. 243 pp.

Perkins, F.O. 1976. Fine structure of lower marine and estuarine fungi. In Gareth Jones, E.B. (ed.): Recent advances in aquatic mycology, pp. 279-312. Elek Scientific Books, Ltd., Oxford, England.

Peters, G.A. 1978. Blue-green algae and algal associations. Bioscience **28**(9):580-585.

Petersen, R.H. (ed.). 1971. Evolution in the higher basidiomycetes. University of Tennessee Press, Knoxville. 562 pp.

Peterson, D.H., and H.C. Murray. 1952. Microbiological oxygenation of steroids at carbon-11. J. Am. Chem. Soc. **74**:1871-1872.

Peveling, E. 1970. Die Darstellung der Oberflächenstrukturen vol Flechten mit dem Rasterelektronenmikroskop. Ber. Duet. Bot. Ges. **4**:89.

Pfiester, L.A. 1975. Sexual reproduction of *Peridinium cinctum* f. *ovoplanum* (Dinophyceae). J. Phycol. **11**:258-265.

Pfitzer, E. 1871. Untersuchungen über Bau und Entwicklung der Bacillariaceen. In Honstein, Bot. Abhandl. a. d. Gebiet der Morph u. Physiol. Heft. 2.

Phaff, H.J. 1981. Industrial microorganisms. Sci. Am. **245**(3):77-89.

Phaff, H.J., M.W. Miller, and E.M. Mrak. 1966. The life of yeasts. Harvard University Press, Cambridge, Mass. 181 pp.

Phatak, S.C., D.R. Sumner, H.D. Wells, D.K. Bell, and N.C. Glaze. 1983. Biological control of yellow nutsedge with indigenous rust fungus *Paccinia canaliculata*. Science **219**:1446-1447.

Phillips, R. 1981. Mushrooms and other fungi of Great Britain and Europe. Pan Books, Ltd., London. 288 pp.

Pickett-Heaps, J.D. 1975. Green Algae: Structure, reproduction and evolution in selected genera. Sinauer Associates, Inc., Sunderland, Mass. 606 pp.

Pickett-Heaps, J.D., and H.J. Marchant. 1972. The phylogeny of the green algae: a new proposal. Cytobios. **6**:255-264.

Pielou, E.C. 1975. Ecological diversity. John Wiley & Sons, Inc., New York. 165 pp.

Pingree, R.D., P.M. Holligan, G.T. Mardell, and R.N. Head. 1976. The influence of physical stability on spring, summer and autumn phytoplankton blooms in the Celtic Sea. J. Mar. Biol. Assoc. (U.K.) **56**:845-873.

Playford, P.E. 1980. Australia's stromalite stronghold. Natural History **89**(10):58-61.

Poelt, J. 1973. Classification. In Ahmadjian, V., and M.E. Hale (eds.): The lichens, pp. 599-632. Academic Press, Inc., New York.

Poff, K.L., W.F. Loomis, Jr., and W.L. Butler. 1974. Isolation and purification of the photoreceptor pigment associated with phototaxis in *Dictyostelium discoideum*. J. Biol. Chem. **249**:2164-2167.

Pohl, R. 1949. Diurnal rhythm in phototactic behavior in *E. gracilis*. Zeitschr. Naturforsch. **3b**(9/10):367-374.

Pontecorvo, G. 1956. The parasexual cycle in fungi. Annu. Rev. Microbiol. **10**:393-400.

Pontecorvo, G., and J.A. Roper. 1952. Genetic analysis without sexual reproduction by means of polyploidy in *Aspergillus nidulans*. J. Gen. Microbiol. **6**:vi-vii (abstract).

Porkorny, K.L. 1967. *Labyrinthula*. J. Protozool. **14**:697-708.

Porter, D. 1972. Cell division in the marine slime mold, *Labyrinthula* sp., and the role of the bothrosome in extracellular membrane production. Protoplasma **74**:427-448.

Porter, L. 1917. On the attachment organs of the common corticolous Ramalinae. Proc. Roy. Irish Acad., Sect. B. **34**:17-32.

Powell, J.H., and B.J.D. Meeuse. 1964. Laminarin in some Phaeophyta of the Pacific Coast. Econ. Bot. **18**:164-166.

Prakash, A. 1975. Dinoflagellates—an overview. In Lo-Cicero, V.R. (ed.): Proceedings, First International Conference on Toxic Dinoflagellate Blooms, pp. 1-6. Massachusetts Science and Technology Foundation, Wakefield, Mass.

Prasertphon, S., and Y. Tanada. 1969. Mycotoxins of entomophthoraceous fungi. Hilgardia **39**:581-600.

Pratt, D.M. 1966. Competition between *Skeletonema costatum* and *Olisthodiscus luteus* in Narragansett Bay and in culture. Limnol. Oceanog. **11**(4):447-455.

Prescott, G.W. 1978. How to know the freshwater algae. Ed. 3. Wm. C. Brown Co., Publishers, Dubuque, Iowa. 293 pp.

Prézelin, B.B. 1976. The role of peridinin–chlorophyll *a*-proteins in the photosynthetic light adaptation of the marine dinoflagellate, *Glenodinium* sp. Planta. **130**:225-233.

Pringsheim, E.G., and O. Pringsheim. 1952. Experimental elimination of chromatophores and eye-spot in *Euglena gracilis*. New Phytol. **51**:65-76.

Proctor, M.C.F. 1979. Surface wax on the leaves of some mosses. J. Bryol. **10**:531-538.

Provasoli, L., S.H. Hunter, and A. Schatz. 1948. Streptomycin-induced chlorophyll-less races of *Euglena*. Proc. Soc. Exper. Biol. Med. **69**:279-282.

Puckett, K.J., E. Nieboer, M.J. Gorzynski, and D.H.S. Richardson. 1973. The uptake of metal ions by lichens: a modified ion-exchange process. New Phytol. **72**:329.

Pueyo, G. 1960. Recherches sur la nature el l'évolution des glucides solubles chez quelgues lichens du bassin Parisien. Année. Biol. **36**:117-169.

Raddum, G.G., and D.A. Saether. 1981. Chironomid communities in Norwegian lakes with different degrees of acidification. Verh. Int. Verein. Limnol. **21**:367-373.

Ragan, N.A., and D.J. Chapman. 1978. A biochemical phylogeny of the protista. Academic Press, Inc., New York. 317 pp.

Ramsbottom, J. 1953. Mushrooms and toadstools. Collins Clear-Type Press, London. 306 pp.

Ramus, J. 1969. Pit connection formation in the red alga *Pseudogloiophloea*. J. Phycol. **5**:57-63.

Ramus, J. 1971. Properties of septal plugs from the red alga *Griffithsia pacifica*. Phycologia **10**:99-103.

Ramus, J., S.I. Beale, and D. Mauzerall. 1976a. Correlation of changes in pigment content with photosynthetic capacity of seaweeds as a function of water depth. Mar. Biol. **37**:231-238.

Ramus, J., S.I. Beale, D. Mauzerall, and K.L. Howard. 1976b. Changes in photosynthetic pigment concentrations in seaweeds as a function of water depth. Mar. Biol. **37**:223-229.

Rao, D.N., and F. LeBlanc. 1966. Effect of sulfur dioxide on the lichen alga, with special reference to chlorophyll. Bryol. **69**:69-75.

Rao, D.N., and F. LeBlanc. 1967. Influence of an iron-scintering plant on corticolous epiphytes in Wawa, Ontario. Bryol. **70**:141-157.

Raper, J.R. 1966. Life cycles, basic patterns of sexuality and sexual mechanisms. In Ainsworth, G.C., and A.S. Sussman (eds.): The fungi, Vol. 2, pp. 473-511. Academic Press, Inc., New York.

Raper, J.R. 1968. On the evolution of fungi. In Ainsworth, G.C., and A.S. Sussman (eds.): The fungi, Vol. 3, pp. 677-693. Academic Press, Inc., New York.

Raper, J.R., and A.S. Flexer. 1970. The road to diploidy with emphasis on a detour. Symp. Soc. Gen. Microbiol. **20**:401-432.

Raper, K.B. 1973. Acrasiomycetes. In Ainsworth, G.C., F.K. Sparrow, and A.S. Sussman (eds.): The fungi: an advanced treatise, Vol. IVB, pp. 9-36. Academic Press, Inc., New York.

Raymont, J.E.G. 1980. Plankton and productivity in the oceans. Ed. 2. Vol. I: Phytoplankton. Pergamon Press, Ltd., Oxford, England. 489 pp.

Reid, D.A. 1976. *Inonotus obliquus* (Pers. ex Fr.) Pilat in Britain. Trans. Brit. Mycol. Soc. **67**:329-332.

Reid, G.K., and R.D. Wood. 1976. Ecology of inland waters and estuaries. Ed. 2. D. Van Nostrand Co., New York. 485 pp.

Reid, I.D. 1974. Properties of conjugation hormones (erogens) from the basidiomycete *Tremella mesenterica*. Can. J. Bot. **52**:521-524.

Reid, P.C. 1975. Large scale changes in North Sea phytoplankton. Nature **257**:217-219.

Reidel, V., G. Gerisch, E. Müller, and H. Beug. 1973. Defective cyclic adenosine-3′, 5′-phosphate-phosphodiesterase regulation in morphogenetic mutants of *Dictyostelium discoideum*. J. Mol. Biol. **74**:573-585.

Reinhard, M.R., A. Armengaud, and J. Dupaquier. 1968. Histoire générale de la population mondiale. Ed. 3. Montchrestien Sàrl, Paris. 708 pp.

Remy, W. 1982. Lower Devonian gametophytes: relation to the phylogeny of land plants. Science **215**:1625-1627.

Rhodes, F.M. 1978. Growth, production, litterfall and structure in populations of the lichen *Lobaria oregana* (Tuck.) Mull. Arg. in canopies of old-growth Douglas fir. Doctoral thesis. University of Oregon, Eugene. 150 pp. Diss. Abs. 7901091.

Richardson, D.H.S. 1973. Photosynthesis and carbohydrate movement. In Ahmadjian, V., and M.E. Hale (eds.): The lichens, pp. 249-288. Academic Press, Inc., New York.

Richardson, D.H.S. 1974. The vanishing lichens. Macmillan Publishing Co., Inc., New York. 231 pp.

Richardson, D.H.S. 1981. The biology of mosses. John Wiley & Sons, Inc. New York. 220 pp.

Richardson, D.H.S., and J. Puckett. 1971. Abstract, First International Mycology Congress, p. 79.

Richardson, W.N. 1970. Studies on the photobiology of *Bangia fuscopurpurea*. J. Phycol. **6**:216-219.

Richardson, W.N., and P.S. Dixon. 1968. Life history of *Bangia fuscopurpurea* (Dillw.) Lyngb. in culture. Nature **218**:496-497.

Rieth, A. 1959. Periodizität beim Ausschlüpfen der Schwärmsporen von *Vaucheria sessilis*. De Candolle. Flora **147**:35-42.

Rippon, J.W. 1982. Medical Mycology. Ed. 2. W.B. Saunders Co., Philadelphia. 842 pp.

Rishbeth, J. 1963. Stump protection against *Fomes annosus* III. Inoculation with *Peniophora gigantea*. Ann. Appl. Biol. **52**:63.

Rizzo, P.J., and L.D. Noodén. 1974. Partial characterization of dinoflagellate chromosome proteins. Acta Biochim. Biophys. **349**:415-427.

Rizzo, P.J., and E.R. Cox. 1977. Histone occurrence in chromatin from *Peridinium balticum*, a binucleate dinoflagellate. Science **198**:1258-1260.

Roberts, D.A., and C.W. Boothroyd. 1972. Fundamentals of plant pathology. W.H. Freeman Co., San Francisco. 402 pp.

Rodricks, J.V., C.W. Hesseltine, and M.A. Mehlman (eds.). 1977. Mycotoxins and human and animal health. Pathotox, Park Forest South, Ill. 807 pp.

Roper, J.A. 1966. Mechanism of inheritance. 3: The parasexual cycle. In Ainsworth, G.C., and A.S. Sussman (eds.): The fungi: an advanced treatise, Vol. 2, pp. 589-617. Academic Press, Inc., New York.

Rose, A.H. 1981. The microbiological production of food and drink. Sci. Am. **245**(3):126-138.

Rosen, S.D., R.W. Reitherman, and S.H. Barondes. 1975. Distinct lectin activities from six species of cellular slime molds. Exp. Cell. Res. **95**:159-166.

Rosowski, J.R., and B.C. Parker (eds.). 1982. Selected papers in phycology, II. Phycological Society of America, Lawrence, Kansas. 866 pp.

Rosowski, J.R., and R.L. Willey. 1975. *Colacium libellae* sp. nov. (Euglenophyceae), a photosynthetic inhabitant of the larval damselfly rectum. J. Phycol. **11**:310-315.

Ross, I.K. 1979. Biology of the fungi. McGraw-Hill, Inc., New York. 499 pp.

Ross, I.K., J.C. Pommerville, and D.L. Damm. 1976. A highly infectious "mycoplasma" that inhibits meiosis in the fungus *Coprinus*. J. Cell Sci. **21**:175-191.

Round, F.E. 1971. The taxonomy of The Chlorophyta, II. Br. Phycol. J. **6**:235-264.

Round, F.E. 1981. The ecology of algae. Cambridge University Press, New York. 653 pp.

Royce, D., and L.C. Schisler. 1980. Mushrooms: their consumption, production, and culture development. Interdisc. Sci. Rev. **5**(4):324-332.

Rudolph, E.D. 1967. Lichen distribution. In Bushell, V.C., (ed.): Terrestrial life of Antarctica, pp. 9-11. Antarctic Map Folio Ser. 5. Amer. Geophys. Soc., New York.

Rumack, B.H., and E. Salzman (eds.). 1978. Mushroom poisoning: diagnosis and treatment. CRC Press, Boca Raton, Fla. 263 pp.

Rusmin, S., and T.J. Leonard. 1975. Biochemical induction of fruiting bodies in *Schizophyllum commune:* bioassay and its application. J. Gen. Microbiol. **90**:217-227.

Russell-Hunter, W.D. 1970. Aquatic productivity. Macmillan Publishing Co., Inc., New York. 306 pp.

Ryther, J.H. 1959. Potential productivity of the sea. Science **130**:602-608.

Ryther, J., W. Dunstan, K. Tenore, and J. Huguenin. 1972. Controlled eutrophication—increasing food production from the sea by recycling human wastes. BioScience **22**:144-152.

Sakagami, Y., A. Isogai, A. Suzuki, S. Tamura, E. Tsuchiya, and S. Fukui. 1978. Isolation of a novel sex hormone, Tremerogen A-10, controlling conjugation tube formation in *Tremella mesenterica*. Fries. Agr. Biol. Chem. **42**: 1093-1094.

Sakai, A., and S. Yoshida. 1967. Survival of plant tissue at superlow temperatures. IV. Effects of cooling and rewarming rates on survival. Plant Physiol. **42**:1695-1701.

Salaman, R.N. 1949. The history and social influence of the potato. Cambridge University Press, London. 685 pp.

Salas, J.A., and J.G. Hancock. 1972. Production of the perfect stage of *Mycena citricolor* (Berk. and Curt.) Sacc. Hilgardia **41**:213-234.

Salisbury, F.B. 1962. Martian biology. Science **136**:17-26.

Sanders, F.E., B. Mosse, and P.S. Tinker. 1976. Endomycorrhizas. Academic Press, Inc., New York. 626 pp.

Santelices, B. 1974. Gelidioid algae, a brief resume of the pertinent literature. Marine Agronomy U.S. Sea Grant Program, Hawaii. Tech. Report No. 1. 111 pp.

Santesson, R. 1952. Foliicolous lichens I: A revision of the taxonomy of the obligatory foliicolous, lichenized fungi. Symb. Bot. Upsal. **12**(1):1-590.

Sarjeant, W.A.S. 1974. Fossil and living dinoflagellates. Academic Press, Inc., New York. 182 pp.

Sauer, H.W. 1973. Differentiation in *Physarum polycephalum*. In Ashworth, J.M., and J.E. Smith (eds.): Microbial differentiation, pp. 375-405. Cambridge University Press, London.

Saunders, G.W. 1972. Potential heterotrophy in a natural population of *Oscillatoria aghardhii* var. *isothrix* skuja. Limnol. Oceanogr. **17**:704-711.

Savile, D.B.O. 1976. Evolution of the rust fungi (Uredinales) as reflected by their ecological problems. Evol. Biol. **9**:137-207.

Savile, D.B.O. 1978. Paleoecology and convergent evolution in rust fungi (Uredinales). Bio. Sys. **10**:31-36.

Scagel, R.F., R.J. Bandoni, J.R. Maze, G.E. Rouse, W.B. Schofield, and J.R. Stein. 1982. Nonvascular plants: An evolutionary survey. Wadsworth Publishing Co., Belmont, Calif. 570 pp.

Scagel, R.F., R.J. Bandoni, G.E. Rouse, W.B. Schofield, J.R. Stein, and T.M.C. Taylor. 1965. An evolutionary survey of the plant kingdom. Wadsworth Publishing Co., Belmont, Calif. 658 pp.

Schantz, E.J., V.E. Ghazarossian, H.K. Schnoes, F.M. Strong, J.P. Springer, J.O. Pezzanite, and J. Clardy. 1975. Paralytic poisons from marine dinoflagellates. In LoCicero, V.R. (ed.): First International Conference on Toxic Dinoflagellate Blooms, pp. 267-274. Massachusetts Science and Technology Foundation, Wakefield, Mass.

Scheirer, D.C., and I.J. Goldklang. 1977. Pathway of water movement in hydroids of *Polytrichum commune* Hedw. (Bryopsida). Am. J. Bot. **64**:1046-1047.

Scheuer, P.J. 1977. Marine toxins. Accounts Chem. Res. **10**:33-39.

Schiff, J.A. 1962. Sulfur. In Lewin, R.A. (ed.): Physiology and biochemistry of algae, pp. 239-246. Academic Press, Inc., New York.

Schiff, J.A. 1978. Photocontrol of chloroplast development in *Euglena*. In Akoyunoglou, G., and J.H. Argyroudi-Akoyunoglou (eds.): Chloroplast development. Elsevier North-Holland, New York.

Schindler, D.W. 1978. Factors regulating phytoplankton production and standing crop in the world's freshwaters. Limnol. Oceanogr. **23**:478-486.

Schmidt, R.J., V.D. Gooch, A.R. Loeblich III, and J.W. Hastings. 1978. Comparative study of luminescent and nonluminescent strains of *Gonyaulax excavata* (Pyrrophyta). J. Phycol. **14**:5-9.

Schmitter, R. 1979. Temporary cysts of *Gonyaulax excavata:* effects of temperature and light. In Taylor, D.L. and H.H. Seliger (eds.): Toxic dinoflagellate blooms, pp. 123-126. Elsevier North-Holland, New York.

Schmitz, K., and C.S. Lobban. 1976. A survey of translocation in Laminariales (Phaeophyceae). Mar. Biol. **36**:207-216.

Schmitz, K., K. Lüning, and J. Willenbrink. 1972. CO_2 fixierung und Stofftransport in benthischen marinen Algen. II: Zum Ferntransport ^{14}C markierter Assimilate bei *Laminaria hyperborea* und *Laminaria saccharina*. Z. Pflanzenphysiol. **67**:418-429.

Schmitz, K., and L.M. Srivastava. 1975. On the fine structure of sieve tubes and the physiology of assimilate transport in *Alaria marginata*. Can. J. Bot. **53**:861-876.

Schneck, N.C. (ed.). 1982. Methods and principles of mycorrhizal research. American Phytopathological Society, St. Paul, Minn. 256 pp.

Schopf, J.W. 1968. Microflora of the Bitter Springs formation; late Precambrian, Central Australia. J. Paleont. **42**:651-688.

Schopf, J.W., and E.S. Barghoorn. 1967. Alga-like fossils from the early Precambrian of South Africa. Science **156**:508-512.

Schramm, J.R. 1966. Plant colonization studies on black wastes from anthracite mining in Pennsylvania. Trans. Am. Phil. Soc. **56**:194.

Schuster, R.M. 1966, 1969, 1974, 1981. The Hepaticae and Anthocerotae of North America east of hundredth meridian. Vols. 1-4. Columbia University Press, New York. Vol. I, 802 pp.; vol. II, 1062 pp.; vol. III, 880 pp.; vol. IV, 1344 pp.

Schwimmer, M., and D. Schwimmer. 1968. Medical aspects of phycology. In Jackson, D.F. (ed.): Algae, man, and the environment, pp. 279-358. Syracuse University Press, Syracuse, New York.

Scott, G.D. 1973. Evolutionary aspects of symbiosis. In Ahmadjian, V., and M.E. Hale (eds.): The lichens, pp. 581-598. Academic Press, Inc., New York.

Seagrave, S. 1981. Yellow rain. M. Evans and Co., New York. 316 pp.

Searles, R.B. 1980. The strategy of the red algal life history. Am. Nat. **115**(1):113-120.

Seliger, H.H., M.E. Loftus, and D.V. Subba Rao. 1975. Dinoflagellate accumulations in Chesapeake Bay. In Lo-Cicero, V.R. (ed.): Proceedings, First Interational Conference on Toxic Dinoflagellate Blooms, pp. 181-205. Massachusetts Science and Technology Foundation, Wakefield, Mass.

Setchell, W.A. 1918. Parasitism among the red algae. Proc. Am. Phil. Soc. **57**:155-172.

Shaffer, R.L. 1975. The major groups of basidiomycetes. Mycologia **67**:1-18.

Shank, R.C. 1981. Mycotoxins and N-nitroso compounds: environmental risks. Vols. I and II. CRC Press, Inc., Boca Raton, Fla. Vol. I, 296 pp.; Vol. II, 248 pp.

Shapiro, J. 1973. Blue-green algae: why they become dominant. Science **179**:182-184.

Shea, K.P. 1972. Captan and folpet. Environ. **14**(1):22-32.

Shear, C.L., and B.O. Dodge. 1927. Life histories and heterothallium of the red bread-mold fungi of the *Monilia sitophila* group. J. Agr. Res. **34**:1019-1042.

Sheath, R.G., and K.M. Cole. 1980. Distribution and salinity adaptations of *Bangia atropurpurea* (Rhodophyta), a putative migrant into the Laurential Great Lakes. J. Phycol. **16**(3):412-420.

Sheath, R.G., M. Munawar, and J.A. Hellebust. 1975. Fluctuations of phytoplankton biomass and its composition in a subarctic lake during summer. Can. J. Bot. **53**(19):2240-2246.

Shelford, V.E. 1913. Animal communities in temperate America. University of Chicago Press, Chicago.

Shimizu, Y., and M. Yoshioka. 1981. Transformation of paralytic shellfish toxins as demonstrated in scallop homogenates. Science **212**:547-549.

Shimony, C., and J. Friend. 1975. Ultrastructure of the interaction between *Phytophthora infestans* and leaves of two cultivars of potato (*Solanum tuberosum* L.) orion and majestic. New Phytol. **74**:59-65.

Shovlin, F.E., and R.E. Gillis. 1974. Microorganisms and diseases of the oral cavity. In Briody, B.A., and R.E. Gillis (eds.): Microbiology and infectious disease, pp. 377-401. McGraw Hill Book Co., New York.

Showman, R.E. 1981. Lichen (*Parmelia caperata*) recolonization following air quality improvement. Bryol. **84**(4):492-497.

Sideman, E.J., and D.C. Scheirer. 1977. Some fine structural observations on developing and mature seive elements in the brown alga *Laminaria saccharina*. Am. J. Bot. **64**:649-657.

Siegel, M.R., and H.D. Sisler (eds.). 1977. Antifungal compounds. Vols. I. and II. Marcel Dekker, Inc., New York. Vol. 1, 600 pp; Vol. 2, 674 pp.

Sievers, A., and K. Schröter. 1971. Versuch einer, Kausalanalyse der geotropischen reaktionskette im *Chara*-rhizoid. Planta **96**:339-353.

Silver-Dowding, E. 1955. *Endogone* in Canadian rodents. Mycologia **47**:51-57.

Simenstad, C.A., J.A. Estes, and K.W. Kenyon. 1978. Aleuts, sea otters and alternate stable-state communities. Science **200**:403-411.

Simon, R.D. 1973. Measurement of the cyanophycin granule polypeptide contained in the blue-green alga *Anabaena cylindrica*. J. Bact. **116**:1212-1216.

Sinden, J.W. 1971. Ecological control of pathogens and weed-molds in mushroom culture. Annu. Rev. Phytopath. **9**:411-432.

Singer, R. 1975. The Agaricales in modern taxonomy. Ed. 3. J. Cramer, Weinheim, Germany. 912 pp.

Singer, S.J., and G.L. Nicolson. 1972. The fluid-mosaic model of the structure of cell membranes. Science **175**:720-731.

Sinha, U., and J.M. Ashworth. 1969. Evidence for the existence of elements of a para-sexual cycle in the cellular slime mold *Dictyostelium discoideum*. Proc. R. Soc. Br. Ser. B. **173**:531-540.

Sinsabaugh, R.L. III, E.F. Benfield, and A.E. Linkins III. 1981. Cellulase activity associated with the decomposition of leaf litter in a woodland stream. Oikos **36**:184-190.

Skuja, H. 1938. Comments on freshwater Rhodophyceae. Bot. Rev. **4**:665-676.

Skye, E. 1958. The influence of air pollution on the fruticolous and foliaceous lichen flora around the shale-oil works at Kvarntorp in the province of Närke. (In Swedish; English summary). Sv. Bot. Tidskr. **52**:131-177.

Skye, E., and I. Hallberg. 1969. Changes in the lichen flora following air pollution. Oikos **20**:547-552.

Slocum, R.D., V. Ahmadjian, and K.C. Hildreth. 1980. Zoosporogenesis in *Trebouxia gelatinosa* (Chlorophyta, Chlorococcales): ultrastructure, potential for zoospore release and implication for the lichen association. Lichenol. **12**(2):173-187.

Smirnov, N.N. 1958. Some data about the food consumption of plant production of bogs and fens by animals. Verh. Int. Venein. Theor. angew. Limnol. **13**:363-368.

Smith, A.L. 1921. Lichens. Cambridge University Press, London. 464 pp.

Smith, D.C. 1962. The biology of lichen thalli. Biol. Rev. Cambridge Phil. Soc. **37**:537-570.

Smith, D.C., and E.A. Drew. 1965. Studies in the physiology of lichens. V: Translocation from the algal layer to the medulla in *Peltigera polydactyla*. New Phytol. **64**:195-200.

Smith, F.A. 1968. Rates of photosynthesis in characean cells. II: Photosynthetic $^{14}CO_2$ fixation and ^{14}C bicarbonate uptake by characean cells. J. Exp. Bot. **19**:207-217.

Smith, G.M. 1950. The freshwater algae of the United States. McGraw-Hill Book Co., Inc., New York. 719 pp.

Smith, G.M. 1955. Cryptogamic Botany. Vol. II: Bryophytes and Pteridophytes. Ed. 2. McGraw-Hill Book Co., Inc., New York. 399 pp.

Smith, G. 1968. Fungi under domestication. In Ainsworth, G.C., and A.S. Sussman (eds.): The fungi, Vol. 3, pp. 273-286. Academic Press, Inc., New York.

Smith, J.E., and D.R. Berry (eds.). 1976. The filamentous fungi. Vol. 2: Biosynthesis and metabolism. Edward Arnold, Ltd., London. 520 pp.

Smith, J.E., and D.R. Berry (eds.). 1978. The filamentous fungi. Vol. 3: Developmental mycology. John Wiley & Sons, Inc. New York. 464 pp.

Snell, W.H., and E.A. Dick. 1970. The boleti of northeastern North America. Verlag von J. Cramer. 3301 Lehre. 115 pp.

Soeder, C.J., and D. Maiweg. 1969. Einfluss pilzlicher parasiten auf unsterile massenkulturen von *Scenedesmus*. Arch. Hydrobiol. **66**:48-55.

Soeder, C., and E. Stengel. 1974. Physico-chemical factors affecting metabolism and growth rate. In Stewart, W.D.P. (ed.): Algal physiology and biochemistry, pp. 714-740. University of California Press, Berkeley.

Spanos, N.P., and J. Gottlieb. 1976. Ergotism and the Salem village witch trials. Science **194**:1390-1394.

Sparrow, F.K. 1958. Interrelationships and phylogeny of the aquatic phycomycetes. Mycologia **50**:797-813.

Sparrow, F.K. 1960. Aquatic Phycomycetes. Ed. 2. University of Michigan Press, Ann Arbor. 1187 pp.

Sparrow, F.K. 1973. Mastigomycotina (zoosporic fungi). In Ainsworth, G.C., F.K. Sparrow, and A.S. Sussman (eds.): The fungi, Vol. IVB, pp. 61-73. Academic Press, Inc., New York.

Sparrow, F.K. 1976. The present status of classification of biflagellate fungi. In Gareth Jones, E.B. (ed.): Recent advances in aquatic mycology, pp. 213-222. Elek Scientific Books, Ltd., London.

Spector, D.L., L.A. Pfiester, and R.E. Triemer. 1981. Ultrastructure of the dinoflagellate *Peridinium cinctum* f. *ovoplanum*. II: Light and electron microscopic observations on fertilization. Am. J. Bot. **68**(1):34-43.

Stakman, E.C. 1955. Progress and problems in plant pathology. Ann. Appl. Biol. **42**:22-33.

Staley, J.T., F. Palmer, and J.B. Adams. 1982. Microcolonial fungi: common inhabitants of desert rocks? Science **215**:1093-1095.

Starr, R.C. 1972. Control of differentiation in *Volvox*. Develop. Biol. (Suppl.) **4**:59-100.

Starr, R.C., and L. Jaenicke. 1974. Purification and characterization of the hormone initiating sexual morphogenesis in *Volvox carteri* f. *nagariensis* Iyengar. PNAS (U.S.) 71:1050-1054.

Starr, T.J., E.F. Deig, K.K. Church, and M.B. Allen. 1962. Antibacterial and antiviral staining activities of algal extracts studied by acridine orange. Tex. Rep. Biol. Med. **20**:271-278.

Steemann-Nielsen, E. 1952. The use of radioactive carbon (C^{14}) for measuring organic production in the sea. J. Cons. Perm. Int. Explor. Mer. **18**:117-140.

Steidinger, K.A. 1975. Basic factors influencing red tides. In LoCicero, V.R. (ed.): Proceedings, First International Conference on Toxic Dinoflagellate Blooms, pp. 153-162. Massuchusetts Science and Technology Foundation, Wakefield, Mass.

Steidinger, K.A., and E.R. Cox. 1980. Free-living dinoflagellates. In Cox, E.R. (ed.): Phytoflagellates, pp. 407-432. Elsevier North-Holland, New York.

Steidinger, K.A., and K. Haddad. 1981. Biological and hydrographic aspects of red tides. BioScience **31**(11):814-819.

Steinbiss, H.H., and Schmitz, K. 1973. CO_2-Fixierung und stofftransport in benthischen marinen Algen. Zur autoradiographischen Lokalisation der Assimilattransportbahnen im Thallus von *Laminaria hyperborea*. Planta (Berlin) **112**:253-263.

Stevens, F.L. 1913. The fungi which cause plant disease. Macmillan, Inc., New York. 469 pp.

Stewart, K.D., and K.R. Mattox. 1975. Comparative cytology, evolution and classification of the green algae with some consideration of the origin of other organisms with chlorophylls *a* and *b*. Bot. Rev. **41**:104-135.

Stewart, K.D., and K.R. Mattox. 1980. Phylogeny of phytoflagellates. In Cox, E.R. (ed.): Phytoflagellates, pp. 433-462. Elsevier North-Holland, New York.

Stewart, W.D., J.L. Danto, and S. Maddin. 1978. Dermatology: diagnosis and treatment of cutaneous disorders. Ed. 4. The C.V. Mosby Co., St. Louis. 634 pp.

Stewart, W.D.P. (ed.). 1974a. Algal physiology and biochemistry. Botanical Monographs, Vol. 10. University of California Press, Berkeley. 989 pp.

Stewart, W.D.P. 1974b. Blue-green algae. In Quispel, A. (ed.): The biology of nitrogen fixation, pp. 202-237. American Elsevier, New York.

Stewart, W.D.P. 1977. Blue-green algae. In Hardy, R.W.F., and W.S. Silver (eds.): A treatise on dinitrogen fixation. Sect. III: Biology, pp. 36-121. John Wiley & Sons, Inc., New York.

Stewart, W.D.P., and G.A. Codd. 1975. Polyhedral bodies (carboxysomes) of nitrogen fixing blue-green algae. Brit. Phycol. J. **10**:273-278.

Stoloff, L. 1977. Aflatoxins—an overview. In Rodricks, J.V., C.W. Hesseltine, and M.A. Mehlman (eds.): Mycotoxins in human and animal health, pp. 7-28. Pathotox Publishers, Park Forest, Ill.

Strathern, J.N., E.W. Jones, and J.R. Broach (eds.). 1981-82. The molecular biology of the yeast *Saccharomyces*. Monograph II A. 1981. Life cycle and inheritance, 751 pp. Monograph II B. 1982. Metabolism and gene expression, 672 pp. Cold Spring Harbor Laboratory, Cold Spring Harbor, N.Y.

Strobel, G.A., and G.N. Lanier. 1981. Dutch elm disease. Sci. Am. **242**(2):56-66.

Sturch, H.H. 1926. *Choreocolax Polysiphoniae* Reinsch. Ann. Bot. **40**:585-605.

Suberkropp, K., and M.J. Klug. 1977. Extracellular hydrolytic capabilities of aquatic Hyphomycetes on leaf litter, p. 639. Abstracts, Second International Mycological Congress, Tampa, Fla.

Subrahmanyan, R. 1954. On the life-history and ecology of *Hornellia marina* gen. et. sp. nov., (Chloromonadineae), causing green discoloration of the sea and mortality among marine organisms off the Malabar Coast. Indian J. Fish. **1**:182-203.

Sudman, M.S. 1974. Prototothecosis. A critical review. Am. J. Clin. Pathol. **61**:10-19.

Sumich, J.L. 1980. An introduction to the biology of marine life. Ed. 2. Wm. C. Brown Co., Publishers, Dubuque, Iowa. 359 pp.

Sunhede, S. 1974. Studies in the Myxomycetes. II: Notes on *Fuligo septica*. Svensk. Bot. Tidskr. **68**:397-400.

Sussman, M., and R. Brackenbury. 1976. Biochemical and molecular genetic aspects of cellular slime mold development. Annu. Rev. Plant Phys. **27**:229-265.

Sutton, A., G.E. Harrison, T.E.F. Carr, and D. Barltrop. 1971. Reduction in the absorption of dietary strontium in children by an alginate derivative. Int. J. Radiat. Biol. **19**:79-85.

Sweeney, B.M., and J.W. Hastings. 1962. Rhythms. In Lewin, R.A. (ed.): Physiology and biochemistry of algae, pp. 687-700. Academic Press, Inc., New York.

Syers, J.K., and I.K. Iskandar. 1973. Pedogenetic significance of lichens. In Ahmadjian, V. and M.E. Hale (eds.): The lichens, pp. 225-248. Academic Press, Inc., New York.

Taggart, R.E., and L.R. Parker. 1976. A new fossil alga from the Silurian of Michigan. Am. J. Bot. **63**(10):1390-1392.

Tamiya, H. 1957. Mass culture of algae. Annu. Rev. Plant Physiol. **8**:309-334.

Tamm, C., and W. Breitenstein. 1980. The biosynthesis of trichothecene mycotoxins. In Steyn, P.S. (ed.): The biosynthesis of mycotoxins: a study in secondary metabolism, pp. 69-104. Academic Press, Inc., New York.

Tarapchak, S.J. 1972. Studies on the Xanthophyceae of the Red Lake wetlands, Minnesota. Nova Hedwigia **23**:1-43.

Tatum, E.L. 1959. A case history in biological research. Science **129**:1711-1715.

Tatum, L.A. 1971. The southern corn leaf blight. Science **171**:1113-1116.

Taylor, D.L., and H.H. Seliger (eds.). 1979. Toxic dinoflagellate blooms. Elsevier North-Holland, New York. 505 pp.

Taylor, F.J.R. 1978. Problems in the development of an explicit hypothetical phylogeny of the lower eukaryotes. Biosystems **10**:67-89.

Taylor, W.R. 1957. Marine algae of the northeastern coast of North America. University of Michigan Press, Ann Arbor. 509 pp.

Taylor, W.R. 1960. Marine algae of the eastern tropical and subtropical coasts of the Americas. University of Michigan Press, Ann Arbor. 870 pp.

Teal, J.M. 1980. Primary production of benthic and fringing plant communities. In Barnes, R.S.K., and K.H. Mann (eds.): Fundamentals of aquatic ecosystems, pp. 67-83. Blackwell Scientific Publications, Ltd., Oxford.

Tedders, W.L. 1981. *In vitro* inhibition of the entomopathogenic fungi *Beauveria bassiana* and *Metarhizium anisopliae* by six fungicides used in pecan culture. Environ. Entomol. **10**(3):346-349.

Tel-Or, E., and W.D.P. Stewart. 1975. Manganese and photosynthetic oxygen evolution by algae. Nature **258**:715-716.

Tett, P.B., and M.G. Kelly. 1973. Marine bioluminescence. Oceanogr. Mar. Biol. Annu. Rev. **11**:89-173.

Tiffney, B.H., and E.S. Barghoorn. 1974. The fossil record of the fungi. Bot. Harv. Unit 7. Occas. Pap. Farlow Herb. Cryptogam:1-42.

Tilson, D.L., F.E. Palmer, and J.T. Staley. 1977. Nitrogen fixation in lakes of the Lake Washington drainage basin. Water Res. **11**:843-847.

Timberlake, W.E., L. McDowell, J. Cheney, and D.H. Griffin. 1973. Protein synthesis during the differentiation of sporangia in the water mold *Achlya*. J. Bacteriol. **116**:67-73.

Ting, I.P. 1982. Plant Physiology. Addison-Wesley Publishing Co., Inc., Reading, Mass. 642 pp.

Tippo, O., and W.L. Stern. 1977. Humanistic botany. W.W. Norton & Co., Inc., New York. 605 pp.

Tomas, C.R. 1980a. IV: *Olisthodiscus luteus* (Chrysophyceae): effects of light intensity and temperature on photosynthesis and cellular composition. J. Phycol. **10**(2):149-156.

Tomas, C.R. 1980b. V: *Olisthodiscus luteus* (Chrysophyceae): its occurrence and dynamics in Narragansett Bay, Rhode Island. J. Phycol. **16**(2):157-166.

Tomas, R.N., and E.R. Cox. 1973a. Observations on the symbiosis of *Peridinium balticum* and its intracellular alga. I: Ultrastructure. J. Phycol. **9**:304-323.

Tomas, R.N., and E.R. Cox. 1973b. The symbiosis of *Peridinium balticum* (Dinophyceae). I: Ultrastructure and pigment analysis. J. Phycol. **9** (Suppl.):16.

Tomas, R.N., E.R. Cox, and K.A. Steidinger. 1973. *Peridinium balticum* (Levander) Lemmermann, an unusual dinoflagellate with a mesokaryotic and an eukaryotic nucleus. J. Phycol. **9**:91-98.

Trainor, F.T. 1978. Introductory phycology. John Wiley & Sons, Inc., New York. 525 pp.

Trench, R.K., M.E. Trench, and L. Muscatine. 1970. Utilization of photosynthetic products of symbiotic chloroplasts in mucus synthesis by *Placobranchus ianthobapsus* (Gould), Opisthobranchia, Sacoglossa. Comp. Biochem. Physiol. **37**:113-117.

Trench, R.K., M.E. Trench, and L. Muscatine. 1972. Symbiotic chloroplasts; their photosynthetic products and contribution to mucus synthesis in two marine slugs. Biol. Bull. **142**:335-349.

Trieff, N.M., V.M.S. Ramanujam, M. Alam, S.M. Ray, and J.E. Hudson. 1975. Isolation, physio-chemical and toxicologic characterization of toxins from *Gymnodinium breve* Davis. In LoCicero, V.R. (ed.), Proceedings, First International Conference on Toxic Dinoflagellate Blooms, pp. 309-323. Massachusetts Science and Technology Foundation, Wakefield, Mass.

Tubaki, K. 1975a. Notes on the Japanese Hyphomycetes. VI: *Candelabrum* and *Beverwykella* gen. nov. Trans. Mycol. Soc. Japan **16**:132-140.

Tubaki, K. 1975b. Notes on the Japanese Hyphomycetes. VII: *Cancellidum*, a new Hyphomycetes genus. Trans. Mycol. Soc. Japan **16**:357-360.

Tuominen, Y., and T. Jaakkola. 1973. Absorption and accumulation of mineral elements and radioactive nuclides. In Ahmadjian, V., and M.E. Hale (eds.): The lichens, pp. 185-223. Academic Press, Inc., New York.

Turner, C.H., and L.V. Evans. 1978. Translocation of photoassimilated ^{14}C in the red alga *Polysiphonia lanosa*. Br. Phycol. J. **13**:51-55.

Twarog, B., and E. Gilfillan. 1975. Session summary—pharmacology. In LoCicero, V.R. (ed.): Proceedings, First International Conference on Toxic Dinoflagellate Blooms, pp. 334-336. Massachusetts Science and Technology Foundation, Wakefield, Mass.

Ueno, Y. 1977. Trichothecenes: overview address. In Rodricks, J.V., C.W. Hesseltine, and M.A. Mehlman (eds.): Mycotoxins in human and animal health, pp. 189-207. Pathotox Publishers, Park Forest, Ill.

Uno, I., and T. Ishikawa. 1974. Effect of glucose on the fruiting body formation and adenosine 3′, 5′-cyclic monophosphate levels in *Coprinus macrorhizus*. J. Bacteriol. **120**:96-100.

Uraguchi, K., and M. Yamazaki (eds.). 1978. Toxicology, biochemistry and pathology of mycotoxins. John Wiley & Sons, Inc., New York. 288 pp.

van der Meer, J.P. 1983. The domestication of seaweeds. BioScience **33**(3):172-176.

van Geel, B., and T. van der Hammen. 1978. Zygnemataceae in Quartenary Colombian sediments. Rev. Paleobot. Palynol. **25**:377-392.

Van Valkenburg, S.D. 1980. Silicoflagellates. In Cox, E.R. (ed.): Phytoflagellates, pp. 335-350. Elsevier North-Holland, New York.

Vartia, K.O. 1973. Antibiotics in lichens. In Ahmadjian, V., and M.E. Hale (eds.): The lichens, pp. 547-561. Academic Press, Inc., New York.

Verrett, M.J., M.K. Mutchler, W.F. Scott, E.F. Reynaldo, and J. McLaughlin. 1969. Teratogenic effects of captan and related compounds in the developing chicken embryo. Ann. NY Acad. Sci. **160**:334-343.

Viala, G. 1966. L'astaxanthine chez le *Chlamydomonas nivalis* Wille. Compt. Rend. Itebd. Séances Acad. Sci. **263**:1383-1386.

Volesky, B., J.E. Zajic, and E. Knettig. 1970. Algal products. In Zajic, J.E. (ed.): Properties and products of algae, pp. 49-82. Plenum Press, New York.

Waddell, D.R. 1982. A predatory slime mold. Nature **298** (5873):464-466.

Wade, N. 1981a. Toxin warfare charges may be premature. Science **214**:34.

Wade, N. 1981b. Yellow rain and the cloud of chemical war. Science **214**:1008-1009.

Walker, J.C. 1969. Plant pathology. Ed. 3. McGraw Hill-Book Co., New York. 819 pp.

Walker, J.D., R.R. Colwell, and Z. Vaituzis. 1975. Petroleum-degrading achlorophyllous alga *Prototheca zopfii*. Nature **254**:423-424.

Walker, L.M., and K.A. Steidinger. 1979. Sexual reproduction in the toxic dinoflagellate *Gonyaulax monilata*. J. Phycol. **15**:312-315.

Walne, P.L. 1967. The effects of colchicine on cellular organization in *Chlamydomonas*. II: Ultrastructure. Am. J. Bot. **54**:564-577.

Walne, P.L. 1980. Euglenoid flagellates. In Cox, E.R. (ed.): Phytoflagellates, pp. 165-212. Elsevier North-Holland, New York.

Walsby, A.E. 1972. Structure and function of gas vacuoles. Bact. Rev. **36**:1-32.

Walsby, A.E. 1977. The gas vacuoles of blue-green algae. Sci. Am. 90-97.

Walsh, J. 1981. Genetic vulnerability down on the farm. Science **214**:161-164.

Walters, A.H., and J.J. Elphick (eds.). 1968-71. Biodeterioration of materials. 2 vols. Elsevier, Amsterdam, Netherlands.

Wardlaw, C.W. 1955. Embryogenesis in plants. John Wiley & Sons, Inc., New York. 380 pp.

Wardle, W.J., S.M. Ray, and A.S. Aldrich. 1975. Mortality of marine organisms associated with offshore summer blooms of the toxic dinoflagellate *Gonyaulax monilata* Howell at Galveston, Texas. In LoCicero, V.R. (ed.): First International Conference on Toxic Dinoflagellate Blooms, pp. 257-263. Massachusetts Science and Technology Foundation, Wakefield, Mass.

Wareing, P.F., and I.D.J. Phillips. 1978. The control of growth and differentiation in plants. Ed. 2. Pergamon Press, Inc., New York. 347 pp.

Wasson, R.G. 1967. Soma, divine mushroom of immortality. Harcourt Brace Jovanovich, Inc., New York. 381 pp.

Wasson, R.G. 1978. Soma brought up to date. Harvard Univ. Bot. Mus. Leaflet **26**(6):211-223.

Wasson, R.G., and V.P. Wasson. 1957. Mushrooms, Russia, and history. 2 vols. Pantheon Books, Inc., New York. 432 pp.

Watanabe, I., K.K. Lee, B.V. Alimagno, M. Sato, D.C. del Rosario, and M.R. de Gusman. 1977. Biological nitrogen fixation in paddy field studied by *in situ* acetylene-reduction assays. IRRI Research Paper Series, No. 3. The International Rice Research Institute, Manila, Philippines.

Waterhouse, G.M. 1973a. Plasmodiophoromycetes. In Ainsworth, G.C., F.K. Sparrow, and A.S. Sussman (eds.): The fungi: an advanced treatise. Vol. IVB, pp. 75-82. Academic Press, Inc., New York.

Waterhouse, G.M. 1973b. Entomophthorales. In Ainsworth, G.C., F.K. Sparrow, and A.S. Sussman (eds.): The fungi: an advanced treatise. Vol. IVB, pp. 219-229. Academic Press, Inc., New York.

Watley, F.R. 1976. Nitrogen, oxygen and manganese. Nature **259**:11.

Watson, P.L., R.J. Barney, J.V. Maddox, and E.J. Armbrust. 1981. Sporulation and mode of infection of *Entomophthora phytonomi*: a pathogen of the alfalfa weevil. Environ. Entomol. **10**(3):305-306.

Webber, J. 1981. A natural biological control of Dutch elm disease. Nature **292**:449-451.

Webster, J. 1980. Introduction to fungi. Ed. 2. Cambridge University Press, London. 669 pp.

Webster, J., and E. Descals. 1981. Morphology, distribution and ecology of conidial fungi in freshwater habitats. In Cole, G.T., and B. Kendrick (eds.): Biology of conidial fungi, Vol. 1., pp. 295-355. Academic Press, Inc., New York.

Wei, C., and C.S. McLaughlin, 1974. Structure-function relationship in the 12, 13-epoxytrichothecenes: novel inhibitors of protein synthesis. Biochem. Biophys. Res. Commun. **57**:838-844.

Weichold, B. (ed.). 1962. Bioenvironmental features of the Ogotoruk Creek area, Cape Thompson, Alaska. U.S. Atomic Energy Commission, Rep. TID-17226.

Weinberg, E. 1971. The roles of trace metals in toxigenesis. In Hemphill, D. (ed.): 4th Annual Conference on Trace Substances in Environmental Health, pp. 233-240. University of Missouri, Columbia.

Weiser, J. 1982. Fungus pathogens as pesticides. In Kurstak, E. (ed.): Microbial and viral pesticides, Marcel Dekker, Inc., New York. 712 pp.

Werner, D. 1977a. Silicate metabolism. In Werner, D. (ed.): The biology of diatoms, pp. 110-149. University of California Press, Berkeley.

Werner, D. (ed.). 1977b. The biology of diatoms. University of California Press, Berkeley. 498 pp.

Wernick, R. 1983. From ewe's milk and a bit of mold: a fromage fit for a Charlemagne. Smithsonian 13(11):56-63.

Wessels, J.G.H. 1965. Morphogenesis and biochemical processes in Schizophyllum commune Fr. Wentia 13:1-113.

West, J.A. 1968. Morphology and reproduction of red alga Acrochaetium pectinatum in culture. J. Phycol. 4:89-99.

Wetzel, R.G. 1975. Limnology. W.B. Saunders Co., Philadelphia. 743 pp.

Whisler, H.C., S.L. Zebold, and J.A. Shemanchuk. 1975. Life history of Coelomomyces psorophorae. PNAS (U.S.) 72:693-696.

White, A.W. 1981. Marine zooplankton can accumulate and retain dinoflagellate toxins and cause fish kills. Limnol. Oceanogr. 26(1):103-109.

Whitehead, D.R., S.E. Reed, and D.F. Charles. 1981. Late glacial and post glacial pH changes in Adirondack Lakes. Bull. Ecol. Soc. Amer. 62:154.

Whittaker, R.H. 1969. New concepts of kingdoms of organisms. Science 163:150-160.

Whittaker, R.H. 1975. Communities and ecosystems. Ed. 2. Macmillan, Inc., New York. 385 pp.

Whittaker, R.H., and G.E. Likens. 1975. The biosphere and man. In Leith, H., and R.H. Whittaker (eds.): The primary productivity of the biosphere, pp. 305-328. Springer-Verlag, New York.

Wiedeman, V.E. 1970. Heterotrophic nutrition of waste-stabilization pond algae. In Zajic, J.E. (ed.): Properties and products of algae, pp. 107-114. Plenum Press, New York.

Wilce, R.T. 1959. The marine algae of the Labrador peninsula and northwest Newfoundland (ecology and distribution). Nat. Mus. Canada Bull. 158:1-103.

Wilcox, H.A. The ocean food and energy farm project. J. Marine Educ., Summer 1976.

Wilcox, H.E. 1971. Morphology of ectendomycorrhizae in Pinus resinosa. In Hocskaylo, E. (ed.): Mycorrhizae: Proceedings, First North American Conference on Mycorrhizae, pp. 54-68. USDA For. Ser. Misc. Publ. 1189.

Willoughby, L.G. 1969. Salmon disease in Windermere and the River Leven: the fungal aspect. Salmon and Trout Magazine 186:124-130.

Willoughby, L.G. 1977. Freshwater biology. Pica Press, New York. 167 pp.

Wilson, J. 1976. Immune response in fish to fungal disease. In Gareth Jones, E.B. (ed.): Recent advances in aquatic mycology, pp. 573-601. Elek Scientific Books, Ltd., London.

Woelkerling, W.J. 1975. Observations on Batrachospermum (Rhodophyta) in southeastern Wisconsin streams. Rhodora 77:467-477.

Wood, H.A., R.F. Bozarth, J. Adler, and D.W. Mackenzie. 1974. Proteinaceous virus-like particles from an isolate of Aspergillus flavus. J. Virol. 13:532-534.

Wood, K.G. 1975. Photosynthesis of Cladophora in relation to light and CO_2 limitation: $CaCO_3$ precipitation. Ecology 56(2):479-484.

Wood, R.D. 1965. Monograph of the Characeae. In Wood, R.D., and K. Imahori: A revision of the Characeae, Vol. I, 904 pp. J. Cramer, Weinheim, Germany.

Wright, B.E. 1973. Critical variables in differentiation. Prentice-Hall, Inc., Englewood Cliffs, N.J. 109 pp.

Wright, B.E., A. Taiz, and K.A. Killick. 1977. Fourth expansion and glucose perturbation of the Dictyostelium kinetic model. Eur. J. Biochem. 74:217-225.

Wurster, C. 1968. DDT reduces photosynthesis by marine phytoplankton. Science 159:1474-75.

Wynne, M.J. 1969. Life history and systematic studies of some Pacific North American Phaeophyceae (brown algae). Univ. Calif. Pub. Bot. 50:1-88.

Wynne, M.J., and S. Loiseaux. 1976. Recent advances in life history studies of the Phaeophyta. Phycologia 15:435-452.

Yentsch, C.M., and L. Incze. 1980. Accumulation of algal biotoxins in mussels. In Lutz, R. (ed.): Mussel culture in North America, pp. 223-246. Elsevier North-Holland, New York.

Yentsch, C.M., C.M. Lewis, and C.S. Yentsch. 1980. Biological resting in the dinoflagellate Gonyaulax excavata. BioScience 30(4):251-254.

Yentsch, C.S. 1963. Primary production. Oceanogr. Mar. Biol. Annu. Rev. 1963. 1:157-175.

Young, P. 1981. Thick layers of life blanket lake bottoms in Antarctica valleys. Smithsonian 12(8):52-61.

Youngman, P.J., P.N. Adler, T.M. Shinnick, and C.E. Holt. 1977. An extracellular inducer of plasmodium formation in Physarum polycephalum. PNAS (U.S.) 74:1120-1124.

Zaar, K., and H. Kleinig. 1975. Spherulation of Physarum polycephalum. I: Ultrastructure. Cytobios 10:306-328.

Zahl, P.A. 1974. Algae: the life givers. Nat. Geog. 145:361-367.

Zilinskas, B.A., L.S. Greenwald, C.L. Bailey, and P.C. Kahn. 1980. Spectral analysis of allophycocyanin I, II, III and B from Nostoc sp. phycobilisomes. Acta Biochim. et Biophys. 592:267-276.

Zingmark, R.C. 1970. Sexual reproduction in the dinoflagellate Noctiluca miliaris Suriray. J. Phycol. 6:122-126.

A

References to illustrations are printed in **boldface** type. Numbered entries in *italics* refer to citations in tables.